A TEXTBOOK OF
GENERAL PHYSIOLOGY

Volume two

A TEXTBOOK OF
GENERAL
PHYSIOLOGY

by

HUGH DAVSON

D.Sc. (Lond.)

Scientific Staff, Medical Research Council;
Honorary Research Associate and Fellow of
University College London

FOURTH EDITION

Volume two

With 876 illustrations including 79 plates

J. & A. CHURCHILL
104 GLOUCESTER PLACE, LONDON
1970

Dedicated to the memory of
SIR CHARLES LOVATT EVANS, F.R.S.
Emeritus Professor of Physiology in the University of London

First Edition . . .		1951
,, ,, *reprinted* . .		1952
,, ,, ,, .		1954
Japanese translation . .		1955
Second Edition . .		1959
,, ,, *reprinted* .		1960
Third Edition . .		1964
,, ,, *reprinted* . .		1966
Italian translation . .		1969/70
Fourth Edition . . .		1970

I.S.B.N. 0.7000.1491.8

Printed in Great Britain

VOLUME TWO CONTENTS

4. *Characteristics of Excitable Tissue*

v

5. The Mechanism of Contraction of Muscle

6. Light: Its Effect on, and its Emission by, the Organism

Volume one contains the following chapters:

4. *Characteristics of Excitable Tissue*

CHAPTER XV

EXCITABILITY AND PROPAGATION OF THE IMPULSE

IN an earlier chapter (p. 564) we have discussed the nature of the potential difference between the inside and outside of certain cells, in particular of nerve and muscle fibres, without, however, attempting to elucidate their functional significance. In the present section we shall be concerned with the problems of excitation and transmission, phenomena that are associated with modifications of these bioelectric potentials. In the complex organism the effects of a stimulus are transmitted by way of specialized cells—*neurones*; transmission is not peculiar to these, however, since a muscle, for example, may exhibit the spread of an excited state in an essentially similar manner to that found in nerve, and even unicellular organisms exhibit responses to local stimulation that are not necessarily confined to the stimulated point. The spread of excitation may thus be a general characteristic of living tissues, attaining, however, a maximal degree of efficiency and speed in the specialized nervous cells. The spread of excitation over a muscle fibre is probably related to the uniform development of contraction, and is thus an important element in its behaviour; nerve and muscle therefore represent specialized tissues in which we may expect to find the properties of excitability and transmission highly developed, so that in this section we may confine attention almost exclusively to them.

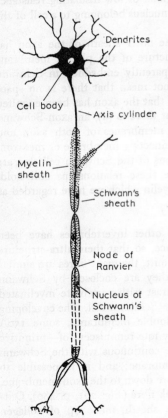

FIG. 519. Typical motor neurone.

The Neurone

A typical vertebrate nerve consists of a bundle of hundreds of fibres, each one of which is an extension of a single nerve cell, or neurone; a motor neurone, Fig. 519, is essentially a cell with a series of processes, one of which, the *axon*, is much longer than the remainder, the *dendrites*. The axon leaves the cell body (soma or perikaryon) as the axon hillock, and eventually makes connection with an effector organ—in this case a number of muscle fibres—or another neurone; the dendrites connect with the axons of other nerve fibres of the central nervous system.* As we shall see, these junctional regions, the *nerve muscle junction* and the nerve-nerve *synapse*, are specialized regions permitting a close association of the two elements; nevertheless there is reason to believe that the fundamental continuity of the plasma membrane over the protoplasmic surface of the nerve cell is maintained even in these regions, *i.e.*, there is probably no cytoplasmic fusion, so that the neurone is a true unit in transmission. As classically represented, the axon consists, from without inwards, of an outermost sheath or *neurilemma* (*Schwann's sheath*); a *myelin sheath* of mainly lipoid material which acts as an insulating layer, and which may be thick in the typical *myelinated* (or *medullated*) *nerves* of vertebrates or very thin in the *non-myelinated* nerves typically found in invertebrates and in some parts of the vertebrate nervous system. The limiting plasma membrane separates the internal *axoplasm* from the outer layers. The thick myelin sheath of the medullated nerve is interrupted at regular intervals—about 1 mm. in man and 3 mm. in the frog—to give constrictions, or *nodes of Ranvier*, which are therefore to be regarded as localized regions of low insulating resistance (p. 1070). In each internodal region there is a nucleus belonging to a cell of the Schwann sheath or neurilemma.

Ultra-Structure. The ultra-structure of the outer sheath of the axon has been discussed earlier in relation to the structure of the plasma membrane (p. 477). We may recall that the axon is apparently embedded in Schwann cytoplasm but, as Gasser showed, this does not mean that there is no space between the two cells. Fig. 261 (p. 480) shows that the axon has been engulfed by an invagination of the Schwann cell so that there is (*a*), an axon-Schwann double membrane consisting of the plasma membranes of both axon and Schwann cell, and (*b*), the so-called surface-connecting membrane or mesaxon, representing the space between the two portions of the Schwann cell that are in close apposition. In the myelinated nerve these relationships still hold, but are obscured by the numerous layers of myelin which we have regarded as essentially spirally wound Schwann membrane.

Invertebrate Fibres

The giant fibres of the squid and certain other invertebrates have been studied extensively because of their convenience, so that their ultra-structure is of some interest. According to Geren & Schmitt, the giant fibres are similar to vertebrate non-myelinated fibres in that they are enclosed by Schwann cells, but the nuclei indent the axon, by contrast with vertebrate myelinated nerve where the nuclei lie in the surface of the Schwann cell. The enveloping Schwann cytoplasm contains some 3 to 6 double membranes, some 150A thick, which would appear to be—or are at any rate reminiscent of—primitive myelin sheath. These membranes appear to be continuous with the Schwann surface membrane and the axon plasma membrane; and it is possible to follow spaces leading from the surface of the fibre down to the axon membrane, similar to the mesaxons described in the Remak fibre (Fig. 261, p. 480). Outside the Schwann layer is a connective-tissue sheath, made up of collagen fibrils well orientated in the axial direction.

The giant axon is surrounded by several Schwann cells some $0 \cdot 1 - 0 \cdot 2\mu$ thick near their ends and some $0 \cdot 8 - 0 \cdot 9\mu$ near the nucleus; these are in a single row around the axon, to form a layer that is crossed by tortuous intercellular channels.

---

* The distinction between axon and dendrite is made on both morphological and functional grounds; it is sufficient for our purpose to note that the neurone is so disposed that impulses pass normally into the dendrites through the cell body and thence along the axon. The long processes, constituting the sensory nerves of the spinal ganglia (p. 1127), are often described as axons, though functionally they behave as dendrites.

External to the Schwann cells is a thick basement membrane ($0\cdot 2\mu$ thick), and beyond this are alternate layers of connective-tissue cells and fibrils forming the endoneurium (Villegas & Villegas, 1968). Sections through the Schwann sheath indicate the presence of some three to six membranes apparently running parallel to the axon (Geren & Schmitt) and analogous with the myelin layer of myelinated nerve. Careful examination of these in the electron-microscope has shown that

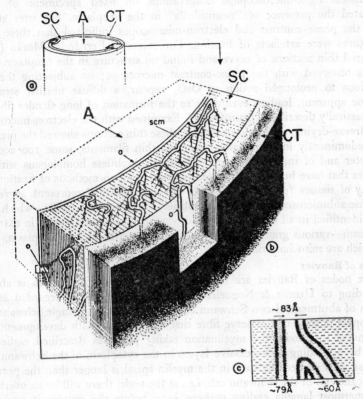

FIG. 520. Three-dimensional diagram of the giant nerve fibre of the squid. (a) A segment of the nerve fibre showing the axon (A) covered by the Schwann cell (SC). The latter is covered by the connective tissue (CT). (b) Enlarged portion of the fibre in which channels (ch) are shown as slits crossing the Schwann cell from outer surface to axonal surface. Note the openings (o) of the channels (ch). It may be seen that some of the channels, which in ultra-thin sections appear ending in a blind alley, are found to be continuous at different levels, as has been observed in serial electron micrographs. (c) Highly enlarged view of one of the channel openings (o) in which the continuity of the channel walls with the Schwann cell membrane (scm) is demonstrated. The fine structure of the axolemma (a) is shown. No difference can be appreciated between this structure (a), the Schwann cell membrane (scm), and the channel wall structure (ch). (Villegas & Villegas. *J. gen. Physiol.*)

they consist of double-edged osmiophilic layers, whilst the less dense zones between the edges are channels connecting the outside of the Schwann cell to the space surrounding the axon as illustrated by Fig. 520. Studies on the permeability to water by Villegas & Villegas gave a value of $1\cdot 42.10^{-4}$ cm./sec, and from this they deduced that the fraction of the total area of the axon available for diffusion was some $0\cdot 23$ per cent., and this would correspond with the area of the channels connecting the connective tissue to the space between axolemma and Schwann cell membrane.*

The arthropod axon is surrounded by a lemnoblast and is described by Edwards, Ruska & de Harven as *tunicated*, since it is loosely mantled by several cytoplasm-enclosing membranes of the lemnoblast.

Neurofibrils

Classical light-microscopical observations on fixed specimens of nerve indicated the presence of "neurofibrils" in the axoplasm, but later studies with the phase-contrast and electron-microscopes indicated that these gross structures were artefacts of fixation. For example, Fernández-Morán (1952) prepared thin sections of nerve and found no structure in the axoplasm when it was observed with the phase-contrast microscope; on subjecting the preparations to prolonged action of OsO_4 vapour, a diffuse fibrillar structure became apparent, leading eventually to the formation of long slender fibrils— the classically described "neurofibrils". Examined with the electron-microscope after freeze-drying, or simple air-drying, these thin sections showed the presence of predominantly longitudinally orientated thin filaments, some 100–200A in diameter and of indefinite length. These are doubtless homologous with the tubules that have been demonstrated by more modern methods of fixation in a variety of tissues (p. 44). Presumably the "neurofibrils" represent aggregates of these submicroscopic "neuroprotofibrils". Besides these fibrils—which have been identified in all the axons studied whether in thin sections or in extruded axoplasm—various granular bodies have been described in the axoplasm, some of which are mitochondria.†

Nodes of Ranvier

The nodes of Ranvier are regions in which the myelin sheath is absent. According to Uzman & Nogueira-Graf, the node must be regarded as the region of abutment of two Schwann cells. On this basis, a single Schwann cell envelops a portion of the nerve fibre that will become, with development, the myelinated internode, the myelination taking place, as described earlier (p. 479), by a winding of successive layers in the cytoplasm of the Schwann cell. Because each succeeding turn in the myelin spiral is longer than the previous one, at the end of the Schwann cell, *i.e.*, at the node, there will be an overhang, the innermost lamella ending perhaps $\frac{1}{2}$–1μ before the outermost one (Fig. 521a). At the node, where the Schwann cells end, the axon is thus covered only by Schwann cytoplasm. According to Robertson, the Schwann cytoplasm in

---

* The measurement was one of the diffusion of tritiated water and passage was presumably largely determined by the migration along the tortuous channels shown in Fig. 520; responses to osmotic gradients, on the other hand, must have been determined by the axolemma; a filtration permeability coefficient of $7\cdot8.10^{-10}$ ml./cm². sec. cm. H_2O pressure was derived from the osmotic measurements. By applying Solomon's theoretical treatment (p. 416), an equivalent pore radius of 4·25A was deduced for the axolemma (Villegas & Barnola, 1961; Villegas, Caputo & Villegas, 1962) which may thus be regarded as the main seat of diffusional resistance to transport across the axon. In a preliminary note, Villegas & Villegas (1962) have pointed to the possibility of a third component, in addition to axolemma and Schwann sheath; this is the layer of endoneurium cells external to the basement membrane of the Schwann cells.

† Maxfield (1953) has isolated, by fractional centrifugation, a protein from extruded squid axoplasm which he calls "axon filaments"; a dried solution appeared under the electron-microscope as fibrils of indefinite length, 90–160A in diameter. These filaments could be reversibly broken down to smaller particles by alkaline K-phosphate, in which case dried specimens showed no fibrils that could be resolved. Elfvin (1961) has described thin (100A) and thick (300A) filaments in unmyelinated nerve. More recently Peters & Vaughan (1967) have described the microtubules and filaments in the developing nerve—at early stages the microtubules of 233–260A diameter dominate the picture; but later the number of neurofibrils, with 90–100A diameter, increases; these, in cross section, also appear as tubules. The suggestion is made that the filaments come from microtubules.

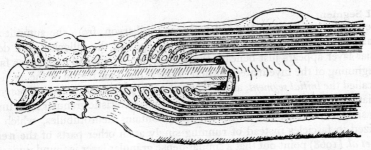

FIG. 521a. Diagrammatic representation of cutaway view of early node of Ranvier illustrating relationship between myelin lamellar endings, Schwann cell cytoplasm, and axon. The myelin lamellæ are cross-hatched; cut surfaces of the Schwann cell cytoplasm are stippled and include outlines of formed cytoplasmic organelles. (Uzman & Nogueira-Graf. *J. biophys. biochem. Cytol.*)

the node breaks up into a number of processes, so that, in effect, the axon membrane is naked in this region.

Schmidt-Lantermann Clefts

The myelin sheath in each Schwann segment is interrupted by several oblique partitions, the *incisions* or *clefts of Schmidt-Lantermann*, easily recognizable in polarized light, and first identified in the electron-microscope by Rozsa, Morgan, Szent-Gyorgyi & Wyckoff. According to Robertson, these

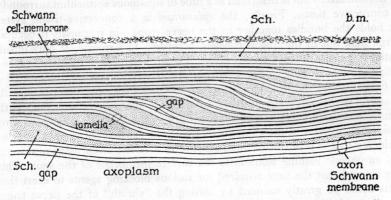

FIG. 521b. Diagram of a Schmidt-Lantermann cleft in a segment of myelin in longitudinal section. The outer Schwann cell membrane is shown above as two lines, making a unit about 75A across. Two such units are seen below, making the axon-Schwann membrane. Each myelin lamella is composed of two such units in contact, and the heavy dense lines are produced where adjacent lamellæ are in contact. In the cleft the lamellæ may separate to give a gap. The basement membrane (b.m.) is indicated above, but not below. (Robertson. *J. biophys. biochem. Cytol.*)

clefts arise by a staggered separation of the myelin lamellæ, as indicated in Fig. 521b, so that, in effect, the gaps are not empty but traversed by successive elements in the concentrically wound myelin lamellæ. If the myelin lamellæ are formed in Schwann cytoplasm, it may be supposed that these spaces are filled with this cytoplasm, and a consideration of the three-dimensional aspect of the clefts will show that they constitute a helical layer of Schwann cytoplasm, leading eventually to the surface of the axon.

Initial Segment

The axon emerges from the cell body, or perikaryon, from the summit of a conical projection, the *axon hillock*. Near the apex of the hillock a thin dense granular layer appears just beneath the plasma membrane and extends as far as the beginning of the myelinated portion of the axon. This region of the axon has been called the *initial segment*, and seems to be that part where an action potential is initiated (p. 1191). In addition to having this characteristic granular layer under its plasma membrane, the initial segment contains neurotubules, which are organized in fascicles instead of running singly as in other parts of the nerve. Palay *et al.* (1968) point out that a similar dense granular layer is found under the plasma membrane at the node of Ranvier. They hazard the suggestion that the fasciculation of the neurotubules subserves a contractile function, perhaps providing the motive force for the protoplasmic streaming from parikaryon to axon.

Outer Membranes. The individual fibres of a nerve trunk are bound together by three connective-tissue systems, namely the *endoneurium, perineurium*, and *epineurium*. Essentially, the endoneurium constitutes a covering for the individual fibres, and it may be described as a connective-tissue tube made up of two layers, the *inner endoneurium*, or sheath of Plenk & Laidlaw, made up of circularly orientated argyrophil fibrils, and an outer endoneurium, or sheath of Key & Retzius.* The *perineurium* is a connective-tissue sheath surrounding a bundle of nerve fibres which, in mammals, is a laminated capsule, each lamella being covered with endothelial cells. In the frog, on the other hand, it is not lamellated but is described as a tube of squamous epithelium surrounded by connective tissue. Finally, the *epineurium* is a connective-tissue system that holds the bundles together in the nerve trunk—in mammals it occupies a large bulk but in the frog it consists, according to Krnjević, of only unsubstantial fragments of loose areolar tissue. The arrangement of the membranes is illustrated schematically by Fig. 522 from Shanthaveerappa & Bourne (1962): as the figure shows, the innermost layer of the perineurium is described as the *perineurial epithelium*; it is a stratified squamous structure, five layers thick in the mammalian sciatic nerve and two layers thick in the frog. They have acquired considerable interest to the experimenter with the isolated nerve since it is now considered that at least one of them constitutes a barrier to diffusion of material from an outside bathing solution to the surface membrane of the axon. Thus Feng showed that the time required for various blocking agents to exert their effects could be greatly reduced by slitting the "sheath" of the nerve trunk, or removing it. As a result of further studies, for example those of Crescitelli, of Huxley & Stämpfli, and most recently of Krnjević, there can be little doubt that the "connective-tissue sheath" of nerve does indeed constitute a barrier to diffusion which may reduce the effective diffusion coefficient of K^+ by a factor of 30 or more, whilst with lipid-soluble substances, such as urethane, little or no "barrier action" is observed. The only point in dispute seems to be the anatomical identification of the barrier; most investigators have described this as the epineurium, and have characterized the "desheathed nerve" as essentially a nerve without epineurium, but with perineurium and endoneurium intact.

---

* The neurilemma, or Schwann membrane of classical light-microscopy, may be described as the innermost portion of the tubular wall that surrounds the nerve fibre, *i.e.*, the innermost portion of the endoneurium. In the electron-microscope it is a fine granular membrane, only a few hundred Ångstrom units thick, reinforced by a network of submicroscopic fibrils. The fibrils have an axial repeat of 600–660A and are apparently collagen; the longitudinal fibrils interweave with circular fibrils to form a network which constitutes the *inner endoneurium* of Plenk & Laidlaw.

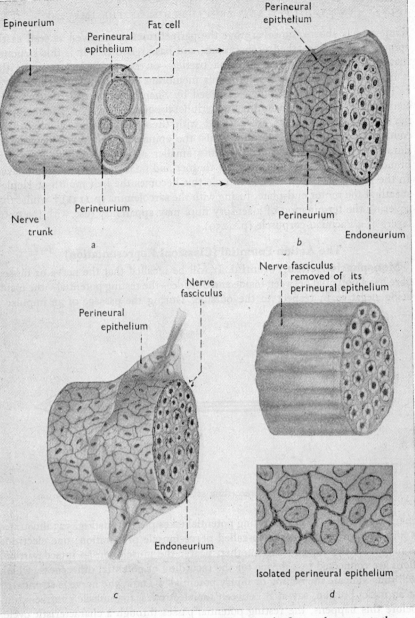

FIG. 522. The outer membranes of peripheral nerve; the figures demonstrate the various stages of isolation of this "perineural epithelium" under a binocular dissection microscope.

(a) The nerve trunk as a whole, with many fasciculi along with their connective tissue components, the epi-, peri-, and endo- neurium. The perineural epithelium is shown in the diagram surrounding each nerve fasciculus, lying under the perineurium.

(b) One nerve fasciculus is removed along with the perineurium, perineural epithelium and endoneurium. Part of the perineurium is removed to show the multiple layered perineural epithelium, lying immediately under the perineurium and on the entire surface of the nerve fasciculus.

(c) The nerve fasciculus removed from its entire perineural connective tissue layer, leaving the perineural epithelium covering the nerve fasciculus. The multiple layered nature of the perineural epithelium is indicated.

(d) Nerve fasciculus removed of its perineurium and perineural epithelium, leaving the nerve fibres of the fasciculus and the endoneurium intact. The isolated perineural epithelium can be seen lying flat on the glass slide. A capillary network in this perineural epithelium is also shown. (Shantha-veerappa & Bourne. J. Anat.)

Krnjević has shown, however, that the perineurium is removed as well as the epineurium in desheathed preparations, and has argued that it is this structure that is responsible for the diffusion barrier—on morphological grounds this is likely since it contains a continuous membrane of closely apposed cells.*

This view has been strongly endorsed by Shanthaveerappa & Bourne, who consider that the perineurium is the peripheral equivalent of the pia-arachnoid; the emerging nerve roots are covered with these membranes which become continuous with the perineurium, whilst the epineurium is the analogue of the dura. As the nerve fasciculus becomes smaller and smaller, with successive branching, the perineurium becomes thinner and finally becomes a single layer. In the motor fibre to a skeletal muscle this becomes the bell mouth of Henle's sheath at the motor end-plate, fusing with the sarcolemma (p.1131),† whilst that covering the termination of a sensory fibre may apparently fuse with the outer layers of a Pacinian corpuscle (p. 1259).

The Action Potential (Classical Representation)

Monophasic Action Potential. It will be recalled that the nerve or muscle fibre has a potential between inside and outside—the resting potential—the inside being negative in respect to the outside. During the passage of an impulse a

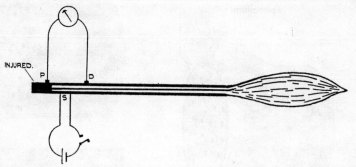

FIG. 523. Stimulation and recording set-up for nerve-muscle preparation.

characteristic change in the resting potential takes place; thus Fig. 523 illustrates a muscle with its nerve, the so-called nerve-muscle preparation; one electrode is placed on the cut end and another, nearer the muscle, on an intact portion. Both are connected to a device for the recording of potential differences, which will thus indicate the resting or injury potential. If, now, the nerve is stimulated at an intact portion, say at S between the electrodes, the muscle contracts but, before this happens, the resting potential passes through a characteristic cycle, falling rapidly to zero and returning to its original value, the whole cycle lasting perhaps only a few thousandths of a second (msec.). If the difference of potential between P and D is plotted against time, we will obtain a curve similar to that in Fig. 524 (a), the potential of the electrode D, in respect to P, being first 30 mV positive, falling to zero, and then rising to its initial value. It is more usual, however, in making records, to balance out the resting potential

* Shanes has drawn attention to a 40 per cent. increase in volume that occurs on soaking a desheathed frog's nerve in Ringer's solution; it would seem to be a gain in extracellular fluid; in the toad nerve no such swelling occurred.

† According to Shanthaveerappa & Bourne, capillaries pass through the perineurium to supply the individual fibres of the fasciculus; it would thus be interesting to determine whether there is a "blood-axon barrier"; if the capillaries are similar fundamentally to those of the central nervous system this would be the case (p. 728).

with a potentiometer, so that, before stimulation, the recorded potential between the electrodes is zero. Stimulation results in a change of potential such that D now becomes negative in respect to P, and the action-potential record, obtained by plotting the potential of D against time, is turned upside down as in Fig. 524 (*b*), potentials above the abscissa being called *negative*. The detailed nature of the action potential and its full interpretation will be dealt with later; for the moment we may note that the record is called a *negative spike*, and that the transmission of the effect of stimulation seems to be associated with a transient fall in the resting potential to zero.* This fall could happen

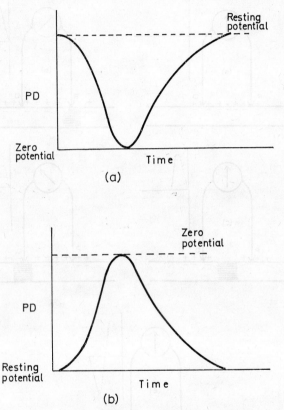

Fig. 524. Monophasic action potential, classically regarded as a falling to zero of the resting potential.

if the effect of the propagated disturbance were to abolish the potential difference between inside and outside of the cell at the point D, *i.e.*, if the electrical condition of the point P were unaffected. That this is the truth of the matter is shown by the fact that the time at which the spike occurs depends on the distance of the point of stimulation from D, and not on its distance from P. Thus the membrane is said to be depolarized by the passage of the propagated disturbance.

An essentially similar type of action potential is obtained by stimulation of the plant cell, *Nitella*, but in this case the process is very much slower, lasting some 15 seconds.

* As we shall see, the action potential consists of something more than a *fall* in the resting potential; the sign is actually reversed.

Diphasic Action Potential. If the conducted effect of the stimulus—the propagated disturbance—consists essentially of the abolition of the resting potential across the wall of the cell, then, since this disturbance travels at a measurable rate, we may expect to find characteristic changes in the difference of potential between two electrodes on intact nerve as the disturbance passes first under one electrode and then under the other. We may expect the changes illustrated in Fig. 525. At (a) the disturbance has not reached either electrode: both of these are on intact nerve so that their potential difference is zero. At

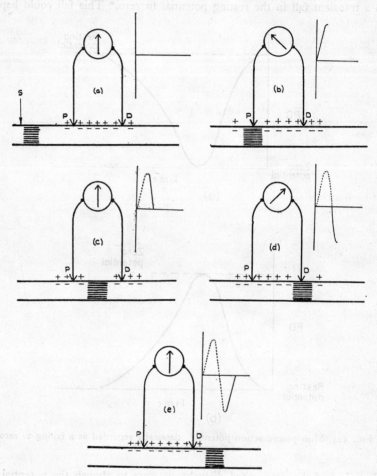

FIG. 525. Schematic illustration of the diphasic action potential.

(b) the disturbance has reached the proximal electrode, P, to abolish its resting potential; consequently P becomes negative in respect to the distal electrode, D, and positive current flows through the recording instrument from D to P. At (c) the disturbance is between the electrodes, and the resting potential at P has been re-established; there is thus no difference of potential between the electrodes —the *isoelectric phase*. At (d) the disturbance has reached the distal electrode, abolishing its resting potential. D thus becomes negative in respect to P and current now flows in the external circuit, *i.e.*, through the recording system, in the reverse direction. Finally, at (e) the disturbance has passed on and the resting potential at D has been re-established; there is now no difference of potential between the two electrodes. If, now, we plot the potential of the proximal electrode, P, in respect to the other electrode, against time we obtain a *diphasic record*, as shown in Fig. 525, where negativity of P is marked by points above the horizontal axis. The shape of the diphasic record will clearly depend on the distance between the electrodes, and the velocity of transmission, among other factors; as the electrodes are brought closer together the isoelectric phase tends to disappear; when the electrodes are very close, the development of negativity at P is followed so closely by the development of negativity at D that the effect is to produce what is described as the "first differential" of the monophasic action potential, *i.e.*, the record exhibits the variation in the *rate of change* of potential under the first active electrode. By integrating this record it is possible to deduce the form of the monophasic action potential.

Electrical Properties of the Membrane

The resting potential depends on the unequal distribution of ions across the fibre membrane; the propagated disturbance, the action potential, is a decrease in this potential—a depolarization. Since the maintenance of the unequal distribution of ions depends on the permeability characteristics of the membrane, the simplest explanation of the depolarization is that it results from a localized change in permeability which, if it were maintained for long enough, would permit the high internal concentration of K^+ to fall and the low concentration of Na^+ to rise. Such a change in permeability should be reflected in a change in such electrical properties of the membrane as its resistance and capacity; let us therefore consider the significance of these characteristics.

Impedance. The bioelectric potentials have been described in terms of the mobilities of certain ions through the plasma membrane of the cell. The selectivity displayed by the plasma membrane should be reflected in other characteristics besides the potential across it. Thus on passing a current through the membrane, *e.g.*, by placing electrodes inside and outside a cell, the flow of current should encounter an ohmic resistance, compounded of that due to the resistance of the solution within the cell, the resistance of the membrane, and the resistance of the extracellular fluid. (By *ohmic* resistance is meant a resistance such that the current is always a definite fraction of the applied voltage.) The membrane, representing as it does a layer, or several layers, of lipoid material, is made of poorly conducting material; in this event the concept of electrical resistance must be extended to include the effects of *specific inductive capacity* or *dielectric constant*. When a potential difference is established across a non-conductor, an electric charge is stored in the material and the electric, or static, capacity is given by the ratio of the quantity of electricity stored over the applied voltage. Thus, if the material is completely non-conducting, a certain amount of current flows during the charging process; when this is complete, the flow ceases. In an alternating current, on the other hand, the applied potential changes its sign repeatedly; during each cycle current flows in a charging and discharging process, so that it is permissible to speak of a flow of alternating current *through* the condenser although it is non-conducting. The resistance to flow of the alternating current is therefore made up of a simple ohmic resistance plus a "capacitative" resistance, the two together producing what is called the "*impedance*" of the system. The cell membrane allows a constant current to flow through it if a voltage is applied across it; during the establishment of the steady flow, however, the fluctuations of current intensity reveal the presence of a capacitative element in the circuit, *i.e.*, the membrane behaves as a condenser. The course of the flow of current, after the sudden application of a potential difference, is described as a *direct current transient*, and from its characteristics important information regarding the capacity and resistance of the system can be deduced. On applying an alternating potential difference to the cell membrane, the impedance may be measured; this varies with the frequency of alternation and thereby reveals the presence of a capacitative element. The pioneering studies of Fricke, and of Cole & Curtis, on the impedance of suspensions of cells and of nerve fibres indicate that the cell membrane may be represented, to a first approximation, as a condenser with a resistance in parallel as in Fig. 526; such a system might be called a "leaky" condenser since current can flow through the shunt resistance. On applying a potential difference across the cell membrane we may expect, on this basis, that current will flow rapidly at first until the condenser charges; when the maximum charge is attained, the flow will be entirely through

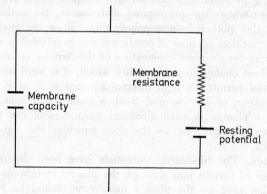

FIG. 526. The membrane regarded as a leaky condenser.

the shunt resistance. The process of charging the condenser consists, in reality, of the accumulation of ions of opposite charges at the surfaces of the membrane, and may be described as a *polarization of the membrane*, the potential on the two sides of the membrane building up to oppose the applied potential.

Length and Time Constants*

The determination of the membrane capacity and resistance usually involves the measure of the spatial and temporal growth or decay of an applied electrotonus, whether this be applied at the surface of the excitable tissue, as in Hodgkin & Rushton's (1946) analysis, or by a microelectrode across the cell membrane as in many more recent studies. The spatial decay or growth is characterized by λ, the *length-constant* or *characteristic length* of Rushton,

in the equation:—

$$\sqrt{\frac{r_m}{r_1 + r_2}},$$

$$P = P_0 \; exp - \frac{x}{\lambda}$$

P_0 being the peak potential at the electrode, and P that at a point distant x from

* Cole (1968) in a fascinating, though formidably technical, account of the development of modern ideas on excitation and propagation has discussed in some detail the methods of determining membrane parameters.

the electrode, r_m the resistance times unit length of membrane (ohm.cm) and r_1 and r_2 the resistances per unit length of outside and inside medium respectively (ohm.cm^{-1}). The temporal decay or growth of an applied electrotonus is defined by a *time constant*, τ_m, which is related to the membrane capacity, c_m, and resistance, r_m, by: $c_m = \tau_m \times r_m$. According to Hodgkin & Rushton's analysis, τ_m is equal to the time to reach *erf* 1 or 85 per cent. of the steady state value, *erf* 1 being given by,

$$erf\; 1 = \frac{2}{\pi} \int_0^1 e^{-w^2} dw = 0.8425.$$

By using transmembrane current through a microelectrode, the input resistance across the membrane may be measured directly, and from the cell dimensions the membrane resistance, r_m, may be calculated. With r_m measured from the growth or decay of the applied electrotonus, the membrane capacity is computed from the relation: $c_m = \tau_m \times r_m$.

Some values of resistance and capacity are shown in Table LXXIII. The results indicate in general that the capacity of the membrane is remarkably constant at about 1 microfarad per sq. cm., whether we consider different nerve fibres, the large plant cell, *Nitella*, or the erythrocyte. The transverse resistance, on the

TABLE LXXIII

Electrical Constants of Certain Cells (after Katz, 1948, and Cole, 1942).

	RESISTANCE			CAPACITY
	Internal ohm.cm.	Outside Medium ohm.cm.	Membrane ohm.cm.^{2}	Membrane μF/cm.^{2}
Axon of *Homarus vulgaris* (75μ)	60	20	600–7,000	1·3
Axon of *Carcinus Mænas* (30μ)	90	20	2,000–16,000 Mean: 8,000	1·1
Giant axon of Squid. *Loligo.* (500μ)	29	20	400–1,100 Mean: 700	1·1
Nitella	58		2·5.10^5	0·94
Erythrocyte . . .	140	—	—	0·8
Frog muscle. (Bundles of add. magnus 75μ) . .	200	87	1,500	6
Frog muscle. (Extensor dig. IV, 45μ) . . .	260	87	4,000	4·5
Frog egg			170	2·0
Arbacia egg. resting . .	180		>100	1·1
fertilized .	210		>100	2·8

other hand, is highly variable, being some 250,000 ohm.cm.^{2} in *Nitella*, 8,000 ohm.cm.^{2} in crab single fibres, and only 700 ohm.cm.^{2} in the squid giant axon; moreover, small differences in technique seem to bring about large changes in the resistance; thus Cole & Curtis' study on impaled axons of the squid gave a value as low as 23 ohm.cm.^{2}. This variability suggests that it is in its transverse resistance, as opposed to its capacity, that physiological changes are reflected. Thus the resistance is essentially a measure of the power of electrolytes to penetrate the membrane whilst the capacity represents its ability to accumulate charge. The appearance of quite a small area of membrane (in comparison with the total) permitting free diffusion would very considerably modify the electrical resistance whilst the change in capacity might be barely measurable. Hodgkin (1947) has calculated that a layer of sea-water as thick as the plasma membrane (say 100A) would have a resistance of only 2.10^{-5} ohm.cm.^{2}, *i.e.*, less than one ten-millionth of that of the squid axon; if only 1 per cent. of the area became freely conducting, the resistance of the membrane as a whole would fall to less than a thousandth of an ohm.cm.^{2}.

Muscle

Katz has applied the method of Hodgkin & Rushton to determine the electrical constants of muscle fibres; the theoretical treatment and the experimental findings were consistent, and suggested a fundamental similarity between nerve and muscle, viewed as sources of potential and conducting units. Quantitatively there were differences; thus the membrane capacity was 4·5–6μF/cm.^{2}, considerably higher than the values found for all other membranes studied; the membrane resistance was 1,500–4,000 ohm.cm.^{2} and the internal and external resistivities were 200–267 and 87 ohm.cm. respectively.*

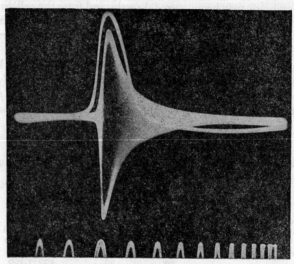

FIG. 527. Superimposed records of the action potential and change in transverse impedance of the giant axon of the squid. (Cole & Curtis. *J. gen. Physiol.*)

More recent studies, e.g. those of Adrian & Freygang and Falk & Fatt, have shown that the simple circuit represented by Fig. 526 is inadequate to describe the electrical parameters of muscle owing to the complexities introduced by the sarcotubular system; this work will be discussed later (p. 1415).

Change in Impedance with Activity. Cole & Curtis have measured simultaneously the action potential and the impedance of the membrane of

---

* The different terms used to describe the electrical resistance of biological structures and fluids, and the units in which they are expressed, are likely to cause confusion. The measured resistance varies inversely as the area, and directly as the thickness of the material being examined, *i.e.*, Resistance = Rl/A. When A is 1 cm.^{2}, and l is 1 cm., and the resistance is measured in ohms, the constant, R, becomes the *specific resistance* or *resistivity*; the units are: ohms × cm.^{2}/cm. = ohm.cm. Frequently it is useful to employ the resistance per unit length of the structure, *e.g.*, the axoplasm. If this is regarded as a cylinder, the specific resistance, R, will be given by: R = $\pi\rho^2 r$, where ρ is the radius and r is the resistance per unit length of axoplasm, with units of ohm.cm.$^{-1}$ Where a membrane of unknown thickness is concerned, the quantity, l, cannot be measured, hence the transverse membrane resistance must be expressed in units independent of this quantity; the *specific transverse resistance* of the membrane, R$_m$, is equal to the specific resistance of the membrane multiplied by its thickness in cm., and its units are ohm.cm. × cm. = ohm.cm.^{2}. Finally, the transverse resistance of unit length of the membrane, r_m, is often used; since the resistance decreases with increasing length of membrane, the quantity required is the resistance times the length, and the unit is ohm.cm. With a cylindrical membrane, r_m is related to the specific transverse resistance, R$_m$, by: $2\pi\rho r_m$ = R$_m$.

Nitella, and of the squid giant axon. The action potential, we will remember, may be considered to be a falling off of the normal resting potential. The latter, we have seen, depends on the permeability characteristics of the membrane, and thus on the electrical impedance. A decrease in the resting potential could be brought about by a permeability of the membrane to previously non-penetrating organic anions, or by an increased permeability to Na^+; both of these effects should be reflected in a decrease in membrane resistance. The record of Fig. 527 shows unmistakably that the transverse impedance decreases as the wave of action potential passes between the impedance-recording electrodes; as we shall see, the time at which the impedance change occurs corresponds to the point on the action potential curve where it is thought that the critical change of membrane potential takes place. A single measurement of a change of impedance is not sufficient to permit of an analysis into a change of capacity and change of resistance; by repeating the experiment at different frequencies of alternating current, however, the analysis could be made, and it was shown that the main change consisted in a fall of resistance to about 28 ohm.cm.^{2}, *i.e.*, a fall of nearly 97 per cent. The change in membrane capacity was very small, only 2 per cent. In *Nitella* essentially similar results were obtained, the capacity decreasing by only about 15 per cent., whilst the resistance decreased from about 10^5 ohm.cm.^{2} to about 500.

Effect of Ions on Membrane Conductance. If the nerve membrane is more permeable to K^+ than to Na^+, as the theory of the resting potential demands, we must expect current to be carried through the membrane, as the result of an applied potential, mainly by K^+ and Cl^-. Increasing the concentration of K^+ in the external medium should therefore, by increasing the concentration of available carriers of electricity, increase the conductance of the membrane. Qualitative effects of this kind have been observed in *Nitella* by Blinks and Osterhout. Hodgkin (1947) has made a detailed study of the effects of ions on the membrane conductance of isolated axons of *Carcinus mænas*; Rb^+, K^+, and Cs^+ increase the membrane conductance; the greater effect of Rb^+ being probably due to the fact that the nerve membrane is more permeable to this ion than to K^+. Na^+ and Li^+ produced only very small increases in membrane conductance when added to the Ringer solution. A study of the membrane capacity revealed very small, if any, effects of K^+.

Effect of Stimulation on Electrolyte Content

The passage of an action current along a nerve is associated with a transient collapse of the fibre's polarization; if this is due to a transient change in the ionic permeability relationships, whereby Na^+ may diffuse into the fibre and K^+ leak out, we must expect to be able to measure a leakage of K^+ from a nerve repeatedly stimulated. Cowan found that stimulation of crab nerve for five minutes at 40 to 140 shocks per second caused a measurable escape of K^+. A. C. Young observed a similar leakage in the leg nerve of *Limulus*, and Arnett & Wilde a definite but limited loss of K^+ from medullated nerve. More recently Keynes has shown, with the aid of radioactive isotopes, that Na^+ penetrates the fibre during repetitive stimulation. We have referred to the increased membrane conductance resulting from raising the concentration of K^+ in the external medium of an isolated axon; Hodgkin & Huxley (1947) have used this change as an index to the loss of K^+ resulting from stimulation. When the isolated axon is surrounded by a layer of sea-water only a few microns thick, any escape of K^+ as a result of stimulation must cause a significant increase in membrane conductance. An increase of four-fold in this quantity was actually observed, following one minute of stimulation; the conductance returned smoothly to its normal value in about five minutes, a recovery presumably due to reabsorption of the lost K^+ by the axon; if the axon was placed in a large volume of sea-water immediately after stimulation, the recovery of normal conductance was almost instantaneous, due this time to the washing away of the K^+ from the surface of the fibre. The loss-per-impulse was computed to be about 1.10^{-12} moles per cm.^{2} of membrane; this actually represents only about $1/100,000$ of the total quantity of K^+ in the fibre. If the fibre is treated as a condenser with a capacity of $1·35$ microfarads per cm.^{2}, across which is the resting potential of 61 mV, it is possible to calculate the loss of K^+ necessary to discharge the condenser; it was computed by Hodgkin & Huxley that the actual loss measured was twice that necessary. With the aid of ^{24}Na and ^{42}K, Rothenberg confirmed that K^+ was lost during stimulation ($2·6$–$6·5.10^{-12}$ meq./cm.^{2}); moreover, Na^+ penetrated at the same time. This was confirmed by Keynes & Lewis who found a loss of K^+ amounting to 3–4.10^{-12} meq./cm.^{2} per impulse, and a net gain* of Na^+ amounting to $3·5$–$3·8.10^{-12}$ meq./cm.^{2}.

Resting and Action Potentials. Quantitative Considerations

On the basis of a simple view of the action potential as a depolarization of the membrane, we may expect it to be at most equal to the resting potential, *i.e.*, the proximal recording lead, used for obtaining a diphasic response, should develop a negativity at most equal to the negativity of the lead on the cut end of a nerve, or muscle fibre. For many years this was thought to be true, although the correct magnitudes of the resting and action potentials were a matter of some doubt owing to the short-circuiting that must have taken place with the usual recording devices. However, the more recent measurements on single fibres of the nerve (Hodgkin & Huxley, 1939, Curtis & Cole, 1942) and muscle (Graham & Gerard, and Nastuk & Hodgkin) have provided accurate quantitative data not only on the resting potential but also on the action potential. As we have seen, the resting potential of the squid giant axon, measured directly across the membrane by means of an inserted micro-electrode, is some 61 mV, and that of isolated fibres of the frog sartorius about 88 mV; the action potential, measured under the same conditions by Curtis & Cole, turned out to be considerably greater, varying between 77 and 168 mV with an average value of 108 mV (Fig. 528). Hodgkin & Huxley obtained essentially similar results on the axon of the British squid, *Loligo forbesi*, although the absolute magnitudes of their potentials were consistently smaller. Fig. 529 illustrates a typical finding; the scale on the record indicates the potential of the electrode inside the fibre in relation to a similar electrode in sea-water, the potential of the latter being put equal to zero. The resting potential was thus 44 mV. On the classical view the spike should rise to the zero level on the scale, at which point the potential of the internal electrode becomes equal to that of the

---

* Passage of an impulse causes a transient increase in influx of Ca^{++} in the squid axon, efflux being unaffected. It seems likely that the influx accompanies the high permeability to Na^+ rather than that to K^+ since a sustained depolarization—due to raised external K^+—has a much smaller effect. With a sustained depolarization, the Na^+-permeability mechanism is largely inactivated, but not that for K^+ (Hodgkin & Keynes, 1957). Keynes & Ritchie (1965) have measured the resting and extra influxes and effluxes of K^+ and Na^+ in non-myelinated nerve fibres, and the results are compatible with the development of an action potential considerably higher than that actually taking place. The amount of K^+ lost per impulse is $1·0$ pmole/cm^2, less than for any other nerve so far examined.

external one in sea-water. In fact, the potential reverses in sign before the spike reaches its maximum height, the magnitude of this *positivity* being variable and sometimes as high as 45 mV. It will be noted also that the action potential of the squid axon is diphasic in the sense that, after falling to zero, the potential

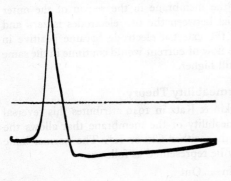

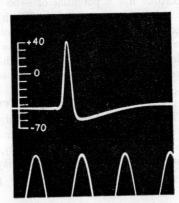

FIG. 528. Membrane action potential of the squid axon. The two horizontal traces are 50 mV apart; the resting potential was 58 mV. Thus the upper horizontal line approximately represents zero potential difference across the membrane, the lower line the resting potential (outside positive), and the action potential, starting from the resting potential, swings to 110 mV (outside negative). Time intervals at the bottom are 0·2 msec. (Curtis & Cole. *J. cell. comp. Physiol.*)

FIG. 529. Similar to Fig. 528; action potential recorded between inside and outside of squid giant axon. The vertical scale indicates the potential of the internal electrode in mV, the sea-water outside being taken as zero potential. Time marker, 500 cyc./sec. (Hodgkin & Huxley. *J. Physiol.*)

difference between inside and outside reverses its sign; this *positive after-potential* amounts to some 15 mV and is typical of all measurements on the squid axon.

In their studies of impaled single muscle fibres Graham & Gerard were unable to record accurately individual action-potentials; nevertheless they were

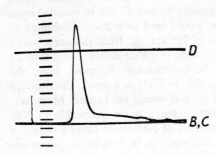

FIG. 530. Resting potential and action potential of muscle fibre. The ordinate scale indicates steps of 10 mV. Records B and C were obtained with the micro-electrode inside the fibre in the resting and stimulated conditions; D, with it outside at the end of the experiment, *i.e.*, D represents zero potential between the electrodes. (Nastuk & Hodgkin. *J. cell. comp. Physiol.*)

able to compute the probable magnitude of a given spike, and showed that it was some 30 mV greater than the resting potential. This was confirmed by Nastuk & Hodgkin who found, at 10° C, an average resting potential of 88 mV and an action potential of 119 mV (Fig. 530).

Reversal of Membrane Potential

An action potential greater than the resting potential must mean that the membrane is not only depolarized but is repolarized with a potential orientated in the opposite direction; thus if the resting potential is 60 mV, the internal electrode has a potential of −60 mV compared with the external electrode on the intact outer surface. When the membrane in the region of the outer electrode is depolarized, the potential between the two electrodes is zero and the spike height is 60 mV; if now the external electrode became positive in relation to the internal electrode, the flow of current would continue in the same direction and the spike would rise still higher.

Sodium Permeability Theory

The theory put forward by Hodgkin & Katz in 1949 attributes this reversal of polarity to a change in the permeability of the membrane that allows the Na^+-ion to penetrate more rapidly than either Cl^- or K^+. The normal ionic distributions, it will be recalled, may be represented as follows:—

$$
\begin{array}{c c}
\text{In} & \text{Out} \\
- \mid & + \\
K^+ - \mid & + Na^+ \\
- \mid & + \\
- \mid & + \\
A^- - \mid & + Cl^- \\
- \mid & +
\end{array}
$$

and the resting potential is due to the high internal concentration of K^+, associated with impermeability to A^- and an "effective impermeability" to Na^+; the potential difference arises through the tendency for K^+-ions to escape from the inside, and the effect is to hasten the inward passage of K^+ and to retard its outward passage, so that the fluxes remain equal in spite of the large difference in concentrations. Similarly the Cl^--ions are accelerated outwards and retarded inwards. If, suddenly, the membrane became equally permeable to all ions the potential would fall to zero, since there would be no restraint on the movement of ions; the system would run down without any appreciable flow of current, the outward migration of K^+ and A^- being associated, on the average, with inward migration of Na^+ and Cl^-. The action potential would therefore be a brief fall to zero of the resting potential; the re-establishment of the normal permeability relationships would lead to a re-establishment of the potential. The result would be a loss of K^+ and A^- from the nerve and a gain of Na^+ and Cl^-. This "blanket" change in permeability would thus be unable to account for a reversal of the polarization. Suppose, now, the membrane suddenly became highly permeable to Na^+; as a result, the restraint on the loss of K^+ would be removed and the p.d. would fall to zero. Moreover, if the membrane in this active state became much more permeable to Na^+ than to K^+ and Cl^-, this increased inward flux of Na^+ could cause a diffusion potential directed so that the inside became positive, in fact, if the permeabilities to K^+ and Cl^- could be neglected, the magnitude of this would be given by the Nernst equation:—

$$ E = \frac{RT}{F} \ln \frac{[Na]_{out}}{[Na]_{in}}. $$

Effect of External Sodium

Qualitatively there is no doubt that this assumption could account for the reversal of polarity that occurs, the inward flow of Na^+, in excess of the outward flow of K⁺ and inward flow of Cl⁻, causing an inward ionic current. If the theory is correct, the height of the action potential should depend critically on the external concentration of Na⁺. Moreover, since propagation of an impulse depends on the action potential's acquiring a sufficient height (p. 1062), we should expect to find a decrease in excitability of the nerve following a reduction of the concentration of Na⁺ in the external medium, this reduction being independent, to some extent at least, of the resting potential. Overton in 1902 showed that a muscle became inexcitable in solutions containing less than 10 per cent. of the normal plasma concentration of Na⁺, and Kato and Erlanger & Blair have more recently demonstrated similar losses of excitability with nerve. Hodgkin & Katz have shown, in support of their theory, that reducing the concentration of Na⁺ in sea-water, by addition of increasing proportions of isotonic dextrose solution, progressively decreases the height of the action potential (recorded by their micro-electrode technique with one electrode inside the squid axon), whilst the size of the resting potential (which depends

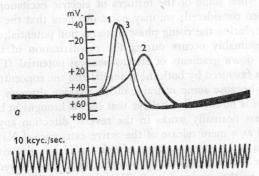

FIG. 531. Action of sodium-deficient solution on the resting and action potentials. 1, response in sea-water; 2, after 16 minutes in 33 per cent. sea-water, 67 per cent. isotonic dextrose; 3, 13 minutes after re-application of sea-water. The scale gives the potential difference across the nerve membrane (outside—inside) with no allowance for the junction potential between the axoplasm and the sea-water in the micro-electrode. (Hodgkin & Katz. *J. Physiol.*)

essentially on the inside concentration of K⁺ and should not be greatly changed by these substitutions) remained effectively constant. Some typical effects are shown in Fig. 531. On a simple theory, the magnitude of the reversed potential difference, *i.e.*, of the positive component of the action potential, or "overshoot" as it is called, should depend on the logarithmic ratio: log $\frac{[\text{Na}]_{\text{test}}}{[\text{Na}]_{\text{sea-water}}}$. On plotting the change in action potential against this ratio, a straight line should be obtained; for both nerve and muscle the agreement between theory and experiment was very satisfactory. The effect of *increasing* the concentration of Na⁺ should be the reverse; however, such an increase must cause such rapid and profound changes in the osmotic relationships that it is questionable whether experiments using hypertonic solutions have much significance; it is of interest, nevertheless, that quite definite, though transient, increases in the height of the action potential were observed with Na⁺-rich solutions, increases which were of the order of magnitude predicted on the basis of simple membrane theory. In further confirmation of the theory it was found that the *rate of rise* of the action potential depended on the outside concentration of Na⁺; thus in Fig. 531 it is quite evident that the action potential is not only reduced in height but also that its rate of rise is reduced when the external concentration is made equal to 33 per cent. of normal. Essentially similar results were obtained with impaled frog muscle fibres by Nastuk & Hodgkin.*

It may be argued that the effects observed are due not so much to dilution of the Na^+ but to reducing the concentration of Cl^- or of some other ion; that Cl^- was not a significant factor was shown by substituting choline chloride for NaCl in sea-water; a solution made up of equal parts of isotonic choline chloride and sea-water gave a positive phase of the action potential equal in magnitude to that found with a mixture of equal parts of sea-water and isotonic dextrose solution. Finally, the concentration of Na^+ in muscle fibres may be increased, as we have seen, by soaking in Ringer for some time; Desmedt has shown that, under these conditions, the "overshoot" of the action potential varies inversely with the internal concentration, as we should expect.

Catalysed Permeability

Further experimental work on the nature of the action potential will be described later when some of the features of electric excitation of nerve and muscle have been considered; we may point out here that the rapid influx of Na^+ that occurs during the rising phase of the action potential, and the escape of K^+ that presumably occurs during the repolarization of the membrane, both take place down gradients of electrochemical potential. Thus, the penetration of Na^+ is favoured by both the potential and the concentration gradient, and takes place because some restraint on its passage through the membrane is released; there is no reason to believe that Na^+ is brought in by active transport—this process normally works in the reverse direction anyway—nor can it be considered as a mere release of the active extrusion of Na^+, because this is a slow process and a mere cessation of this would not give rise to the rapid influx that must occur if the resting potential is to be reversed. The change may be likened to the catalysed or facilitated permeability to Na^+ described by Davson & Reiner in the cat erythrocyte.

Potassium Efflux

The loss of K^+ that occurs also takes place down a gradient of electrochemical potential. Thus, in the inactive state, the K^+-ion is in approximate or complete equilibrium, the electrochemical potential being equal on both sides of the membrane; with a fall in the resting potential, however, the system is not in equilibrium any longer and the electrochemical potential becomes higher inside; consequently, as the resting potential falls and reverses, the force driving K^+ out of the fibre increases.

We may note that the actual loss of K^+ and gain of Na^+ associated with a single action potential will be too small to be detected but, as we have seen, repeated stimulation leads to measurable changes. Ultimately, then, the system should run down, the internal concentration of Na^+ rising and that of K^+ falling, both effects leading to a fall in resting potential and in action potential. In fact, continuous stimulation does, indeed, have these effects, but under normal conditions the restitutive processes, namely the extrusion of Na^+ associated with accumulation of K^+ (p. 630), keep pace with this running down.

Opacity Changes

Finally, we may note that exchanges and transfers of electrolytes are likely to be associated with changes in the volume of an axon owing to the osmotic shifts of water that must take place. D. K. Hill & Keynes observed a reversible

* Grundfest, Shanes & Freygang (1953) showed that the impedance changes associated with the spike were affected by external K^+ and Na^+ concentrations in a manner predictable from the Hodgkin-Katz theory.

decrease in the opacity of the crustacean nerve trunk during stimulation, and subsequent work of D. K. Hill made it very probable that this change was a result of an increase in volume of the axons. The simple exchange of K^+ for Na^+ might, at first thought, be expected to result in no change in water content; however, the smaller hydration of the K^+-ion—3·8 molecules of H_2O per ion compared with 8 molecules for the Na^+-ion—must result in an increase in the water content of the axon when K^+ exchanges for Na^+; excitation, moreover, is accompanied by a definite penetration of Na^+ and Cl^-, in addition to the K^+—Na^+ exchange, consequently a marked increase in volume is to be expected. According to Hill, the changes in volume of the axons, deduced from the changes in opacity, are of the correct order to be accounted for by ionic movements, although more recent and extensive studies of Shaw & Tobias and Bryant & Tobias indicate that the phenomenon is generally more complicated.

Recording the Potentials

The use of an internal electrode to record the resting and action potentials has revolutionized the experimental study of excitability, if only because it has permitted measurements of potential that are not vitiated by short-circuiting losses. With very fine fibres the insertion of an internal electrode is not practicable,

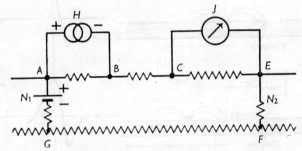

FIG. 332. Diagram to show principle of method of measuring resting potential. ABCE, fluid surrounding fibre; FG, axoplasm; N_1, normal node of Ranvier; N_2, depolarized node; H, adjustable source of current; J, amplifier. (Huxley & Stämpfli. *J. Physiol.*)

so that external records, in which short-circuiting is cut to a minimum, have been exploited with great advantage. One such method is that of Huxley & Stämpfli illustrated in circuit form by Fig. 532; the node of Ranvier of an isolated mye-linated fibre, N_1, is in Ringer's solution whilst the next node, N_2, is in KCl which completely depolarizes it; in the absence of current flow the difference of potential between A and E would be the resting potential. In fact, an injury current flows in the circuit ABCEFGA and is detected by the potential drop measured across CE where the resistance in the external fluid is made high. Current is passed between A and B from a source H, making A positive to B and thus opposing the injury current; H is adjusted so that no current is detected at C and E. Under these conditions the potential between A and G—the resting potential— is equal to that between A and B which can easily be measured. The resting potential found was 71 mV and the action potential 116 mV.

Other devices have consisted in increasing artificially the longitudinal resistance of the sheath between two nodes either by causing a stream of non-electrolyte solution to flow over the internodal region (the sucrose-gap technique of Stämpfli)* or by placing the internode in an air-gap whilst the two adjacent nodes are bathed in Ringer's solution; the drying of the internode increases the resistance of this region, and since the short-circuiting factor is given by $\dfrac{r_1}{r_1 + r_2}$, where r_1 is the

---

* König has applied the sucrose-gap technique to frog muscle, using the extensor longus digiti which consists of some 20–60 parallel fibres. The recorded potential changes were some 60–80 per cent. of the actual.

external longitudinal resistance and r_2 the internal longitudinal resistance, the factor becomes close to unity (see, for example, Tasaki & Frank, 1955).

Volume Conductor. Where the actual size of the recorded action potential is immaterial, an external recording may be made with a fine metallic electrode. When used in conjunction with an intracellular electrode the external recording may add considerably to our information as to the sum total of electrical changes taking place in the neighbourhood of the excitable cell in which we are interested. The interpretation of the extracellular records is by no means easy except in the rather simple situation we have so far discussed when the conducting fibres are all well aligned and when, experimentally, the conducting tissue may be surrounded

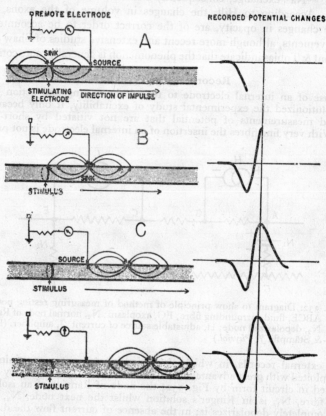

FIG. 533. Illustrating the recording of electrical changes taking place in an excitable cell and propagated in a conducting medium. The recording electrode is on the surface of the cell, whilst the second is on inactive tissue having the properties of an electrolytic conductor. The excursions of the recording instrument are shown on the right. (Brazier, *The Electrical Activity of the Nervous System*. London: Pitman Medical.)

by paraffin oil in order to increase the external resistance and so reduce short-circuiting to a minimum. When the recording electrode is not close to the origin of the electrical changes it is supposed to record, then the resulting changes of potential between this and an indifferent electrode, *i.e.*, one remote from the electrical changes, will be a very attenuated and highly distorted version of what actually happened. Under these conditions we are said to be recording from a *volume conductor*, a situation discussed in the now classical paper by Bishop. The state of affairs is illustrated by Fig. 533 from Brazier, where we imagine that a propagated impulse is passing from left to right and approaching the active or exploring electrode; as the disturbance approaches, the electrode is said to become a source of positive current which passes into the sink represented by the active region; with respect to the remote electrode, then, the exploring electrode is positive and the potential change is represented by a downward movement; when the activity comes immediately under the exploring electrode the latter now becomes a sink for positive current so that the record shows an upward spike. When the activity has passed on, the exploring electrode becomes once again a source of positive current and so the record passes through another positive phase before finally the two electrodes adopt the same potential. The record is thus triphasic, and its shape may vary considerably according to its position in relation to the direction of movement of the propagated electrical change that one is trying to record.

ELECTRICAL EXCITATION

The primary facts of electrical stimulation of nerve and muscle are, in effect, simple; nevertheless the development of mathematical theories to describe the phenomena has prompted such exhaustive investigations, in which practically every possible variable has been altered, that the newcomer into the field is presented with a bewildering mass of facts and a highly specialized terminology. In the present section it must be our aim to throw the main facts into prominence and to indicate the mode in which the modern investigator chooses to interpret them, bearing in mind, always, that the final aim must be to relate the electrical phenomena with the distribution of electrolytes across the plasma membrane of the excitable tissue, the restraints placed on ionic movement across this membrane, and the relationship of metabolic processes to these factors.

The Stimulus

Let us consider two electrodes, say 1 cm. apart, on the intact surface of a nerve, connected to a source of e.m.f. by way of a morse-key. On closure of the circuit the nerve is stimulated, and a wave of action potential passes in both directions along the nerve. The wave starts at the negative electrode and, if the applied current is strong enough, may be extinguished when it reaches the positive electrode—there is an *anodic depression* of excitability; consequently, the more effective stimulus to a nerve-muscle preparation is obtained when the negative electrode, or *cathode*, is the nearer to the muscle, *i.e.*, when the stimulating current is *descending*, since with the electrodes in the *ascending* position the action potential may be blocked at the anode in many, if not all, of the nerve fibres. Only one action potential is usually obtained on making the circuit, so that we may think of the nerve as having *accommodated* itself to the new conditions imposed by the passage of a current through it. If the nerve consists of a bundle of fibres, their thresholds will be different; consequently, on raising the applied voltage, successively larger responses may be obtained—either measured as the height of the action potential or as the muscular contraction in a nerve-muscle preparation—until all the fibres are stimulated. Raising the voltage above this point has no further effect on the magnitude of the response, a phenomenon that has given rise to the expression "*All-or-None Law*", by which is meant that a given nerve or muscle fibre responds with a fixed size of action potential (or muscular contraction) which depends on its physiological condition but not on the strength of the applied stimulus. The action potential recorded from a bundle of nerve fibres is the summated effect of changes occurring in many individual fibres; as we have seen, short-circuiting results in losses of potential at the recording leads, so that the measured action potential is only a fraction of the true electrical change; with a large number of fibres activated, the recorded action potential is greater than with a small number, just because of these short-circuiting losses. With a single fibre, on the other hand, if the all-or-none law holds, the action potential should be of fixed height independently of the strength of stimulus; this is indeed found to be the case, and the same is true of single muscle fibres and of the wave of action potential in *Nitella*.

Electrotonus

On breaking the circuit, provided that the current has been strong enough, a new response is elicited from the nerve; this time it begins at the anode. These simple phenomena suggest that it is the *establishment* of a definite flow of current, rather than the flow itself, that is the effective stimulus; in fact we have seen that the propagated action potential may be suppressed by the anode when the current is flowing. If the stimulus consists in the establishment of the action potential, which then propagates along the nerve, we may treat the action potential at any given point along the nerve as the effective stimulus to an immediately adjoining point, which will make *it* active and pass on its activity to a more remote point, and so on. The suppression of the action potential by the anodal current may therefore be regarded as an expression of some tendency working against stimulation, *i.e.*, the *excitability* may be said

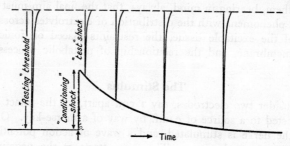

FIG. 534. Illustrating *latent addition*. As a result of a conditioning shock, the threshold to the test shock is reduced. (Katz. *Electric Excitation in Nerve*. O.U.P.)

to be lowered in the region of the anode. The effects of the passage of a constant current through a nerve on its excitability were described long ago by Pflüger, and gave rise to the concept of *electrotonus*; in general, it is found that the excitability of nerve is greatest in the region of the cathode and least around the anode during the flow of a constant current. This phenomenon we may call "*Pflüger's electrotonus*", since the term electrotonus is now used in a rather more special sense to describe the changes in potential in the region of the stimulating electrode during the first few milliseconds of the application of an external electromotive force. The Pflüger electrotonus is a rather more complex phenomenon than this simple account suggests, since it changes its sign on increasing the intensity of the constant current sufficiently, *i.e.*, the nerve in the region of the anode may become the more excitable; according to Rosenblueth its sign and magnitude also depend on the distance between the electrodes.*

* The changes occurring with constant currents of long duration—the Pflüger electrotonus—are closely related to the phenomenon of "cathodic depression", the depression of excitability at the cathode following lengthy application of a potential difference. Thus, to say that the polarity of the electrotonus is reversed is rather misleading, since what probably happens is that the nerve under the cathode becomes relatively inexcitable, in which case excitation is easier at the anode. The cathodic depression is due, presumably, to a decrease in membrane resistance resulting from the continued outflow of K^+, which produces a condition essentially similar to that pertaining in the refractory state.

Latent Addition. If the applied stimulus is too weak to excite, *i.e.*, to set up a propagated disturbance, this does not mean that it has no influence on the excitability of the tissue; a second stimulus, of the same subliminal strength, falling within a few msec., may excite—the phenomenon of *subliminal summation* or *latent addition*. Thus, on plotting the strength of stimulus necessary to excite at different times after the first, *conditioning*, stimulus, against time after the conditioning stimulus, we obtain the change in excitability with time as in Fig. 534. Alternatively we may estimate the excitability, at a fixed time following a conditioning stimulus, at different points in the regions of both electrodes; it is found that excitability is greatest in the region of the cathode and least around the anode; between the electrodes is a neutral point at which the increased *cathodic excitability* is just balanced by the *anodic depression*.

Strength-Duration Curve. To elicit a response in a nerve, the stimulus must be applied for a finite time; the weaker the current the longer the time

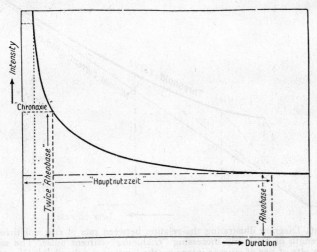

Fig. 535. Diagram of the strength-duration curve, illustrating the relation between the threshold strength and the duration of a rectangular current pulse. (Katz. *Electric Excitation in Nerve*. O.U.P.)

required for stimulation. By plotting the current-strength against the duration required to excite, we get the well-known *strength-duration curve* shown in Fig. 535. We may note that if the strength is below a certain minimum quantity —the *rheobase*—the current may flow indefinitely without exciting. The time necessary for stimulation on applying a current of twice the rheobase is called by Lapique the *chronaxie*, and is said to be a measure of the excitability of a given tissue. With very brief pulses of current it was found by Gildemeister that the total *quantity* of current required to excite was constant, *i.e.*, $i \times t = k$. This *constant-quantity* relationship clearly does not apply to comparatively long times of current-flow, since otherwise the strength-duration curve would be a rectangular hyperbola.

Minimal Current Gradient. One further characteristic of the electrical stimulus, of great theoretical importance, is the phenomenon of a *minimal current gradient*. It is not sufficient to state that a current of a definite magnitude, flowing for a definite time, will excite; if the current is established slowly, *i.e.*, if the applied voltage is built up gradually, it may not excite, whereas the sudden application of the same voltage may excite. For any strength of current capable of exciting, therefore, there will be a minimum rate of growth below which the current will be ineffective (Fig. 536). This phenomenon reveals an adaptive, or *accommodative*, change in nerve, which may be regarded as a progressive increase in the threshold of excitability during the process of stimulation.

Nernst Theory. Much mathematical ingenuity has been displayed in interpreting the strength-duration curve and related phenomena of electrical stimulation; the most significant contributions in this direction have been the *membrane theory* of Nernst and the *core-conductor theory* of Hermann, in that they represent definite attempts to relate the phenomena of electrical excitation with the known physico-chemical characteristics of excitable tissue. Nernst was impressed with the fact that alternating currents of very high frequency failed to stimulate at all. He argued that the effect of passing an electric

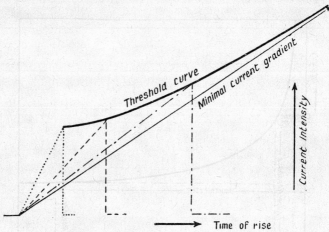

Fig. 536. Diagram illustrating the relation between rate of rise and threshold strength of a linearly increasing "triangular" current pulse. The steeper the rate of rise of current intensity, the lower is the threshold. (Katz, after Fabre. *Electric Excitation in Nerve.* O.U.P.)

current through a tissue would be to cause a local accumulation of ions in the region of the electrodes; this accumulation could reach significant proportions if there were membranes, capable of slowing the passage of ions, interposed in the path of the current through the tissue. Such an accumulation would require an appreciable time to attain any specified magnitude, so that, if a certain critical accumulation of electrical charge at a membrane were necessary for excitation, too rapid an alternation of the current would be useless for stimulating. Nernst developed an equation to describe the accumulation of ions at a membrane, transversely across the axis of a tube of fluid; as a result of the applied electromotive force, ions of one charge would be accelerated across the membrane whilst those of the opposite charge would accumulate on the other side. Because of diffusion, however, this accumulation would tend to be dissipated, and for any current-strength a steady-state would be achieved with a definite concentration of ions in the region of the membrane. If, now, the requirement for a current to excite be that it produce a critical accumulation, Nernst showed that the product: $i \times \sqrt{t}$ should be constant. Over a certain range of current-strengths this prediction accorded remarkably well with the experimental results. The phenomenon of "rheobase", however, namely the existence of a minimal current-strength, below which an indefinitely long period of current-flow would still fail to excite, put a limit to the applicability of the equation, so that Nernst invoked an accommodative process, during the flow of current, that raised the threshold for stimulation. Too slow a development of the current, or too low a current-strength, failed to excite because the accumulation of ions could not keep pace with the progressive increase of threshold. Empirically a good equation, covering the results of stimulation under a variety of conditions, is given by that derived by Lapicque in 1907:—

$$I = I_{Rh}/(1 - {}^{-}e^{t/\tau})$$

where I_{Rh} and τ are constants with dimensions of current and time respectively.

Hill's Treatment. A better fit of the experimental facts of stimulation with the predictions of theory has been achieved by dropping all attempts at a physical explanation of the underlying phenomena. Instead, it is merely postulated (*e.g.*, by Hill; see also similar theories of Rashevsky and Monnier) that the passage of a current through a nerve increases a vaguely-defined "*local potential*" (V_0) above its normal resting value, *i.e.*, V_0 rises at the cathode to a new value at any time, t, equal to V. At the same time the original, resting, threshold, U_0, rises at the cathode at a rate dependent on the rise in local potential. By making certain assumptions about the rate of growth and decay of both the local potential and the change in threshold, a simple equation describing the strength-duration curve was formulated; the equations, in general, embodied two "time constants", one, represented by the symbol k, indicating the rate of decay of local potential on cessation of the current, and the other, λ, representing the rate of return of the threshold. A current is said, on the basis of this theory, to excite when the local potential, V, reaches the threshold, U, which, of course, changes during the excitation process. The accommodative changes are known to take place much more slowly than the changes in local potential, consequently, by choosing the conditions of stimulation suitably, it is possible to isolate this factor and estimate the magnitude of λ. Similarly, by using currents of short duration, during which the accommodative factor barely enters, the excitation constant, k, may be isolated and shown to be simply related to the chronaxie. In general, the theory of Hill, and the essentially similar theories of Rashevsky and Monnier, provide equations which describe many of the details of electrical stimulation of nerve and muscle. The concept of local potential has not been strictly defined; and the fundamental equations contain four arbitrary constants, so that a failure of these equations to fit the experimental results might be more surprising than the actual fit. In so far, however, as the theory has led to predictions that have been subsequently verified experimentally, its value cannot be seriously questioned; moreover, the theory co-ordinates a vast number of the facts of electrical stimulation and allows them to be assessed in terms of the two time-constants, k and λ.

Core-Conductor Theory

A more realistic interpretation of the electrical events leading to excitation at an applied electrode is provided by the core-conductor theory, developed from the original concepts of Hermann and Matteucci by Cremer and later workers. On applying electrodes to the surface of a nerve, potentials develop at, and in the neighbourhood of the electrodes, *electrotonic potentials*, which indicate that there is a flow of current along certain specific lines as in Fig. 537. The potential at any point on the surface of the nerve, in relation to the potential at a point a long distance away from the electrodes, is called the electrotonic potential; in the region of the cathode it is negative or *catelectrotonic*, and in the region of the anode it is positive or *anelectrotonic*. The potential change is greatest immediately underneath the electrodes, and falls off exponentially in the extrapolar region; it requires a measurable time to develop, both at the electrodes and in their immediate neighbourhood. The development of these potentials, and their associated currents, can best be accounted for if the nerve fibre is treated as a *cable* or *core-conductor*, *i.e.*, as a cylinder of conducting fluid, the axoplasm, surrounded by a layer of conducting medium

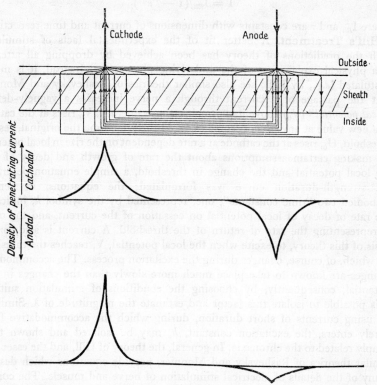

FIG. 537. Diagram of current distribution in a core-conductor. Upper figure: The direction of current flow through outside, sheath, and inside is indicated by arrows. Lower figure: The lateral spread of the traversing current lines is illustrated, by plotting current density against length of nerve. (Katz. *Electric Excitation in Nerve.* O.U.P.)

and separated from this layer by a sheath or polarizable membrane of high electrical resistance (Fig. 538*a*).* This sheath is thought to impose a capacity on the system, so that the nerve fibre may be viewed as a condenser capable of being charged on application of a potential difference across it. To account for the flow of current that takes place across the sheath, on applying a difference of potential under steady-state conditions, the condenser is said to be

---

* As Katz (1939) has pointed out, it would be unwise to attempt an exact correlation between the histological structures of the myelinated nerve fibre and these abstract entities: *core, outer layer,* and *sheath*. In the non-myelinated axon the core is the axoplasm, the outer layer is the enveloping fluid plus connective tissue, whilst the plasma membrane constitutes the highly resistant and polarizable "sheath". The myelin sheath of a medullated nerve must contribute to r_m.

leaky; and it may be treated as a condenser plus a shunting resistance, R, in parallel (Fig. 538b). To account for the time characteristics of the build-up of electrotonic potential, it is necessary to treat the nerve as a series of units, as in Fig. 538c, but the essential principle is the same as in the simpler case shown in Fig. 538b. When a potential is applied across two electrodes, side-by-side on the nerve, the flow of current charges the condensers, with the result that the potential at and near the electrodes builds up and tends to oppose the further flow of current. The form of the curve of development of electrotonus with time may be predicted on the basis of such a model, if the values of the capacity and resistance of the membrane, and the resistances of sheath and core, are known. When the steady-state has been attained, moreover, the

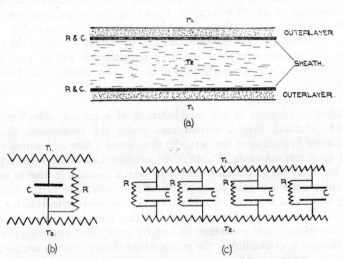

FIG. 538. Illustrating the core-conductor treatment of the axon. R and C are the transverse resistance and capacity of the sheath respectively.

distribution of electrotonic potentials around the electrodes can also be computed. Thus, according to Rushton's analysis, the potential, P, at any distance, x, from the electrode, during the passage of a weak current is given by:—

$$P = P_0\, exp -\left(\frac{x}{\sqrt{r_m/r_1 + r_2}}\right)$$

P_0 being the peak potential at the electrode, r_m the resistance times unit length of nerve membrane, and r_1 and r_2 the resistances per unit length of outside and axoplasm respectively. The quantity $\sqrt{\dfrac{r_m}{r_1 + r_2}}$ is called the *characteristic*

length of nerve; and it has been shown by Rushton that the equations describing the strength-interpolar distance relationship of stimulation, and also the resistance-length relationship, all contain this length-factor which is, essentially, an index to the degree of electrotonic spread produced by a given potential. Fig. 537 indicates the flow of current during the steady-state. If, now, we accept the hypothesis that the essential feature of an effective stimulus is the build-up of a sufficiently high current density directed outwards through the membrane,* or, more precisely, the build-up of a sufficiently high electrotonic potential at a given point, a great many of the phenomena of stimulation may be explained, at least on qualitative grounds. An all-embracing quantitative treatment of electrical stimulation is, however, no more possible on the basis of cable theory than on the basis of Nernst's membrane hypothesis, and until the physico-chemical mechanisms involved in accommodation are elucidated, it is unlikely that progress will be made in this direction; we may merely note that the equation developed by Cremer to describe the strength-duration curve has a similar form to that developed on the basis of the membrane theory. In the hands of Rushton, Cole & Curtis, Rosenberg, and Hodgkin, the cable theory has been developed to explain the detailed characteristics of the build-up of the electrical stimulus and its propagation; in this respect (as opposed to the development of an all-embracing theory of the quantitative aspects of electrical stimulation) the theory has been remarkably successful, as the following pages will testify. Before entering into the details of some of the modern studies of the excitation process and its propagation, it will be profitable to indicate certain qualitative explanations of the simpler phenomena of electrical excitation.

Qualitative Considerations

The effective stimulus is the establishment of a critical reduction of the membrane potential, over a certain area under the stimulating electrode; with excitation beginning at the cathode, this means a flow of positive current from the axoplasm outwards. A subliminal stimulus causes too small a flow of current, but since the electrotonic potentials take a measurable time to subside, a new subliminal shock may summate its effects with those of the first, conditioning shock and so excite. A region of catelectrotonus is therefore a region of increased excitability, and it may be shown that there is a perfect correlation between the electrotonic potentials caused by an applied electromotive force and the changes in excitability at the points where these potentials are measured. At the anode the current-flow is in the opposite direction (anelectrotonus) so that a stimulus, to excite, must drive current against an opposing potential difference; the applied shock must therefore be greater, and the excitability is less. If the stimulating electrodes are very close together, the cathodic and anodic electrotonus will interfere, and the threshold stimulus should be larger than when the separation is greater; Rushton showed that this was the case.

THEORY OF EXCITATION AND PROPAGATION

On the basis of the core-conductor theory, the effective stimulus is the establishment of a region of catelectrotonus under the stimulating electrode; i.e., the surface of the nerve fibre is made negative in respect to other portions. It will be recalled that the normal resting potential is directed so that the inside of the fibre is negative in respect to the outside; electrical stimulation, therefore, consists of the neutralization, or partial neutralization, of the resting potential in the region of the cathode. The fibre is said to be *depolarized* by the stimulus. As a result of the stimulus, provided that it is great enough, this local electric

---

* Direct confirmation of this is provided by "transmembrane" stimulation; Graham & Gerard inserted an electrode directly into the muscle fibre and placed another, above it, on the outside of the fibre. Passage of a current from inside to outside, i.e., when the internal electrode was positive, acted as a stimulus, a quantity of electricity as low as 1.10^{-10} coulombs being effective; on reversing the polarity no excitation could be obtained, even when the current flow was increased a hundred-fold over that required with the internal anode.

effect is propagated as an action potential, which consists of a similar, but more extreme, form of depolarization, involving a definite breakdown of the membrane. An exact analysis of the electrical changes taking place at the anode and cathode, as a result of a brief stimulus, can therefore throw a great deal of light on the nature of the primary excitation process and its propagation.

Local Subthreshold Responses

We will suppose that the shock is applied, and the recording made, as indicated in Fig. 539. A and B are the stimulating electrodes, B being the cathode from which the propagated disturbance will travel; the effects of the shock are recorded at B and C, *i.e.*, the electrode B is used both for stimulation and recording. On applying a brief thyratron shock lasting only, say, 60 μsec. (Hodgkin, 1938), B becomes progressively negative, as a result of the passive build-up of electrotonic potential which may be likened, as we have seen, to the charging of a condenser. If the stimulus is well below threshold, the

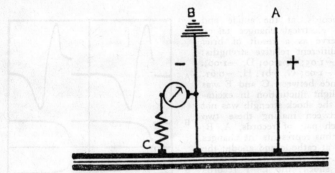

FIG. 539. Stimulating and recording set-up for study of electrotonic and propagated potentials. A and B are stimulating electrodes, whilst recording is from B and C. (After Hodgkin. *Proc. Roy. Soc.*)

potential passes through a maximum and falls to its original value, the rate of fall being slower than the rate of rise since the fall depends on the discharge of a leaky condenser through a high resistance. The course of the electrotonic changes at the cathode is therefore given by the record shown in Fig. 540G from Hodgkin. At the anode quite symmetrical changes are observed (Fig. 540H). The shock has thus produced local changes in electrical potential that are easily predictable from an equivalent electrical circuit.*

Action Potential

If the stimulus is above threshold, the same electrotonic changes occur, but are succeeded by electrical events that reveal a more fundamental alteration in the state of the nerve; the action potential, in effect, makes itself manifest, as in Fig. 540A. The "negativity" of electrode B, in relation to C, builds up to a far greater extent than can be accounted for by the primary electrotonic effect of the applied shock, and is due to the complete breakdown and reversal of the resting potential (p. 1045). Moreover, this change is propagated, so that C eventually becomes negative in respect to B, and the change is diphasic. If the stimulus is reduced in intensity, but maintained above threshold, the transition

* That these changes are, nevertheless, a characteristic of living tissue is shown by experiments on dead nerve, in which the electrotonic potentials were only a small fraction of the values obtained on living nerve and were due to some electrode polarization.

from the passive, electrotonic, to the active condition, as revealed by the development of the action potential proper, is more protracted; we may therefore speak of a *latent-period* of development of the action potential which, in the record of Fig. 540C, amounts to about 1/3 msec. but is much shorter with stronger stimuli (Fig. 540A). Thus Figs. 540A and C represent the primary, electrotonic, plus the true action, potentials at the cathode resulting from a brief shock; since only electrotonic changes occur at the anode, and since there is every reason to believe that they are the mirror images of those taking place at the cathode (Fig. 540G and H), we may allow for the electrotonic effects at the cathode by merely taking the algebraic sum of the cathodic and anodic potentials; the action potential, uncomplicated by the primary electrotonic effects, is thus obtained, as in Fig. 541.

Local Subthreshold Changes

For a long time it was thought that the transition from a purely electrotonic polarization to the active state was sharp, so that a nerve could be considered either to have been activated or not, according as the strength of the stimulus

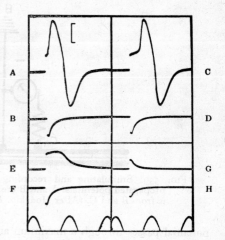

FIG. 540. Records, at the anode and cathode, of electrical changes taking place in nerve as a result of brief stimuli of different relative strengths: A, 1·05; B, −1·05; C, 1·00; D, −1·00;. E, 1·00; F, −1·00; G, 0·61; H, −0·61. The difference between C and E was due to a slight fluctuation in excitability, since the shock strength was not changed between making these two records. Each pair of records, A, B; C, D, etc., thus represents the changes taking place at cathode and anode; the A—B and C—D records result from a supraliminal shock; the E—F records result from a shock high enough to produce a local response, whilst the G—H records indicate the purely electrotonic responses to a definitely subthreshold stimulus. Scale, 15 mV; time, 1 msec. (Hodgkin. *Proc. Roy. Soc.*)

was great enough to excite or not. In 1937 Katz, by studying the changes in excitability of nerve in the neighbourhood of the cathode, immediately after a just subliminal shock, obtained evidence for the existence of local changes which were too great to be accounted for on the basis of electrotonus, and yet were too small, and were not propagated far enough, to be described as typical action potentials. Following on this, Hodgkin, using the stimulating and recording set-up illustrated in Fig. 539, showed that electrical stimuli of strength greater than half the threshold could produce local changes in potential at the recording lead, which were quite distinct from the purely electrotonic ones induced by the stimulus. Thus in Fig. 540 the electrical changes at the cathode are shown for a definitely subthreshold shock (G), and for one very close to the threshold (E); the first record is one of a typical electrotonic change and is the mirror image of that obtained at the anode (Fig. 540H); the potential in the second record, on the other hand, lasts much longer and is quite different from the record at the anode (Fig. 540F). By adding the cathodic and anodic changes algebraically, as in Fig. 542, the local excitatory disturbance, uncomplicated by the electrotonic effects of the stimulus, may be obtained. In this case, the local response had a rising phase of 270 μsec. and a total duration of 1,500 μsec.

The gradual transition from the subliminal to the threshold response (leading to a propagated action potential) is shown in Fig. 543; it appears from this figure that it is only when the local potential reaches a certain critical size that it develops into a true propagated disturbance; if it fails to achieve this it dies out as a localized monophasic wave—it is said to *decrement* to extinction. It is interesting, moreover, that the latency of the propagated impulse is, in essence, determined by the time required for the local disturbance to build up; this is

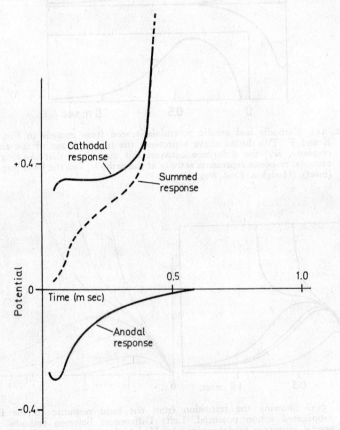

FIG. 541. Showing how the response at the cathode, after allowing for purely electrotonic effects, may be determined. The plotted curve is the algebraic sum of the cathodal and anodal responses. Ordinates: Fraction of the spike height. Abscissæ: Time in msec. (After Hodgkin.)

evident from Fig. 543. In general, if a given fibre has a long latency, the local build-up of its sub-threshold potential is slow and *vice versâ*.

Spread of Potential

The local disturbance, following a subliminal stimulus, is not confined to the region of the nerve immediately under the electrode; in other words, there is some propagation to neighbouring regions. Since the effects do not extend far, however, the study of the spread is difficult (allowance being made, of course, for the spread of electrotonic potential).*

* In fibres with a long latency, the subliminal response may spread over appreciable lengths of the nerve before dying away. If, now, the spreading subliminal response passes into a region of greater excitability, it may grow into a true spike; in this case, as Hodgkin has shown, a spike may originate at some distance from the cathode.

34*

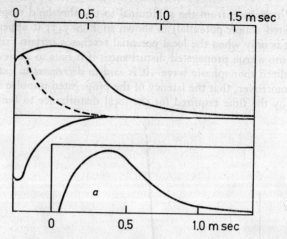

FIG. 542. Cathodic and anodic potentials, traced from records in Fig. 540 E and F. The dotted curve represents the mirror image of the anodal response, and the difference between this curve and that showing the cathodal response represents activity at the cathode, *i.e.*, the local response (inset). (Hodgkin. *Proc. Roy. Soc.*)

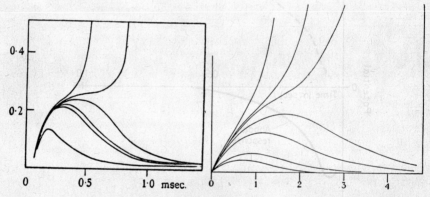

FIG. 543. Showing the transition from the local response to the fully propagated action potential. Left: Differences between cathodal and anodal responses as determined by Hodgkin's technique. Right: Curves derived from excitability changes following threshold and subthreshold shocks by Katz. (Hodgkin, *Proc. Roy. Soc.*, and Katz, *Proc. Roy. Soc.*)

Growth to Action Potential

These studies of Katz and Hodgkin on the subliminal response are very suggestive; they indicate that the first element in the series of events leading to the production of a propagated action potential is the establishment of an electrotonic potential of a certain magnitude; the flow of current across the membrane, thus established, appears to release some mechanism which continues the process of depolarization. As a result, the local potential, in the region of the cathode, becomes higher than the electrotonic potential. If the conditions are suitable, this process goes to completion, *i.e.*, if the initial electrotonic potential is high enough, or the excitability sufficiently great, the local disturbance grows into an "all-or-none" type of action potential which is propagated away from the cathode. If the conditions are unsuitable, the local disturbance fails to reach the minimum height necessary for propagation as a true action potential; it travels a little way, decrementing to extinction.*

Refractory State. Immediately after stimulation, or the passage of a wave of action potential, excitable tissue passes through a stage of complete inexcitability, the *absolutely refractory period*; following this, there is a period of reduced excitability, the *relatively refractory period*; before returning to its normal resting condition the tissue often passes, finally, through a stage of hyperexcitability, the *supernormal phase*. During the relatively refractory period, the local response to a subliminal stimulus is very considerably reduced, whilst the electrotonic polarization is virtually unchanged; during the supernormal phase the local response is accentuated. If, now, we compare the heights to which a local response must rise in order to grow into "orthodox spikes" in the relatively refractory, resting, and supernormal phases we find that they decrease in this order, a large local response, and therefore of course a large stimulus, being necessary to cause a propagated spike in the refractory condition. In this condition, indeed, the difference between a local response which fails to propagate, and a propagated spike, tends to become vanishingly small; in resting nerve the difference in size may be about five-fold.

Local Circuit Theory

These facts indicate that the ability of a local disturbance, produced at the cathode, to propagate depends on its size in relation to the state of excitability of the tissue. In other words, in order that the locally developed action potential may *move away* from its seat of origin, it must be able to *stimulate adequately an adjoining region*. The rationale for such a stimulation by an adjoining region of tissue is provided by the *local-circuit theory* of propagation. This concept follows directly from the core-conductor theory of stimulation; essentially it is argued that, during the passage of a wave of action potential, the surface of the active, or depolarized, nerve, being at a different potential from that in an adjoining region, causes a local flow of current which is adequate to activate the adjoining region. The analogy between nerve, with a cathodic electrotonic potential applied externally, and a nerve during the passage of an action potential is illustrated in Fig. 544 from Hodgkin. The cathode causes electric circuits through the nerve fibre to be made as in Fig. 544, and the electrotonus in the region of this electrode is such as to increase excitability in neighbouring regions, because the effective stimulus is the establishment of a critical potential across the membrane. Now let us consider the state of affairs with active nerve. The surface of the active region becomes negative in respect to the inactive region; consequently positive current tends to flow from the inactive to the active region of the surface. Within the core, the active region becomes more positive (or less negative) in relation to an adjoining inactive region, so that positive current flows, within the core, from active to inactive regions. A local circuit is thus established causing a flow of current which will be by no means confined to the junction between active and inactive regions; consequently the condition at the point X may be essentially similar in the two conditions in that, in each case, it shares in the electrotonic spread from an active region. In the immediate neighbourhood of the active region, however, the current density through the membrane is sufficiently great to

---

* The question whether subliminal responses may be elicited in medullated nerve has been discussed by Katz (1947), who has brought forward convincing evidence in favour of it. More recently Huxley & Stämpfli (1951) and del Castillo-Nicolau & Stark (1951), have recorded local responses from single medullated fibres.

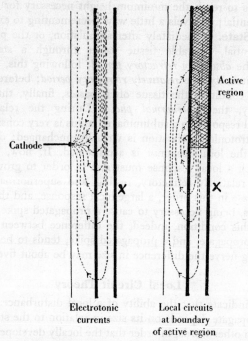

Cathode

Active region

X   X

Electrotonic currents

Local circuits at boundary of active region

FIG. 544. Illustrating the analogy between the electrotonic currents due to an applied cathode and the local circuits at the boundary of an active region. (Hodgkin. *J. Physiol.*)

excite the adjoining, inactive, region which then excites the next point, and so on; meanwhile the electrotonus extends farther and farther along the nerve.

Spread Beyond a Block

On the basis of this theory a number of important deductions may be made. Thus Osterhout & Hill argued that if a tissue is made locally inexcitable, *e.g.*, by treating a portion of its surface with chloroform, it should be possible for

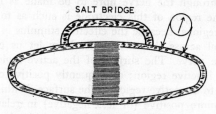

SALT BRIDGE

NARCOTISED REGION

FIG. 545. The spread of excitation through a salt-bridge. (After Osterhout & Hill. *Cold Spring Harbor Symposia*.)

the wave of activity to be continued around the block by a salt-bridge as in Fig. 545. and they demonstrated that this could indeed occur in *Nitella*. This is a most convincing piece of evidence in favour of the essentially electrical nature of the transmission process, since it is quite inconceivable that any chemical substance, liberated proximally to the block, could diffuse through the salt-bridge within a short time and affect the distal region of the cell's surface. Again, Hodgkin has pointed out that, since the blockage of a nerve by cold or a narcotic* does not affect appreciably its longitudinal electrical resistance, the action potential, when it reaches the block, may influence regions beyond. This point is illustrated in Fig. 546. Hodgkin found in a nerve blocked with cold that, whereas a maximal stimulus applied to A produced no action potential beyond the block (at C), a sub-threshold stimulus, applied to B 1–2 msec. after the maximal shock at A, produced an action potential at C. Hence the propagated disturbance from the first stimulus, although it failed to pass through the block, increased the excitability of the nerve in the region immediately distal to the block. Changes in threshold down to 10 per cent. of the normal resting value could be obtained under appropriate conditions

Extrinsic Potentials. Clearly, if the active region of nerve proximal to the block, *i.e.*, on the side nearest A in Fig. 546, is modifying the excitability of the

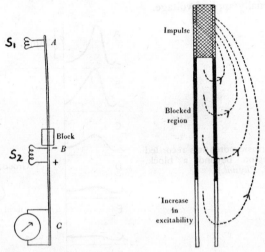

FIG. 546. Illustrating Hodgkin's study of the spread of electrotonic potential beyond a blocked region. Left: General scheme of the stimulating and recording set-up. Right: Showing how the currents, due to local circuits, may modify the excitability beyond a block. (Hodgkin. *J. Physiol.*)

region distal to the block, some electrical effects should be observable beyond the block. In Fig. 547 some records of Hodgkin are shown of the action potential proximal to the block (A) and the electrotonic potentials at increasing distances beyond this region (B-F). These potentials are much smaller than

---

* Conduction through a nerve may be blocked by a number of experimental procedures, *e.g.*, by narcotics applied locally. If the narcotization is not too severe, however, the action potential, although depressed in height, continues to travel through the region of depressed excitability and re-acquires its former height on emerging into the normal stretch of nerve. For some time it was thought that the nerve impulse experienced *decremental conduction*, *i.e.*, that it was propagated for considerable distances through regions of low excitability with a gradually decreasing spike-height. Such a conclusion has been shown to be wrong (*e.g.*, by Kato and by Davis *et al.*), so that the probable course of events, if the narcotization is sufficient to suppress excitability completely, is an abrupt cessation of activity at the boundary between normal and narcotized nerve. In the region where excitability is only depressed, and not abolished, we may expect that the propagated disturbance will decrement rapidly to a lower level, at which it will remain constant until it passes back to normal nerve, when it will rise rapidly, but not instantaneously, to normal height. The depression in height of the action potential may be severe; Davis *et al.*, working with a mixed nerve, found a decrease of 75 per cent., and that this was not due to complete blocking of a significant proportion of the fibres was indicated by the fact that the emerging action potential regained 99 per cent. of its original height.

the true action potential and, instead of propagating at the same height, they decay logarithmically to $1/e$ of their value in 2 mm. They are called *extrinsic potentials*, since they are due entirely to electrical activity in a region beyond the recording electrodes, being electrotonic effects of the action potential. The strong analogy between the extrinsic potential and the electrotonic effect of an externally applied cathodic potential was shown by further experiments of Hodgkin; a potential was applied with the cathode in the blocked region, and it was shown that, in the normal region beyond the block, the response obtained could be made almost identical with that caused by a propagated disturbance which died away in the block. Finally, Hodgkin showed that the time-courses of development of the change in excitability beyond the block, and of the rise in extrinsic potential, were identical; the efficiency of an extrinsic potential in raising excitability was found, however, to be some 1·4–2·3 times greater than that of an externally applied voltage.

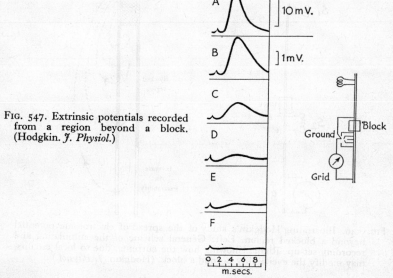

FIG. 547. Extrinsic potentials recorded from a region beyond a block. (Hodgkin. *J. Physiol.*)

Propagation

This concept, of an activated region of nerve acting as a stimulus for an adjoining region, permits of a ready interpretation of the details of the propagated action potential. This potential, it will be recalled, is an expression of the changes of potential taking place at an electrode on the intact surface of the fibre at different times during the approach of a propagated disturbance. When the approaching disturbance has reached a certain distance from the electrode, the electrotonic effects, spreading ahead of the active region, increase the negativity at the electrode. The initial deflection on the recording instrument, the *foot* of the action potential, is thus due to an *extrinsic potential* and is not a sign of "activity" under the electrode. At a certain point in time, when the active region has come sufficiently close, the region under the electrode is excited, and at this point the negativity increases rapidly to a maximum. With the re-establishment of polarization, which must take a measurable time, the negativity falls.*

* The foot of the action potential represents an extrinsic potential, so that the shape and time constants of the rising phase will be largely determined by the velocity of propagation and the characteristics of electrotonic spread. The "wavelength" of the action

Point of Inflection

The point of inflection on the rising phase of the action potential, where the rate of rise of negativity is greatest, represents the moment at which the region of nerve under the electrode becomes fully active, as a result of the excitatory influence of the electrotonic potential. We have already seen (p. 1043) that Cole & Curtis have recorded simultaneously the changes in potential and impedance at a point on the nerve during the passage of a propagated disturbance, and it is of the utmost interest that they observed that the change of impedance did indeed occur at the point of inflection of the action potential curve (Fig. 548).

After-Potentials. The action potential does not fall simply to zero; studies in the schools of Gasser and Erlanger have shown that the sequence of events is usually: *action potential, negative after-potential, positive after-potential.* The relative prominence of these after-potentials varies with the type of nerve fibre studied; the negative after-potential is manifest as a failure of the record to return to its base-line, so that its beginning is generally difficult to identify; the positive after-potential, on the other hand, represents a reversal in sign of the record. Associated with these states of residual negativity and positivity

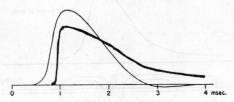

Fig. 548. Membrane conductance increase (heavy line), after approximate corrections for electrode length and bridge amplifier response, and monophasic action potential (light line) obtained from the first derivative, after approximate correction for action potential amplifier response. (Cole & Curtis. *J. gen. Physiol.*)

are found conditions of hyper- and hypo-excitability, as one should expect. We may defer the explanation of the positive after-potential, occurring typically in the squid axon, till later (p. 1085). It was suggested by Shanes in 1949 that the negative after-potential was due to the accumulation of K^+ on the outside of the fibre membrane following the action potential, and he backed this suggestion by a great deal of evidence, showing that the effect could be reduced by increasing the bulk of medium surrounding the fibre, that veratrine, which increased the after-potential, increased the loss of K^+ on stimulation, and so on. At first thought, however, this explanation might be considered unlikely, in view of the minute amount of K^+ liberated by a single impulse, but a consideration of the anatomical relations of the squid giant axon, for example, shows that very small amounts may nevertheless lead to appreciable concentrations. Thus, the electron-microscope has shown that these non-myelinated fibres are enveloped in Schwann cytoplasm, being separated by the axon-Schwann double membrane. It is essentially the separation between these two

---

potential is given by the velocity of transmission divided by the "frequency", *i.e.*, the reciprocal of the duration. A crab fibre, with a velocity of 3–5 m./sec., and a duration of 0·8–1·0 msec., has a wavelength of about 4 mm. Myelinated fibres of the frog have a wavelength of about 40 mm. The difference is largely due to the myelin sheath with its high longitudinal resistance which permits a far greater electrotonic spread. Thus in medullated nerve an applied electrotonic potential falls to $1/e$ of its value in 2–3 mm., whereas in single fibres of crab non-medullated nerve it falls to the same extent in only 0·5 mm.

limiting membranes that constitutes the immediate outside medium, so that if diffusion away from here is restricted, high local concentrations can be built up. Frankenhaeuser & Hodgkin stimulated isolated axons repetitively and observed that the positive after-potential decreased with successive stimuli, due, apparently, to the summing of residual negative after-potentials. These effects could always be imitated by an appropriate external concentration of K+, the rise after one impulse being equivalent to about 1·6 mM. For the escaped K+ to be present in this concentration it had to be assumed that there was a barrier to diffusion of this K+ away from the axon surface, presumably the Schwann membrane; moreover, the actual space between axon and this barrier would have had to be only some 300A. It is still not clear whether the K+ lost from the fibre would escape by crossing into the Schwann cytoplasm or alternatively by passing up the mesaxons or surface connecting membranes (p. 480).*

Interaction Between Fibres. A further confirmation of the primarily electrical changes associated with nervous conduction is provided by the work of Katz & O. H. Schmitt on the interaction of single nerve fibres in a bundle.

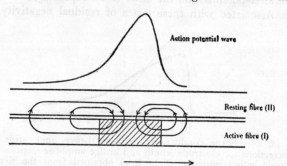

FIG. 549. Local circuit diagram illustrating the penetration of Fibre II by the action currents of Fibre I. The shaded area indicates the "active region" of Fibre I. Note that the direction of the penetrating current reverses twice. (Katz & O. H. Schmitt. *J. Physiol.*)

A study of the local circuits during a propagated disturbance, illustrated in Fig. 549, shows that positive current tends to flow, within the fibre, from the active to the inactive region and penetrates the membrane, in the inactive region, from within → out. This type of penetration, we have argued, is equivalent to catelectrotonus and should increase excitability. If a resting fibre is lying alongside an active one in close apposition, as in Fig. 549, some of the current tends to flow from the inactive region of the active fibre into the resting fibre in the direction: out → in, *i.e.*, it should depress excitability; in the region of activity, however, the current-flow through the resting fibre is from in → out, a condition that must increase its excitability. The excitability at a point on a resting fibre must therefore pass through a cycle of changes, during the passage of an impulse along an adjacent fibre, because of the alternate anelectrotonic and catelectrotonic influences of the active region as it passes by. By stimulating only one of a pair of adjacent isolated fibres of a non-myelinated crab nerve, and measuring the change in excitability of the other, results were obtained indicating influences of one fibre on the other that were

* Although Shanes' view of the cause of the negative after-potential seems to be applicable to the squid axon under conditions of repetitive stimulation, there is some doubt as to whether the effects of anoxia and drugs on the potential are completely explicable in terms of K+-loss. This is especially manifest in myelinated fibres (see Frankenhaeuser & Hodgkin, 1956, p. 372).

qualitatively predictable on the basis of the local-circuit theory of propagation. Quantitatively, the threshold lowering amounted to about 20 per cent., *i.e.*, it was sufficiently small to render negligible the possibility of one fibre activating another. Two impulses passing in separate adjacent fibres should interfere, either by summating or cancelling their electrotonic effects, the exact nature of the interaction depending on the phase relationships of the impulses; these electrotonic interactions should be reflected in changes in velocity of propagation, since this must depend on electrotonic spread (p. 1072). The experimental analysis of such possibilities, carried out by Katz & O. H. Schmitt, has shown that if one impulse slightly precedes, it accelerates the rate of conduction of the lagging action potential, an effect leading to a synchronization of impulses in adjacent fibres if their speeds differ only slightly.*

Safety-Margin. If it is the potential difference between adjacent portions of the surface of the nerve fibre that provides the condition for propagation, it follows that the height of the action potential should be equal to, or exceed, the adequate stimulus for the nerve fibre. It is known, moreover, that under conditions approaching nerve block, or during the relatively refractory period, the threshold for stimulation may be increased some five-fold. Under the same conditions the height of the propagated action potential may be reduced to one-half, consequently, during propagation through relatively refractory nerve, the effective stimulus provided by the action potential is only one half as great as normal, and the excitability is only one fifth. The normal action potential should therefore be some 10 times higher than necessary if it is to have a *safety-margin* to permit propagation under adverse conditions. An exact quantitative comparison between the potential required to stimulate and the height of the action potential, is not easy; according to Hodgkin the smallest extrinsic potential in a mixed nerve trunk was 2–3 mV, comparing with an action potential, recorded under the same conditions, of 20–30 mV, *i.e.*, there was a safety-margin of about 10. Moreover, unless the action potential, initiated at the cathode, has a considerable reserve of potential, the propagated disturbance should invariably be wiped out by the electrotonic influence of the anode. It is known, of course, that a propagated impulse *can* be blocked at the anode if the applied voltage is high enough, but a quantitative study reveals that the anodic voltage necessary to block an impulse is some 10 to 15 times the cathodic voltage required to initiate one, *i.e.*, if the anodic polarization, necessary to cause block, is equivalent to the height of the action potential, the safety-margin is indeed more than 10-fold.

Saltatory Conduction. We have so far supposed the spread of electrotonus along the nerve fibre to take place continuously; in the non-myelinated fibres of invertebrates this seems to be true, but Erlanger & Blair in 1934 showed that, in medullated nerve, the blocking effects of anodic polarization showed discontinuities which could most reasonably be attributed to the fact that the polarizing current entered the nerve sheath only through the nodes of Ranvier. The evidence for this cannot be presented here, but it is convincing and indicates that a polarizing current, passed through a nerve, enters mainly, if

---

* Rosenblueth (1944) has discussed the possibility that fibre interaction is a factor in *recruitment, i.e.,* the systematic increase in the number of nerve fibres stimulated by the successive shocks of a sub-maximal train; recruitment is not exclusively due to local conditions at the stimulating electrode, and the observation that stimulation of fibres other than those tested gives recruitment certainly suggests that interaction of spike potentials is a factor. Rosenblueth infers that interaction is partly electrical and partly due to the escape of K^+; it is considerably enhanced by veratrine. Under strictly normal conditions *in vivo*, Esplin (1962) was unable to find any interaction of nerve fibres so far as their conduction velocities were concerned.

not exclusively, at the nodes of Ranvier where the myelin sheath is thinnest. If this is true of externally applied currents, it should also be true of the spread of electrotonic currents from the advancing wave-front of the propagated disturbance. Evidence has been accumulating that this is indeed true* and in 1949 Huxley & Stämpfli developed methods of recording the action potentials from very short lengths of nerve fibre so that records could successively be taken from at least three points within one internode, a distance of

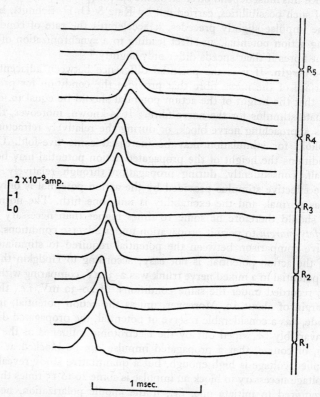

FIG. 550. Saltatory conduction. Tracings of records obtained at a series of positions along a single medullated fibre. Diagram of fibre on right-hand side shows positions where each record was taken. (Huxley & Stämpfli. *J. Physiol.*)

only 3 mm., on the average, in frog nerve. It was pointed out by Hodgkin that, if current enters the axoplasm only through nodes, the flow of current through any internode, at any moment during the approach of an impulse, must be the same at all points along the internode; in other words, the recorded action potential should have the same time relations at all points in the internode.†

* The earlier work has been summarized by Huxley & Stämpfli and, more fully, by von Muralt (1946); credit for the establishment of this point is due particularly to the Japanese worker, Tasaki, and his colleagues; Pfaffman, using an isolated fibre, obtained larger action potentials from the nodes than from internodal regions. The monophasic action potential, recorded at the internode, was different from that at the node, and suggested that the recorded potential was due to an out-of-phase spread of electrotonic potential originating at the two nodes.

† An apt analogy would be that provided by water flowing through a tube; if there are leaks in the wall the flow will decrease progressively along the tube; if the leaks are confined to definite nodes, the flow between the nodes will be uniform, although from one segment to another the flow will decrease in accordance with the loss at the nodes.

A series of records at different points on the internodes of a nerve are shown in Fig. 550, and it will be seen that the groups of records from any internode are practically synchronous, whilst the different groups are displaced in time. This result certainly suggests that the nerve fibre becomes active only at the nodes, or at any rate that the effects of its activity are only propagated through these discontinuities in the sheath. We may note that the records were taken with two electrodes on intact nerve (*i.e.*, the recording would be described as "diphasic") so that the rise and fall of potential are mainly due to activity occurring first at one node and then at the other; the interval is so small however (the internodal distance is only 3 mm. and the velocity of propagation about 23 m./sec., with the result that the time-interval is only 0·13 msec.) that the second phase of the action potential is ill defined, the rising limb of the first phase representing activity mainly at the proximal node and the falling phase activity mainly at the distal node; according as the leads are taken from points close to the proximal or distal node we may expect the time relations of the different phases of the action potential to change; this is clear from the graphs of Fig. 550, and an elaborate experimental and theoretical analysis of the effects of the distance of the recording leads from the node tends to confirm the saltatory interpretation of the phenomena, although a certain amount of capacity current-flow does take place through the myelin sheath of the internodes.

Resistance at the Node

If conduction is saltatory in myelinated nerve it is because current flows more easily across the nodal membrane than across the internodes; at the node, the myelin is missing, so that it follows that the electrical resistance of the myelin sheath must contribute greatly to the transverse resistance of the fibre. Direct measurements of the resistance and capacity at node and internode have been made recently by Tasaki, who found the following:—

				Internode	Node
Resistance	.	.	.	10^5 Ohm.cm.^{2}	8–20 Ohm.cm.^{2}
Capacity	.	.	.	0·005 μF/cm.^{2}	3–7 μF/cm.^{2}

Thus, the resistance of the node is low (cf. Table LXXIII, p. 1042), whilst that of the internode is some 10,000 times greater; the capacity of the node is of the same order as that of other cell membranes whilst that of the myelin is very low. It was considered at one time that the myelinated fibres of the central nervous system had no nodes of Ranvier; there seems little doubt now, however, that conduction in these fibres is saltatory (Tasaki, 1952) and that there are regions of high permeability to methylene blue at intervals of some 500 to 1,000μ along the fibres, presumably corresponding to regions in which myelin is absent (Hess & Young).

Velocity of Propagation

Diameter of the Fibre. It follows, from Rushton's analysis of the implications of the core-conductor theory, that the velocity of propagation should increase with the diameter of the nerve fibre, other things being equal. The diameters of the medullated fibres in a mixed nerve, such as a frog's sciatic, vary over a wide range, between 3 and 29μ. We must therefore expect the action potential, recorded from a mixed nerve such as this, to represent the summated effects of the action potentials, moving with different velocities, in hundreds of fibres; at the point of initiation of the impulse we may expect the disturbances to be, to some extent, in phase, but as the propagated disturbance travels along the nerve they should fall "out of step" so that at a few centimetres' distance we may expect to record, first, the spikes of the most rapidly conducting fibres; the spikes of more slowly conducting fibres adding their contributions later. The shape of the action potential, recorded from a bundle of nerve fibres, may therefore be expected to change significantly with distance travelled. The complete action potential of the bull-frog's peroneal nerve,

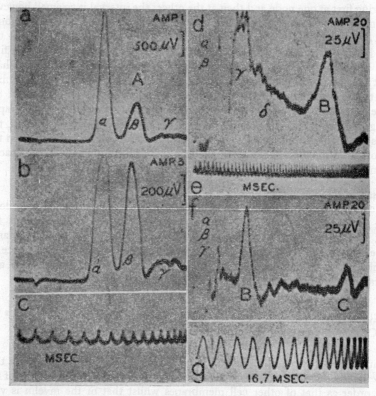

Fig. 551. Complete action potential of bullfrog's peroneal nerve after conducting through a distance of 13·1 cm. The time falls off logarithmically from left to right; scale c applies to records a and b, whilst scale e applies to record d, and g to record f. Note different amplifications. (Erlanger & Gasser. *Electrical Signs of Nervous Activity*. U. of Pennsylvania Press, Philadelphia.)

after travelling 13 cm., is shown in Fig. 551; it contains a number of spikes, representing the activities of groups of fibres of different conduction velocities; the fastest fibres produce an A-wave which may be analysed into α-, β-, and γ-elevations, corresponding to groups of fast fibres with conduction velocities decreasing in this order.* The waves due to the slower groups are named B and C. The differences in size of the elevations, due to the three main groups

* There is not complete agreement as to the theoretical effect of internodal distance on velocity of conduction; practically there is no influence (Sanders & Whitteridge); theoretical considerations are given by Offner *et al.*, and by Huxley & Stämpfli; according to the last authors, theory predicts that the velocity will have a maximum value over a certain range of internodal distances, a range that might be sufficiently wide to make the velocity, in practice, independent of this variable.

of fibres, were so great that the records had to be amplified to different extents; thus in the records (d) and (f) the amplification was 20 times that in (a); moreover, separate time-scales were necessary. A painstaking analysis of the action potential records, correlated with measurements of the distribution of fibre diameters,* by Erlanger & Gasser has provided the classification of fibre-types shown in Table LXXIV. More recently Gasser & Grundfest have analysed the components of the cat and rabbit saphenous nerves as in Table LXXV. It should be noted that there is considerable overlap between the velocities of the A- and B-groups; their classification is not, however, exclusively on a basis of conduction velocity. The duration of the spike, the characteristics of the positive after-potential, the refractory period, etc., are all taken into account. Thus the slowest fibres of the A-group in the rabbit's and cat's saphenous nerves are described as the δ-elevation; they are actually slower than B-fibres, but differ sufficiently in other characteristics to warrant their inclusion in the A-group.

TABLE LXXIV

Classification of Frog Fibres According to Fibre Diameter and Conduction Velocity.

Elevation		Fibre Diameter	Velocity
A { α	. . . . .	18·5 μ	42 m./sec.
β	. . . . .	14·0 μ	25 m./sec.
γ	. . . . .	11·0 μ	17 m./sec.
B	. . . . .	—	4·2 m./sec.
C	. . . . .	2·5 μ	0·4–0·5 m./sec.

TABLE LXXV

Analysis of Components of Cat and Rabbit Saphenous Nerves (Grundfest, 1940).

Group	A	B	C
Diameter of fibre (μ)	20–1	3	—
Conduction velocity (m./sec.) . .	100–5	14–3	2
Duration of action potential (msec.) .	0·4–0·5	1·2	2·0
Absolute refractory period (msec.) .	0·4–1·0	1·2	2·0

Fibre Size and Function

Attempts at establishing a correlation between the size of the fibre and its function were, at first, hopeful; thus, in general, the motor root contains only the α-elevation of the A-group, and the C-elevation, the latter being due to very fine non-myelinated fibres; the sensory root contains the whole A- and C-groups; pre-ganglionic (myelinated) nerves of the autonomic system give the B-type of elevation. The finding of four groups of fibres in sensory nerves suggested that different groups conducted different sensations. Although

---

* The mathematical relationship between velocity and diameter was originally thought by Blair & Erlanger to be of the form: $V \alpha D^2$; however, the most recent work of Gasser & Grundfest has established that the relationship is approximately linear in A-fibres of the cat and rabbit saphenous nerves, with diameters ranging from 10–2 μ. In the giant axons of the squid Pumphrey & Young found the relationship $V \alpha D^{0·6}$. It should be noted that Gasser & Grundfest obtained the best linear fit for their results by using the axon, as opposed to the fibre, diameter. The proportion of sheath to whole fibre becomes larger as the fibres become smaller; thus the ratio: axon diameter/fibre diameter is about 0·75 for fibres of 8μ and larger, but may fall as low as 0·2 for the thinnest myelinated fibres (Taylor; Gasser & Grundfest). Hodgkin (1954) concludes from a theoretical analysis that, when a nerve or muscle fibre is in a large volume, the velocity should vary as the square root of diameter; in a small volume, it should be inversely proportional to the square root of the sum of the external and internal resistances per unit length. In myelinated fibres, with saltatory conduction this analysis is not applicable.

some correlation between the extinction of a given elevation and the disappearance of a given type of sensation has been found, the most recent work on the subject makes it unlikely that there is a strict segregation of the modalities of sensation into fibre-groups. The fact that the fibres in a given bundle can differ in their conduction rates by a factor of about 100 is doubtless significant from the point of view of central co-ordination; thus it is possible for impulses to reach the highest centres before others, travelling along thin fibres, reach the spinal cord.

Thickness of Sheath. Lest it should be thought that conduction velocity is entirely determined by the diameter of the fibre it should be emphasized that, when the fibres of different phyla are compared, there is no strict correlation. Several authors have drawn attention to the thickness of the myelin sheath; thus the largest squid fibres, with diameters of the order of 700μ, conduct at about 23 m./sec. whilst the fastest A-fibres of the frog, with a diameter of only 18μ, conduct about twice as fast. In the squid, the myelin layer represents only some 1 per cent. of the thickness of the fibre, whilst it

TABLE LXXVI

Showing Relative Constancy of Product of Birefringence of Sheath Times Diameter of Fibre in Various Nerve Fibres (Taylor, 1942).

	Fibre diam. (μ)	Axon diam. (μ)	Axon diam. / Fibre diam.	Birefringence	Velocity (ca.) (m./sec.)	Birefringence × Fibre diam.
Squid giant . . .	650	637	0·98	−0·0001	25	—
Earthworm giant . .	100	90	0·90	0·0010	25	0·10
Shrimp giant . . .	50	43	0·87	0·0024	25	0·12
Frog sciatic . . .	10	7·5	0·75	0·0105	25	0·105
Cat saphenous (calc.) 20° C.	8·7	6·6	0·76	0·014	25	0·12
Catfish	8·8	5·8	0·58	0·012	25	0·105

may be as high as 25 per cent. in vertebrates. It would appear, therefore, that a greater relative thickness of the myelin sheath, combined with the presence of nodes, accounts for the higher velocity in the frog fibre. This view is supported by Table LXXVI; in it the birefringences of a number of fibres, all with approximately the same conduction rate, are shown together with the thicknesses of the whole fibres and of the axons. It will be seen that, as the fibre diameter decreases, the proportionate thickness of sheath increases. If the lipid content of the sheath is the important factor, it will be the birefringence that determines its value in aiding conduction; and it is interesting that the product of the birefringence times the fibre diameter is approximately constant for all the fibres considered. It should be noted, however, that when fibres in a single nerve are compared, the axon diameter seems to be the determining factor (Gasser & Grundfest). In the invertebrate nervous system, the importance of fibre diameter is strikingly indicated by the giant fibres of the squid; J. Z. Young showed that these fibres, acting as the motor pathway for contraction of the mantle, were developed by the fusion of numerous smaller nerve cells into a syncytium; this fusion is to be regarded as an adaptive process to enable rapid conduction to the mantle, which is used in propulsion.

External Resistance. The velocity of propagation, according to Rushton's analysis, should depend on the length factor, L (p. 1041), which, as we have seen, is a function of the resistances of the outside and inside media of the fibre; increasing L means that the extrinsic potential will extend farther at any given moment and so activate a given point, ahead of the action potential, sooner. Since L increases with decreasing external resistance, we should expect a nerve to conduct more rapidly in sea-water than in oil. With isolated fibres of the squid giant axon, Hodgkin demonstrated an increase of 80–140 per cent. in conduction velocity on transferring a nerve fibre from oil to sea-water. It is interesting, in this connection, that Auger increased the velocity of propagation of the wave of action potential in *Nitella* by covering the cell with a strip of filter paper moistened with saline solution. Conversely Katz has shown that the conduction velocity in a single nerve fibre is reduced by replacing some of the salt in its outside medium by non-electrolyte.*

Spike-Height. The greater the height of the action potential, the greater will be the extent of the extrinsic potential, and hence the greater the conduction velocity. In a narcotized region the action potential propagates at a lower height, so that narcotization should be reflected in a lower velocity of propagation. According to Adrian's experiments this is true, the velocity falling, under suitable conditions, to less than a third of normal.

The Ionic Currents

In this discussion of electrical excitation, the nerve fibre has been treated as an electrical analogue; this treatment, as we have seen, has been successful in explaining and ordering most of the experimental facts. In so far, however, as the postulated changes in potential and impedance, and the associated flows of current, are essentially ionic events, it is evident that a satisfactory explanation of the phenomena of excitation, and propagation of the impulse, must ultimately be given in terms of movements of ions under their appropriate gradients of electrochemical potential. It will be recalled that the changes taking place during the passage of an action potential were tentatively described in terms of a sudden increase in permeability to Na^+, leading to an influx of this ion, followed by a loss of K^+ associated with the repolarization of the membrane. From the preceding pages it has become clear that the event leading up to this sudden change is a partial depolarization of the cell membrane, a reduction of the potential across this membrane by a critical amount—of the order of 15 mV—leading to a further all-or-none change. If, then, it could be shown that partial depolarization induced the required changes in permeability, and if, moreover, experiments could be carried out permitting a measurement of the time-courses and magnitudes of these changes, the possibility would be opened for explaining many of the phenomena of excitation and propagation in terms of permeability and ionic fluxes. Such, indeed, has been the approach of Hodgkin and his collaborators (Katz, Huxley, Keynes), an approach that has been successful in so far as it has enabled the calculation of the shape of the action potential, for example, in terms of experimentally determined parameters.

Voltage-Clamp Technique. Hodgkin, Huxley & Katz introduced two long electrodes down the length of a squid axon; one measured the potential across the membrane, and the other the current passing through. Any required potential could be established across the membrane, and maintained at this value, by a "negative feed-back", *i.e.*, the circuit was such that any spontaneous change in membrane potential caused an output current to flow in a direction that restored the potential to its original value. The system was arranged so that no current passed through the nerve in the resting condition; at the required

---

* The conduction velocity in regenerated nerve fibres is less than normal even 16 months after the original axotomy; the decrease is partly accounted for by a reduction in average diameter, but not completely (Cragg & Thomas, 1964).

moment a rectangular pulse of current was fed in, raising the membrane potential abruptly to a new level, which was maintained by the feed-back amplifier.

When the potential was increased, *i.e.*, the inside was made more negative, there was a small inwardly directed current of some $30\mu A/cm.^2$. This is understandable, the membrane behaving like an ohmic resistance; in terms of ionic movements it could be stated that the rise in resting potential above the equilibrium value for K^+ raised the electrochemical potential of K^+ outside, and thus caused an increased inward flux of this ion.

Effect of Depolarization

When the potential was decreased, for example by 65 mV, the state of affairs was entirely different; if the membrane had behaved like a simple ohmic resistance the flow of current should have been outward, and been represented by a flow of K^+ adapting itself to the new potential. Instead, the initial change was an inward current, which had a maximal value of $600\mu A/cm.^2$. Such a current would, of course, have depolarized the membrane, but the voltage-clamp maintained the potential at the same value; from the known capacity of the membrane it could be calculated just how rapidly this influx of current would have depolarized the membrane; it came out at 750 V/sec., agreeing well with the maximal rate of rise of the spike. The phase of inward current changed fairly rapidly into a prolonged period of outward current which, in the absence of a feed-back, would have repolarized the membrane. This outward current seemed to be maintained indefinitely, by contrast with the transient nature of the inward current. The initial inward current failed to appear if the depolarization of the membrane, induced by the voltage-clamp, was less than about 10 mV, or if it was greater than about 100 mV. If it was due to an influx of Na^+ this meant that the depolarization of the membrane had to exceed 10 mV in order that the rise in permeability to Na^+ should manifest itself in a sodium-current; it meant also that, when the membrane was depolarized by more than 100 mV, the orientation of the potential difference (outside now negative) was such as to reduce the electrochemical potential gradient for Na^+ to such an extent that net influx of Na^+ became negligible.*

Sodium Current

To prove that the initial inward current was carried by Na^+, voltage-clamps were established in solutions with concentrations of Na^+ varying from zero to normal, by replacement of Na^+ with choline. In the absence of Na^+, there was no inward current whilst the late outward current, due to K^+, was hardly affected. With intermediate Na^+ concentrations the initial current was smaller, and could be predicted with some accuracy on the assumption that the rate of inward movement of Na^+ was determined by the deviation of the actual potential difference across the membrane from the so-called "sodium-potential"—the potential that would be present were it determined entirely by the difference in concentration of Na^+ across the membrane, *i.e.*, by:—

$$E_{Na} = \frac{RT}{F} \ln \frac{[Na]_{In}}{[Na]_{Out}}$$

* If the potential at which the early component of ionic current reverses is determined by the concentration-ratios: $[Na]_{Out}/[Na]_{In}$, then at constant external Na^+-concentration, variations in this reversal potential will be due to variations in internal Na^+-concentration. This point was exploited by Moore & Adelman and Adelman & Moore, who showed that the net gain of Na^+ of an axon in sea-water was some 40 pmole/cm.^{2}/sec. Reduction of the external concentration of Ca^{++} by a factor of 10 increased the net influx of Na^+ twofold.

Thus, the initial inward current could be unequivocally associated with an influx of Na+; the later outward current, being unaffected by the outside concentration of Na+, was probably due to an outflow of K+; and by making some reasonable assumptions it was possible to separate the magnitudes of these two currents under any given experimental conditions. As a result, it could be shown that the outflow of K+ was due to a steep *increase* in permeability to this ion resulting from the depolarization, this change in permeability being delayed to give an S-shaped curve of development.* These changes in permeability induced by an applied depolarization are obviously suited by their timing to bring about the initial reversal of potential, followed by a rapid restoration to its original value; if they occurred simultaneously, on the other hand, they would be ineffective since Na+—K+ exchanges would occur and the necessary flow of current, required to reverse the potential, would be impossible.

Time-Course of Permeability Change

The increase in permeability to Na+ and K+ must be of brief duration, if the nerve is to be able to recover rapidly from the effects of a stimulus or the passage of an action potential, *i.e.*, to be repolarized and to be able to respond to a new stimulus, or allow the passage of a new action potential. The time-course of the changes in permeability to Na+ is therefore of great interest, and an accurate knowledge of this is necessary for a quantitative treatment of the problem of conduction. The changes were studied from two aspects, namely the return of the permeability, as measured by the Na+-current, to normal when the resting potential was restored to its original value after a depolarization, and the changes of permeability that occurred when a given degree of depolarization was maintained for a long time. On re-establishing a normal resting potential, permeability to Na+ returns very rapidly to its normal value, the rate depending on the actual value at the time, being expressed by a simple exponential equation of the type:—

$$\frac{P_{Na}}{P_{Na0}} = e^{-b_{Na}t}$$

The rate-constant, b_{Na}, varied with temperature and with the degree of depolarization, and was of the order of 15 msec.$^{-1}$, indicating a half-life of the order of 0·05 msec.

Inactivation

When a given degree of depolarization was maintained by the voltage-clamp, it was found that the Na+-permeability steadily decreased, suggesting a process of inactivation. This inactivation process is illustrated in Fig. 552 which shows the effects of an initial depolarization of 8 mV on the Na+-current induced by a subsequent depolarization of 44 mV. The depolarization of 8 mV was too small to produce any measurable Na+-current, but if it was allowed to persist as long as 20 msec. it reduced the Na+-current caused by the 44 mV depolarization by as much as 40 per cent. The figure shows that the inactivation takes time to develop, and it was found that the process followed a simple logarithmic curve with a rate-constant of some 0·14 msec.$^{-1}$. The inactivation process was thus very slow compared with the rapid onset of Na+-permeability when the depolarization was established. If the initial, conditioning, potential was a hyperpolarization, *i.e.*, if the inside was made more negative, the Na+-current

* Hodgkin & Huxley (1953) estimated the outward flux of ^{42}K from nerve, with a potential applied, and showed that this was of comparable magnitude with the outwardly directed current.

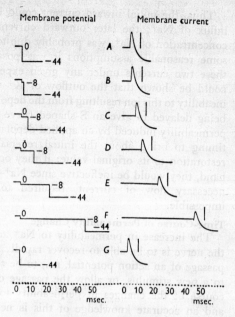

FIG. 552. Development of "inactivation" during constant depolarization of 8 mV. Left-hand column: time-course of membrane potential (the numbers show the displacement of the membrane potential from its resting value in mV). Right-hand column: time-course of membrane current-density. (Vertical lines show the sodium-current expected in the absence of the conditioning step.) (Hodgkin & Huxley. *J. Physiol.*)

subsequently evoked by the 44 mV depolarization was increased; the effect, once again, depended on the duration of the applied hyperpolarization, and if greater than 15 msec. an increase of as much as 70 per cent. could be achieved by a hyperpolarization of 31 mV (Fig. 553). The resting potential of isolated

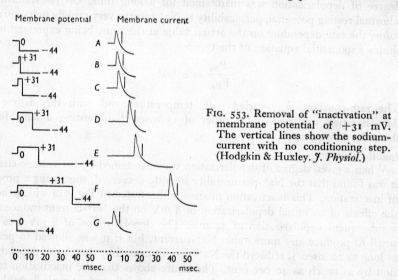

FIG. 553. Removal of "inactivation" at membrane potential of +31 mV. The vertical lines show the sodium-current with no conditioning step. (Hodgkin & Huxley. *J. Physiol.*)

nerve is low, in the sense that it is considerably below the theoretical Nernst K^+-potential; it is therefore probably in a state of partial depolarization, so that some inactivation of Na^+-permeability is present as a steady-state condition; raising the resting potential by a voltage-clamp may therefore be considered to be a restoration of the nerve to its ideal state in which there is no inactivation.

Thus, if I_{Max} is the maximum Na^+-current that can be evoked from the nerve in its ideal state, and if I is the actual Na^+-current when a given degree of polarization is applied, then the ratio of these, h, will be less than unity in the isolated axon; actually it is about 0·6, indicating that, in its partially depolarized state, resulting from isolation and maintenance in sea-water, the power to increase its Na^+-permeability is decreased to some 60 per cent. of its ideal capacity, *i.e.*, when hyperpolarized. Fig. 554, illustrates the variation of h with applied voltage. These results, and a number of others, indicate quite strongly that the process that leads to an increase in Na^+-permeability is followed by an opposite process, called *inactivation*, that restores the permeability to its low value and, moreover, precludes a successive depolarization from inducing a new increase in Na^+-permeability. It is established during a maintained depolarization and so is different from the much more rapid fall in Na^+-permeability that is associated with repolarization and which may be regarded simply as a reversal of the initial change. Essentially, then, this slow

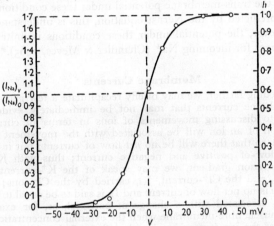

FIG. 554. Influence of membrane potential on "inactivation" in the steady state. Abscissa: Displacement of membrane potential from its resting value during conditioning step. Ordinate: Sodium-current during test step relative to sodium-current in unconditioned test step (left-hand scale) or relative to maximum sodium-current (right-hand scale). The smooth curve has been calculated. (Hodgkin & Huxley. *J. Physiol.*)

inactivation is the basis of the refractory period; it takes time to establish and a corresponding time to decay. The K^+-permeability shows no inactivation since, as we have seen, it is maintained at a steady value when voltage-clamps are maintained at a constant value. On removal of a depolarization the K^+-permeability decays, but persists for some time after the fall in the spike. Thus, immediately after the spike, the nerve is in a state of partial inactivation of the Na^+-permeability mechanism, whilst the K^+-permeability is still increased; both of these factors militate against the development of a new action potential, and so the refractory period may be accounted for.

Equilibrium Potential

If the axon is behaving, during the period of maximum Na^+-permeability, like a sodium-electrode, *i.e.*, behaving like a system governed only by the permeability to Na^+, then the potential that can be developed will be given by the Nernst equation:—

$$E = RT/F \ln \frac{[Na]_o}{[Na]_i}$$

With a ratio of activities of, say, 10, we may expect a value of V_e, the equilibrium potential, of some 58 mV. Experimentally, the value of V_e is given by the internal potential of the axon that just prevents the initiation of the early inward current under voltage-clamp conditions. By varying the external concentration of Na^+, Julian, Moore & Goldman (1962) obtained reasonable agreement between the predicted and measured values of V_e. When the internal concentration of Na^+ was varied, by employing the perfused squid axon, Adelman & Fok (1964) obtained the following values:—

Na_i	Na_o	V_e Calc.	V_e Exptl.
200	430	31·1	31·7
100	430	47·3	46·3
0	430	—	—

With zero internal concentration of Na^+ they failed to obtain an early outward current, and if the trans-membrane potential under these conditions was indeed "infinite", as predicted by the Nernst equation, this is understandable. As we shall see, however, the potential under these conditions is finite because the K^+-ion exchanges for incoming Na^+ (Chandler & Meves, 1965).*

Membrane Currents

Before proceeding farther, we may briefly recapitulate a few general principles regarding membrane currents that may not be immediately evident to those unaccustomed to discussing movements of ions in terms of current-flow. In general, diffusion of an ion will be associated with the movement of an ion of opposite charge, so that there will be no *net* flow of current; but it is customary to speak of a flow of positive and negative current; thus with KCl diffusing down a concentration gradient, we may speak of the K^+-current, I_K, carried by the K^+-ions, and the Cl^--current, I_{Cl}, carried by the Cl^--ions; under these conditions there is no net flow of current and I_K is said to be equal to $- I_{Cl}$. When the diffusion is associated with a change of potential, as, for example, when K^+ leaves the muscle fibre membrane when the external concentration is altered, the movement of the ion is associated with a flow of electrical current because the potential changes, and we may say that the flow of current is equivalent to charging or discharging the condenser constituted by the cell membrane separating the internal and external solutions; it will be given by the product of the capacity, C, and the rate of change of potential, dV/dt. When the membrane potential has achieved its new value, the flows of positive and negative ions become equal, and the *net* current becomes zero. According to the extent to which the current is carried by a given ion we may speak of the Na^+-current, K^+-current and so on. Conversely, we may apply a potential across a membrane different from its resting potential, and in this case the flow of current will correspond to the movement of ions along the gradient of electrical potential.

Let us consider, then, the case where a steady potential from an external battery is imposed on the existing resting potential of a cell; this has been represented by the circuit of Fig. 555a, the external battery E has been connected across the membrane, and in this case it is hyperpolarizing it, making the outside more positive. Positive current is shown flowing *inwards*, and we may expect to measure a membrane conductance, given by the ratio of current over applied voltage. In general, if the system considered corresponds to nerve, the inward current will be carried by K^+-ions, moving into the axon along the new electrochemical potential gradient, and by Cl^--ions moving outwards; but under conditions where the permeability of the membrane to K^+ greatly exceeds that to the anion, we may equate the membrane current to the K^+-current, I_K, and refer to the

---

* Tasaki and his collaborators (see, for example, Tasaki & Luxoro, 1964) described overshoots in the internally perfused axon that were considerably larger than the theoretically possible ones, on the basis of the Nernst equation; thus with 50 mM Na^+ inside and 500 mM outside, the theoretical V_e is about 50 mV, yet overshoots of 100–140 mV were obtained. Chandler & Hodgkin (1965) showed that these were most probably due to an electrode artefact; their own experiments showed that the overshoot never exceeded the theoretical potential based on the Goldman equation and assuming a P_{Na}/P_K ratio of 5.

K^+-conductance, g_K. This conductance is obviously related to the permeability of the membrane to this ion and is in fact given by the relationship:

$$g_K = F^2 \times \text{Unidirectional Flux}/RT$$

If the responses to applied potential are purely passive, we may expect the current-voltage relationship to be linear in accordance with Ohm's Law, and with hyperpolarizing currents this is, indeed, found. When the membrane is depolarized by an external potential, as in Fig. 555b, passive flow of current will be in the reverse direction, so that positive current will flow *outwards*; so long as the currents are passive, we may expect to measure a membrane conductance, as before. The conductance of the membrane to a given ion need not be the same

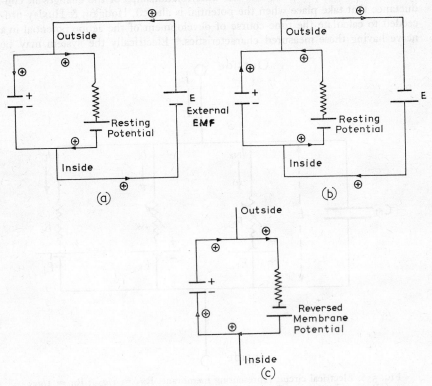

FIG. 555. Showing effects of applied electromotive forces to the excitable cell membrane. (a) A hyperpolarizing e.m.f. is applied; (b) a small depolarizing e.m.f.; (c) a larger depolarization has taken place leading to the development of the Na potential so that now there is an inward flow of positive current by contrast with the outward flow that occurs during passive depolarization shown in (b).

in both directions, even though the membrane is behaving passively; this *rectification* is a simple consequence of the unequal distribution of ions. Thus, the slopes of the current-voltage curves will be different according as current passes inwards or outwards.

With membranes that are electrically excitable, we have seen that, at a certain critical depolarization, the flow of current in the external circuit reverses; thus in Fig. 555c the flow of current has reversed, and this is made possible by a change in the nature of the membrane's own source of electromotive force. Under passive conditions, this source was the resting potential, but now, under these regenerative conditions, it has changed sign and has become the sodium-potential, determined by the difference of concentration of Na^+ across the membrane, which, in the squid axon, is some 40 mV, inside positive. The onset of the active state, during which positive current flows inwards, is equivalent, therefore, to switching off the resting electromotive force and replacing it with the "sodium-battery" of opposite polarity. Positive current, represented by a net flux of Na+-ions, will flow from outside to inside and will thus change the polarity of the charge across the membrane-condenser, making the inside positive. It will be clear that when the membrane becomes active the current-voltage relationship will not be simple, regenerative activity being equivalent to the resistance of the membrane becoming *negative*. See, for example, Fig. 575, p. 1105.

Predicted Action Potential. Having analysed, by the voltage-clamp technique, the dependence of the Na+- and K+-permeabilities, or conductances, on the membrane potential, and the time-relationships of the changes in conductance that take place when the potential is altered, Hodgkin & Huxley proceeded to calculate the time-course of development of the action potential in a nerve having these measured characteristics.* Electrically the system may be

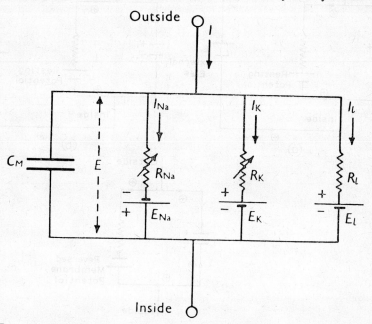

FIG. 556. Electrical circuit representing membrane. $R_{Na} = 1/g_{Na}$; $R_K = 1/g_K$; $R_l - 1/g_l$. R_{Na} and R_K vary with time and membrane potential; the other components are constant. (Hodgkin & Huxley. *J. Physiol.*)

represented by Fig. 556 where three batteries represent the potentials that the membrane can develop under appropriate conditions; thus at the phase of high Na+-permeability the Na+-battery may be considered to be switched on, and so on.

The fundamental equation describing the membrane current, I, is as follows:

$$I = C_m(dV/dt) + \bar{g}_K n^4(V-V_K) + \bar{g}_{Na} m^3 h(V - V_{Na}) + \bar{g}_l(V - V_l) \ldots (66)$$

The first term represents the passive charging or discharging of the membrane

* The calculation was applied to a relatively simple condition where it was imagined that the membrane was excited simultaneously along its whole length, so that throughout the sequence of changes the potential at any point was always equal to that at any other point; there was thus no flow of current longitudinally along the axon as would, of course, occur were the action potential propagating. The action potential obtained in this way is described as a "*membrane action potential*"; the only current-flow would be devoted to charging or discharging the membrane-condenser, so that $C_m dV/dt$ is equal to the sum of the Na+, K+ and leakage currents, *i.e.*, I in Equation (66) may be put equal to zero.

capacity in response to a change of potential; the second term describes the K$^+$-current as the product of the partial K$^+$-conductance,* $\bar{g}_K$ (which as we have seen in an index to K$^+$-permeability) and the potential causing a K$^+$-current which will be the difference between the actual membrane potential, V, and the equilibrium Nernst potential, V_K, computed from the K$^+$-concentrations. The term n^4 describes the variations of g_K with potential in accordance with empirical equations derived from the voltage-clamp experiments; thus

$$dn/dt = \alpha_n(1 - n) - \beta n$$

$$\alpha_n = 0\cdot01(V + 10)/\left(\exp\frac{(V + 10)}{10} - 1\right)$$

and

$$\beta_n = 0\cdot125 \exp (V/80)$$

α_n and β_n being described as the rate-constants determining the variation of K$^+$-conductance with time under an applied potential. The third term describes the Na$^+$-current, which, besides the term m has an extra factor, h, which, as we have seen, indicates the inactivation process; m is the factor determining the dependence of the Na-conductance on potential and is described, like n, by a set of differential equations:

$$dm/dt = \alpha_m(1 - m) - \beta_m$$

$$\alpha_m = 0\cdot1(V + 25)/\left(\exp\frac{V + 25}{10} - 1\right)$$

$$\beta_m = 4 \exp (V/18)$$

whilst h is similarly determined by a set of equations:

$$dh/dt = \alpha_h(1 - h) - \beta_h$$

$$\alpha_h = 0\cdot07 \exp (V/20)$$

$$\beta_h = 1/\left(\exp\frac{V + 30}{10} + 1\right)$$

The final term represents a leak current which takes into account the fact that the membrane has a finite permeability to anions, so that the full theoretical values of the K$^+$- and Na$^+$-batteries are not developed; the leak current becomes zero when the membrane potential, V, is equal to the leak potential, V_l, which is given experimentally by the condition when the membrane is depolarized by some 10·6 mV.

As Fig. 557 shows, there is reasonable agreement between the calculated and

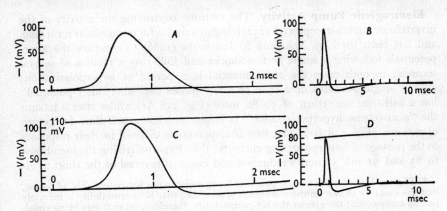

FIG. 557. A and B: Calculated action potential on two different time-scales. C and D recorded action potentials from squid axons on approximately the same vertical and horizontal scales as those employed for the calculated action potentials. (Hodgkin & Huxley. *J. Physiol.*)

* $\bar{g}_{Na}$ and $\bar{g}_K$ are the maximal conductances developed, and for the squid axon are 120 and 36 mmho.cm.$^{-2}$ respectively at 6·3° C.

recorded spikes, when allowance is made for the difference of temperature at which the measurements were made.

Positive After-Potential. The positive after-potential, which makes the spike in the squid axon diphasic, turns out to be a necessary consequence of the persistence of the raised K^+-permeability. In an ideal axon, *i.e.*, one in which the resting potential was equal to the K^+-potential, a change in K^+-permeability should not affect the resting potential, this being simply determined by the concentrations of the ion, and not its mobility. The isolated axon, however, has a resting potential considerably less than its K^+-potential, so that it represents an unstable system that is running down; in this case, the value of the potential depends on the other ions in the system besides K^+ and the equation describing it contains the mobility of K^+ (p. 571); increasing this tends to increase the resting potential, bringing it closer to a pure K^+-potential*; thus the rise in the resting potential, which appears as a positive after-potential,† may be explained.

Post-Tetanic Hyperpolarization

In connexion with the positive after-potential following a single spike we may mention, here, the more prolonged effects of repetitive stimulation of nerve fibres, manifest in a phase of hyperpolarization; this might be explained on the same basis, namely a prolonged high permeability to K^+, but it could be due to changes in the concentration of K^+ outside the nerve fibre opposite to those normally expected, due to a very rapid re-accumulation of K^+, as suggested by Ritchie & Straub (1956), or finally it could be due to the operation of an electrogenic Na^+-pump as indicated in Chapter X (p. 635). Hyperpolarization due to excessive reabsorption of K^+ was invoked to account for the situation in small C-fibres; reabsorption would deplete the small spaces between the fibre and its Schwann cell (p. 480). Poisoning the active-transport mechanism with DNP or ouabain, or exposure of the bundle to solutions containing deficient K^+, abolished the hyperpolarization, as we should expect on the basis of this explanation. That the phenomenon is only observed in very small fibres is presumably due to the large value of the quotient: Area/Volume, permitting very large increases in internal Na^+-concentration, and it is thought that these large changes would stimulate the active-transport mechanisms; at any rate, Hodgkin & Keynes (1956) showed that a raised internal concentration of Na^+ increased the efflux of this ion from the giant axon of the squid.

Electrogenic Pump Activity. The neurone controlling the activity of the invertebrate stretch-receptor (p. 1252) is highly suitable for intracellular recording, and has been used by Nakajima & Takahashi (1966) to compare the after-potentials following a single or few shocks and following a tetanus of several seconds; in both cases the after-potential is a period of hyperpolarization; after a single stimulus or a train of a few pulses the "short after-potential" has a half-time for return of 50–80 msec (Fig. 558 A1) whilst after a tetanus the "post-tetanic hyperpolarization" is larger and has a half-time for return of 2–6 sec. (Fig. 558 B1). The two after-potentials differed in their responses to the passage of hyperpolarizing currents; thus hyperpolarizing the membrane to 83 and 91 mV caused a reduction and eventual reversal of the short after-

---

* The condition for the resting potential being equal to the K^+-potential is, of course, that the mobility of Na^+ in the membrane is zero, *i.e.*, that K^+-permeability is infinitely fast by comparison; the greater the K^+-permeability, therefore, other things being equal, the nearer will the resting potential be to the K^+-potential. This is the third mechanism concerned with the production of after-potentials, the other two being accumulation of K^+ outside the fibre or the over-activity of the accumulation process (p. 1069 and *above*); Ritchie (1961) has given an excellent discussion of all aspects of the production of after-potentials in nerve fibres.

† In the recording of the spike, positivity inside is indicated upwards; an increase in the resting potential means a decrease in positivity inside, and so is represented downwards.

potential (Fig. 558 A2, A3) whilst the same treatment only increased the post-tetanic hyperpolarization (B2, B3). The reversal point for the short after-potential was around -82 mV; such a reversal would be expected were the after-potential due to delayed rectification, *i.e.* increased permeability to K^+; thus hyperpolarizing the membrane beyond the Nernst K^+-potential would lead to a potential after the stimulus less than this hyperpolarized potential, *i.e.* the record would finish above the base-line and give a "negative after-potential".

The enhancement of the post-tetanic after-potential suggests a different mechanism, and its abolition by replacement of external Na^+ by Li^+ suggests that it is closely related to pumping of Na^+ out of the cell; this could be due to reduction of the outside concentration of K^+, as suggested by Ritchie & Straub, or to the electrogenic nature of the Na^+-pump. Nakajima & Takahashi decided

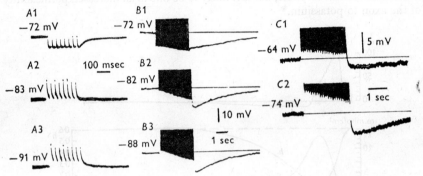

FIG. 558. Effects of conditioning hyperpolarization on the after-potentials of the stretch-receptor neurone of the cray-fish. *A:* short after-potential. *B* and *C:* post-tetanic hyperpolarization (PTH). *A* and *B:* records from the same cell. *C:* another cell. A_1, B_1, and C_1 show the after-potentials at the membrane potential without conditioning current. With conditioning hyper-polarizations, the polarity of the short after-potential was reversed (A_2 and A_3). but PTH became larger (B_2, B_3, C_2). The membrane potentials are indicated over each base line. Tetanus was induced by intra-cellularly applied short current pulses of 31/sec in B, and 50/sec in C. The spikes are off the trace at the amplification used. (Nakajima & Takahashi. *J. Physiol.*)

in favour of the latter hypothesis because, at the height of the hyperpolarization, there seemed to be little or no change in the equilibrium K^+-potential, E_K. Thus, if we assume that at the height of the short after-potential the membrane has such a high K^+-permeability that the membrane potential is about equal to E_K, then we should expect the magnitude of this short after-potential to be increased by a reduction of outside K^+ concentration; in fact measurement of this after-potential during different degrees of post-tetanic hyperpolarization showed that it was the same as that obtained during an artificial hyperpolarization of equal amount. If we accept the implied assumption that this artificial hyperpolarization does not affect the concentration of K^+ outside the fibre appreciably, then this agreement between the two sets of measurements is good evidence favouring the "electrogenic pump" hypothesis.*

* A state of hyperpolarization means a state of reduced excitability, since a greater reduction in membrane potential is necessary to initiate a spike; Gage & Hubbard (1966) have used the decreased excitability of mammalian motor neurones, measured by their responses to antidromic stimulation at their junctions with the muscle, as a measure of post-tetanic hyperpolarization. From the effects of variations in the external concentrations of K^+ and Na^+, of de- and hyper-polarizing currents, and of application of metabolic and

Mammalian Non-Myelinated Fibres

Experiments on non-myelinated C-fibres by Rang & Ritchie (1968) have led to the same conclusion; the after-potential is normally only one or two millivolts, but this can be increased to as high as 35 mV by replacing Cl^- in the medium by the slowly penetrating isethionate ion. The effect takes some 15 minutes to develop, so it seems that it is due to washing out the Cl^- from the fibre; the extrusion of Na^+, if not associated with the intake of a K^+-ion, will be much more highly electrogenic now that it can no longer take place in company with a relatively permeable anion like Cl^-. Ouabain, or replacement of external Na^+ by Li^+, both inhibit the development of the isethionate-after potential; furthermore, K^+, applied to the nerve after stimulation in K^+-free medium, causes a hyper-polarization rather than the normal depolarization, showing that the hyper-polarization cannot be due to a fall in periaxonal concentration of K^+ resulting from its active reabsorption; nor is it likely to be due to an increased permeability of the axon to potassium.[*]

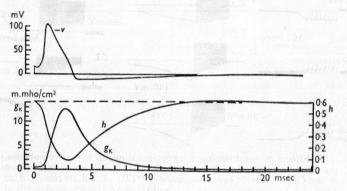

Fig. 559. Calculated action potential (upper graph) with curves (lower graph) showing the time-course of the potassium-conductance (g_k) and the parameter h indicating inactivation of the sodium-mechanism. (Hodgkin & Huxley. *J. Physiol.*)

Refractory Period. Not only could the shape of the action-potential be predicted but also the velocity of propagation and the time-course of the change of impedance, and once again excellent agreement between theory and experiment was obtained. The refractory period is due, as we have indicated, to the persisting inactivation of the Na^+-permeability system, associated with persistence of the raised permeability to K^+. The duration of these effects, during and after an action potential, is illustrated in Fig. 559; it will be seen that the power of the membrane to increase its permeability to Na^+, indicated by the parameter h (p. 1080), is at a minimum some 3 msec. after the stimulus, and it remains impaired for as long as 13 msec. The K^+-permeability remains higher than normal for 25 msec. or more. Calculation showed that, up to 5-6 msec. after the stimulus, the nerve should be absolutely refractory; after this, small spikes should be evoked, and even up to 10 msec. they should be of less than normal height. In practice this was found.

active-transport inhibitors, they concluded that the hyperpolarization was essentially the result of an increased K^+-permeability, *i.e.*, the delayed rectification that is at the basis of the positive after-potential following the spike. On this basis, then, both "short after-potential" and post-tetanic hyperpolarization have the same cause.

[*] Rang & Ritchie point out that the large effect of anion substitution indicates that the mammalian C-fibre has a relatively high permeability to Cl^-, being thus similar to muscle.

Effects of Temperature. By employing a digital computer to the solution of the equations defining the action potential, the work was enormously reduced so that it became practicable to deduce the effects of temperature on the essential features of the travelling spike. The results, based on the primary assumption that the main effects of temperature were rather on the variations of permeability with time and potential (the α's and β's of the equations), are illustrated by Fig. 560 where the upper records are the responses of the squid giant axon at

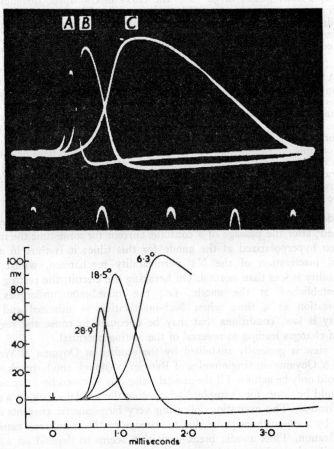

FIG. 560. Top: Superimposed records of action potentials at (A) 32·5°C, (B) 18·5°C and (C) 5°C. The respective amplitudes are 74·5, 99, and 108·5 mV; time marks are 1 msec.
Bottom: Propagated action potentials computed for the squid axon at 3 temperatures allowing only for the effect of temperature in increasing the rates of change of the permeabilities. Each spike is drawn with a conduction time (measured at 30 per cent. of the height of the spike) inversely proportional to its conduction velocity. (Huxley. *Ann. N. Y. Acad. Sci.*)

three temperatures whilst the lower curves are calculated responses; it will be seen that spike-height increases with decreasing temperature whilst the velocity, indicated roughly by conduction time, decreases. Rate of rise of the spike increases with increasing temperature. An interesting feature in the examination of the effects of temperature on conduction velocity was that the latter should theoretically reach a maximum at about 33° C, and this was indeed found by Hodgkin & Katz (1949), propagation being abolished above about 38° C and returning when the temperature was lowered. The reason for this is essentially that the effect of a rise of temperature on the recovery processes—inactivation of Na⁺-permeability and increase in K⁺-permeability—is more pronounced than on the initial process, namely the increase in Na⁺-permeability.*

Threshold

The effects of temperature on threshold depend on the duration of the stimulus; with 100 msec. stimuli there is a steady increase with increasing temperature, whilst with short pulses of 50 μsec. the curve is U-shaped, threshold passing through a minimum at about 15 °C (Guttman, 1962). Fitzhugh (1966) has applied the Hodgkin–Huxley equations, on the basis of Moore's experimental finding that the membrane conductance increases by 4 per cent. of its value at 15 °C for each degree centigrade. The ionic conductances were multiplied by the factor:—

$$A \left[1 + B(T - 6·3) \right]$$

where A is the ratio between the ionic conductance at 6·3 °C and the value used by Hodgkin & Huxley, which they assumed independent of temperature. A depends on the physiological condition of the axon. B is determined by the rate of change of conductance with temperature; with Moore's value of 4 per cent. per degree, this is 0·061. The solutions to the equations predicted the actual effects with both long and short pulses. Thus with short pulses, at low temperatures, the decrease in relaxation time of Na⁺-activation (m) relative to the electrical relaxation time (RC) favours activation and decreases threshold; at higher temperatures the effect on m saturates, but the decreasing relaxation times of Na⁺-inactivation (h) and of K⁺-activation (n) favour accommodation and increase the threshold, hence the U-shape of the threshold-temperature curve.

Anode-Break. The phenomenon of anode-break stimulation may also be explained; after the passage of a constant current for some time the membrane has been hyperpolarized at the anode for this time; it is thus in a state of minimal inactivation of the Na⁺-permeability mechanism, whilst the K⁺-permeability is less than normal. On breaking the circuit, the resting potential is re-established at the anode, *i.e.*, the membrane undergoes a rapid depolarization at a time when Na⁺-inactivation is minimal and K⁺-permeability is low, conditions that may be adequate to cause the regenerative series of changes leading to reversal of the resting potential.

This view is generally sustained by the studies of Ooyama & Wright and Wright & Ooyama on single nodes of Ranvier, although anode-break excitation here could only be achieved if the anodal pulse was followed by a depolarization. This could be done, for example, by imposing the anodal shock on a sustained depolarization. The excitation following very large anodic currents was considered by these authors to be due to damage to the membrane causing local depolarization. Thus anodic break excitation seems to depend on a tendency towards depolarization, which may be induced artificially, or may be associated with deterioration of the condition of the tissue.

Accommodation. Accommodation, the increase in threshold associated with passage of a current, is also explicable in terms of changes in the activity of the Na⁺-permeability system. Passage of a constant current causes the K⁺-permeability to rise, and the inactivation of the Na⁺-system to increase at the cathode; both factors raise the threshold and so give rise to accommodation. The phenomenon of minimal current gradient (p. 1054) has, as its basis, the phenomenon of accommodation. It will be recalled that a

---

* Cole and his collaborators have carried out a valuable analysis of the possible errors in the voltage-clamp technique and have examined minutely the variation in K⁺-conductance over a wide range of membrane potential. In general, the Hodgkin-Huxley approach emerges unscathed, although the equation relating K⁺-conductance to membrane potential is modified slightly (Cole & Moore, 1960; Cole, 1961; Taylor, Moore & Cole, 1961).

given applied potential may stimulate if developed quickly, but fail to stimulate if allowed to grow very slowly; for example, a depolarization of 30 mV rapidly established causes an action potential; if a sub-threshold depolarization of 5 mV were established and maintained there would be a partial inactivation of the Na⁺-mechanism and an increased K⁺-permeability; on increasing the depolarization by another 5 mV, because of these effects, the depolarization would still be inadequate to initiate the regenerative response and, once again, the Na⁺-mechanism would be further inactivated and the K⁺-permeability increased; and so on.

Variations in Accommodation

Nerve fibres vary considerably in their power of accommodation, and this is of great physiological significance since the power to accommodate determines, in effect, whether a sustained stimulus will produce more than a single action potential (Chapter XVII); an analysis of the excitability parameters of nerve fibres, showing different accommodation, will be of value in showing which of these are important. On theoretical grounds, variations in the rate-constants, α and β, as well as the permeability constants, P_{Na}, P_K and P_p,* could all affect this feature of the excitable system. Vallbo (1964) examined single myelinated fibres under voltage-clamped conditions when linearly rising currents were applied, whilst Frankenhaeuser & Vallbo (1965) made computations of the effects of variations in the basic

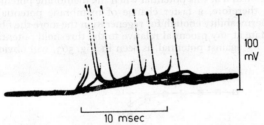

FIG. 561. Membrane potential records from a single node when stimulated by linearly rising currents of various slopes. Seven superimposed records. (Vallbo. *Acta. physiol. scand.*)

parameters, on the basis of a model fibre and using the appropriate modifications of the Hodgkin–Huxley equations to myelinated nerve (p. 1096). Fig. 561 illustrates seven superimposed records resulting from applying linearly increasing currents of decreasing slope, dI/dt,; in six, the linearly increasing current produced action potentials and in one, the most slowly rising current, there was only a hump with delayed rectification. Regenerative activity started at about the same potential regardless of the slope of the stimulating current, so that essentially the *threshold* was unchanged. It will be seen that the action potentials decreased in size progressively as the slope, dI/dt, was decreased; and failure was sudden, so that it was not difficult to assess the critical value of dI/dt required to stimulate. With individual fibres this varied† from 4·5 to 75·7 mA.cm⁻².sec⁻¹. In Fig. 562 are the computed responses to linearly rising currents on the basis of the model nerve, and the qualitative agreement between theory and experiment is satisfactory. Fig. 563 shows the steady-state inactivation curves of three fibres; A had 10 times the critical slope of C whilst B had an intermediate value; it is clear that, for a given degree of depolarization, the strongly accommodating fibre (A) has a greater degree of steady-state inactivation than that of the less strongly accommodating fibres. During a slowly rising current there will be a continuous decrease in h before regeneration occurs, and the slower the rise in membrane potential, the closer will h be to its steady-state value, h_∞, at any potential. The threshold is approximately the minimum potential at which regenerative activity will begin. The lower the

* For the definition of P_p, in relation to myelinated nerve, see p. 1096.
† Vallbo points out that a better measurement of the critical slope, when comparing one fibre with another, is given by rheobases/sec.; in this case the slopes varied from 35 to 256 rheobases/sec.

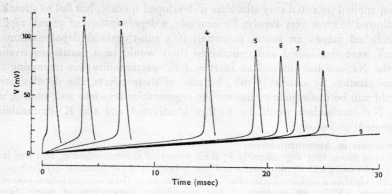

FIG. 562. Illustrating accommodation in nerve. Computed membrane potential changes in response to step-current stimulation (1) and linearly rising currents (2) to (9) with rates of rise varying from 125mA/cm²/sec (2) to 21·5 mA/cm²/sec (8) and 21·2 mA/cm²/sec (9). (Frankenhaeuser & Vallbo. *Acta. physiol. scand.*)

steady-state value of h at a given potential in relation to threshold, the lower will be the actual value of h at this potential when the membrane potential changes with a certain rate; therefore, a faster change of membrane potential is required to increase Na⁺-permeability enough for regeneration the more the fibre is inactivated in the steady-state at any ptotential relative to the threshold potential. Thus a shift of the curve of h_∞ against potential, as seen in Fig. 563, will obviously change the

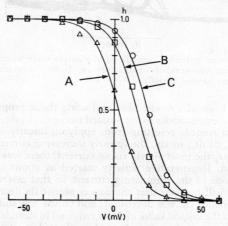

FIG. 563. Steady-state inactivation of the sodium mechanism, h, has been plotted as ordinate against membrane potential as abscissa. Axon A is represented by the triangles; axon B by squares and C by circles. A had the steepest critical slope, i.e. accommodated fastest, and has the greatest degree of inactivation for a given value of the membrane potential. (Vallbo. *Acta. physiol. scand.*)

critical slope, dI/dt, required to excite. The degree of inactivation depends on the rate-constants α_h and β_h, the equations predicting that a shift along the potential axis corresponds to a higher value of α_h and/or a lower value of β_h. Analysis of the voltage-clamp data indicated that β_h hardly varied, whereas α_h varied by a factor of 7·7. Frankenhaeuser & Vallbo (1965) have, as indicated above, examined the effects of varying all of the significant parameters in the Hodgkin–Huxley equations and have shown that almost every one of these alters, to some extent, the critical slope of dI/dt; a change that decreased the outward ionic current or increased the inward current at potentials close to the threshold potential making the accommodation slower, and *vice versa*. Their analysis showed that, as found experimentally, the inactivation of the sodium mechanism had a marked effect on the critical slope; in addition, the turning on of the K^+-permeability, indicated by n, had some effect, as well as the permeability constants for Na^+ and K^+, whilst the leak conductance did not affect the critical slope.*

Threshold. It is obvious from the phenomenon of rheobase that a simple definition of threshold for excitation of a tissue is only feasible under very limited conditions; of more significance is the strength-duration curve, but here, again, the "constants" defining the curve depend greatly on the mode of stimulation, whilst their theoretical significance is limited by the essentially empirical nature

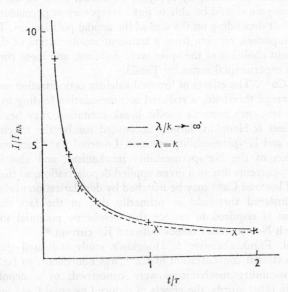

FIG. 564. Strength-duration curve for a uniformly polarized membrane (x) and a point-polarized cable (+) whose membrane ionic currents obey the Hodgkin—Huxley equations. The lines indicate the range of prediction of Hill's model as the ratio λ/k is varied from 1 to ∞. (Noble & Stein. *J. Physiol.*)

of the equations defining them. Cooley, Dodge & Cohen (1965) and Noble† & Stein (1966) have applied the Hodgkin–Huxley concepts to this particular aspect of excitability; thus Fig. 564 gives the dimensionless predicted strength-duration curve in terms of I/I_{Rh} and t/τ, where I is the exciting current and I_{Rh} is the rheobasic current, t is the duration of the stimulus and τ is defined by: $\tau = \underset{t\to\infty}{Lim}\ I.t/I_{Rh}$.

The points are predicted on the basis of current-voltage membrane relations defined by the Hodgkin–Huxley equations, and the lines are the limits of the predictions derived from Hill's equation when λ/k, as defined by Hill, varies from 1 to infinity.

* Frankenhaeuser & Vallbo's analysis indicated that the rheobasic current was nearly proportional to the leak conductance; this is understandable since the existence of rheobase, *i.e.*, of a current that will not excite even though of indefinite duration, can be explained by a "leak" such that the applied current fails to establish the required depolarization.

† Noble (1966) has reviewed the application of Hodgkin–Huxley equations to a variety of excitable tissues.

Delayed Rectification. Since the K^+-conductance of the membrane increases when the membrane is depolarized, and decreases when it is hyperpolarized, we may expect rectification of an alternating current on this ground. To distinguish this from the *anomalous rectification* that is manifest in skeletal muscle and cardiac muscle in particular (p. 575), this form of rectification is referred to as delayed rectification. Cole & Curtis noted that the squid giant axon behaved as a rectifier, the outwardly directed current associated with a steady depolarization being greater than the inwardly directed current due to a hyperpolarization of the same magnitude.

Abolition of the Spike. The rising phase of the spike depends on a specific increase in the Na^+-permeability whose magnitude is determined at any moment by the membrane potential; by applying an anodal pulse to the nerve during the rising phase we should be able to put a temporary or permanent stop to the spike, the effect depending on the size of the anodal polarization. Theoretically, a series of responses, ranging from a transient repolarization of the membrane to a permanent abolition of the spike were deduced, and these compared quite well with an experimental series by Tasaki.

Effect of Ca^{++}. The effects of lowered calcium concentration on excitability, namely a lowered threshold, a reduced accommodation leading to spontaneous firing and a tendency towards anodic break excitation, may be, according to Frankenhaeuser & Hodgkin, largely accounted for by the effects of this ion on the Na^+- and K^+-permeability systems. Lowered Ca^{++} increases the degree of inactivation of the Na^+-permeability mechanism, and also increases the Na^+- and K^+-currents due to a given applied depolarization, so that, in general, the effects of lowered Ca^{++} may be matched by depolarization of the membrane. Thus, the lowered threshold is primarily due to the fact that a smaller depolarization is required to reduce the membrane potential to the critical value at which Na^+ current exceeds outward K^+ current.*

In general, Frankenhaeuser & Hodgkin's study indicated that a fivefold decrease in external concentration of Ca^{++} was equivalent, so far as the Na^+- and K^+-permeability mechanisms were concerned, to a depolarization of 10-15 mV. In other words, the effects of reduced external Ca^{++} on the electric responses of the squid axon should be predictable by very simple alterations in the parameters governing these permeabilities (the α's and β's of the Hodgkin–Huxley equations). As we shall see (Chapter XVII), the oscillatory type of response, leading to bursts of spikes in response to a single stimulus, is a predictable consequence of this alteration of the permeability parameters (Huxley, 1959).†

Myelinated Nerve

Huxley & Stämpfli (1951) demonstrated the importance of external Na^+ for the development of the action potential at a single node of Ranvier in myelinated nerve, whilst Frankenhaeuser and Dodge & Frankenhaeuser applied

---

* Removal of *all* the Ca^{++} from squid nerve causes a state of block, due to a high K^+-conductance and almost complete inactivation of the Na^+-permeability mechanism; the blockage of the myelinated frog nerve by complete removal of Ca^{++} is very rapidly and reversibly achieved; the removal of Ca^{++} must be very complete, the amounts of this ion normally present in reagent grades of Na^+ being sufficient to maintain excitability.

† The studies of Dalton & Adelman (1960) and of Adelman & Dalton (1960) on the effects of Ca^{++} and K^+ on the resting and action potentials of lobster giant axons may well be found to be completely consistent with the Hodgkin–Huxley theory when the effects of the Ca^{++}-ions on the appropriate parameters of Na^+- and K^+-conductance have been determined with precision. Heene (1962) has examined the effects of pH on the resting and action potentials; acid solutions have the same effects as those of a high concentration of Ca^{++}, and may well be related to their influence on the activity of this ion at the membrane.

the voltage-clamp technique to the study of the Na+ and K+ currents at different stages of depolarization of a single node. In general, their studies showed a remarkable agreement, both qualitative and quantitative, with those described by Hodgkin & Huxley for the squid giant axon.

Sodium Currents. Thus Fig. 565 shows the peak initial currents following stepwise depolarizations of the node, the membrane potential being maintained at the degree of depolarization by clamping. It will be seen that the current passes through a maximum and becomes zero at a depolarization of just over

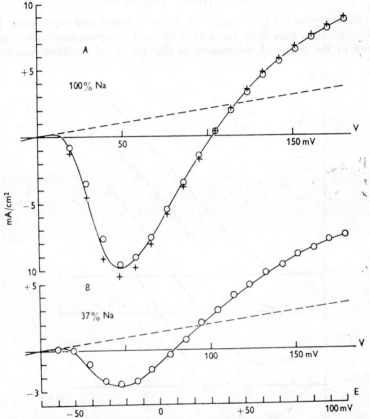

FIG. 565. Peak initial (sodium) currents for voltage-clamped myelinated axon plotted against V, and E in normal Ringer's solution (A) and in Ringer's solution containing 37 per cent. of the normal Na+-concentration (B). Here E is the absolute value of the membrane potential and V is the value in relation to the resting potential, *i.e.* V = E − E$_r$. Interrupted line represents anodal polarization. (Dodge & Frankenhaeuser. *J. Physiol.*)

100 mV which, on the basis of the Hodgkin–Huxley theory, represents the Na+-potential. In B of Fig. 565 we have the same curve for the condition where the external concentration of Na+ has been reduced to 37 per cent. of normal. Not only has the current decreased but also the reversal point has fallen to about 80 mV and this is reasonably close to that calculated from the Nernst equation. The Na+ mechanism showed a similar inactivation process when a depolarization was sustained and, in general, the Na+-permeability, calculated from the peak currents, could be described by similar equations relating it to membrane potential and time as those determined by Hodgkin & Huxley.

35*

Potassium Currents. In spite of the claim of Tasaki & Freygang to the contrary, the myelinated axon also showed a delayed current that could be ascribed to an increased permeability to K^+; this was demonstrated by applying the Goldman equation to calculate the K^+-permeability from the current, I_K, and the membrane potential, E. Thus under two conditions of external concentration $[K]_0$ and $[K']_0$ the ratio of the currents carried by K^+ should be given by:—

$$\frac{I'_K}{I_K} = \frac{[K']_0 - [K]_i \exp EF/RT}{[K]_0 - [K]_i \exp EF/RT}$$

Fig. 566 illustrates the good agreement between theory and experiment. Here the depolarization caused by the altered external K^+-concentration was counteracted by the feed-back mechanism so that the initial condition was one at

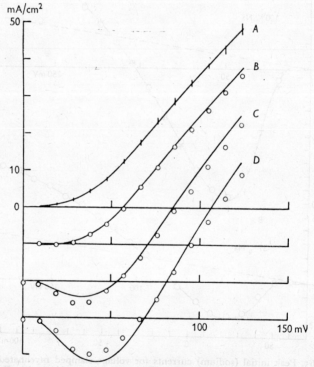

FIG. 566. Experimental and calculated delayed feed-back currents following step-wise depolarization of the myelinated fibre, when the external concentration of K^+ was 2·5 mM (A), 12·5 mM (B), 62·5 mM (C) and 114·5 mM (D). Ordinate scale is for A. The origin for B, C, and D has been successively shifted 10 mA/cm² downwards. Bars in A indicate extreme values for five runs. Continuous lines of B, C, and D have been calculated from the Goldman equation employing the smooth line drawn through the experimental points in A. (Frankenhaeuser. *J. Physiol.*)

which the membrane potential was equal to the resting potential. The curves show the currents induced by stepwise depolarizations from this resting condition. An interesting feature of this figure is that, at high external K^+-concentrations, the effect of a depolarization is to cause an inward current, *i.e.*, to increase the effect of the passive depolarization. As indicated earlier, the passive response to an imposed depolarization is an outward flow of positive current, in accordance with Ohm's Law; this passive flow tends to oppose the depolarization process;

the *active* or regenerative response to a depolarization, on the other hand, is manifest as an inward flow of positive current that continues the depolarization. Thus, in high external K^+-concentrations, a depolarization may have a regenerative action, leading to a minor form of action potential. Hence the apparently anomalous action potentials described by Mueller (1958) and Moore (1959) are predictable from the ionic hypothesis provided that the current is predominantly carried by the K^+-ion.*

Computed Action Potential. Quantitative studies of the voltage and time-dependence of the potassium currents revealed an essentially similar situation to that described for the voltage-clamped squid axon, but the repolarizing, or

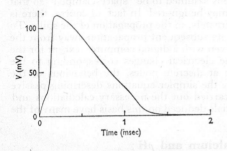

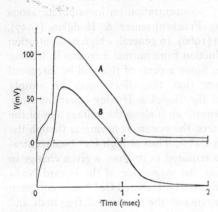

FIG. 567. Above: Computed action potential for myelinated nerve of toad. Below: Recorded action potentials from two axons. Ordinates for *B* shifted, 50 mV down. Axon in *A* polarized continuously with a weak anodal current in order to show the slowly decaying after-potential in this situation. (Frankenhaeuser & Huxley. *J. Physiol.*)

delayed, current could not be described entirely in terms of a permeability to K^+ since in its later stage there was evidence that some of it was mainly carried by Na^+ rather than K^+, so that, in order to define the system, extra parameters, P_p (permeability constant for the ions carrying the current) and α_p and β_p had to be measured and defined (Frankenhaeuser, 1963). Having measured the necessary parameters the next step was to compute the theoretical action potential; this was done by Frankenhaeuser & Huxley (1964) and the result is illustrated by Fig. 567; the agreement with an experimental record is sufficiently striking; the absence of a period of hyperpolarization (positive after-potential, p. 1085)

* The lobster giant axon has too small a diameter (*ca.* 100μ) to make insertion of a long internal electrode, necessary for voltage-clamping, practicable. Being unmyelinated, an isolated node cannot be employed for voltage clamping. Julian, Moore & Goldman (1962) have made an artificial node by exposing a small region to sea-water whilst the adjoining regions were exposed to isotonic sucrose. Thus the sheath resistance is raised sufficiently high that the potential measured across the inside of the axon and the surface exposed to sea-water is very close to the true membrane potential. With this technique they were able to demonstrate the essential similarity of this axon to that of the squid, so far as the currents under voltage-clamp conditions were concerned.

distinguishes the myelinated fibre from that of the squid, and is due to the prompt return of the phase of high K^+-permeability to its resting condition. In a later paper Frankenhaeuser (1965) computed some of the excitability properties of the theoretical membrane; thus subthreshold, and just threshold, stimuli gave predicted responses comparable with those described by Hodgkin (p. 1061). Other features, such as the strength-duration curve, and the effects of temperature and external Na^+ concentration on the spike, were also dealt with.

Impulse Initiation. The computed action potentials of Fig. 557 and 567 have been calculated for the simplified situation in which the potential, current and other variables relating to the state of the membrane vary with time but not with distance along the fibre; in other words, the fibre is assumed to be "space-clamped" so that interactions between adjacent portions may be ignored. In fact, of course, there is the spatial variation that is manifest, ultimately, in the propagation of the spike. To compute the growth of an impulse and its subsequent propagation away from the point of stimulation is not practicable, even with a digital computer, except for the case of the myelinated axon, where the electrical changes corresponding to the Hodgkin–Huxley equations occur only at discrete nodes, the behaviour of the intervening stretches being governed by the simpler equations describing passive cable properties. Fitzhugh (1962) has carried out the necessary calculations and shown that the propagated action potentials deduced on this basis have many of the features described experimentally.

Effects of Calcium and pH

Calcium. The effects of altered Ca^{++}-concentration on invertebrate axons have been measured and discussed by Frankenhaeuser & Hodgkin (1957), Huxley (1959) and Blaustein & Goldman (1966). In general, a high concentration of Ca^{++} decreases excitability and a reduction from normal increases it, leading to spontaneous and repetitive excitation. Some aspects of this will be discussed in Chapter XVII; here we may note that the effects may be simply described in terms of the parameters of the Hodgkin–Huxley equations since they can be considered to result from a simple shift along the voltage axis of the values of the voltage-dependent parameters, the system behaving as though the membrane potential were raised by high Ca^{++}. Thus at high Ca^{++}-concentration a greater degree of depolarization is required to produce a given change in m_∞ or h_∞, the parameters that determine the magnitude of the inward Na^+-current. Hille (1968) has measured the effects of both altered Ca^{++}-concentration and altered pH on the Na^+- and K^+-currents of the myelinated frog node and found, so far as Ca^{++} is concerned, essentially the same effects, which could be analysed in terms of alterations of the parameters in the Hodgkin–Huxley equations as indicated in Table LXXVII, where the effective shifts in voltage produced by an e-fold change of Ca^{++}-concentration are given.

TABLE LXXVII

Voltage Shifts of the Relation between Parameters of Na^+-and K^+- Activation and Voltage due to an e-Fold Change in Concentration of Ca^{++} or H^+ (Hille 1968).

Cation	Voltage Shift					
	m_∞	τ	h_∞	τh	n_∞	τn
Ca^{++} (0·45–22 mM)	8·7	8·4	6·5	8·4	1·6	0·0
H^+ (pH 4·1–5·5)	13·5	—	13·5	—	—	—
H^+ (pH 5·5–10·1)	1·3	—	1·3	—	—	4·0

Here the time-constants, τ_m and τ_n, are given by: $\tau_n = 1/(\alpha_n + \beta_n)$, $\tau_h = 1/(\alpha_h + \beta_h)$ and so on.

pH. The effects of altered *p*H, when on the acid side of neutrality, were striking; Fig. 568 shows the effects on $\bar{g}_{Na}$, the relative amplitudes of the Na$^+$-conductance during the spike, and the voltage shifts of m_∞ and h_∞. The curve of $\bar{g}_{Na}$ versus *p*H is reminiscent of a dissociation curve for a weak acid of pK_a 5·2, and suggests that the carrier-mechanism for Na$^+$ involves a weakly acidic group that becomes ineffective when it is protonated. A similar sensitivity of the carrier-mediated Na$^+$-permeability of the cat erythrocyte to *p*H changes on the acid side of neutrality has already been described (p. 461).[*]

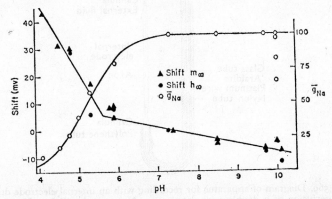

FIG. 568. The effects of *p*H on sodium parameters of the myelinated frog node. The shifts of the voltage-dependence of the parameters, m_∞ and h_∞, and the relative amplitudes of the sodium conductance, $\bar{g}_{Na}$, are plotted as a function of the *p*H of the medium. (Hille. *J. gen. Physiol.*)

The Perfused Giant Axon

In Chapter X we referred briefly to the possibility of extruding the axoplasm from the squid's giant axon and subsequently perfusing solutions of known composition through the remaining tube, consisting of the axon membrane and sheath. The experimental arrangement that permitted recording of resting and action potentials with an internal electrode is illustrated in Fig. 569, whilst action potentials obtained from the normal axon and from one perfused with isotonic K$_2$SO$_4$, are shown in Fig. 570. The similarity is sufficiently striking to justify the use of this preparation in studies on the effects of variations of the internal concentration on the electrical properties. Increases of the internal Na$^+$-concentration reduced the overshoot in a reversible manner; this was not due to the reduction of internal K$^+$-concentration, since replacement of the K$_2$SO$_4$ with isotonic sucrose actually increased the overshoot. Replacement of internal K$^+$ by Li$^+$ had about the same effect as Na$^+$, in that the overshoot was reduced and the membrane was depolarized.

Delayed Rectification. In an intact axon, as we have seen, depolarization causes a delayed rise in conductance, which has been attributed to an increase in permeability to K$^+$, and this accounts for the delayed rectification; if this interpretation is correct, delayed rectification should disappear when internal

---

[*] Hille (1968) points to the similarity of these effects of external Ca^{++} and *p*H with those obtained by decreasing the ionic strength inside the perfused axon (p. 1102), and attributes them to changes in the concentration or effectiveness of fixed charges in the membrane. Blaustein & Goldman (1968) have examined the effects of other divalent, and some trivalent cations on the voltage-clamped lobster axon. So similar are the effects to those of Ca^{++} that they are best interpreted in terms of interactions with the same anionic membrane groups. An increased influx of Ca^{++} occurs on stimulating the squid giant axon (Tasaki, Watanabe & Lerman, 1967).

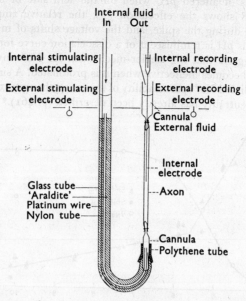

FIG. 569. Diagram of apparatus for recording with an internal electrode during perfusion of a doubly cannulated axon. Not to scale. (Baker, Hodgkin & Shaw. *J. Physiol.*)

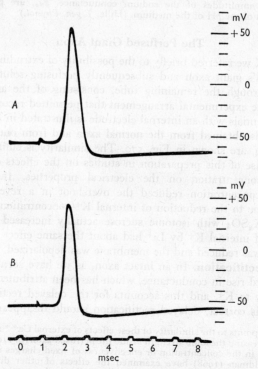

FIG. 570. A. Action potential recorded with internal electrode from axon when its axoplasm was extracted and replaced with isotonic potassium sulphate. B. Action potential of an intact axon with same amplification and time-scale. (Baker, Hodgkin and Shaw. *J. Physiol.*)

K^+ is replaced by Na^+. To test this, the effects of inward and outward currents on the membrane potential were measured with normal and K^+-free internal solutions; when an inward current depolarized the fibre by 50 mV, an outward current of the same magnitude depolarized it by only 4 mV when the internal fluid contained K_2SO_4; when this was replaced by Na_2SO_4, the resting potential fell by 35 mV and the rectification practically disappeared.

Internal Cations. The perfused axon preparation provides the opportunity for systematic variations of the internal composition of the axon; we have already (p. 1080) briefly referred to the changes in the equilibrium Na^+-potential following changes in internal Na^+-concentration. When this is zero, the Nernst equation predicts an infinitely positive internal potential; in fact Chandler & Meves (1965) obtained an experimental value of 69 mV for the potential required to reduce the early Na^+-current to zero under voltage-clamp conditions. The evidence

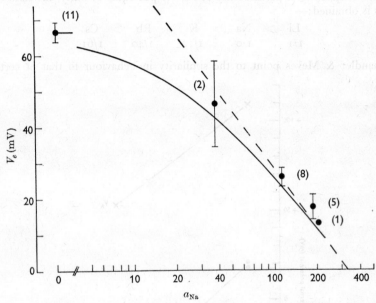

Fig. 571. Effect of internal sodium on equilibrium potential, V_e. Ordinate: Equilibrium potential, V_e. Abscissa: Internal sodium activity, a_{Na} (m-equiv./kg H_2O). a_{Na} was altered by mixing 300 mM-KCl + sucrose in various proportions with 300 mM-NaCl + sucrose. The points are averages from measurements on different axons; number of experiments indicated. The interrupted curve is drawn on the basis of the simple Nernst relation whilst the continuous curve is drawn on the basis of Equation 67 with $P_{Na}/P_K = 11\cdot5$. (Chandler & Meves. *J. Physiol.*)

suggested that, under these conditions, the internal potassium was carrying some of the current, in the sense that K^+ was exchanging for Na^+; according to the terminology of Chandler & Meves, K^+ was sharing the same channel as Na^+ in the phase of increased Na^+-permeability. In this event, the application of the Goldman equation gives:—

$$V_e = RT/F \ln \frac{P_{Na}\,[Na]_o}{P_K\,[K]_i + P_{Na}\,[Na]_i} \tag{67}$$

and, as Fig. 571 shows, the estimated equilibrium potential fits the experimentally determined values; the permeability ratio, P_{Na}/P_K, assumed to give the best fit, was $11\cdot5$.

More unequivocal evidence favouring the view that there was, indeed, an increase in K^+-permeability coinciding with that of Na^+ was provided by studying the early *outward* current when the axon was so strongly depolarized that its internal potential exceeded the equilibrium potential. With a normal internal composition, this outward current is carried by Na^+; with no internal Na^+ there is, indeed, an early outward current; furthermore, when the variation of this outward current with potential is studied, an identical "inactivation curve" is obtained to that found on the same axon when the early current was inward, and must have been carried by Na^+.

Discrimination between Cations. When the internal solution of the perfused axon was varied by using different cations, and the external solution was varied by substituting Na^+ by Li^+, relative permeabilities for the different cations in the active state could be obtained from estimates of V_e and application of the Goldman equation. If permeability of Na^+ is put equal to unity, the following series is obtained:—

$$Li > Na > K > Rb > Cs$$
$$1\cdot1 \quad 1\cdot0 \quad 1/12 \quad 1/40 \quad 1/61$$

Chandler & Meves point to the similarity in behaviour to that of certain

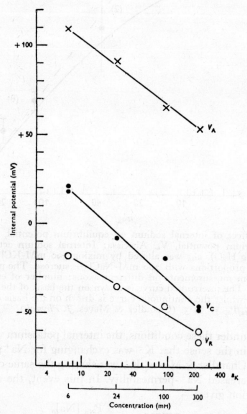

FIG. 572. Effect of varying the concentration of KCl inside the squid axon on resting potential, critical potential and overshoot. Abscissa, upper scale: Internal potassium activity (m-equiv/kg H_2O); lower scale, internal concentration of KCl; $\bigcirc$, resting potential, V_R; $\bullet$, critical potential, V_C; $\times$, potential at crest of spike, V_A. The external solution was K-free sea water. (Baker, Hodgkin & Meves. *J. Physiol.*)

ion-selective glass membranes employed experimentally as cation-selective electrodes in the determination of pH, Na^+-activity, and so on. The theory of such selectivity (Eisenman 1962) presupposes the presence in the membrane of fixed negative charges, and an important element is the field strength of these charges; if it is very high, only the hydrogen ion can permeate; thus to obtain significant permeability to the larger cations some shielding of the fixed charges is important. It is easy to envisage the development of an active state of a membrane permitting selective permeability to one special ion on this basis.

Action Potentials. A surprising finding during the early work on perfused axons was that large action potentials could be obtained from axons perfused with isotonic sugar solutions, although there was little or no resting potential; an axon depolarized in this fashion should be incapable of developing the high permeability to Na^+ because of the high degree of inactivation (Fig. 554, p. 1081), so it must be assumed, if the Hodgkin–Huxley theory is valid, that there has been a shift in the inactivation curve in the direction of more positive internal potentials; associated with this there must be a similar shift in the relation between sodium-conductance and membrane potential. To analyse the situation completely, a knowledge of the resting potential, the threshold depolarization required to induce a spike—critical potential—and the height of the crest of the action potential were measured. As Fig. 572 shows, all these parameters, when plotted as a function of internal potassium concentration, exhibit a logarithmic relation. Examination of the Figure shows, for example, that at very low internal concentrations of K^+ the low resting potential of about 10 mV requires, for initiation of an action potential, a reversal of membrane potential to +25 mV, and the spike rises to give a positivity of some 110 mV compared with an "overshoot" of only 50 mV at normal internal potassium.

Inactivation

Indirect evidence indicated that there had been a shift in the inactivation curve on reduction of internal K^+, and voltage-clamp studies confirmed this; Fig. 573, from Moore, Narahashi & Ulbricht (1964), illustrates this; thus, whereas the Na^+-mechanism is normally 50 per cent. inactivated at an internal potential of —50 mV, when the internal K^+ is only 11 mM, approximately the same degree of inactivation is given by complete depolarization. Figure 574 from Chandler, Hodgkin & Meves (1965) illustrates the same phenomenon; here the internal solutions were 300 mM KCl, 50 mM KCl + 250 mM NaCl, and 50 mM KCl, isotonicity being maintained with sucrose. Only when KCl was replaced by the non-electrolyte was there the shift in the inactivation curve, showing that it is the reduced ionic strength rather than the internal concentration of K^+ that is responsible. A possible explanation for this effect of ionic strength was suggested by Chandler *et al.* on the basis of the presence of fixed negative charges on the inside of the axon membrane; these would attract counterions to give the classical double layer, or rather "atmosphere", of counterions; the lower the ionic strength of the medium, the fewer the available counterions, with the result that the electrical potential, due to this system of fixed ions and mobile counterions, is greater. This potential will not affect the thermodynamic resting potential, since this is determined by the difference of concentration in accordance with the Nernst relation, but the activated state of the membrane may well be influenced by it, in addition to the thermodynamic resting potential, in which event, with low ionic strength the membrane will be effectively hyperpolarized, so that even with no resting potential the Na^+-mechanism will be active. Chandler *et al.* applied the Debye–Hückel theory to the situation, and on the assumption of a plausible density of fixed charges they were able to predict shifts of the

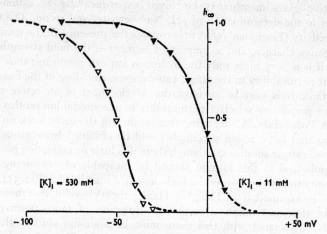

FIG. 573. Sodium inactivation curves for high-potassium and low-potassium perfused giant axons. Here h_∞ is the peak sodium current, during a pulse bringing the membrane to a fixed potential, as a fraction of the maximum current. Abscissa: Steady membrane potential preceding the pulse. (Moore, Narahashi & Ulbricht. *J. Physiol.*)

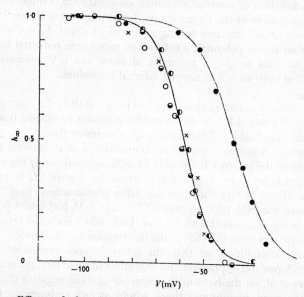

FIG. 574. Effects of altered internal ionic composition on the steady-state relation between h_∞ and the membrane potential of perfused squid giant axons. Here h_∞ is measured as the sodium-current associated with a test depolarization (following a 70 msec prepulse), expressed in units of maximum current. Abscissa is the prepulse potential, V. Internal solutions: ◑, 300 mM-KCl; ●, 50 mM-KCl; ○, 300 mM-KCl; ×, 50 mM-KCl + 250 mM-NaCl; ◓, 300 mM-KCl. Isotonicity maintained with sucrose. External solution: K-free artificial sea-water. The curves are theoretical. (Chandler, Hodgkin & Meves. *J. Physiol.*)

Na⁺-inactivation curve of comparable magnitude with those actually observed On the basis of this theory, these authors were able to explain the effects on resting potential of reduced ionic strength; earlier studies (Baker *et al.* 1964) had shown that the resting potential at low ionic strength could be accounted for by a relatively lower permeability to anions; lowered ionic strength increases the negativity of the membrane and thus will reduce anion permeability.

Heart-like Action Potentials

Narahashi (1963) observed that when the axon was perfused with K^+-free medium and bathed in a medium likewise containing NaCl but no K^+, the falling phase of the action potential was prolonged to give a record reminiscent of those obtained from Purkinje fibres of the heart (p. 1279). On the basis of the Hodgkin–Huxley theory, the falling phase is determined by an outward flow of K^+-current; in the absence of this ion, repolarization would rely entirely on the subsidence of the Na^+-permeability, but Baker, Hodgkin & Meves (1964) were of the opinion that a prolongation to give action potentials lasting 1–5 sec. must require something more. Perhaps, in the absence of K^+, the K^+-channels that normally provide the high permeability to this ion during the falling phase of the spike become available to Na^+.

Effects of Some Drugs

Tetrodotoxin. This non-polypeptide, low molecular weight neurotoxin, extracted from the Japanese puffer fish, *Sphoeroides rubripes*, is chemically identical with that extracted from the eggs of the California newt, *Tarachia torox*, and called *tarichotoxin*.* The toxin blocks transmission along the axon without affecting its resting potential, and thus bears a superficial resemblance, in its action, to the local anæsthetics, such as procaine (Taylor, 1959).†

Effect on Membrane Currents

Takata *et al.* (1966) examined the effect of tetrodotoxin on the voltage-clamped axon, and confirmed the generally held belief that it blocked the early Na^+-current of the action potential; careful examination indicated that the late K^+-current was unaffected.

This is illustrated by Fig. 575 from a more recent study of Narahashi & Moore (1968); the curves are the typical current-voltage relations for the lobster giant axon when the currents recorded are the "peak transient", representing the phase of activated Na^+-permeability, and the steady-state current to which the voltage-clamped axon settles down and representing the K^+-current. The full curves represent controls; on the abscissæ are plotted the membrane potentials to which the axon is clamped, positive values indicating a reversal of the potential; on the ordinates are plotted current, a positive value indicating movement of positive current out of the axon. With the steady-state K^+-current, there is almost a linear relation with applied voltage, and it is unaffected by tetrodotoxin. The curve for Na^+-current shows the characteristic minimum as depolarization is increased, and the reversal in sign of the current at the equilibrium Na^+-potential of 40 mV. Tetrodotoxin almost completely abolishes the Na^+-currents.

There was no effect on the steady-state inactivation, h; very little on the time-constant, τ_h, for removal of inactivation, and very little on m, determining the way the Na^+-current is turned on, and measured as the time to reach half-peak of the spike. When the toxin was applied internally in the perfused axon it had no effect on any parameter of the excitability system and thus allowed the action potential to develop normally (Narahashi, Anderson & Moore, 1967).

---

* Mosher, Fuhrman, Buchwald & Fischer (1964) showed that the two were identical chemically; they give an interesting history of the discovery of these potent poisons. Saxitoxin, isolated from mussels, is different chemically from these two.

† Barbiturates and alcohols are similar to procaine in blocking both the early Na^+-permeability and late K^+-permeability of the spike (Armstrong & Binstock, 1964; Blaustein, 1968).

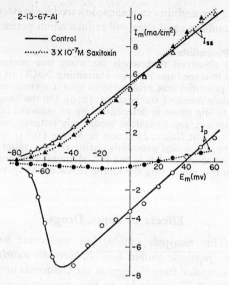

FIG. 575. Current-voltage relations for peak transient current (I_p) and for steady-state current (I_{ss}), before and during external application of 3×10^{-7} M saxitoxin, in a lobster giant axon. I_m, membrane current; E_m, membrane potential. Note that saxitoxin almost completely blocks the inward current carried by Na^+ (I_p) but has little effect on the delayed current carried by K^+ (I_{ss}). (Narahashi & Moore. *J. gen. Physiol.*)

Procaine

Takata *et al.*, and later Narahashi *et al.* compared the action with that of procaine; internally this was just as effective as when applied externally, and the difference between the two was probably the lipid-solubility of the procaine that permitted the internally applied molecule to reach the external sites that are presumably involved in the Na^+-current mechanism. Procaine inhibited both the early Na^+-current and the late steady-state K^+-current to the same extent. It was suggested that the presence of the guadinium group in the tetrodotoxin molecule was at the basis of the block; thus guadinium ion will replace Na^+ in supporting the action potential in myelinated nerve (Lüttgau, 1958), whilst substituted guanidines only gave local non-propagated responses (Deck, 1958), so it was argued that tetrodotoxin, passing along the Na^+-channel, blocked it. On this basis, the blockage would only occur on the way inwards. Narahashi, Moore & Poston (1966) have argued that guanidine is not necessarily the responsible group, since a tetrodotoxin derivative failed to show blocking action. The toxin presumably reacts irreversibly with active sites responsible for the facilitated transfer of Na^+ described as "activated permeability"; its failure to act internally is probably largely due to the extremely low concentrations employed (nanomolar vs millimolar quantities with procaine); if there are non-specific adsorptive sites within the axon that take up the toxin before it can reach the specific sites, which may well be concentrated on the outside, there will be insufficient to block these specific sites.

Early K⁺-Current

It will be recalled that Chandler & Meves showed that, in the axon perfused with a Na^+-free medium, some of the early current was carried by K^+, or when the inside of the axon was made strongly positive so that the early current became outward, with no Na^+ inside the axon, this early current was carried by K^+.

Tetrodotoxin blocks this early K$^+$-current* so that its action seems to be on this early increase in cation permeability, as such, rather than specifically on Na$^+$.†

Snake Venoms. Certain venoms, notably that of the cottonmouth moccasin snake, block conduction; the process has been studied by Rosenberg, and apparently depends on an increased permeability of the nerve resulting from a splitting off of lysophosphatides from the membrane by a phospholipase-A present in the venom (Condrea, Rosenberg & Dettbarn, 1967). As a result of what seems to be a purely non-specific increase in permeability, substances like curare and acetylcholine are able to penetrate the squid axon and to exert effects on its excitability that are not shown in the normal axon (Rosenberg & Podlewski, 1963). It is interesting that, in order to obtain an effect on the giant fibre, some adhering myelinated fibres must be left attached, and it is the lysophosphatides split off from their myelin that affects the giant axon (Martin & Rosenberg, 1968).

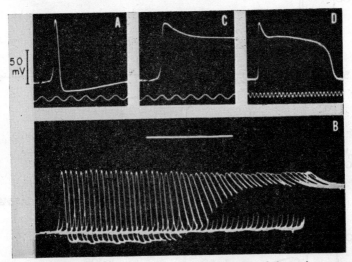

FIG. 576. Records showing changes in the time-course of the action potential caused by intracellular injection of tetraethylammonium chloride. Record A, before injection. The white line in record B represents the time during which the injecting pipette was moved from one end of the axon to the other; shock interval, 1 sec. Records C and D, after injection. Time marker, 1 msec. 22°C. (Tasaki & Hagiwara. *J. gen. Physiol.*)

Action of Tetraethylammonium. Tasaki & Hagiwara observed that the falling phase of the action potential of the squid giant axon was remarkably prolonged, to give a plateau very similar to that normally seen in single Purkinje fibres of the heart, when tetraethylammonium (TEA) was injected into the axoplasm (Fig. 576). This plateau could be abolished by an appropriate anodic (hyperpolarizing) pulse; *i.e.*, the action potential could be brought to an abrupt end by such a pulse. These authors regarded their results as in conflict with the Hodgkin-Huxley theory, and considered that the membrane was able to adopt two stable states and could jump from one to another. Thus, when the axon was depolarized by a high external concentration of K$^+$, they found that an anodal pulse would lead to a regenerative spike in which the fibre became hyperpolarized, and they considered that under these conditions the membrane was jumping to a second stable state. Other studies, involving the behaviour of axons and other excitable tissues in Na$^+$-free media and in the presence of high concentrations of alkali-earths such as Ca^{++} and Ba^{++}, have likewise been interpreted by Spyropoulos & Tasaki as being in contradiction to the ionic hypothesis. We must hesitate a

* Personal communication of Chandler & Meves to Moore *et al.* (1967).
† Watanabe, Tasaki, Singer & Lerman (1967) showed that tetrodotoxin would block the spike in the perfused axon in the absence of NaCl inside or outside, e.g. with hydrazinium chloride and CaCl$_2$ outside and CsF and Cs phosphate inside, and concluded that there was no specificity in the action so far as Na$^+$ was concerned.

long time, however, before abandoning the basic concepts of the Hodgkin-Huxley treatment for several obvious reasons, notably the sound empirical basis of the fundamental equations, and the success in interpreting so many aspects of excitation.

Application of Hodgkin-Huxley Equations

Thus, Fitzhugh and George & Johnson have independently examined the application of the Hodgkin–Huxley equations to the case of the TEA-injected axon, making use of Tasaki & Hagiwara's observation that the effect of this injection is to decrease the rate of onset of the K^+-permeability mechanism. If we assume that this is the only effect, then we may simply substitute α_n/k and β_n/k for α_n and β_n

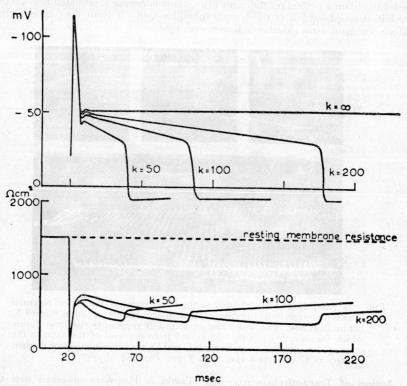

FIG. 577. Effect of slowing down the rate constants involved in the potassium-carrying mechanism by a factor k.

Upper figure: Action potential wave-forms for retardation factors (k) of 50, 100, 200 and ∞.

Lower figure: Time-course and magnitude of membrane resistivity during the action potentials for $k = 50$, 100 and 200 shown in the upper figure. (George & Johnson. *Austral. J. Exp. Biol. Med. Sci.*)

in the fundamental equations, where k is a factor by which α and β for the K^+-permeability mechanism are reduced. Fig. 577 from George & Johnson shows some calculated action potentials when k varied between 50 and infinity, and it is seen that there is a remarkable similarity between the theoretical action potentials and those obtained by Tasaki & Hagiwara from the TEA-injected axon. The plateau is brought to a sudden stop by a repolarizing process that leads to an undershoot, and this is, indeed, observed experimentally. The effects of applied pulses of current during the plateau were also calculated, and it was shown that under appropriate conditions they could abolish the plateau permanently. We may conclude, therefore, that the effects of TEA are not inconsistent with the Hodgkin–Huxley equations but rather the remarkable coincidence of the calculated with the experimental action potentials further strengthens the basis of these equations.

¶ In a similar way we may examine the effects of raised external K^+-concentration; under these conditions the resting potential is reduced, so that in the equations a new value of V_K must be substituted. Examination of the theoretical effects of pulses of current on the membrane potential and resistance showed that, above a critical pulse strength, there was a discontinuity of membrane resistance that led to a hyperpolarizing response, *i.e.*, to the reversed type of action potential that had been described by Segal (1958) and Tasaki (1959) in the squid axon, and by Stämpfli in the node of Ranvier of the frog axon.

Upper and Lower Stable States

A general analysis of the membrane currents that occur at any given membrane potential is illustrated by Fig. 578, where the steady-state Na^+-current is indicated by the curve marked I_{Na}; the K^+-current has been calculated on the assumption that the K^+-conductance remains constant at its resting value and has been lumped

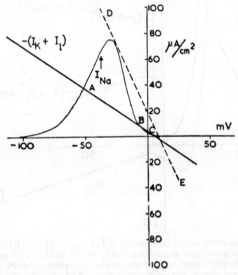

FIG. 578. Graphs of the various currents for the steady state and where the potassium conductance is constant at the resting value. The solution of the Hodgkin-Huxley equation for this case is given by the intersection of the two full lines, *i.e.*, where the inward and outward currents are equal. It is seen that two stable states are possible, A and C. For values of potassium conductance greater than the limiting case shown by the interrupted line, only one stable state is possible. (George & Johnson. *Austral. J. Exp. Biol. Med. Sci.*)

with the leakage current, I. For an equilibrium, the solution to the fundamental equation is given by the intersection of this straight line, describing K^+- and leakage-currents, with the Na^+-current curve, since here inward and outward currents just balance, and this is the condition for equilibrium.

There are three points of intersection corresponding to voltages of -52, -6 and $+4$ mV; of these states A and C are stable, since an increase of potential leads to an increase in current; at B we have the opposite condition, an increase in current causing the point B to move in the negative direction. The two stable states probably correspond to the upper and lower stable states described by Tasaki, and are possible because the K^+-conductance has been held at its resting value. The solutions to the fundamental equation are given by the intersections of the straight line with the Na^+-current curve; if the K^+-conductance is given a higher value we have a new straight line, and at a certain value this line will just graze the Na^+-curve (the line DE of Fig. 578), and now there is only one solution, namely a membrane potential of about 4 mV above the resting potential. We may say, therefore, that the upper stable state of Tasaki corresponds to the plateau of the TEA-action potential, where the K^+-conductance is low; if the K^+-conductance slowly rose this would mean a clockwise rotation of the straight line, and when the position DE was reached only one stable state would be possible and the membrane potential would return sharply to above the resting value.

Hyperpolarizing Spike

A corresponding analysis applied to the effects of high external K^+ indicated that there was only one stable state, the point A of Fig. 579; when a hyperpolarizing current is imposed on the membrane, the curve representing K^+- and leakage-currents is shifted upwards and it is seen that beyond a certain value of this hyperpolarizing current there is no intersection with the Na^+-current, so that the only stable state becomes the point C where all currents are zero, *i.e.*, at a state of high membrane hyperpolarization. This corresponds then to the hyperpolarizing spike in high external K^+.

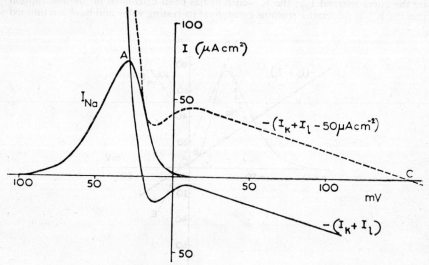

FIG. 579. The separate membrane currents I_{Na} and $- (I_K + I_l)$ in the presence of raised extracellular potassium ($[K]_o = 46$ mM). The solution of the Hodgkin-Huxley equation is given by the intersection of the two solid lines (point A) and this is the only stable state. The situation for added hyperpolarizing current of $50~\mu A$. cm. $^{-2}$ is shown by the interrupted line. For I equal to or greater than this value only one stable state is possible corresponding to the hyperpolarized state, point C. (George & Johnson. *Austral. J. Exp. Biol. Med. Sci.*)

Blockage of Delayed K^+-Current

The blockage of the delayed K^+-current by TEA has been examined in some detail in several laboratories (Armstrong & Binstock, 1965; Armstrong, 1967; Hille, 1967; Koppenhöfer, 1967; Schmidt & Stämpfli, 1966). The phenomenon is comparable with the blocking of the early Na^+-current by tetrodotoxin, but the concentration required is of the order of 5 mM as opposed to nanomolar quantities with tetrodotoxin. In the squid it only manifests itself after injection into the axon whereas at the node of Ranvier it occurs with external application; with the latter preparation both inward and outward movements of K^+ were blocked, whereas in the squid the inhibition applied only to the outward current, thus producing anomalous rectification. The reversal potential for the inward Na^+-current at the node of Ranvier is less than the Na^+-potential, so that K^+ probably passes through the Na^+-channel; Koppenhöfer found no effect of TEA on this reversal potential whereas, if K^+-permeability had been affected during this phase of the spike, there should have been an increase of some 19 mV. In general it would seem that TEA combines reversibly with the specific carriers concerned with the delayed K^+-permeability, thereby blocking access by K^+.

Action of DDT. Hille (1968) has compared the actions of tetrodotoxin and saxitoxin with that of the insecticide, DDT (1,1,1-trichloro-2,2-*bis*(*p*-chlorophenyl)-ethane. The last named compound prolongs the inward Na^+-current of the spike to give a potential "tail" of current after a test pulse. This tail does not occur in the absence of Na$^+$ so that, to use the current terminology, we may say that DDT holds open the Na$^+$-channels by contrast with tetrodotoxin which closes them. The effect of DDT is not just to prevent inactivation of Na$^+$-permeability; thus, during the action potential, the "turning off" of Na$^+$-permeability depends on two processes, namely a decrease in h when the nerve has become depolarized (inactivation) and a decrease in m when the nerve is repolarized or hyperpolarized, giving the typical tail of Na$^+$-current. In DDT-treated nodes Hille showed that the persisting Na$^+$-current did not turn off at the normal rate during depolarization and hyperpolarization, so that both processes, described by m and h, appear to be overridden when DDT (or veratrine, Ulbricht, 1965) is present. Thus the total Na$^+$-current could be divided into two components, that which persisted and the remainder, which behaved more orthodoxly, and it was found that the decay of the orthodox (transient) component had, within the experimental error, the same time-constants for rise and decay as did the Na$^+$-currents in the unpoisoned node.

Action of Caesium. Baker *et al.* (1962) found that perfusion of the squid axon with Rb$^+$ prolonged the declining phase of the spike, suggesting that Rb$^+$ did not

TABLE LXXVIII

Effluxes of Rb$^+$ and Cs$^+$ During Stimulation of Squid Giant Axon (Sjodin, 1966).

Axon Concn. (meq./litre)		Resting Efflux (pmoles./cm.^{2}/sec.)		Extra Efflux/Impulse (pmole.)	
K$^+$	Cs$^+$	K$^+$	Cs$^+$	K$^+$	Cs$^+$
217	180	30·1	12·4	3·24	0·45
232	174	40·9	20·6	5·38	0·0

pass through the activated channels as well as K$^+$. The effects of cæsium were more extreme and led to eventual inexcitability with a reduced resting potential. Later Chandler & Meves (1965) found that replacement of half the internal K$^+$ by Cs$^+$ reduced the delayed outward current to 2–5 per cent. of the control value, showing that Cs$^+$ interfered with the activated K$^+$-permeability; and this was confirmed by tracer studies of Sjodin (1966). As Table LXXVIII shows, the extra efflux of Cs$^+$ per impulse is negligible compared with that of K$^+$, so that the delayed rectification is selective for K$^+$. Sjodin deduced from the measured efflux of K$^+$, and that expected of a normal axon under voltage clamp conditions, that the presence of Cs$^+$ reduced this efflux by about 92 per cent. Voltage-clamp studies of the perfused axon by Adelman & Senft (1966) have confirmed this viewpoint, and have shown that the prolongation of the action potential is due, not only to the decreased K$^+$- conductance, but also to a prolongation of Na$^+$-activation.*

Electric Excitation of Muscle

As we have seen, the action potential, recorded from a single muscle fibre, exhibits the typical overshoot that has been interpreted to be the result of a phase of high Na$^+$-permeability; the effects of reducing the concentration of Na$^+$ in the outside medium are consistent with this view (Nastuk & Hodgkin).

In general, the spike in muscle is slower in development than in nerve and its velocity of propagation is about one-tenth or even less (Göpfert & Schaefer, 1938); for example, in the frog's sartorius the velocity is about 2 m./sec. comparing with a velocity in the motor nerve fibres of over 50 m./sec. at 20–25° C; the total duration of the negative spike is about 2 msec., the crest is attained in 0·75 msec., and the wavelength is some 7 mm., *i.e.*, only about one-fifth that of myelinated motor nerve fibres, a difference due mainly to the smaller electrotonic spread in the low-resistance muscle sheath. It will be

* Adelman & Senft point out that the resting potential of the axon with equal internal concentrations of K$^+$ and Cs$^+$ is about what would be expected of the reduced K$^+$-concentration, so that Cs$^+$ seems not to affect the resting permeability to K$^+$, and this is confirmed by Sjodin's flux measurements which indicate normal resting values but greatly reduced active outflux.

recalled that the wavelength of the action potential in unmyelinated crab nerve is also very small; thus the distances for "half-decay" in medullated nerve, crab nerve and frog sartorius muscle are 2–3 mm., 0·5 mm., and 1 mm. respectively.*

Tetrodotoxin. Tetrodotoxin blocks the action potential with little effect on the resting potential (Narahashi *et al.*, 1960) thus confirming the similarity in spike mechanisms. Narahashi *et al.* pointed out that the delayed rectification in muscle is not obvious in normal muscle fibres, but it becomes manifest in tetrodotoxin-poisoned muscles, presumably because the action potential normally obscures it. Similar effects were observed by Kao & Nishiyama (1965), using saxitoxin from mussels; they concluded that the poison had no effect on K^+- and Cl^+-conductances, either at rest or during the spike.

Membrane Currents. Clamping the voltage over an area of muscle membrane presents greater difficulties than with the squid nerve fibre because it is difficult to isolate a piece of membrane electrically and because of the contraction of the fibre when depolarized. Adrian, Chandler & Hodgkin (1966, 1968) used three intracellular electrodes, one to deliver current and the other two to measure membrane current and voltage. Contraction of the muscle fibre was prevented by using a Ringer's solution made hypertonic with sucrose (p. 1398). Inward going, or anomalous, rectification was demonstrated, in that the current for a hyperpolarizing step was larger than for a corresponding depolarizing step. A sudden depolarization from a resting potential of -100 mV to between -60 mV and $+20$ mV gave a brief inward current, abolished by tetrodotoxin and therefore, presumably, carried by Na^+; its equilibrium potential was some 20 mV positive. The rapid decrease in this inward current indicated an inactivation process, and the steady-state inactivation versus membrane potential curve was similar to that of squid nerve, the value for full activation (h_∞) occurring at -92 mV. The delayed current, following the initial Na^+-current, differed from that in the squid axon in several respects; for example, its equilibrium potential was consistently further from the K^+-equilibrium potential than the resting potential; thus in 2·5 mM K^+, the equilibrium potential for the delayed current was 10–15 mV positive to the resting potential. This means that during this delayed current the K^+-Na^+ selectivity of the membrane is less than that prevailing under resting conditions, in fact the relative permeabilities were 30:1 compared with 100:1 under resting conditions (Hodgkin & Horowicz, 1959).

Negative After-Potential

Muscle exhibits a negative after-potential, *i.e.*, a failure of the membrane potential to return to its baseline after the spike; the squid axon exhibits a positive after-potential, *i.e.*, the membrane becomes hyperpolarized immediately after the spike. The squid's after-potential has been explained in terms of the enhanced K^+-selectivity of the membrane—delayed rectification—that brings the membrane potential closer to the K^+-potential than it is during the resting condition. Thus the squid's resting potential is lower than the K^+-potential with the result that an increased K^+-selectivity makes the membrane potential more negative, and thus gives a positive after-potential. With muscle, during the delayed rectification phase the K^+-selectivity is decreased, so that the membrane potential becomes *less* negative than the resting potential, and the after-potential is negative.

---

* Falk & Landa (1960) have described an interesting prolongation of the negative after-potential of muscle when it is stimulated in a ferrocyanide-Ringer, but its interpretation is not obvious unless repolarization depends on anion-currents to some extent, in which case substitution of a non-permeating anion would prevent repolarization.

Effects of Pulse Duration

Finally, the space between the squid axon and the Schwann cells allows K^+ to accumulate in it during a sustained depolarization; and this means that the equilibrium potential for the delayed current depends on the length of the current pulse (Frankenhaeuser & Hodgkin, 1956). Muscle fibres, lacking this envelopment, would not be expected to show this effect, and in fact Adrian *et al.* found only very small effects of duration of the current pulse.

Calcium. Because changes in the internal concentration of Ca^{++} may well be concerned with the coupling of excitation with mechanical contraction of muscle (p. 1397), the effects of stimulation on the fluxes of this ion are of some interest; they have been examined by Bianchi & Shanes and Shanes & Bianchi; both influx and efflux are increased by stimulation, by contrast with nerve where only influx increased. The situation is complicated, of course, by the strong binding in non-ionic form, of this alkali-earth.

Barnacle Giant Muscle Fibre. Hoyle & Smyth (1963) described giant fibres making up the depressor muscle to the operculum of the barnacle, *Balanus nubilus*; these were some 2 mm. in diameter and several centimetres long and were thus ideal for intracellular recording and perfusion. The internal ionic composition is indicated in Table LXXIX.

TABLE LXXIX

Ionic Composition ($mM/litre\ H_2O$) of Sarcoplasm of Barnacle Giant Muscle Fibre Compared with that of "Barnacle Ringer" (Hagiwara, Chichibu & Naka, 1964).

Ion	Sarcoplasm	Barnacle Ringer
K^+	157	8
Na^+	21	476
Cl^-	32	539

Like other crustacean muscle fibres the usual response to an applied depolarizing current was a small non-propagated response associated with development of tension; sometimes a spike would occur. Hagiwara & Naka (1964) found that injection of alkali earth precipitants or complexers, such as K_2SO_4 or EDTA and EGTA,* converted the response to a large spike; the height of the overshoot increased linearly with the concentration of Ca^{++} in the medium up to about 100 mM, the theoretical increase of 29 mV per 10-fold increase of concentration being obtained; when the concentration was decreased below normal (20 mM) the overshoot decreased. The actual internal concentration of Ca^{++} below which spike activity occurred was 8.10^{-8} M; between this and 5.10^{-7} M there was oscillatory activity (Hagiwara & Nakajima, 1966b).

Initial Inward Current

Voltage-clamp studies indicated that the initial inward current was, indeed, carried by Ca^{++}; this was proved by the increased influx of ^{45}Ca that took place during the spike. When the effects of ions on the resting potential were studied, it became evident that this was largely determined by K^+; furthermore, the falling phase of the spike was probably associated with increased K^+-permeability, raising the concentration of K^+ inside decreasing the resting potential and increasing the overshoot and prolonging the spike. When the height of the overshoot was plotted against the logarithm of the $(Ca^{++})/(K^+)$ ratio, a linear relation with a slope of 29 mV per ten-fold increase in the ratio was obtained; such a simple relation was surprising in view of the difference of valency of the two ions (Hagiwara, Chichibu & Naka, 1964).† Tetrodotoxin blocks the Na^+-current responsible for

---

* EDTA is the abbreviation for ethylene-diamine-tetra-acetate; EGTA is ethylene glycol bis (β-aminoethyl ether) -N,N' -tetra-acetic acid. EDTA binds both Ca^{++} and Mg^{++}, whereas EGTA does not affect Mg^{++} appreciably. Van der Kloot & Glovsky (1965) consider that the mere increase in the concentration gradient of Ca^{++} between sarcoplasm and medium, resulting from complexing by EDTA, is unlikely to be the cause of the overshoot since the concentration gradient is so high normally.

† Hagiwara & Takahashi (1967) have examined the possibility that deviations of the overshoot equilibrium potential from the Nernst relation are due to the fact that the

the rising phase of the axon spike; interestingly it has no effect on the barnacle muscle spike; procaine, likewise, does not block, in fact it causes a small rise in the overshoot perhaps because of an effect on K^+-permeability (Hagiwara & Nakajima, 1966a).

METABOLIC ASPECTS OF EXCITATION

The metabolism of muscle is complicated by the fact that it is concerned not only with the conducting, but also with the contractile, mechanism; so that it would be most profitable to concentrate attention on the metabolism of nerve, where the conduction mechanism has been developed to the greatest extent. The frog nerve, at rest, consumes O_2 and produces heat at rates of the order of 50 mm.^{3}/g.hr. and 0·5 cal./g.hr., respectively.

Effects of Anaerobiosis and Metabolic Inhibitors

In an atmosphere of pure N_2 the resting heat production falls, over a period of two hours, but even after this there is a steady evolution of heat corresponding to one-quarter to one-fifth of the normal resting rate; on re-admitting O_2 the heat production rises rapidly to overshoot the resting level for about an hour. There is thus developed an "oxygen debt", but the extra consumption of O_2 only represents about 15 per cent. of that missed during anaerobiosis. During anaerobiosis the production of CO_2 continues for some time in spite of the fact that, according to Hill's calculation, all the dissolved O_2 in nerve should be lost in a few minutes; the relatively slow fall in heat production is therefore said to be due to the presence of an "oxidative reserve", i.e., the presence within the nerve of a hydrogen acceptor capable of taking the place of molecular O_2. It is thought that, when the nerve in N_2 has settled down to its low rate of heat production, the true anaerobic metabolism, presumably involving the formation of lactic acid, is operating alone. The steady anaerobic heat production amounts to some 10^{-3} cal./g.hr., equivalent to the energy obtained from the formation of 2.10^{-4} g. of lactic acid per gramme of nerve per hour; according to Schmitt & Cori, the actual formation of lactic acid corresponds to 2.10^{-5} g./g.hr.

Conduction Block. In the absence of O_2, the nerve slowly loses its power to conduct an impulse; Gerard placed several recording electrodes at different positions along the frog's sciatic nerve; if the impulse had to pass through a portion of nerve exposed to an atmosphere of N_2, the height of the action potential, recorded from a part of the nerve in O_2, fell progressively during a period of an hour or more. For example, in one experiment there was a 3 per cent. loss in 15 minutes, 13 per cent. in 30 minutes, 49 per cent. in 40 minutes and complete block occurred in one hour. The higher the temperature, or the more rapid the resting O_2-consumption, the more rapidly the block occurred; thus the order of decreasing rates of blockage was: Dog nerve > green-frog > bull-frog, the order of their resting O_2-consumptions. If there is an oxidative reserve, the exhaustion of which causes eventual block, then clearly the rate of block must run parallel with the normal requirements of the nerve. On re-admitting O_2 to the nerve, conduction returned in one minute, the nerve being, for a time, more excitable than in its initial resting condition; in this connection it is of interest that, in the early stages of asphyxia, the response to stimulation actually increases above normal before falling.

Relation to Resting Potential. We need not expect the blocking action of

---

concentration of Ca^{++} at the surface of the muscle fibre is different from the bulk concentration. Sr^{++} will substitute for Ca^{++} so far as the overshoot is concerned.

reduced metabolic activity to be reflected in a changed resting potential, since the influence may well be exerted on the changes of sodium and potassium permeability that are at the basis of the excited state. This is revealed by the effects of local anæsthetics, such as procaine, which block conduction in nerve but have little or no effect on the resting potential (Bishop; Bennett & Chinbury). Studies with the voltage-clamp technique indicated, on the other hand, that the transient inward current, indicating the phase of increased Na^+-permeability, immediately following the depolarizing stimulus, is reduced; the later K^+-current was also reduced, so that the two effects of the local anæsthetic oppose each other, the actual state of block being due to the greater influence of the changed Na^+-permeability (Taylor, 1959; Shanes et al., 1959).* The actual effects of an interference with metabolic activity on the resting potential need not be great; this was revealed by Hodgkin & Keynes' study of the effects of DNP on the squid giant axon (p. 638); although the Na^+-pump was put out of action the resting potential was hardly affected. Again, it by no means follows that the abolition of the resting potential should be immediately followed by a levelling up of the concentrations of K^+ in the fibre and extracellular fluid; the potential depends on a restraint placed on the migration of K^+ by the impermeability of certain anions; the removal of this restraint will rapidly abolish the potential, although it may take a considerable time for the K^+ to leak away from the fibre.†

Anodic Polarization. A striking observation, originally made by Thörner and confirmed and extended by Lorente de Nó (1946/7) and by Gallego, is that the excitability of nerve, destroyed by anoxia or narcotics, may be restored by anodal polarization, i.e., by the passage of a weak constant current through the nerve, the tested region being near the positive pole of the polarizing electrodes. As Lorente de Nó pointed out, this observation means that the excitability of the nerve fibre is determined primarily by its electrical condition, metabolism influencing the excitability in so far as it is necessary to restore the nerve's original (or approximately so) electrical condition after the depolarization which follows a stimulus and propagated disturbance. In view of the dependence of the degree of activation of the Na^+-mechanism on the degree of polarization of the membrane, it is quite likely that the effects of anodal polarization are to be traced to an increase in the activation of this process.

Heat Production During Activity

As a result of a propagated disturbance the nerve fibre is depolarized, and subsequently, during the refractory periods, re-establishes its resting polarized condition; later still, as the consequence of active extrusion of Na^+ and accumulation of K^+, the small changes in ionic distribution resulting from the action potential are corrected. We may expect these events to be associated with changes in the heat content and heat production of the nerve, although the precise nature of these changes would be difficult to predict. Thus, depolarization of the membrane represents the discharge of a condenser and could lead to heat

* Blaustein & Goldman (1966) have compared the actions of Ca^{++} and procaine on the voltage-clamped lobster axon; they have confirmed the antagonistic action of the two described by Aceves & Machne (1963). We have seen that increasing the concentration of Ca^{++} causes a shift of the Na^+ and K^+ conductance curves along the voltage axis so that a larger depolarization is required to produce a given conductance step. The action of procaine is to cause a reduction in the conductances at all potentials rather than to shift the curve.

† According to Furusawa (1929), the resting potential of crustacean nerve is abolished by anaerobiosis. Shanes & Brown (1942) have divided the resting potential into a number of fractions according to their sensitivity to metabolic poisons. Whilst there seems little doubt that the resting potential is affected by metabolism, it is by no means easy to interpret the effects in terms of ionic permeability.

production; repolarization is a recharging of the condenser and could lead to heat absorption. The active processes of restitution of the K^+ and Na^+ concentrations are associated with heat production, and so on.

Tetanus. The absolute amounts of heat involved in the passage of an impulse are very small and they present serious technical problems in their measurement, especially since the rate of liberation is rapid. The history of the analysis of the heat production of nerve activity is thus the history of the development of suitable thermopiles, thermostats, and galvanometers. Space will not permit a description of the apparatus, developed in its most refined form by A. V. Hill, but if it is appreciated that the rise in temperature of nerve, accompanying a high frequency tetanic stimulation lasting one second, is only 6.10^{-6} ° C, the high sensitivity of the thermopile, and the accuracy of the temperature regulation, become evident. Until very recently, the heat liberated as a result of a single shock applied to a nerve was too small to measure, so that the effects of a volley of impulses in several nerves resting on a thermopile were studied and the curve, obtained by plotting the heat liberated against time of stimulation, was analysed into components corresponding to what we may call the *"initial heat"*, i.e., a burst of heat production immediately associated with the passage of the impulse, and a *"delayed heat"*, associated with restitutive reactions. Thus, during a faradic discharge at, say, a frequency of 40 per sec., the heat liberated at any moment represents the sum of the initial heat, due to the impulse at that moment, plus the accumulation of the recovery heat from many of the preceding stimuli; the number of these preceding stimuli contributing to the instantaneous heat production must clearly depend on the duration of the recovery heat; if this lasts for 30 minutes, then the first stimulus must contribute heat to the instantaneous heat production for the first 30 minutes of repetitive stimulation. It will be clear, too, that at this point, i.e., when the stimulation has lasted for the duration of the recovery heat, a steady-state of heat production will be reached.

Analysis of Heat Production

The response to a faradic stimulus lasting for 16 seconds is shown in Fig. 580, the rate of heat production, computed from an analysis of the galvanometer deflection, being plotted against time. There is a rapid rise at the beginning of stimulation and rapid decline at the end, both of which may be attributed to the initial heat. The slower rise in rate of heat production is due to the cumulative effect of the delayed heat of preceding impulses, plus a steady component of initial heat which may be assumed to remain constant. If the frequency of stimulation is not too high, a steady-state may be reached in about 25–40 minutes in frog nerve at 20° C, hence the total duration of the recovery heat lies within this range. If the stimulation is stopped when the steady-state has been reached, there is a sudden fall in heat production, followed by a more gentle fall lasting over the same period as that required to establish the steady-state. The sudden fall is due predominantly to the falling out of the initial heat, and the slower one to the gradual decrease in the recovery heat. A study of the rate of fall of heat production in these circumstances shows that the recovery heat may be divided into a very rapid "A-component", which is half-completed in 2–3 seconds, and a slow "B-component", half-completed in 4–5 minutes. The total heat liberated during the rapid recovery process is about equal to that liberated as initial heat, and is doubtless equivalent to the heat liberated during the rapid re-establishment of excitability during the absolutely refractory phase. The slower phase, lasting half-an-hour, is probably an oxidative process which restores the energy utilized in the rapid phase.

Single Shock. In 1958 Abbott, Hill & Howarth described their results obtained with highly improved instruments that permitted the recording of heat production resulting from a single shock. They showed that the initial heat was composite, being the result of an initial burst of heat production, lasting about 100 msec., followed by a slower absorption of heat—*negative heat production*—of about the same total magnitude, and lasting some 300 msec. at 0° C. Since the bundle of nerve fibres on the thermocouple was at some distance from the point of stimulation, in order to avoid the heating effect of the stimulus, the action potentials arrived at different times in different fibres owing to the variation in conduction velocity; in consequence, the positive and negative heat productions were out of phase, so that what was in effect

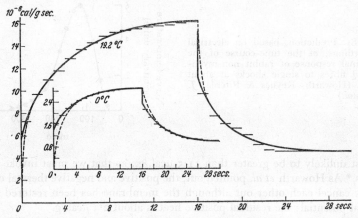

FIG. 580. Extra heat production of nerve during and after a 16-sec. tetanus. The extra heat liberated has been computed from the galvanometer deflection over successive 1-sec. intervals, as indicated by the short horizontal lines. The full curves are drawn through these lines, it being assumed that the initial rise of heat production on stimulation, and the fall on cessation of the stimulus, are sudden. The broken curves are drawn on the assumption that the initial rise and final fall of heat production are continuous, not sudden. (Feng, after Hill. *Ergebn. d. Physiol.*)

measured was the resultant of a partial mutual cancellation. When an approximate allowance had been made for this, the actual heats per impulse were estimated as 14.10^{-6} and -12.10^{-6} cal./g. respectively.

Non-Myelinated Fibres

In non-myelinated mammalian fibres of the rabbit cervical vagus nerve, also, the initial heat is composed of both positive and negative phases; with this preparation, because the range of nerve diameters is restricted, action potentials with only a small temporal dispersion can be obtained; by working at 5° C the time-scale may be expanded and thus a fairly precise resolution of the positive and negative phases of heat production may be obtained (Howarth, Keynes & Ritchie, 1968). Fig. 581 shows the time-course of the predicted thermal changes in nerve on the assumption that heat is released only during the rising phase of the spike and absorbed during the falling phase; a comparison of these curves with experimental parameters, such as the latent period before positive heat can be detected, and the time to beginning of fall in temperature, agreed remarkably well: 33 and 130 msec predicted, versus 34 and 136 msec. Ba^{++} considerably delays the repolarization phase of the spike and therefore should postpone the development of negative heat; since there is considerable overlap between positive and negative heats, owing to the mixed character of the nerve, this means that the positive heat will be greatly increased by the postponement of the negative heat; and this was actually found. A number of other environmental influences that alter the spike altered the course of evolution and absorption of heat in a predictable manner, so that it is safe to conclude that liberation and absorption of heat are, indeed, associated with the depolarization and repolarization phases of the spike respectively. The actual heat liberated was, on average, 7·2 μcal./g. per impulse, but since there is considerable overlap of positive and negative phases, the probable value is some 3·4 times this, namely 24·5 μcal./g. This is much larger than the theoretical energy liberated when a condenser corresponding to the nerve fibre is discharged, which may be as small as 0·75 μcal./g. and

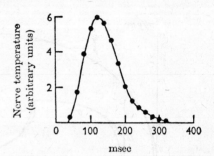

FIG. 581. Prediction, based on electrical recordings, of the time-course of the thermal response of rabbit non-myelinated fibres to single shocks at about 5°C. (Howarth, Keynes & Ritchie. *J. Physiol.*)

is most unlikely to be greater than 11·5 μcal./g., so that we must invoke other factors.* As Howarth *et al.* pointed out, the positive and negative thermal events do not cancel each other out although the membrane has been restored to its original potential; the residual positive heat is about 2·3 μcal./g.

Electric Organ

In the electric organ the negative heat is much more obvious, the tissue actually cooling below its initial temperature after a single shock (Abbott, Aubert & Fessard, 1958). It must be emphasized, when comparing thermal events in a nerve impulse with those in the electric organ, that in the latter the spike lasts only 8 msec. compared with over 200 msec. for the non-myelinated nerve at 5° C when dispersed over 7 mm. of thermopile; in the electric organ, therefore, it is not possible to separate the consequences of depolarization and repolarization; and the cooling that is observed must be equated with the total net initial heat of the nerve, which is invariably positive, and not with the cooling associated with the downstroke of the spike. A plausible explanation for the large negative phase of heat production of the electric organ was put forward as long ago as 1906 by Bernstein & Tschermak, who argued that, if the energy for the discharge consisted of movements of ions down concentration gradients, the process would be analogous with the adiabatic expansion of a gas and thus accompanied by cooling. Aubert & Keynes (1968) examined the thermal events during the discharge of *Electrophorus electricus* with this aspect in mind; they observed three phases, an initial positive, Q_1, followed by a negative Q_2 and a positive Q_3; Q_1 could be dismissed as an artefact; Q_2 lasted some 70 msec. after the electrical discharge whilst Q_3, presumably a recovery heat, had a half-time to completion of some 37 sec. On the adiabatic expansion theory, Q_2 should be closely related to the discharge and should thus vary with the resistance through which it is shunted in a manner predicted by ordinary electrical theory, reaching a maximum when internal and external resistances are equal. The situation proved more complex, and the cooling phase could be resolved into a component behaving like an adiabatic expansion and another, independent of load,

* In earlier discussions the authors considered that dilution of K^+ and Na^+ salts, occurring during the action potential, might account for the thermal changes; however experimental measurements showed that these were usually in the wrong direction. The exchange of ions in solution for those bound at membrane sites might well be accompanied by large thermal changes (Howarth *et al.* 1968).

that persisted on open circuit. To explain this open-circuit cooling it was suggested that the rapid efflux of K^+ under a high electrochemical gradient at the plateau of the action potential might reverse the Na^+-pump and promote synthesis of ATP.

Extra Oxygen Consumption

The lengthy phase of recovery heat corresponds with the restitutive processes that restore the lost K^+ to the fibre and extrude the gained Na^+. These should be reflected in an increased O_2-consumption; in fact the recovery heat should be equivalent to the extra-oxygen consumption if the energy supply is ultimately derived from oxidation. Recent application of the oxygen electrode has permitted the determination of the extra oxygen consumption resulting from relatively few impulses; Ritchie (1967), for example, measured the changes in O_2-content of the Ringer solution flowing over mammalian unmyelinated nerve. The resting consumption was o·24 $\mu M/g./min.$ at $35 \cdot 8°C$ and o·35 μM at $11 \cdot 6°C$, values that were not greatly different from those of other nerves, either myelinated or non-myelinated. The extra consumption per impulse decreased as the frequency of stimulation increased; at some 2·5 shocks/min. the estimated extra consumption per shock was some 900 pmole/g.; this is equivalent to some 100 $\mu cal./g.$ comparing with some 50 $\mu cal./g.$ found experimentally for the same nerve by Howarth et al. (1966).* Since ouabain abolished the extra O_2-consumption we may assume that it is, indeed, devoted to pumping K^+ back into the fibre and extruding Na^+; the O_2-consumption is equivalent to a K^+/O_2 ratio of about 5. It might be expected that the extra O_2-consumption would be influenced by external K^+ since this is required for adequate functioning of the Na^+-pump; the effects of removal of K^+ are not large, but this is probably due to the circumstance that repetitive stimulation leads to accumulation of K^+ outside the fibres so that the medium here is not K^+-free even when the bulk of the medium is (Rang & Ritchie, 1968).

* Ritchie (1967) has compared the "oxygen cost" of conducting an impulse along myelinated and non-myelinated fibres; it turns out to be the same in spite of the fact that, in the myelinated nerve, the breakdown of membrane potential is confined to restricted regions, namely, the nodes of Ranvier. The unmyelinated fibre seems, therefore, to have adapted itself in some way to conduct more economically; for example, the K^+-loss per impulse is smaller than from myelinated nerves (Keynes & Ritchie, 1965).

REFERENCES

ABBOTT, B. C., AUBERT, X., & FESSARD, A. (1958). La production de chaleur associée à la décharge du tissu électrique de la Torpille. *J. Physiol.*, Paris, **50**, 99–102.

ABBOTT, B. C., HILL, A. V., & HOWARTH, J. V. (1958). "The Positive and Negative Heat Production associated with a Nerve Impulse." *Proc. Roy. Soc., B,* **148**, 149–187.

ACEVES, J. & MACHNE, X. (1963). The action of calcium and of local anesthetics on nerve cells, and their interaction during excitation. *J. Pharmacol.*, **140**, 138–148.

ADELMAN, W. J., & DALTON, J. C. (1960). Interactions of calcium with sodium and potassium in membrane potentials of lobster giant axon. *J. gen. Physiol.*, **43**, 609–619.

ADELMAN, W. J. & FOK, Y. B. (1964). Internally perfused squid axons studied under voltage clamp conditions. *J. cell. comp. Physiol.*, **64**, 429–443.

ADELMAN, W. J., & MOORE, J. W. (1961). Action of external divalent ion reduction on sodium movement in the squid giant axon. *J. gen. Physiol.*, **45**, 93–103.

ADELMAN, W. J. & SENFT, J. P. (1966). Voltage clamp studies on the effect of internal cesium ion on Na and K currents in the squid giant axon. *J. gen. Physiol.*, **50**, 279–293.

ADRIAN, E. D. (1914). "The Relation between the Size of the Propagated Disturbance and the Rate of Conduction in Nerve." *J. Physiol.*, **48**, 53.

ADRIAN, R. H., CHANDLER, W. K. & HODGKIN, A. L. (1968). Voltage clamp experiments in striated muscle fibers. *J. gen. Physiol.*, **51**, 188–192S.

ADRIAN, R. H., CHANDLER, W. K. & HODGKIN, A. L. (1966). Voltage clamp experiments in skeletal muscle fibres. *J. Physiol.*, **186**, 51–52P.

ADRIAN, R. H. & FREYGANG, W. H. (1962). The potassium and chloride conductance of frog muscle membrane. *J. Physiol.*, **163**, 61–103.

ADRIAN, R. H. & FREYGANG, W. H. (1962). Potassium conductance of frog muscle membrane under controlled voltage. *J. Physiol.*, **163**, 104–114.

ARMSTRONG, C. M. (1967). Time course of TEA$^+$-induced anomalous rectification in squid giant axons. *J. gen. Physiol.*, **50**, 491–503.

ARMSTRONG, C. M. & BINSTOCK, L. (1964). The effects of several alcohols on the properties of the squid giant axon. *J. gen. Physiol.*, **48**, 265–277.

ARMSTRONG, C. M. & BINSTOCK, L. (1965). Anomalous rectification in the squid giant axon injected with tetraethylammonium chloride. *J. gen. Physiol.*, **48**, 859–872.

ARNETT, V., & WILDE, W. S. (1941). "Potassium and Water Changes in Excised Nerve on Stimulation." *J. Neurophysiol.*, **4**, 572.

AUBERT, X. & KEYNES, R. D. (1968). Temperature changes during and after the discharge of the electric organ in *Electrophorus electricus*. *Proc. Roy. Soc. B.*, **169**, 241–263.

AUGER, D. (1933). "Contribution à l'étude de la Propagation de la Variation Électrique chez les Characées." *C. r. Soc. Biol.*, Paris, **113**, 1437.

BAKER, P. F., HODGKIN, A. L. & MEVES, H. (1964). The effect of diluting the internal solution on the electrical properties of a perfused giant axon. *J. Physiol.*, **170**, 541–560.

BAKER, P. F., HODGKIN, A. L. & SHAW, T. I. (1962). Replacement of the axoplasm of giant nerve fibres with artificial solutions. *J. Physiol.*, **164**, 330–354.

BAKER, P. F., HODGKIN, A. L. & SHAW, T. I. (1962). Effects of changes in internal ionic concentrations on electrical properties of perfused giant axons. *J. Physiol.*, **164**, 355–374.

BENNETT, A. L., & CHINBURY, K. G. (1946). "The Effects of several Local Anæsthetics on the Resting Potential of Isolated Frog Nerve." *J. Pharmacol.*, **88**, 72.

BIANCHI, C. P. & SHANES, A. M. (1959). Calcium influx in skeletal muscle at rest, during activity, and during potassium contracture. *J. gen. Psysiol.*, **42**, 803–815.

BISHOP, G. H. (1932). Action of nerve depressants on potential. *J. cell. comp. Physiol.*, **1**, 177.

BISHOP, G. H. (1937). "La Théorie des Circuits Locaux Permet-elle de Prévoir la Forme du Potentiel d'Action ?" *Arch. int. Physiol.*, **45**, 273.

BLAIR, E. A. & ERLANGER, J. (1933). A comparison of the characteristics of axons through their individual electrical responses. *Amer. J. Physiol.*, **106**, 525–564.

BLAUSTEIN, M. P. (1968). Barbiturates block sodium and potassium conductance increases in voltage-clamped lobster axons. *J. gen. Physiol.*, **51**, 293–307.

BLAUSTEIN, M. P. & GOLDMAN, D. E. (1966). Competitive action of calcium and procaine on lobster axon. *J. gen. Physiol.*, **49**, 1043–1063.

BLAUSTEIN, M. P. & GOLDMAN, D. E. (1968). The action of certain polyvalent cations on the voltage-clamped lobster axon. *J. gen. Physiol.*, **51**, 279–291.

BLINKS, L. R. (1930). Direct current resistance of *Nitella*. *J. gen. Physiol.*, **13**, 495.

BRAZIER, M. (1960). *The Electrical Activity of the Nervous System*. 2nd ed. London: Pitman.

BRYANT, S. H., & TOBIAS, J. M. (1955). "Optical and Mechanical Concomitants of Activity in *Carcinus* Nerve." *J. cell. comp. Physiol.*, **46**, 71–95.

DEL CASTILLO-NICOLAU, & STARK, L. (1951). "Local Responses due to Stimulation at a Single Node of Ranvier." *J. Physiol.*, **114**, 19P.

CHANDLER, W. K. & HODGKIN, A. L. (1965). The effect of internal sodium on the action-potential in the presence of different internal anions. *J. Physiol.*, **181**, 594–611.

CHANDLER, W. K., HODGKIN, A. L. & MEVES, H. (1965) Effect of changing the internal solution on Na inactivation and related phenomena in giant axons. *J. Physiol.*, **180**, 821–836.

CHANDLER, W. K. & MEVES, H. (1965). Voltage clamp experiments on internally perfused giant axons. *J. Physiol.*, **180**, 788–820.

COLE, K. S. (1941). "Rectification and Inductance in the Squid Giant Axon." *J. gen. Physiol.*, **25**, 29.

COLE, K. S. (1942). "Impedance of Single Cells." *Tab. Biol.*, **19** (2), 24.

COLE, K. S. (1961). An analysis of the membrane potential along a clamped squid axon. *Biophys. J.*, **1**, 401–418.

COLE, K. S. (1968). *Membranes, Ions and Impulses*. Berkeley: Univ. of California Press.

COLE, K. S., & CURTIS, H. J. (1936). "Electric Impedance of Nerve and Muscle." *Cold Spr. Harb. Symp. quant. Biol.*, **4**, 73.

COLE, K. S., & CURTIS, H. J. (1938). "Electric Impedance of *Nitella* during Activity." *J. gen. Physiol.*, **22**, 37.

COLE, K. S., & MOORE, J. W. (1960). Potassium ion current in the squid giant axon; dynamic characteristic. *Biophys. J.*, **1**, 1–14.

CONDREA, E., ROSENBERG, P. & DETTBARN, W. D. (1967). Demonstration of phospholipid splitting as the factor responsible for increased permeability and block of axonal conduction induced by snake venom. *Biochim. biophys. acta.*, **135**, 669–681.

COOLEY, J., DODGE, F. & COHEN, H. (1965). Digital computer solutions for excitable membrane models. *J. cell comp. Physiol.*, **66**, Suppl. 2., 99–109.

COWAN, S. L. (1934). "Action of K+ and other Ions on the Injury Potential and Action Current in *Maia* Nerve." *Proc. Roy. Soc., B*, **115**, 216.

CRAGG, B. G. & THOMAS, P. K. (1964). The conduction velocity of regenerated peripheral nerve fibres. *J. Physiol.*, **171**, 164–175.

CRESCITELLI, F. (1951). "Nerve Sheath as a Barrier to the Action of Certain Substances." *Amer. J. Physiol.*, **166**, 229–240.

CURTIS, H. J., & COLE, K. S. (1942). "Membrane Resting and Action Potentials in Giant Fibres of Squid Nerve." *J. cell. comp. Physiol.*, **19**, 135.

DALTON, J. C., & ADELMAN, W. J. (1960). Some relations between action potential and resting potential of the lobster giant axon. *J. gen. Physiol.*, **43**, 597–607.

DAVIS, H., FORBES, A., BRUNSWICK, D., & HOPKINS, A. McH. (1926). "The Question of Decrement." *Amer. J. Physiol.*, **76**, 448.

DAVSON, H., & REINER, J. M. (1942). Ionic permeability: enzyme-like factor concerned in migration of Na through cat erythrocyte membrane. *J. cell. comp. Physiol.*, **20**, 325.

DECK, K. A. (1958). Über die Wirkung des Guanidinhydrochlorids und anderer Substanzen auf das Aktionspotential der Nerveneinzelfaser. *Pflüg. Arch. ges. Physiol.*, **266**, 249–265.

DESMEDT, J. E. (1953). Electrical activity and intracellular sodium concentration in frog muscle. *J. Physiol.*, **121**, 191–205.

DODGE, F. A., & FRANKENHAEUSER, B. (1958). Membrane currents in isolated frog nerve fibre under voltage clamp conditions. *J. Physiol.*, **143**, 76–90.

DODGE, F. A., & FRANKENHAEUSER, B. (1959). Sodium currents in the myelinated nerve fibre of *Xenopus lævis* with the voltage clamp technique. *J. Physiol.*, **148**, 188–200.

EDWARDS, G. A., RUSKA, H. & DE HARVEN, E. (1958). Electron microscopy of peripheral nerves and neuromuscular junctions in the wasp leg. *J. biophys. biochem. Cytol.*, **4**, 107–114.

EISENMAN, G. (1962). Cation selective glass electrodes and their mode of action. *Biophys. J.*, **2**, No. 2, Pt. 2, 259–323.

ELFVIN, L. G. (1961). The ultrastructure of the node of Ranvier in cat sympathetic nerve fibres. *J. Ultrastr. Res.*, **5**, 374–387.

ERLANGER, J., & BLAIR, E. A. (1938). "Action of Isotonic Salt-free Solutions on Conduction in Medullated Nerve Fibres." *Amer. J. Physiol.*, **124**, 341.

ERLANGER, J., & GASSER, H. S. (1937). *Electrical Signs of Nervous Activity*. Philadelphia. Univ. of Pennsylvania Press.

ESPLIN, D. W. (1962). Independence of conduction velocity among myelinated fibres in cat nerve. *J. Neurophysiol.*, **25**, 805–811.

FALK, G. & FATT, P. (1964). Linear electrical properties of striated muscle fibres observed with intracellular electrodes. *Proc. Roy. Soc. B*, **160**, 69–123.

FALK, G., & LANDA, J. F. (1960). Prolonged response of skeletal muscle in the absence of penetrating anions. *Amer. J. Physiol.*, **198**, 289–299.

FENG, T. P. (1936). "The Heat Production of Nerve." *Ergebn. d. Physiol.*, **38**, 73.

FENG, T. P., & LIU, Y. M. (1949). "The Connective Tissue Sheath of the Nerve as Effective Diffusion Barrier." *J. cell. comp. Physiol.*, **34**, 1–16.

FERNÁNDEZ-MORÁN, H. (1952). The submicroscopic organization of vertebrate nerve fibres. *Exp. Cell Res.*, **3**, 282–359.

FITZHUGH, R. (1960). Thresholds and plateaus in the Hodgkin-Huxley equations. *J. gen. Physiol.*, **43**, 867–896.

FITZHUGH, R. (1962). Computation of impulse initiation and saltatory conduction in a myelinated nerve fibre. *Biophys. J.*, **2**, 11–21.

FITZHUGH, R. (1966). Theoretical effect of temperature on threshold in the Hodgkin-Huxley nerve model. *J. gen. Physiol.*, **49**, 989–1005.

FRANKENHAEUSER, B. (1959). Steady state inactivation of sodium permeability in myelinated nerve fibres of *Xenopus lævis*. *J. Physiol.*, **148**, 671–676.

FRANKENHAEUSER, B. (1960). Quantitative description of sodium currents in myelinated nerve fibres of *Xenopus lævis*. *J. Physiol.*, **151**, 491–501.

FRANKENHAEUSER, B. (1962). Delayed currents in myelinated nerve fibres of *Xenopus lævis* investigated with voltage clamp technique. *J. Physiol.*, **160**, 40–45.

FRANKENHAEUSER, B. (1962). Instantaneous potassium currents in myelinated nerve fibres of *Xenopus lævis*. *J. Physiol.*, **160**, 46–53.

FRANKENHAEUSER, B. (1962). Potassium permeability in myelinated nerve fibres of *Xenopus lævis*. *J. Physiol.*, **160**, 54–61.

FRANKENHAEUSER, B. (1963). A quantitative description of potassium currents in myelinated nerve fibres of *Xenopus lævis*. *J. Physiol.*, **169**, 424–430.

FRANKENHAEUSER, B. (1965). Computed action potential in nerve from *Xenopus lævis*. *J. Physiol.*, **180**, 780–787.

FRANKENHAEUSER, B., & HODGKIN, A. L. (1956). "The After-effects of Impulses in the Giant Nerve Fibres of *Loligo*." *J. Physiol.*, **131**, 341–376.

FRANKENHAEUSER, B., & HODGKIN, A. L. (1957). "The Action of Calcium on the Electrical Properties of Squid Axons." *J. Physiol.*, **137**, 218–244.

FRANKENHAEUSER, B. & HUXLEY, A. F. (1964). Action potential in myelinated nerve fibre of *Xenopus lævis* as computed on basis of voltage clamp data. *J. Physiol.*, **171**, 302–315.

FRANKENHAEUSER, B. & VALLBO, A. B. (1965). Accommodation in myelinated nerve fibres of *Xenopus lævis* as computed on basis of voltage clamp data. *Acta physiol. scand.*, **63**, 1–20.

FRICKE, H. (1925). "The Electric Capacity of Suspensions with Special Reference to Blood." *J. gen. Physiol.*, **9**, 137–152.

FURUSAWA, K. (1929). "The Depolarisation of Crustacean Nerve by Stimulation or Oxygen Want." *J. Physiol.*, **67**, 325.

GAGE, P. W. & HUBBARD, J. I. (1966). The origin of the post-tetanic hyperpolarization of mammalian motor nerve terminals. *J. Physiol.*, **184**, 335–352.

GAGE, P. W. & HUBBARD, J. I. (1966). Investigation of post-tetanic potentiation of end-plate potentials at a mammalian neuromuscular junction. *J. Physiol.*, **184**, 353–375.

GALLEGO, A. (1948). On the effect of ethyl alcohol upon frog nerve. *J. cell. comp. Physiol.*, **31**, 97.

GASSER, H. S. (1950). "Unmedullated Fibres originating in Dorsal Root Ganglia." *J. gen. Physiol.*, **33**, 651–690.

GASSER, H. S., & GRUNDFEST, H. (1939). Axon diameter in relation to spike dimension and conduction velocity in mammalian A fibres. *Amer. J. Physiol.*, **127**, 393.

GEORGE, E. P., & JOHNSON, E. A. (1961). Solutions of the Hodgkin–Huxley equations for squid axon treated with tetraethylammonium and in potassium-rich media. *Aust. J. exp. Biol. med. Sci.*, **39**, 275–294.

GERARD, R. W. (1936). "Metabolism and Excitation." *Cold Spr. Harb. Symp. quant. Biol.*, **4**, 194.

GEREN, B. B., & SCHMITT, F. O. (1954). Structure of the Schwann cell and its relation to the axon in certain invertebrate nerve fibres. *Proc. Nat. Acad. Sci., Wash.*, **40**, 863–870.

GÖPFERT, H., & SCHAEFER, H. (1938). "Uber den Direkt und Indirekt Erregten Aktionstrom und die Funktion der Motorische Endplatte." *Pflüg. Arch.*, **239**, 597.

GRAHAM, J., & GERARD, R. W. (1946). "Membrane Potentials and Excitation of Impaled Single Muscle Fibres." *J. cell. comp. Physiol.*, **28**, 99.

GRUNDFEST, H. (1940). Bioelectric potentials. *Ann. Rev. Physiol.*, **2**, 213.

GRUNDFEST, H., SHANES, A. M., & FREYGANG, W. (1953). "The Effect of Sodium and Potassium Ions on the Impedance Change accompanying the Spike in the Squid Giant Axon." *J. gen. Physiol.*, **37**, 25–37.

GUTTMAN, R. (1962). Effect of temperature on the potential and current thresholds of axon membrane. *J. gen. Physiol.*, **46**, 258–266.

HAGIWARA, S., CHICHIBU, S. & NAKA, K.-I. (1964). The effects of various ions on resting and spike potentials of barnacle muscle fibers. *J. gen. Physiol.*, **48**, 163–179.

HAGIWARA, S. & NAKA, K.-I. (1964). The initiation of spike potential in barnacle fibers under low intracellular Ca^{++}. *J. gen. Physiol.*, **48**, 141–162.

HAGIWARA, S. & NAKAJIMA, S. (1966, a). Differences in Na and Ca spikes as examined by tetrodotoxin, procaine, and manganese ions. *J. gen. Physiol.*, **49**, 793–806.

HAGIWARA, S. & NAKAJIMA, S. (1966, b). Effects of intracellular Ca ion concentration upon excitability of the muscle fiber membrane of a barnacle. *J. gen. Physiol.*, **49**, 807–818.

HAGIWARA, S. & TAKAHASHI, K. (1967). Surface density of calcium ions and calcium spikes in the barnacle muscle fiber membrane. *J. gen. Physiol.*, **50**, 583–601.

HEENE, R. (1962). Das Aktionspotential des isolierten Ranvierschen Schnürring bei Erhohung der extracellulären Wasserstoffionen-Konzentration in calciumhaltigen Lösungen. *Pflüg. Arch.*, **275**, 1–11.

HESS, A., & YOUNG, J. Z. (1949). "Correlation of Internodal Length and Fibre Diameter in the Central Nervous System." *Nature*, **164**, 490–491.

HILL, A. V. (1932). Closer analysis of heat production of nerve. *Proc. Roy. Soc., B*, **111**, 106.

HILL, A. V. (1933). Three phases of nerve heat production. *Proc. Roy. Soc., B*, **113**, 345.

HILL, A. V. (1936). Excitation and accommodation in nerve. *Proc. Roy. Soc., B*, **119**, 305.

HILL, D. K. (1950). "Effect of Stimulation on Opacity of Crustacean Nerve Trunk and its Relation to Fibre Diameter." *J. Physiol.*, **111**, 283.

HILL, D. K. (1950). "Volume Change resulting from Stimulation of a Giant Nerve Fibre." *J. Physiol.*, **111**, 304.

HILL, D. K., & KEYNES, R. D. (1949). "Opacity Changes in Stimulated Nerve." *J. Physiol.*, **108**, 278.

HILLE, B. (1967). The selective inhibition of delayed potassium currents in nerve by tetraethylammonium ion. *J. gen. Physiol.*, **50**, 1287–1302.

HILLE, B. (1968). Charges and potentials at the nerve surface. Divalent ions and pH. *J. gen. Physiol.*, **51**, 221–236.

HILLE, B. (1968). Pharmacological modifications of the sodium channels of frog nerve. *J. gen. Physiol.*, **51**, 199–219.

HODGKIN A. L. (1937). "Evidence for Electrical Transmission in Nerve." I & II. *J. Physiol.*, **90**, 183, 211.

HODGKIN, A. L. (1938). "The Subthreshold Potentials in a Crustacean Nerve Fibre." *Proc. Roy. Soc., B,* **126**, 87.

HODGKIN, A. L. (1939). "The Relation between Conduction Velocity and the Electrical Resistance Outside a Nerve." *J. Physiol.*, **94**, 560.

HODGKIN, A. L. (1947). "The Membrane Resistance of Non-medullated Nerve Fibre." *J. Physiol.*, **106**, 305.

HODGKIN, A. L. (1947). "The Effect of Potassium on the Surface Membrane of an Isolated Axon." *J. Physiol.*, **106**, 319.

HODGKIN, A. L. (1954). "A Note on Conduction Velocity." *J. Physiol.*, **125**, 221–224.

HODGKIN, A. L. & HOROWICZ, P. (1959). Movements of Na and K in single muscle fibres. *J. Physiol.*, **145**, 405–432.

HODGKIN, A. L., & HUXLEY, A. F. (1939). "Action Potentials recorded from inside a Nerve Fibre." *Nature,* **140**, 710.

HODGKIN, A. L. & HUXLEY, A. F. (1947). Potassium leakage from an active nerve fibre. *J. Physiol.*, **106**, 341.

HODGKIN, A. L., & HUXLEY, A. F. (1952). Currents carried by sodium and potassium ions through the membrane of the giant axon of *Loligo. J. Physiol.*, **116**, 449–472.

HODGKIN, A. L. & HUXLEY A. F. (1952). "The Components of Membrane Conductance in the Giant Axon of *Loligo.*" *J. Physiol.*, **116**, 473–496.

HODGKIN, A. L., & HUXLEY, A. F. (1952). "The Dual Effect of Membrane Potential on Sodium Conductance in the Giant Axon of *Loligo.*" *J. Physiol.*, **116**, 497–506.

HODGKIN, A. L., & HUXLEY, A. F. (1952). Quantitative description of membrane current and its application to conduction and excitation in nerve. *J. Physiol.*, **117**, 500–544.

HODGKIN A. L., & HUXLEY, A. F. (1953). "Movements of Radioactive Potassium and Membrane Current in a Giant Axon." *J. Physiol.*, **121**, 403–414.

HODGKIN, A. L., HUXLEY, A. F., & KATZ, B. (1952). Measurement of current-voltage relations in membrane of giant axon of *Loligo. J. Physiol.*, **116**, 424–448.

HODGKIN, A. L., & KATZ B., (1949). "Effect of Na⁺ on the Electrical Activity of the Giant Axon of the Squid." *J. Physiol.*, **108**, 37.

HODGKIN, A. L., & KEYNES R. D. (1956). "Experiments on the Injection of Substances into Squid Giant Axons by means of a Microsyringe." *J. Physiol.*, **131**, 592–616.

HODGKIN, A. L., & KEYNES, R. D. (1957). "Movements of Labelled Calcium in Squid Giant Axons." *J. Physiol.*, **138**, 253–281.

HODGKIN, A. L., & RUSHTON, W. A. H. (1946). "The Electrical Constants of a Crustacean Nerve Fibre." *Proc. Roy. Soc., B,* **133**, 444.

HOWARTH, J. V., KEYNES, R. D. & RITCHIE, J. M. (1966). The heat production of mammalian non-myelinated (C) nerve fibres. *J. Physiol.*, **186**, 60–62 P.

HOWARTH, J. V., KEYNES, R. D. & RITCHIE, J. M. (1968). The origin of the initial heat associated with a single impulse in mammalian non-myelinated nerve fibres. *J. Physiol.*, **194**, 755–793.

HOYLE, G. & SMYTH, T. (1963). Neuromuscular physiology of giant muscle fibers of a barnacle, *Balanus nubilus* Darwin. *Comp. Biochem. Physiol.*, **10**, 291–314.

HUXLEY, A. F. (1959). Ionic movements during nerve activity. *Ann. N.Y. Acad. Sci.*, **81**, 221–246.

HUXLEY, A. F., & STÄMPFLI, R. (1949). "Evidence for Saltatory Conduction in Peripheral Myelinated Nerve Fibres." *J. Physiol.*, **108**, 315.

HUXLEY, A. F., & STÄMPFLI, R. (1951). Direct determination of membrane resting potential and action potential in single myelinated nerve fibres. *J. Physiol.*, **112**, 476.

HUXLEY, A. F., & STÄMPFLI, R. (1951). "Effect of K⁺ and Na⁺ on Resting and Action Potentials of Single Myelinated Nerve Fibres." *J. Physiol.*, **112**, 496.

HUXLEY, A. F., & STÄMPFLI, R. (1951). Direct determination of membrane resting potential and action potential in single myelinated nerve fibres. *J. Physiol.*, **112**, 476-508.

JULIAN, F. J., MOORE, J. W., & GOLDMAN, D. E. (1962). Current-voltage relations in lobster giant axon membrane under voltage clamp conditions. *J. gen. Physiol.*, **45**, 1217–1238.

JULIAN, F. J., MOORE, J. W., & GOLDMAN, D. E. (1962). Membrane potentials of lobster giant axon obtained by use of sucrose-gap technique. *J. gen. Physiol.*, **45**, 1195–1216.

KAO, C. Y. & NISHIYAMA, A. (1965). Actions of saxitoxin on peripheral neuromuscular systems. *J. Physiol.*, **180**, 50–66.

KATO, G. (1936). "Excitation, Conduction and Narcotisation of Single Nerve Fibres." *Cold Spr. Harb. Symp. quant. Biol.*, **4**, 43.

KATZ, B. (1937). "Experimental Evidence for a Non-conducted Response of Nerve to Subthreshold Stimulation." *Proc. Roy. Soc., B*, **124**, 244.

KATZ, B. (1939). *Electrical Excitation in Nerve.* Oxford. O.U.P.

KATZ, B. (1947). "Subthreshold Potentials in Medullated Nerve." *J. Physiol.*, **106**, 66.

KATZ, B. (1947). "The Effect of Electrolyte Deficiency on the Rate of Conduction in a Single Nerve Fibre." *J. Physiol.*, **106**, 411.

KATZ, B. (1948). "Electrical Properties of the Muscle Fibre Membrane." *Proc. Roy. Soc., B*, **135**, 506.

KATZ, B., & SCHMITT, O. H. (1940). "Electric Interaction between Two Adjacent Nerve Fibres." *J. Physiol.*, **97**, 471.

KEYNES, R. D., & LEWIS, P. R. (1951). "The Sodium and Potassium Content of Cephalopod Nerve Fibres." *J. Physiol.*, **114**, 151–182.

KEYNES, R. D. & RITCHIE, J. M. (1965). The movements of labelled ions in mammalian non-myelinated nerve fibres. *J. Physiol.*, **179**, 333–367.

KÖNIG, K. (1962). Membranpotentialmessungen am Skeletmuskel mit der "Saccharose-Trennwand" Methode. *Pflüg. Arch.*, **275**, 452–460.

KOPPENHÖFER, E. (1967). Die Wirkung von Tetraäthylammoniumchlorid auf die Membranströme Ranvierscher Schnürringe von *Xenopus lævis*. *Pflüg. Arch. ges. Physiol.*, **293**, 34–55.

KRNJEVIĆ, K. (1954). "The Connective Tissue of the Frog Sciatic Nerve." *Quart. J. exp. Physiol.*, **39**, 55–71.

LORENTE DE NÓ, R. (1946–7). "Correlation of Nerve Activity with Polarisation Phenomena." *Harvey Lectures*, **42**, 43.

LÜTTGAU, H. -C. (1958). Die Wirkung von Guanidinhydrochlorid auf die Erregungsprozesse an isolierten markhaltigen Fasern. *Pflüg. Arch. ges. Physiol.*, **267**, 331–343.

MARTIN, R. & ROSENBERG, P. (1968). Fine structural alterations associated with venom action on squid giant nerve fibers. *J. Cell Biol.*, **36**, 341–353.

MAXFIELD, M. (1953). "Axoplasmic Proteins of the Squid Giant Nerve Fibre with particular reference to the Fibrous Proteins." *J. gen. Physiol.*, **37**, 201–216.

MONNIER, A. M. (1934). "*L'Excitation Electrique des Tissues.*" Paris, Hermann.

MOORE, J. W. (1959). Excitation of the squid axon membrane in isosmotic potassium chloride. *Nature*, **183**, 265–266.

MOORE, J. W., & ADELMAN, W. J. (1961). Electronic measurement of the intracellular concentration and net flux of sodium in the squid axon. *J. gen. Physiol.*, **45**, 77–92.

MOORE, J. W., BLAUSTEIN, M. P., ANDERSON, N. C. & NARAHASHI, T. (1967). Basis of tetrodotoxin's selectivity in blockage of squid axons. *J. gen. Physiol.*, **50**, 1401–1411.

MOORE, J. W., NARAHASHI, T. & ULBRICHT, W. (1964). Sodium conductance shift in axon internally perfused with a sucrose and low-potassium solution. *J. Physiol.*, **172**, 163–173.

MOSHER, H. S., FUHRMAN, F. A., BUCHWALD, H. D. & FISCHER, H. G. (1964). Tarichatoxin-tetrodotoxin: a potent neurotoxin. *Science*, **144**, 1100–1110.

MUELLER, P. (1958). Prolonged action potentials from single nodes of Ranvier. *J. gen. Physiol.*, **42**, 137–162.

NAKAJIMA, S. & TAKAHASHI, K. (1966). Post-tetanic hyperpolarization and electrogenic Na-pump in stretch receptor neurone of crayfish. *J. Physiol.*, **187**, 105–127.

NARAHASHI, T. (1963). Dependence of resting and action potentials on internal potassium in perfused squid giant axons. *J. Physiol.*, **169**, 91–115.

NARAHASHI, T., ANDERSON, N. C. & MOORE, J. W. (1967). Comparison of tetrodotoxin and procaine in internally perfused squid giant axons. *J. gen. Physiol.*, **50**, 1413–1428.

NARAHASHI, T., DEGUCHI, T., URAKAWA, N. & OHKUBO, Y. (1960). Stabilization and rectification of muscle fiber membrane by tetrodotoxin. *Amer. J. Physiol.*, **198**, 934–938.

NARAHASHI, T. & MOORE, J. W. (1968). Neuroactive agents and nerve membrane conductances. *J. gen. Physiol.*, **51**, 93–101 s.

NARAHASHI, T., MOORE, J. W. & POSTON, R. N. (1966). Specific action of tetrodotoxin derivatives on nerve. *Science*, **154**, 425.

NASTUK, W. L., & HODGKIN, A. L. (1950). "The Electrical Activity in Single Muscle Fibres." *J. cell. comp. Physiol.*, **35**, 39.

NERNST, W. (1908). "Zur Theorie des elektrischen Reizes." *Pflüg. Arch.*, **122**, 275.

NOBLE, D. (1966). Applications of Hodgkin–Huxley equations to excitable tissues. *Physiol. Rev.*, **46**, 1–50.

NOBLE, D. & STEIN, R. B. (1966). The threshold conditions for initiation of action potentials by excitable cells. *J. Physiol.*, **187**, 129–162.

OFFNER, F., WEINBERG, A. & YOUNG, G. (1940). Nerve conduction theory. Some mathematical consequences of Bernstein's model. *Bull. Math. Biophys.*, **2**, 89.

OOYAMA, H., & WRIGHT, E. B. (1961). Anode break excitation on single Ranvier node of frog nerve. *Amer. J. Physiol.*, **200**, 209–218.

OSTERHOUT, W. J. V., & HILL, S. E. (1930). "Salt Bridges and Negative Variations." *J. gen. Physiol.*, **13**, 547.

PALAY, S. L., SOTELO, C., PETERS, A. & ORKAND, P. M. (1968). The axon hillock and the initial segment. *J. Cell Biol.*, **38**, 193–201.

PETERS, A. & VAUGHN, J. E. (1967). Microtubules and filaments in the axons and astrocytes of early postnatal rat optic nerves. *J. Cell Biol.*, 32, 113–119.

PFAFFMAN, C. (1940). Potentials in isolated medullated axon. *J. cell. comp. Physiol.*, **16**, 407.

PUMPHREY, R. J., & YOUNG, J. Z. (1938). "The Rates of Conduction of Nerve Fibres of Various Diameters in Cephalopods." *J. exp. Biol.*, **15**, 453.

RANG, H. P. & RITCHIE, J. M. (1968). Electrogenic sodium pump in mammalian non-myelinated nerve fibres and its activation by various external cations. *J. Physiol.*, **196**, 183–221.

RANG, H. P. & RITCHIE, J. M. (1968). The dependence on external cations of the oxygen consumption of mammalian non-myelinated fibres. *J. Physiol.*, **196**, 163–181.

RASHEVSKY, N. (1936). "Physico-mathematical Aspects of Excitation and Conduction in Nerves." *Cold Spr. Harb. Symp. quant. Biol.*, **4**, 90.

RITCHIE, J. M. (1961). Possible mechanisms underlying production of afterpotential in nerve fibres. In *Biophysics of Physiological and Pharmacological Actions*. Ed. Shanes. Washington: Amer. Ass. Adv. Sci., pp. 165–182.

RITCHIE, J. M. (1967). The oxygen consumption of mammalian non-myelinated nerve fibres at rest and during activity. *J. Physiol.*, **188**, 309–324.

RITCHIE, J. M., & STRAUB, R. W. (1956). "The After-effects of Repetitive Stimulation on Mammalian Non-medullated Fibres." *J. Physiol.*, **134**, 698–711.

ROBERTSON, J. D. (1957). "The Ultrastructure of Nodes of Ranvier in Frog Nerve Fibres." *J. Physiol.*, **137**, 8–9P.

ROBERTSON, J. D. (1958). "The Ultrastructure of Schmidt-Lanterman Clefts and related Shearing Defects of the Myelin Sheath." *J. biophys. biochem. Cytol.*, **4**, 39–46.

ROSENBERG, H. (1937). "The Physico-chemical Basis of Electrotonus." *Trans. Farad. Soc.*, **33**, 1028.

ROSENBERG, P. & PODLEWSKI, T. R. (1963). Ability of venoms to render squid axons sensitive to curare and acetylcholine. *Biochim. biophys. acta.*, **75**, 104–115.

ROSENBLUETH, A. (1944). "The Interaction of Myelinated Fibres in Mammalian Nerve Trunks." *Amer. J. Physiol.*, **140**, 656.

ROTHENBERG, M. A. (1950). "Ionic Movements across Axonal Membranes." *Biochim. biophys. Acta*, **4**, 96.

ROZSA, G., MORGAN, C., SZENT-GYÖRGYI, A., & WYCKOFF, R. W. G. (1950). "Electron Microscopy of Myelinated Nerve." *Biochim. biophys. Acta*, **6**, 13.

RUSHTON, W. A. H. (1937). "The Initiation of the Propagated Disturbance." *Proc. Roy. Soc., B*, **124**, 201.

SANDERS, F. K., & WHITTERIDGE, D. (1946). "Conduction Velocity and Myelin Thickness in Regenerating Nerve Fibres." *J. Physiol.*, **105**, 152.

SCHMIDT, H. & STÄMPFLI, R. (1966). Die Wirkung von Tetraäthylammoniumchlorid auf den einzelnen Ranvierschen Schnürring. *Pflüg. Arch. ges. Physiol.*, **287**, 311–325.

SCHMITT, F. O., & CORI, C. F. (1933). "Lactic Acid Formation in Medullated Nerve." *Amer. J. Physiol.*, **106**, 339.

SEGAL, J. R. (1958). Anodal threshold phenomenon in squid giant axon. *Nature*, **182**, 1370.

SHANES, A. M. (1949). "Electrical Phenomena in Nerve. I. Squid Giant Axon." *J. gen. Physiol.*, **33**, 57. "II. Crab Nerve," p. 75.

SHANES, A. M. (1954). "Effects of Sheath Removal on the Sciatic Nerve of the Toad, *Bufo marinus*." *J. cell. comp. Physiol.*, **43**, 87–98.

SHANES, A. M. & BIANCHI, C. P. (1959). The distribution and kinetics of release of radiocalcium in tendon and skeletal muscle. *J. gen. Physiol.*, **42**, 1123–1137.

SHANES, A. M., & BROWN, D. E. S. (1942). "The Effect of Metabolic Inhibitors on the Resting Potential of Frog Nerve." *J. cell. comp. Physiol.*, **19**, 1.

SHANES, A. M., FREYGANG, W. H., GRUNDFEST, H., & AMATNIEK, E. (1959). Anæsthetic and calcium action in voltage clamped squid giant axon. *J. gen. Physiol.*, **42**, 793–802.

SHANTHAVEERAPPA, T. R. & BOURNE, G. H. (1962). The "perineurial epithelium", a metabolically active, continuous, protoplasmic cell barrier surrounding peripheral nerve fasciculi. *J. Anat.*, **96**, 527–537.

SHAW, S. N., & TOBIAS, J. M. (1955). "On the Optical Change associated with Activity in Frog Nerve." *J. cell. comp. Physiol.*, **46**, 53–70.

SJODIN, R. A. (1966). Long duration responses in squid giant axons injected with [134]Cs sulfate solutions. *J. gen. Physiol.*, **50**, 269–278.

SPYROPOULOS, C. S., & TASAKI, I. (1961). Nerve excitation and synaptic transmission. *Ann. Rev. Physiol.*, **22**, 407–432.

STÄMPFLI, R. (1958). Die Strom-Spannungs-Charakteristik der erregbaren Membran eines einzelnen Schnürrings und ihre Abhängigkeit von der Ionenkonzentration. *Helv. Physiol. Acta*, **16**, 127–146.

TAKATA, M., MOORE, J. W., KAO, C. Y. & FUHRMAN, F. A. (1966). Blockage of sodium conductance increase in lobster giant axon by tarichatoxin (tetrodotoxin). *J. gen. Physiol.*, **49**, 977–988.

TASAKI, I. (1939). "The Electro-saltatory Transmission of the Nerve Impulse and the Effect of Narcosis upon the Nerve Fibre." *Amer. J. Physiol.*, **127**, 211.

TASAKI, I. (1952). "Properties of Myelinated Fibres in Frog Sciatic Nerve and in Spinal Cord as examined with Micro-electrodes." *Jap. J. Physiol.*, **3**, 73–94.

TASAKI, I. (1955). New measurements of capacity and resistance of myelin sheath and nodal membrane of isolated frog nerve fibre. *Amer. J. Physiol.*, **181**, 639–650.

TASAKI, I. (1956). Initiation and abolition of the action potential of a single node of Ranvier. *J. gen. Physiol.*, **39**, 377–395.

TASAKI, I. (1959). Demonstration of two stable states of the nerve membrane in potassium-rich media. *J. Physiol.*, **148**, 306–331.

TASAKI, I., & FRANK, K. (1955). Measurement of the action potential of myelinated nerve fibre. *Amer. J. Physiol.*, **182**, 572–578.

TASAKI, I., & FREYGANG, W. H. (1955). The parallelism between the action potential, action current, and membrane resistance at a node of Ranvier. *J. gen. Physiol.*, **39**, 211–223.

TASAKI, I., & HAGIWARA, S. (1957). Demonstration of two stable potential states in squid giant axon under tetraethylammonium chloride. *J. gen. Physiol.*, **40**, 859–885.

TASAKI, I. & LUXORO, M. (1964). Intracellular perfusion of Chilean giant squid axons. *Science*, **145**, 1313–1315.

TASAKI, I., WATANABE, A. & LERMAN, L. (1967). Role of divalent cations in excitation of squid giant axons. *Amer. J. Physiol.*, **213**, 1465–1474.

TAYLOR, G. W. (1942). "The Correlation between Sheath Birefringence and Conduction Velocity with special reference to Cat Nerve Fibres." *J. cell. comp. Physiol.*, **20**, 359.

TAYLOR, R. E. (1959). Effect of procaine on electrical properties of squid axon membrane. *Amer. J. Physiol.*, **196**, 1071–1078.

TAYLOR, R. E., MOORE, J. W., & COLE, K. S. (1960). Analysis of certain errors in squid voltage clamp measurements. *Biophys. J.*, **1**, 161–202.

THÖRNER, W. (1922). "Elektrophysiologische Untersuchungen am Alterierten Nerven." *Pflüg. Arch.*, **197**, 159.

ULBRICHT, W. (1965). Voltage clamp studies of veratrinized nodes. *J. cell. comp. Physiol.*, **66**, Suppl. 2, 91–94.

UZMAN, B. G., & NOGUEIRA-GRAF, G. (1957). "Electron Microscope Studies of the Formation of Nodes of Ranvier in Mouse Sciatic Nerves." *J. biophys. biochem. Cytol.*, **3**, 589–598.

VALLBO, A. B. (1964). Accommodation related to inactivation of sodium permeability in single myelinated nerve fibres from *Xenopus lævis*. *Acta. physiol. scand.*, **61**, 429–444.

VAN DER KLOOT, W. G. & GLOVSKY, J. (1965). The uptake of Ca^{++} and Sr^{++} by fractions from lobster muscle. *Comp. Biochem. Physiol.*, **15**, 547–565.

VILLEGAS, G. M. & VILLEGAS, R. (1968). Ultrastructural studies of the squid nerve fibres. *J. gen. Physiol.*, **52**, 44–60s.

VILLEGAS, R., & BARNOLA, F. V. (1961). Characterization of the resting axolemma in the giant axon of the squid. *J. gen. Physiol.*, **44**, 963–977.

VILLEGAS, R., & VILLEGAS, G. M. (1960). Characterization of the membranes in the giant nerve fibre of the squid. *J. gen. Physiol.*, **43**, Suppl. 73–96.

VILLEGAS, R., & VILLEGAS, G. M. (1962). The endoneurium cells of the squid giant nerve and their permeability to [^{14}C]glycerol. *Biochim. Biophys. Acta*, **60**, 202–204.

VILLEGAS, R., CAPUTO, C., & VILLEGAS, L. (1962). Diffusion barriers in the squid nerve fiber. The axolemma and the Schwann layer. *J. gen. Physiol.*, **46**, 245–255.

WATANABE, A., TASAKI, I., SINGER, I. & LERMAN, L. (1967). Effects of tetrodotoxin on excitability of squid giant axons in sodium-free media. *Science*, **155**, 95–97.

WRIGHT, E. B., & OOYAMA, H. (1961). Anode break excitation and Pflüger's Law. *Amer. J. Physiol.*, **200**, 219–222.

YOUNG, J. Z. (1939). Fused neurons and synaptic contacts in the giant nerve fibres of cephalopods. *Phil. Trans.*, **229**, 465.

CHAPTER XVI

TRANSMISSION OF THE IMPULSE

THE wave of action potential, initiated at any point on the nerve or muscle fibre, travels over its whole surface as a result of the self-propagating characteristic discussed earlier. The transmission of this disturbance from one cell to another, be it from one nerve to another, or from a nerve to a muscle or secretory cell, introduces a new problem, the bridging of a gap in protoplasmic continuity. The existence of regions of functional discontinuity, at least in so far as the nervous organization is concerned, is clearly necessary if a single impulse is not to excite all nerves with which it is related, either directly or indirectly. The problem of transmission is, therefore, twofold in its nature; it is concerned with how an impulse bridges a gap in protoplasmic continuity and why, in some circumstances, it fails to do so. So complex are the phenomena of transmission and inhibition, however, that a lucid and full account is beyond the scope of this book; all that the author can hope to do is to present some of the more striking facts regarding transmission from nerve to muscle and, in certain comparatively simple instances, from nerve to nerve, and to indicate the trends of thought in their interpretation. Before considering, in detail, some of the special problems of transmission, we may first briefly review the general organization of the vertebrate nervous system.

Neuronal Organization

The neurone is the unit of conduction, but only in rare cases, in the higher organisms, is it adequate to initiate and complete any response to an environmental change; in general, such a response is brought about by groups of neurones subserving separate functions, the propagated disturbances initiated in one being passed, by way of the *synapse*, to new, second-order, neurones, which may then pass the impulse on to others, until finally the neurone making relation with the effector organ, *e.g.*, a muscle fibre, is activated. If it is realized that a single neurone may make direct synaptic relations with many other neurones, the complexity of the organization of the nervous system becomes evident, and the difficulties of physiological study, in terms of the passage of impulses, become only too manifest. In the present treatment of the subject we shall therefore confine attention to those experiments in which there is reason to believe that the pathway is simple, so that, for this purpose, it is necessary only to indicate the neuronal organization at its lowest level, the nervous pathway involved in the simple *spinal reflex*, and the motor pathway in the *visceral reflexes* mediated by a sympathetic ganglion. Fig. 582*a* is the time-honoured representation of the reflex arc; a sensory nerve, with its cell body located in the *dorsal root ganglion*, has two processes, the peripheral axon* ramifying at its termination in the skin, the ramifications becoming specialized to respond to a definite stimulus, *e.g.*, of touch, or heat, and called *receptors*. The central end enters the spinal cord (a group of these fibres constituting the *posterior root*) and makes relations with numerous other neurones; in the simplest monosynaptic type of reflex—*e.g.*, the myotatic—the impulses may be effectively transmitted directly to the motor neurones, the cell-bodies

* Actually a dendrite (p. 1031).

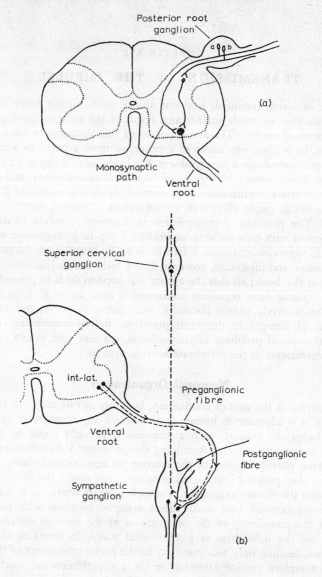

FIG. 582. (a) Illustrating simple reflex arcs; a sensory fibre, with its cell body in the posterior root ganglion, may synapse directly with a motor neurone in the ventral grey matter, or indirectly by way of an intercalary neurone. (b) Illustrating the sympathetic outflow. The axons of sympathetic motor neurones emerge by way of the ventral root and enter a ganglion of the sympathetic chain where they may or may not synapse immediately.

of which are in the *ventral horn*, their axons passing out at the motor, or *anterior*, *root* to end in a number of muscle fibres.

The *autonomic* division of the nervous system is concerned with the activation of effectors not under voluntary control, *e.g.*, the pupil of the eye. The motor pathway of the *sympathetic* division is mediated by motor neurones with their cell bodies in the intermedio-lateral grey matter of the cord; their axons pass out in the anterior root with the voluntary motor fibres already described, but leave these to pass into one of a chain of ganglia, the *sympathetic chain*, where they may or may not relay, *i.e.*, make synaptic relations with post-ganglionic nerve cells. Thus fibres to the eye (Fig. 582b) pass *through* the *stellate ganglion* to synapse in a ganglion higher up, the *superior cervical ganglion*. Fibres to the heart, on the other hand, relay in the stellate ganglion. The motor pathway for the *parasympathetic* is by way of special nerves, most of them emanating directly from the brain, the so-called *cranial nerves*; they relay in ganglia closely related to the organs they innervate, *e.g.*, the *ciliary ganglion* in the orbit of the eye. Hence in the autonomic system we have to distinguish, in the motor path, between *pre-* and *post*-ganglionic fibres, and in any ganglion we must find out which pre-ganglionic fibres synapse and which pass straight through.

The Synapse

The modes of connection between neurones are so various that it is impossible to speak of a typical synapse; as we have seen, the impulses in a neurone are carried by the axon away from the cell-body or soma; the axon

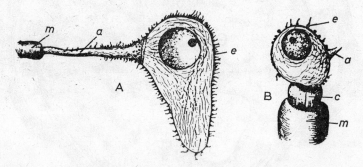

FIG. 583. Synapses. A, large motor-type cell from the reticular formation of the goldfish showing relatively uniform distribution of homogeneous boutons, *e*, on soma and proximal part of axon, *a*. × 960. This type of synaptic system is the most common in the vertebrate nervous system and is characteristic of motor neurones and intermediate neurones. B, cell of reticular formation of the goldfish showing, in addition to small boutons, a single large club ending, *c*, of a myelinated axon, *m*. × 960. (Bodian, after Lorente de Nó. *Physiol. Rev.*)

generally ramifies at its termination, its branches making contact with the dendrites or soma of the post-synaptic neurone. Frequently this contact is made by *terminal boutons*, the fine ramifications ending in swellings which are apposed to the post-synaptic neurone, as in Fig. 583, which gives some idea of the large number of boutons, derived from many axons, that may make relation with a single neurone; so great may be the number on the ventral horn cell that, according to Wyckoff & Young, the boutons may be said to cover the surface of the cell body and dendrites virtually completely so that, in all, perhaps more than 80 per cent. of the neurone's surface is covered. In other cases a single axonic termination may occupy a large proportion of the synaptic surface; this is especially manifest in the calyciform endings on the post-ganglionic neurones of the ciliary ganglion of the chick, illustrated schematically by Fig. 584; the endings of the pre-synaptic neurone have large expansions which extend over a considerable area of the post-synaptic cell surface, often appearing to envelop the cell entirely. In general, we may take it that a post-synaptic neurone receives, or is capable of receiving, impulses transmitted from the axons of several pre-synaptic fibres.

Synaptilemma. Until the application of high-resolution electron-microscopy to the problem, the evidence for a protoplasmic discontinuity at the synapse was mainly physiological, although classical histological studies suggested the

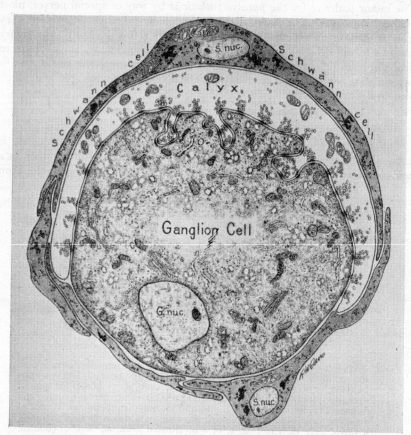

FIG. 584. Highly schematic drawing incorporating the principal details of calyciform endings in the ciliary ganglion of the chick. The calyx is shown in contiguity with a considerable area of ganglion cell surface. In all calyces the appositional membranes exhibit localized dense regions. At these locations, clusters of synaptic vesicles are evident in the presynaptic terminal. The synaptic cleft is uniformly 300 to 400 A wide. Occasionally, a dense line is resolved between the appositional membranes in the synaptic cleft. Schwann cells and their processes invest both the calyx and the ganglion cell (S. nuc. identifies the Schwann cell nucleus and G. nuc., the ganglion cell nucleus). Endoplasmic reticulum, cisternæ, cytoplasmic granules and mitochondria are schematically represented. (de Lorenzo. *J. biophys. biochem. Cytol.*)

presence of a membrane separating pre- and post-synaptic cells—the so-called *synaptilemma*. According to Palay's study of dendrite-axon and axon-perikaryon connections in the central nervous system, this synaptilemma consists of two membranes with a space of some 200A between them (Fig. 585, Pl. LIII). Thus, the surface membrane completely invests the terminal axon to form one membrane, the other being the surface membrane of the dendrite or perikaryon. If the pre-synaptic axon is myelinated, it loses this sheath a short distance

PLATE LIII

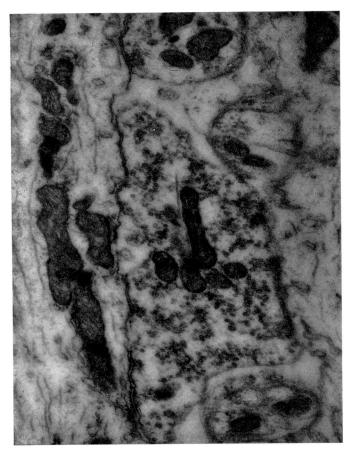

Fig. 585. Electron-micrograph of an elongated axon terminal upon a dendrite in the neuropil of the abducens nucleus. The dendrite, containing mitochondria and endoplasmic reticulum, is left, but is only partly included in the picture. The ending contains mitochondria and numerous synaptic vesicles. Note that many of the mitochondria have longitudinally orientated cristæ. The synaptilemma clearly consists of two separate membranes. × 32,000. (Palay. *J. biophys. biochem. Cytol.*)

PLATE LIV

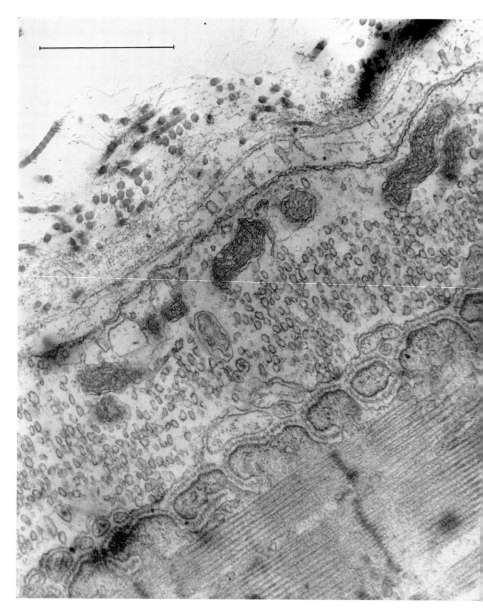

FIG. 588. Longitudinal section through a neuromuscular junction. Fixed OsO_4 and stained in PTA. Some details have been traced in Fig. 589. (Birks, Huxley & Katz. *J. Physiol.*)

from the contact surface, whilst the Schwann cell surrounds the terminal axon right down into the depression on the post-synaptic cell, but it does not, according to de Robertis & Bennett, extend over the very end of the fibre, so that, as indicated above, the protoplasm of the pre- and post-synaptic cells is separated only by plasma membranes. The end-feet, or pre-synaptic terminals, are easily recognizable in the electron-microscope by the high concentration of mitochondria within them, together with minute circular profiles —probably representing sections through spherical vesicles approximately 200–650A in diameter—surrounded by a thin dense line. These bodies may well be the loci in which the chemical transmitter substance is stored, since cutting the axon of the pre-synaptic neurone, *i.e.*, separating it from its cell body, leads to rapid disappearance of these vesicles in association with blockage of transmission; furthermore, changes in activity of the neurones are said to be reflected in changes in the number of vesicles identified in sections (de Robertis).*

The Neuromuscular Junction

Structure. The detailed structure of the voluntary muscle fibre will be discussed later in relation to its contractile properties (p. 1347); we may note here that it is a long cell, containing many nuclei, with fibrillæ embedded in

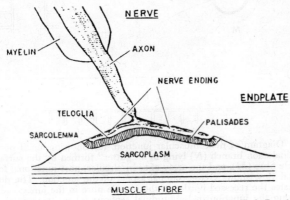

FIG. 586. The neuromuscular junction. (Acheson. *Fed. Proc.*)

a matrix of sarcoplasm, the whole being surrounded by a connective tissue-type of membrane, the sarcolemma, the functional plasma membrane lying beneath this. The junction between the fibre and its motor nerve fibre is described as the *end-plate* (Fig. 586).

Classical light-microscopical studies revealed that, at the nerve terminal, the fibre lost its myelin sheath and that the endoneurial sheath of the nerve fibre (sheath of Key & Retzius or Henle) fused with the endomysium or sarcolemma, whilst the unmyelinated terminal twigs appeared to penetrate the muscle surface and ramify in an accumulation of granular sarcoplasm—the end-plate—lying beneath the sarcolemma and containing numerous nuclei. According to Shanthaveerappa & Bourne (p. 1037), the actual membrane that fuses with the sarcolemma is the perineurium.

---

* The ultrastructure of the synapse, and its variations, have been reviewed by Gray (1969); in general, synapses fall into two main categories according as the membranes of the apposing cells remain separate (Category 1) or become fused (Category 2). The neuromuscular junction belongs to Category 1. Category 3 is constituted by serial junctions where three cells come into close relation (p. 1213).

Junctional Folds

In the electron-microscope the essential discontinuity between nerve fibre, on the one hand, and muscle fibre on the other, was demonstrated conclusively, in the sense that both the fine nerve terminal and the muscle membrane remained separated by a measurable gap. Thus Robertson described how the terminal axon became partially enveloped by a trough or gutter formed by a depression of the surface of the muscle fibre, whilst the muscle membrane, in this gutter, was thrown into a series of "post-junctional folds", some 0·5 μ to 1·0 μ long and 500–1,000A wide. Three views of this arrangement are illustrated schematically by Fig. 587 from Birks, Huxley & Katz, which shows (a) a frontal view, (b) a longitudinal section and (c) a so-called tangential section.* In Fig. 588, Pl. LIV, is shown an electron-micrograph of a longitudinal section, whilst Fig. 589 is a sketch indicating the main features of this micrograph. It will be seen from these figures that the terminal axon is partially

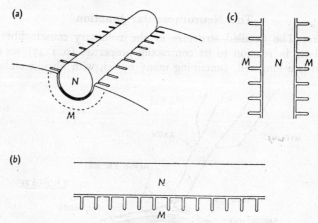

FIG. 587. Diagram of neuromuscular junction. (a) Showing small portion of a terminal axon branch (N) lying in a "gutter" formed by the surface of the muscle fibre (M), and the array of semi-circular, post-junctional folds. The frontal view shows cross-sectional aspect of the junction. The dotted line indicates the recessed border of the muscle fibre in the junctional fold. (b) Viewing the junction in longitudinal section; (c) in "tangential" section. (Birks, Huxley & Katz. J. Physiol.)

invested by the Schwann cell layer, which separates it from the overlying connective tissue; this layer, however, never engulfs the axon completely so that the synaptic space between axon and muscle fibre is free of Schwann cytoplasm, although an occasional finger-like process may invade this space, where it is seen in the longitudinal section of Fig. 589 as an irregular profile (S.F. of Fig. 589). The space of several hundred Ångstrom units between axonal and muscle membranes, and between the surfaces of the junctional folds, constitutes a true extracellular space into which externally applied drugs would be most likely to diffuse.

Schwann Cell

The Schwann cell shows the typical relationship with the axon described earlier, being separated from this by some 150A; on its external surface facing the connective-tissue space there is a dense layer separated from the Schwann cell membrane by some 200–300A; this layer can be followed around the surface of

* It is these junctional folds that, on cross-section, give rise to the appearance of *palisades* described by Couteaux in light-microscope studies of the neuromuscular junction.

the Schwann cell up to the place where it makes contact with the muscle fibre; at this point the external layer fuses with a similar dense layer enveloping the muscle membrane.

Synaptic Vesicles

The cytoplasm of the terminal axon contains numerous "synaptic vesicles". These bodies, some 300–400A in diameter, were first described by de Robertis & Franchi in the cytoplasm of the pre-synaptic nerve ending of a retinal synapse, and by Robertson in the axon of the neuromuscular junction. Section of the

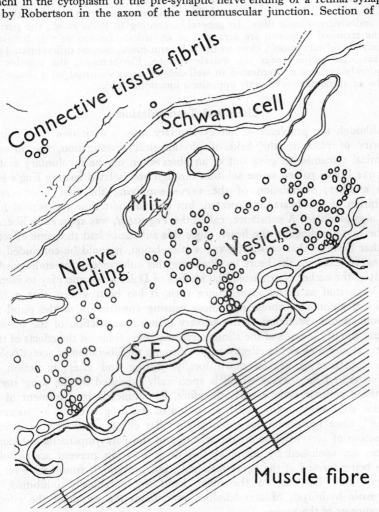

Fig. 589. Diagram illustrating main features of the neuromuscular junction as revealed by the electron-micrograph of Fig. 588, Plate LIV. (Birks, Huxley & Katz. *J. Physiol.*)

nerve, *i.e.*, separation of the axons from their cell bodies, caused a rapid (22–24 hour) disappearance of the vesicles, which are presumably formed in the cell body whence they migrate to the terminals (de Robertis, 1956). Since their original discovery they have been described in all synapses in which there is reason to believe that transmission is brought about by the liberation of a chemical mediator or "transmitter", and it is assumed that they contain this transmitter, which is liberated in response to a nerve impulse. Birks, Huxley & Katz have discussed the possibility that these are in fact cross-sections of a series of tubules, but point out that, if this were so, a given set of vesicles would appear different according to the plane of sectioning, and yet this is not the case; moreover, when serial sections were examined, individual vesicles dropped out in successive slices, whereas true cylindrical structures, such as collagen fibrils, persisted from slice to slice.

Fig. 590, Pl. LV shows vesicles in a synapse in the rat's central nervous system at very high magnification; the triple-layered plasma membranes surrounding the vesicles are easily resolvable.

Mitochondria

The mitochondria are characteristically different in structure from those of the underlying muscle fibre. In general, according to Birks *et al.*, the particles in the terminal axoplasm are arranged in an orderly fashion so that frequently a "mitochondrial region", close to the Schwann cover, may be differentiated from a "vesicular region" near the muscle surface. Furthermore, the vesicles may frequently be seen concentrated in well-defined areas focussed on a dense zone of the axon membrane directly opposite a junctional fold.

Chemical Transmission

Although the problems of nerve-voluntary muscle excitation have enjoyed priority of place in the field of physiological investigation, the ideas of chemical transmission grew out of an observation on the involuntary system. In 1921 Loewi passed saline solution through one isolated beating frog's heart into another; stimulation of the vagus supplying the first heart caused inhibition of this heart, as expected, but after a short time the second heart was also inhibited. A substance, called the *Vagusstoff*, was apparently liberated at the nerve endings in the heart; since this substance had the same effect on another heart as that caused by vagus stimulation, it could be concluded that the nerve acted on the heart by liberating this substance at its endings. As a result of the earlier pharmacological studies of Dale, it became easy to identify the Vagusstoff as *acetylcholine*. Since then, it has been shown that a large group of autonomic fibres, *e.g.*, those causing constriction of the pupil and accommodation, secretion of the salivary glands, constriction of the bronchi, contraction of the walls of the alimentary canal, etc., transmit the effects of their electrical impulses to the effector organ through the liberation of acetylcholine. This substance is broken down locally, during and after its action, by *cholinesterase*,* an enzyme which is specifically inhibited by the drug *eserine*. All these nervous actions may, therefore, be mimicked by treatment of the effector with acetylcholine; such an action of the drug is called a "*muscarine action*", since muscarine has essentially similar effects. With few exceptions, the action of acetylcholine on the effectors of the parasympathetic autonomic system are abolished by atropine; this drug does not prevent acetylcholine from being formed at the nerve terminals, but prevents it from acting on the effector cell. Eserine, on the other hand, by specifically inhibiting the enzymatic hydrolysis of acetylcholine, potentiates the action of the effector substance or of the nerve.

Sympathetic

The classical researches of Elliot and of Cannon showed that another group of effectors, mainly controlled by the sympathetic division of the autonomic system, were stimulated by the liberation of a substance similar to, if not identical with, *adrenaline*, a hormone secreted by the suprarenal gland. For example, the effect of stimulating the accelerator nerves to the heart may be mimicked by intravenous injection of adrenaline. Subsequent work, notably that of Cannon & Rosenblueth, suggested the existence of two sympathetic

---

* Mendel & Rudney have shown that plasma and tissues contain a non-specific esterase, *pseudo-cholinesterase*; true cholinesterase, specific for acetylcholine, may be separated from the non-specific enzyme.

PLATE LV

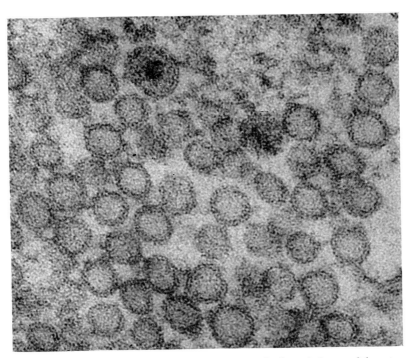

FIG. 590. Synaptic vesicles in a synapse of the anterior hypothalamus of the rat. Observe the triple layered membrane that limits each vesicle, which has an approximate diameter of 500A. Fixation in formaldehyde-osmium. × 200,000. (Courtesy, E. De Robertis.)

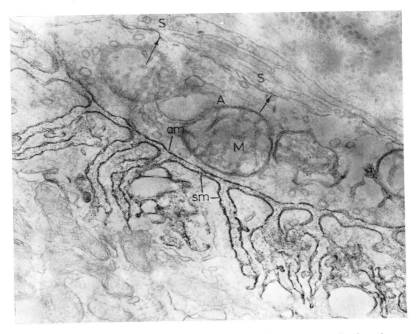

FIG. 591. Localization of acetylcholinesterase at the neuromuscular junction. The prejunctional (am) and post junctional (sm) membranes are distinctly stained. Moderate staining is also present in the axonal plasma membrane (arrows) facing the teloglial Schwann cell (S) covering. M, mitochondrion; A, axon. × 63,500 approx. (Davis & Koelle. *J. Cell Biol.*)

[To face p. 1134.

PLATE LVI

Fig. 610. Teased tortoise muscle fibres stained with gold chloride. *Left*: En plaque termination. *Right*: Clusters of "en grappe" endings. × 1800 approx. (Proske & Vaughan. *J. Physiol.*)

transmitters, one largely concerned with excitatory phenomena—*sympathin-E*, and one concerned with inhibition—*sympathin-I*; and it later materialized that sympathin-E corresponded with *nor-adrenaline*, first extracted from tissues by v. Euler, whilst adrenaline corresponded with sympathin-I. According to v. Euler, nor-adrenaline is the true sympathetic transmitter, whilst adrenaline acts only after it has been liberated from the adrenal gland. Since the original division of the autonomic system into parasympathetic fibres, liberating acetylcholine, and sympathetic fibres, liberating adrenaline or *sympathin*, has turned out to be inaccurate,* it is now customary to describe nerve fibres as either *cholinergic* or *adrenergic*, according to whether their action may be mimicked by acetylcholine or adrenaline.

The responses in these effectors controlled by the autonomic system are slow, so that the concept of a chemical process intervening between the more rapid electrical ones was not considered an insuperable difficulty in the way of acceptance of the chemical mediator hypothesis; moreover, the fact that the effects of a single nerve volley could persist for some time (several seconds in the case of vagal inhibition of the heart), and that this persistence could be extended by the application of eserine, rather favoured the hypothesis.

Nicotine Action

Transmission of the nerve impulse through a synapse or myoneural junction is a very rapid process, the delay-time being of the order of one msec. or less; nevertheless the existence of this delay, together with the specific effects of certain drugs, notably nicotine and curare, on the junctional region, marks the latter off as a possible site for something more than purely depolarization phenomena in the conduction process. Nicotine, in small doses, was shown by Langley to have a stimulating action on ganglionic synapses, and to cause a muscle to contract if applied to the end-plate region, the effect becoming weaker the farther removed the point of application from the end-plate. Recent work has shown that both the ganglionic synapse, and the end-plate of voluntary muscle, are activated by acetylcholine, so that it is customary to speak of the *nicotine action* of acetylcholine on these structures, to distinguish it from the muscarine action on *autonomic effectors*. The demonstration that acetylcholine is normally present in ganglia, and that it is liberated on stimulating the pre-ganglionic fibres (Witanowski; Chang & Gaddum), has placed the position of acetylcholine as an important element in the transmission of impulses through ganglionic synapses beyond question.† The extension of the hypothesis to include neuromuscular transmission in the vertebrate voluntary effectors has met with more opposition, but to-day even those, who at one time were the most ardent supporters of an electrical theory of neuromuscular transmission, concede that one stage in the process involves the liberation of acetylcholine. Dale, Feldberg & Vogt showed that stimulation of the motor nerve to perfused voluntary muscle liberates acetylcholine into the venous effluent. The effects of intravenous injections of acetylcholine are obscured by its rapid destruction by the enzyme cholinesterase in the blood and tissues; however, Brown, Dale & Feldberg showed that if the injection is made by way of the artery immediately supplying the muscle, very small quantities, *e.g.*, 2 μg., will cause a contraction as strong as that obtained by

---

* For example, the sweat glands are innervated by the sympathetic, whilst the nerve-endings apparently liberate acetylcholine.

† According to Bülbring, the superior cervical ganglion may also be activated by low concentrations of adrenaline; larger doses have an inhibitory action.

maximal nerve stimulation.* Curarine, in sufficient doses, completely blocks transmission from nerve to muscle, leaving the excitability of both tissues unaffected; it does not, however, affect the liberation of acetylcholine at the end-plate, so that its action must consist in preventing the liberated acetylcholine from exerting its normal effects on the muscle tissue; it is thought that it competes with acetylcholine for the active groups in the end-plate substance.†￼ The response to injected acetylcholine is essentially similar to that evoked by repeated nerve stimulation, consisting of a brief asynchronous tetanus with typical all-or-nothing spikes at a frequency of some 200/sec. (Brown, 1937); eserine, by inhibiting cholinesterase, may modify the response to a single nerve volley, from a single twitch to a waning tetanus of the muscle fibres; moreover, it causes a muscle to twitch spontaneously.

Cholinesterase. We shall see that a single volley of impulses, that is, a single shock applied to all the nerve fibres, in the vertebrate motor nerve produces only a single action potential, and contraction, in the muscle fibres supplied by it; if transmission is effected by the depolarizing action of the liberated acetylcholine in the end-plate region, in order that it may not cause a second depolarization it must be removed within the refractory period of the muscle fibre, i.e., within about 5 msec. in the frog and about 2 msec. in the mammal. If the acetylcholine theory is correct, therefore, we may expect the rate of hydrolysis of acetylcholine to be rapid. A correlation between the concentration of cholinesterase in any region and the number of end-plates might also be expected; but we must appreciate that the acetylcholine is liberated in highly localized regions, and it could easily be that diffusion alone would be adequate to lower its concentration sufficiently to render it ineffective; furthermore, it need not necessarily be removed by hydrolysis, but could be restored to the inactive form from which it had been liberated if, as is thought, it exists as an inactive complex in the resting nerve terminals.

Localization at End-Plate

The concentration of cholinesterase in muscle as a whole is low; hence, if the local concentration in the nerve endings were the same as in the muscle generally, the necessary rate of hydrolysis of the liberated acetylcholine would be some 50,000 times too slow to permit its complete elimination during the refractory period; thus the concentration at the end-plate must be some 50,000 times higher than elsewhere in the muscle.

The pioneering studies of Marnay & Nachmansohn, of Koelle & Friedenwald, and of Couteaux, left little doubt as to the accumulation of the enzyme in the end-plate region. Thus Koelle & Friedenwald (1949) localized cholinesterase activity by incubating the tissue with a sulphur derivative of acetylcholine, acetylthiocholine, in the presence of a copper salt; the primary reaction product was precipitated as copper sulphide, and their published photographs reveal a dense accumulation of this product at the end-plates. Subsequent improvements have increased the specificity of the method by distinguishing between the so-

---

* Amphibian voluntary muscle and denervated mammalian muscle are very sensitive to acetylcholine, so that treatment with this drug generally causes the muscle to go into a *contracture*, i.e., the activation of the normal contractile mechanism in which tension, heat and lactic acid are produced, but in which conduction of the mechanical response and a wave of action potential are missing (Gasser). This is probably due, as Brown pointed out, to the relatively smaller amounts of cholinesterase in these tissues than in normal mammalian muscle.

† Beani, Bianchi & Ledda (1964) have described a reduced output of acetylcholine from the guinea pig phrenic nerve-diaphragm preparation when it is stimulated at high frequencies in the presence of tubocurarine; Chang, Cheng & Chen (1967), however, were unable to confirm this.

called non-specific cholinesterase, *ChE*, and the esterase that specifically attacks acetylcholine, *AcChE*. This may be achieved by the use of specific inhibitors; *e.g.*, 10^{-7} M DFP specifically inhibits ChE and 10^{-6} ambenonium chloride inhibits the AcChE. Acetylthiocholine and butyrylthiocholine are used as specific substrates for AcChE and ChE respectively in place of the less specific thiol-acetic acid.* For the application of these techniques to electron-microscopy the primary reaction product was precipitated in the form of lead sulphide and this was found on the plasma membrane of the muscle fibre in the space between the axon and muscle—*primary synaptic cleft*—and the space formed by the junctional folds—*secondary synaptic cleft*; in addition, it was found on the plasma membrane of the terminal axon and, surprisingly, within the terminal axoplasm on the vesicles and tubular structures.

HIGHER RESOLUTION

A higher degree of resolution was achieved by Davis & Koelle (1967) by precipitating gold instead of lead, the particles of Au sulphide (from thiolacetic acid) or Au thiocholine phosphate (from acetylthiocholine) being much finer. Fig. 591, Pl. LV illustrates the accumulation of Au in the primary and secondary junctional folds; under conditions where precipitation was not so dense, the main accumulations were at the prejunctional axon membrane and the postjunctional sarcoplasmic membrane. When non-specific cholinesterase was studied this had the same distribution, but there were dense accumulations of reaction product between the Schwann cell membrane and axon. Salpeter (1967) labelled the cholinesterases with ^{3}H-DFP and made quantitative estimates of density by grain counts in autoradiographs. 80–85 per cent. was over the junctional folds, but the resolution was inadequate to determine whether the activity here was associated purely with the subneural apparatus or also with the axonal membrane.†

Marnay & Nachmansohn calculated that at a single end-plate 8.10^9 molecules of acetylcholine could be split during the refractory period; this compares with Krnjević & Miledi's estimate of about 1.10^6 molecules liberated per impulse (p. 1146).

Nerve-Nerve Synapses

Studies on the distribution of cholinesterase in synapses told essentially the same story, the concentration in the nerve being considerably less than that in the ganglion where it made its synapse; degeneration of the pre-ganglionic fibres caused a loss of up to 60 per cent. of the activity of the ganglion, a fact suggesting that this proportion is present in the nerve fibres themselves, the remainder being extracellular, and possibly acting as a barrier to prevent the diffusion of acetylcholine away from the synaptic region.

ELECTRICAL ASPECTS OF NEUROMUSCULAR TRANSMISSION

For a long time the process of neuromuscular transmission was regarded as a purely electrical phenomenon, the condition at the junction being represented as similar to that in a partially blocked nerve. Thus Hodgkin's and Tasaki's studies have shown that a discontinuity, consisting of several millimetres of inactive nerve, can be bridged by the extrinsic potential moving ahead of an active region. The work of the past 20 years, however, has built up a convincing

---

* Thiolacetic acid is not very specific as a substrate by comparison with the thiocholines, but it has the advantage that it penetrates relatively rapidly into the tissue. For electron-microscopic work, the precipitation of copper with $(NH_4)_2S$ may be dispensed with if the tissue is fixed with $KMnO_4$, which oxidizes the copper thiocholine endproduct, making it more dense and less susceptible to removal (Brzin, Tennyson & Duffy, 1966).

† Miledi (1964) described accumulations of acetylthiocholine reaction product in the central elements of the triads of the sarcotubular system (p. 1349), in addition to large amounts in the synaptic clefts and synaptic vesicles.

body of evidence in favour of a chemical mode of transmission at practically all the junctions that have been studied. In general, the sequence of events is as follows: Arrival of nerve impulse at terminal branches—Liberation of acetylcholine from these branches—Depolarization of the end-plate region by acetylcholine—Initiation of propagated spike in muscle fibre. It is important, nevertheless, that we should have an exact picture of the electrical events taking place at the end-plate during transmission. Because of the many factors involved, namely the electrotonic action of the arriving nerve impulse, the chemical reactions involved in the liberation and hydrolysis of acetylcholine, and the depolarization process leading to the initiation of a new spike in the muscle, any measured electrical events are rather more complex than those we have so far considered.

End-Plate Potentials

Eccles & O'Connor, in 1939, studied the effects of two successive stimuli on the action potentials that could be recorded by an electrode on the end-plate region of muscle. It was found that, for 0·7 to 1·8 msec. after the first stimulus, a second volley failed to set up a propagated spike in any muscle fibre—the

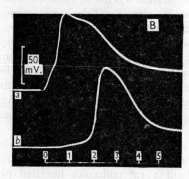

Fig. 592. Action potential at the nerve-muscle junction, recorded from the inside of muscle fibre at the nerve-muscle junction (a), and 2·5 mm. away (b). Time-scale msec. (Fatt & Katz. *J. Physiol.*)

muscle in this region was absolutely refractory for this period. If the delay was greater than 0·7–1·8 msec., however, the second volley caused a small rise and fall of negativity which they called the *end-plate potential*. The greater the delay, the greater the height of this potential until finally, after some 2 msec., an ordinary spike potential was recorded. The time-course of the rise and fall of the end-plate potential was characteristically different from that of the action potential, being altogether slower.

The end-plate potentials, discovered in this way, are characteristic of the refractory condition of the end-plate; nevertheless there is good reason to believe that a similar potential rises during the establishment of any spike in muscle, although its time-course can only be deduced indirectly, since it is swamped by the actual spike. Kuffler's studies on isolated single nerve-muscle preparations, which permitted a close approximation of the recording electrode to the end-plate region, and Fatt & Katz's measurements with an intracellular electrode permitting an even closer approximation to the end-plate, left no doubt as to the normal occurrence of this preliminary potential change, out of which the spike grew; Fig. 592 brings out the difference between the spike, recorded at the end-plate by an intracellular electrode, and that recorded 2·5 mm. away. The end-plate potential forms a step during the rising phase of the record, and after the peak a discrete hump is seen, suggesting that the change in the end-plate is "obtruding itself" on the spike potential of the muscle fibre, *i.e.*, it would seem that the spike is a characteristic of the muscle fibre proper, whilst the electrical change occurring at the end-plate is characteristically different, being slower in onset and more sustained in action. This view of the end-plate potential as something distinct from the action potential is strengthened by the observation of Fatt & Katz that the height of the action potential, recorded at the end-plate, was less than that recorded elsewhere.

Short-Circuit of Membrane Potential

It could be argued that the change at the end-plate was a simple short-circuiting of the membrane, as distinct from a reversal of the resting potential. Such a short-circuiting could initiate an action-potential in adjoining tissue by virtue of the local-circuit currents that would be set up, and if it were sustained during and after the active phase of the muscle fibre, would obtrude itself in this way, maintaining an electronegativity at the end-plate that would delay the fall in the spike. Furthermore, the fact that the spike-height, recorded from the end-plate, was smaller than elsewhere would also be explained; on this basis, the actual reversal of polarization that constitutes the real feature of the spike would belong to the adjoining tissue of the muscle fibre and it would be recorded at the end-plate by virtue of electrotonic spread. The presence of a short-circuit in the region of the end-plate would naturally reduce the positivity measured by the electrode.

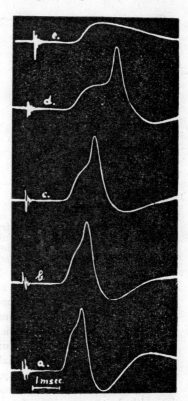

Effect of Curarine. Curarine blocks neuromuscular transmission; its effects on the end-plate potential are therefore of great interest; in Fig. 593 some results of Kuffler on the isolated end-plate are shown; *a* represents the normal end-plate potential, following on to a spike, in an uncurarized preparation; *b* to *e* represent progressive effects of curarine, blockage being reached at *e*; it will be noted that curarine diminishes the initial end-plate potential and progressively lengthens the latent period of the spike. Further increases in dosage of curarine progressively reduce the height of the end-plate potential. By the use of an intracellular electrode in the end-plate region, Fatt & Katz were able to confirm this in a more quantitative fashion; curarine had no effect on the resting potential, whilst increasing doses caused progressively smaller end-plate potentials. If we regard the establishment of a certain critical end-plate potential as a stage in the process of transmission, we may explain the effect of curarine as a progressive lowering of the end-plate potential until it is too small to initiate a propagated impulse in muscle. The fact that block may be obtained without the complete extinction of the end-

FIG. 593. Effects of curarine on the potentials, recorded at the end-plate region of a single-fibre nerve-muscle preparation, in response to nerve stimulus. *a*, Before application of curarine; *b*, *c*, and *d* during progressive curarization, show the diminution of the initial end-plate potential and the progressive lengthening of the spike latent period; *e*, pure end-plate potential, no spike being set up. (Kuffler. *J. Neurophysiol.*)

plate potential makes the curarized nerve-muscle preparation a useful one for the study of the potential, in fact it was first discovered by Göpfert & Schaefer in the curarized frog nerve-muscle preparation; these authors were inclined to regard it as the electrical sign of events following the liberation of acetylcholine.

Transmitter Action

The curarized nerve-muscle preparation provided an excellent opportunity for a detailed study of the end-plate potential, which was described by Eccles, Katz & Kuffler in 1941. In general, the potential, as recorded by an external electrode in the region of the end-plate, could be regarded as an electrotonic spread from a focus at the end-plate itself, similar to that recorded in the region of an applied sub-threshold negative potential. So far as the time-course of the development and decay was concerned, it appeared to result from a brief "transmitter action", which could be described as a highly localized depolarization lasting only a few milliseconds and presumably caused by the

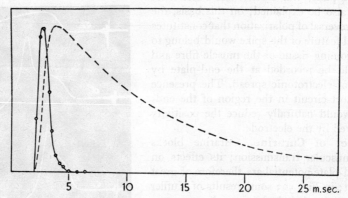

FIG. 594. End-plate potential and transmitter action. Broken line represents the end-plate potential of frog's sartorius at 17·5° C., with a time-constant for decay of 9·8 msec. Full line represents the probable course of the "transmitter action". Ordinates in arbitrary units. Abscissæ: Time after nerve stimulus. (Eccles, Katz & Kuffler. *J. Neurophysiol.*)

liberation of acetylcholine. The actually recorded potentials would result from electrotonic spread and electrotonic decay, processes that would depend on the passive cable properties of the muscle membrane (Fig. 594).

End-Plate Current. This view is confirmed by Takeuchi & Takeuchi's study on the voltage-clamped junction; by clamping the potential across the end-plate at the resting potential, the spread of current could be prevented, so that the feed-back current required to clamp the membrane potential during the nerve impulse was a measure of the current that would have flowed during the brief transmitter phase. In Fig. 595A the end-plate potential recorded in a curarized junction is shown, whilst in B the *end-plate current* is shown. In C the end-plate current is compared with a "calculated end-plate potential" deduced on the assumption that the electrotonic spread from the active region takes place in accordance with the cable properties of muscle. The end-plate current lasts about 5 msec, in good agreement with the duration of the active phase deduced by less direct means (Eccles, Katz & Kuffler).

Effect of Eserine. Eserine, by poisoning the enzyme cholinesterase, should prolong the action of acetylcholine at the end-plate and therefore, if liberation of this substance does represent the transmitter process, should lengthen the end-plate potential. It was found by Brown, Dale & Feldberg that eserine

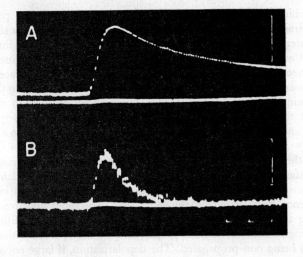

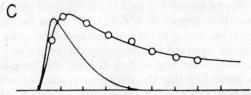

FIG. 595. A. End-plate potential recorded intracellularly without feed-back.
B. End-plate current from same end-plate with feed-back. C. Superimposed
tracings of end-plate potential and current; circles indicate potential change
calculated from end-plate current on assumption of membrane time-constant of 25 msec. (Takeuchi & Takeuchi. *J. Neurophysiol.*)

causes repetitive discharges in cat's muscle in response to single nerve stimuli,
whereas in normal preparations a single impulse in the nerve produces only a
single response in any fibre. Unless the nerve-muscle preparation is blocked in
some way, it is not easy to study the effects of eserine on the end-plate potential,
owing to the intervention of the spike; in the curarized preparation there is an

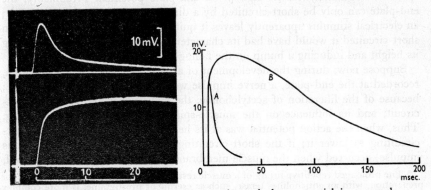

FIG. 596. The effect of prostigmine on the end-plate potential in a muscle
fibre blocked by a low-Na$^+$ medium (4/5 of Na$^+$ replaced by sucrose).
Left: End-plate potential in Na$^+$-deficient muscle (upper record) and
the same after addition of prostigmine (lower record). Right: Superimposed tracings. A: Na$^+$-deficient muscle. B: After addition of prostigmine. (Fatt & Katz. *J. Physiol.*)

undoubted increase in the size and duration of the end-plate potential, as recorded intracellularly, but perhaps the most unequivocal effect is shown when block is achieved by bathing the preparation in a low-sodium medium; as Fig. 596 shows, the end-plate potential rises to a plateau which is sustained for some time—an effect best explained by the greater life of the liberated acetylcholine, due to the poisoning of the cholinesterase at the end-plate. Applying the voltage-clamp technique, Takeuchi & Takeuchi observed an increase in the average duration of the rising phase of the end-plate current from 0·77 msec. to 1·28 msec., and of the peak to half-decline time from 1·08 msec. to 2·01 msec.*

End-Plate Potential as a Short-Circuit. We may now draw a tentative picture of the events taking place at the neuromuscular junction when an impulse in the nerve arrives. As a result of the liberation of acetylcholine from the nerve terminals, the end-plate becomes highly permeable to ions and is, in consequence, depolarized, in the sense that it becomes negative in respect to adjoining tissue. The depolarization, however, is distinctly different from that occurring in the muscle fibre proper, and we may postulate that it differs from this in being non-propagated. The depolarization, if large enough, causes local circuits of sufficient intensity to initiate a propagated spike in the adjoining tissue of the muscle fibre proper. It is difficult to assess the magnitude of the end-plate depolarization, owing to the intervention of the spike, but a minimal value can probably be obtained by measuring the height of the step leading to the action potential. According to Fatt & Katz, the "step-height" is some 33 mV, and it is interesting that the minimum amount of cathodic depolarization required to initiate a spike in the muscle amounts to 36 mV,† so that it is very likely that the primary depolarization at the end-plate is, indeed, sufficiently large to initiate a spike.

Interaction of Spikes. This view of the end-plate potential as a depolarization, as opposed to the reversal of the resting potential that occurs during the spike, is consistent with several observations, in particular the interaction of end-plate potential changes, due to direct and indirect stimulation. Fatt & Katz showed that the potential changes, recorded with an intracellular electrode at the end-plate, were different according as the nerve or the muscle was electrically stimulated. With nerve stimulation, the typical end-plate step initiated the spike, whereas with direct stimulation of the muscle a pure spike was obtained with no sign of an end-plate potential. Presumably, the end-plate can only be short-circuited by a direct action of acetylcholine on it; an electrical stimulus apparently leaves it quite unaffected since, if it had been short-circuited it would have had its characteristic effect on the spike, reducing its height and inducing a hump on the falling limb.

Suppose now, during the development of a muscle-induced action potential, recorded at the end-plate, a nerve impulse were timed to reach the end-plate; because of the liberation of acetylcholine, the end-plate would act as a short-circuit, and its influence on the muscle-spike would depend on the timing. Thus, when the action potential was at its height, we should expect the short-circuiting to lower it; if the short-circuiting (*i.e.*, the arrival of the nervous impulse) occurred when the muscle membrane was just completely depolarized,

* The prolonged repetitive firi ng of a muscle resulting from treatment of the animal, or preparation, with an anticholines terase, such as eserine or prostigmine, is more complex than at first sight and will be disc ussed later; it seems that the liberated acetylcholine acts on the nerve terminals causing the nerve to fire repetitively and antidromically, so that spikes may be recorded from the motor root of the nerve (Masland & Wigton, 1940).

† Nastuk (1953) found a critical depolarization of 48 mV to be necessary to induce a spike in a nerve-muscle preparation, the depolarization being measured with an intracellular electrode.

the end-plate and adjoining tissue would have the same potential and there would be no interaction; finally, if the nerve impulse arrived when the muscle was only partially depolarized, *e.g.*, if the resting potential had fallen to 40 mV, then a complete short-circuiting of the end-plate would favour the further depolarization of the muscle membrane, and the end-plate potential would add to the action potential. These predictions were amply verified by Del Castillo & Katz, as Fig. 597 shows, with the proviso that the potential difference across the end-plate, approached during nervous stimulation, was not zero—as we would expect were there a true short-circuiting—but some 15–20 mV, the

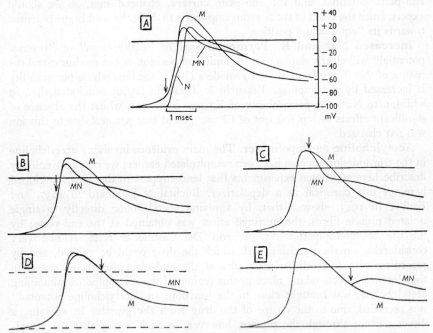

FIG. 597. Series of records from a single end-plate illustrating the interaction of the muscle- and nerve-induced spikes when the latter arrives later and later. M is the muscle-induced spike alone; N the nerve-induced spike, and MN the record obtained when both interact, the beginning of the N-response being indicated by an arrow. Note that the liberation of transmitter, *i.e.*, the arrival of the N-spike at the end-plate, causes a fall of the muscle-spike if it occurs during the positive phase, so that the end-plate potential, in this case, may be said to be reversed in sign. At a certain point on the falling limb of the muscle-spike, the transmitter delays the fall and may actually cause a hump on the record (E). (Del Castillo & Katz. *J. Physiol.*)

inside being negative. Actually, this may well be the potential difference that one would expect of a system that suddenly became indiscriminately permeable to all ions, since a diffusion potential would be established owing to the influx of Na^+ and anions and the efflux of K^+, and calculations show that this might be of the order of 15 mV.*

* If the end-plate potential really arises as a short-circuiting it should never exceed that which may theoretically be obtained by short-circuiting the resting potential. If we take the resting potential as about 100 mV, and the short-circuit potential as 15 mV, then the maximal end-plate potential is $100 - 15 = 85$ mV. It is difficult to assign a maximal value experimentally since in normal circumstances an end-plate potential of more than about 35 mV leads to an action potential. Indirect evidence, obtained from a relatively inexcitable muscle, certainly indicates that the end-plate potential approaches a maximum within this limit.

Reversal Potential. The voltage-clamp technique is ideally adapted to studying the relationship between the membrane potential and the effect of a nerve stimulus at the end-plate; if the equilibrium position obtained on complete short-circuiting of the membrane were zero potential, then plotting end-plate current, measured as the feed-back current required to maintain the potential constant during the nervous impulse, against applied potential, should give us a line passing through the origin; in fact, however, extrapolation to zero end-plate current gave a membrane potential of about 15 mV, in agreement with that deduced by Fatt & Katz. If the membrane was further depolarized, then the end-plate potential, and the end-plate current, changed sign, as we should expect, since the effect of the nervous impulse is to drive the membrane potential towards its "equilibrium position".

Increased Na⁺ and K⁺ Permeability. The "equilibrium" or "reversal potential" will clearly depend on the ionic composition of the medium; and the nature of this dependence should provide a clue to the ions whose permeability is increased by the impulse. Takeuchi & Takeuchi (1960) concluded that, in addition to Na^+, the permeability of K^+ was increased, whilst the absence of significant effects of replacement of Cl^- suggested that permeability to this ion was not changed.

Acetylcholine as Depolarizer. The main evidence involving acetylcholine in the transmission process has been recapitulated earlier; we may conveniently describe here additional experiments that lend support to the view, developed here, of this chemical as a depolarizer. Buchtal & Lindhard in 1937, and Kuffler in 1943, showed that, by applying acetylcholine directly to single isolated muscle fibres, the maximal effect was obtained at the end-plate; by the use of micropipettes Nastuk, and Del Castillo & Katz, refined very considerably on the precision with which the drug could be applied, and by recording with intracellular electrodes at the end-plate they could ascertain the electrical effects taking place in this region. If the micropipette, containing acetylcholine, was brought close to the junction, an "acetylcholine potential" was recorded, due to the escape of the drug from the pipette; by applying a direct current through the pipette, however, in such a way that the acetyl-choline ions were moved away from the tip electrolytically, the pipette could be brought quite close to the end-plate without producing any electrical change. Brief pulses of electricity in the opposite direction now permitted the escape of controlled quantities of acetylcholine at the end-plate. In this way it was found that the minimum quantity required to excite the junction, *i.e.*, to initiate a spike in the muscle, was 10^{-15}—10^{-16} moles.

Action of Acetylcholine on Nerve

Nachmansohn (see, for example, his monograph, 1959) has argued that propagation of the impulse along the axon involves the liberation of acetylcholine from binding sites in the axon; this is said to depolarize the axon and to be destroyed by cholinesterase. In apparent support of this hypothesis is the finding of Armett & Ritchie that high concentrations, applied to unmyelinated fibres, depolarize them, reducing the spike-height, slowing conduction, and increasing the size and speed of onset of the positive after-potential. A study of the ionic requirements for the effect (Armett & Ritchie, 1963) suggested that the phenomenon could be accounted for by an increased permeability to Na^+ and divalent cations. The pharmacology of the effect, however, ruled out the intervention of acetylcholine in the conducting mechanism (Ritchie & Armett, 1963); for example, hexamethonium and prostigmine failed to block conduction although they, along with a large number of compounds, are able to block the acetylcholine effect on the axon. In a similar way, the blocking of axonal conduction by high concentrations of a cholinesterase inhibitor, such as DFP, may well be due to a non-specific narcotic action of the drug since it can be shown to block the facilitated transfer of Na^+ across the cat erythrocyte membrane (Davson & Matchett, 1951). The effects of drugs, such as acetylcholine, on the axon are doubtless the result of interaction with specific groupings in the membrane that affect the main parameters of excitability (see, for example, Ehrenpreis, 1964). As we shall see, acetylcholine can excite the non-junctional muscle membrane, but once again all the evidence indicates that transmission along the muscle fibre occurs by virtue of the potential changes in accordance with the Hodgkin-Huxley principles*.

Equivalent Circuit.† The short-circuiting of the resting potential may be represented by the closing of the key in an equivalent circuit (Fig. 598). Here, the right-hand element (E.R.) is the ordinary ion pathway through the membrane, E being the resting potential, and R the transmembrane resistance; transmitter action opens up a new pathway represented by the ϵ, r system, where ϵ is the equilibrium potential of the transmitter and r is the resistance of the short-circuit region. If ϵ is less than E, a current, I, will flow in the direction of the arrow. By the application of Ohm's Law, I will be given by: $I = E - \epsilon/R + r$. The reduction in membrane potential, e, which is given by the current times the resistance, R, is therefore given by: $e = R(E - \epsilon)/(R + r)$.

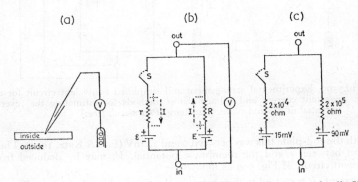

FIG. 598. (a) Intracellular electrode inserted at junctional region of cell: V, potential difference measured between inside and outside. (b) and (c), Equivalent circuit for transmitter action at the end-plate. Current I flows only when key, S, is shut. (Ginsborg. *Pharmacol. Rev.*)

I corresponds to the flow of positive ions outwards through the ordinary channels (or negative ions inwards) and the equal net flow of positive ions inwards through the channels opened up by the transmitter. The reduction of potential, e, is equivalent to depolarization of the cell and, if sufficient, will cause an action potential. Typical figures for the frog's end-plate are represented by Fig. 598c, with a resting potential of 90 mV and resistance of 2.10^5 ohms, an equilibrium potential, ϵ, of 15 mV and a resistance at site of transmitter of 2.10^4 ohms. When the end-plate is activated the current will be:—

$$\frac{(90 - 15)\ mV}{(2.10^5 + 2.10^4)\ Ohm} = 3\cdot4.10^{-7}\ amp$$

and this will depolarize the muscle fibre by $3\cdot4.10^{-7} \times 2.10^5$ ohm $= 68$ mV unless an action potential intervenes.

Reversal or Equilibrium Potential

On the basis of the equivalent circuit just discussed we may define ϵ as an *equilibrium potential*, measured experimentally by fixing the potential difference across the cell membrane at such a value that pressing the key has no effect; the fixed potential must be ϵ. We may predict, moreover, that by holding the potential below ϵ we may reverse the sign of the transmitter action so that the equilibrium

* Cholinesterase may be identified on the surface of the unmyelinated axon, deposits of reaction-product being seen in the mesaxon and space between axon and Schwann cell. In myelinated nerve, however, it cannot always be demonstrated, and never at the node of Ranvier (Schlaepfer & Torack, 1966). Brzin et al. (1965) have computed the density of cholinesterase per unit area of squid axon membrane.

† The reader may be referred to an excellent review by Ginsborg (1967) for an exposition of the equivalent circuitry of junctional transmission.

potential may be defined as the potential at which transmitter action reverses its sign, the *reversal potential*. A commonly applied method of determining this equilibrium potential is illustrated by Fig. 599; the membrane potential is recorded with one intracellular electrode, and a second is used to pass different steady-state currents across the membrane to displace its potential; both electrodes are in the end-plate region. The amplitude of the end-plate potential, e, was increased when the membrane was hyperpolarized, and the straight-line relation between e and the membrane potential indicated that the membrane was, indeed, behaving as an ohmic resistance; extrapolation to a zero value of e gives the equilibrium potential,

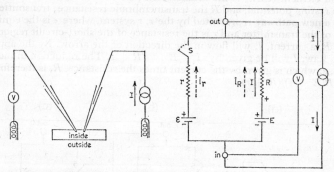

FIG. 599. Experimental arrangement and simplified equivalent circuit for the constant current and voltage clamp methods for estimating the reversal potential of the end-plate. (Ginsborg. *Pharmacol. Rev.*)

V; with the end-plate this was between o and 14 mV (Fatt & Katz, 1951). The linear relation between e and the membrane potential, V, may be deduced from the equivalent circuit of Fig. 599 as follows:—

A current, I, is caused to flow between the terminals with the key, S, open; the resulting membrane potential, V, is given by:—

$$V = E - RI \qquad (1)$$

With the key shut, provided the total current is held constant, part of I now flows through the left element, i_r, and part through the right, i_R:—

$$I = i_r + i_R \qquad (2)$$

If the potential between the terminals, as a result of closing the key, changes to V', then,

$$V' = E - R i_R \qquad (3)$$
$$\text{and} \quad V' = \epsilon - r i_r \qquad (4)$$

Equations 2, 3, and 4 are solved for V' in terms of I, and from (1) we obtain:—

$$V' - V = [R/r + R] (V - \epsilon)$$

Thus the depolarization e, which is equal to $V' - V$, is linearly related to V. It is clear from this formulation that if the membrane potential, V, is made less negative than ϵ, the transmitter effect will reverse in sign. This is not possible in normal muscle since an action potential occurs, but in the electroplaque which, as we shall see, is analogous with an end-plate, such a reversal is seen (Bennett, Wurzel & Grundfest, 1961).[*]

Minimum Amount of Transmitter Liberated. Krnjević & Miledi (1958) found that the minimum amount of acetylcholine required to induce an end-plate potential in rat diaphragm was even smaller than that found by Del Castillo & Katz, namely 10^{-17} to 10^{-16} moles. In view of the relatively disadvantageous

[*] In some cases the membrane at which transmitter action occurs does not behave like a simple ohmic resistance but has rectifying qualities, the resistance being a function of the applied potential (see, for example, the study of Kandel & Tauc, 1966, on giant cells of the sea slug, *Aplysia*).

position that the externally applied acetylcholine would have, by comparison with an equal quantity liberated from the nerve terminals directly into the synaptic clefts, it would not be surprising if the experimentally determined amount liberated at a single synapse, in response to a single nerve stimulus, were less than this. Estimates of this quantity have been made by Straughan, and by Krnjević & Mitchell, who stimulated the nerve to a rat diaphragm repetitively and estimated the amount of acetylcholine liberated into the surrounding medium. With repetitive stimulation the amount liberated per impulse decreases rapidly, because the amount released at a terminal is reduced for some 3 sec. after a single stimulus (Lundberg & Quilish). Thus the repetitive stimulation, which is necessary to enable the experimenter to obtain enough transmitter for his analysis, tends to give too low a value for the amount that would be liberated by a single impulse. Krnjević & Mitchell obtained the greater yield, which corresponded, at a stimulation rate of some 5/sec., to about 1.10^{-13} moles per impulse; the phrenic nerve contained some 10,000 fibres and presumably there were the same number of end-plates, so that about 1.10^{-17} moles were liberated per impulse, *i.e.*, some 6 million molecules.* This is a very small fraction of the total amount present.

Spontaneous Miniature End-Plate Potentials. With an intracellular electrode, Fatt & Katz (1952) observed, in normal frog muscle, a series of spontaneously occurring end-plate potentials, being on average 0·5 mV, compared with an estimated 50 mV for the normal spike-producing potential, varying between limits of 0·1 and 3 mV, and occurring at frequencies of from 0·1 to 100 per sec. Curarine decreased the amplitude of the potentials, whilst the anticholinesterase, prostigmine, increased both the amplitude and duration; in fact, it could cause a summation of successive miniature potentials which occasionally led to the growth of an action potential with resultant contraction of the fibre. The frequency with which the potentials occurred was not affected by either drug. The authors considered that these miniature end-plate potentials were the result of the spontaneous release of small quantities of acetylcholine from nerve endings; a statistical test indicated that the events were random, and a study of the effects of Ca^{++} and Mg^{++} strongly suggested that the events were quantal in nature, *i.e.*, that the acetylcholine was released in packets of the same amount, so that the end-plate potentials could be considered as multiples of a unit. Thus, on reducing the Ca^{++} and raising the Mg^{++} concentrations in the Ringer solution, the neuromuscular junction could be blocked (p. 1159) and the end-plate potentials, resulting from nervous stimulation, could be reduced till finally they were equal in height to that of a miniature spontaneous potential. With slightly less blocking, the end-plate potential produced by a nervous stimulus varied in height in a random fashion, with an average value some 4–5 times that of the spontaneous miniature potential. Under these conditions it could be argued that a stimulus led to the liberation of a randomly variable number of discrete quanta of acetylcholine; if this were true the variations in the response to a stimulus should be predicted by Poisson's Law, and a study of the distribution of responses showed this to be approximately true.

---

* An interesting finding of Del Castillo & Katz (1955) was that acetylcholine, injected into the muscle fibre, had no influence; this is reasonable, as we should expect the active groupings with which the drug reacts to be on the outside of the end-plate; in order to have an effect, the acetylcholine would have to diffuse across the muscle fibre membrane; presumably this would be a slow process so that an adequate concentration would not be built up. The possibility that the acetylcholine was hydrolysed within the fibre was investigated and rejected.

Localized Active Spots

When recordings at the end-plate were made extracellularly with fine focal electrodes, the miniature spontaneous potentials were found to be highly localized, so that moving the electrode some $50\,\mu$ caused sufficient attenuation to make them disappear. Fatt & Katz (1952) suggested that the miniature end-plate potentials did, in fact, arise from such localized areas scattered throughout the junctional region; the effects spread electrotonically and thus are recorded intracellularly without much spatial discrimination. A comparison of intra- and extra-cellular recording confirmed this hypothesis (Del Castillo & Katz, 1956); Fig. 600 illustrates the probable flow of current when there has been a local depolarization; current intensity would be highest at A but this is inaccessible so that it is the condition at B that is relevant, and it is here that the extracellular electrode picks up the most intense current. Katz & Miledi (1965) have analysed

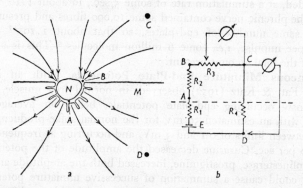

FIG. 600. Schematic drawing of active spot illustrating the probable flow of current when there has been a local depolarization. N: cross-section of a nerve terminal; M: interior of muscle fibre. Various possible electrode positions are indicated by points A–D. (*b*) ‘ Equivalent circuit ’ diagram to illustrate conditions of external recording (*e.g.* between B and C) and internal recording (between C and D). (Del Castillo & Katz. *J. Physiol.*)

the localization in greater detail by placing one electrode on an active spot and moving another in relation to it; if the two electrodes were close enough they picked up the same focus of activity; when the second was moved transversely across the muscle, a movement of as little as 7–$10\,\mu$ could be enough to extinguish the record in this. When electrodes were $50\,\mu$ apart the spontaneous potentials were definitely from different active spots, bearing no time relation to each other.

Quantal Content of Impulse

This aspect was studied in some detail by Boyd & Martin, and by Liley, on mammalian miniature end-plate potentials, which were the same in most essential respects as those of the frog. When the junction was blocked by Mg^{++}, the end-plate potentials in response to a nerve stimulus varied in a random fashion, as with the frog. In Fig. 601 the frequencies with which responses of a given magnitude occurred have been represented on a histogram, and it is easy to see that the most probable values occur as multiples of 0.4 mV; as the inset shows, moreover, the spontaneous miniature potentials fluctuated about the mean of 0.4 mV. If we accept this mean value of 0.4 mV as the most probable response of a single "unit" in the neuromuscular junction, and if we assume that the neuromuscular junction contains a large number of these units, each with its probability of responding to a nervous impulse, then the fluctuations in response to a nervous impulse may be predicted when this probability is low, as in the presence of Mg⁺⁺. Thus the mean quantum content of the end-plate potential in response to an impulse will be given by the mean amplitude of the potentials measured in the 200 observations described by Fig. 601, divided by 0·4 mV, the mean of the spontaneous miniature potentials. If this mean quantal content is represented by m, then the numbers of end-plate potentials with numbers of quanta in them given by x (0, 1, 2, 3, etc.) are calculated from the Poisson distribution: $e^{-m}m^x/x!$. The smooth line in Fig. 601 has been drawn through the predicted frequencies and the agreement is remarkably good.* On the assumption that the same statistical treatment was applicable to normal muscle, where the probability of a quantal response is very much higher, Boyd

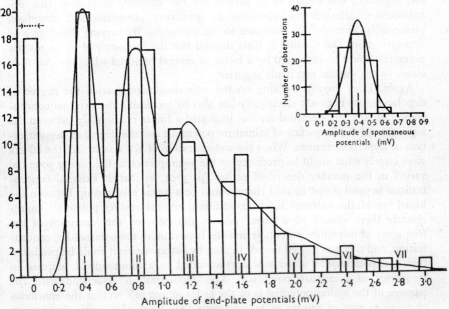

FIG. 601. Histograms of e.p.p. and spontaneous potential amplitude distributions in a fibre in which neuromuscular transmission was blocked by increasing the external magnesium concentration. Peaks in e.p.p. amplitude distribution occur at 1, 2, 3 and 4 times the mean amplitude of the spontaneous miniature potentials. Gaussian curve is fitted to spontaneous potential distribution and used to calculate theoretical distribution of e.p.p. amplitude. (Boyd & Martin. *J. Physiol.*)

& Martin computed that the mean number of quanta in a normal end-plate potential was some 310, corresponding to an end-plate potential of about 44 mV when allowance is made for a non-linear summation.

Nerve Depolarization

It was considered at first that the spontaneous potentials resulted from spontaneous nervous discharges at the terminals; however, Del Castillo & Katz (1955) showed that spontaneous changes in conductance at the end-plate, which reflected the release of acetylcholine, could occur even with the muscle immersed in isotonic K_2SO_4, *i.e.*, when both nerve and muscle were completely

* The continuous line in Fig. 601 has actually been drawn through a series of Gaussian distribution curves; thus the Poisson equation gives the numbers of responses to be expected containing 0, 1, 2, etc., quanta. From the measured variance of the spontaneous potentials, σ^2, a normal distribution curve about x, as a mean, may be calculated; and this has been done for $x = 1$, variance $= \sigma^2$; $x = 2$, variance $2\sigma^2$, and so on.

depolarized. The spontaneous activity is therefore the consequence of the random liberation of small parcels or quanta of acetylcholine. Del Castillo & Katz (1954) showed that the frequency of the spontaneous miniature potentials could be greatly increased by depolarization of the nerve to the muscle; by placing two electrodes on the nerve, near its junction, they modified its resting potential by passing a steady current through it; an intracellular electrode in the muscle fibre under the end-plate failed to record this steady potential, thus conclusively proving that there was no protoplasmic continuity across the neuromuscular junction. It could have been argued, however, that the failure to record this steady depolarization was a consequence of the failure of the nerve depolarization to spread electrotonically to the very terminals of the nerve, but this argument was shown to be invalid by the demonstration that this depolarization affected the spontaneous miniature potentials and therefore, presumably, extended its influence to the terminals. By varying the degree of depolarization, Del Castillo & Katz showed that the frequency of the miniature potentials could be increased by a factor of several hundred when the electrode closer to the muscle was made negative.

Again, Liley (1956), working on the rat's diaphragm, varied the degree of depolarization, not only electrically but also by increasing the external concentration of K^+. His electrical studies indicated a linear relationship between the logarithm of the frequency of miniature potentials and the degree of depolarization or hyperpolarization. When the concentration of K^+ was varied, the effects were largely what might be predicted on the assumption that the resting potential varied in the manner described earlier (p. 570), so that at external concentrations beyond about 10 mM there would be a linear relationship between the logarithm of the external K^+ concentration and the resting potential; consequently there should be a linear relationship between the logarithm of the frequency of miniature potentials and the logarithm of the external K^+ concentration, and this was indeed found. In fact, by extrapolation it could be predicted that a depolarization of 60 mV corresponded to a thousandfold increase in frequency of miniature potentials. The depolarization associated with the passage of the action potential is of the order of 100 mV so that the enormous increase in rate of release of acetylcholine that takes place with the passage of the nervous impulse is quite consistent with these findings. We may conclude, then, that the normal end-plate potential, *i.e.*, the response to a single nerve impulse, is made up of the synchronization, within a fraction of a millisecond, of a few hundred miniature potentials that would otherwise have occurred spontaneously over a much longer period of time.

Effect of Hypertonicity

Fatt & Katz observed that the most effective way of altering the frequency of the miniature end-plate potentials was to increase the tonicity of the bathing medium; Furshpan (1956) showed that the essential factor was the difference of osmotic pressure, since sucrose and NaCl were much more effective than glycerol and ethanol. They showed that the mean quantal content of the response to a nerve stimulus was unaffected, and this presents a stumbling block to accepting Liley's suggestion that the response to a nerve stimulus is the result of a momentary acceleration of spontaneous release of transmitter. Hubbard, Jones & Landau (1968) thought that the different effects on spontaneous release and on the response to a stimulus might be due to the circumstance that *m*, the mean quantal content in response to a nerve stimulus, is measured when the junction is depolarized, by contrast with the measurement of the frequency of spontaneous potentials; they found that, when the neuromuscular junction was depolarized with 40 mM KCl, a change in osmolality of the medium had little or no effect on the frequency of miniature potentials. As they point out, this does remove the stumbling block to acceptance of Liley's hypothesis, but the explanation of the effect of hypertonicity remains to be found; the effect may be resolved into an initial large increase in frequency of miniature potentials, which subsides to a smaller sustained one; the initial transient effect could be due to osmosis of water carrying with it synaptic vesicles against an adverse potential gradient; depolarization, by reducing this adverse potential, would abolish the effect of hypertonicity.

Tetrodotoxin

Katz & Miledi (1967d) profited by the circumstance that tetrodotoxin abolishes spike activity in nerve and muscle but leaves the end-plate unaffected. The end-plate could thus be stimulated by electrotonic spread along the nerve, or by application of a focal stimulating electrode. Increasing the strength of the stimulus increased the end-plate potential, but the quantal nature was unaffected, the numbers of responses of given magnitudes following the Poisson distribution as in normal muscle. The increase was non-linear, in the sense that there was a steep rise in response as the current-density of the electrotonic pulse increased.* Interestingly, the duration of the depolarization was important, and the increase in response with increasing duration was also non-linear, the response rising steeply with increasing duration. This indicates a temporal facilitation, the effect of the first time-step potentiating that of the next, and so on (Katz & Miledi, 1967b).

Fibre-Size

In a given muscle, the size of the spontaneous end-plate potentials may vary by a factor of more than 10; this could be due to a variation in the size of the quantum liberated or, alternatively, to variation in the electrical characteristics of the fibres; for example, if the transverse resistance varied from fibre to fibre the effects of a quantum of acetylcholine would vary—the greater the resistance, the bigger the potential. Since the resistance varies inversely with fibre-size, we would expect an inverse correlation between fibre-size and spontaneous end-plate potential; in fact, this is what Katz & Thesleff found; for example, fibres of 140 μ diameter had a mean miniature end-plate potential of 0·26 mV, whilst fibres of 10 μ had one of 5·6 mV.

Frequency and Amplitude

In general, then, we may conclude that the frequency of the miniature potentials is controlled entirely by the conditions of the pre-synaptic membrane, whilst their amplitude is determined by the properties of the post-synaptic element, in this case the muscle fibre membrane (Katz, 1962).†

Neuromuscular or Synaptic Delay. A shock applied to the nerve of a muscle evokes an action potential after a delay which, in the frog, may amount to 1–2 msec.; this delay is compounded of the time for propagation along the nerve and its fine terminals in the end-plate region, the chemical events resulting from the arrival in the endings, and the consequent development of the end-plate potential. It has been argued that transmission along the fine non-myelinated terminal twigs is not of the all-or-none type, but electrotonic. However, Hubbard & Schmidt (1963) used a focal electrode in the isolated nerve-diaphragm preparation to identify the origin of miniature end-plate potentials; having located an active spot in this way, they stimulated the nerve and obtained an extracellularly recorded nerve spike of the expected shape, followed, after the

---

* Since the duration of the depolarization affects the quantal content of the end-plate potential, prolonging the action potential, by treatment of the preparation with Cs^+ (p. 1110), should increase the quantal content. Ginsborg & Hamilton (1968) found an increased amplitude of the evoked end-plate potential with no change in the size of the miniature potentials. After prolonged exposure, however, the resting potential and action potential of the nerve declined, and this was accompanied by a reduced end-plate potential, and final block.

† Elmqvist & Quastel (1965) have described a fairly exhaustive study of the isolated human nerve-muscle preparation—intercostal muscle obtained at thoracotomy; the results are consistent with earlier studies on the frog and guinea-pig preparations, and the main interest of the paper is the discussion of the results in terms of a simple model describing transmitter synthesis and release.

synaptic delay, by the end-plate potential. Passing a stimulus through the microelectrode caused a propagated antidromic spike in the nerve; thus proving that the terminal fibre from which they had recorded a spike was capable of propagating an all-or-none type of response.* Katz & Miledi (1965b), using fine microelectrodes to record at two different points along the course of the fine terminal, established that propagation was active, and took place with a velocity of some 0·3 m./sec. which, for non-myelinated nerves, agrees reasonably with the predicted velocity on the basis of the diameter, capacity and resistance (Katz, 1948). With a recording electrode immediately over an active spot, the interval between the arrival of the nerve spike and the beginning of the end-plate potential was measured. The preparation was in a Ca^{++}-free Ringer to block the muscle action potential, and the recording electrode contained Ca^{++}, which could be expelled by an iontophoretic pulse. In this way only the active spot in the immediate neighbourhood of the electrode gave rise to end-plate potentials, so

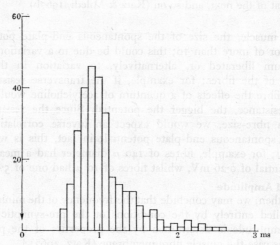

FIG. 602. Distribution of synaptic delays at the neuromuscular junction. Zero time (marked by arrow) corresponds to negative peak of focal nerve spike. The most frequent delay was 0·9 msec. (Katz & Miledi. *Proc. Roy. Soc.*)

that the record was uncomplicated by activity in neighbouring spots. Fig. 602 illustrates the frequency with which delays of varying magnitude occurred; the minimum was 0·5 msec., and the most frequent was 0·9 msec. These fluctuations in latency are the necessary consequence of a system in which the nerve impulse increases the *probability* of an augmented number of quantal discharges.

The minimum delay cannot be accounted for by diffusion time, since calculations indicate that this is not likely to be greater than 50 μsec; similarly, the time for acetylcholine to affect the postsynaptic membrane is probably not a significant factor if one may extrapolate from microinjection experiments, so that it seems safe to say that the real delay is that between the arrival of the electrical change at the terminal and the increase in probability of acetylcholine release. If this view is correct, we may look on Fig. 602 as representing the initial delay and subsequent time-course of the transmitter release. This, of course, describes events at a single point; with a terminal some 150 μ long the process will begin at

* It seems that propagation along the terminals in the crayfish preparation is electrotonic (Dudel, 1965). Braun & Schmidt (1966) confirmed that spike activity is propagated along the terminals of the frog end-plate at some 30 cm./sec.

the proximal end some 0·5 msec. before it reaches the distal end, and at each point the process will be dispersed over a few msec. The theoretical course of events at the whole end-plate will therefore be that indicated by Fig. 603.

Role of the Synaptic Vesicles. When first discovered, it was thought that the miniature potentials might result from random impacts of single transmitter molecules leaking out of the axon, but this was ruled out by quantitative studies, which showed that a single miniature potential must result from the synchronous discharge of a substantial packet of acetylcholine molecules with little time for diffusion from distances greater than about 1 μ. The suggestion naturally arose, therefore, that the spontaneous emptying of a single synaptic vesicle might be the cause of the miniature potential; and evidence supporting this was obtained by Birks, Katz & Miledi, who carried out parallel studies on the electron-microscopical appearance of the junction and the spontaneous electrical activity.

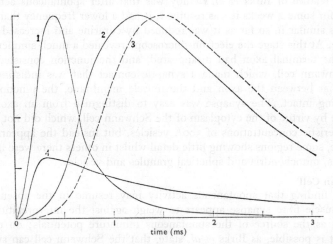

FIG. 603. Computed time-course of release and action of transmitter. Curve 1: Distribution of synaptic delays at one spot. Curve 2: Averaged end-plate current at that spot. Curve 4: Sample of unitary end-plate response. Curves 1 and 2 show, respectively, the average time-course of transmitter release (1) and transmitter action (2) at one point of the nerve-muscle junction. Curve 3 shows the time-course of transmitter action dispersed over the whole junction. (Katz & Miledi. *Proc. Roy. Soc.*)

Nerve Section

On section of the nerve supplying a muscle there is a loss of indirect excitability, which in the frog takes about 3–4 days to manifest itself, if the animal is kept at 20° C, but as long as 30–40 days if kept at 5° C. The generally accepted view of the cause of this slow loss of indirect excitability is that it is due to the failure of the degenerating axons to produce transmitter.* In accordance with this view, it was found that the spontaneous miniature potentials ceased in parallel with the loss of indirect excitability; when individual junctions were studied, it was found that not all of them failed at once; moreover, the failure to transmit, or to exhibit spontaneous miniature potentials, occurred suddenly over a period of a few hours with no very well defined transitional state of reduced

* Since the original study of MacIntosh in 1938, subsequent work has thoroughly established the proposition that the block of transmission through a sympathetic ganglion is due to failure of synthesis of acetylcholine resulting from the loss of choline acetylase (Banister & Scrase, 1950).

excitability.* Thus, after a preliminary period of apparently normal behaviour, neuromuscular transmission stops abruptly within a few hours, and spontaneous activity also stops. Since the junction is still normally sensitive to acetylcholine applied locally at this stage, it is difficult to escape the conclusion that failure is due to loss of transmitter substance at the nerve terminals.

Electron-Microscopy of Nerve Terminals

In the electron-microscope the junction showed no obvious changes during the stage of normal indirect excitability; during the next phase (3–4 days after denervation at 20° C) there were extensive degenerative changes at the junction, most characteristic being the appearance of "honeycomb" structures, which may have been masses of agglutinated vesicles; at other, and probably later stages, islets of these honeycombs were completely enclosed in Schwann cytoplasm, which now established "synaptic contact" with the muscle fibre. A surprising feature of Birks et al.'s study was that after spontaneous activity had ceased for some 2 weeks it was resumed; it had a lower frequency than normal, but was similar in so far as it was reduced by curarine and increased by prostigmine. At this stage the electron-microscope revealed a much simpler picture, since the terminal axon had disappeared, and the junction consisted of only the Schwann cell, which made a synaptic contact that was indistinguishable from that between the axon and the muscle membrane, the junctional folds remaining intact. The synapse was easy to distinguish from an axon-muscle synapse by virtue of the cytoplasm of the Schwann cell, which did not show the characteristic concentrations of 500A vesicles, but instead the appearance was variable, some regions showing little detail whilst in others there were numerous particles, mitochondria and spherical granules and vesicles.

Schwann Cell

This finding that spontaneous activity may resume in the absence of an axon-muscle fibre synapse appears to weigh against the interpretation of the vesicles as the source of the spontaneous miniature potentials; on the other hand, it is possible, as Birks et al. state, that the Schwann cell can synthesize acetylcholine, which may be concentrated in the vesicles which are certainly present, although in not so regular a manner as in the axon. In this event the "silent period" of a week or more before the reappearance of spontaneous activity might be the manifestation of a delay in developing the synthetic apparatus for producing acetylcholine.†

Emptying the Vesicle

In the normal end-plate we might expect to find vesicles in electron-micrographs at the moment of emptying their contents into the "synaptic space" between axon and muscle membranes; in fact, this appearance has not been described with any frequency but it is easy to calculate that the probability that such a situation would be seen is very small indeed; thus at the highest peak of activity only a few hundred quanta are released at one junction, whilst the total length of arborization is of the order of 1 mm; hence one emptying would occur

---

* The actual transition from activity to inactivity was followed in some cases; the most striking feature was a decline in rate of spontaneous discharge, which was sometimes preceded by a period of high-frequency discharge such as could have resulted from depolarization of the axon terminal; the average amplitude diminished with time.

† Miledi & Slater (1968) have carried out a similar study on the rat neuromuscular junction. Denervation resulted usually in a permanent abolition of the miniature potentials. In the electron microscope it was clear that the Schwann cells usually failed to take up a synaptic relationship with the synaptic cleft, but disappeared. Occasionally, 1–3 weeks after denervation, miniature potentials were observed, and this may have been associated with the presence of a Schwann cell overlying the cleft.

at points 3 μ apart, so that the chances of finding an emptying vesicle in a thin section are indeed remote (Birks *et al.*).

Isolated Vesicles

The nerve-nerve synapse is more amenable to direct chemical study since high concentrations of these may be obtained by homogenizing nervous tissue, for example the superficial layers of the cerebral cortex. The pioneering studies of Whittaker in this respect led to the separation of a fraction with high acetyl-choline activity which, on examination in the electron-microscope, was shown to consist of synaptic vesicles enveloped in larger particles, which were apparently broken-off nerve terminals. By more vigorous homogenization de Robertis *et al.* were able to prepare a fraction containing many intact vesicles which had been liberated from the nerve endings. The transmitter at the adrenergic nerve ending is nor-adrenaline; by squashing splenic nerves and extraction, v. Euler & Lishajko have separated granules at 50,000 g containing the transmitter, which is released by detergents or acids.*

Regeneration of the Neuromuscular Junction

Phase of Increased Excitability. It is well known that structures deprived of their normal innervation become supersensitive to their transmitter substance (see, for example, Cannon & Rosenblueth, 1949); thus the sensitivity of a denervated skeletal muscle to applied acetylcholine may well be 100 times that of a normal muscle. The effects of denervation described above, in relation to the onset of miniature end-plate potentials, occur long before this phase of supersensitivity in the end-plates.

Spread of Chemical Sensitivity. The problem of the supersensitivity of denervated muscle has been examined by Axelsson & Thesleff in the cat, and by Miledi in the frog. The striking feature of denervation, discovered by Axelsson & Thesleff, was that the whole length of a given muscle fibre became sensitive to local application of acetylcholine, in marked contrast to the normal fibre where sensitivity to minute applications (electrophoretic), was confined to discrete spots. There was no evidence that the end-plate region became supersensitive, so that the increased sensitivity was due entirely to this spread from the end-plate region. Studies of the equilibrium potential showed that the denervated fibre now behaved like an end-plate, giving a value for this of about 10 mV. Isotopic studies showed that permeability of the fibre membrane to K^+ and Na^+ became high when acetylcholine was applied; and this increase in permeability occurred even when the fibre was depolarized in a high-K^+ medium. The permeability to chloride was barely affected† (Klaus *et al.*, Jenkinson & Nichols). In general, then, the whole fibre behaved like an end-plate with the exception that its acetylcholine-sensitivity was not prolonged by anticholinesterases such as edrophonium; consequently the denervated muscle fibre presumably has no high concentration of cholinesterase, such as is found at the end-plate.

Electrical and Chemical Excitability

It has been argued by Grundfest (1957) that chemically excitable tissue, such as the post-synaptic membranes of the neuromuscular junction or nerve-nerve

---

* Cholinesterase is associated with a microsome fraction of brain; this fraction is rich in membranous vesicles whilst a fraction rich in granules has low activity. It would appear, therefore, that this activity is associated with the membranous part of the endoplasmic reticulum (Toschi, 1959; Hanzon & Toschi, 1959).

† The permeability to Ca^{++} was also increased on treatment with acetylcholine, and it may be that the contracture that develops under these conditions is due to the influx of this ion (p. 1403).

synapse, are electrically inexcitable, so that passing an electric current through the membrane leads to no regenerative response such as that typically seen in the propagated action potential. The denervated muscle fibre membrane constitutes one of several exceptions to this generalization since Axelsson & Thesleff were able to obtain propagated action potentials along their denervated fibres.

Frog Fibres

Miledi measured the activity of single fibres with internal microelectrodes in muscles that had been completely and partially denervated. Because of the multiple innervation of the sartorius muscle of the frog, partial denervation frequently leads to the presence of single fibres with one innervated and one denervated end-plate; when such fibres are found, the one end-plate may act as a control for the other. At a time when the sensitivity of a denervated muscle to applied acetylcholine was 100 times normal, it was found that the amplitude of the spontaneous miniature end-plate potentials was no greater than normal;

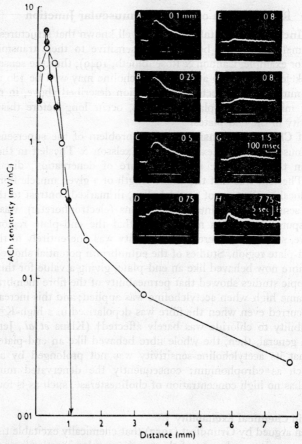

FIG. 604. Distribution of acetylcholine sensitivity in a muscle fibre 66 days after complete denervation (O) and at a normal end-plate in the control muscle (●). Abscissæ, distance along fibre; ordinates, sensitivity to iontophoretic pulses of ACh, log. scale. Inset; upper traces, examples of ACh potentials of denervated fibre at specified distances; lower traces, records of current flowing through acetylcholine pipette. The muscle supersensitivity is expressed as the ratio, control: denervated, of minimal ACh concentrations that evoke muscle twitches. (Miledi. *J. Physiol.*)

moreover, in the region of the end-plate the sensitivity to acetylcholine, applied locally by the iontophoretic method, was normal. As Axelsson & Thesleff had shown in the cat, the reason for the increased sensitivity of the muscle as a whole was the spread of sensitivity from the end-plate, where it is normally localized, to the whole length of the muscle fibre (Fig. 604). When muscle fibres containing both innervated and denervated end-plates were studied, it was shown that the sensitivity of the innervated one, both to applied acetyl-choline or to nervous impulses, was quite unaffected by the denervation of the other. A completely denervated muscle fibre atrophies rapidly, but the spread of acetylcholine sensitivity is not due to this since the partially denervated fibre remains completely normal from this aspect yet its sensitivity spreads from the denervated end-plate region.

Inhibitory Action of Nerve

Thus the presence of the nerve terminals in the end-plate in some way inhibits the development of "extra-junctional receptors" as Miledi calls them. This, as we shall see, is not the only example of a control, by a nerve fibre, of the functional state of a muscle fibre (p. 1179); it is apparently not exerted through the liberation of transmitter since Miledi showed that prolonged immersion of the denervated muscle in high concentrations of acetylcholine did not prevent the spread of sensitivity.*

Botulinum Toxin

On the other hand, the finding of Thesleff that botulinum toxin, which prevents the release of acetylcholine from the nerve terminals whilst leaving them otherwise intact, causes the whole muscle fibre to become sensitive to acetylcholine, strongly suggests that it is by releasing acetylcholine that the nerve terminals maintain their control over the muscle fibre's acetylcholine-sensitivity.

Neuromuscular Regeneration. In the recently denervated muscle the amplitude and frequency of the miniature end-plate potentials are far below normal, as we have seen. Later, there develops the spread of sensitivity to applied acetylcholine. Miledi has followed the reversal of these changes during the process of regeneration of the nerve fibres up to, and beyond, the point at which neuromuscular transmission is resumed. The first stage is shown by an increase in the amplitude and frequency of the miniature potentials, whilst the curve of distribution of amplitudes becomes less skew; it is as though the axon were taking over from the Schwann cell by pushing it away from its synaptic contact, or by inhibiting it from releasing its packets of acetylcholine. Only when the frequency and amplitude of the miniature potentials have returned to normal is neuromuscular transmission restored, and associated with this there is a narrowing of the acetylcholine-sensitive region on the fibre. An interesting feature is the all-or-none nature of the resumption of transmission in so far as the end-plate potential, in response to a nervous stimulus, is concerned; before transmission is established, a nerve impulse produces no end-plate potential—in other words it is unable to cause the liberation of acetylcholine. Thus, as soon as an end-plate potential is evoked by a nervous stimulus, this end-plate potential is large enough to evoke a spike in the muscle fibre.†

* By studying end-plate regions in the rat, which are much more highly circum-scribed, being only some 30 μ wide, Miledi showed that the sensitivity to acetylcholine extended well beyond this width, up to some 500 μ, whilst the region over which miniature end-plate potentials could be recorded was confined to the 30 μ area. Thus the sensitivity of the muscle fibre to acetylcholine extends well beyond the end-plate proper (Diamond & Miledi).

† The neuromuscular junction is not completely normal when transmission has been re-established, showing an increased latency, a more ready presynaptic failure of propagation and a pronounced post-tetanic facilitation (Miledi, 1960).

Mechanism of Chemical Excitation

It is not sufficient to say that acetylcholine is liberated at a junctional region; we must ask ourselves how this liberation initiates a new impulse in the effector cell, *i.e.*, how a chemical substance, or a chemical reaction, can produce the same results as those of an applied electrotonus. In the first place we may note that the phenomena of sensory excitation, to be discussed briefly later, provide many instances of chemical excitation; the end organs of smell and taste are merely specialized nerve endings adapted to respond to minute changes in the chemical environment; the retinal receptors are specialized cells capable of responding electrically to a photochemical reaction taking place at their surface; and so on. It will therefore be nothing new to demonstrate that a substance such as acetylcholine, even in minute concentrations, can initiate impulses in an effector cell.

Depolarization. We have described electrical excitation, and the propagation of the action potential, as a *depolarization* phenomenon, and the action of acetylcholine as a *depolarizing action*; let us review the significance of these statements. An applied cathodal potential causes current to flow as in Fig. 537 (p. 1057), *i.e.*, in such a way that positive current flows across the membrane from inside to the outside: this flow of current causes the breakdown in impedance of the membrane that gives the full action potential; and since the flow of current during excitation is such as to break down the normal resting potential, the excitation process is described as one of depolarization. Propagation is due to the flow of current from the active to the inactive region as indicated in Fig. 544 (p. 1065); this flow is such as to depolarize the adjoining portion of the membrane, and therefore to excite it. Any chemical reagent that is capable of reducing the resting potential at a given point on the surface of an excitable tissue must set up demarcation currents in the same direction as those that normally are responsible for the propagation of the impulse. Now KCl, applied to a nerve or muscle, reduces the resting potential for reasons that are quite clear (p. 570); a localized application of KCl to the nerve should, and actually does, initiate a propagated disturbance, whilst immersion of the whole tissue in a solution with a high concentration of KCl abolishes conduction. Similarly, any reagent that breaks down the selective ionic permeability of the membrane should depolarize it and therefore induce a propagated disturbance. For example, a high concentration of alcohol, by its destructive action on the cell membrane, probably makes it permeable to Na^+ and other ions; applied locally to nerve it initiates an impulse (Arvanitaki & Fessard); in fact the literature is well stocked with examples of chemical compounds which, even in low concentrations, depolarize the membranes of excitable tissues. Thus the dialkylsuccinylsulphates have been shown by Höber to depolarize frog muscle in concentrations in the region of 1.10^{-4} M. In general, we may distinguish two types of depolarization; the first, typified by K^+, is due to a reduction of the resting potential on simple thermodynamic grounds; its effect can be mimicked by other cations to which the membrane is permeable, *e.g.*, Rb^+, Cs^+, NH_4^+ and probably certain organic derivatives of the ammonium ion (Wilbrandt; Cowan & Walter); to the other class belong substances such as alcohol, bile salts, saponin, etc., whose influence is directly on the cell membrane, breaking down selective permeability when applied in sufficiently high concentration, and causing reversible block in lower concentration. In general, the effects of substances of the first class are reversible whilst those of the second class are only reversible under restricted conditions, the effects on the membrane being often too severe for recovery on removal of the agent.

Acetylcholine as Cation

We may now ask: How does acetylcholine depolarize? It is a quaternary ammonium base of the formula:—

$$\text{(CH}_3)_3\text{N-CH}_2.\text{CH}_2\text{OOCCH}_3$$
$$\overset{\displaystyle \text{OH}}{\underset{|}{}}$$

Quaternary ammonium bases are strong bases, so that acetylcholine exists in solution, under physiological conditions, as a cation. If the end-plate tissue were permeable to this cation, and if its local concentration were effectively high, depolarization could occur as a result of a thermodynamic effect rather similar to that of K^+. The local concentration, however, would have to be exceedingly high in comparison with the bulk concentrations that are effective; and a simple quantitative analysis of the amount of acetylcholine that would be necessary to maintain the prolonged depolarization observed in the eserinized preparation convinced Fatt & Katz that this mechanism was quite unfeasible—there simply could not be enough acetylcholine in a nerve terminal to cause even a single end-plate response, let alone the large numbers that may be obtained from repetitive stimulation.

Change of Membrane Permeability

The alternative hypothesis, that acetylcholine depolarizes by virtue of a reversible change in membrane permeability, is easier to reconcile with the facts, since it is known that permeability may be affected by quite low concentrations of certain ions, e.g., Ag^+, or by certain surface-active organic substances. Presumably the end-plate membrane has a specific chemical grouping on its surface capable of reacting with acetylcholine; as a result of this reaction, it could be supposed that the properties of the membrane would be altered so as to cause a breakdown in the normal resting permeability relationships to give the depolarized condition. As we have indicated earlier, this depolarization may well be different from the reversal of polarity that occurs during the rising phase of the propagated nerve or muscle spike, different in the sense that it is less specific and represents, instead of an ordered sequence of increase in Na^+ and K^+ permeability, an indiscriminate increase in permeability to all ions, or at any rate to Na^+, K^+, Cl^- and HCO_3^-.

As to the nature of the "receptive substance" with which acetylcholine reacts, little is known. Chagas and his colleagues profited by the fact that the large electric organ (p. 1223) consists of little more than neuromuscular junctions and thus should provide a good source of the material to which acetylcholine attaches itself. By injecting the eel with a radioactive curare analogue in order to render the receptive substance insoluble, Chagas et al. were able to prepare extracts that remained in their dialysis sacs after prolonged dialysis. Refinements of the procedure by Ehrenpreis have shown a good correlation between the power of binding of various drugs by the preparation and the activity of these drugs at the neuromuscular junction or electric organ. On the other hand, Trams et al. conclude from their studies that the precipitation of protein by curarine is not highly specific so that the precipitated material might well not be the "receptive substance".

Neuromuscular Block

Magnesium and Calcium. Mg^{++}, in concentrations of some 5–10 mM,* causes a reversible block in neuromuscular transmission, without affecting

---

* The normal plasma concentration is about 1 mM.

appreciably conduction along nerve or muscle. The effect is antagonized by Ca^{++}, in the sense that raising the concentration of this ion requires that the blocking concentration of Mg^{++} be higher. The effect of Mg^{++} has some similarity with that of curarine, in that the end-plate potential is reduced—the greater the concentration of Mg^{++} the smaller the potential. However, the effect of curarine is completely accounted for by its effect on the "post-synaptic membrane", i.e., on the end-plate substance, preventing acetylcholine from acting by competition for specific reactive groups. Curarine therefore inhibits the action of directly applied acetylcholine. Mg^{++} does, indeed, reduce the effect of applied acetylcholine but, according to Del Castillo & Engbaek, this effect is too small to account completely for its blocking action. Thus, when Mg^{++} has reduced the sensitivity to acetylcholine by 25 per cent., the end-plate potential following a nervous stimulus is reduced by 75 per cent., and it would seem that the main action of Mg^{++} is to reduce the amount of acetylcholine liberated by the nerve impulse.*

The action of Ca^{++} is not confined to antagonizing the effects of excess Mg^{++}; increasing the concentration of Ca^{++} in a normal Ringer actually increases the size of the end-plate step in response to a nerve stimulus; presumably it increases the amount of acetylcholine liberated at the junction.

Frequency of Miniature Potentials

So far as miniature end-plate potentials were concerned, neither Ca^{++} nor Mg^{++} seemed to affect their frequency in amphibian muscle, although Boyd & Martin described an increase when the concentration of Ca^{++} was raised in the mammalian preparation. A more exact study by J. I. Hubbard on the rat diaphragm has shown, however, that both ions can affect the frequency, when the base-line employed is the frequency in the complete absence of Ca^{++}. Thus even in the presence of a chelating agent and the absence of Ca^{++} in the medium, there is a residual "fixed fraction" of end-plate activity; adding Ca^{++} increased the frequency of this activity, the increase being proportional to the logarithm of the concentration; at a given concentration of Ca^{++}, the frequency was decreased by increasing concentrations of Mg^{++}; in the absence of Ca^{++}, Mg^{++} had no effect at all. Apart from these rather small effects, the main influences of the alkali earth ions are on the release in response to nervous activity, i.e., in response to depolarization of the nerve terminal; and it is considered by Hubbard that the spontaneous activity, over and above the "fixed fraction", is really determined by the state of polarization of the nerve, and certainly the effect of Mg^{++} was much more pronounced when some depolarization had been induced by raising the external concentration of K^+.

Calcium Hypothesis. A somewhat *ad hoc* hypothesis to explain the Ca^{++}-Mg^{++} antagonism could be that these ions competed for sites on some complexing molecule in the presynaptic terminal; the complex with Ca^{++}, Ca_a, would lead in some way to release of acetylcholine, whereas the Mg-complex would be inactive in this respect. Jenkinson (1957) varied the Ca^{++} and Mg^{++} concentrations over a wide range, and concluded that their effects on the end-plate potential were

---

* Direct proof on a perfused muscle has not so far been carried out; however, in the superior cervical ganglion Mg^{++} blocks synaptic transmission and it has been shown by Hutter & Kostial that the amount released by the perfused ganglion during repetitive stimulation is reduced in these circumstances, an effect that can be reversed by Ca^{++}. Hoyle has drawn attention to the very high concentrations of Mg^{++} that may occur in insect hæmolymph; the action of Mg^{++} on the insect junction is qualitatively similar to that on the frog's, but it would seem that the insect junction is more tolerant to the ion. This is mainly due to the circumstance that very small end-plate potentials may cause contraction of muscle fibres in the invertebrates, without producing propagated spikes. In the stick insect, *Carausius morosus*, there is an undoubted tolerance of the neuromuscular junction to very high Mg^{++} concentrations (Wood, 1957).

reasonably consistent with a Michaelis–Menten type of competition. More recently Dodge & Rahamimoff (1967) have examined the relation between quantal release and divalent cation concentration in greater detail, especially in the region of low concentrations of Ca^{++}. On the basis of simple competition, the concentration of calcium complex should be proportional to:—

$$\frac{[Ca^{++}]}{1 + \dfrac{[Ca^{++}]}{K_1} + \dfrac{[Mg^{++}]}{K_2}}$$

and if quantal release is proportional to the concentration of complex, the quantal content of the end-plate potential should be proportional to this too. In fact, a fourth power relation gave a better fit (Fig. 605):—

$$Quantal\ Content = K \left(\frac{[Ca^{++}]}{1 + \dfrac{[Ca^{++}]}{K_1} + \dfrac{[Mg^{++}]}{K_2}} \right)^4$$

This could mean a stoicheiometric relation between complex anion and calcium, or it could mean a stochastic relation, *i.e.*, that four sites over a small area of membrane had to be occupied to obtain a maximal effect.

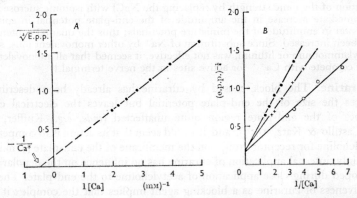

FIG. 605. Left: Lineweaver-Burk plot for relationship between [CaX⁴] and end-plate response. Ordinate: Reciprocal of fourth root of end-plate potential. Abscissa: Reciprocal of [Ca]. Right: A similar plot for a different end-plate exposed to different concentrations of Mg. Note that the straight lines have approximately the same intercept. ○, 0·5 mM-Mg; +, 2·0 mM-Mg; ●, 4·0 mM-Mg. (Dodge & Rahamimoff. *J. Physiol.*)

Influx of Ca⁺⁺

The importance of Ca^{++} in transmission suggests that this ion penetrates the nerve terminal as a result of an increased permeability caused by the action potential. By applying Ca^{++} micro-iontophoretically to the end-plate at different times, Katz & Miledi (1967c) showed that this ion was only effective over a brief interval, namely 50 to 300 μsec. before the depolarizing pulse; application after the pulse was without effect. The inhibitory effects of Mg^{++}-pulses were also confined to this brief interval before the depolarizing pulse. By applying tetrodotoxin, thereby blocking the nerve spike, Katz & Miledi (1967a) were able to vary the duration of the stimulating pulse—conducted electrotonically along the nerve. They found that the latency for the end-plate potential was more prolonged, the longer the pulse, but the time from the end of the pulse to the beginning of the end-plate potential became shorter as the pulse lengthened, so that eventually, after 4–5 msec., the end-plate response occurred while the depolarizing pulse was still on. The longer the pulse, the greater was the end-plate potential. A possible interpretation of this delaying effect of the depolarizing pulse is that it delays electrically the passage of Ca^{++} into the nerve, although in this event we should expect the end-plate response to be delayed for as long as the depolarizing pulse lasts, but this is not so.

Very Low Ca^{++} Concentrations

Hubbard, Jones & Landau (1968) studied the frequency of spontaneous potentials over a wide range of external Ca^{++}, in order to determine whether there was a limiting concentration below which spontaneous activity ceased; if this were so, then Ca^{++} could be said to initiate the process, rather than to accelerate a process that carried on in its absence. To achieve the low concentrations, a buffer system made up of EDTA and EGTA was employed, and the actual concentration under any particular conditions was calculated with the aid of a computer; in this way the concentration was controlled over a range extending from 10^{-10} to 10^{-2} M. Hubbard *et al.* found that there was, indeed, a basic frequency of spontaneous potentials that was unaffected by Ca^{++} concentrations from zero to 10^{-5} M. At concentrations of 10^{-4} M and above there was an acceleration of the basic rate; and the effects of mixtures of Ca^{++} and Mg^{++} suggested competitive action. When the concentration of Ca^{++} was low, *e.g.*, 10^{-7} M, Mg^{++} was found to increase the basic frequency. Miledi (1966) had shown that micro-iontophoretic application of Sr^{++} could cause end-plate potentials, although less effectively than Ca^{++}, so that it would appear that, under appropriate conditions, the three alkali earth metals, Mg^{++}, Ca^{++}, and Sr^{++}, have the same action.*

Effect of Sodium

Kelly (1968) found that when the concentration of Ca^{++} was low (0·23 mM), reduction of the ionic strength by replacing the NaCl with isotonic sucrose caused an immediate increase in the amplitude of the end-plate potential in spite of a reduction in amplitude of the miniature potentials; thus the quantal content must have been increased. Since substitution of Na^{+} by other monovalent ions, such as methylammonium or lithium, was not effective, it seemed that all monovalent ions could compete with Ca^{++} for active sites in the nerve terminal.†

Curarine. The block caused by curarine has already been described; it reduces the size of the end-plate potential but leaves the electrical characteristics of the end-plate region quite unaffected (*vide, e.g.,* Kuffler, 1945; Del Castillo & Katz, 1957), and it would seem that its action is to compete with acetylcholine for receptor groups on the membrane of the end-plate (Jenkinson). Certainly, internal application of curarine has no influence on the depolarization developed during close application of acetylcholine to the end-plate. The great effectiveness of curarine as a blocking agent implies that the complex it forms at the end-plate is more stable than the acetylcholine complex, and some evidence favouring this view is given by the observation that the effect of a brief "pulse" of curarine, delivered electrolytically by a micropipette at the end-plate, decays some 35 times more slowly than does the effect of a brief pulse of acetylcholine. Prostigmine has a curarizing action, and the reason why it is so much less efficient than curarine may be that the complex it forms is much less stable; at any rate its effect decays some 10 times more rapidly (Del Castillo & Katz, 1957).

Prostigmine

Eserine prolongs the action of acetylcholine (p. 1140) because it inhibits the enzyme cholinesterase. Prostigmine has the same effect but, as just mentioned, it may also behave like curarine, inhibiting the action of acetylcholine. At first thought this is puzzling, but it becomes intelligible when it is appreciated that

---

* Elmqvist & Feldman (1965) were able to abolish all spontaneous activity at the mammalian end-plate by soaking in EDTA solutions; Ba^{++} and Sr^{++}, in addition to Ca^{++}, restored activity abolished by EDTA.

† Colomo & Rahamimoff (1968) found a similar effect and suggested a competition of Ca^{++}, Mg^{++} and Na^{+} for the same site; the predicted relations on the basis of Michaelis–Menten kinetics did not conform perfectly with the experimental, but this could be due to other effects of low Na^{+}, namely depressed synthesis of acetylcholine (Birks, 1963) and a change in the acetylcholine equilibrium potential at the end-plate. Gage & Quastel (1966) have examined the relations between Na^{+}, K^{+}, Ca^{++} and Mg^{++} concentrations on the frequency of spontaneous end-plate potentials in some detail, and analysed them in terms of competition for a binding molecule which is active when combined with one Ca^{++}-ion.

acetylcholine must react (*a*) with some constituent of the end-plate membrane so as to depolarize it and (*b*) with the enzyme, cholinesterase, to be hydrolysed. Both the end-plate "receptive substance" and cholinesterase may therefore depend, for their activity, on similar chemical groupings—groupings that would permit some sort of combination with acetylcholine. It is therefore likely that a substance that competes with acetylcholine for the end-plate (curarizing) will also compete with acetylcholine for cholinesterase (eserinizing) and, finally, it may also mimic acetylcholine by depolarizing. Some of these predictions are verified in the behaviour of choline and decamethonium which, under appropriate conditions, can produce a weak depolarization, a potentiation of acetylcholine (eserine-like action) and an inhibition of acetylcholine in the presence of prostigmine, *i.e.*, a curarizing activity (Del Castillo & Katz, 1957).

Acetylcholine Block. It was early discovered that neuromuscular block could be caused by an excess of acetylcholine at the neuromuscular junction; and it was considered that this block was the consequence of a permanent state of depolarization of the end-plate, similar to the neuromuscular block that may be caused by high concentrations of K$^+$ in the external medium (*e.g.*, Jenerick & Gerard, 1953). This view was strengthened, moreover, by the observation that synthetic substances like "decamethonium" blocked neuro-muscular transmission, but also depolarized the neuromuscular junction. A careful examination of the potential at the neuromuscular junction with an intracellular electrode by Thesleff showed, however, that the block that developed on placing frog muscle in a bath containing 0·5–2·0 . 10^{-5} M acetyl-choline was quite unconnected with the fall in resting potential since, within 7–15 minutes, the resting potential had returned to normal, yet block was maintained as long as acetylcholine remained in the outside solution, and was only overcome by washing the muscle for 15–45 minutes in normal Ringer; in fact, complete neuromuscular block was only obtained when the resting potential had returned to normal. There seems little doubt from this that acetylcholine block—and the similar block caused by nicotine—is due to a progressively developing insensitivity of the "post-synaptic membrane" to the chemical mediator. So far no explanation for this "desensitizing action" of acetylcholine and other depolarizing agents has been put forward.*

Membrane Resistance

Manthey (1966) showed that the desensitization of frog muscle to applied carbamylcholine was associated with a return of the transmembrane resistance to normal; increasing the concentration of Ca^{++} from 0 to 10 mM gave a seven-fold increase in rate of desensitization, measured by the change of transmembrane resistance; the effect was antagonized by Na$^+$. Manthey suggested that the effect of Ca^{++} was on end-plate membrane permeability to ions, decreasing this and thus reducing the effect of acetylcholine; at any rate Takeuchi (1963) found that a high concentration of Ca^{++} shifted the reversal point for acetylcholine-produced end-plate currents to more negative values and decreased the end-plate currents.

Botulinum Toxin. This toxin, liberated by the organisms responsible for food-poisoning (*Clostridium*), causes a neuromuscular block that is different from that caused by curare, in that the end-plate is still sensitive to acetylcholine (Burgen, Dickens & Zatman); the end-plate potential in response to nerve stimulation is abolished, and this suggests that the block is due to the failure of the nerve terminals to liberate acetylcholine. This view is strengthened by

* In the mammal, the repolarization following application of acetylcholine is not very marked. However, the depolarization that remains is not sufficient to cause the observed neuromuscular block; thus, the resting potential should be reduced to 50–55 mV for this to happen, whereas Thesleff found that acetylcholine and other blocking agents of this type reduced it to only 60–70 mV.

Brooks' observation that the frequency of miniature end-plate potentials is diminished, although the amplitude of individual ones is unaffected. As indicated earlier (p. 1157), the effect of the toxin is analogous with that of denervation in that the whole muscle fibre becomes sensitive to acetylcholine (Thesleff).

Tetrodotoxin. The non-polypeptide neurotoxins, such as tetrodotoxin and saxitoxin, exert their action, as we have seen, on the Na^+-activation process and therefore block the propagated action potential in both nerve and muscle. Since the end-plate response does not involve this specific increase in Na^+-permeability we may expect its excitability to be unaffected; and this is, indeed, true so that acetylcholine readily depolarizes the end-plate (Kao & Nishiyama, 1965) whilst transmission from nerve to muscle can be achieved by electrotonically conducted depolarizations (Katz & Miledi, 1967).*

Depression and Facilitation

Fatigue. With repetitive stimulation of a muscle through its nerve, the tension developed gradually falls off and finally attains a negligible value; the failure is due to transmission block since action potentials are still carried by the nerve, and direct stimulation of the muscle reveals that it is still excitable. The essential feature of fatigue is therefore a neuromuscular block,† and, theoretically at least, it may be due either to failure of the nerve terminals to produce acetylcholine sufficiently rapidly to keep pace with the demands of repetitive stimulation, or to diminished sensitivity of the end-plate to acetylcholine, or to both of these factors. All the evidence indicates that the dominant factor in this synaptic depression is a failure of the nerve terminals to liberate sufficient transmitter.

Quantitative aspects of production and release during stimulation indicate that failure of transmission is not due to complete exhaustion of the transmitter in the terminals; far from it, only about 1/500 of the total being *available* for liberation in response to a nerve action potential (Krnjević & Mitchell, 1961); furthermore, in response to a given stimulus, only about one-third of the immediately available transmitter is, in fact, released. If this is true, it is not possible to detect, by chemical methods, whether the block following repetitive stimulation is due to depletion of readily available transmitter, so that essentially indirect methods have been used to reach this conclusion.

Examination of the events taking place at the end-plate with successive, or just single, shocks reveals both synaptic depression and facilitation, a second stimulus producing a smaller or larger end-plate potential than that produced by the first, conditioning, shock. The ratio of the sizes of the end-plate potentials, or better still, the ratio of the quantal contents, as estimated from fluctuations in the magnitudes of the end-plate potentials on the basis of statistical theory, is used as a measure of facilitation or depression.

Reduced Number of Quanta

By fatiguing a normal nerve-muscle preparation to the point where block occurred in the fibre examined, Del Castillo & Katz found that this block was associated with a fall in the end-plate potential, resulting from a nerve stimulus, to a few millivolts, whilst spontaneous activity increased markedly.

---

* Katz & Miledi (1968) have shown, however, that the electrotonic spread from the action potential is insufficient, in the frog neuromuscular preparation, to cause sufficient transmitter release to permit muscle activation; thus local electrophoretic application prevents the production of end-plate potentials distal to the site while they are produced normally at proximal sites of the same terminal arborization. This means simply that the spread of current during the spike is insufficient to cause release of transmitter.

 † The block associated with repetitive stimulation is usually described as *Wedensky inhibition.*

The reduction in the end-plate potential could have been due to a reduction in the size of the quantum of acetylcholine released, or to a reduction in the number of quanta released by a nerve impulse; inspection of the size of the spontaneous potentials showed that, although they may have been reduced a little, the reduction could not have accounted for the fall in size of the end-plate potential following a nerve stimulus, which must therefore be attributed to a diminished liberation by the nerve impulse, *i.e.*, by a reduced number of quanta.

By reducing the concentration of Ca^{++} and increasing that of Mg^{++}, neuro-muscular block is achieved; a block which, unlike that caused by curare, is due to failure of the terminals to liberate acetylcholine in response to a nerve impulse; as a result, there is no diminution in the size of the spontaneous miniature end-plate potentials. When the concentration of Ca^{++} is increased, then a given stimulus that just causes depression at the low concentration will now produce a well-marked depression; since Ca^{++} increases the liberation of transmitter per impulse, this effect is understandable on the basis of a reduction in the available amount of transmitter caused by the first stimulus (Lundberg & Quilisch, 1953).

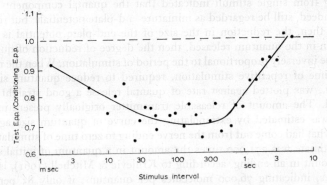

FIG. 606. Illustrating depression at the end-plate following a single stimulus. The relative amplitudes of test end-plate potentials are plotted as ordinates against the interval between conditioning and test stimuli as abscissae. (Thies. *J. Neurophysiol.*)

Recently Thies (1965) has examined the changes in excitability of the mammalian end-plate, as revealed by the quantal content of the end-plate potential, immediately following a single stimulus; as others had found, there is usually an early period of facilitation, and this is followed by depression which may last for several seconds (Fig. 606); he found a good correlation between the estimated quantal content of the end-plate potential and the size of the latter, indicating that the sensitivity of the end-plate membrane to transmitter had not been altered, thus confirming earlier experiments in which acetylcholine had been applied directly to the end-plate during the depression following stimulation (Krnjević & Miledi; Otsuka, Endo & Nonomura, 1962). By varying the proportions of Ca^{++} and Mg^{++} in the Ringer solution, Thies varied the quantal content of the conditioning end-plate potential, and thus the amount available for release by the test stimulus, and there was an excellent correlation between the degree of depression and the quantal content of the conditioning stimulus. The mechanism for this post-stimulatory depression presumably resides, then, in the failure of the transmitter replacement mechanism to operate rapidly enough, so that the probability that a given number of quanta will be released by an action potential is reduced.

Hemicholinium

The synthetic action of choline acetylase is blocked by hemicholinium (HC-3), so that we may expect fatigue to manifest itself more rapidly in a preparation stimulated in the presence of this drug; this is true of low doses, the block in transmission after lengthy tetanic stimulation being associated with a reduction in the acetylcholine content of the cat's muscle by some 80 per cent. (Thies, 1962).* With high doses ($>10^{-5}$ M) there is a rapid block of transmission long before reserves of transmitter could have been used up; and this *post*-synaptic action shares many of the features of curare-action (Martin & Orkand, 1961). The *pre*-synaptic action has been studied in some detail by Elmqvist & Quastel (1965), who measured the end-plate potentials resulting from stimulating the nerve of a nerve–muscle junction in the presence of HC-3, and the alterations in the size of the spontaneous miniature potentials. By working at a concentration of 4.10^{-6} M, the *post*-synaptic action was held at a low level, revealed by an unchanged response to directly applied carbachol.

The striking feature was the reduction in size of the miniature potentials, as repetitive stimulation continued; statistical analysis of the end-plate potentials resulting from single stimuli indicated that the quantal components of these could, indeed, still be regarded as miniature end-plate potentials, but reduced in size. If, then, the reduction in the size of the end-plate potential is due to a reduction in the quantum released, then the degree of reduction of quantal size should be inversely proportional to the period of stimulation. When the reciprocal of the time of repetitive stimulation, required to reduce quantum size by 50 per cent., was plotted against rate of quantal release a good straight line was obtained. The amount of releasable transmitter originally present in a nerve ending was estimated by extrapolating the curve of quantum size against the sum of what had come out from the nerve ending to zero time of stimulation. The mean estimate was 271,000 times the amount in a quantum of initial size. The total amount in an ending, according to Krnjević & Mitchell (1961), is $1.8.10^{10}$ molecules, indicating 76,000 molecules per quantum; if only 85 per cent. is releasable, this reduces to 57,000, which compares with the estimate of 45,000 as the number that could be packed into a synaptic vesicle of 500A diameter (Canepa, 1964).

Orbeli's Phenomenon

Transmission through the fatigued neuromuscular junction is facilitated by sympathetic stimulation—*Orbeli's phenomenon*. Hutter & Loewenstein (1958) showed that the effect was associated with an increased amplitude of the end-plate potential, and Jenkinson, Stamenovic & Whitaker (1968) have shown that the effect of noradrenaline—the sympathetic transmitter—is essentially pre-junctional, the sensitivity of the end-plate to applied acetylcholine being unaffected whilst the frequency, but not the amplitude, of the miniature spontaneous potentials, was increased. The coefficient of variation in the mean quantal content (m) of responses to repeated constant nerve stimuli is given, for a Poisson distribution, by $1/\sqrt{m}$; thus the variability in response should be smaller the larger the mean quantal content; Jenkinson *et al.* found that this variability was decreased after application of noradrenaline.†

---

* The reduction in the guinea pig muscle is much less and may be no greater than in the unpoisoned preparation.

† The neuromuscular block of botulinum toxin may be overcome by tetanic stimulation (Brooks, 1956); this finding supports the belief that the phenomenon of post-tetanic potentiation at the neuromuscular junction is a feature of the nerve terminals rather than of the end-plate; the probability of release of acetylcholine is reduced by the toxin (reduced frequency of miniature potentials) whilst tetanic stimulation increases the probability, as

Facilitation. Immediately following the conditioning stimulus there is usually a period of facilitation in which the end-plate potential is higher than in the conditioning response; the extent of this facilitation varies with the species and the conditions of the experiment; for example, it is not so easy to see in the curarized preparation, and this is presumably because the facilitatory, or potentiating, action is obscured by the effects of transmitter depletion; so as Hubbard (1963) emphasized, the phenomenon is best studied in the Mg^{++}-blocked preparation; since here the block is due to the failure of the nervous impulse to liberate sufficient acetylcholine from the nerve terminals; in addition, the confusing effects of muscle action potentials and twitches are avoided, whilst propagation along the nerve is not impaired. The period of facilitation following

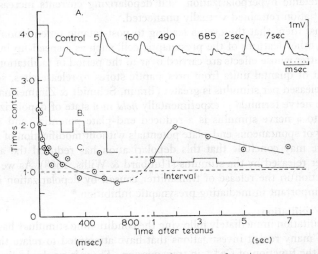

FIG. 607. The effect of 50 impulses at a frequency of 200/sec upon e.p.p. amplitude and m.e.p.p. frequency in a Mg-paralysed preparation. A shows records of testing e.p.p.s. elicited at the indicated interval (msec or sec) after the tetanus. In the graph C the average amplitude is plotted as a multiple of the control e.p.p. amplitude found in the absence of tetanic stimulation. Note the amplitude of testing e.p.p.s. was potentiated at intervals up to 400 msec after the tetanus, was depressed at longer intervals up to 1 sec, and thereafter potentiated again. The plot B is the probability of m.e.p.p. occurrence after 50 impulses at 200/sec, as a multiple of the control probability. Note the break in the abscissal scale between 800 msec and 1 sec and the similar breaks in the graphs. (Hubbard. *J. Physiol.*)

one or a few conditioning stimuli, and lasting some 100 msec. in the mammalian junction, is described as *primary facilitation* by Hubbard, in contrast with the classical *post-tetanic potentiation*, revealed, for example, by the overcoming of a neuromuscular block by repetitive stimulation. The two types of facilitation are illustrated by Fig. 607; the duration of the post-tetanic potentiation lasts seconds compared with the milliseconds of primary facilitation.

Post-tetanic Potentiation

Like the primary facilitation, this post-tetanic potentiation is presynaptic, the sensitivity of the end-plate membrane being unaffected (Hutter, 1952); and it was considered by Hubbard that the increased transmitter released in this condition was due to a hyperpolarization of the nerve terminals, since Hubbard & Schmidt (1963) showed that applied hyperpolarizing currents did, in fact, produce large

shown especially by the increased frequency of the miniature potentials immediately following a tetanus (Brooks, 1956).

increases in the end-plate potentials along with an increased spike-amplitude in the nerve. Evidence from other sources likewise supports the notion that hyperpolarization of the presynaptic terminals increases the availability of transmitter for release by a subsequent depolarizing spike. A number of studies have, however, cast doubt on Hubbard's hypothesis; thus Braun & Schmidt found that, after a tetanus, the size of the presynaptic action potential was reduced; in spite of this, the end-plate potential was increased. Again, Gage & Hubbard (1966) made precise measurements of both post-tetanic hyper-polarization (p. 1085) and the post-tetanic potentiation, and concluded that the differences in duration were sufficiently great to cast doubt on the existence of any close relation between the two; moreover, hyperpolarizing currents decreased the post-tetanic hyperpolarization, and depolarizing currents increased it, yet the potentiation remained virtually unaffected.

It seems, then, that the events leading to post-tetanic potentiation are to be sought in other features of the presynaptic cell, such as a speed-up in synthesis of transmitter whose effects are carried over to the period of facilitation, and/or a movement of quantal units from presynaptic stores to release-sites, so that the amount released per stimulus is greater (Braun, Schmidt & Zimmerman, 1966). When the nerve terminal is experimentally *held* in a state of depolarization, the response to a nerve stimulus is a reduced end-plate potential, and a reduced frequency of spontaneous end-plate potentials without modification of their size, so that we may conclude that this depolarization has reduced the amount of transmitter released by the stimulus (Hubbard & Willis, 1968). As we shall see, this reduction in the release of transmitter caused by depolarization is physio-logically important in mediating presynaptic inhibition.*

Primary Facilitation

The facilitation immediately following the conditioning stimulus† has been the subject of many recent investigations that have attempted to relate the pheno-menon to the function of Ca^{++} in transmission. There is no doubt that it is not the consequence of an increase in the presynaptic action potential, since the phenomenon may be observed when this is smaller than the initial, conditioning, impulse (Hubbard & Schmidt, 1963; Katz & Miledi, 1967).

Statistical Analysis

According to Del Castillo & Katz, facilitation is essentially an increase in the probability that a nerve impulse will release a quantum of acetylcholine from a terminal unit, so that if a given unit fails to respond to the first impulse its chances of responding to the second will be greater. As a result, the quantal content of the second end-plate potential will be increased, although the size of the quantum is unaffected. Facilitation is thus also revealed in the frequency of occurrence of spontaneous potentials. A further interesting feature is that facilitation may occur even though, because of high Mg^{++} at the end-plate, the conditioning stimulus has not caused any release of transmitter, suggesting to Hubbard that the effect is rather on the *availability* of acetylcholine for release

---

* Gage & Hubbard consider that some other hypotheses are also ruled out by their experiments, such as the postulated effects of increased intracellular Na^+ in the terminals, or increased K^+ outside; an increased supply of available transmitter is unlikely, since the amount is actually decreased (Liley & North, 1953; Elmqvist & Quastel, 1965), the actual augmentation of the end-plate potential being due to the release of a larger fraction of the depleted store. Cardiac glycosides and K^+-free solutions abolished post-tetanic potentia-tion (Gage & Hubbard, 1966). According to Elmqvist & Feldman (1965) ouabain increases the frequency of miniature potentials by a factor of 200 in the otherwise normal preparation.
† Even this primary facilitation seems to have two components; according to Mallart & Martin (1967) the decay of facilitation follows an initial rapid, and then a slow time-course suggesting two processes, with decay constants of 35 and 250 msec. respectively.

than on the ability of the impulse to release, *i.e.*, the effect is to mobilize acetylcholine from previously unavailable supplies.*

Facilitation and Depression

Mallart & Martin (1968) have compared facilitation in the Mg^{++}-blocked and curarized preparations; in the former instance release of transmitter in response to nerve stimulation is very small, whereas with the curarized preparation it is closer to normal, so that the effects of depletion may manifest themselves. They confined their study to the first rapid component of facilitation (p. 1168, footnote) and showed that, when the quantal content of the conditioning stimulus was varied, by varying the concentration of Mg^{++}, the facilitation decreased with increasing quantal content to reach a plateau at 10 quanta. In a curarized preparation the depressive effects of transmitter depletion became manifest, so that the effects of a train of stimuli could be resolved into facilitatory and depressive actions, and the interaction between the two could be predicted on the basis that facilitation was due to an increased probability of release of transmitter whilst depression was due to depletion. Thus the results oppose the theory that this early facilitation is due to an increased availability of transmitter since in this case a simple subtractive rule would govern interaction between facilitation and depression, the time-course of the amplitude changes in the end-plate potential predicted on the basis of subtraction not agreeing with that obtained experimentally.

Function of Calcium

Katz & Miledi (1965) pointed to the similarity between facilitation and calcium action; a single stimulus increased the probability of quantal discharge and so did application of Ca^{++}; this was very neatly demonstrated by applying a focal pipette containing Ca^{++} under iontophoretic control, close to a sensitive spot of a junction in Ca^{++}-free Ringer at the end-plate. By ejecting Ca^{++}, large focal end-plate potentials, containing as many as 10 quanta, were obtained; since a focal end-plate potential represents only 1–10 per cent. of the whole, this means that very large end-plate potentials would be developed by the whole junction. Since it seems likely that the release of acetylcholine requires the formation of some Ca^{++}-complex in the presynaptic membrane, facilitation may represent the presence of a residual amount of this complex persisting after the first stimulus. A critical test for the importance of Ca^{++} in facilitation was carried out by Katz & Miledi (1968) by the simple procedure of applying Ca^{++} iontophoretically at the end-plate by a single pulse just before the application of the conditioning stimulus. The end-plate receives two stimuli, N_1 and N_2; N_2 is the test stimulus and is applied first without N_1 and then with. The ratio of the responses gives the facilitation. If N_1, the conditioning stimulus, was applied first with a low external Ca^{++} and again when a high concentration was present, the facilitation under the latter condition was some twice that when there was a low external Ca^{++}. Presumably more Ca^{++} entered the presynaptic terminal and left a greater residue of the complex to facilitate the test response. When the intervals between test and conditioning pulses were reduced below 100 msec., the facilitation diminished, probably because the spike in the nerve terminal had not recovered, and at 10 msec. the system was absolutely refractory. When the impulses were eliminated by tetrodotoxin, a very high degree of facilitation at short time-intervals was revealed; thus with a pulse-duration of 0·9 msec., the mean quantal content was only 0·023; with a duration of 1·7 msec. the mean quantal content was 1·2, indicating a facilitation of 51. Katz & Miledi interpreted their results in terms of the calcium hypothesis along the lines indicated in Fig. 608; (a) indicates three depolarizing impulses, and (b) the corresponding rises in

* This primary facilitation tends to summate so that, if the interval between several successive stimuli is not too long, the facilitation following a burst of, say, 50 stimuli is considerably greater than after a single one. In the mammalian preparation of Hubbard, the duration of facilitation following a single shock was about 100 msec., consequently primary facilitation was not observed at frequencies below 10/sec.

calcium conductance leading to penetration of the ion into the nerve terminal; (c) shows the build-up of active Ca^{++}, Ca_a, responsible for transmitter release. The degree of facilitation with two pulses in rapid succession depends on the resulting accumulation of Ca_a, which depends on the decline of g_{Ca} after the first pulse and the removal of Ca_a from the critical sites. If the interval is very short, g_{Ca} will not have decayed to the base-line, and the calcium conductance during the second pulse will be greater than in the first, so that the amount of Ca_a may be more than doubled. Thus in Fig. 608 c, doubling the pulse width has trebled the amount of Ca_a and made it act for longer. Now the rate of the reaction, following the formation of Ca_a, that causes increased probability of transmitter release apparently varies as a power of the concentration of Ca_a; if this power factor were 4 very large facilitating factors would be obtained.*

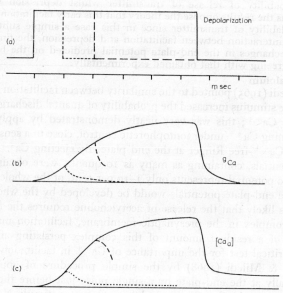

FIG. 608. Schematic presentation of "calcium hypothesis". *a*: Depolarizing pulses of 1, 2 and 5·5 msec duration. *b*: Rise and fall of "Ca conductance". *c*: Rise and fall of "active calcium" concentration. (Katz & Miledi. *J. Physiol.*)

Synaptic Vesicles

The presence of vesicles within the nerve terminals is a feature of all synapses, and there is little doubt that these vesicles contain the transmitter—acetylcholine in the case of the neuromuscular end-plate. It is tempting to speculate that it is the release of the contents of a single vesicle that determines the quantal packet, but as Hubbard & Kwanbunbumpen (1968) have emphasized, this is only speculation since attempts to relate synaptic function with alterations in the vesicle population have not been successful.

Relation to Junctional Folds. In their own study they examined with care the relation of the vesicles to the junctional folds. Thus, when they counted the

* On the basis of the calcium hypothesis, at high external concentrations of Ca^{++} the number of sites occupied by Ca^{++} after an impulse would be high, so that the possibility of facilitation should be reduced, if facilitation is, indeed, a sign of residual occupation of sites. To some extent this prediction was fulfilled by experiment, namely when the time-intervals between conditioning and test stimuli were short; at longer intervals the opposite was found. This was due to the altered time-course of facilitation at high Ca^{++} (Rahami-moff, 1968).

vesicles touching or actually fused with the axoplasmic membrane, the number opposite junctional folds was very significantly higher, 11·25 vesicles/μ compared with 1·7 away from the fold.* By bathing the preparation in Ringer containing 20 mM KCl the frequency of miniature end-plate potentials was increased, and this decreased the number of touching or fused vesicles opposite the junctional folds from a control value of 11·16/μ to 8·4/μ. MgCl$_2$ prevented this action of KCl. However, the decrease in vesicles was not peculiar to those opposite the folds since corresponding counts applied to large areas showed the same quantitative difference. In so far as an increased rate of firing of spontaneous end-plate potentials is accompanied by a change in the distribution of vesicles in the axon terminal this work supports the concept that transmitter is held in vesicles but, as the authors point out, it is not possible to equate the vesicle with the commonly employed parameters of transmission, namely n, the total number of available quanta, or with p, the probability of release.

Synaptosomes. Whittaker (see, for example, 1966) has separated a fraction from brain homogenates that he calls *synaptosomes*, membrane-bound bodies containing the synaptic vesicles and other inclusions. These could be burst osmotically to release the synaptic vesicles, which were collected as a separate fraction. The acetylcholine within the vesicles can be released from its "bound" form by heating, or treatment with acid; they also contain choline acetylase (Caman *et al.* 1965) and thus, most probably, the synthesizing system for the transmitter as well as the transmitter itself. It is interesting that the membrane of the synaptic vesicle contains a lower concentration of cholesterol and ganglioside than the membrane of the presynaptic nerve ending, so it is unlikely that the vesicle is formed by pinching off this membrane; it seems more likely that it is derived from the endoplasmic reticulum.

Membrane Fusion. Blioch *et al.* (1968) have argued that Ca^{++} might favour release of transmitter by reducing the ionic charge on the synaptic vesicles thereby favouring fusion of their membranes with that of the presynaptic terminal and thus permitting exocytosis. In this event the effect of hypertonic solutions, in increasing the frequency of spontaneous miniature potentials, observed by Fatt & Katz (1952) and confirmed by Blioch *et al.* in greater detail, would be attributable to the increased concentration of Ca^{++} within the terminal. Again, other divalent ions, such as Sr^{++} and Ba^{++}, and trivalent ions such as La^{+++}, might be expected to be effective if, like Ca^{++}, their influx into the terminal were increased during depolarization. These ions did, indeed, increase the frequency of spontaneous potentials occurring after a nerve stimulus, and they showed that the times for complete fusion of artificial lipid membranes, after being brought into contact, were inversely related to the valency of the added cation; thus in 100 mM K$^+$ the time was 16 seconds, in 0·1 mM Ca^{++} it was 5·2 seconds and in 0·01 mM La^{+++} it was 0·6 seconds. On this basis we should expect Mg^{++} to favour release of transmitter, yet it does the opposite in so far as it blocks the response to nerve stimulation; in the complete absence of Ca^{++}, however, Blioch *et al.* found that 5–20 mM Mg^{++} could replace Ca^{++} by restoring, to some extent, the rise in miniature potential frequency following a stimulus.

Invertebrate Junctions

Crustaceans. We may note that facilitation in the vertebrate neuromuscular junction may only be demonstrated in the blocked preparation; in the normal

---

* A further feature was the location of characteristic densely staining areas of axoplasmic membrane opposite the junctional folds; these areas had a strong probability of having a vesicle in contact with, or fusing with, them.

preparation a single impulse always causes a propagated spike in the muscle. An entirely different state of affairs is found in certain invertebrate systems, *e.g.*, in the anemone *Calliactis parasitica* (Pantin) or the leg muscles of crabs (Pantin; Katz, 1936). A study of the "excitatory junctional potentials" in the crab muscles by Katz and Kuffler has shown that summation occurs, such that, with repetitive stimulation, a potential five times as large as that evoked by a single stimulus may be obtained; moreover, whereas a single stimulus rarely induces an action potential in a muscle fibre, repetitive stimulation does so; the crustacean nerve-muscle system is therefore similar to the curarized vertebrate system, or to the normal vertebrate synapse (p. 1181). Wiersma (1941) suggested that this type of response was due to the presence of nerve endings widely distributed along the surface of the muscle fibres, so that the potential, recorded from the whole muscle, might well be the summated effect of a series of non-propagated junctional potentials. By recording from the inside of a single crab muscle fibre, Fatt & Katz showed that this interpretation was correct, the "end-plate potential" being obtained at any position of the electrode in the fibre, and its magnitude being barely affected by position.*

Effects of Alkali Earths. Fatt & Katz observed that the graded responses of crayfish muscles were converted into all-or-none spikes if Ba^{++} or Sr^{++} was substituted for Na^+ in the outside medium; it was considered possible, therefore, that these ions might carry the inward current necessary for the regenerative activity. More recently Werman, McCann & Grundfest and Werman & Grundfest have examined the effects of partial and complete replacements of Na^+ by these alkali earths on the behaviour of insect and lobster muscles. In both cases it was only necessary to replace about 10 per cent. of the Na^+ by Ba^{++} in order to convert the graded non-propagated response to an electrical stimulus into a propagated spike response. When all the Na^+ was replaced by Ba^{++}, spike activity still occurred, so that it does appear that the inward current may be carried by this ion; nevertheless, in solutions containing both Na^+ and Ba^{++} there was little doubt that the inward current was carried at least partly by Na^+. With relatively high concentrations of Ba^{++} very prolonged action potentials, lasting several seconds, were obtained, with a long plateau reminiscent of the Purkinje fibre (p. 1279); since the alkali earths increased the membrane resistance, it was concluded that a prominent element in their effect in converting graded to all-or-none responses was a decreased K^+-permeability or, as these authors preferred to call it, a K^+-inactivation process. Such a decrease would prolong the falling phase of the action potential leading to a plateau. Evidence indicating that Ba^{++} decreases K^+-permeability was given by the finding that the muscle fibre behaved less like a K^+-electrode in a high concentration of this ion.

Miniature Potentials. As with the vertebrate muscle fibre, spontaneous miniature potentials, of amplitude about 0·5 mV and occurring on the average about once per second, may be recorded from an intracellular electrode in the crayfish muscle fibre (Dudel & Kuffler; Reuben & Grundfest). The spontaneous potentials could be recorded from all positions of the microelectrode in the fibre; however, this was not so much because of the diffuse innervation but because of the much more extensive spread of depolarization in consequence of the small

---

* Hoyle & Wiersma (1958) have examined the excitatory responses in single fibres of a number of crustacean muscles; their results emphasize the diversity of response, although the general thesis that the junctional potentials build up as a result of recruitment of individual small responses is sustained. Hoyle & Wiersma consider that separate transmitter substances must be postulated for the nerve endings on the "fast" and "slow" fibres, *i.e.*, the fibres that respond with a twitch and sustained contraction respectively. The actual differences between the two fibres in terms of their electrical responses were difficult to define; spike activity is certainly not characteristic of the fast fibre.

space-constant of the crustacean muscle fibre. In order to localize their origin, therefore, an extracellular electrode was necessary; with this it was found that the spontaneous potential could only be recorded from those spots where an excitatory junctional potential was obtained in response to nerve stimulation. At a given site, the frequency of the spontaneous potentials was only about one per minute, compared with one per second with an internal electrode; this indicates that there are some 60 nerve-muscle junctions on a given fibre since, as indicated above, a discharge at any point on the muscle fibre will be recorded from an internal electrode. The time-course of the miniature potential was essentially the same as that of an excitatory junctional potential (e.j.p.), whilst the smallest externally recorded e.j.p. in response to a nerve stimulus was equal in amplitude to that of a spontaneous potential. A statistical study indicated, moreover, that the e.j.p. could be regarded as being made up of variable numbers of a single quantum, corresponding to a single spontaneous potential.

Procaine Spike. Fatt & Katz (1953), when studying the effects of cation substitutions on the electrical responses of the crustacean muscle to direct stimulation, found that replacement of Na^+ by choline gave a propagated action potential with an overshoot, whereas it was very difficult to obtain propagated responses in normal Ringer's solution. Tetraethyl ammonium (TEA) was even more effective, giving action potentials as high as 118 mV. Finally procaine, usually regarded as an anæsthetic and blocker of impulses, also permitted the production of propagated action potentials. According to Takeda (1967) this procaine-spike is unaffected by tetrodotoxin so that the inward current is apparently not carried by Na^+; raising the concentration of Ca^{++} increased the overshoot; soaking in a K^+-free medium also increased the overshoot, probably because the activation of K^+-permeability is responsible for the repolarizing phase.

Procaine blocks the invertebrate junction by reducing the size of the junctional potential; it also changes its shape so that it has a fast spike-like component followed by a plateau; according to Maeno (1966), the effects can be explained on the assumption of a reduction in the magnitude of the peak Na^+-conductance, so that the increased K^+-conductance repolarizes more rapidly; the increased Na^+-conductance is sustained for a longer time, however, and this leads to the plateau.

Facilitation. The facilitation described above, whereby the post-synaptic potential set up by one impulse is greater if it is preceded by one or more preceding (conditioning) stimuli, seems also to be the consequence of the liberation of more quanta at the nerve terminals in response to a nerve impulse, *i.e.*, the phenomenon is *pre-synaptic*. Thus Dudel & Kuffler analysed the e.j.p.'s obtained during stimulation by the statistical methods already outlined, and they showed that the larger facilitated potentials could still be described as multiples of the same unit quantum as in the smaller e.j.p.'s. Thus facilitation increases the probability that a terminal will discharge a quantum of transmitter. On this basis we might expect the frequency of the spontaneous potentials to increase after a conditioning nervous discharge; under certain conditions this increase could indeed be observed, although this was not easy because of the readiness with which a "fatigue" of spontaneous discharge sets in.

Insect Junctions. The skeletal muscle fibres of insects have anatomical and physiological features in common with those of the crustacea; both are innervated multiterminally and frequently polyneuronally; they are usually gradedly responsive. It was considered that the insects differed, in that their neuromuscular control did not involve peripheral inhibition (Hoyle, 1955). From the excitatory aspect we may note that Usherwood (1963) observed spontaneous miniature junctional potentials occurring randomly, whose frequency and amplitude were depressed by Mg^{++} and accentuated by Ca^{++}. As with crustacea, "cholinergic" drugs were without effect on transmission and on the miniature potentials. The inhibition has been studied in some detail by Usherwood & Grundfest (1965); only the slow muscle fibres receive inhibitory innervation, and this was revealed physiologically as the appearance of hyperpolarizing IPSP's; when these occurred before, or during, the response to the excitatory nerve, the EPSP was reduced (Fig. 609). The effects of inhibitory stimulation could be mimicked by locally applied GABA (p. 1219), and could be attributed to an increased permeability to chloride*.

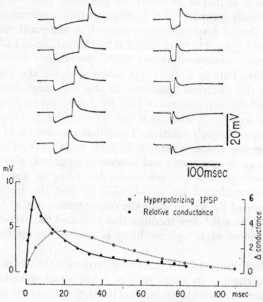

FIG. 609. Illustrating the duration of inhibitory action at the locust neuro-muscular junction. *Above:* An EPSP was evoked at various intervals after an IPSP was initiated. Inhibition was most marked when the EPSP arose slightly before the peak of the hyperpolarization caused by the IPSP (lowest record on right). *Below:* Comparisons of the time course of hyper-polarization during a single IPSP (open circles, *ordinate* on left) with the time course of the increase in membrane conductance. The increase is expressed in arbitrary units (*ordinate* on right) as the decrease in amplitude of a testing pulse which was caused by an intracellularly applied brief current of constant amplitude delivered at different times during the IPSP. (Usherwood & Grundfest. *J. Neurophysiol.*)

Electron-Microscopy. The minute anatomy of the arthropod nerve-muscle system has been described in detail by Edwards, Ruska & de Harven; the axon is surrounded by a lemnoblast, equivalent to the vertebrate Schwann cell, and this exhibits the typical mesaxon (p. 480): the outer surface of this lemnoblast has a thick basement membrane of a laminated cuticular material; at the nerve-muscle junction the basement membrane fuses with the muscle sarcolemma and the tracheoblast, whilst the nerve and its accompanying tracheole run longitudinally in a groove in the surface of the muscle fibre between the merged basement membranes and the plasma membrane of the muscle cell; at the synaptic regions the membranes are separated by only 120A, and the axon here contains vesicles of about 250A diameter and many mitochondria. The thick lemnoblast base-

---

* Picrotoxin is known to inhibit the action of vertebrate inhibitory nerves if these act presynaptically (Eccles, 1963), and thereby causes convulsions. Usherwood & Grundfest found that picrotoxin prevented the action of inhibitory impulses, and of GABA, so far as the IPSP was concerned.

ment membrane and the fused basement membranes at the junction probably constitute an efficient diffusional barrier that restricts exchanges between the extracellular fluid and the surrounding hæmolymph, a necessary condition in view of the enormous changes in ionic composition that may take place in many arthropod hæmolymphs (Hoyle, 1962).

Twitch and Slow Fibres

The distinction between invertebrate and vertebrate neuromuscular activity is not so sharp as was at one time thought; in the frog—but apparently not in the mammal—there are two types of muscle fibre, characterized by different types of innervation. The one most commonly studied is called the *"large-nerve"* or *"twitch"* system, and corresponds to the system already discussed; the nerve fibres supplying the muscle fibre are large and rapidly conducting, and terminate at typical end-plates; the effect of a single nerve stimulus is an all-or-none spike followed by a twitch. These fibres are responsible for fast contractions. The other type, called the *"small-nerve"* or *"slow"* system is, as Tasaki & Mizutani were the first to show, analogous to the invertebrate type, giving rise to slow contractions, and potentials lasting relatively much longer. Presumably they are concerned with the maintenance of posture.

Histology. Krüger (1952) emphasized the structural differences between the two fibres, the slow fibres lacking the M-band, having thicker Z-lines and a more diffuse fibrillar structure and described as possessing, in consequence, a *Felderstrucktur* by contrast with the *Fibrillenstruktur* of the fast fibre. Peachey & Huxley (1962) and Page (1965) compared the two in the electron-microscope; in the twitch fibres the myofibrils were less than 1μ across, with well delineated sarcoplasmic reticulum and numerous mitochondria. In the slow fibres the fibrils were less regular and fused together to form a more or less continuous mass in which isolated areas of sarcoplasm were interposed. The innervation of the two types of fibre is characteristically different; in the slow fibres, the motor fibres are small and end in characteristic bulbous structures on the surface of the muscle; these are densely grouped in some areas so that as many as ten could be seen within a length of 80 μ. These are the endings that have been described by the light-microscopist as "terminaisons en grappe" distributed along the length of the fibre by contrast with the fast end-plate confined to a localized region (Fig. 610, Pl. LVI). Thus the fast fibre relies for activation of the whole fibre on the spread of the action potential, whilst the slow fibre depends on the simultaneous activation of numerous grape-like endings, no one of which can induce a propagated disturbance. A feature of the slow ending is the virtual absence of junctional folds (Salpeter, 1967) and weak cholinesterase activity, this latter feature accounting for the prolonged action of applied acetylcholine.

Small-Junction Potentials. The potentials occurring as a result of small-nerve stimulation were studied with an intracellular electrode by Kuffler & Williams; because of the multiple innervation of the single fibres, the junctional potential recorded in the fibre was usually composite, rising in steps as the successive contributions from different junctions reached the electrode. The rising phase of a single step occupied some 2–3 msec., and it was followed by a restitution of the membrane potential with a half-life of some 23–39 msec., leading on to a phase of hyperpolarization (*i.e.*, positive after-potential) which finally subsided. The whole electrical change represented by this *"small-junction potential"* lasted some 0·4 sec. With repetitive stimuli the successive potentials summed, to give a plateau representing a maximal depolarization of some 30–35 mV; no propagated spike arose out of the small-junction potential. An interesting feature of the slow muscle fibre is that its resting potential is lower, being distributed about a mean of 60 mV compared with one of 95 mV for the twitch fibre. Again, the response to acetylcholine was characteristically different, being a slow and maintained contracture (p. 1136) compared with the transient contraction and subsequent neuromuscular block found with the twitch system.*

Equilibrium Potential

The slow muscle fibre seems, then, to be inexcitable, in the sense that a self-propagating disturbance cannot be initiated in it. This view is borne out by the work of Burke & Ginsborg, who inserted both stimulating and recording electrodes into a single slow fibre; they found that they could depolarize the fibre completely without inducing an action potential, the imposed catelectrotonus decaying in accordance with simple cable theory (p. 1042). When the membrane was hyperpolarized, the small-junction potential resulting from a nerve stimulus was greater than normal; when the membrane was partially depolarized the potential was reduced and at a certain limit—*the reversal level* —the small-junction potential was annulled. Beyond this, the junctional potential changed sign, *i.e.*, the effect of a stimulus was to repolarize the membrane. Thus, as with the end-plate, a stimulus tends to drive the membrane potential towards a certain equilibrium level which, according to Burke & Ginsborg, lies between 10 and 20 mV, an effect that could be achieved by a non-specific increase in permeability to all the ions of the system (p. 1143).

Miniature Potentials

As with the twitch fibres, intracellular recording reveals the presence of spontaneous miniature potentials which arise from nervous activity, since they are abolished by *d*-tubocurarine or atropine. According to Burke, a small-junction potential can be viewed as the result of a nearly simultaneous discharge of units that are spontaneously active but normally not synchronized. Because a slow fibre has at least 70 nerve-muscle junctions, the actual number of these units normally activated may be as little as five.

Mammalian Fast and Slow Muscles. The differences between twitch and slow fibres described above are peculiar to amphibian muscles. In mammals, however, the muscles have been classed as either fast or slow in accordance with their speed of response to a nerve stimulus, and consequently the frequency of stimulation necessary to provoke a tetanus (p. 1389). The slow muscles are relatively red because of their high concentration of myoglobin, and they are concerned with sustained postural contractions, by contrast with the "twitch-type" of muscle concerned in phasic movements.

Histochemical Differences. The difference between the two main types has prompted careful histological and histochemical studies of the constituent muscle fibres as well as of the mechanical responses of these. In general, two main types of fibre may be distinguished by virtue of their enzyme contents; Type 1 fibres are rich in oxidative enzymes such as succinate dehydrogenase, $NADH_2$ and diaphorase and probably depend mainly on the Krebs cycle for energy; these correspond to the slow fibres of red muscle. Type 2 fibres are rich in the enzymes of glycolysis, such as α-glycerophosphate dehydrogenase. These correspond to the fast fibres of white muscle.

Stein & Padykula (1962) separated the muscle fibres into three types, A, B and C on the basis of their succinase dehydrogenase activity and glycogen content. Types B and C with high activity and low glycogen content correspond to slow red fibres, C being differentiated from B by its large accumulation of succinic dehydrogenase activity under the sarcolemma. The fibres could also be distinguished on the basis of diameter, A being largest and C smallest. Romanul (1964), using a wider variety of histochemical tests, distinguished 8 types of fibre in gastrocnemius but only three in soleus; at one end of the spectrum was the white fibre with high capacity to utilize glycogen, low lipid and oxidative metabolism and myoglobin content, and

* Acetylcholine is assayed biologically by measuring the tension developed in the rectus abdominis muscle; it is because this muscle contains a large number of slow fibres that this is practicable, the intensity of the sustained contraction being proportional to the concentration of the applied acetylcholine.

at the other the red fibre with the opposite features. In general, as Henneman & Olsom (1965) have emphasized, the slow, red fibres are used more often and for longer times; they exhibit intense ATPase activity by contrast with the fast, white fibres; and the slowness with which they develop their tension is a reflection of the economical use of energy. The white fibres utilize glycogen and a predominantly anaerobic metabolism, fatigue easily but are capable of rapid and forceful activity. This latter feature follows not only from their individual sizes but from their organization into large "units"; being supplied by large motor fibres, these can branch more profusely and innervate more muscle fibres than the smaller fibres that innervate the red muscle fibres.

Motor-Units

Corresponding with these two or three types of muscle fibre, Close has identified in the skeletal muscles of the rat three types of motor unit, *i.e.*, mechanical responses to selective stimulation of the motor nerve fibres. Thus earlier work on the difference between slow and fast mammalian muscles was based on the assumption that these could be considered to be made up of a uniform population of a single fibre type, but Close (1967), by careful selective stimulation of the motor nerves, was able to distinguish in the "slow" soleus muscle some three out of thirty units that responded with an intermediate contraction-time of 18 msec., compared with 38 msec. for the long times of the remaining typical slow units. In the extensor digitorum longus, a

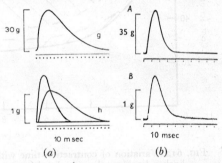

Fig. 611. (*a*) Isometric twitch of whole muscle (*g*) and low-threshold units (*h*) of the slow soleus muscle. (*b*) Isometric twitch of whole muscle (*A*) and a motor unit (*B*) of the fast muscle extensor digitorum longus. (Close. *J. Physiol.*)

fast muscle, the population was uniform, being made up of some forty fast units with an average contraction-time of 11 msec. It seems likely that the three types correspond with the three identified by Stein & Padykula, by histochemical techniques, their Type A being the typical fast fibre and the Types B and C being two kinds of slow or red fibre. Thus Close estimated that the two components of the slow soleus muscle contributed some 80 and 20 per cent. to development of the total tetanic tension whilst Stein & Padykula's section of the same muscle suggested that the C fibres constituted some 20 per cent. of the total cross-sectional area. An example of the difference in response of the whole muscle and of slow and inter-mediate units of the soleus is illustrated in Fig. 611, *a*. The homogeneity of the fast extensor digitorum longus is revealed by the corresponding records for whole muscle and a single unit (Fig. 611, *b*).*

Change during Development. The longer period of hyperpolarization of the slow nerves confines their activity to low-frequency discharges, and since the fusion frequency required for a tetanus in slow muscle is less than for a fast muscle, this means, in essence, that the nerve is matched to the muscle (Eccles, Eccles & Lundberg, 1958). The different innervation raises the problem as to whether the type of innervation determines the muscular response or *vice versa*. Buller, Eccles & Eccles examined the changes occurring during growth in new-born kittens; at birth all the muscles were slow, with a contraction-time of some 80 msec. After some 4–6 weeks the contraction-time of the fast muscles, such as lateral gastrocnemius, had

---

* A comparable study of the motor units of the cat's gastrocnemius and soleus muscles is that of Burke (1967); by intracellular stimulation of the motor neurones belonging to the units, Burke was able to characterize the features of the nerve impulse as well as the mechanical features of the muscle unit; thus the motor neurones innervating fast muscle units had faster axonal conduction velocity, shorter post-spike hyperpolarization and lower input resistances, although no combination of motor neurone properties alone was sufficient to separate the two types of motor unit unequivocally. In general, the fast units develop greater tension in a twitch but their tetanus/twitch tension ratios are smaller.

shortened to about 25 msec. whilst that of the slow muscles had passed through a minimum and finally, at about 10 weeks, reached a maximum of about 70 msec (Fig. 612). Essentially similar results were described later in a more complete study (Buller & Lewis, 1965) on the typically slow soleus and fast flexor hallucis longus; in addition to speed of contraction, the tetanus/twitch tension ratio and the rate of tension development in an isometric tetanus were studied, in an endeavour to isolate the mechanical features that are concerned in these ontogenetic changes. Whereas the maximum rate of development of tetanic tension increased to a maximum by 6 weeks, the tetanus/twitch ratio remained at the adult value from birth.

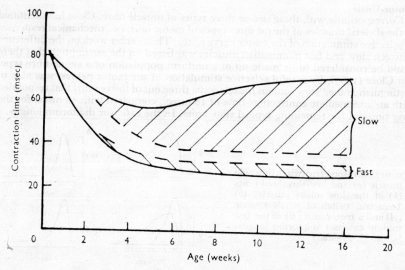

FIG. 612. Variation of contraction-time with age for slow and fast muscles. In broken lines are the postulated curves for these muscles in the absence of neural influences, so that the hatched areas indicate the effects of nervous activity. (Buller, Eccles & Eccles. *J. Physiol.*)

Responses during Regeneration

The behaviour of the nerve-muscle system during regeneration after cutting the nerve may have some bearing on this aspect, since the responses to nerve stimulation in the early stages are essentially the responses of newly developed nerve-muscle junctions. Maclagan & Vrbová (1966) found a marked increase in sensitivity to depolarizing blocking drugs such as decamethonium; in the newborn kitten, on the other hand, sensitivity is less than in the adult, so that this greater sensitivity is not due to the "newness" of the junction *per se*, hence the argument that the denervated muscle reverts to a more primitive type, and resembles foetal or neonatal muscle, may well be unsound.

Abolition of Nervous Influences. By cutting off all nervous influx to the cord, and therefore virtually abolishing all activity in the motor nerves to the muscles, the changes taking place in the fast muscles were barely affected, whilst the slow muscles became fast muscles, as indicated by the dashed lines in Fig. 612. Thus the influence of the nervous supply is chiefly manifest in maintaining slowness of contraction.

Cross-Innervation. By cross-innervation, so that the nerve to a fast muscle was made to grow into a slow muscle during regeneration, the slow muscle, *e.g.*, soleus, was transformed into a fast one so long as the cord was not deprived of afferent impulses; similarly, slow neurones converted fast muscles into slow ones. It could be argued that the bombardment of the muscle by, say, high frequency impulses over a period of time might be the cause of its adapting itself to respond to a high frequency of nervous discharge; the evidence, however, is against this mechanism since the cutting off of all impulses might be expected to act as a slowing mechanism to a fast muscle, yet this did not happen. It seems most likely, therefore, that the nerve, during activity, liberates some substance at the junction that modifies the muscle's contractile characteristics.*

Biochemical Changes

Examination of the biochemistry of the muscle fibre showed a reversal of enzymatic profiles; thus the slow soleus muscle of the rat or cat after cross-innervation by a fast nerve showed a preponderance of fibres with the high α-glycerophosphate dehydrogenase activity that is characteristic of fast fibres (Romanul & Van der Meulen, 1966). Again, Dubowitz (1967) cross-innervated the fast flexor hullucis longus of the cat with the slow soleus nerve; in every case there were regions on the muscle showing the usual pattern for this muscle, namely a chequer-board pattern of Types 1 and 2 fibres; in addition there were regions composed entirely of the slow Type 1 fibres and indistinguishable from those seen in normal soleus. The soleus muscle, cross-innervated by a fast nerve, was always completely normal, however. In kittens and young rabbits it was possible to find the fast pattern of muscle fibres on the slow soleus muscle after cross-innervation. Finally, Guth & Watson (1967) extracted the proteins from cross-innervated muscles and examined their electrophoretic profiles. Denervation *per se* caused the profile of the fast plantaris muscle of the rat to change to that of the slow soleus, a process that was reversed by re-innervation. Denervation of the soleus had no effect. When the soleus was cross-innervated by nerve that predominantly supplied fast muscles its protein profile came to resemble that of the fast muscles. There is little doubt, then, that the nerve can influence the biochemistry of the muscle fibre it supplies, but the mechanism remains obscure.†

Sub-Mammalian Species

In lower orders, such as birds and amphibia, the effects of cross-innervation are harder to demonstrate. Hník *et al.* (1967) cross-innervated the slow anterior latissimus dorsi of the chicken with the nerve to the fast posterior latissimus dorsi, and *vice versa*. The end-plates on the cross-innervated fast muscle had a distinct *en grappe* appearance with weak cholinesterase activity, but there was genuine multiple innervation with densely distributed endings in only two out of 155 fibres. Associated with this morphological difference it was found that the muscle would not now respond to a single stimulus unless this had been potentiated by tetanic stimulation; the mechanical response, when it occurred, was typically fast, so that the alteration is essentially presynaptic and due to the smaller amount of transmitter liberated by the new type of junction. The slow muscle innervated by fast nerve exhibited characteristic fast-type end-plates, with high concentration of cholinesterase; propagated action potentials occurred which were restricted, however, to small regions of the fibre. In the amphibian, the effects of cross-innervation are still more limited. Miledi & Orkand (1966) cross-innervated the mixed ileofibularis muscle of the frog with the fast sartorius nerve and found that acetylcholine now only produced a transient contracture compared with the prolonged one that occurs normally, but there was considerable variation in response. In the electron microscope the slow fibres were unchanged. Close & Hoh (1968) cross-innervated the fast sartorius of the toad and the mixed posterior semitendinosus but were unable to distinguish any alteration in the contractile properties; the contracture of the

---

* According to Vrbová (1963), mere cutting the tendon of the slow soleus muscle converts it into a fast muscle after some 3 weeks; the observed change was due to the abolition of reflex nervous activity and not to disuse, *per se*; thus the electromyograph indicates continuous impulse activity in this postural muscle. A critical examination of the changes in soleus following tenotomy by Buller & Lewis (1965) indicates, however, that whereas the twitch contraction of soleus is, indeed, speeded by tenotomy, it would be quite wrong to characterize the muscle in this state, as fast, as judged by several criteria. Vrbová (1963) observed that the fast muscles, such as tibialis anterior, could be slowed by continual use; she draws attention to an earlier observation that repetitive stimulation of a muscle will, after an initial decrease, cause the glycogen content to rise; this only occurs with stimulation through the nerve.

† The subject has been critically discussed by Buller & Lewis (1965) who emphasize that the change from one type of muscle to another is far from complete so that it might be best to characterize cross-innervated muscles as "hybrid". The effects of innervating a slow muscle with a fast nerve might be the result of decreasing the duration of its active state (p. 1394); this, other things being equal, would decrease the twitch tension and shorten the development and decay of tension. The effects of innervating a fast muscle by a slow nerve cannot be so simply explained, however.

cross-innervated fast muscle in response to acetylcholine was, indeed, larger and longer but retained its transitory character.*

CHELONIAN MUSCLE

The chelonian retractor capitis muscle is composed mainly, if not exclusively, of fast fibres with high acetylcholine sensitivity confined to spots. In the tortoise this sensitivity spreads to a greater extent than in the terrapin or amphibian fast muscles, but the difference is quantitative rather than qualitative. Thus it is possible, in the tortoise, to detect a level of acetylcholine sensitivity everywhere on the membrane of the innervated fast muscle fibre, so that we must describe the reduction in chemosensitivity that runs parallel with innervation during embryogenesis as being quantitatively less in the tortoise than in the frog. On this basis Levine (1966) concluded that the membrane of a twitch fibre was composite, being made up of both electrically and chemically excitable elements; and it is the proportion of the latter that is regulated by the motor neurone.†

LIZARD MUSCLE

Proske & Vaughan (1968) have made a combined histological and electro-physiological study of skeletal muscle of the blue tongue lizard, *Tiliqua nigrolutea*. About two-thirds of the impaled fibres responded with action potentials on nerve stimulation, whilst the remainder gave only a junctional potential. Subthreshold junctional potentials could also be obtained from the fibres responding with a spike, but these were characteristically faster than those of the non-spike fibres. Direct stimulation of fibres showing a resting potential greater than 65 mV resulted in action potentials, whereas those showing resting potentials of 45–55 mV failed to give an action potential even if they were pre-polarized to 80 mV. Histologically the fibres fell into two groups, the majority exhibiting the typical *terminaison en plaque* and the remainder the *en grappe* endings distributed over the fibre, the numbers varying from three to sixteen (Fig. 610, Pl. LVI).‡

Extraocular Muscles. Mammalian extraocular muscles, *e.g.*, those of the cat and man, contain typical large fibres with end-plate innervation, but in addition there are many fine fibres multiply innervated by the grape-like endings seen on amphibian slow fibres. These last give slow contractions and Hess & Pilar (1963) were unable to obtain propagated spike potentials in them; Bach-y-Rita & Ito (1966), however, were able to obtain spikes with marked overshoot in both types of fibre, and they attributed the "wave of depolarization" observed by Hess & Pilar to damage of the fine fibre as a result of electrode insertion.

Buckley & Heaton (1968) have estimated the cholinesterase activity at the two types of junction in rat and guinea pig extraocular muscles; here the distinction between end-plate and grape-like terminations is not so obvious, but the fibres fall into two categories with either focal (*i.e.*, single) or multiple innervation. The cholinesterase activity at the multiply innervated fibres was much less. It is considered that the slow fibre is concerned with movements of vergence, whilst the fast fibres are concerned with the rapid saccadic type of movement. It is these slow fibres that presumably give the extraocular muscle its sensitivity to acetylcholine (Duke-Elder & Duke-Elder, 1930).

---

* Close & Hoh point out that denervated slow muscle shows poorly developed acetyl-choline contracture and suggest that the transient contracture in the ileofibularis found by Miledi & Orkand might have been the consequence of denervation rather than of re-innervation by the fast sartorius nerve.

† Levine has discussed his findings on the chemosensitivity of the chelonian muscle fibre in the light of Nachmansohn's hypothesis that views acetylcholine as the transmitter in the propagated impulse over the whole surface of muscle or nerve. He rejects the hypothesis on the grounds that the muscle spike is the same wherever recorded, in spite of differences in sensitivity ranging over a factor of $10^5:1$; that the action potential recorded the end-plate is markedly different from that recorded elsewhere; that acetylcholine at receptors on non-end-plate regions are readily inhibited by curare but this does not affect the electrically excited action potential; and finally that there is no cholinesterase at non-junctional sites despite the fact that curare reaches the receptor structure.

‡ On the basis of this different innervation we may expect to find differences in ultra-structure of the muscle fibres; the authors state that preliminary studies indicate that the expected differences will be harder to detect than in snake muscle.

THE NERVE-NERVE SYNAPSE

The synaptic relationship between nerves is usually a more complex one than that between nerve and muscle although, as we shall see, it may be simpler. Thus, with the vertebrate "twitch-fibre" system, the synapse constitutes essentially the mechanism for transmission of impulses from nerve to muscle, and may be regarded simply as a transducing mechanism whereby the energy of the nervous impulse is transformed into the energy of the muscle impulse. The effect is normally all-or-none; either the impulse gets through or it does not.

Integration. In slow vertebrate muscle fibres, or in the invertebrate fibre, the junction is usually rather more than a transmitter, since it has the power of integrating the incoming impulses, building up a large or small junctional potential—and consequently inducing a weak or powerful contraction—in accordance with the frequency of the input from the motor nerve. Moreover, a given junction may be influenced by an inhibitory junction in its neighbourhood so that the integrative process takes note of both excitatory and inhibitory inputs. In nerve-nerve synapses the integrative process shown by the post-synaptic cell may be highly complex, in so far as a single post-synaptic neurone may make synaptic contact with many hundreds of pre-synaptic neurones, both excitatory and inhibitory, so that its condition of excitability at any moment is determined by streams of impulses arriving at different synapses some of which, by virtue of their position on the neurone, may be better able to influence the excitatory state than others. Thus, where these complex synaptic arrangements are met, we need not expect to find any obvious or subtle relationship between the input of a given pre-synaptic nerve fibre and the discharge from the post-synaptic neurone.

Autonomic Ganglion. The vertebrate autonomic ganglia contain synapses of a relatively simple kind, and because of their accessibility and the fact that they can be isolated and perfused with artificial fluids, they have been studied in detail. In so far as they illustrate the phenomenon of chemical transmission and the elementary integrative process of facilitation they may be briefly touched upon.

Facilitation. The stellate ganglion of the cat was studied by Bronk; here the post-ganglionic neurones make synapses with axons from several neurones derived from different nerve bundles, so that it is possible to observe the responses of these post-ganglionic neurones to stimuli reaching them at different synapses at the same or different times. Spatial summation is illustrated by Fig. 613 in which root A was stimulated alone, and together with stimulation of root B. The increased response when both roots were stimulated means that, whereas stimulation of root A or B alone failed to excite all the neurones in the ganglion, it nevertheless increased the excitability of those that failed to respond, with the result that the two together caused maximal discharge. The maximum facilitation is obtained when the stimuli are simultaneous, but the state of increased excitability may last as long as 200 msec., passing after this into a state of diminished excitability. If impulses from one group of fibres can facilitate the passage of impulses initiated in another, we can expect that successive impulses in one group of pre-ganglionic fibres may gradually build up an increasing response, *i.e.*, *recruitment* or *temporal summation* may occur. This does indeed happen, but there is a limit to which this build-up can extend; if the frequency of pre-ganglionic stimulation exceeds about 20/sec. in the stellate ganglion, the height of the post-ganglionic action potential decreases. A study of the action

potentials in single fibres in these circumstances shows that, when the preganglionic stimulation exceeds about 20/sec., the frequency of response ceases
to follow the frequency of stimulation; instead, the single fibres respond at
variable rates and so fail to give a *synchronized* response in the whole nerve;
this dispersion of the individual action potentials causes a diminished height
of the recorded action potential from the whole nerve.

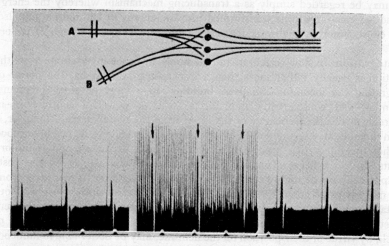

FIG. 613. *Above.* Schematic representation of the innervation of ganglion cells
by fibres from different roots. *Below.* First record: Postganglionic responses
to preganglionic volleys in root A. Second record: Arrows indicate
responses to similar volleys during concurrent repetitive stimulation of
root B. Last record: Stimulation of root A alone. Time: 0·2 msec. (Bronk.
J. Neurophysiol.)

Post-Tetanic Facilitation. As with the neuromuscular junction, facilitation
may be induced by tetanic stimulation; Larrabee & Bronk (1947) proved that the
increased synaptic excitability was a property of the presynaptic fibres, since
repetitive stimulation of one set of preganglionic fibres did not alter the response
of the same post-ganglionic neurones when activated by another set of preganglionic fibres. They showed, also, that occluding the blood-supply to the
ganglion, *i.e.*, interference with the nutrition of the post-ganglionic neurones, did
not affect the time-course of facilitation.

Chemical Transmitter. The chemical phenomena associated with transmission through an autonomic ganglion, such as the superior cervical ganglion,
are essentially similar to those described for the neuromuscular junction; and it
is because this ganglion can be studied in isolation, perfused with artificial
media, that a great deal of our knowledge on chemical transmission was acquired
in the first place. Thus, as we have already seen, stimulation of the pre-ganglionic
fibres leads to the appearance of acetylcholine in the perfusate. The artificially
circulated system studied by MacIntosh and his colleagues illustrated the
importance of the synthesis of acetylcholine, catalysed by the enzyme *choline
acetylase.* Thus, when perfusing with a saline solution, the release of transmitter,
measured per impulse, decreased during excitation of the pre-ganglionic fibres,
but if an appropriate supply of metabolites, such as glucose and choline, was
added, the synthetic process would keep pace with loss due to stimulation.
Of particular interest was the effect of hemicholinium which caused ganglionic
block because it inhibited the synthetic process, so that soon the stores in the

pre-synaptic terminals were exhausted, as revealed by the absence of acetylcholine in the perfusate.

Quantitative Aspects

Brown & Feldberg (1936) found that, with repetitive stimulation of the
preganglionic nerve, the amount of acetylcholine released into the perfusion
medium from the perfused ganglion fell progressively, to reach a plateau which
could be only some 20 per cent. of the original output.* Analysis of the ganglion
failed to reveal significant losses, so that resynthesis was apparently keeping pace
with release and the problem remained as to why the amount released should fall
to this extent. It appeared that the resynthesized acetylcholine was not "readily
available" for release by nerve impulses (Perry, 1953). Birks & MacIntosh (1961)
have made a careful quantitative study of the acetylcholine economy of the
ganglion. When breakdown was inhibited by eserine, the amount in the ganglion
steadily rose, indicating that the ganglion normally synthesizes continuously to
produce a surplus that is broken down. This surplus, when allowed to accumulate,
is not readily releasable since the amount released per stimulus is unaltered;
moreover, as soon as the inhibition of cholinesterase is overcome, the surplus
disappears. We may think of binding sites at which acetylcholine is held, ready
for release; if these are all full, then the synthesized material is surplus and is
removed by cholinesterase; presumably these sites are immune to the action of
the enzyme, *e.g.*, the sites may be within the vesicles. The basic synthesis was
never greater than 20 per cent. of that obtained during stimulation, so that clearly
the stimulus triggers both the release of acetylcholine and the process of its
synthesis. When synthesis of acetylcholine is blocked by hemicholinium,
prolonged stimulation leads to depletion of the stores of acetylcholine but there
is always a residue, unreleasable by stimulation, representing some 15 per cent.
of the total; this *stationary* acetylcholine may thus be distinguished from the rest,
which may be called *depot* acetylcholine. Analysis of the time-course of the release
during repetitive stimulation showed that this depot was composed of two
fractions, one smaller and more readily released than the other; the two appeared
to be "in series", in the sense that the transmitter lost by the first was replaced
by the second. Of the 260 nanograms in the ganglion, 40 consisted of stationary,
and 50 of readily available; each volley discharged about 1/1,200 of the smaller
depot fraction.

Presynaptic Action of Acetylcholine

Neuromuscular Junction. We have seen that anticholinesterases, such as
eserine or prostigmine, applied to the end-plate or given intra-arterially, potentiate
the action of acetylcholine liberated in response to a nerve stimulus and give rise
to an enhanced end-plate potential and to repetitive firing of the muscle revealed
as "eserine twitching" (p. 1141). It was considered that the effect was entirely
post-synaptic in the sense that the liberated acetylcholine acted for a longer time
and was able to induce end-plate potentials outlasting the refractory period of the
muscle. However, Masland & Wigton (1940) showed that during eserine twitching action potentials could be recorded from the motor nerve; these were not
affected by cutting the motor root and so could not have been due to activity in
the spinal cord. When time was allowed for degeneration of the motor nerve,

---

* More recently Nishi, Soeda & Koketsu (1967) have estimated the release during
repetitive stimulation; at 1/sec. $1\cdot6.10^{-16}$ g. of acetylcholine were collected per volley per
single ganglion cell; at 5, 10 and 100/sec. the amounts were $0\cdot9$, $0\cdot6$ and $0\cdot19.10^{-16}$ g. per
volley respectively. From the anatomy of the synapse it was deduced that a single synaptic
knob would liberate 2 or 5 quanta.

however, the potentials ceased, and Masland & Wigton suggested that the recorded action potentials were antidromically conducted spikes initiated in the nerve terminals on the muscle fibre; it was suggested that these terminals were sensitive to the acetylcholine liberated in response to the orthodromic discharge, and in this sense were analogous to the end-plate in their chemical sensitivity. Subsequent work of Werner (1960) supported this concept; thus the antidromic activity was greatly enhanced by tetanic stimulation of the nerve, in fact it could be initiated, under these conditions, without an anticholinesterase. Again, the latency could be so long that it was unlikely that the discharge was due to back-firing from the muscle. Finally, agents that depressed nerve activity leaving transmission relatively unaffected, such as procaine, could block the antidromic activity leaving transmitted responses unimpaired; Riker *et al.* (1959) likened the presynaptic response to a generator potential (p. 1252).

Autonomic Ganglia. Autonomic ganglia also exhibit a prolonged after-discharge after preganglionic stimulation, or after injection of an anticholinester-ase or of acetylcholine (Volle & Koelle, 1961), and it has been argued that the presynaptic nerve endings are responsible for this (Koelle, 1961) so that, in

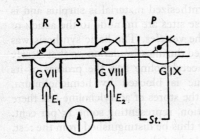

FIG. 614. Schematic drawing of the experimental arrangement for the sucrose-gap method employed for bullfrog sympathetic ganglia chain. R, S, T and P are Ringer solution, sucrose (224 mM) solution, test solution (Ringer solution containing certain drugs) and paraffin, respectively. E_1 and E_2 are calomel electrodes. St. represents the stimulator. G VII, G VIII and G IX represent the 7th, 8th and 9th sympathetic ganglion, respectively. (Koketsu & Nishi. *J. Physiol.*)

general, we may speak of the dual role of acetylcholine, activating the post-synaptic cell but also exerting an influence on presynaptic cells, causing them to release transmitter, which need not necessarily be the same as that primarily released. As Koketsu & Nishi (1968) have pointed out, however, in the normal ganglion no evidence for antidromic discharges, analogous with those in the motor fibres, has been produced. They recorded from the isolated paravertebral sympathetic chain of the frog, and the superior cervical ganglion of the rat, using a sucrose-gap technique designed to permit recording of preganglionic activity, that due to post-ganglionic cells being excluded (Fig. 614). Acetylcholine, applied through the test-compartment, T, caused a depolarization that was due to the preganglionic terminals rather than the axonal part of the fibres; this primary acetylcholine depolarization was followed by a slow repolarization. The significance of this presynaptic depolarization is not clear, since we shall see that after-discharges in these ganglia are probably triggered by post-synaptic potentials (p. 1200).*

---

* Repetitive stimulation of preganglionic fibres leads to post-tetanic facilitation, just as with the neuromuscular junction (p. 1167); Larrabee & Bronk (1947) showed that this was due to changes at the presynaptic terminals; and it is interesting that repetitive presynaptic stimulation profoundly affects the responses to intra-arterial acetylcholine (Volle, 1962), thus, once again, emphasizing a possible presynaptic locus for the action of acetylcholine in addition to its postsynaptic one. There was no antidromic discharge in the preganglionic fibres, in spite of the enhancement of the action of injected acetylcholine by the post-tetanic facilitation. It is worth recalling that Dempshire & Riker (1957) invoked a presynaptic action of acetylcholine to explain the antidromic discharges from the sympa-thetic ganglion of pseudo-rabies infected animals.

Electrical Events in Synaptic Transmission

Giant Synapse. The giant synapse of the stellate ganglion of the squid provides a chance for intracellular recording of the changes taking place in both post- and presynaptic cells, by contrast with other synaptic systems in which only the postsynaptic events can be recorded intracellularly.

The anatomy of the synapse was described by Young (1939), and the relevant physiological anatomy has been discussed more recently by Miledi (1967); there is one preganglionic nerve containing several giant fibres; and emerging from the stellar ganglion there are 8–11 postganglionic nerves each containing either one or two giant axons. One of the preganglionic giant fibres gives off a sequence of collateral branches, each of which enters into synaptic contact with one of the postganglionic giant fibres, forming a unique set of giant synapses; and it is these synapses that have been studied.

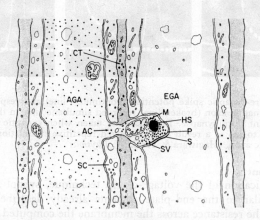

Fig. 615. Schematic drawing summarizing the fine structure of the squid giant synapse. *AGA*, afferent giant axon; *EGA*, efferent giant axon; *AC* collateral process from the afferent giant axon; *HS*, homogeneous dense substance; *M*, mitochondria; *P*, presynaptic membrane; *S*, postsynaptic membrane; *SV*, synaptic vesicles; *CT*, connective tissue sheath; *SC*, Schwann cell sheath. (Castejón & Villegas. *J. ultrastr. Res.*)

The relation between pre- and postsynaptic fibres is illustrated schematically by Fig. 615 from Castejón & Villegas (1964); as indicated, the presynaptic process contains numerous vesicles that presumably contain transmitter; the vesicles are of two types, one similar to those described by De Robertis & Bennett of 500–800A diameter, and the other of 800–1,000A diameter containing dense cores that are probably equivalent to those described in adrenergic terminals (p. 1155).

Synaptic Potentials

The early studies with extracellular recording by Bullock, and with an intracellular electrode in the post-synaptic neurone by Bullock & Hagiwara, demonstrated the existence of a synaptic potential similar to the end-plate potential of muscle, with a synaptic delay of some 1–2 msec. With electrodes in both pre-synaptic axon and in the post-synaptic axon immediately under the synaptic region, Hagiwara & Tasaki were able to demonstrate the complete insulation of the two neurones from each other; the response to the pre-synaptic impulse being an action potential that grew out of a synaptic potential which began after a definite synaptic delay. This growth of the synaptic potential is easiest to recognize in the fatigued preparation as illustrated in Fig. 616, where it is seen that there is no evidence of an electrically transmitted effect of the action potential in the post-synaptic neurone; as the pre-synaptic neurone becomes more and more fatigued a point is reached where transmission fails, in the sense that the post-synaptic response is only a synaptic potential; the highest potential that failed to induce an action potential was some 10–15 mV, and this corresponds to the threshold depolarization necessary to stimulate the post-synaptic axon directly through an intracellular electrode. The insulation between the two neurones was further demonstrated by applying large hyperpolarizations to the pre-synaptic neurone, or antidromic impulses to the post-synaptic neurone, and in no case was there evidence of spread of current from the one to the other.

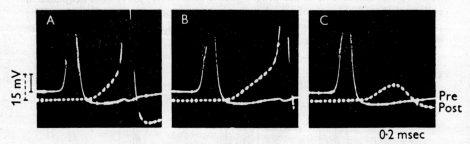

FIG. 616. Pre-synaptic spike potentials (continuous line) and responses of the post-synaptic axon (broken line) recorded simultaneously in the vicinity of the giant fibre synapse. Records were taken when synaptic transmission started to fail as a result of a prolonged repetitive stimulation of the pre-synaptic axon. (Hagiwara & Tasaki. *J. Physiol.*)

Synaptic Current

By the application of the voltage-clamp technique, synaptic currents were measured similar to the end-plate currents already described, and from a knowledge of the resistance across the membrane the computed synaptic potential that would have occurred agreed well with that actually measured under unclamped conditions. Thus the synaptic current passing inward at the synapse, as illustrated by Fig. 617, is adequate to stimulate the neighbouring axon membrane by causing an outward flowing current there.

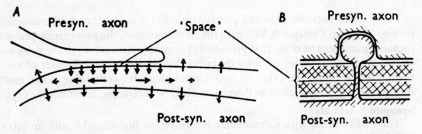

FIG. 617. A: Distribution of electric currents in the vicinity of the giant fibre synapse at the time when a synaptic potential is produced in the post-synaptic axon. B: simplified drawing showing the structure of the squid synapse; one axoplasmic process extending from the post-synaptic axon towards the pre-synaptic axon is shown. "Space" indicates the region where the d.c. potential is close to that of the surrounding sea water. (Hagiwara & Tasaki. *J. Physiol.*)

Equilibrium Potential

By applying different polarizing currents across the post-synaptic membrane, the latter could be hyperpolarized to different extents, and it was found that the size of the synaptic potential increased with degree of hyperpolarization in a linear fashion; this, of course, is to be expected if the effect of the transmitter is to produce an increased conductance to Na^+, K^+, Cl^-, etc., as postulated for the end-plate, the bigger the potential to begin with the bigger the fall when it is short-circuited. On extrapolating to zero synaptic potential, a value for the membrane potential of -60 mV was obtained, approximately equal to the resting potential. Thus the equilibrium potential of this synaptic system is in the region of zero (p. 1143).

Synaptic Delay

By carefully pushing the internal electrode in the post-synaptic axon into the synaptic region it could be made to emerge in the synaptic cleft, *i.e.*, the space between the two axons. Under these conditions the recording electrode now picked up a reversed synaptic potential,* and the electrical changes occurring at the post-synaptic membrane were recorded with the minimum of delay; the delay between the beginning of this synaptic potential and the peak of the pre-synaptic spike was between 0·3 and 0·6 msec.

Later Miledi & Slater (1966) recorded the arriving presynaptic spike and post-synaptic potential extracellularly, after localizing their electrode to a sensitive spot (p. 1148); the delay measured in this way was 0·5 to 0·8 msec.

Facilitation

Facilitation of the response to a stimulus by a preceding stimulus was a characteristic feature of transmission through the squid giant synapse (Takeuchi & Takeuchi); by varying the interval between the two stimuli, larger and larger junction potentials could be recorded, and these correlated well with the size of the action potential in the pre-synaptic axon, thus suggesting that the amount of transmitter liberated was proportional to this latter parameter. However, Miledi & Slater, (1966), whilst they confirmed the facilitation, were unable to correlate this with an increased presynaptic spike amplitude, in fact facilitation could be observed with a smaller amplitude of presynaptic spike. Bryant (1958) found that Ca^{++} facilitated, and Mg^{++} depressed, transmission across the synapse, and these ions were found to increase and decrease respectively the synaptic current measured by the voltage-clamp technique; they had no effect on the presynaptic action potential so that if their effect was to modify the liberation of transmitter it was exerted more directly. Miledi & Slater showed that in Ca^{++}-free Ringer the synaptic potential finally disappeared; using a Ca^{++}-filled micropipette to eject the ion iontophoretically, these authors found discrete sensitive spots on the postsynaptic axon where release of Ca^{++} permitted the appearance of synaptic potentials in response to presynaptic stimulation. Interestingly, injection of Ca^{++} into the presynaptic fibre failed to induce a synaptic potential, so that mere entry of Ca^{++} into the axoplasm is not sufficient to activate the transmitter-release mechanism.

Quantal Release of Transmitter

Miniature spontaneous synaptic potentials were not observed by Takeuchi & Takeuchi, but they pointed out that this might merely have been due to the size of

* The sign of the electrical changes will be reversed because of the different position of the active recording electrode in the two cases. When it is internal it is negative in comparison with the external one in the bathing medium, and becomes positive in relation to this as soon as the synaptic potential operates; the flow of positive current is thus from the outside neutral electrode *to the* internal recording one. When the active electrode is in the synaptic space both are at the same zero potential; with the onset of the synaptic potential, positive current flows *away from* the recording active electrode into the post-synaptic axon, *i.e.*, current flow is in the reverse direction. In the first case we should say the recording electrode became positive in relation to the neutral one in the bathing medium; in the latter case we should say it became negative, or a *sink* for positive current.

the postsynaptic fibre since work on the end-plate had shown that the recorded size of the end-plate potential was inversely proportional to fibre diameter, so that by extrapolating to the squid axon it could be computed that potentials of only 10–50 μV would be recorded, *i.e.*, close to the background noise of the instruments. Miledi (1967) used the smaller axons of the last stellar nerve, and was able to record spontaneous miniature potentials; they were usually very small—20–50 μV—whilst the largest were 0·5–1·0 mV. Presynaptic stimulation increased their frequency whilst blocking the axon with tetrodotoxin failed to influence them. Iontophoretically applied glutamate, which depolarizes crayfish muscle when applied to junctional spots (Takeuchi & Takeuchi, 1964), also depolarized the giant postsynaptic axon, the finding of local regions of high sensitivity suggesting a junctional action; and it is possible that glutamate is, indeed, the synaptic transmitter. At the giant synapse,

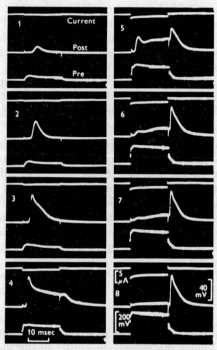

FIG. 618. Effect of prolonged depolarization of the presynaptic fibre of the giant synapse on the postsynaptic potential. Records 1 to 8 are with increasing pulse intensity. Note gradual change from ON- to OFF-response. (Katz & Miledi. *J. Physiol.*)

a nerve impulse can generate a synaptic potential of some 50 mV, suggesting that some 10^4 quanta are released at the synapse by the arriving impulse; this is high compared with the hundred or so at the end-plate, but in view of the different sizes, the density of release may well be comparable. When the axons of the synapse were blocked by tetrodotoxin, Katz & Miledi (1967d) found synaptic potentials in response to the electrotonically transmitted impulses whose heights were proportional to the degree of presynaptic depolarization.

However, when the depolarization of the presynaptic fibre was increased and prolonged* there was a striking delay and falling off of the postsynaptic potential, so that with long pulses of high amplitude the response was negligible until the pulse ceased, when the synaptic potential appeared as an OFF-effect (Fig. 618). On the basis of the Ca^{++}-hypothesis, the effect of the prolonged depolarization

* The prolongation was made possible by injecting TEA into the presynaptic axon; this reduced the delayed rectification and thus the outward current that would otherwise have occurred, which the microelectrodes would have been unable to sustain.

was predictable, since it would delay the entry of Ca^{++} into the axon by virtue of the internal electropositivity (p. 1161).*

Spinal Motor Neurones

Ventral Root Potential. The development of negativity in the spinal motor neurones was deduced from the pioneering experiments of Barron & Matthews who recorded with electrodes on the axons emerging from the spinal cord carrying impulses from the motor neurones (Fig. 582, p. 1128); after these neurones had been activated by stimulating the dorsal root, "ventral root potentials" were recorded, which appeared to be derived, by electrotonic spread, from the motor neurones; when these built up to a critical value, spike potentials appeared.

Synaptic Potentials. Intracellular recording has amply confirmed this view of the ventral root potential. Thus Brock, Coombs & Eccles inserted a pipette-

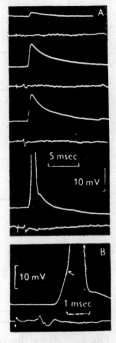

FIG. 619. Intracellular potentials set up in biceps-semitendinosus neurone by various sizes of volley in the afferent nerve (lower records). A shows synaptic potentials (upper records) of graded size, the largest setting up a spike. In B a faster record of this response is given, showing spike arising at arrow from initial synaptic potential. (Brock, Coombs & Eccles. *J. Physiol.*)

electrode directly into the soma of the motor neurone, and recorded the effects of reflex and antidromic excitation. Penetration of the motor neurone was recognized by the development of a resting potential which amounted, on the average, to 70 mV, although values as high as 80 mV were sometimes obtained. Weak stimulation of the pre-synaptic neurones caused the appearance of a synaptic potential (Fig. 619); on increasing the strength of this stimulus, the synaptic potential became larger until, eventually, when the depolarization was of the order of 8–15 per cent. of the spike action potential, *i.e.*, 6–14 mV, a spike was initiated. Presumably the membrane of the motor neurone is depolarized locally at the synaptic terminals by the liberation of some transmitter substance;

* Kusano, Livengood & Werman (1967) increased the postsynaptic potential at the squid giant synapse by injecting TEA into the presynaptic axon; this was presumably the result of the prolongation of the spike and increase in its amplitude, due to the reduced delayed rectification caused by TEA. Kusano (1968) has examined the relation of the pre- to the postsynaptic potentials at this synapse in some detail; the minimum presynaptic potential required to produce a postsynaptic one was lower than 25 mV.

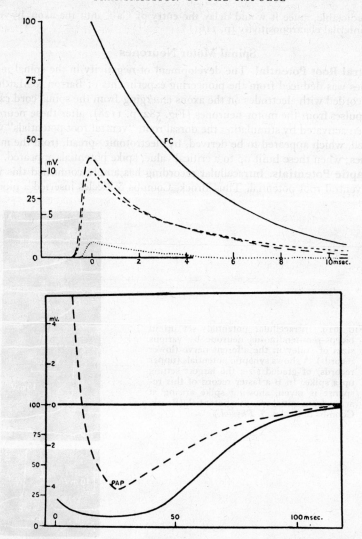

FIG. 620. *Above*: Broken and dotted lines show three post-synaptic potentials recorded from a motor neurone intracellularly. A standard facilitation curve, FC, taken from Lloyd's work, is also drawn with ordinates as percentage of maximum facilitation, which is at zero interval between the condition and testing volleys. *Below*: Broken line gives typical positive after-potential, PAP, recorded intracellularly when a motor neurone is activated by an impulse at zero on the time-scale. Ordinates: potential change in mV relative to resting potential. Continuous line gives measure of excitability of the motor neurones, following their anti-dromic activation, plotted on same time-scale with ordinates the size of testing monosynaptic reflex spike as a percentage of its control size. (Eccles. *Cold. Spring Harbor Symposia*.)

a weak stimulus to the nerve excites only a few sensory fibres so that the total area of depolarization is insufficiently large to cause a propagated spike, and the depolarization decays electrotonically, *i.e.*, in the same way as an applied negative charge would decay by virtue of the core-conductor properties of the neurone. With a stronger stimulus, the increased number of activated fibres permits sufficient spatial summation of the localized depolarizations to give a spike. Following the spike, there is a prolonged state of hyperpolarization—or positive after-potential—and it is interesting that this corresponds, in its time-course, to the period of reduced excitability of the motor neurones as determined by their response to a second stimulus (*vide, e.g.*, Lloyd, 1946). The rise and fall of the synaptic potential are slow processes compared with the spike, so that a first stimulus, which itself fails to excite, may facilitate a second stimulus by addition of synaptic potentials. Fig. 620 shows the correlation between the time-course of the synaptic potential and that of the process of reflex facilitation; in the same figure the correlation between the after-potential following a spike and the period of reduced reflex excitability is shown.

IS and SD Spikes

An analysis of the course of the action potential recorded by an internal electrode in the motor neurone has shown it to be complex, and to be determined by the differing excitabilities of two main zones, namely the initial unmedullated segment of the axon including the axon hillock (IS) and the soma and dendrites (SD). The complexity was first revealed in the antidromic spike, elicited

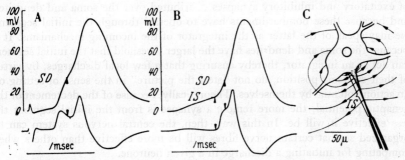

FIG. 621. Intracellularly recorded spike potentials evoked by antidromic (A) and monosynaptic (B) stimulation of a motor neurone. The lower traces show the electrically differentiated records. C shows diagrammatically the lines of current flow that occur when a synaptically induced depolarization of the soma-dendritic membrane electrotonically spreads to the initial segment. (Eccles. *Ergebn. d. Physiol.*)

by stimulating the motor axon, which passed back to the body of the neurone; as Fig. 621 shows, there is an initial sharp rise, the IS-spike, which is followed by a slower rise, the SD-spike; when excitability is reduced, only the first step may appear and it was argued by Brock *et al.* that the spread of excitation antidromically would have a low safety-factor—and would therefore be subject to block easily—at the junction between the unmedullated portion of the axon and the soma of the neurone; they therefore considered that the initial step on the spike was due to activity in this non-medullated region.

Flow of Current

Subsequent studies on the normally activated potential, when recorded at sufficient speed, showed that here, too, there was an inflexion on the rising phase corresponding to the IS-spike (Fig. 621). Thus, the course of events during normal activation of a motor neurone is probably as follows: the arrival of the afferent impulses at the synaptic knobs on the soma and dendrites develops a sufficiently high synaptic potential to act as a sink for current-flow in the directions indicated by Fig. 621, the flow being from the initial segment towards the depolarized regions; because the initial segment has a much lower threshold for spike activity than the soma and dendrites, this outward flow of current from the initial segment induces the IS-action potential, so that this region rapidly becomes more negative than the soma and dendrites; current-flow is now in the reverse direction to that indicated by Fig. 621 and this induces the SD-spike in soma and dendrites. By applying voltage-clamp techniques, along the lines indicated earlier for the end-plate, Araki & Terzuolo have confirmed the general thesis that the SD-spike does, indeed, arise from the soma where the threshold for excitation is high. Thus, by clamping the somal membrane at its resting potential, and then imposing a transient depolarization, a current passed through the clamp, after a latency, that corresponded exactly with what would be expected were the initial segment to have fired and the soma not to have reached threshold.

The timing of the impulse in the medullated axon indicated that it originated just before the onset of the IS-spike and thus was presumably initiated at the first node by a flow of current comparable with that flowing from the initial segment* (Coombs, Curtis & Eccles, 1957b).

Initial Segment as Integrator

The motor neurone is subjected to bombardment through large numbers of excitatory and inhibitory synapses distributed over the soma and dendrites, and because these bombardments have to operate through the initial segment, we may speak of the latter as the integrator of the incoming mechanisms. It is because the soma and dendrites have the larger threshold that the initial segment can act as an integrator, thereby ensuring that a few local discharges, by virtue of their strategic position, do not "steal the picture" in the sense of setting off an action potential by themselves. Theoretically, because of the decrement of the synaptic potential, the more remote a synapse is from the initial segment the less effective it will be. In this way, then, the central nervous system can be organized so that certain nerve fibres will be more effective than others when competing for initiating a discharge in a given neurone.

In general, the studies of Coombs, Curtis & Eccles on the cat's motor neurones, and of Machne, Fadiga & Brookhart on frog motor neurones, have shown that the initial segment has the lower threshold; in the cat the IS-threshold was 5–18 mV, with a mean of 10·6 mV, and for the SD-spike 19–37 mV, with a mean of 25 mV; in the frog, mean values were 9 mV and 26 mV respectively. With many other types of neurone, however, the transition from IS- to SD-spike is not so obvious, and in some there is no evidence of a transition at all (see, for example, Eccles, Eccles & Lundberg, 1960).

As indicated above, the excitatory synaptic potential has an indirect action in initiating a spike, since it arises on the soma and dendrites; in *Aplysia*, Tauc has shown that the impulse arises in the axon rather than the soma not only because of its lower threshold but also because the synaptic endings are there.

Nature of the Synaptic Potential

In line with studies on other synaptic systems, we may treat the excitatory synaptic potential of the motor neurone as a short-circuiting of the resting potential. This view is substantiated by the thorough investigation of Coombs, Eccles & Fatt, who inserted a double-barrelled electrode; current was passed

---

* The induction of the SD-spike reveals that the somal and dendritic membranes are electrically excitable, allowing very high transmembrane currents; moreover, as Fatt (1957) and later Nelson, Frank & Rall (1960) have shown, antidromic impulses pass from the axon over the soma and along the dendrites of motor neurones. Since these membranes are excited by chemical transmitters, they would appear to be exceptions to Grundfest's generalization to the effect that membranes excited chemically are electrically inexcitable (p. 1155). Whilst there is little doubt that the soma and dendrites have a higher threshold for electrical excitation than the axonal segment, there is no justification for assuming an absence of electrical excitability (Eccles, 1961).

through one, thereby altering the membrane potential, whilst recording of the synaptic potential was done through the other. Depolarization of the membrane decreased the synaptic potential following nerve stimulation; if the membrane potential was reversed, then the nerve stimulus caused a reversal of the synaptic potential, *i.e.*, the internal electrode, instead of becoming less negative became more negative. The level of membrane potential at which this reversal in sign of the synaptic potential occurred was about zero, and this represents the equilibrium potential. The absence of specific effects of ions injected into the neurone was in accord with the view of the synaptic potential as a generalized increase in membrane permeability.*

Transmitter Action

On this view, then, excitatory impulses, reaching the synaptic terminals, cause a brief "transmitter action" which consists of a short-circuiting of the post-synaptic membrane; it probably does not last much longer than 1 msec., but its effects persist as a synaptic potential. According to the frequency with which impulses reach a given synapse, and according to the spatial distribution of the synapses that are activated, this synaptic potential may build up to a sufficient height—about 10–15 mV—to initiate an IS-spike which will invade the soma as an SD-spike, whilst a spike is propagated along the axon to the muscle fibre.

"Spontaneous Activity"

In the absence of afferent stimulation to the motor neurones, Brock *et al.* observed a background of small potentials, described by Eccles (1961) as "synaptic noise". These miniature synaptic potentials could be the analogue of those at the end-plate, representing release of a single quantum of transmitter at a single ending, although Brock *et al.* considered that a single quantal event taking place in the motor neurone, with its large number of synaptic knobs, might well be too small to be detected; and it seemed more likely that the background was due to synaptic bombardment by discrete propagated impulses reaching the neurone from its multiplicity of central connexions. Katz & Miledi (1963) blocked afferent activity in the isolated frog spinal cord by treating it with Mg^{++} or high concentrations of K^+, and found that spontaneous activity still occurred, the recorded potentials having an amplitude of less than 2 mV and occurring in random sequence; the variety in amplitude and time-course was far greater than at the end-plate, presumably because of the great variety of locations of the synapses. By blocking presynaptic activity more effectively with tetrodotoxin, Colomo & Erulkar (1968) again found no appreciable reduction in the frequency, average amplitude, and amplitude histogram of the spontaneous miniature potentials. Thus, so far as the frog is concerned, synaptic noise is due to spontaneous release of transmitter rather than to the bombardment of the motor neurones with presynaptic impulses. In cats, Kuno (1964) measured the responses of a motor neurone to repeated just-threshold stimuli through a nerve making a monosynaptic Ia connexion, and analysed the magnitudes of these responses by the statistical methods already employed at the end-plate. The analysis was not so accurate, since an independent estimate of the quantum of activity could not be made; thus at the end-plate this is the smallest miniature potential recorded, but there is no certainty that spontaneous potentials are the result of activity in the particular nerve fibre that is being stimulated. However, the mean number of quanta, *m*, calculated from the failures of responses on the basis of a Poisson series, showed a linear relation with the mean amplitude of the observed EPSP's, and the entire distribution of the sizes of the EPSP responses was in good agreement with Poisson's Law. Thus the monosynaptic EPSP can be regarded as being made up of units, each with a small probability of responding to an afferent impulse. The unit

* Later studies agree in showing that the peak amplitude of the excitatory synaptic potential does not increase with hyperpolarization (Nelson & Frank, 1967), whilst there is no clear-cut equilibrium potential with strong depolarization; this is probably because the distance of the recording electrode (in the soma) from the site of the synaptic potential (on the dendrites) is high (Smith, Wuerker & Frank, 1967; Ginsborg, 1967); an additional factor is undoubtedly the anomalous rectification that causes membrane resistance to decrease with hyperpolarization (Nelson & Frank, 1967).

size lay between 0·12 and 0·24 mV. Using this value, and a critical firing level of 10 mV, Kuno computed that some 50 to 100 fibres would be required to produce a reflex discharge in a motor neurone. When impulse activity in the cat's cord was blocked by local application of tetrodotoxin, Hubbard, Stenhouse & Eccles (1967) reported a virtually complete cessation of "spontaneous activity" and concluded that the synaptic noise was not due to spontaneous release of transmitter. Blankenship & Kuno (1968), using both intravenous* and local application of the toxin, confirmed that spontaneous activity was reduced, but they did find a residual activity which might well have been due to spontaneous release analogous with that at the end-plate and at the frog's motor neurone; at any rate its amplitude-distribution was that of a nearly Gaussian type with a mean of 0·2 mV. Thus "spontaneous" potentials with relatively large amplitude (greater than 0·4 mV) would seem to be the result of synaptic nerve activity, whilst the residual activity, after poisoning with tetrodotoxin, is the equivalent of the spontaneous potentials at the end-plate.

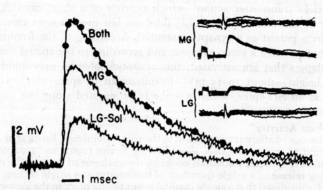

FIG. 622. Illustrating summation of EPSP's evoked in the motor neurone. Superimposed tracings of EPSP's evoked in a gastrocnemius motor neurone by volleys in the LG-Sol and MG nerves separately, and then together simultaneously. Each trace is the average of 25 successive responses, averaged in a CAT 400B computer and plotted on an X-Y plotter. Large dots denote the algebraic sum of the LG-Sol and MG responses. (Burke. *J. Neurophysiol.*)

Composite Nature of the EPSP

Normally the motor neurone will receive impulses from different nerve fibres, and it is interesting to examine their single and combined effects. Burke (1967) showed that when records were taken from a gastrocnemius motor neurone, and the afferent monosynaptic inputs were from the lateral-gastrocnemius-soleus and medial gastrocnemius nerves, there was a remarkably linear summation, as illustrated by Fig. 622. This apparent absence of interaction could be attributed to the large spatial separation of the evoked potentials (Rall, 1967). The absence of recorded impedance change with more than half the EPSP's (Smith, Wuerker & Frank, 1967) supports this assumption since the recording electrode is necessarily in the soma of the motor neurone, and if the EPSP is evoked on a dendrite the effect on impedance will not be measurable although the current-flow, giving rise to the EPSP, will.† An additional obscuring factor was anomalous rectification of the type described by Kandel & Tauc (1966) in the molluscan neurone; here a failure of the EPSP to increase in amplitude with hyperpolarization is due to a decrease in membrane resistance as the membrane hyperpolarizes; such an anomalous rectification in the motor neurone was demonstrated by Nelson & Frank (1967).

---

* Slow intravenous injection into the cat to give a final dose of 20–80 μg./kg. was employed; apparently vital centres are spared, whilst activity in motor and sensory neurones is blocked. Colomo & Erulkar (1968) have suggested that Blankenship & Kuno had not completely blocked impulse activity, so that their residual synaptic potentials were not spontaneous; it could be, as they argue, that spontaneous activity in the cat is of such low amplitude as not to be resolved from the instrumental background noise.

† Smith *et al.* found impedance changes with all inhibitory postsynaptic potentials, suggesting that these arose on the soma.

Autonomic Synapse

The essential features of the autonomic ganglion as a relay system have been already analysed; we may now consider the events taking place in the post-synaptic cell as a result of the arrival of the presynaptic impulse; in general, these bear a very strong analogy with the events in the end-plate region of the muscle fibre, so that a full description of the many experiments would mean a tedious repetition of principles already established.

Synaptic Structure. The synaptic arrangements in the superior cervical ganglion of the cat have been described by Elfvin (1963); here the ganglion cells have complex dendritic processes round which the presynaptic fibres wind, forming several synaptic contacts; at these points the nerve fibre widens and here its axoplasm is seen to contain numerous synaptic vesicles; small ones of 300–500A diameter and larger ones, 700–1,000A, with a central dense core. The membranes of pre- and postsynaptic cells are separated by a space of some 60–80A. In addition to these "classical" synaptic contacts Elfvin saw regions of apparent membrane fusion suggestive of electrotonic contacts (p. 1216). In the amphibian, the ganglion cells are simpler, having no dendrites, so that synapses are with the axon and soma; the presynaptic fibre winds round the axon and cell body making frequent synaptic contacts (Taxi, 1965). As with the mammalian ganglion, the presynaptic terminal contains numerous vesicles both small and large, the latter containing a dense core and considered, *e.g.*, by Uchizono (1964), to contain the adrenergic transmitter noradrenaline.* Within the cytoplasm of the postsynaptic ganglion cell there are specializations in the form of dense plates of material parallel with the synaptic membranes. Using the usual acetylcholinesterase localization techniques, Brzin, Tennyson & Duffy (1966) showed that reaction-product is probably first deposited on the internal surfaces of the membranes of the endoplasmic reticulum and projections of this into the cisterns; later the product appears in the synaptic cleft, but resolution was not good enough to determine with which membrane it was associated.

Avian Ciliary Ganglion

In the bird's ciliary ganglion there are two types of synaptic contact; a calyx embracing the ganglion cell near the axonal pole and sending out large finger-like processes extending for considerable distances over the cell body (Fig. 584, p. 1130); in other cells the presynaptic fibres end in diffusely branching terminal networks making a "bouton" type of contact (Martin & Pilar, 1963). As we shall see, the former, caliciform, synapse is associated with an electrotonic, as opposed to chemical, type of transmission.

Electrical Events. R. M. Eccles (1955) inserted microelectrodes into the cells of the rabbit's isolated superior cervical ganglion; she observed resting potentials in the range 65–80 mV which, however, were maintained for only a few minutes, possibly because of the leakage of KCl from the electrode into these small cells (25 μ diameter). The response to preganglionic stimulation was a slowly rising synaptic potential, out of which the spike emerged; this was said to be followed by periods of hyperpolarization, depolarization and hyperpolarization, succeeding each other. Usually, the threshold synaptic depolarization leading to a spike was about 15 mV; in a few cases synaptic potentials, uncomplicated by a spike, could be obtained, in which event the synaptic potential was some 30 mV, lasting some 20–60 msec.†

* Taxi (1965) failed to find any change in these large vesicles in the reserpinized animal so prefers to call the vesicles neurosecretory rather than specifically adrenergic.
† Erulkar & Woodward (1968) have succeeded in maintaining electrode penetration of the rabbit's ganglion cells for long periods, so that many further features of their activity could be examined. It is interesting that, whereas preganglionic stimulation gave rise to multiple spike responses, direct stimulation of the ganglion cells gave only single spike responses; this suggests that the repetitive response is due to the arrival of successive impulses dispersed in time, perhaps through internuncial connections, although such have not been demonstrated histologically. Alternatively, the dendritic character of many of the synapses, described by Elfvin, may be responsible for multiple responses to a single presynaptic volley.

In the frog, Nishi & Koketsu (1960) studied the action potentials in impaled postsynaptic cells, following both antidromic and orthodromic stimulation; and the difference between the two responses could be attributed to the action of a transmitter substance, reducing the membrane resistance, *i.e.*, increasing its permeability to ions, in the case of orthodromic stimulation. These results, and those of Blackman, Ginsborg & Ray (1963), were sufficiently similar to those obtained by Katz and his colleagues on the end-plate to make a detailed recapitulation unnecessary. The successive changes in the membrane potential of the postsynaptic cell during the evolution of the spike are shown schematically in Fig. 623; the spike rises from a synaptic potential analogous with the end-plate potential; following the spike there is a wave of hyperpolarization, which passes into a period of negative after-potential; this last is apparently due to persistence of transmitter action, and its long duration may be due to the absence of adequate amounts of strategically placed cholinesterase in the cleft, although this is by no means proved. The hyperpolarization, or positive after-potential, may be

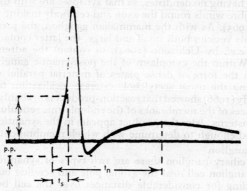

FIG. 623. Illustrating the successive phases of the orthodromic response of a postsynaptic cell of the sympathetic ganglion of the frog. p.p., peak positive phase; S, synaptic step; t_s duration of synaptic step; t_n, time to peak of negative wave. (Blackman, Ginsborg & Ray. *J. Physiol.*)

analogous with the delayed rectification of the squid axon, and due to a persistence of high permeability to K^+ that brings the membrane potential close to the K^+-potential. If this is true, *i.e.*, that the resting potential is less negative than the K^+-potential, the reversal potential, *i.e.*, the membrane potential at which the effect of transmitter changes sign, should be negative; in fact Nishi & Koketsu found a mean value of 14·8 mV when they measured the synaptic potential as a function of the membrane potential of the postsynaptic cell.

The fact that curare blocks synaptic transmission through the autonomic ganglion suggests that the transmitter is acetylcholine; and Blackman *et al.* were able to evoke synaptic potentials by its iontophoretic application.

Spontaneous Potentials

Nishi & Koketsu observed spontaneous miniature potentials, and Blackman *et al.* (1963c, d) examined these along the lines employed by Katz, Liley, and others on the end-plate; their frequency was increased strikingly by raising the concentration of K^+, and statistical analysis of the responses to preganglionic stimulation under conditions of partial block (*e.g.*, with high Mg^{++}- and low Ca^{++}-concentrations) suggested that, as at the end-plate, these could be compounded of multiple miniature potentials or quanta. When successive stimuli were applied at short intervals, the response to the second stimulus was reduced, and if the interval was short enough only a synaptic potential developed (Hunt & Riker, 1966).

Denervation

We have seen that denervation of the muscle leads to block of the junction, with disappearance of spontaneous end-plate potentials within 4–5 days; this is associated with degeneration of the nerve terminals on the end-plate (p. 1154). In the autonomic ganglion the block occurs sooner—24 to 48 hours—with a similar disappearance of spontaneous miniature potentials (Hunt & Nelson, 1965); even after 30 days, however, there was no reappearance of these potentials, by contrast with the frog end-plate. In the electron microscope, block was seen to be associated with clumping and disappearance of the vesicles and enlargement of the mitochondria. Early in the course of degeneration, the Schwann cells came into approximation with the postsynaptic cell and appeared to engulf the residues of the degenerated nerve terminals. The changes in the postsynaptic cell after axotomy have been studied by Hunt & Riker (1966); the most striking feature is the post-tetanic potentiation that becomes manifest; this normally cannot be shown, since a presynaptic impulse nearly always evokes a postsynaptic spike; the decrease in synaptic potential associated with axotomy now permits potentiation.

Double Type of Transmission

The ciliary ganglion relays impulses entering along the oculomotor nerve to postsynaptic neurones emerging as the ciliary nerves. In the bird, as we have seen, there

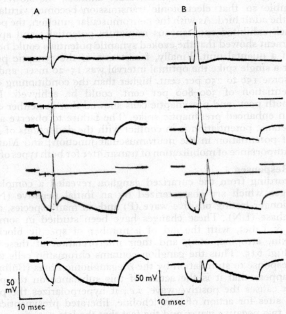

FIG. 624. Illustrating transmission at ciliary ganglion of the chicken. Response of two ganglion cells (*A* and *B*) to preganglionic stimulation. Upper records with no current pulse, subsequent records with hyperpolarizing current pulses of increasing amplitude. In the lower record of *A* spike fails, leaving only e.p.s.p. In *B*, e.p.s.p. is preceded by a faster potential change (coupling potential) which at resting membrane potential initiates spike. (Martin & Pilar. *J. Physiol.*)

are two types of synaptic contact; and corresponding with these there are two types of transmission, electrotonic, mediated by the calyciform synapse, and humoral, mediated by the bouton type of contact. Martin & Pilar (1963–64) inserted microelectrodes into the isolated ganglion and were able to make insertions into both the ganglion cell and its calyciform presynaptic terminal. The type of penetration was recognized by responses to ortho- and antidromic stimulation. When a ganglion cell was penetrated, preganglionic stimulation could have one of two effects, as illustrated by Fig. 624. In A, the spike is initiated by a synaptic potential; by hyperpolarizing the cell before stimulating, the synaptic potential became more pronounced, and finally it was inadequate to initiate a spike when the hyper-polarization had become too great. In B, the synaptic potential was preceded by a rapid depolarization, which normally fired off the spike; if sufficient hyper-polarization preceded the stimulus, this initial depolarization was inadequate to fire the spike, which was then fired by the synaptic potential. Finally, with high enough hyperpolarization, the two depolarizations were distinguishable without a spike. The initial change had all the features of an electrotonic coupling potential resulting from the arrival of the action potential; thus it occurred some $1 \cdot 8$ msec. before the later synaptic potential, an interval corresponding with the synaptic delay and hence it coincided with the arrival of the action potential at the junction; its magnitude, unlike that of the later synaptic potential (which was increased), was unaffected by hyperpolarization; finally it was unaffected by curare. The coupling between pre- and postganglionic cells was two-way, as revealed by a hump of negativity after the action potential had passed, and this was due to the initiation of a synaptic potential by the presynaptic terminals, *i.e.*, the depolarization of the ganglion cell had been transmitted to the preganglionic cell and thus had caused release of transmitter; treatment with curare blocked this late hump. In the intact isolated ganglion, taken from 3–5 day-old chicks, stimulation of the preganglionic nerve leads to a bimodal response in the postganglionic nerve, with a delay of $1 \cdot 5$ to $3 \cdot 0$ msec. between the two responses; treatment with curare blocks the second response. When ganglia were taken from older animals, the slower component became negligible so that electrotonic transmission becomes virtually the sole mechanism in the adult bird. As with the neuromuscular junction, the post-synaptic cell of the chick exhibited spontaneous miniature potentials; and application of statistical treatment showed that the evoked synaptic potentials could be considered as multiples of a quantal unit. Finally, facilitation of the synaptic potential was observed; after a single spike the optimal interval was 15–20 msec. and the second potential was some 150 to 250 per cent. higher than the conditioning one; after a tetanus a potentiation of 300–800 per cent. could be achieved. Intracellular recordings in both post- and presynaptic cells showed that in neither case was the effect due to an enhanced presynaptic spike. The failure to observe an enhanced spike in post-tetanic potentiation is in conflict with the deductions of Hubbard as to the cause of potentiation in the neuromuscular junction; and Martin & Pilar emphasize the importance of mobilization of transmitter for both types of facilitation.

Multi-Phasic Responses

External recording from the curarized ganglion revealed a complex series of potential changes which are characterized by an initial negative (N) potential followed by a longer-lasting positive wave (P) and, in many species, a prolonged late negative phase (LN). These changes have been studied in some detail by R. M. Eccles & Libet, with the aid of a number of specific blocking agents (botulinum toxin, atropine, etc.), and their interpretation of these changes is illustrated by Fig. 625. Thus the ganglion contains chromaffin cells that liberate adrenaline in response to stimulation by the preganglionic fibres (Bülbring, 1944), and it would appear that it is the action of this substance on the postsynaptic membrane that causes the positive wave, *i.e.*, it hyperpolarizes the postsynaptic cells. The two sites for action of acetylcholine, liberated presynaptically, would account for the two negative waves and the fact that the late negative wave (LN) is blocked by atropine, so that this site partakes of the character of a muscarine-sensitive membrane similar to smooth muscle and cardiac fibres.

INTRA-ARTERIAL ACETYLCHOLINE

Subsequent work has confirmed this somewhat speculative picture of events in the autonomic ganglion. Thus intra-arterial injection of acetylcholine in moderate doses could elicit a triphasic response corresponding to the early and late responses or, with a smaller dose, only the late hyperpolarizing and depolarizing responses; atropine blocked the hyperpolarizing and late depolarizing responses (Takeshige & Volle, 1964).

LATE NEGATIVE POTENTIAL

Again, the existence of the late negative potential in the normally conducting, *i.e.*, uncurarized, ganglion was confirmed by Libet (1964) who argued that, if this really occurred, it should impose itself on the long-lasting ganglionic after-potentials that occur after preganglionic stimulation. The slow potentials may be blocked by atropine, leaving transmission unaffected, so that the effects of this slow potential can be deduced by subtraction. In fact, the change produced by atropine on the after-positivity and other post-stimulatory features of the ganglion gave unequivocal evidence of the existence of a slow negative potential similar to that observed in the curarized preparation. The evidence indicated that the potential occurred in the same ganglion cells as those generating the postganglionic discharge. Thus the LN has all the features of an EPSP and may be described as the slow EPSP. Later, Libet & Tosaka (1966) used intracellular recording in the paravertebral ganglion of the frog, and the rabbit's superior cervical ganglion, to confirm that the slow negative and positive potentials were postsynaptic; the slow responses were blocked by atropine as in whole-ganglion recording. In the frog ganglion, Tosaka, Chichibu & Libet (1968) found, moreover, that the same postsynaptic ganglion cell could give rise to a nicotinic EPSP as well as to a muscarinic slow EPSP indicating the existence of receptive sites for the two types of cholinergic activity.

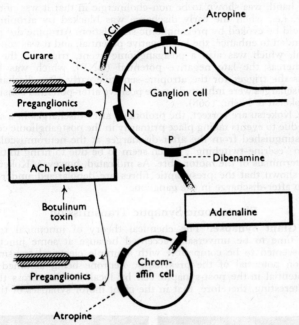

FIG. 625. Diagram of theory suggested to explain the synaptic origins and blockade of the N, P and LN potentials. Receptor sites for the transmitter substances are shown as small blocks on cell membranes. (Eccles & Libet. *J. Physiol.*)

LONG SYNAPTIC DELAYS

Libet (1968) remarked on the extremely long synaptic delays in the initiation of the slow EPSP and IPSP, namely 200–300 msec. and 35 msec. respectively; these are vastly greater than the common run of monosynaptic responses, but it may be that this is a feature of cholinergic synapses sensitive to atropine and mimicked by muscarine; thus the transmission of vagal inhibitory effects has a long latency, and the responses of many central neurones (which are sensitive to atropine) to applied acetylcholine develop slowly, delays of 2–30 sec. being recorded for thalamic relay neurones (Andersen & Curtis, 1964).*

LN AND LLN

Nishi & Koketsu (1966) described two negative after-potentials in the frog para-vertebral sympathetic ganglion; both of these are due to activity in the postsynaptic

* Sanghvi, Murayama, Smith & Unna (1963) found that subthreshold doses of muscarine, acting presumably on late-negative receptors, potentiated postganglionic discharges elicited by preganglionic stimulation, thus emphasizing the importance of the muscarinic action of acetylcholine on autonomic ganglionic transmission. Ambache was the first to demonstrate the excitatory effect of muscarine on the perfused autonomic ganglion.

cell, exhibiting equilibrium potentials; thus the authors refer to the late negative (LN) and late-late negative (LLN) potentials; unlike the LN, the LLN is not abolished by atropine and is presumably initiated by a non-cholinergic receptive site on the ganglion cell different from the cholinergic site on the same cell giving rise to the late negative potential (LN) or late-EPSP as it is called. The LLN lasts for as long as 2–5 min., compared with some 20 sec. for the LN.

Relation to After-Discharges

It is well established that the autonomic ganglion gives rise to after-discharges succeeding tetanic stimulation; in the amphibian sympathetic ganglion studied by Nishi & Koketsu (1968) this could be resolved into two independently initiated discharges called early (EAD) and late (LAD) respectively. The early after-discharge was abolished by atropine and therefore reminiscent of the muscarinic late negative potential evoked in the postganglionic cell; the later after-discharge, on the other hand, was shown to be non-cholinergic in that it was not evoked by acetylcholine, *i.e.*, when the early discharge was blocked by atropine an after-discharge could be evoked by preganglionic stimulation. Atropine did not abolish, but rather tended to enhance, the late negative potential, and it was concluded that this potential, which was also a postganglionic event, triggered the late after-discharge whereas the late negative potential (LNP), which was blocked by atropine, was the trigger for the atropine-sensitive early after-discharge (EAD). Both after-discharges were inhibited by the positive after-potential considered to be an IPSP (Koketsu & Nishi, 1968).

If Nishi & Koketsu are correct, the prolonged after-discharges in a sympathetic ganglion are due to events taking place primarily in the postganglionic cell and thus are to be distinguished from the after-discharges at the neuromuscular junction giving rise to "eserine-twitching", which seem to be due to firing initiated in the presynaptic terminals of the motor fibre. As indicated, however, Koketsu & Nishi (1968) have shown that the presynaptic fibres are depolarized under conditions giving rise to after-discharge in the ganglion.

Electrotonic Synaptic Transmission

Crayfish Giant Synapse. The chemical theory of junctional transmission took a long time to be universally accepted because at some junctions many of the facts seemed to be compatible with a simple electrical transmission, the arriving action potential of the presynaptic neurone being assumed to induce an action potential in the post-synaptic cell by local circuits across the synaptic cleft. It is interesting, therefore, that in the giant motor synapses of the crayfish,

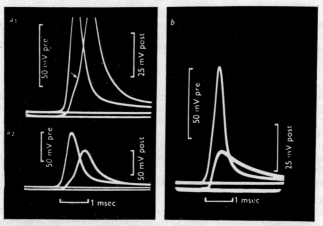

FIG. 626. Transmission across the giant motor synapse of the crayfish recorded by electrodes in pre- and post-synaptic fibres. Upper trace, pre-synaptic potential; lower trace, post-synaptic. a_1 and a_2 recorded from same synapse at different amplifications. The p.s.p. at about the point indicated by the arrow evoked a spike. A sub-threshold p.s.p. is shown in *b*. (Furshpan & Potter. *J. Physiol.*)

which are anatomically very similar to those of the squid, where chemical transmission prevails, Furshpan & Potter have described a form of one-way transmission that seems to rely only on the local circuits at the junction. With microelectrodes in both pre-synaptic and post-synaptic fibres (lateral giant, and giant motor, fibres respectively), stimulation of the pre-synaptic fibre gave rise to propagated action potentials in both, as illustrated in Fig. 626. The post-synaptic spike began as a slower post-synaptic potential, reminiscent of other junctional potentials, and if excitability was low only a subthreshold synaptic potential was obtained (Fig. 626, b). The synaptic delay, corrected for conduction time, was very short, only 0·12 msec. If transmission were electrical, across the synaptic cleft, we should expect relatively small changes in potential of the pre-synaptic fibre to be reflected in electrotonic changes of potential of the post-synaptic fibre, and this should apply to both depolarizations and hyperpolarizations. The fact that transmission is one way would suggest that corresponding changes in the post-synaptic fibre might not be reflected in the pre-synaptic fibre. Experiments showed that neither of the two suppositions was completely correct; depolarizations of the pre-synaptic fibre were indeed reflected in depolarizations of the post-synaptic fibre, but hyperpolarizations were not; by contrast, depolarizations of the post-synaptic fibre had no influence on the pre-synaptic fibre whilst hyperpolarizations did.

Rectifier Effect

Furshpan & Potter concluded from these observations that the synaptic membrane was polarized, in the sense that it permitted positive current to flow inwards from pre- to post-synaptic fibre, but not in the reverse direction, *i.e.*, that it behaved as a rectifier. The four situations described above are illustrated in Fig. 627, whence it is seen that orthodromic depolarization (*a*) and antidromic hyperpolarization (*d*) produce identical situations, so far as local current-flow is concerned. It was not possible to measure directly the current traversing the synapse under both of these conditions, so that the rectifier characteristics could not be directly

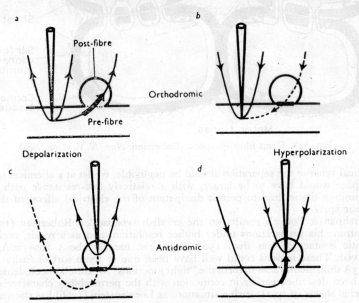

FIG. 627. Synaptic rectifier hypothesis. The post-fibre is shown in transverse section at the point at which it crosses over the pre-fibre; the junction is indicated by the dotted line or a heavy bar, representing a low or high synaptic resistance, respectively. The arrows give the direction of (positive) current entering or leaving the current-passing micro-electrode; dashed lines indicate negligible current flow due to high synaptic resistance. Diagrams *a* and *d*, corresponding to the two situations in which trans-synaptic effects were observed, show that in both cases current would cross the junction in the same direction (indicated by the heavy arrows). (Furshpan & Potter. *J. Physiol.*)

determined, but a study of the current-voltage relationships when current was passed through pairs of microelectrodes in both pre- and post-synaptic elements permitted the calculation of the current-voltage curves for the synapse under conditions of orthodromic depolarization and antidromic hyperpolarization. A comparison of these two calculated curves showed them to be the same within experimental error, so that it was concluded that a synaptic rectifier with the calculated current-voltage characteristics accounted for the diverse results of current-passing experiments.*

Morphology of the Electrotonic Synapse

In the electron-microscope the early studies provided little to differentiate the chemically transmitting squid synapse from the electrically transmitting crayfish one (Robertson); in both, intimate contact was brought about by processes from the post-synaptic fibre pushing into the cytoplasm of the pre-synaptic fibre (Fig. 628). In these regions the two cytoplasms were separated by as little as 150–300A, and since there are two 75A thick membranes in this space, the actual separation could have been very small indeed. For maximal efficiency at an

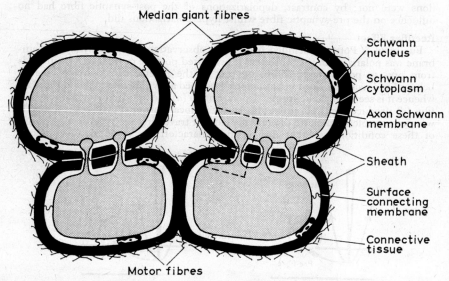

Median giant fibres

Schwann nucleus

Schwann cytoplasm

Axon Schwann membrane

Sheath

Surface connecting membrane

Connective tissue

Motor fibres

Fig. 628. Giant fibre synapses. (Robertson. *Ann. N. Y. Acad. Sci.*)

electrical synapse the separation should be negligible, whilst at a chemical synapse the space would have to be larger, with a relatively low-resistance path to the surrounding tissue fluid to permit dissipation of the electrical effects of the pre-synaptic spike.

Furshpan & Potter's results on the crayfish synapse led Robertson (1961) to re-examine his preparations under higher resolution, and as a result, regions of synaptic contact of less than 150A thickness, namely about 100–120A, were observed. These regions could well have been due to some sort of fusion of the two 75A unit-membranes to form the "tight junctions", or zonulae occludentes, that have been described earlier in connexion with the permeability characteristics of epithelial sheets (p. 492); such a situation, as Loewenstein (1966) has shown, gives rise to an electrotonic coupling, in the sense that a potential applied to one cell is reflected in a corresponding potential change in the coupled cell.

---

* Rectification may also occur across a chemically sensitive synaptic membrane, notably in the giant metacerebral cells of the snail, *Aplysia*. Kandel & Tauc (1966) have shown that, at potentials near the resting level, there is a decrease in conductance with depolarization, and an increase with hyperpolarization, so that the EPSP actually decreases with hyperpolarization instead of increasing. This anomalous rectification may well be physiologically significant, since the changes in membrane resistance are such that a small increase in background synaptic bombardment, leading to a small depolarization, will increase synaptic efficacy and so magnify the effect of this small increase in excitatory drive.

Electrotonic Coupling. This coupling might well permit transmission of the action potential in the presynaptic cell to the postsynaptic cell, provided the attenuation is not too great. Since this study, electrotonic transmission has been established in a number of other synapses; and in all cases the electron microscope has revealed regions of membrane fusion. Thus the supramedullary neurones of the puffer fish have large cell bodies, greater than 200 μ in diameter; they fire synchronously when activated synaptically, whilst an impulse directly evoked in one cell spreads to all. In the electron microscope Bennett, Nakajima & Pappas (1967) observed that pairs of axons frequently had regions of fusion to give a combined membrane thickness of 150A; in addition, typical synapses, with vesicles in the presynaptic axon, were seen. By inserting microelectrodes in pairs of cells, coupling could be demonstrated, in so far as the current reaching the unstimulated cell was greater than if the same stimulus had been applied extracellularly.*

Electrical Requirements for Coupling

The electrical requirements for coupling between cells, on the basis of simple equivalent circuits, have been analysed by Bennett (1966); the *coupling coefficient*, k, is defined as the ratio of the potential produced on one side by a given potential applied to the other; thus on passing from cell (1) to cell (2) $k_{12} = V_2/V_1$, and in the reverse direction, $k_{21} = V_1/V_2$ (Fig. 629). These coefficients may be shown to be given by:—

$$k_{12} = \frac{1}{1 + r_o/r_2}$$

$$k_{21} = \frac{1}{1 + r_o/r_1}$$

k_{12} tends to unity as r_o/r_2 tends to zero, and k_{21} tends to zero as r_1/r_o tends to zero. Thus perfect coupling in one direction can be achieved with zero coupling in the opposite direction, so that the electrotonic synapse may, indeed, be unidirectional.

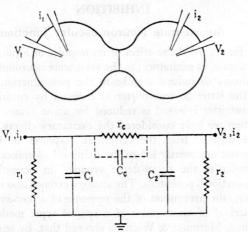

FIG. 629. Illustrating the model for an electrotonic junction *Above:* Diagram of the cells. *Below:* An equivalent circuit for the cells. The junctional capacity is connected in by dotted lines to signify that it is usually neglible. (Bennett. *Ann. N.Y. Acad. Sci.*)

This asymmetry is due to the relation between resistances of presynaptic and postsynaptic structures, and these depend on surface area and areal resistivity, *e.g.*:—

$$k_{12} = \frac{1}{1 + \dfrac{\rho_o \, A_2}{\rho_2 \, A_o}}$$

where ρ is the areal resistivity and A the area. Where passage of impulses is

* In the frog's spinal cord, Charlton & Gray (1965) have described axo-dendritic contacts in which the synaptic cleft is obliterated to give the typical 3-lined complex some 150A thick; these presumably correspond with a form of electrical transmission.

concerned, a further asymmetry may be provided by the pre- and postjunctional time-constants and thresholds. Finally, we have the rectifying characteristics of the junctional membrane discussed above, but this appears to be peculiar to the squid giant synapse.

Physiological Significance. The electromotor neurones of mormyrid fish exhibit a great symmetry in their action, which seems to be brought about by electrotonic coupling between the spinal electromotor neurones; and the electron-microscopical evidence of Bennet *et al.* (1967) strongly suggests this. Again, the electric organ of *Malapterus electricus* is controlled by a pair of giant neurones, one on each side; in this case there is no evidence that these are coupled, but their afferent neurones seem to be coupled to both, as seen by areas of fusion between them and the soma and dendrites. In both these examples we have neurones that, to be effective, must discharge the individual elements of the electric organ, the electroplaques, as synchronously as feasible; and it may well be that electrotonic coupling is the most efficient method of achieving this.

Two-Way Transmission. The colossal cells of the leech ganglion show a simple two-way electrical transmission that is by no means confined to the action potential, small non-propagated electrotonic potentials being transmitted whether these be de- or hyperpolarizations; Hagiwara & Morita, who have described the phenomena, have also summarized the literature on this form of transmission.

The Ephapse. The findings of Furshpan & Potter recall the theoretical studies of Arvanitaki on the possibilities of direct excitation of one fibre by the action potential in another. She has applied the general considerations on the changes of potential in the vicinity of an advancing wave of action potential, developed by Bishop (p. 1051), to the special case of the *ephapse*—the region of contact between two laterally apposed axons—and has deduced that, given an appropriate geo-metrical arrangement, electrical transmission is possible.

INHIBITION

Invertebrate Neuromuscular Junction

We have so far regarded the effect of an impulse at a junction, be it neuro-muscular or synaptic, as excitatory; in the vertebrate neuromuscular preparation the effect is always to induce a spike in the post-junctional cell, unless the sensitivity of the latter to transmitter is reduced by curarine, or unless the amount of transmitter released is reduced for some cause. In the crustacean and insect system we have considered the excitatory effects of nerve impulses too; they have differed from the vertebrate response in that the junctional potentials have not necessarily led to the firing off of spikes. The invertebrate junction is similar to the nerve-nerve synapse, in permitting the temporal summation of junctional potentials. The analogy extends also to the phenomenon of inhibition, *i.e.*, the prevention of the response of a post-synaptic cell due to the previous arrival of an impulse of a special type, mediated by a special nerve fibre. Thus, Marmont & Wiersma showed that, by stimulating one fibre —the inhibiting fibre—to a crab muscle, a complete relaxation of the muscle could be induced in spite of the concurrent stimulation of another fibre—the excitatory fibre.

Possible Mechanism of Inhibition. Before going into further details we may briefly consider how an electrical impulse may inhibit a synapse. Essen-tially, the phenomenon is revealed by the interference of this inhibitory effect with the normal excitatory process which, at a synapse, is a depolarization due to an increase in permeability to certain ions, possibly Na^+, K^+ and Cl^-; the essential feature of this depolarization is that Na^+ may penetrate the fibre, and the magnitude of the depolarization will depend on the relative permea-bilities of the membrane to Na^+, K^+ and Cl^-, as we have seen in discussing the action potential in nerve. Thus an increased permeability to Na^+ causes depolarization provided that the permeability to K^+ is not increased too, since we have seen that an increased permeability to K^+ will act to repolarize the nerve. Consequently, if an excitatory influence on the junction consists primarily of an increased permeability to Na^+, we may inhibit its effects if, at another nearby point, we produce a marked increase in the permeability of the membrane to K^+. This inhibitory influence, acting by itself, need have no effect on the resting potential, in strong contrast to the excitatory effect, which must depolarize. Alternatively, the inhibitory effect might actually cause a hyperpolarization; thus we have seen that the resting potential may be described as being determined either by the K^+ distribution or by the Cl^- distribution; if these concentration-ratios conform to a strict Nernst equilibrium, of course the potential will be the same whichever ion is considered; if, because of active processes or for other reasons, the resting potential is not equal to the Nernst potential appropriate to the K^+ concentrations, say, but conforms more closely to that for Cl^- concentrations, then an increase in permeability to K^+ will tend to impose the Nernst K^+-potential on the system instead of that determined by the prevailing Cl^- distribution. The direction in which the resting potential changes will obviously depend on whether the Nernst K^+-potential is greater, or less than, the resting potential; if it is greater than this, then increasing the K^+-permeability will raise the resting potential, *i.e.*, it will hyperpolarize the membrane. If the Nernst potential is less than the resting potential the inhibitory impulse will depolarize, whilst if the resting potential and equilibrium potential are equal we may expect the effect to be zero.

Point of Reversal

Theoretically, on a superficial view, this condition will be a point of reversal, in the sense that the discharge will hyperpolarize when the resting potential is below this value and will depolarize when it is above. Because of complicating factors, the reversal potential need not necessarily equal the K^+-potential (supposing that inhibition does consist in a specific increase in K^+-permeability), but its measurement, which may be carried out by voltage-clamp techniques, or by passing current across the membrane, certainly provides a clue as to the character of the inhibitory and, as we have seen, of the excitatory stimulus. So far as the crustacean neuromuscular synapse is concerned, this reversal has been elegantly demonstrated by Dudel & Kuffler, who varied the degree of polarization across the junction, by passing current through an internal electrode, and recorded the effects of an inhibitory train of impulses, as shown in Fig. 630. Starting with a resting potential of 80 mV the inhibitory junction potentials became negligibly small when this was reduced to 72 mV, and reversed their direction with additional depolarization. The reversal potential has been described by Kuffler as the *inhibitory equilibrium potential*, and, as the records show, it is approximately the potential to which the junction is driven by the inhibitory discharge.*

Potential Changes. Fatt & Katz showed that the inhibitory impulses always had an electrical effect on the excitatory junctional potential, causing it to subside more rapidly than in its absence. Since the muscular contraction depends on the summation of the local junctional potentials distributed through-

* Excitation across the giant crayfish synapse is, as we have seen, electrical; it is interesting that, in addition to the fast spikes that travel across this synapse, Furshpan & Potter observed slow potentials whose sign depended on the level of polarization of the post-synaptic axon; these were clearly inhibitory in action and were presumably mediated by a chemical transmitter since the reversal of sign of the response, in accordance with the degree of polarization of the post-synaptic membrane, cannot be explained by electrical transmission. The effect of nerve stimulation was mimicked by GABA (p. 1219).

out the fibre, an increase in rate of subsidence of these potentials must reduce the degree of summation, and thence the force of contraction. This effect of the inhibitory impulse was associated with a 20–50 per cent. increase in membrane conductance, *i.e.*, an apparently similar effect to that of an excitatory impulse.

In the absence of a previous excitatory impulse, the inhibitory impulse had, usually, little or no effect on the resting potential, although sometimes a small hyperpolarization, *i.e.*, an increase in the internal negativity, was observed. If the membrane was partially depolarized by a cathodic current, then the inhibitory impulse caused an invariable rise in resting potential; if, on the other hand, the membrane was hyperpolarized by an anodic current, the inhibitory impulse caused a decrease in the membrane potential. In other words, the inhibitory impulse brought the membrane potential towards its

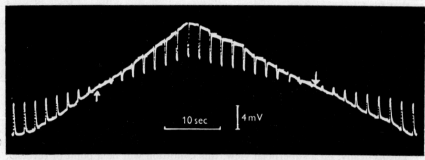

FIG. 630. Inhibitory junctional potentials with varying resting potential beginning with 80 mV and falling to 61 mV by passing current through a second intracellular microelectrode. Arrows indicate the reversal potential at 72 mV. (Dudel & Kuffler. *J. Physiol.*)

normal resting value. Fatt & Katz suggested that this could be due to a specific increase in permeability of the fibre-membrane to K+ and Cl−.

Transmitter Substances

It was concluded by Fatt & Katz that this junctional change was not the most important effect of the inhibitory impulse—in fact, the large decrease in the junctional potential (it could be reduced to one tenth of its normal value) could not have been due to this postulated increase in permeability to K+. It was suggested, therefore, that both the excitatory and inhibitory impulses released separate transmitter substances which competed for active groupings in the postsynaptic membrane; it would be this competition that reduced the junctional potential.

Synaptic Contacts

In the electron-microscope Atwood (1968) was able to differentiate two types of junction according as the synaptic vesicles were uniformly spherical or irregular in shape; the former were seen exclusively in muscles where no inhibitory innervation was present and so could be identified as excitatory junctions. It was remarkable that junctions of dissimilar type were frequently very close to each other and this would, presumably, be a factor favourable to inhibition of the activity induced by release of excitatory transmitter. Examination of the excitatory terminals showed that these frequently contained axo-axonal synapses formed by an inhibitory nerve-ending, recognized by its synaptic vesicles, and these would be the morphological basis for presynaptic inhibition.

Ionic Substitutions

The electrical effect of an inhibitory impulse was investigated in more detail by Boistel & Fatt, who argued that the effects of ionic substitutions should indicate which ion or ions became specifically more permeable. By replacing the external Cl⁻ by a relatively impermeant anion, an increase in permeability to Cl⁻ should now cause the resting potential to shift towards a new chloride-potential determined by the relative concentrations of Cl⁻ inside and outside the fibre which, because of the removal of external Cl⁻, would be such as to depolarize the fibre; this did, indeed, happen, whilst alteration of the external concentration of K⁺, although it modified the resting potential, did not appreciably affect the direction of the inhibitory potential change. Boistel & Fatt concluded, therefore, that the inhibitory impulse caused a specific increase in permeability to Cl⁻. Subsequent studies from Takeuchi's laboratory, in which inhibitory activity was induced by the local application of the presumptive inhibitor, γ-aminobutyric acid (GABA, p. 1219), have confirmed the importance of anion permeability in the initiation of the inhibitory junctional potential (Takeuchi & Takeuchi, 1966, 1967).

Equivalent Circuit

A simple equivalent circuit for a postjunctional membrane, subject to both excitatory and inhibitory actions, is illustrated by Fig. 631 from Ginsborg (1967); ϵ is the equilibrium potential for excitation and ϵ' that for inhibition. It will be noted that ϵ' has been given a value less negative than the resting potential of 75 mV so

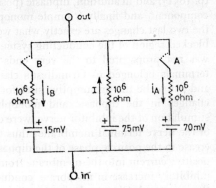

FIG. 631. Illustrating the equivalent circuit for interaction between inhibitory (A) and excitatory (B) transmitter effects on the postjunctional membrane. (Ginsborg. *Pharmacol. Rev.*)

that, acting alone, it also would tend to depolarize the membrane. When both excitatory and inhibitory transmitters act together, the displacement of the membrane potential will not be a simple algebraic sum of their effects when acting alone. We may assume that the channels opened by excitatory and inhibitory transmitters are entirely distinct. The action of the inhibitory transmitter is given by closing switch A, which produces a depolarization of 2·5 mV $(R(E - \epsilon')/R + r)$. The excitatory transmitter, indicated by switch B, causes a depolarization of 30 mV. When A and B are both shut, the membrane depolarization, e, is given by:—

$$(10^6 \text{ ohm}) \times (I \text{ mAmp}), \text{ where } I = i_A + i_B.$$
$$75 - 10^6 . I = 70 + 10^6 .i_A$$
$$\text{and } 75 - 10^6 . I = 15 + 10^6 .i_B$$

Whence e is approximately 22 mV, so that A has inhibited B by reducing its normal depolarization from 30 mV to 22 mV. This calculation emphasizes that it is not necessary for an inhibitor to hypolarize a junctional membrane; in general, the excitatory transmitter will be inhibited if its equilibrium potential is more negative than the threshold for excitation.

Pre-Synaptic Inhibition

Dudel & Kuffler confirmed the experiment of Boistel & Fatt, and pointed out that inhibition might occur in two other ways, namely by competition of an inhibitory transmitter for the excitatory transmitter's receptor sites, as suggested by Fatt & Katz, or it might prevent the release of transmitter by the excitatory nerve terminals—*pre-synaptic inhibition*. That this last mechanism must be operative in the crayfish junction was demonstrated by choosing conditions of stimulation such that the effects of the inhibitory increase of membrane conductance would be to increase, instead of antagonize, the excitatory junctional potential. Under these conditions the increase in membrane conductance would be unable to influence the externally recorded potentials, yet it was found that this response was in fact reduced in amplitude, whilst failures to produce any response to a nerve impulse increased. A statistical treatment suggested that the effect of inhibition was to reduce the average number of quanta in a response whilst the individual quanta retained their normal average size; furthermore, the frequency of the spontaneous potentials decreased.

It could well be, as Dudel & Kuffler point out, that this pre-synaptic inhibition is mediated by a primary electrical change acting through a transmitter on the excitatory terminal; if this reduced the effect of the nerve impulse on the liberation of quanta of acetylcholine its inhibitory action would be accounted for.* Later, Dudel (1963–1965) made extracellular recordings of the early electrical changes occurring in the junctional regions when the excitatory nerve was stimulated. The *excitatory nerve terminal potentials* recorded in this way were of three kinds; the expected triphasic (positive, negative, positive) change, (p. 1051); and in addition, diphasic (positive-negative) with a prominent positive component, and finally a simple monophasic positive wave. Dudel argued that the two last changes are exactly what would be expected of records taken over a blocked region of the conducting system, and he concluded that the nerve spike was not propagated to the very ends of the nerve terminals, *i.e.*, the nerve terminals belonged to Grundfest's class of electrically inexcitable membranes and would be the presumptive sites for the action of a chemical inhibitor. The changes in the diphasic and monophasic records, resulting from previous stimulation of the inhibitor nerve, were consistent with an increased conductance of the nerve terminal membrane; thus the most pronounced feature was an increase in the positive phase of the diphasic potential, due to the drawing of more positive current into the membrane from the approaching action potential. If the inhibitory increase in membrane conductance is similar to that associated with the increased Cl^- conductance of the postsynaptic membrane, then it will presumably interact with an excitatory depolarization, blocking this, in the same way as at the post-junctional membrane.†

Inhibition of Motor Neurones

The synaptic excitation of the motor neurones of the spinal cord, discussed earlier, was brought about by stimulation of the sensory fibres from a muscle;

---

* All these effects are in essence attenuations of the electrical events associated with excitation and have been classed as α-inhibition, in contrast to a possible direct action of the impulse on the muscle, say by uncoupling the process interposed between membrane potential changes and contraction of muscle substance (β-inhibition). The relative extents of these two types of inhibition have been studied by Hoyle & Wiersma in a number of different crustacean muscles.

† The involvement of excitatory and inhibitory activity in the nerves supplying crustacean muscles during natural movements, such as opening and closing of the claws, have been analysed by Bush (1962a, b) and Wilson & Davis (1965). Thus stroking the inside surface of the claw elicits a strong closing reaction, and this is associated with high E-activity in the closer and high I-activity in the antagonistic opener. Again, inhibitory activity, induced by the stretch-receptor attached to the dactylopodite, is involved in a negative feed-back circuit tending to stabilize the position of the claw (Wilson & Davis, 1965).

these fibres were derived from the muscle spindles and formed the basis of the stretch-reflex, *i.e.*, the contraction of a muscle that results from stretching of the muscle or its tendon. If the sensory nerve from an antagonist muscle is stimulated, this stretch-reflex may be inhibited; *e.g.*, the excitatory effect of stimulating the nerve from the biceps-semitendinosus may be inhibited by a previous stimulation of the afferent fibres from the quadriceps. The excitatory effect of the stretch reflex is mediated by a two-neurone arc, as illustrated by Fig. 582*a*, where the afferent neurone makes direct synaptic contact with the motor neurone—the pathway is said to be *monosynaptic*. It seems unlikely that the same neurone would also exert an inhibitory action, and it has now been conclusively shown that the inhibition of the motor neurone is brought about by an interneurone that presumably liberates its own kind of transmitter at its synaptic knobs on the soma and dendrites.*

Inhibitory Post-Synaptic Potential. By means of an intracellular electrode, Brock, Coombs & Eccles were able to show that the electrical effects of an inhibitory volley on the motor neurones was characteristically different from that of an excitatory volley; in the latter case, as we have seen, a depolarization

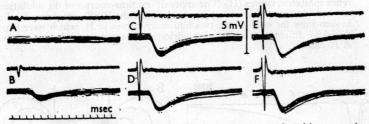

FIG. 632. Lower records give intracellular responses of a biceps-semiten-dinosus motor neurone to a quadriceps volley of progressively increasing size as is shown by the upper records which are taken from the L 6 dorsal root by a surface-electrode (downward deflection signalling negativity). Note three gradations in the size of the inhibitory post-synaptic potential: from A to B, from B to C and from D to E. Voltage scale gives 5 mV for intracellular records, downward deflections indicating membrane hyper-polarization. (Coombs, Eccles & Fatt. *J. Physiol.*)

occurred—the excitatory post-synaptic potential (EPSP)—whilst with the inhibitory volley there was a characteristic hyperpolarization, amounting to some 2–3 mV—the *inhibitory post-synaptic potential* (IPSP)—as shown in Fig. 632.

Relation to Reflex Inhibition

That this potential reflected true inhibitory activity was made very probable by comparing its time-course with the degree of inhibition of the motor neurones, measured either by examining their excitability to direct stimulation through an intracellular electrode or, more naturally, by examining the reflex discharge of the motor neurones in response to an excitatory afferent volley at different times after an inhibitory volley had entered the cord. In Fig. 633 the course of the inhibitory synaptic potential has been plotted on the same time-scale with the course of decay of inhibition of reflex discharge brought about by an

---

* There has been some disagreement with Eccles' assumption of this inhibitory inter-neurone, because of the claim of Lloyd & Wilson (1959) and Lloyd (1960) that the long latency of the inhibitory potentials recorded in motor neurones, attributed by Eccles and his co-workers to synaptic delay in the inhibitory neurone, were artefactual; these workers maintained that an afferent neurone might excite through one collateral and inhibit through another, both in monosynaptic pathways. The point was finally settled in favour of the extra neurone by Araki, Eccles & Ito (1960).

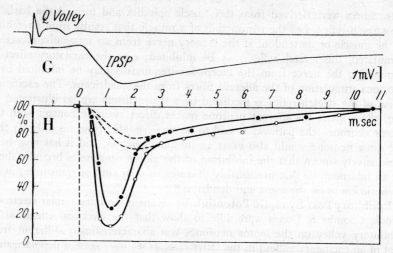

FIG. 633. Illustrating the inhibitory post-synaptic potential (G) and its associated reflex inhibition curves (H). The approximate time-courses of the inhibition attributable directly to the hyperpolarization of the IPSP's are shown by the broken lines for each of the two inhibitory curves in H, which were produced by different strengths of quadriceps afferent volley. (Eccles. *Ergebn. d. Physiol.*)

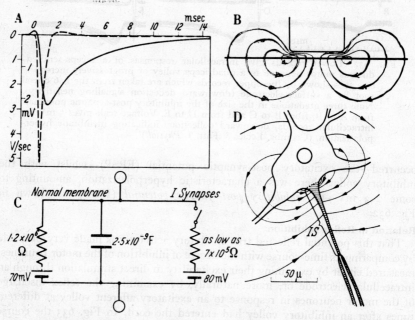

FIG. 634. A. The recorded IPSP is shown in full line whilst the brief flow of current is indicated by the broken line. B shows the flow of post-synaptic currents at a synapse. C is a formal circuit-diagram of the membrane of a motor neurone; on the right, in addition, is a representation of the sub-synaptic areas of the membrane that are activated in producing the IPSP. Maximum activation of these areas will be given by closing the switch. D, Diagrammatic representation of current that flows as the IPSP generated in the soma-dendritic membrane spreads electrotonically to hyperpolarize the initial segment (IS), which is the site of initiation of impulses discharged from the motor neurone. (Eccles. *Ergebn. d. Physiol.*)

inhibitory volley. In general, this inhibition of reflex activity shows an initial intense phase and a prolonged tail; and the duration of this tail accords well with the duration of the IPSP. The reason for the intense period is to be sought in the electrical events that take place at the synapse, and of which the IPSP is a reflexion.

Transmitter Action

It will be recalled that the end-plate potential could be represented as the effect of a brief transmitter action, *i.e.*, an intense end-plate current lasting only for a few msec., which produced a local depolarization that decayed over a longer period by virtue of the cable properties of the muscle membrane. From a knowledge of the end-plate current it was possible to calculate the form of the end-plate potential, or from a measurement of the end-plate potential, the phase of intense transmitter action could be deduced. Fig. 634, A illustrates Eccles' computation of the course of the transmitter action by comparison with the inhibitory synaptic potential of the motor neurone. It is this intense phase of current-flow that accounts for the intense inhibitory effect of the afferent impulses; the inhibitory current, acting by itself, will cause a current-flow at the initial segment as illustrated by Fig. 634, D, *i.e.*, in the opposite direction from that caused by an excitatory synapse (Fig. 621 C, p. 1191), and this constitutes the intense inhibitory effect of the current-flow. The inhibitory potential, which decays so much more slowly, will also exert an effect, but not so strong, tending to reduce the excitatory potentials built up in its neighbourhood. The broken lines in Fig. 633 represent the contribution of this component of inhibitory action on the post-synaptic membrane to reducing reflex activity, as calculated by Araki, Eccles & Ito.

Latencies

The latency of an inhibitory volley, so far as its ability to influence an excitatory volley is concerned, is some 0·65 msec., *i.e.*, if the inhibitory volley precedes the excitatory volley by 0·65 msec. or more it exerts an inhibitory action; if it precedes it by less than this, there is no action. The inhibitory synaptic potential begins some 1·5 msec. after the inhibitory volley has entered the cord, hence, if this volley is given a 0·65 msec. start over the excitatory volley, its synaptic potential will arise 1·5 − 0·65 = 0·85 msec. after the excitatory volley has entered. An action potential will arise out of a synaptic potential some 1·05 msec. after the excitatory volley has entered the cord, hence, under these limiting conditions, there is a period of 0·2 msec. during which the inhibitory synaptic potential may exert its effect. There is nothing in the latencies of the inhibitory and excitatory effects that is inconsistent with this view of the hyperpolarization as an antagonist to the excitatory synaptic depolarization.

The Ionic Basis of the IPSP

The size of the IPSP increases with depolarization of the post-synaptic membrane and decreases with hyperpolarization; in fact, when the membrane potential is set at a sufficiently high value there is a reversal to give an equilibrium potential of about 80 mV at which the IPSP is zero; this represents some 6 mV of hyperpolarization of the resting potential and, arguing along the lines developed earlier, we may say that the inhibitory change at the membrane causes the membrane potential to shift towards the Nernst potential corresponding to either the Cl^-- or K^+-distributions, or of both, on the assumption that the resting potential is held at a different level from these. Studies on the effects of injecting anions into the motor neurone strongly indicated that these had a dominating influence; thus, increasing the internal chloride concentration caused a reversal in sign such as would be expected were the potential, during inhibition, determined by the new value of $[Cl]_i/[Cl]_o$. To employ a formal electrical diagram, inhibition would be represented by closing the switch representing the I-synapses (Fig. 634,C).*

Pre-Synaptic Inhibition. The inhibition due to the IPSP is a post-synaptic event; in the crustacean neuromuscular junction we have seen that inhibition may be brought about by an influence of the inhibitory fibre on the pre-synaptic fibre. In the vertebrate nervous system, pre-synaptic inhibition has also been described. For example, Frank & Fuortes (1957) observed that the gastrocnemius motor neurones could be inhibited up to 50 per cent. by a preceding volley in the hamstring muscle nerve, although stimulation of this nerve alone had no influence on the post-synaptic neurone, *i.e.*, it produced no IPSP, nor did it alter the electrical excitability as tested by stimulation through internal microelectrodes. The

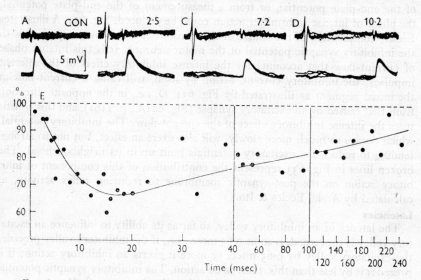

FIG. 635. Illustrating pre-synaptic inhibition. In A to D the lower records give the EPSP's reflexly evoked by maximal Group Ia volleys. The upper records are from the dorsal root of L7 as it enters the cord, negativity being recorded downwards. A is the control record, whilst the responses in B and D are those that occur after previous stimulation of another sensory group, namely Group I of the posterior biceps and semitendinosus. If the interval between stimuli is small, as in B, the depression of the EPSP is small. In E the time-course of depression in EPSP is shown, when the abscissæ are the intervals after the conditioning posterior biceps and semitendinosus volley. (Eccles, Eccles & Magni. *J. Physiol.*)

phenomenon was repeated in some detail by Eccles, Eccles & Magni, and Fig. 635 from their paper shows the depression of the EPSP in the motor neurones of the gastrocnemius-soleus nerve by conditioning volleys in another muscle-nerve, namely that of posterior biceps and semitendinosus—volleys that themselves had no influence on the membrane potential of the gastrocnemius-soleus motor neurones. The total duration of this inhibitory action is of the order of 200 msec. By the selective stimulation of the afferent fibres, it could be shown that the effective ones belonged to Groups Ia and Ib, Groups II and III being ineffective.†

* Araki, Ito & Oscarsson (1961) and Ito, Kostyuk & Oshima (1962) have examined the effects of injecting many different anions, and if their ability to influence the IPSP is a measure of their ability to cross the post-synaptic membrane during the IPSP, it can be concluded that, during this inhibitory transmitter action, there is an increase in permeability that allows ions of diameter some 1·14 times that of K^+ to pass rapidly, whilst those with diameters not much greater are unable to pass. Eccles (1961) has discussed the possible role of increased permeability to K^+.

† The afferent fibres from a muscle have been divided, on the basis of diameter, into four groups: Group I with diameters from 13 to 20 μ; Group II from 4 to 12 μ; Group III

Dorsal Root Reflexes

The clue to the phenomenon was given by the study of the so-called *dorsal root reflexes*, the appearance of a discharge in an *afferent* nerve when another afferent nerve is stimulated, which commonly occurs when a cutaneous afferent nerve is stimulated but also, as shown by Brooks (cf. Brooks & Koizumi, 1956), occurs from muscle-afferent to muscle-afferent when the spinal cord is cooled. In some way, cooling increased the excitability of this neurone-arc, so that, as Eccles, Kozak & Magni argued, in the normal animal only a subthreshold depolarization of the afferent neurones to one muscle would occur when those to another were stimulated. In the monosynaptic stimulation of motor neurones employed in studies of motor neurone excitability, it is the Group I sensory fibres that are stimulated so that if, before a given set of Group I fibres were stimulated, another set, belonging to a different muscle, were stimulated, we would expect a long-lasting depolarization of the former fibres. If, as seems likely from Hagiwara & Tasaki's study on the synapse, the size of the excitatory synaptic potential depends on the degree of polarization of the pre-synaptic axon, then clearly a depolarization will decrease the EPSP, presumably by liberating less transmitter. Studies on the depolarization of the afferent neurones with intracellular electrodes confirmed this view of the mechanism of pre-synaptic inhibition, which is probably produced by a prolonged depolarizing action on the branches of the afferent neurone in the spinal cord, probably in the motor neurone and intermediate nuclei. The depolarization is brought about by synaptic contacts with intermediate neurones since there is a considerable synaptic delay in the initiation of this depolarization (4 to 10 msec.). Because of the much greater duration of pre-synaptic inhibition (200 or more msec.) by comparison with the post-synaptic inhibitory and excitatory effects of an afferent stimulus, we may expect to find that stimulation of the Group I fibres to any muscle will have a pre-synaptic inhibitory effect on the motor neurones to all the muscles of a limb as soon as the post-synaptic events have subsided; according to Eccles, Schmidt & Willis (1962) this is indeed the case, especially when the afferent nerves to flexor muscles are stimulated.

Serial Synapses

The morphological basis for presynaptic inhibition is, we may assume, the serial synapse through which the presynaptic terminal of one neurone receives a synaptic knob from another neurone. Such synapses have been described in the spinal cord (Charlton & Gray, 1966) and in the crustacean muscle, another site of presynaptic inhibition (Atwood, 1968).

Electrical Inhibition. The post-synaptic inhibition that we have so far considered has represented a change in the permeability characteristics of the post-synaptic membrane, such that the active depolarization associated with the action potential is mitigated or prevented. The change is essentially a clamping of the resting potential and may be manifest as a hyperpolarization, a depolarization, or it may be isopotential. An extrinsic source of potential, such as a positively charged electrode, could also act in an inhibitory manner by simply building up the membrane potential in the immediate neighbourhood of the electrode, causing an inward flow of current that hyperpolarizes the membrane locally. Such an extrinsic source of anodic current could possibly be mimicked by an arrangement of nerve terminals in the neighbourhood of a part of an excitable cell, the arrangement being such that the action potentials did not invade completely these terminals, stopping short to give a region of positivity; and it would seem that in the Mauthner cells of the goldfish, studied by Furshpan and Furukawa, we have an example of this electrical type of inhibition.

Mauthner Cells

The medulla of the goldfish contains two remarkable neurones, one on each side, called Mauthner cells and illustrated by Fig. 636; because of the size of

---

from 1 to 4 μ; and Group IV, unmyelinated fibres. Largely as a result of Hunt's (1954) work, these groups may be related to function; thus Group I are sensory fibres from the muscle spindles and Golgi tendon organs (Ia and Ib respectively).

the dendrites this cell is of especial value for the study of the site of origin of the spike and the extent to which it may invade the soma and dendrites. As Fig. 636 shows, the cell is abundantly supplied with pre-synaptic endings of several types, many of them derived from collaterals of the Mauthner cell of the opposite side, or from interneurones with which this makes connexion, since the Mauthner cell of one side may be fired when the contralateral one is excited. The Mauthner cells cannot be seen by the investigator, so that the position of the

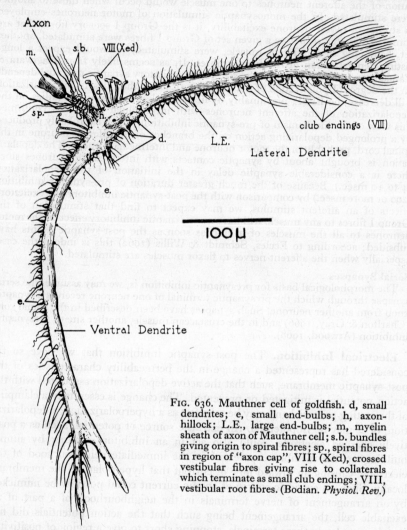

FIG. 636. Mauthner cell of goldfish. d, small dendrites; e, small end-bulbs; h, axon-hillock; L.E., large end-bulbs; m, myelin sheath of axon of Mauthner cell; s.b. bundles giving origin to spiral fibres; sp., spiral fibres in region of "axon cap", VIII (Xed), crossed vestibular fibres giving rise to collaterals which terminate as small club endings; VIII, vestibular root fibres. (Bodian. *Physiol. Rev.*)

recording electrode has to be deduced for any given recording, and in a brilliant study Furshpan & Furukawa were able to correlate the different potential changes with the position of the external electrode as subsequently established by electrolytic deposition of iron salts from the tip of the steel electrode. Thus, the very prominent negative spike could be localized to a circumscribed region corresponding to the axon hillock; on moving away from here the negative spike became less prominent, whilst positive waves became more noticeable, so that in the more distant regions of the dendrites the external record was a simple positive wave. It was concluded that the axon hillock was the site of intense regenerative activity, acting as a sink for current from more remote regions that were probably not invaded by the spike, and therefore gave a simple positive wave of potential, acting as a source (p. 1051).

EXTRINSIC HYPERPOLARIZING POTENTIAL

Inhibition was revealed by the difference in effect of stimulating the ipsi- and contralateral VIIIth nerves; only the ipsilateral nerve led to firing of the Mauthner cell, whilst the contralateral nerve could suppress firing. This inhibi- tory activity was studied by Furukawa and Furshpan. As Fig. 637 shows, the negative spike following antidromic stimulation of the Mauthner cell axon is followed by positive waves, which have been described as the *extrinsic hyper- polarizing potential* (EHP), because they differ fundamentally from a later potential change that may be identified as the typical inhibitory post-synaptic potential and called by Furukawa & Furshpan *late collateral inhibition* (LCI).

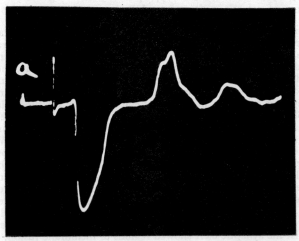

Fig. 637. Extracellular recording from vicinity of a Mauthner cell axon hillock in response to stimulus applied to spinal cord. The negative antidromic spike (measured downwards) is followed by positive waves constituting the extrin- sic hyperpolarizing potential. (Furukawa & Furshpan. *J. Neurophysiol.*)

The inhibitory character of the EHP was demonstrated by its ability, when it exceeded some 10 mV, to prevent the spike from invading the axon hillock, and to raise the threshold of the Mauthner cell to a suitably timed excitatory VIIIth nerve volley. All the evidence suggested that the site of action of the EHP was the axon cap, and intracellular recording from the soma strongly indicated that the block was a failure of electrotonic spread, such as could be induced by the application of an electrotonus locally. Thus the hyper- polarization of the EHP was fundamentally different from what would be expected were it the result of a change in membrane permeability such as that described for the IPSP; in this case the intracellular recording would indicate a phase of negativity corresponding to the extracellular record of positivity; by contrast, the internal and external electrodes both recorded positivity during the hyperpolarization. This hyperpolarization is therefore equivalent to the application of an external anode; and by inserting a micro- electrode into the axon cap, the spike potential could, indeed, be blocked by passing a hyperpolarizing current. As indicated earlier, this extrinsic anodal current could be achieved by fibres that enter the axon cap region, and if the action potential failed to invade the terminals this particular part of the Mauthner cell would be exposed to such a current; since it is here that the spike potential probably originates, the EHP would effectively block excitation.*

INHIBITORY NERVE ENDINGS

Antidromic excitation of the Mauthner cell leads to characteristic IPSP's called *late collateral inhibition*; these are due to the excitation of collaterals from the axon of the Mauthner cell that feed back to their own dendrites or soma, presumably by an interneurone, analogous with the axon collaterals of the spinal motor neurone operating through Renshaw cells (p. 1217). Normally this activity does not appear as a clearly marked depolarization, but it is converted to this by the KCl leaking out of the microelectrode. Asada showed that injected anions increased the size of the IPSP, in fact such large depolarizations could be achieved as to induce spikes. Unlike the motor neurone, the Mauthner cell was unaffected by injections of K^+, no hyperpolarization being obtained even with large injections of potassium citrate or sulphate.

Electrotonic Excitatory Transmission. Furshpan & Furukawa (1962) noticed a very short-latency response to stimulation of the ipsilateral N VIII, suggesting electrotonic transmission at an excitatory synapse. Later Furshpan (1964) recorded intracellularly from a lateral dendrite, the cell body, and the axon; finely graded responses with a latency of 0.1–0.15 msec., and a duration of 1.5–2.0 msec., were obtained; they could evoke a spike and be diminished by an inhibitory input, so they were similar to an ordinary EPSP except for the time-relations. The most probable site of this response was the myelinated club ending, because of its size; and this was confirmed by the finding that the responses were largest in the distal half of the lateral dendrite where these endings are concentrated. When axons of N VIII were penetrated, some produced very short-latency responses in the Mauthner cell; in this event they were invaded by antidromic stimuli, whereas those giving long-latency responses were not. Invasion by an antidromic stimulus from the Mauthner cell means passage across the synapse in the "wrong direction" and thus suggests an electrotonic type of synapse. Earlier, Robertson, Bodenheimer & Stage (1963) had suggested an electrotonic transmission through the club endings since these, in contrast with the *bouton* terminals, contained very few synaptic vesicles, whilst the synaptic cleft showed regions of membrane fusion.

Transmitter Substances

Excitatory Substances. So far as excitatory systems are concerned, acetylcholine is the substance to which a transmitter action in the vertebrate central nervous system can be unequivocally ascribed, and this is the excitation of Renshaw cells by collaterals from the motor neurones (Fig. 638).† This was demonstrated by Curtis & Eccles (1958) by iontophoretic application of acetylcholine‡ in the immediate environment of the cells, and more satisfactorily by Kuno & Rudomin (1966), who showed that antidromic stimulation of the motor

---

* By its very nature, such an inhibitory effect must be restricted to a small region of the cell, by contrast with a change in the electrical characteristics of the membrane which may spread for some distance. As Furukawa & Furshpan emphasized, the appearance of an extrinsic "source" must be associated with the appearance elsewhere of a "sink" that will favour excitation; in fact, under appropriate conditions, regions of extracellular negativity, probably corresponding to these "sinks", could be localized.

† In the ganglion of the invertebrate, *Aplysia*, acetylcholine seems to act both as excitatory and inhibitory transmitter, according to the cell being examined; in the former case it depolarizes, whilst in the latter it hyperpolarizes, thereby mimicking the normal inhibitory synaptic activity (Tauc & Gerschenfeld).

‡ Because of the blood-brain barrier that prevents acetylcholine from diffusing easily from blood to the surrounding nervous tissue (p. 722) the pharmacological agents had to be applied locally; even by this route, however, relatively high concentrations must be employed presumably because the synapses are protected by another barrier, perhaps the Schwann sheaths of the terminal axons.

nerves of the cat would bring about liberation of acetylcholine into the perfused spinal cord; the fact that this was increased by post-tetanic potentiation strongly suggests that the acetylcholine was liberated from presynaptic terminals on the Renshaw cell, and the observed parallelism between Renshaw cell activity and acetylcholine release provided further support. Thus the motor neurone, at its main axonal terminal on the muscle fibre, liberates acetylcholine, and also at the terminal of its collateral on a Renshaw cell. This is consistent with Dale's view that a given neurone will liberate the same transmitter at all its terminals, having, presumably, the synthetic apparatus for only the one transmitter (in this case choline acetylase, coenzyme A, choline and acetate; see, for example, Birks & MacIntosh, 1957).

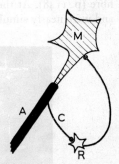

FIG. 638. Illustrating the activation of a Renshaw cell, R, by the collateral, C, of a motor neurone, M. Through this circuit the motor neurone inhibits its own excitability.

Amines

Noradrenaline was originally recognized as a peripheral transmitter, conveying the effects of nerve impulses to their effector cells, *e.g.*, in smooth muscle. Recently more and more evidence has accumulated implicating noradrenaline, and other amines, in central nervous transmission (see, for example, Fuxe & Gunne, 1964). The evidence is mainly morphological and biochemical; the presence of noradrenaline, dopamine, 5-hydroxytryptamine (5-HT), etc. in neurones being demonstrated histochemically, whilst synaptic vesicles have been shown to contain noradrenaline (Whittaker, 1964).*

Amino Acids

Glutamate occurs in large amounts in nervous tissue, and certainly it excites a variety of vertebrate neurones; thus Curtis, Phillis & Watkins applied this acidic amino-acid, as well as aspartic and cysteic acids, ionto-phoretically on to individual neurones of the spinal cord and found that interneurones, Renshaw cells and motor neurones were excited so that if, for example, a neurone was firing at a low frequency, application of one of these amino acids caused a large increase in frequency of firing. These authors concluded, however, that the effect was a non-specific increase in excitability of the post-synaptic membrane, perhaps by mimicking the action of the true transmitter. The action of glutamate is real, nevertheless, and may well be concerned in pathological convulsive states; the fact that it was active in concentrations of less than $M/100$, whilst its normal overall concentration in the mammalian central nervous system is about $M/100$, indicates that a great deal of this ion is intracellular. In invertebrate systems, including the neuromuscular

* Noradrenaline seems not to require an enzyme to remove it from its site of action after liberation by the nerve impulse; instead, there is a mechanism for its return to the nerve fibres where it is stored for subsequent release (see, for example, Gillespie, 1966; Axelrod, 1966). The enzymes *catechol oxidase* and *O-methyl transferase* both inactivate the transmitter, but their function seems to be that of preventing excessive accumulation, either intracellularly (catechol oxidase) or extracellularly (transferase). Axelrod & Kopin (1969) have reviewed the metabolism of noradrenaline in sympathetic nerves.

junction, acetylcholine is agreed *not* to be an excitatory transmitter (Ellis, Thienes & Wiersma; Grundfest, Reuben & Rickles; Wood). Robbins observed that certain amino-acids, notably L-glutamic acid, would excite the crustacean neuromuscular junction in low concentrations—of the order of 5.10^{-4} M— whilst at higher concentrations it would reversibly block conduction; a more extensive study of Van Harreveld & Mendelson showed that the same effects could be found in a number of species of crustacea, and that application led to a depolarization of the muscle fibre lasting about 0·1 sec.

By simultaneously recording with intra- and extracellular electrodes, Takeuchi & Takeuchi (1964) were able to localize very sharply "active spots" on the muscle, equivalent to those described by Del Castillo & Katz on the vertebrate muscle fibre (p. 1148). At these spots the recorded junctional potentials were maximal and were nearly simultaneous with the externally recorded presynaptic spike. As

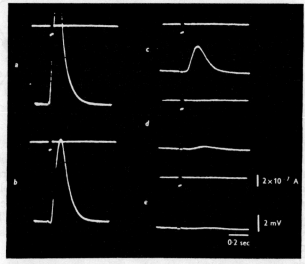

Fig. 639. The effects of glutamate on crayfish muscle. Lower traces, L- gluta- mate induced depolarizations. Upper traces, monitored injection current. *a*, L-glutamate filled pipette was located near a focus. From *a* to *e* the L-glutamate pipette was moved along muscle fibre in 8μ steps. (Takeuchi & Takeuchi. *J. Physiol.*)

Fig. 639 shows, these spots are highly sensitive to iontophoretically applied glutamic acid; movement of the micropipette by $8\,\mu$ steps results in progressive diminution of the response. These responses interacted with the normally evoked excitatory junctional potentials, and they represented a direct action on the muscle since they occurred in the denervated preparation; furthermore, the glutamate did not affect the frequency of the spontaneous miniature junctional potentials. Unlike the situation described by Curtis & Watkins (1960), the D-isomer was inactive.* According to Kerkut's studies, other invertebrate nerve-muscle junctions are sensitive to glutamate, *e.g.*, those of the cockroach, *Periplaneta*, of the snail, *Helix*, and the crab, *Carcinus*. By passing a Ringer's

---

* Tetrodotoxin and saxitonin had no effect on the response of the crustacean neuro-muscular junction to applied glutamate. The procaine-treated muscle permits the propaga-tion of spikes, but these are not affected by the poisons (Ozeki, Freeman & Grundfest, 1960). These findings are consistent with the notion that tetrodotoxin only affects processes mediated by an activated Na^+-transport; the procaine spike is probably the result of altered permeability to Ca^{++}.

solution over the exposed nerve-muscle preparation, Kerkut & Walker (1966) and Kerkut *et al.* (1965) showed that, in the cockroach and snail, stimulation of the excitatory nerve led to the appearance of glutamate in the outflowing fluid, the amount collected being proportional to the number of stimuli.*

Factor S

Van der Kloot (1960) extracted a substance from the crayfish, which he called Factor S, that stimulated contraction of its closer muscle. It appeared in perfusates from stimulated, but not from unstimulated, claws. Chemically it seemed to be similar to a catechol derivative—*catechol-4*—isolated by Östlund (1954) from a number of invertebrate species. A similar substance has been extracted by Cook (1967) from a number of insects; it, too, is excitatory when applied to muscle.

Inhibitory Transmitter. Florey in 1954 prepared an extract of mammalian brain which he called Factor I, and which had a powerful inhibitory action on the invertebrate stretch-receptor (p. 1252); when topically applied to the vertebrate spinal cord it inhibited the monosynaptic tendon reflex (Florey & McLennan) and subsequent studies have shown it to be effective on a number of other systems including the mammalian cerebral cortex, where it alters the configuration of the potentials recorded from the surface in response to afferent stimulation. Bazemore, Elliott & Florey considered that the active principle of Factor I was γ-aminobutyric acid (GABA), and this substance certainly mimics the action of Factor I in a number of situations; thus Edwards & Kuffler have shown that inhibition of the crayfish stretch receptor (p. 1252) may be brought about by application of the compound in a concentration as low as 10 μg./ml., the effect being to drive the activated cell membrane towards its resting potential, above the neurone's firing level, just as with the nervous impulse.

When studied on the invertebrate neuromuscular system it mimicked not only the inhibitory action, as revealed by the cessation of contraction in response to excitation, but also the electrical events so that, for example, the application of GABA forced the resting potential to the same equilibrium potential as that found with maximal inhibitory stimulation (Dudel & Kuffler); moreover it had the same pre-synaptic inhibitory action.

Extracts of the lobster central nervous system contained, in addition to other amino acids and amines, GABA, taurine and betaine, which all had a strong blocking action on the junction (Dudel *et al.*, 1963); the same substances were extracted from the motor-inhibitor bundle to the leg muscle and, if the inhibitor and excitor nerves were separated, it was found that GABA was confined to the inhibitor nerve whilst the other blocking compounds, such as taurine and betaine, were found in both (Kravitz *et al.*, 1963a, b).

Actions of GABA

Takeuchi & Takeuchi (1966) showed that the action of iontophoretically applied pulses of GABA was highly localized to active spots, so that when the amplitudes of GABA- and glutamate-induced potentials were plotted against distance from a given point on the muscle, the peaks of activity coincided accurately; corresponding with this it has been shown histologically that the excitatory and inhibitory nerves run parallel courses and often end almost at the same place (see, for example, Atwood, 1968). There is no doubt, now, that the two substances do not act on the same receptors; thus a decreased sensitivity of the muscle, due to protracted application of glutamate, had no effect on the GABA-potentials. Takeuchi & Takeuchi (1967) measured the membrane conductance of the crayfish muscle fibre on treatment with GABA when Cl^- was substituted by other anions; in normal Ringer's solution GABA changes the slope of the voltage-current relation indicating increased conductance; when Cl^- was replaced by successively greater proportions

---

* Florey & Woodcock (1968) have discussed the possibility that glutamate may act presynaptically at invertebrate junctions.

of methyl sulphate, the size of the conductance increase was diminished until, with no Cl⁻, it was abolished. In general, the anions could be placed in a series with Br⁻ at one end and BrO_3^- at the other; Br⁻ actually enhanced the effect of GABA whilst the remaining anions decreased it.* Inhibition at the crustacean junction is both pre- and postsynaptic; Dudel (1965b) has measured the effects of GABA and other drugs on the muscle membrane conductance, *i.e.*, their postsynaptic effects, and on the amount of transmitter released by the nerve terminal, as deduced from statistical analysis of the quantum contents of the junctional potentials. GABA and a number of related compounds mimicked the inhibitory nerve in increasing membrane conductance, and it reduced the amount of transmitter released per stimulus, thus inhibiting both post- and presynaptic activities. Fig. 640 illustrates schematically this dual action of GABA.

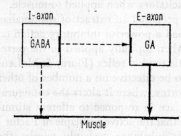

FIG. 640. Diagrammatic illustration of synaptic contacts in the crayfish opener muscle. Inhibitory transmission, dotted arrows; excitatory transmission, solid arrows; GABA, γ-aminobutyric acid; GA, glutamic acid. (Atwood. *Experientia.*)

Transmitter Release

Direct proof of the action of a transmitter is best provided by its collection from the perfused system after stimulation of the nerve; with GABA this is complicated by the spontaneous release of this amino acid that normally takes place; however, if the exposed neuromuscular preparation is washed for some hours, the amount entering the washing fluid settles down to a steady value. As Fig. 641 shows, stimulation of the inhibitory nerve caused a marked increase in the output, but not stimulation of the excitatory nerve; furthermore, the amount released increased with frequency of stimulation, whilst with low Ca^{++} in the

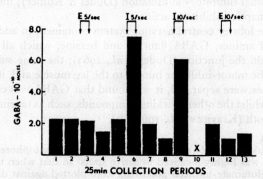

FIG. 641. Illustrating release of GABA into the fluid perfusing the opening muscle of the crusher claw of the lobster, in response to I-nerve stimulation. The preparation was washed with saline medium for 4 hr prior to the first resting collection (no. 1). GABA was assayed in 25-min collection samples during rest and during E- and I-nerve stimulation. In most experiments the assay failed on one or more samples, indicated by X. (Otsuka *et al.* *Proc. Nat. Acad. Sci.*)

* The essential characteristic of the anions that favours permeability through the inhibitory junction membrane is not easy to determine, so that no single factor is responsible. Certainly, as Takeuchi & Takeuchi have shown, hydrated ion size, "naked ion" size, or hydration energy of the ion cannot be directly correlated with permeability; lipid-solubility is likewise not determining, since acetate is much less effective than formate, at any rate in the Mauthner cell (Asada, 1963).

medium stimulation no longer produced inhibitory junctional potentials and no longer increased output of GABA, although action potentials were recorded from the nerve. The amount released was some $1-4.10^{-14}$ moles per impulse compared with Takeuchi & Takeuchi's (1965) estimate of 4.10^{-15} moles required for iontophoretic stimulation.

Vertebrate Central Nervous System. In the vertebrate central nervous system the function of GABA seems not so well substantiated, and its depressant action may well be non-specific on all responses—the EPSP, the IPSP and spike potential being all depressed, most probably by causing an increase in chloride-conductance (McLennan, 1957; Curtis, Phillis & Watkins, 1959; Curtis & Watkins, 1960). This non-specific depression of neuronal activity is well demonstrated by the effects of topical application to the cerebral cortex where it progressively abolishes the responses to afferent stimulation to greater and greater depths as the drug slowly diffuses from the surface into the deeper structures (Bindman, Lippold & Redfearn, 1962).*

Although GABA may not be the essential principle of Factor I, it is very likely that the extract of this name does contain an inhibitory transmitter; thus Florey & McLennan, as indicated above, reversibly blocked the reflex pathway through the cord by topical application of the extract, whilst the effect was abolished by strychnine, a convulsant drug that exerts its activity by blocking the inhibitory pathway impinging on the monosynaptic reflex arc in the cord.† (Curtis 1962).

Mauthner Cell

The electrical excitatory and inhibitory actions taking place in the Mauthner cell have already been alluded to; there is no doubt that the electrical inhibitory potential is followed by inhibitory postsynaptic potentials—the late collateral inhibition—and its inhibitory character may be demonstrated by the diminished height of a second antidromic impulse initiated in the cell. Fig. 642 shows records taken by Diamond (1968) with an electrode whose position was subsequently identified as being some $250-300$ μ away from the axon-hillock, *i.e.*, from the point at which the antidromically evoked spike was initiated. It is in this region that the club-ending type of synapse is collected, and since these are definitely excitatory it follows that the inhibitory endings are interspersed with these and may well be the small endings that are distributed all over the cell, but are fairly highly concentrated in this part of the lateral dendrite. By applying a micropipette loaded with GABA to different regions of the Mauthner cell, Diamond showed that the lateral dendrite was sensitive, in so far as GABA diminished the height of the antidromically

* GABA alters the configuration of the succession of potential changes recorded from the surface of the cortex in response to an afferent stimulus, so that the predominant effect now becomes one of positivity; Purpura, Girado & Grundfest (1957) considered that GABA left intact the inhibitory component of the potential complex, namely, a hyperpolarization which, according to them, would be recorded as a phase of positivity on the surface. Bindman *et al.* showed, however, that all activity at the surface was non-specifically blocked, so that the surface was behaving as a "source" to "sinks" of activity in the depths of the tissue and thus becoming positive.

† McLennan (1959) has shown that chromatographic separation of Factor I gives evidence not only of GABA but of guanidino-propionic acid and γ-guanidino-butyric acid, which both have similar effects to that of GABA on the crayfish stretch-receptor and mammalian cerebral cortex. γ-Amino-butyrylcholine has been reported to be present also and to be some 500–1,000 times more active on the cerebral cortex than GABA (Takahashi, Nagashima & Koshino, 1958). Honour & McLennan (1960) have examined these substances and shown that Factor I cannot be mimicked by any one of them. The effects of catecholamines, such as adrenaline, have been examined by McLennan (1961); 3-hydroxytyramine, topically applied to the cord, reversibly abolishes the knee-jerk like Factor I, but the concentration required was very large—50 g./litre—and Curtis (quoted by McLennan) failed to find any effect of iontophoretic application to motor neurones in the cord. When tested for inhibitory activity on the crayfish stretch-receptor, a large number of compounds, containing acidic and basic groups separated by some 4A, were effective; outstanding was hydroxytyramine (McGeer, McGeer & McLennan, 1961).

initiated spike,* and it is indeed remarkable that inhibitory action, initiated some 300 μ away from the axon hillock, can affect the size of the spike; an analysis of the variation of the dose-response curve with distance from the axon hillock showed, however, that the effects were consistent with cable-theory if the inhibitory transmitter was, indeed, producing a local increase in membrane conductance.

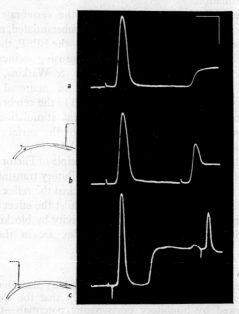

Fig. 642. Records of antidromic spike, and depolarization during late collateral inhibition of the Mauthner cell. (a) and (b), recording from branch of lateral dendrite, 300 μ from the axon hillock; the beginning of the late depolarization is seen following the antidromic spike in (a), and in (b) a second antidromic spike, fired at the beginning of the depolarization, is seen to be diminished in size. (c) a typical cell-body recording (from another preparation) showing the much earlier onset of the late depolarization, and a second diminished spike during it. Calibrations; 10 mV, 2 msec. (Diamond. *J. Physiol.*)

Glycine

In their early studies on the effects of micro-injections of amino acids in the spinal cord, Curtis & Watkins observed hyperpolarizing responses and depression of activity when glycine was injected in the neighbourhood of motor neurones. Later Curtis *et al.* (1967) showed that the hyperpolarization was blocked by strychnine, a drug that causes convulsions by virtue of its blockage of inhibitory circuits. By contrast, it had no effect on the depression caused by GABA. Later microinjection studies of Werman, Davidoff & Aprison (1968) established a firm parallelism between the effects of glycine and those of inhibitory afferent stimulation; thus the equilibrium potentials for the hyperpolarizations were the same; glycine produced blockade of antidromic spike invasion of the motor neurone soma and reduced excitatory synaptic potentials in precisely the way that inhibitory afferent stimulation did. These facts, taken in conjunction with the finding of high concentrations of glycine in the ventral horn grey matter, and a diminution of this when anoxic lesions had caused a great reduction in in-

---

* Earlier Diamond had found that the sensitivity of the lateral dendrite to GABA fell off strongly with distance, but this was because of the attenuation of its effect when recorded from the axon hillock region; when the recording electrode was inside the distal region of the lateral dendrite, marked effects could be recorded. The reader is referred to Diamond's paper for an interesting discussion of the factors involved in remote inhibitory action.

hibitory reflex activity (Davidoff *et al.* 1967) lend strong support to this amino acid as a candidate for inhibitory transmitter in the mammalian central nervous system.

Convulsants

The identification of the inhibitory transmitter in the mammalian central nervous system permits a more precise determination of the mechanisms of action of convulsant agents. Strychnine abolishes the reflexly evoked IPSP of a motor neurone but has no effect on the EPSP. Strychnine also blocks the IPSP induced by microelectrophoretic application of glycine, hence strychnine competes with glycine for postsynaptic receptor sites, rather than by reducing the amount of transmitter released. By contrast, it would seem that tetanus toxin acts presynaptically, preventing the release of the inhibitory transmitter. Thus Curtis & De Groat (1968) studied the inhibition of Renshaw cells following stimulation of an afferent nerve; locally applied tetanus toxin caused a gradual increase in spontaneous firing of the Renshaw cell and reduced the inhibitory effect of an afferent stimulus, whilst it had no effect on the depression of activity caused by local injection of glycine. It is very likely, therefore, that tetanus toxin inhibits the release of transmitter from the afferent inhibitory terminals on the Renshaw cell. Strychnine has no effect on presynaptic inhibition, but the latter is blocked by picrotoxin (Eccles, Schmidt & Willis, 1963) so that the convulsant action of this chemical is presumably due to the presynaptic action; as we should expect on this basis, picrotoxin has no effect on the IPSP's developed by glycine (Curtis, 1969).

Electric Organs

Structure. The electric organs of certain fishes have been studied in some detail in the hope that they would cast some light on the mechanism of neuromuscular transmission, since at one time it was considered that the unit of this organ was essentially an end-plate. The electric organ is highly developed in *Electrophorus electricus*, representing a very large proportion* of the animal's tissue. The four organs in this species have been accurately described by Couceiro & Akerman (1948), and more recently by Bloom & Barrnett (1966). They are large masses of tissue divided into compartments, or *electroplaxes*, by connective-tissue laminae running longitudinally and transversely (Fig. 643), with dimensions of 30 $\times$ 1·8 $\times$ 0·1 to 14 $\times$ 2 $\times$ 1·9 mm. according to their position. Each electroplaque is a multinucleate syncytium, and is enclosed within a dense collagenous rectangular compartment whose long axis runs medio-laterally (Fig. 644). Across the caudal third of each compartment the electroplaque extends as a plate or band, with large papillæ extending from its anterior face, and shorter and more numerous papillæ from its posterior face. The electroplaque is activated by a nerve fibre which ramifies on its posterior aspect, so that we may speak of the innervated and non-innervated faces (Fig. 644). The surfaces of both the innervated and non-innervated portions of the electroplaque have multiple caveolæ extending into the most peripheral 100 to 1,000 mμ of the plaque cytoplasm, being more developed on the papillary projections than at the base. Myelinated axons approach the plaque surface; when close they lose their glial and myelin sheaths, which are replaced by a thin glial layer that usually just covers the surface not apposed to the electroplaque; one fibre may have several adjacent zones of apposition. Each zone shows accumulation of synaptic vesicles 400–600A in diameter packed in the nerve terminal; the junctional clefts are of the order of 150–300A thick. Cross-sections revealed neurotubules and neurofilaments as well as synaptic vesicles. At all junctional areas the caveolæ could be frequently seen to be open. Bloom & Barrnett examined the localization of cholinesterase within the electroplaque in electron-microscopical preparations; invariably the majority of precipitated reaction product was deposited on or near

---

* All the viscera of the animal are under the bones of the skull and the first 20 vertebræ, whilst the electric organs occupy the space under the remaining 230 vertebræ; it is therefore possible to keep an *Electrophorus* alive after amputation of its posterior four-fifths.

39*

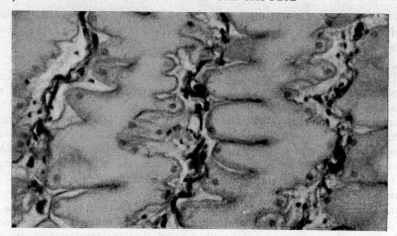

Fig. 643. The electric organ of *Electrophorus electricus*. High power micro-photograph, showing part of two electroplaxes which are cut transversely. × 500. (Couceirs & Akerman. *Anais Acad. Bras. de Ciencias.*)

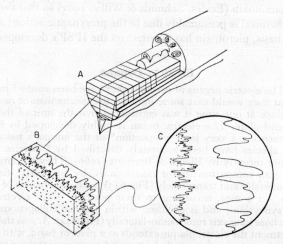

Fig. 644. Diagrammatic sketch of *Electrophorus* cut away dorsally to reveal electric organ (*A*) *in situ*. *B* indicates posterolateral view of one electroplaque within its mucoid-filled compartment, and *C* shows edge of plaque with innervated (left) and non-innervated (right) sides. (Bloom & Barrnett. *J. Cell Biol.*)

the external innervated surface of the plaque and within the tubulovesicular organelles, or caveolæ, opening on to this innervated surface; reaction product was also present, as with Miledi's study on the neuromuscular junction, in the synaptic vesicles, but this appeared when the preparation was treated with the cholinesterase inhibitor physostigmine.

Electric Discharge. On open circuit the electric eel can produce a discharge of some 600 volts, with a maximal power-output of some 100 watts. In *Torpedo* the voltage is lower, some 30–60 volts, whilst in the ray it is only 1–3 volts; as Nachmansohn & Meyerhof showed, these voltages correlated well with the number of electroplaxes in the respective organs, being 5,000–6,000 in the electric eel, 400–500 in the torpedo and only 60–80 in the ray.

Mechanism of Discharge. If each electroplax behaved like a muscle fibre or an end-plate, having a resting potential of, say, 80–90 mV, and giving an action- or end-plate potential consisting of a reversal of polarization or, if it behaved like an end-plate, just a depolarization, then we could expect to develop a summated response, leading to a high voltage discharge, only if the electroplaxes had some electrical asymmetry. If the whole surface of the electroplax behaved uniformly, then the effects of a depolarization, or action potential, would be similar to those occurring in a muscle or nerve bundle, the maximal potential difference developed being no greater than the depolarization, or action potential. Definite proof that the high voltage actually developed was due to an asymmetry across the electroplax was provided by Keynes & Martins-Ferreira, working in Chagas' laboratory. They inserted microelectrodes into individual electroplaxes and recorded the effects of stimulation.

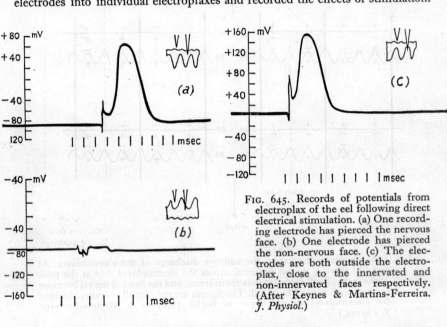

Fig. 645. Records of potentials from electroplax of the eel following direct electrical stimulation. (a) One recording electrode has pierced the nervous face. (b) One electrode has pierced the non-nervous face. (c) The electrodes are both outside the electroplax, close to the innervated and non-innervated faces respectively. (After Keynes & Martins-Ferreira. *J. Physiol.*)

On the average, the resting potential across the electroplax surface was 85 mV, and this corresponded with the distribution of K$^+$ between electroplax and extracellular fluid found by Davson & Lage, and Nishie & Harris. Stimulation of the organ caused a reversal of polarization, the outside becoming some 65 mV negative in respect to the inside, thereby giving an action potential of some 150 mV. As to whether or not an action potential could be recorded, depended on the relative positions of the electrodes with regard to the anterior and posterior faces of the electroplax. In Fig. 645(a) the electrodes were placed so as to record the potential difference across the posterior, or innervated, face, and the large action potential recorded showed that a reversal of polarization occurred across this face. In (b) the potential across the non-nervous, or anterior, face was recorded, and it will be seen that a negligible change of potential took place. In (c) is shown the effect of a stimulus on the potential difference between the two electrodes placed so that the tip of neither was within the electroplax, the one being close to the outside of the anterior surface, the other close to the outside of the posterior surface. At rest, the potential difference was zero and, if the electroplax were quite symmetrical in its response to stimulation, this zero potential difference would be maintained after an electrical stimulus. As the record shows, however, an action potential was recorded. This time it is imposed on the zero-potential base-line, and is obviously due to a reversal of polarity across the nervous face. Schematically, then, the development of a high potential across the organ, leading to a high-voltage discharge, may be represented by Fig. 646; the reversal of polarization across only one face of the electroplax leads to what is, in effect, a series connection of the electromotive forces developed by the individual electroplaxes, by contrast with the parallel connection in a muscle or nerve bundle.

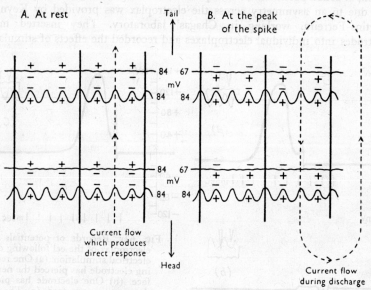

FIG. 646. Diagram illustrating the additive discharge of the electroplaxes. At rest (A) there is no net potential across the electroplaxes, but at the peak of the spike (B) all the potentials are in series, and the head of the eel becomes positive with respect to its tail. The figures are the overall averages obtained for electroplaxes in the organ of Sachs. (Keynes & Martins-Ferreira. *J. Physiol.*)

Direct and Indirect Stimulation. The electroplaxes in a lump of excised organ of the electric eel can be stimulated directly, by simply establishing a potential difference across the block of tissue (Albe-Fessard, Chagas & Ferreira), or indirectly through the nerve. To excite, the current must pass in the antero-posterior direction, *i.e.*, in such a way as to depolarize the nervous face. The critical degree of depolarization required to activate the discharge was of the same order, namely 12 mV, as that required to excite the squid membrane. The indirect response, obtained by stimulation of the nerve, had an irreducible latency of some 1·5 msec., compared with only 0·1 msec. for direct stimulation; and there seems little doubt from the studies of Chagas and his collaborators that nervous excitation is similar to nerve-muscle excitation, being mediated by acetylcholine and blocked by curare. Keynes & Martins-Ferreira found little evidence, however, of a synaptic potential comparable with the end-plate potential, although Albe-Fessard & Chagas found a definite pre-potential that grew into a spike when the stimulus-strength was great enough.

Mechanism of the Spike. The mechanism of the spike has been discussed recently by Nakamura, Nakajima & Grundfest (1965); voltage-clamp measurements indicated an intense early current, presumably carried by Na^+, which was reversed in direction at $+62$ mV and abolished by tetrodotoxin. A delayed outward current, corresponding to the delayed rectification of the squid axon, was missing, however, so that the decline of the spike apparently depends on the cutting off of the Na^+-activation process alone. Altramirano in 1955 had observed that the resistance of the electroplaque increased 2–3-fold when it was depolarized by applied currents above a certain threshold; this increase was evident in the falling phase of the spike and persisted as long as the current pulse, namely 20 msec. It also occurred when the electroplaque was in a high concentration of K^+ and the spike was abolished. Nakamura *et al.* confirmed this, and attributed it to a K^+-inactivation process that, according to Grundfest, seems to be a feature of many excitable systems; thus the anomalous rectification of frog muscle (p. 575) could be explained on the basis of a depolarizing inactivation of K^+-permeability and a hyperpolarizing activation of K^+-permeability.

Torpedo and Raia

By contrast with that of the electric eel, the electric organs of the torpedo and ray were found to be inexcitable by directly applied shocks, and thus their electroplaxes might be likened to end-plates rather than muscle fibres.

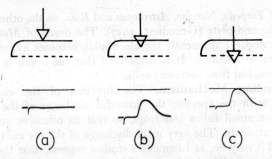

(a)   (b)   (c)

Fig. 647. Records from the electroplaxes of the torpedo; the innervated face is represented by the full line. (a) the recording electrode is outside the cell close to the innervated face; (b) the electrode is within the electroplax. (c) the electrode has passed through the electroplax. Note that the records in (a) and (c) are different although in both cases the electrode is outside the electroplax. (After Bennett, Wurzel & Grundfest. *J. gen. Physiol.*)

By employing microelectrodes along the lines developed by Keynes & Martins-Ferreira, Bennett, Wurzel & Grundfest have confirmed the direct electrical inexcitability of the electroplaxes not only of *Torpedo nobiliana* but also of *Narcine brasiliensis* and of the stargazer *Astroscopus y-graecum*. They have demonstrated, too, the essential asymmetry of the electroplaque in the torpedo, only the innervated face showing any response to a nervous stimulus. This is revealed by the records of Fig. 647. At *a* the recording electrode was just outside the innervated face, and a stimulus caused only a small negativity; at *b* the electrode had entered and recorded a resting potential of 50–70 mV. The stimulus now caused an internal positivity which more than brought the resting potential to zero, *i.e.*, there was some overshoot. At *c* the electrode had passed right through the electroplax, as indicated by the absence of a resting potential, but, because of the asymmetry of the system, there was still a large positivity of the recording electrode on stimulating. Thus the uninnervated face has not responded to the stimulus. However close the stimulating electrodes were to the electroplax there was always a delay in the response of 1–3 msec., suggesting a synaptic type of transmission. Furthermore, the innervated face showed no evidence of active propagation of a locally induced depolarization, as with nerve or muscle; instead, the electrical response could have different amplitudes and forms in closely adjacent regions. All attempts to excite the electroplax by direct currents failed, the membrane potential changing linearly with applied current over a large range, and thus showing no evidence of the regenerative activity that is such a pronounced feature of the axon membrane. The response to drugs was essentially what would be expected of a system activated by acetylcholine, so that curare, for example, blocked the electrical response just as with the electric eel.

In the ray, although not so elaborately studied, Brock, Eccles & Keynes showed that the action potential is only a depolarization and not a reversal of polarity; it lasted some 20 msec. compared with 3–4 msec. for *Electrophorus* and was thus analogous with an end-plate potential.

If the analogy is sound, therefore, we may characterize the electric eel's electroplax as one in which the end-plate and the electrically excitable muscle membrane have been retained, only the contractile machinery having been lost; this permits both electrical and neural stimulation, with sensitivity to acetylcholine; the neurally and acetylcholine-stimulated responses being blocked by curare.

The organs of *Torpedo*, *Narcine*, *Astrocopus* and *Raia*, on the other hand, have retained only the end-plate (Grundfest, 1957). The organ of *Malapterus* falls within neither category; apparently the electroplax becomes hyperpolarized as a result of neural stimulation. It is considered that the organ is not derived from muscle tissue but from secretory cells.

Direction-Finding Mechanism. The function of the electric organ probably varies with the species; the powerful discharge of the electric eel can certainly stun small fishes and frogs, so that its offensive and defensive values are without doubt. The very weak discharge of the ray calls for another explanation, and it may be, as Lissmann's studies suggest, that the discharges subserve a direction-finding mechanism. Lissmann observed that *Gymnarchus niloticus* emitted an uninterrupted series of impulses; if any change in the electrical field near the fish was artificially superimposed—for example, by placing a U-shaped copper wire in the water—the fish was immediately aware of it and showed characteristic escape-reactions. Again, Coates, Altamirano & Grundfest observed a continuous low-intensity discharge at some 65–300 per sec. from the knife-fish, which may have had a similar function. Finally, the organ of Sachs is separate from the main electric organ of *Electrophorus electricus*; it gives rise to a much weaker discharge and it has been suggested that it is responsible for a similar direction-finding mechanism (Coates, 1950).

Sensory Mechanism

The mechanism whereby receptors of *Gymnarchus niloticus* are able to convert the short pulses at a frequency of some 300 cycles per sec. into a nervous discharge that will indicate, by its frequency, the intensity of the pulse, has been discussed by Machin & Lissmann, whilst the actual sensory responses, recorded from single fibres of the lateral line nerve, have been described by Hagiwara, Kusano & Negishi. The actual end-organs that receive the electrical pulses may well be the so-called *mormyroblasts*; they are found in the skin and are composed of several sensory cells aggregated around a small cavity which connects to the outside through a jelly-filled canal.

Feedback Mechanism

In a recent study Larimer & MacDonald (1968) have confirmed the interesting finding of Watanabe & Takeda (1963), to the effect that, when the gymnotid *Eigenmannia* is exposed to a frequency of discharge of a slightly altered frequency from that which it emits, the fish alters the frequency of its discharge away from that of the imposed signal; thus it responds with a higher frequency if the signal has a lower frequency than that of its original discharge, and *vice versa*. If the imposed signal has approximately the same frequency as that of the discharge, there is no alteration in the frequency of the latter. Thus there is a sensitive feedback mechanism that presumably enables the fish to alter the frequency of its own discharge when there is a likelihood of confusion through the discharges of other fishes in the neighbourhood.

REFERENCES

ALBE-FESSARD, D., & CHAGAS, C. (1954). "Mise en évidence d'un potentiel de jonction par dérivation intra-cellulaire dans une électroplaque de l'organe de Sachs du Gymnote." *C. R. Acad. Sci.*, Paris, **239**, 1682–1684.

ALBE-FESSARD, D., CHAGAS, C., & MARTINS-FERREIRA, H. (1951). "Sur l'existence d'un système directement excitable dans l'organe électrique de l'*Electrophorus electricus*, L." *C. R. Acad. Sci.*, Paris, **232**, 1015–1017.

ANDERSEN, P. & CURTIS, D. R. (1964). The pharmacology of the synaptic and acetyl-choline-induced excitation of ventrobasal thalamic neurones. *Acta physiol. scand.*, **61**, 100–120.

ARAKI, T., ECCLES, J. C., & ITO, M. (1960). Correlation of the inhibitory post-synaptic potential of motoneurons with the latency and time course of inhibition of mono-synaptic reflexes. *J. Physiol.*, **150**, 354–377.

ARAKI, T., ITO, M., & OSCARSSON, O. (1961). Anion permeability of the synaptic and non-synaptic motoneurone membrane. *J. Physiol.*, **159**, 410–435.

ARAKI, T., & TERZUOLO, C. A. (1962). Membrane currents in spinal motoneurones associated with the action potential and synaptic activity. *J. Neurophysiol.*, **25**, 772–789.

ARMETT, C. J. & RITCHIE, J. M. (1963). The ionic requirements for the action of acetyl-choline on mammalian non-myelinated fibres. *J. Physiol.*, **165**, 141–159.

ARVANITAKI, A., & FESSARD, A. (1935). "Sur les Potentiels Retardés de la Réponse du Nerf de Crabe. Action de la Vératrine et de la Privation d'Électrolytes." *C. r. Soc. Biol.*, **118**, 419.

ASADA, Y. (1963). Effects of intracellularly injected anions on the Mauthner cells of gold-fish. *Jap. J. Physiol.*, **13**, 583–598.

ATWOOD, H. L. (1968). Peripheral inhibition in crustacean muscle. *Experientia.*, **24**, 753–763.

AXELROD, J. (1966). Methylation reactions in the formation and metabolism of catechol-amines and other biogenic amines. *Pharmacol. Rev.*, **18**, 95–113.

AXELROD, J. & KOPIN, I. J. (1969). The uptake, storage, release and metabolism of nor-adrenaline in sympathetic nerves. *Progr. Brain Res.*, **31**, 21–32.

AXELSSON, J., & THESLEFF, S. (1959). A study of supersensitivity in denervated mammalian skeletal muscle. *J. Physiol.*, **149**, 178–193.

BACH-Y-RITA, P. & ITO, F. (1966). *In vivo* studies on fast and slow muscle fibres in cat extraocular muscles. *J. gen. Physiol.*, **49**, 1177–1198.

BANISTER, J., & SCRASE, M. (1950). Acetylcholine synthesis in normal and denervated sympathetic ganglion of the cat. *J. Physiol.*, **111**, 437–444.

BARRON, D. H. & MATTHEWS, B. H. C. (1938). The interpretation of potential changes in the spinal cord. *J. Pshyiol.*, **92**, 276.

BAZEMORE, A., ELLIOTT, K. A. C., & FLOREY, E. (1956). "Factor I and γ-Aminobutyric Acid." *Nature*, **178**, 1052–1053.

BEANI, L., BIANCHI, C. & LEDDA, F. (1964). The effect of tubocurarine on acetylcholine release from motor nerve terminals. *J. Physiol.*, **174**, 172–183.

BENNETT, M. V. L. (1966). Physiology of electrotonic junctions. *Ann. N.Y. Acad. Sci.*, **137**, 509–539.

BENNETT, M. V. L., NAKAJIMA, Y. & PAPPAS, G. D. (1967). Physiology and ultrastructure of electrotonic junctions. I. Supramedullary neurons. *J. Neurophysiol.*, **30**, 161–179.

BENNETT, M. V. L., PAPPAS, G. D., ALJURE, E. & NAKAJIMA, Y. (1967). Physiology and ultrastructure of electrotonic junctions. II. Spinal and medullary electromotor nuclei in mormyrid fish. *J. Neurophysiol.*, **30**, 180–208.

BENNETT, M. V. L., PAPPAS, G. D., GIMÉNEZ, M. & NAKAJIMA, Y. (1967). Physiology and ultrastructure of electrotonic junctions. IV. Medullary electromotor nuclei in gymnotid fish. *J. Neurophysiol.*, **30**, 236–300.

BENNETT, M. V. L., & WURZEL, M., & GRUNDFEST, H. (1961). Properties of electroplaques of *Torpedo nobiliana*. *J. gen. Physiol.*, **44**, 757–804.

BINDMAN, L. J., LIPPOLD, O. C. J., & REDFEARN, J. W. T. (1962). The non-selective blocking action of γ-aminobutyric acid on the sensory cerebral cortex of the rat. *J. Physiol.*, **162**, 105–120.

BIRKS, R. I. (1963). The role of sodium ions in the metabolism of acetylcholine. *Canad. J. Biochem. Physiol.*, **41**, 2573–2597.

BIRKS, R., HUXLEY, H. E., KATZ, B. (1960). The fine structure of the neuromuscular junction of the frog. *J. Physiol.*, **150**, 134–144.

BIRKS, R., KATZ, B., & MILEDI, R. (1960). Physiological and structural changes at the amphibian myoneural junction, in the course of nerve degeneration. *J. Physiol.*, **150**, 145–168.

BIRKS, R. I., & MACINTOSH, F. C. (1957). Acetylcholine metabolism at nerve-endings. *Brit. med. Bull.*, **13**, 157–161.

BIRKS, R. & MACINTOSH, F. C. (1961). Acetylcholine metabolism of a sympathetic ganglion. *Canad. J. Biochem. Physiol.*, **39**, 787–827

BLACKMAN, J. G., GINSBORG, B. L. & RAY, C. (1963, a). Synaptic transmission in the sympathetic ganglion of the frog. *J. Physiol.*, **167**, 355–373.

BLACKMAN, J. G., GINSBORG, B. L. & RAY, C. (1963, b). Some effects of changes in ionic concentration on the action potential of sympathetic ganglion cells in the frog. *J. Physiol.*, **167**, 374–388.

BLACKMAN, J. G., GINSBORG, B. L. & RAY, C. (1963, c). Spontaneous synaptic activity in sympathetic ganglion cells of the frog. *J. Physiol.*, **167**, 389–401.

BLACKMAN, J. G., GINSBORG, B. L. & RAY, C. (1963, d). On the quantal release of the transmitter at a sympathetic synapse. *J. Physiol.*, **167**, 402–415.

BLANKENSHIP, J. E. & KUNO, M. (1968). Analysis of spontaneous subthreshold activity in spinal motoneurons of the cat. *J. Neurophysiol.*, **31**, 195–209.

BLIOCH, Z. L., GLAGOLEVA, I. M., LIBERMAN, E. A. & NENASHEV, V. A. (1968). A study of the mechanism of quantal transmitter release at a chemical synapse. *J. Physiol.*, **199**, 11–35.

BLOOM, F. E. & BARRNETT, R. J. (1966). Fine structural localization of acetylcholinesterase in electroplaque of the electric eel. *J. Cell Biol.*, **29**, 475–495.

BODIAN, D. (1942). "Cytological Aspects of Synaptic Function." *Physiol. Rev.*, **22**, 146.

BOISTEL, J., & FATT, P. (1958). Membrane excitability change during inhibitory transmitter action in crustacean muscle. *J. Physiol.*, **144**, 176–191.

BOYD, I. A., & MARTIN, A. R. (1956). Spontaneous subthreshold activity at mammalian neuromuscular junctions. *J. Physiol.*, **132**, 61–73.

BOYD, I. A., & MARTIN, A. R. (1956). The end-plate potential in mammalian muscle. *J. Physiol.*, **132**, 74–91.

BRAUN, M. & SCHMIDT, R. F. (1966). Potential changes recorded from the frog motor nerve terminal during its activation. *Pflüg. Arch. ges. Physiol.*, **287**, 56–80.

BRAUN, M., SCHMIDT, R. F. & ZIMMERMAN, M. (1966). Facilitation at the frog neuromuscular junction during and after repetitive stimulation. *Pflüg. Arch. ges. Physiol.*, **287**, 41–55.

BROCK, L. G., COOMBS, J. S., & ECCLES, J. C. (1952). "The Recording of Potentials from Motoneurones with an Intracellular Electrode." *J. Physiol.*, **117**, 431–460.

BROCK, L. G., COOMBS, J. S., & ECCLES, J. C. (1953). "Intracellular Recording from Antidromically Activated Motoneurones." *J. Physiol.*, **122**, 429–461.

BROCK, L. G., ECCLES, R. M., & KEYNES, R. D. (1953). "The Discharge of Individual Electroplates in *Raia clavata*." *J. Physiol.*, **122**, 4–6 P.

BRONK, D. W. (1939). "Synaptic Mechanism in Sympathetic Ganglia." *J. Neurophysiol.*, **2**, 380.

BROOKS, C. McC., & KOIZUMI, K. (1956). Origin of the dorsal root reflex. *J. Neurophysiol.*, **19**, 61–74.

BROOKS, V. B. (1956). An intracellular study of the action of repetitive nerve volleys and of botulinum toxin on miniature end-plate potentials. *J. Physiol.*, **134**, 264–277.

BROWN, G. L. (1937). "Action Potentials of Normal Mammalian Muscle. Effects of Acetylcholine and Eserine." *J. Physiol.*, **89**, 220.

BROWN, G. L., DALE H. H., & FELDBERG, W. (1936). "Reactions of the Normal Mammalian Muscle to Acetylcholine and to Eserine." *J. Physiol.*, **87**, 394.

BROWN, G. L. & FELDBERG, W. (1936). The acetylcholine metabolism of a sympathetic ganglion. *J. Physiol.*, **88**, 265–283.

BRYANT, S. H. (1958). Transmission in squid giant synapses. The importance of oxygen supply and the effects of drugs. *J. gen. Physiol.*, **41**, 473–484.

BRZIN, M., DETTBARN, W. D., ROSENBERG, P. & NACHMANSOHN, D. (1965). Cholinesterase activity per unit surface area of conducting membrane. *J. Cell Biol.*, **26**, 353–364.

BRZIN, M., TENNYSON, V. M. & DUFFY, P. E. (1966). Acetylcholinesterase in frog sympathetic and dorsal root ganglia. *J. Cell Biol.*, **31**, 215–242.

BUCHTAL, F., & LINDHARD, (1942) J.. Transmission of impulses from nerve to muscle fibre. *Acta physiol., scand.*, **4**, 136.

BUCKLEY, G. A. & HEATON, J. (1968). A quantitative study of cholinesterase in myoneural junctions from rat and guinea-pig extraocular muscles. *J. Physiol.*, **199**, 743–749.

BULBRING, E. (1944). The action of adrenalin on transmission in the superior cervical ganglion. *J. Physiol.*, **103**, 55.

BULLER, A. J., ECCLES, J. C., & ECCLES, R. M. (1960). Differentiation of fast and slow muscles in the cat hind limb. *J. Physiol.*, **150**, 399–416.

BULLER, A. J., ECCLES, J. C., & ECCLES, R. M. (1960). Interaction between motoneurones and muscle in respect of the characteristic speeds of their responses. *J. Physiol.*, **150**, 417–439.

BULLER, A. J. & LEWIS, D. M. (1965). Further observations on the differentiation of skeletal muscle in the kitten hind limb. *J. Physiol.*, **176**, 355–370.

BULLER, A. J. & LEWIS, D. M. (1965). Some observations on the effects of tenotomy in the rabbit. *J. Physiol.*, **178**, 326–342.

BULLER, A. J. & LEWIS, D. M. (1965). Further observations on mammalian cross-innervated skeletal muscle. *J. Physiol.*, **178**, 343–358.

BULLOCK, T. H. (1948). Properties of a single synapse in the stellate ganglion of the squid. *J. Neurophysiol.*, **11**, 343.

BULLOCK, T. H., & HAGIWARA, S. (1957). Intracellular recording from the giant synapse of the squid. *J. gen. Physiol.*, **40**, 565–577.

BURGEN, A. S. V., DICKENS, F. & ZATMAN, L. J. (1949). The action of botulinum toxin on the neuromuscular junction. *J. Physiol.*, **109**, 10–24.

BURKE, R. E. (1967). Composite nature of the monosynaptic excitatory postsynaptic potential. *J. Neurophysiol.*, **30**, 1114–1137.

BURKE, R. E. (1967). Motor unit types of cat triceps sural muscle. *J. Physiol.*, **193**, 141–160.

BURKE, W. (1957). Spontaneous potentials in slow muscle fibres of the frog. *J. Physiol.*, **135**, 511–521.

BURKE, W., & GINSBORG, B. L. (1956). The electrical properties of the slow muscle fibre membrane. *J. Physiol.*, **132**, 586–598.

BURKE, W., & GINSBORG, B. L. (1956). The action of the neuromuscular transmitter on the slow fibre membrane. *J. Physiol.*, **132**, 599–610.

BUSH, B. M. H. (1962, a). Peripheral reflex inhibition in the claw of the crab, *Carcinus maenas* (L.) *J. exp. Biol.*, **39**, 71–88.

BUSH, B. M. H. (1962, b). Proprioceptive reflexes in the legs of *Carcinus maenas* (L.) *J. exp. Biol.*, **39**, 89–105.

CANEPA, F. G. (1964). Acetylcholine quanta. *Nature*, **201**, 184–185.

CANNON, W. B., & ROSENBLUETH, A. (1937). *Autonomic Neuroeffector Systems*. New York: Macmillan.

CANNON, W. B., & ROSENBLUETH, A. (1949). *The Supersensitivity of Denervated Structures*. New York: Macmillan.

CASTEJÓN, O. J. & VILLEGAS, G. M. (1964). Fine structure of the synaptic contacts in the stellate ganglion of the squid. *J. Ultrastr. Res.*, **10**, 585–598.

CHAGAS, C., PENNA-FRANCA, E., NISHIE, K., & GARCIA, E. J. (1958). A study of the specificity of the complex formed by gallamine triethiodide with a macromolecular constituent of the electric organ. *Arch. Biochem. Biophys.*, **75**, 251–259.

CHANG, C. C., CHENG, H. C. & CHEN, T. F. (1967). Does d-tubocurarine inhibit the release of acetylcholine from motor nerve endings? *Jap. J. Physiol.*, **17**, 505–515.

CHANG, H. C., & GADDUM, J. H. (1933). Choline esters in tissue extracts. *J. Physiol.*, **79**, 255.

CHARLTON, B. T. & GRAY, E. G. (1965). Electron microscopy of specialized synaptic contacts suggesting possible electrical transmission in frog spinal cord. *J. Physiol.*, **179**, 2–4 P.

CHARLTON, B. T. & GRAY, E. G. (1966). Comparative electron microscopy of synapses in the vertebrate spinal cord. *J. Cell Sci.*, **1**, 67–80.

CLOSE, R. (1967). Properties of motor units in fast and slow skeletal muscles. *J. Physiol.*, **193**, 45–55.

CLOSE, R. & HOH, J. F. Y. (1968). Effects of nerve cross-union on fast-twitch and slow-graded muscle fibres in the toad. *J. Physiol.*, **198**, 103–125.

COATES, C. W. (1950). Electric fishes. *Electr. Eng.*, **69**, 47–50.

COATES, C. W., ALTAMIRANO, M., & GRUNDFEST, H. (1954). Activity in electrogenic organs of knifefishes. *Science*, **120**, 845–846.

COLOMO, F. & ERULKAR, S. D. (1968). Miniature synaptic potentials at frog spinal neurones in the presence of tetrodotoxin. *J. Physiol.*, **199**, 205–221.

COLOMO, F. & RAHAMIMOFF, R. (1968). Interaction between sodium and calcium ions in the process of transmitter release at the neuromuscular junction. *J. Physiol.*, **198**, 203–218.

COOK, B. J. (1967). An investigation of factor S, a neuromuscular excitatory substance from insects and crustacea. *Biol. Bull.*, **133**, 526–538.

COOMBS, J. S., CURTIS, D. R., & ECCLES, J. C. (1957a). The generation of impulses in motoneurones. *J. Physiol.*, **139**, 232–249.

COOMBS, J. S., CURTIS, D. R., & ECCLES, J. C. (1957b). The interpretation of spike potentials of motoneurones. *J. Physiol.*, **139**, 198–231.

COOMBS, J. S., ECCLES, J. C., & FATT, P. (1955). Excitatory synaptic action in motor neurones. *J. Physiol.*, **130**, 374–395.

COOMBS, J. S., ECCLES, J. C., & FATT, P. (1955). The electrical properties of the moto-neurone membrane. *J. Physiol.*, **130**, 291–325.

COOMBS, J. S., ECCLES, J. C., & FATT, P. (1955). The specific ionic conductances and the ionic movements across the motoneuronal membrane that produce the inhibitory post-synaptic potential. *J. Physiol.*, **130**, 326–373.

COOMBS, J. S., ECCLES, J. C., & FATT, P. (1955). The inhibitory suppression of reflex discharges from motoneurones. *J. Physiol.*, **130**, 396–413.

COUCEIRO, A., & AKERMAN, M. (1948). Sur quelques aspects du tissu électrique de l'*Electrophorus électricus* (Linnæus). *Anais da Acad. Bras. de Ciencias*, **20**, 383.

COUTEAUX, R. (1958). Morphological and cytochemical observations on the post-synaptic membrane at motor end-plates and ganglionic synapses. *Exp. Cell Res.*, Suppl. 5, pp. 294–322.

COWAN, S. L., & WALTER, W. G. (1937). "The Effects of Tetraethyl Ammonium Iodide on the Electrical Response and the Accommodation of Nerve." *J. Physiol.*, **91**, 101.

CURTIS, D. R. (1962). The depression of spinal inhibition by electrophoretically administered strychnine. *Int. J. Neuropharmacol.*, **1**, 239–250.

CURTIS, D. R. (1969). The pharmacology of spinal postsynaptic inhibition. *Progr. Brain Res.*, **31**, 171–189.

CURTIS, D. R. & DEGROAT, W. C. (1968). Tetanus toxin and spinal inhibition. *Brain Res.*, **10**, 208–212.

CURTIS, D. R., & ECCLES, R. M. (1958). The excitation of Renshaw cells by pharmacological agents applied electrophoretically. *J. Physiol.*, **141**, 435–445.

CURTIS, D. R., HÖSLI, L., JOHNSTON, G. A. R. & JOHNSTON, I. H. (1967). Glycine and spinal inhibition. *Brain Res.*, **5**, 112–114.

CURTIS, D. R., PHILLIS, J. W., & WATKINS, J. C. (1959). The depression of spinal neurones by γ-amino-n-butyric acid and β-alanine. *J. Physiol.*, **146**, 185–203.

CURTIS, D. R., PHILLIS, J. W., & WATKINS, J. C. (1960). The chemical excitation of spinal neurones by certain acidic amino acids. *J. Physiol.*, **150**, 656–682.

CURTIS, D. R., & WATKINS, J. C. (1960). The excitation and depression of spinal neurones by structurally related amino acids. *J. Neurochem.*, **6**, 117–141.

DALE, H. H., FELDBERG, W., & VOGT, M. (1936). "Release of Acetylcholine at Voluntary Motor Nerve Endings." *J. Physiol.*, **86**, 353.

DAVIDOFF, R. A., SHANK, R. P., GRAHAM, L. T., APRISON, M. H. & WERMAN, R. (1967). Is glycine a neurotransmitter? *Nature*, **214**, 680–681.

DAVIS, R. & KOELLE, G. B. (1967). Electron microscopic localization of acetylcholin-esterase and non-specific cholinesterase at the neuromuscular junction by the gold thiocholine and gold thiolacetic acid methods. *J. Cell Biol.*, **34**, 157–171.

DAVSON, H., & LAGE, H. V. (1953). "The Extra-cellular Space and the Internal Potassium Concentration of the Electric Organ of *Electrophorus electricus*." *Anais da Acad. Bras. de Ciencias*, **25**, 303–307.

DAVSON, H. & MATCHETT, P. A. (1951). The non-specific narcotic action of DFP. *J. cell. comp. Physiol.*, **37**, 1–3.

DEL CASTILLO, J., & ENBAEK, L. (1954). The nature of the neuromuscular block produced by magnesium. *J. Physiol.*, **124**, 370–384.

DEL CASTILLO, J., & KATZ, B. (1953). The failure of local-circuit transmission at the nerve-muscle junction. *J. Physiol.*, **123**, 7–8 P.

DEL CASTILLO, J., & KATZ, B. (1954). The effect of magnesium on the activity of motor nerve endings. *J. Physiol.*, **124**, 553–559.

DEL CASTILLO, J., & KATZ, B. (1954). Quantal components of the end-plate potential. *J. Physiol.*, **124**, 560–573.

DEL CASTILLO, J., & KATZ, B. (1954). Statistical factors involved in neuromuscular facilitation and depression. *J. Physiol.*, **124**, 574–585.

DEL CASTILLO, J., & KATZ, B. (1954). Changes in end-plate activity produced by pre-synaptic polarization. *J. Physiol.*, **124**, 586–604.

DEL CASTILLO, J., & KATZ, B. (1954). The membrane change produced by the neuro-muscular transmitter. *J. Physiol.*, **125**, 546–565.

DEL CASTILLO, J., & KATZ, B. (1955). On the localization of acetylcholine receptors. *J. Physiol.*, **128**, 157–181.

DEL CASTILLO, J., & KATZ, B. (1955). Local activity at a depolarized nerve-muscle junction. *J. Physiol.*, **128**, 396–411.

DEL CASTILLO, J. & KATZ, B. (1956). Localization of active spots within the neuromuscular junction of the frog. *J. Physiol.*, **132**, 630–649.

DEL CASTILLO, J., & KATZ, B. (1957). A study of curare action with an electrical micro-method. *Proc. Roy. Soc., B*, **146**, 339–356.

DEMPSHIRE, J. & RIKER, W. K. (1957). The role of acetylcholine in virus-infected sym-pathetic ganglia. *J. Physiol.*, **139**, 145–156.

DIAMOND, J. (1968). The activation and distribution of GABA and L-glutamate receptors on goldfish Mauthner neurones; an analysis of dendritic remote inhibition. *J. Physiol.*, **194**, 669–723.

DIAMOND, J., & MILEDI, R. (1962). A study of foetal and new born rat muscle fibres. *J. Physiol.*, **162**, 393–408.

DODGE, F. A. & RAHAMIMOFF, R. (1967). Co-operative action of calcium in transmitter release at the neuromuscular junction. *J. Physiol.*, **193**, 419–432.

DUBOWITZ, V. (1967). Cross-innervated skeletal muscle: histochemical, physiological and biochemical observations. *J. Physiol.*, **193**, 481–496.

DUDEL, J. (1963). Presynaptic inhibition of the excitatory nerve terminal in the neuro-muscular junction of the crayfish. *Pflüg. Arch. ges. Physiol.*, **277**, 537–557.

DUDEL, J. (1965, a). Potential changes in the crayfish motor nerve terminal during repetitive stimulation. *Pflüg. Arch. ges. Physiol.*, **282**, 323–337.

DUDEL, J. (1965, b). Presynaptic and postsynaptic effects of inhibitory drugs on the cray-fish neuromuscular junction. *Pflüg. Arch. ges. Physiol.*, **283**, 104–118.

DUDEL, J. (1965, c). The mechanism of presynaptic inhibition at the crayfish neuro-muscular junction. *Pflüg. Arch. ges. Physiol.*, **284**, 66–80.

DUDEL, J., GRYDER, R., KAJI, A., KUFFLER, S. W. & POTTER, D. D. (1963). Gamma-aminobutyric acid and other blocking compounds in crustacea. I. Central nervous system. *J. Neurophysiol.*, **26**, 721–728.

DUDEL, J., & KUFFLER, S. W. (1961). The quantal nature of transmission and spon-taneous miniature potentials at the crayfish neuromuscular junction. *J. Physiol.*, **155**, 514–529.

DUDEL, J., & KUFFLER, S. W. (1961). Mechanism of facilitation at the crayfish neuro-muscular junction. *J. Physiol.*, **155**, 530–542.

DUDEL, J., & KUFFLER, S. W. (1961). Presynaptic inhibition at the crayfish neuromuscular junction. *J. Physiol.*, **155**, 543–562.

DUKE-ELDER, W. S. & DUKE-ELDER, P. M. (1930). The contraction of the extrinsic muscles of the eye by choline and nicotine. *Proc. Roy. Soc., B*, **107**, 332–343.

ECCLES, J. C. (1952). "The Electrophysiological Properties of the Motoneurone." *Cold Spr. Harb. Symp. quant. Biol.*, **17**, 175–183.

ECCLES, J. C. (1961). The mechanism of synaptic transmission. *Ergebn. Physiol.*, **51**, 299–430.

ECCLES, J. C. (1961). The nature of central inhibition. *Proc. Roy. Soc. B*, **153**, 445–476.

ECCLES, J. C. (1963). Postsynaptic and presynaptic inhibitory actions in the spinal cord. *Progr. Brain. Res.*, **1**, 1–22.

ECCLES, J. C., ECCLES, R. M., & KOZAK, W. (1962). Further investigations on the influence of motoneurones on the speed of muscle contraction. *J. Physiol.*, **163**, 324–339.

ECCLES, J. C., ECCLES, R. M., & LUNDBERG, A. (1958). The action potentials of the alpha motoneurones supplying fast and slow muscles. *J. Physiol.*, **142**, 275–291.

ECCLES, J. C., ECCLES, R. M., & MAGNI, F. (1961). Central inhibitory action attributable to presynaptic depolarization by muscle afferent volleys. *J. Physiol.*, **159**, 147–166.

ECCLES, J. C., KATZ, B., & KUFFLER, S. W. (1941). "Nature of the 'End-Plate Potential' in Curarized Muscle." *J. Neurophysiol.*, **5**, 211.

ECCLES, J. C., KOZAK, W., & MAGNI, F. (1961). Dorsal root reflexes of muscle group I afferent fibres. *J. Physiol.*, **159**, 128–146.

ECCLES, J. C., & O'CONNOR, W. J. (1939). "Responses which Nerve Impulses evoke in Mammalian Muscles." *J. Physiol.*, **97**, 44.

ECCLES, J. C., SCHMIDT, R. F., & WILLIS, W. D. (1962). Presynaptic inhibition of the spinal monosynaptic reflex pathway. *J. Physiol.*, **161**, 282–297.

ECCLES, J. C., SCHMIDT, R. F. & WILLIS, W. D. (1963). The mode of operation of the synaptic mechanism producing presynaptic inhibition. *J. Neurophysiol.*, **26**, 523–538.

ECCLES, R. M. (1955). "Intracellular Potentials recorded from a Mammalian Sympathetic Ganglion." *J. Physiol.*, **130**, 572–584.

ECCLES, R. M., & LIBET, B. (1961). Origin and blockade of the synaptic responses of curarized sympathetic ganglia. *J. Physiol.*, **157**, 484–503.

EDWARDS, C., & KUFFLER, S. W. (1957). "Inhibitory Mechanisms of Gamma Amino-butyric Acid on an Isolated Nerve Cell." *Fed. Proc.*, **16**, 34.

EDWARDS, G. A., RUSKA, H., & DE HARVEN, E. (1958). Electron microscopy of peripheral nerves and neuromuscular junctions in the wasp leg. *J. biophys. biochem. Cytol.*, **4**, 107–114.

EHRENPREIS, S. (1960). Isolation and identification of the acetylcholine receptor protein of electric tissue. *Biochem. biophys. Acta*, **44**, 561–577.

EHRENPREIS, S. (1964). Acetylcholine and nerve activity. *Nature*, **201**, 887–893.

ELFVIN, L. G. (1963). The ultrastructure of the superior cervical ganglion of the cat. I. and II. *J. Ultrastr. Res.*, **8**, 403–440; 441–476.

ELLIS, C. H., THIENES, C. H., & WIERSMA, C. A. G. (1942). "The Influence of Certain Drugs on the Crustacean Nerve-muscle System." *Biol. Bull. Wood's Hole*, **83**, 334–352.

ELMQVIST, D. & FELDMAN, D. S. (1965). Calcium dependence of spontaneous acetyl-choline release at mammalian motor nerve terminals. *J. Physiol.*, **180**, 487–497.

ELMQVIST, D. & FELDMAN, D. S. (1965). Effects of sodium pump inhibitors on spon-taneous acetylcholine release at the neuromuscular junction. *J. Physiol.*, **181**, 498–505.

ELMQVIST, D. & QUASTEL, D. M. J. (1965). Presynaptic action of hemicholinium at the neuromuscular junction. *J. Physiol.*, **177**, 463–482.

ELMQVIST, D. & QUASTEL, D. M. J. (1965). A quantitative study of end-plate potentials in isolated human muscle. *J. Physiol.*, **178**, 505–529.

ERULKAR, S. D. & WOODWARD, J. K. (1968). Intracellular recording from mammalian superior cervical ganglion *in situ*. *J. Physiol.*, **199**, 189–203.

v. EULER, U. S. (1959). Autonomic neuroeffector transmission. In *Handbook of Physiology*, Vol. I, pp. 215–237.

v. EULER, U. S., & LISHAJKO, F. (1961). Noradrenaline release from isolated nerve granules. *Acta physiol. scand.*, **51**, 193–203.

FATT, P. (1957). "Electric Potentials occurring around a Neurone during its Antidromic Activation." *J. Neurophysiol.*, **20**, 27–60.

FATT, P., & KATZ, B. (1951). "An Analysis of the End-plate Potential recorded with an Intra-cellular Electrode." *J. Physiol.*, **115**, 320–370.

FATT, P., & KATZ, B. (1952). "Spontaneous Subthreshold Activity at Motor Nerve Endings." *J. Physiol.*, **117**, 109–128.

FATT, P., & KATZ, B. (1952). "The Effect of Sodium Ions on Neuromuscular Trans-mission." *J. Physiol.*, **118**, 73–87.

FATT, P., & KATZ B. (1953). "The Effect of Inhibitory Nerve Impulses on a Crustacean Muscle Fibre." *J. Physiol.*, **121**, 374–389.

FATT, P., & KATZ, B. (1953). "Distributed 'End-plate Potentials' of Crustacean Muscle Fibres." *J. exp. Biol.*, **30**, 433–439.

FATT. P. & KATZ, B. (1953). The electrical properties of crustacean muscle fibres. *J. Physiol.*, **120**, 171–204.

FLOREY, E. (1954). "An Inhibitory and an Excitatory Factor of Mammalian Central Nervous System, and their Action on a Single Sensory Neuron." *Arch. int. Physiol.* **62**, 33–53.

FLOREY, E., & McLENNAN, H. (1955). "Effects of an Inhibitory Factor (Factor I) from Brain on Central Synaptic Transmission." *J. Physiol.*, **130**, 446–455.

FLOREY, E. & WOODCOCK, B. (1968). Presynaptic excitatory action of glutamate applied to crab nerve-muscle preparations. *Comp. Biochem. Physiol.*, **26**, 651–661.

FRANK, K., & FUORTES, M. G. F. (1957). Presynaptic and postsynaptic inhibition of monosynaptic reflexes. *Fed. Proc.*, **16**, 39-40.

FURSHPAN, E. J. (1956). The effects of osmotic pressure changes on the spontaneous activity at motor nerve endings. *J. Physiol.*, **134**, 689-697.

FURSHPAN, E. J. (1964). " Electrical transmission " at an excitatory synapse in a vertebrate brain. *Science*, **144**, 878-880.

FURSHPAN, E. J., & FURUKAWA, T. (1962). Intracellular and extracellular responses of the several regions of the Mauthner cell of the goldfish. *J. Neurophysiol.*, **25**, 732-771.

FURSHPAN, E. J., & POTTER, D. D. (1959). Transmission at the giant motor synapses of the crayfish. *J. Physiol.*, **145**, 289-325.

FURSHPAN, E. J., & POTTER, D. D. (1959). Slow post-synaptic potentials recorded from the giant motor fibre of the crayfish. *J. Physiol.*, **145**, 326-335.

FURUKAWA, T., & FURSHPAN, E. J. (1963). Two inhibitory mechanisms in the Mauthner neurons of goldfish. *J. Neurophysiol.*, **26**, 140-176

FUXE, K. & GUNNE, L. M. (1964). Depletion of the amine stores in brain catecholamine terminals on amygdaloid stimulation. *Acta. physiol. scand.*, **62**, 493-494.

GAGE, P. W. & HUBBARD, J. I. (1966). The origin of the post-tetanic hyperpolarization of mammalian motor nerve terminals. *J. Physiol.*, **184**, 335-352.

GAGE, P. W. & HUBBARD, J. I. (1966). An investigation of the post-tetanic potentiation of end-plate potentials at a mammalian neuromuscular junction. *J. Physiol.*, **184**, 353-375.

GAGE, P. W. & QUASTEL, D. M. J. (1966). Competition between sodium and calcium ions in transmitter release at mammalian neuromuscular junctions. *J. Physiol.*, **185**, 95-123.

GASSER, H. S. (1930). "Contractures of Skeletal Muscle." *Physiol. Rev.*, **10**, 35.

GILLESPIE, J. S. (1966). Tissue binding of noradrenaline. *Proc. Roy. Soc., B*, **166**, 1-10.

GINSBORG, B. L. (1967). Ion movements in junctional transmission. *Pharmacol. Rev.*, **19**, 289-316.

GINSBORG, B. L. & HAMILTON, J. T. (1968). The effect of caesium ions on neuromuscular transmission in the frog. *Quart. J. exp. Physiol.*, **53**, 162-169.

GRAY, E. G. (1969). Electron microscopy of excitatory and inhibitory synapses: a brief review. *Progr. Brain Res.*, **31**, 141-155.

GRUNDFEST, H. (1957). "The Mechanisms of Discharge of the Electric Organs in Relation to General and Comparative Electrophysiology." *Progr. Biophys.*, **7**, 1-85.

GRUNDFEST, H., REUBEN, J. P., & RICKLES, W. H. (1959). The electrophysiology and pharmacology of lobster neuromuscular synapses. *J. gen. Physiol.*, **42**, 1301-1324.

GUTH, L. & WATSON, P. K. (1967). The influence of innervation on the soluble proteins of slow and fast muscles in the rat. *Exp. Neurol.*, **17**, 107-117.

HAGIWARA, S., KUSANO, K., & NEGISHI, K. (1962). Physiological properties of electro-receptors of some gymnotids. *J. Neurophysiol.*, **25**, 430-449.

HAGIWARA, S., & MORITA, H. (1962). Electrotonic transmission between two nerve cells in leech ganglion. *J. Neurophysiol.*, **25**, 721-731.

HAGIWARA, S., & TASAKI, I. (1958). A study on the mechanism of impulse transmission across the giant synapse of the squid. *J. Physiol.*, **143**, 114-137.

HANZON, V., & TOSCHI, G. (1959). Electron microscopy on microsomal fractions from rat brain. *Exp. Cell Res.*, **16**, 256-271.

HENNEMAN, E. & OLSON, C. B. (1965). Relations between structure and function in the design of skeletal muscles. *J. Neurophysiol.*, **28**, 581-598.

HESS, A. & PILAR, G. (1963). Slow fibres in the extraocular muscles of the cat. *J. Physiol.*, **169**, 780-789.

HNIK, P., JIRMANOVÁ, I., VYKICKÝ, L. & ZELENÁ, J. (1967). Fast and slow muscles of the chick after nerve cross-union. *J. Physiol.*, **193**, 309-325.

HÖBER, R., ANDERSH, M., HÖBER, J. & NEBEL, B. (1939). Influence of organic electrolytes and non-electrolytes on the membrane potentials of muscle and nerve. *J. cell. comp. Physiol.*, **13**, 195.

HONOUR, A. J., & MCLENNAN, H. (1960). The effects of γ-aminobutyric acid and other compounds on structures of the mammalian nervous system which are inhibited by Factor I. *J. Physiol.*, **150**, 306-318.

HOYLE, G. (1955). The anatomy and innervation of locust skeletal muscle. *Proc. Roy. Soc., B*, **143**, 281-292.

HOYLE, G. (1962). Neuromuscular physiology. In *Advances in Physiology and Biochemistry*. **1**, 177-216.

HOYLE, G., & WIERSMA, C. A. G. (1958). Excitation at neuromuscular junctions in crustacea. *J. Physiol.*, **143**, 403-425.

HOYLE, G., & WIERSMA, C. A. G. (1958). Inhibition at neuromuscular junctions in crustacea. *J. Physiol.*, **143**, 426–440.

HOYLE, G., & WIERSMA, C. A. G. (1958). Coupling of membrane potential to contraction in crustacean muscle. *J. Physiol.*, **143**, 441–453.

HUBBARD, J. I. (1961). The effect of calcium and magnesium on the spontaneous release of transmitter from mammalian motor nerve endings. *J. Physiol.*, **159**, 507–517.

HUBBARD, J. I. (1963). Repetitive stimulation at the mammalian neuromuscular junction, and the mobilization of transmitter. *J. Physiol.*, **169**, 641–662.

HUBBARD, J. I., JONES, S. F. & LANDAU, E. M. (1968). On the mechanism by which calcium and magnesium affect the spontaneous release of transmitter from mammalian motor nerve terminals. *J. Physiol.*, **194**, 355–380.

HUBBARD, J. I. JONES, S. F. & LANDAU, E. M. (1968). An examination of the effects of osmotic pressure changes upon transmitter release from mammalian motor nerve terminals. *J. Physiol.*, **197**, 639–657.

HUBBARD, J. I. & KWANBUNBUMPEN, S. (1968). Evidence for the vesicle hypothesis. *J. Physiol.*, **194**, 407–420.

HUBBARD, J. I. & SCHMIDT, R. F. (1963). An electrophysiological investigation of mammalian motor nerve terminals. *J. Physiol.*, **166**, 145–167.

HUBBARD, J. I., STENHOUSE, D. & ECCLES, R. M. (1967). Origin of synaptic noise. *Science*, **157**, 330–331.

HUBBARD, J. I. & WILLIS, W. D. (1968). The effects of depolarization of motor nerve terminals upon the release of transmitter by nerve impulses. *J. Physiol.*, **194**, 381–405.

HUNT, C. C. (1954). Relation of function to diameter in afferent fibres of muscle nerves. *J. gen. Physiol.*, **38**, 117–131.

HUNT, C. C. & NELSON, P. G. (1965) Structural and functional changes in the frog sympathetic ganglion following cutting of the presynaptic nerve fibres. *J. Physiol.*, **177**, 1–20.

HUNT, C. C. & RIKER, W. K. (1966). Properties of frog sympathetic neurons in normal ganglia and after axon section. *J. Neurophysiol.*, **29**, 1096–1114.

HUTTER, O. F. (1952). Post-tetanic restoration of neuromuscular transmission blocked by D-tubocurarine. *J. Physiol.*, **118**, 216–227.

HUTTER, O. & KOSTIAL, K. (1954). Effect of magnesium and calcium ions on the release of acetylcholine. *J. Physiol.*, **124**, 234–241.

HUTTER, O. F., & LOEWENSTEIN, W. R. (1955). "Nature of Neuromuscular Facilitation by Sympathetic Stimulation in the Frog." *J. Physiol.*, **130**, 559–571.

ITO, M., KOSTYUK, P. G., & OSHIMA, T. (1962). Further study on anion permeability of inhibitory post-synaptic membrane of cat motorneurones. *J. Physiol.*, **164**, 150–156.

JENERICK, H. P., & GERARD, R. W. (1953). "Membrane Potential and Threshold of Single Muscle Fibres." *J. cell. comp. Physiol.*, **42**, 79–102.

JENKINSON, D. H. (1957). The nature of the antagonism between calcium and magnesium ions at the neuromuscular junction. *J. Physiol.*, **138**, 434–444.

JENKINSON, D. H., & NICHOLS, J. G. (1961). Contractures and permeability changes produced by acetylcholine in depolarized denervated muscle. *J. Physiol.*, **159**, 111–127.

JENKINSON, D. H., STAMENOVIĆ, B. A. & WHITAKER, B. D. L. (1968). The effect of noradrenaline on the end-plate potential in twitch fibres of the frog. *J. Physiol.*, **195**, 743–754.

KANDEL, E. R. & TAUC, L. (1966). Anomalous rectification in the metacerebral giant cells and its consequences for synaptic transmission. *J. Physiol.*, **183**, 287–304.

KAO, C. Y. & NISHIYAMA, A. (1965). Actions of saxitoxin on peripheral neuromuscular systems. *J. Physiol.*, **180**, 50–66.

KATZ, B. (1936). "Neuromuscular Transmission in Crabs." *J. Physiol.*, **87**, 199.

KATZ, B. (1948). Electrical properties of the muscle fibre membrane. *Proc. Roy. Soc., B*, **135**, 506.

KATZ, B. (1962). The transmission of impulses from nerve to muscle, and the subcellular unit of synaptic action. *Proc. Roy. Soc. B*, **155**, 455–477.

KATZ, B. & MILEDI, R. (1963). A study of spontaneous miniature potentials in spinal motoneurones. *J. Physiol.*, **168**, 389–422.

KATZ, B. & MILEDI, R. (1965, a). Propagation of electric activity in motor nerve terminals. *Proc. Roy. Soc., B*, **161**, 453–482.

KATZ, B. & MILEDI, R. (1965, b). The effect of calcium on acetylcholine release from motor nerve terminals. *Proc. Roy. Soc., B*, **161**, 496–503.

KATZ, B. & MILEDI, R. (1967, a). Tetrodotoxin and neuromuscular transmission. *Proc. Roy. Soc. B.*, **167**, 8–22.

KATZ, B. & MILEDI, R. (1967, b). The release of acetylcholine from nerve endings by graded electric pulses. *Proc. Roy. Soc., B*, **167**, 23–38.

KATZ, B. & MILEDI, R. (1967, c). The timing of calcium action during neuromuscular transmission. *J. Physiol.*, **189**, 535–544.

KATZ, B. & MILEDI, R. (1967, d). A study of synaptic transmission in the absence of nerve impulses. *J. Physiol.*, **192**, 407–436.

KATZ, B. & MILEDI, R. (1968). The effect of local blockage of motor nerve terminals. *J. Physiol.*, **199**, 729–741.

KATZ, B. & MILEDI, R. (1968). The role of calcium in neuromuscular facilitation. *J. Physiol.*, **195**, 481–492.

KATZ, B., & THESLEFF, S. (1957). "On the Factors which determine the Amplitude of the Miniature End-plate Potential." *J. Physiol.*, **137**, 267–278.

KELLY, J. S. (1968). The antagonism of Ca^{2+} by Na^+ and other monovalent ions at the frog neuromuscular junction. *Quart. J. exp. Physiol.*, **53**, 239–249.

KERKUT, G. A., LEAKE, L. D., SHAPIRA, A., COWAN, S. & WALKER, R. J. (1965). The presence of glutamate in nerve-muscle perfusates of *Helix*, *Carcinus* and *Periplaneta*. *Comp. Biochem. Physiol.*, **15**, 485–502.

KERKUT, G. A. & WALKER, R. J. (1966). The effect of L-glutamate, acetylcholine and gamma-aminobutyric acid on the miniature end-plate potentials and contractures of the coxal muscles of the cockroach, *Periplaneta Americana*. *Comp. Biochem. Physiol.*, **17**, 435–454.

KEYNES, R. D., & MARTINS-FERREIRA, H. (1953). "Membrane Potentials in the Electroplates of the Electric Eel." *J. Physiol.* **119**, 315–351.

KLAUS, W., KUSCHINSKY, G., LÜLLMAN, H. & MUSCHOLL, E. (1959). Über den Einfluss von Acetylcholin auf die Kaliumpermeabilität der denervierten Muskelmembran im polarisierten und depolarisierten Zustand. *Med. exp.*, **1**, 8–11.

KOELLE, G. B. (1961). A proposed dual neurohumoral role of acetylcholine: its functions at the pre- and post-synaptic sites. *Nature*, **190**, 208–211.

KOELLE, G. B., & FRIEDENWALD, J. S. (1949). "A Histochemical Method for Localising Cholinesterase Activity." *Proc. Soc. exp. Biol., N.Y.*, **70**, 617.

KOKETSU, K. & NISHI, S. (1968). Calcium spikes of nerve cell membrane: role of calcium in the production of action potentials. *Nature*, **217**, 468–469.

KOKETSU, K. & NISHI, S. (1968). Cholinergic receptors at sympathetic preganglionic nerve terminals. *J. Physiol.*, **196**, 293–310.

KRAVITZ, E. A., KUFFLER, S. W. & POTTER, D. D. (1963, b). Gamma-aminobutyric acid and other blocking compounds in crustacea. III. Their relative concentrations in separated motor and inhibitory axons. *J. Neurophysiol.*, **26**, 739–751.

KRAVITZ, E. A., KUFFLER, S. W., POTTER, D. D. & VAN GELDER, N. M. (1963, a). Gamma-aminobutyric acid and other blocking compounds in crustacea. II. Peripheral nervous system. *J. Neurophysiol.*, **26**, 728–738.

KRNJEVIĆ, K., & MILEDI, R. (1958). Acetylcholine in mammalian neuromuscular transmission. *Nature, Lond.*, **182**, 805–806.

KRNJEVIĆ, K., & MITCHELL, J. F. (1961). The release of acetylcholine in the isolated rat diaphragm. *J. Physiol.*, **155**, 246–262.

KRÜGER, P. (1952). *Tetanus und Tonus der quergestreiften Skelettmuskel der Wirbeltiere und des Menschen*. Leipzig: Acad. Verlag, Geest & Portig.

KUFFLER, S. W. (1942). "Electrical Potential Changes at an Isolated Nerve-muscle Junction." *J. Neurophysiol.*, **5**, 18.

KUFFLER, S. W. (1945). "Electric Excitability of Nerve-muscle Fibre Preparations." *J. Neurophysiol.*, **8**, 75.

KUFFLER, S. W., & WILLIAMS, E. M. V. (1953). "Small Nerve Junctional Potentials. The Distribution of Small Motor Nerves to Frog Skeletal Muscle, and the Membrane Characteristics of the Fibres they Innervate." *J. Physiol.*, **121**, 289–317.

KUFFLER, S. W., & WILLIAMS, E. M. V. (1953). "Properties of the 'slow' Skeletal Muscle Fibres of the Frog." *J. Physiol.*, **121**, 318–340.

KUNO, M. (1964). Quantal components of excitatory synaptic potentials in spinal motoneurones. *J. Physiol.*, **175**, 81–99.

KUNO, M. & RUDOMIN, P. (1966). The release of acetylcholine from the spinal cord of the cat by antidromic stimulation of motor nerves. *J. Physiol.*, **187**, 177–193.

KUSANO, K. (1968). Further study of the relationship between pre- and post-synaptic potentials in the squid giant synapse. *J. gen. Physiol.*, **52**, 326–345.

KUSANO, K., LIVENGOOD, D. R. & WERMAN, R. (1967). Correlation of transmitter release with membrane properties of the presynaptic fiber of the squid giant synapse. *J. gen. Physiol.*, **50**, 2579–2601.

LARIMER, J. L. & MacDONALD, J. A. (1968). Sensory feedback from electroreceptors to electromotor pacemaker centers in gymnotids. *Amer. J. Physiol.*, **214**, 1253–1261.

LARRABEE, M. G. & BRONK, D. W. (1947). Prolonged facilitation of sympathetic excitation in sympathetic ganglia. *J. Neurophysiol.*, **10**, 139–154.

LEVINE, L. (1966). An electrophysiological study of chelonian skeletal muscle. *J. Physiol.*, **183**, 683–713.

LIBET, B. (1964). Slow synaptic responses and excitatory changes in sympathetic ganglia. *J. Physiol.*, **174**, 1–25.

LIBET, B. (1968). Long latent periods and further analysis of slow synaptic responses in sympathetic ganglia. *J. Neurophysiol.*, **30**, 494–514.

LIBET, B. & TOSAKA, T. (1966). Slow postsynaptic potentials recorded intracellularly in sympathetic ganglia. *Fed. Proc.*, **25**, 270.

LILEY, A. W. (1956). "An Investigation of Spontaneous Activity at the Neuromuscular Junction of the Rat." *J. Physiol.*, **132**, 650–666.

LILEY, A. W. (1956). "The Quantal Components of the Mammalian End-plate Potential." *J. Physiol.*, **133**, 571–587.

LILEY, A. W. (1956). The effects of presynaptic polarization on the spontaneous activity at the mammalian neuromuscular junction. *J. Physiol.*, **134**, 427–443.

LILEY, A. W. & NORTH, K. A. K. (1953). An electrical investigation of effects of repetitive stimulation on mammalian neuromuscular junctions. *J. Neurophysiol.*, **16**, 509–527.

LISSMANN, H. W. (1951). "Continuous Electrical Signals from the Tail of a Fish, *Gymnarchus niloticus* Cuv." *Nature*, **167**, 201–202.

LLOYD, D. P. C. (1946). "Facilitation and Inhibition of Spinal Motoneurons." *J. Neurophysiol.*, **9**, 421–438.

LLOYD, D. P. C. (1960). Spinal mechanisms involved in somatic activities. In *Handbook of Physiology*, Section I, Vol. II. Washington: Amer. Physiol. Soc., pp. 929–949.

LLOYD, D. P. C., & WILSON, V. J. (1959). Functional organization in the terminal segments of the spinal cord with a consideration of central excitatory and inhibitory latencies. *J. gen. Physiol.*, **42**, 1219–1232.

LOEWENSTEIN, W. R. (1966). Permeability of membrane junctions. *Ann. N.Y. Acad. Sci.*, **137**, 441–472.

DE LORENZO, A. J. (1960). The fine structure of synapses in the ciliary ganglion of the chick. *J. biophys. biochem. Cytol.*, **7**, 31–36.

LUNDBERG, A., & QUILISCH, H. (1953). "Presynaptic Potentiation and Depression of Neuromuscular Transmission in Frog and Rat." *Acta physiol. scand.*, **30**, Suppl. 111, 111–120.

MACHIN, K. E., & LISSMANN, H. W. (1960). The mode of action of the electric receptors in *Gymnarchus niloticus*. *J. exp. Biol.*, **37**, 801–811.

MACHNE, X., FADIGA, E., & BROOKHART, J. M. (1959). Antidromic and synaptic activation of frog motor neurones. *J. Neurophysiol.*, **22**, 483–503.

MacINTOSH, F. C. (1959). Formation, storage, and release of acetylcholine at nerve endings. *Canad. J. Biochem. Physiol.*, **37**, 343–356.

MACLAGAN, J. & VRBOVÁ, G. (1966). A study of the increased sensitivity of denervated and re-innervated muscle to depolarizing drugs. *J. Physiol.*, **182**, 131–143.

MAENO, T. (1966). Analysis of sodium and potassium conductances in the procaine end-plate potential. *J. Physiol.*, **183**, 592–606.

MALLART, A. & MARTIN, A. R. (1967). An analysis of facilitation of transmitter release at the neuromuscular junction of the frog. *J. Physiol.*, **193**, 679–694.

MALLART, A. & MARTIN, A. R. (1967). Two components of facilitation in the neuromuscular junction in the frog. *J. Physiol.*, **191**, 19–20 P.

MALLART, A. & MARTIN, A. R. (1968). The relation between quantum content and facilitation at the neuromuscular junction of the frog. *J. Physiol.*, **196**, 593–604.

MANTHEY, A. A. (1966). The effect of calcium on the desensitization of membrane receptors at the neuromuscular junction. *J. gen. Physiol.*, **49**, 963–976.

MARMONT, G., & WIERSMA, C. A. G. (1938). "On the Mechanism of Inhibition and Excitation of Crayfish Muscle." *J. Physiol.*, **93**, 173.

MARNAY, A., & NACHMANSOHN, D. (1938). "Choline Esterase in Voluntary Muscle." *J. Physiol.*, **92**, 37.

MARTIN, A. R. & ORKAND, R. K. (1961). Postsynaptic effects of HC-3 at the neuromuscular junction of the frog. *Canad. J. Biochem. Physiol.*, **39**, 343–349.

MARTIN, A. R. & PILAR, G. (1963). Dual mode of synaptic transmission in the avian ciliary ganglion. *J. Physiol.*, **168**, 443–463.

MARTIN, A. R. & PILAR, G. (1964). Quantal components of the synaptic potential in the ciliary ganglion of the chick. *J. Physiol.*, **175**, 1–16.

MARTIN, A. R. & PILAR, G. (1964). Presynaptic and post-synaptic events during post-tetanic potentiation and facilitation in the avian ciliary ganglion. *J. Physiol.*, **175,** 17–30.

MASLAND, R. L. & WIGTON, R. S. (1940) Nerve activity accompanying fasciculation produced by prostigmin. *J. Neurophysiol.*, **3,** 269–275.

McGEER, E. G., McGEER, P. L., & McLENNAN, H. (1961). The inhibitory action of 3-hydroxytyramine, gamma-aminobutyric acid (GABA) and some other compounds towards the crayfish stretch receptor neuron. *J. Neurochem.*, **8,** 36–49.

McLENNAN, H. (1957). "A Comparison of some Physiological Properties of an Inhibitory Factor from Brain (Factor I) and of γ-Aminobutyric Acid and Related Compounds." *J. Physiol.*, **139,** 79–86.

McLENNAN, H. (1959). The identification of one active component from brain extracts containing factor I. *J. Physiol.*, **146,** 358–368.

McLENNAN, H. (1961). The effect of some catecholamines upon a monosynaptic reflex pathway in the spinal cord. *J. Physiol.*, **158,** 411–425.

MENDEL, B., & RUDNEY, H. (1943). " Studies on Cholinesterase. I. Cholinesterase and Pseudo-Cholinesterase." *Biochem. J.*, **37,** 59.

MILEDI, R. (1960). The acetylcholine sensitivity of frog muscle fibres after complete or partial denervation. *J. Physiol.*, **151,** 1–23.

MILEDI, R. (1960). Properties of regenerating neuromuscular synapses in the frog. *J. Physiol.*, **154,** 190–205.

MILEDI, R. (1964). Electron-microscopical localization of products from histochemical reactions used to detect cholinesterase in muscle. *Nature*, **204,** 293–295.

MILEDI, R. (1966). Strontium as a substitute for calcium in the process of transmitter release at the neuromuscular junction. *Nature*, **212,** 1233–1234.

MILEDI, R. (1967). Spontaneous synaptic potentials and quantal release of transmitter in the stellate ganglion of the squid. *J. Physiol.*, **192,** 379–406.

MILEDI, R. & ORKAND, P. (1966). Effect of a " fast " nerve on " slow " muscle fibres in the frog. *Nature*, **209,** 717–718.

MILEDI, R. & SLATER, C. R. (1966). The action of calcium on neuronal synapses in the squid. *J. Physiol.*, **184,** 473–498.

MILEDI, R. & SLATER, C. R. (1968). Electrophysiology and electron-microscopy of rat neuromuscular junctions after nerve degeneration. *Proc. Roy. Soc. B*, **169,** 289–306.

NACHMANSOHN, D. (1959). *Chemical and Molecular Basis of Nerve Activity.* N.Y.: Academic Press.

NACHMANSOHN, D., & MEYERHOF, B. (1941). "Relation between Electrical Changes during Nerve Activity and Concentration of Choline Esterase." *J. Neurophysiol.*, **4,** 348.

NAKAMURA, Y., NAKAJIMA, S. & GRUNDFEST, H. (1965). Analysis of spike electrogenesis and depolarizing K inactivation in electroplaques of *Electrophorus electricus*, (L). *J. gen. Physiol.*, **49,** 321–349.

NASTUK, W. L. (1953). "Membrane Potential Changes at a Single Muscle End-plate produced by Transitory Application of Acetylcholine with an Electrically Controlled Microjet." *Fed. Proc.*, **12,** 102.

NASTUK, W. L. (1953). "The Electrical Activity of the Muscle Cell Membrane at the Neuromuscular Junction." *J. cell. comp. Physiol.*, **42,** 249–272.

NELSON, P. G. & FRANK, K. (1967). Anomalous rectification in cat spinal motoneurons and effect of polarizing currents on excitatory postsynaptic potential. *J. Neurophysiol.*, **30,** 1097–1113.

NELSON, P. G., FRANK, K., & RALL, W. (1960). Single spinal motoneuron extracellular potential fields. *Fed. Proc.*, **19,** 303.

NISHIE, K., & HARRIS, E. J. (1955). "Electrolytes in the Tissues of *Electrophorus electricus* L." *J. cell. comp. Physiol.*, **45,** 484–486.

NISHI, S. & KOKETSU, K. (1960). Electrical properties and activities of single sympathetic neurones in frogs. *J. cell comp. Physiol.*, **55,** 15–30.

NISHI, S. & KOKETSU, K. (1966). Late after-discharge of sympathetic postganglionic fibres. *Life Sci.*, **5,** 1991–1997.

NISHI, S. & KOKETSU, K. (1968). Analysis of slow inhibitory postsynaptic potential of bullfrog sympathetic ganglia. *J. Neurophysiol.*, **31,** 717–728.

NISHI, S., SOEDA, H. & KOKETSU, K. (1967). Release of acetylcholine from sympathetic preganglionic nerve terminals. *J. Neurophysiol.*, **30,** 114–134.

OSTLUND, E. (1954). The distribution of catechol amines in lower animals and their effect on the heart. *Acta physiol. scand.*, **31,** Suppl. 112.

OTSUKA, M., ENDO, M. & NONOMURA, Y. (1962). Presynaptic nature of neuromuscular depression. *Jap. J. Physiol.*, **12**, 573–584.

OTSUKA, M., IVERSEN, L. L., HALL, Z. W. & KRAVITZ, E. A. (1966). Release of gamma-aminobutyric acid from inhibitory nerves of lobsters. *Proc. Nat. Acad. Sci., Wash.*, **56**, 1110–1115.

OZEKI, M., FREEMAN, A. R. & GRUNDFEST, H. (1966). The membrane components of crustacean neuromuscular systems. I. II. *J. gen. Physiol.*, **49**, 1319–1334; 1335–1349.

PAGE, S. G. (1965). A comparison of the fine structures of frog slow and twitch fibres. *J. Cell Biol.*, **26**, 477–497.

PALAY, S. L. (1956). "Synapses in the Central Nervous System." *J. biophys. biochem. Cytol.*, **2**, Suppl., 193–201.

PANTIN, C. F. A. (1936). "On the Excitation of Crustacean Muscle. II. Neuromuscular Facilitation." *J. exp. Biol.*, **13**, 111.

PEACHEY, L. D. & HUXLEY, A. F. (1962). Structural identification of twitch and slow striated muscle fibers of the frog. *J. Cell Biol.*, **13**, 177–180.

PERRY, W. L. M. (1953). Acetylcholine release in the cat's superior cervical ganglion. *J. Physiol.*, **119**, 439–454.

PROSKE, U. & VAUGHAN, P. (1968). Histological and electrophysiological investigation of lizard skeletal muscle. *J. Physiol.*, **199**, 495–509.

PURPURA, D. P., GIRADO, M., & GRUNDFEST, H. (1957). Selective blockade of excitatory synapses in the cat brain by γ-aminobutyric acid (GABA). *Science*, **125**, 1200–1202.

RAHAMIMOFF, R. (1968). A dual effect of calcium ions on neuromuscular facilitation. *J. Physiol.*, **195**, 471–480.

RALL, W. (1967). Distinguishing theoretical synaptic potentials computed for different soma-dendritic distributions of synaptic input. *J. Neurophysiol.*, **30**, 1138–1168.

REUBEN, J. P., & GRUNDFEST, H. (1960). Inhibitory and excitatory miniature post-synaptic potentials in lobster muscle fibres. *Biol. Bull.*, **119**, 335–356.

RIKER, W. K., WERNER, G., ROBERTS, J. & KUPERMAN, A. (1959). Pharmacologic evidence for the existence of a presynaptic event in neuromuscular transmission. *J. Pharmacol.*, **125**, 150–158.

RITCHIE, J. M. & ARMETT, C. J. (1963). On the role of acetylcholine in conduction in mammalian nonmyelinated nerve fibers. *J. Pharmacol.*, **139**, 201–207.

ROBBINS, J. (1958). The effects of amino acids on the crustacean neuro-muscular system. *Anat. Rec.*, **132**, 492–493.

DE ROBERTIS, E. (1956). Submicroscopic changes of the synapse after nerve section in the acoustic ganglion of the guinea pig. *J. biophys. biochem. Cytol.*, **2**, 503–512.

DE ROBERTIS, E., & BENNETT, H. S. (1955). "Some Features of the Submicroscopic Morphology of Synapses in Frog and Earthworm." *J. biophys. biochem. Cytol.*, **1**, 47–58.

DE ROBERTIS, E., & FRANCHI, C. M. (1956). Electron microscope observations on synaptic vesicles in synapses of the retinal rods and cones. *J. biophys. biochem. Cytol.*, **2**, 307–317.

DE ROBERTIS, E., DE IRALDI, A. P., RODRIGUEZ, G., & GOMEZ, C. J. (1961). On the isolation of nerve endings and synaptic vesicles. *J. biophys. biochem. Cytol.*, **9**, 229–235.

ROBERTSON, J. D. (1955). "Recent Electron Microscope Observations on the Ultra-structure of the Crayfish Median-to-Motor Giant Synapse." *Exp. Cell Res.*, **8**, 226–229.

ROBERTSON, J. D. (1956). "The Ultrastructure of a Reptilian Myoneural Junction." *J. biophys. biochem. Cytol.*, **2**, 381–394.

ROBERTSON, J. D. (1961). Ultrastructure of excitable membranes and the crayfish median-giant synapse. *Ann. N.Y. Acad. Sci.*, **94**, 339–389.

ROBERTSON, J. D., BODENHEIMER, T. S. & STAGE, D. E. (1963). The ultrastructure of Mauthner cell synapses and nodes in goldfish brains. *J. Cell Biol.*, **19**, 159–199.

ROMANUL, F. C. A. (1964). Enzymes in muscle. I. *Arch. Neurol.*, **11**, 355–368.

ROMANUL, F. C. A. & VAN DER MEULEN, J. P. (1966). Reversal of the enzyme profiles of muscle fibres in fast and slow muscles by cross-innervation. *Nature*, **212**, 1369–1370.

SALPETER, M. M. (1967). The distribution of acetylcholinesterase at motor end-plates of a vertebrate twitch muscle. *J. Cell Biol.*, **32**, 379–389.

SANGHVI, I., MURAYAMA, S., SMITH, C. M. & UNNA, K. R. (1963). Action of muscarine on the superior cervical ganglion of the cat. *J. Pharmacol.*, **142**, 192–199.

SCHLAEPFER, W. W. & TORACK, R. M. (1966). The ultrastructural localization of cholinesterase activity in the sciatic nerve of the rat. *J. Hist. Cyt.*, **14**, 369–378.

SHANTHAVEERAPPA, T. R. & BOURNE, G. H. (1962). The "perineurial epithelium", a metabolically active, continuous, protoplasmic cell barrier surrounding peripheral nerve fasciculi. *J. Anat.* **96**, 527–537.

SMITH, T. G., WUERKER, R. B. & FRANK, K. (1967). Membrane impedance changes during synaptic transmission in cat spinal motoneurons. *J. Neurophysiol.*, **30**, 1072–1096.

STEIN, J. M. & PADYKULA, H. A. (1962). Histochemical classification of individual skeletal muscle fibers of the rat. *Amer. J. Anat.*, **110**, 103–115.

STRAUGHAN, D. W. (1960). The release of acetylcholine from mammalian motor nerve endings. *Brit. J. Pharm.*, **15**, 417–424.

TAKAHASHI, H., NAGASHIMA, A., & KOSHINO, C. (1958). Effect of γ-aminobutyrylcholine upon the electrical activity of the cerebral cortex. *Nature*, **182**, 1443–1444.

TAKEDA, K. (1967). Permeability changes associated with the action potential in procaine-treated crayfish abdominal muscle fibers. *J. gen. Physiol.*, **50**, 1049–1074.

TAKESHIGE, C. & VOLLE, R. L. (1964). A comparison of the ganglion potentials and block produced by acetylcholine and tetramethylammonium. *Brit. J. Pharmacol.*, **23**, 80–89.

TAKEUCHI, A., & TAKEUCHI, N. (1959). Active phase of frog's end-plate potential. *J. Neurophysiol.*, **22**, 395–411.

TAKEUCHI, A., & TAKEUCHI, N. (1960). On the permeability of end-plate membrane during the action of transmitter. *J. Physiol.*, **154**, 52–67.

TAKEUCHI, A., & TAKEUCHI, N. (1962). Electrical changes in pre- and postsynaptic axons of the giant synapse of *Loligo*. *J. gen. Physiol.*, **45**, 1181–1193.

TAKEUCHI, A. & TAKEUCHI, N. (1964). The effect on crayfish muscle of iontophoretically applied glutamate. *J. Physiol.*, **170**, 296–317.

TAKEUCHI, A. & TAKEUCHI, N. (1965). Localized action of gamma-aminobutyric acid on the crayfish muscle. *J Physiol.*, **177**, 225–238.

TAKEUCHI, A. & TAKEUCHI, N. (1966). A study of the inhibitory action of γ-aminobutyric acid on neuromuscular transmission in the crayfish. *J. Physiol.*, **183**, 418–432.

TAKEUCHI, A. & TAKEUCHI, N. (1967). Anion permeability of the inhibitory post-synaptic membrane of the crayfish neuromuscular junction. *J. Physiol.*, **191**, 575–590.

TAKEUCHI, N. (1963). Effects of calcium on the conductance change of the end-plate membrane during the action of the transmitter. *J. Physiol.*, **167**, 141–155.

TASAKI, I., & MIZUTANI, K. (1944). "Comparative Studies on the Activities of the Muscle evoked by Two Kinds of Motor Nerve Fibres." *Jap. J. med. Sci.*, **10**, 237–244.

TAUC, L. (1960). The site of origin of the efferent action potential in the giant nerve cell of *Aplysia*. *J. Physiol.*, **152**, 36–37P.

TAUC, L., & GERSCHENFELD, H. M. (1962). A cholinergic mechanism of inhibitory synaptic transmission in a molluscan nervous system. *J. Neurophysiol.*, **25**, 236–262.

TAXI, J. (1965). Contribution à l'étude des connexions des neurones moteurs du système nerveux autonome. *Ann. Sci. Nat. Zool. Paris*, 12 Sér., **7**, 413–674.

THESLEFF, S. (1955). "The Mode of Neuromuscular Block caused by Acetylcholine, Nicotine, Hexamethonium and Succinylcholine." *Acta physiol. scand.*, **34**, 218–231.

THESLEFF, S. (1955). "The Effects of Acetylcholine, Decamethonium and Succinylcholine on Neuromuscular Transmission." *Acta physiol. scand.*, **34**, 386–392.

THIES, R. E. (1962). Depletion of acetylcholine from muscles treated with HC-3. *Physiologist*, **5**, 220.

THIES, R. E. (1965). Neuromuscular depression and the apparent depletion of transmitter in mammalian muscle. *J. Neurophysiol.*, **28**, 427–442.

TOSAKA, T., CHICHIBU, S. & LIBET, B. (1968). Intracellular analysis of slow inhibitory and excitatory postsynaptic potentials in sympathetic ganglia of the frog. *J. Neurophysiol.*, **31**, 396–409.

TOSCHI, G. (1959). A biochemical study of brain microsomes. *Exp. Cell Res.*, **16**, 232–255.

TRAMS, E. G., IRWIN, R. L., LAUTER, C. J. & HEIN, M. M. (1962). Properties of electroplax protein. *Biochim. biophys. Acta*, **58**, 602–604.

UCHIZONO, K. (1964). On different types of synaptic vesicles in the sympathetic ganglia of amphibia. *Jap. J. Physiol.*, **14**, 210–219.

USHERWOOD, P. N. R. (1963). Spontaneous miniature potentials from insect muscle fibres. *J. Physiol.*, **169**, 149–160.

USHERWOOD, P. N. R. & GRUNDFEST, H. (1965). Peripheral inhibition in skeletal muscle of insects. *J. Neurophysiol.*, **28**, 497–518.

VAN HARREVELD, A., & MENDELSON, M. (1959). Glutamate-induced contractions in crustacean muscle. *J. cell. comp. Physiol.*, **54**, 85–94.

VAN DER KLOOT, W. G. (1960). Factor S—a substance which excites crustacean muscle. *J. Neurochem.* **5**, 245–252.

VOLLE, R. L. (1962). Enhancement of postganglionic responses to stimulating agents following repetitive preganglionic stimulation. *J. Pharmacol.*, **136**, 68–74.

VOLLE, R. L. & KOELLE, G. B. (1961). The physiological role of acetylcholinesterase (AChE) in sympathetic ganglia. *J. Pharmacol.*, **133**, 223–240.

VRBOVÁ, G. (1963). The effect of motoneurone activity on the speed of contraction of striated muscle. *J. Physiol.*, **169**, 513–526.

WATANABE, A. & TAKEDA, K. (1963). The change of discharge frequency by a.c. stimulus in a weak electric fish. *J. exp. Biol.*, **40**, 57–66.

WERMAN, R., DAVIDOFF, R. A. & APRISON, M. H. (1968). Inhibitory action of glycine on spinal neurons in the cat. *J. Neurophysiol.*, **31**, 81–93.

WERMAN, R., McCANN, F. V., & GRUNDFEST, H. (1961). Graded and all-or-none electrogenesis in arthropod muscle. *J. gen. Physiol.*, **44**, 979–995.

WERMAN, R., & GRUNDFEST, H. (1961). Graded and all-or-none electrogenesis in arthropod muscle. *J. gen. Physiol.*, **44**, 997–1027.

WERNER, G. (1960). Neuromuscular facilitation and antidromic discharges in motor nerves: their relation to activity in motor nerve terminals. *J. Neurophysiol.*, **23**, 171–187.

WHITTAKER, V. P. (1959). The isolation and characterization of acetylcholine containing particles from brain. *Biochem. J.*, **72**, 694–706.

WHITTAKER, V. P. (1964). Investigations on the storage sites of biogenic amines in the central nervous system. *Progr. Brain Res.*, **8**, 90–117.

WHITTAKER, V. P. (1966). Some properties of synaptic membranes isolated from the central nervous system. *Ann. N.Y. Acad. Sci.*, **137**, 982–998.

WIERSMA, C. A. G. (1941). "The Efferent Innervation of Muscle." *Biol. Symp.*, **3**, 259–289.

WILBRANDT, W. (1937). "Effect of Organic Ions on the Membrane Potential of Nerves." *J. gen. Physiol.*, **20**, 519.

WILSON, D. M. & DAVIS, W. J. (1965). Nerve impulse patterns and reflex control in the motor system of the crayfish claw. *J. exp. Biol.*, **43**, 193–210.

WITANOWSKI, W. R. (1925). "Uber humorale Ubertragbarkeit der Herznervenwirkung." *Pflüg. Arch.*, **208**, 694.

WOOD, D. W. (1957). "The Effect of Ions upon Neuromuscular Transmission in a Herbivorous Insect." *J. Physiol.*, **138**, 119–139.

WYCKOFF, R. W. G., & YOUNG, J. Z. (1956) "The Motoneuron Surface." *Proc. Roy. Soc., B*, **144**, 440–450.

YOUNG, J. Z. (1939). Fused neurons and synaptic contacts in the giant nerve fibres of cephalopods. *Phil. Trans.*, **229**, 465.

CHAPTER XVII

THE SENSORY RESPONSE

ELECTRICAL excitation is a valuable tool in the study of general mechanisms, on account of the great accuracy with which the stimulus may be controlled; but of course the natural external stimuli are either mechanical or chemical, and are applied to the sensory nerve terminals usually through the inter-mediary action of a specialized receptor. This receptor may take the form of a mechanical transmitter of the impulse, as for example in the Pacinian corpuscle which mediates pressure sensations, or it may be a specialized cell in which an electrical charge is initiated and transmitted to the sensory nerve fibre through a synapse, as in the case of the rods and cones of the vertebrate retina.

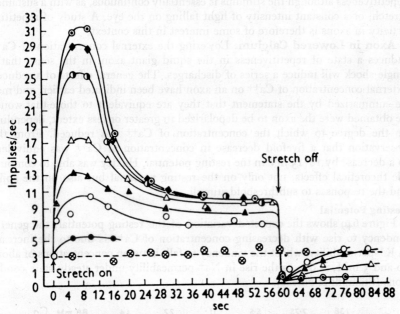

FIG. 648. Rates of discharge of a single stretch fibre of frog skin with different degrees of stretch. Impulse rate of successive experiments at 8 min. interval in which the preparation is abruptly stretched by 6 per cent., ●——●; 9 per cent., O——O; 10 per cent., ▲——▲; 14 per cent., △——△; 18 per cent., ◑——◑; 23 per cent., ◆——◆; 28 per cent., ◉——◉; starting from its initial minimal length (100 per cent.). ⊗——⊗, frequency at minimal length. (W. R. Loewenstein. *J. Physiol.*)

Discharge in Sensory Nerve. The result of stimulating the receptor is the propagation of a series of spikes along the sensory fibre; they are typically all-or-none, maintaining the same characteristics in spite of wide variations in intensity of stimulus, the only effect of this variable being a change in the frequency of discharge. Again, when fibres mediating different sensations are compared, no differences in the nature of the spikes have been definitively correlated with the type of sensation mediated, so that it is generally con-sidered that the nature of the sensation is determined by the connections the sensory fibre ultimately establishes in the central nervous system. In Fig. 648

the frequency of discharge in a single fibre, innervating a stretch receptor in the skin of the frog, has been plotted against the time during which the stretch was maintained. The greater the degree of stretch, the greater the frequency of discharge, and, for a given degree of stretch, it is found that the frequency tends to fall off with time to reach a steady level—the phenomenon of *adaptation* or *accommodation*. As a result of this, the response of a receptor may be divided into a *dynamic* and a *static* phase; the former is the immediate discharge following the initiation of a sustained stimulus, whilst the latter is the discharge finally attained after adaptation. In some receptors, for example the muscle spindle, this static phase is not much different in intensity from the dynamic phase; in the Pacinian corpuscle, on the other hand, adaptation is so great that there is no static phase.

Repetitive Responses

The remarkable feature of the response in a receptor's sensory neurone is its repetitiveness although the stimulus is essentially continuous, as with a sustained stretch, or a constant intensity of light falling on the eye. A study of repetitive activity in axons is therefore of some interest in this context.

Axon in Lowered Calcium. Lowering the external concentration of Ca^{++} induces a state of repetitiveness in the squid giant axon, in the sense that a single shock will induce a series of discharges. The general effects of a reduced external concentration of Ca^{++} on an axon have been indicated earlier, and may be summarized by the statement that they are equivalent to those that would be obtained were the axon to be depolarized to greater or less extent, depending on the degree to which the concentration of Ca^{++} was reduced. From the observation that a fivefold decrease in concentration of Ca^{++} was equivalent to a decrease by 10–15 mV in the resting potential, Huxley was able to compute the theoretical effects, not only on the resting potential but also on the spike and the responses to subthreshold stimuli.

Resting Potential

Figure 649 shows the expected variation of the resting potential; the general tendency to rise with decreasing concentration of Ca^{++} is due to the increase in K^+-permeability, whilst the tendency to fall, between concentrations of about 20 and 5 mM, is due to the rise in Na^+-permeability under steady-state conditions (Stämpfli & Nishie).

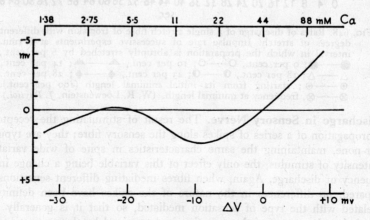

FIG. 649. Variation of squid axon resting potential with concentration of Ca^{++} in the external medium. (Huxley. *Ann. N.Y. Acad. Sci.*)

Oscillatory Responses

Both theoretically and practically it may be shown that the response of the squid axon to a single subthreshold stimulus is not a simple return from its depolarized state, but that it consists of a series of oscillations; at normal Ca^{++}-concentration these are heavily damped but, as Fig. 650 shows, the damping, measured by the natural logarithm of the ratio of the height of each peak to that of its successor, decreases with decreasing $[Ca^{++}]$ until at 16·2 mM it reaches zero; beyond this it becomes negative, indicating that the amplitudes of successive oscillations increase and may thus, presumably, lead to the firing off of a spike. We may begin by considering the effect of progressively reduced $[Ca^{++}]$ on the spike response; between 44 mM (normal) and 19·9 mM there is a progressive fall in the threshold, whilst the oscillations that follow the spike are increased in amplitude; otherwise the predicted behaviour is not far from normal. Qualitative differences may be expected just below 19·9 mM, when a single stimulus gives rise to a small number of spikes of decreasing amplitude,

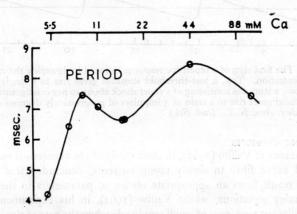

Fig. 650. Calculated period of oscillations of resting potential of very small amplitude as a function of the concentration of Ca^{++} in the external medium. (Huxley. *Ann. N.Y. Acad. Sci.*)

provided that the stimulus is well above threshold (Fig. 651). This happens over such a narrow range of Ca^{++}-concentrations that it is unlikely to be observed experimentally. At lower concentrations, down to 17·7 mM, the threshold response is a series of spikes that continue indefinitely, whilst a just-subthreshold stimulus still causes a damped oscillation.

At still lower concentrations—between 17·7 and 16·2 mM—the predicted effects of a cathodal stimulus are of three kinds depending on the strength: (*a*) a weak shock causes only damped oscillations; (*b*) a shock of sufficient size sets off directly a train of spikes; (*c*) an intermediate condition in which an oscillation of increasing amplitude, *i.e.*, negative damping, is established leading to a spike and, thereafter, a train of spikes. In this range, an anodal shock is very nearly as effective as a cathodal one.

At concentrations of Ca^{++} below 16·2 mM the membrane potential is unstable, so that any deviation produces oscillations that build up to spikes; with still lower concentrations the frequency of the discharge increases, whilst the amplitude tends to fall, so that the difference between oscillations and spikes tends to vanish. At the very lowest concentrations nothing but damped oscillations can be elicited. In general, these theoretical deductions are amply confirmed experimentally on the squid axon.

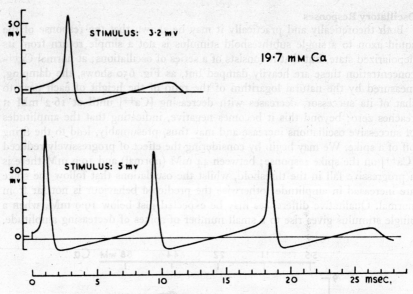

FIG. 651. The first sign of a repetitive response induced by lowering the calcium
concentration. Top, a just-threshold stimulus gives rise to a single spike.
Bottom, a stimulus consisting of a short shock about 50 per cent. greater than
threshold gives rise to a train of 3 impulses of progressively decreasing size.
(Huxley. *Ann. N. Y. Acad. Sci.*)

SLOWLY RISING CURRENTS

Frankenhaeuser & Vallbo (1965), in their study of the theoretical responses of
a myelinated nerve fibre to slowly rising currents, concluded that repetitive
firing would result from an appropriate choice of parameters in the modified
Hodgkin–Huxley equations, whilst Vallbo (1964), in his experimental study,
observed oscillatory responses of small amplitude when the rate of rise in current,
dI/dt, was below a certain value.

Single Spike as Special Case

Thus the single spike response to a single depolarization-step may be regarded
as the result of a special relationship between the Na^+- and K^+-permeabilities
and the factors determining their variations. Repetitive responses to a single
or maintained* depolarization are therefore not the surprising events that they
were before the analysis of the action potential carried out by Hodgkin & Huxley.
This is not to suggest, nevertheless, that the sensory axon has a special repetitiveness; as we shall see, it comes into relationship with a specialized region that
develops a generator potential, and it may simply be the maintenance of this that
acts as a stimulus leading to repetitive discharge.

RESPONSE IN INVERTEBRATE EYE

The Eye of *Dytiscus*

The repetitive response is, of course, not peculiar to sense organs; the
motor neurones of the spinal cord give a repetitive response to a constant
current applied to the cord (Barron & Matthews) and the postganglionic

* Essentially, the single response to a sustained applied current may be described as
resulting from accommodation, and the factors determining this, as we have seen in
Chapter XV (p. 1089), are largely the state of activation of the Na^+-permeability at a given
value of the membrane potential and the turning on of the K^+-permeability.

sympathetic fibres, already considered, give a repetitive discharge with frequencies which bear no simple relationship to the frequencies in the pre-ganglionic fibres. In these cases we have seen that the spikes apparently discharge as a result of a build-up of electrotonic potential, and it is of interest that Bernhard has described a similar phenomenon in the compound eye of the water-beetle *Dytiscus*. The compound eye of the insect and many other invertebrates is essentially a group of unit eyes, or *ommatidia* as they are called, each being provided with a refracting system which focuses an image on a group of sensitive *retinula cells*. In *Dytiscus*, nerve fibres leading from the excitable retinula cells soon end in the optic ganglion or optic lobe, a pear-shaped structure tapering off into an optic nerve which runs to the supra-œsophageal ganglion, the whole distance from retinula cell to supra-œsophageal ganglion being only 2·5 mm. The optic lobe consists of neurones which pre-sumably synapse with the fibres of the retinula cells.

Generator Potential. By placing electrodes at different points on this sensory pathway, as indicated schematically in Fig. 652, Bernhard was able to record the electrical events taking place at different stages. With leads 3 and 5, *i.e.*, on the optic ganglion and supra-œsophageal ganglion, the record consisted in

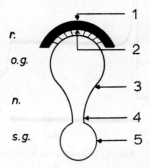

FIG. 652. Schematic drawing of eye of *Dytiscus* prepared for an experiment, showing receptorial layer, *r*, optic ganglion, *o.g*, optic nerve, *n*, and supra-œsophageal ganglion, *s.g*. The numbers 1–4 show the different loci used for the active electrode, whereas 5 is the constant locus of the reference electrode on the supra-œsophageal ganglion. (Bernhard. *J. Neurophysiol.*)

a slow rise in potential persisting as long as the light-stimulus, the optic ganglion being negative; superimposed on this wave of negativity were typical spikes; the farther up the optic nerve the active electrode was placed, the smaller the negativity; and the record became eventually a typical repetitive discharge of spikes (leads 4 and 5). If the optic ganglion and central parts were severed from the eye, leads 1 and 2 recorded only a slow wave of negativity without spikes. This observation suggests that the first effect of the light-stimulus is the develop-ment of an electrotonic negative potential in the receptors (retinula cells), and that this spreads to the optic ganglion and initiates spike potentials in the neurones there. The potential in the receptor would act as a generator—*the generator potential*—behaving like a constant current that induces repetitive discharges in the crustacean axon.

The Eye of *Limulus*

An even more striking demonstration of the significance of the electrotonic generator potential is provided by the study of the responses of another invertebrate eye, namely the lateral eye of the horse-shoe crab, *Limulus*. The ommatidium contains some 8–20 retinula cells on which the light is focused; these cells are continued out of the eye as long fibres to form the optic nerve, which relays synaptically in the lamina ganglionaris. The development of a generator potential, spreading electrotonically to the ganglion layer, would be out of the question here, since the distance is too great, and it was thought that the retinula cells developed both generator potential and propagated spikes, since action potentials could be recorded from a single fibre following illumination of the ommatidium from which it was derived. However, the studies of MacNichol, Wagner & Hartline, and of Waterman & Wiersma, have made it very likely that action potentials do *not* develop in the fibres leading from the retinula cells but only in the fibres from another type of cell— the *eccentric cell*—which at its distal end has a sharp pointed process that projects into the canal formed by the light-sensitive retinula cells, whilst proximally it extends as a fibre-like process in the optic nerve with the retinula fibres. This pointed process, because of its intimate association with the retinula cells, would be likely to be activated by their generator potentials.*

Recordings with intracellular electrodes have generally confirmed this view; thus the retinula cells are coupled with each other and also with the eccentric cell, so that an electrical event occurs simultaneously in all; in response to light the retinula cells only developed a slow wave of depolarization, whilst the

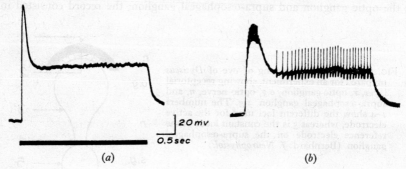

20 mv
0.5 sec

(a)   (b)

Fig. 653. Responses in retinula cell (a) and eccentric cell (b) to light stimulus.
(Dowling, *Nature*.)

eccentric cell developed a slow wave with large spikes as illustrated by Fig. 653 taken from Dowling (1968). The morphological studies of Lasansky (1967) confirm the probable electrotonic coupling in so far as they show that the cells make intimate contact with each other through interdigitating microvilli; these regions of contact are characterized by very close apposition of the cell membranes to give the typical quintuple-layered structure although the absence of appreciable thinning of the combined membrane-pair raises the question as to whether the junction is a true zonula occludens or a mere apposition of membranes as considered by Moody & Robertson (1960).†

* In the vertebrate eye, too, it would seem that the light-sensitive cells, rods and cones, only develop electrotonic generator potentials (Tomita & Funaishi, 1952); intracellular recordings from cones, carried out by Svaetichin, indicate that the cones respond by a *positive* generator potential, *i.e.*, the cell is hyperpolarized for the duration of the light-stimulus.

† Dowling (1968) has examined the potentials recorded from retinula and eccentric cells under conditions of very low illumination in the dark-adapted eye; under these conditions the responses are remarkably similar, and consist of small waves of depolarization—*quantum bumps*—interspersed with larger regenerative responses. As the intensity of the light is increased, the quantum bumps fuse and the response in the retinula cell becomes, finally, the sustained generator potential. In the eccentric cell, the response at high light intensity is also a sustained depolarization but, in addition, this is accompanied by a nervous spike discharge; thus at low intensities of stimulus, in the dark-adapted eye, it seems that the responses are not adequate to produce a sustained generator potential in the eccentric cell.

Inhibition. The studies of Hartline, Wagner & Ratliff on responses in single fibres from the eccentric cells of single ommatidia of *Limulus* have demonstrated that excitation of one ommatidium is associated with inhibition of the simultaneous response in an adjoining one. Two ommatidia were studied simultaneously by separately illuminating them, and recording from their fibres. Fig. 654 illustrates the mutual inhibition of one ommatidium by another; thus, the top record shows the response to illumination of ommatidium A alone, giving a frequency of discharge of 53/sec., whilst the bottom record is that obtained from a nearby ommatidium, B, which alone gives a frequency of 46/sec. When both are illuminated together, A and B have frequencies of 43 and 35/sec. respectively. The degree of inhibition exerted by a neighbouring ommatidium on another depended on the distance between the two—the

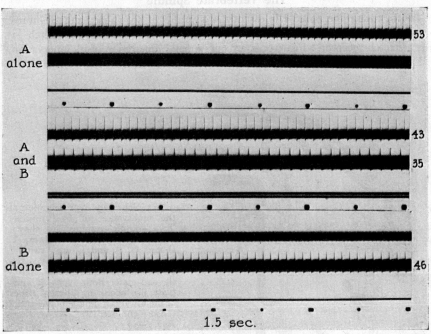

FIG. 654. Illustrating mutual inhibition of two neighbouring ommatidia of the eye of *Limulus*. (Hartline & Ratliff. *J. gen. Physiol.*)

greater the distance the smaller the inhibitory effect—and on the relative intensities of illumination of the two ommatidia. The latter factor is illustrated in Fig. 654, since ommatidium A, with the stronger discharge, is reduced by 10/sec. whilst ommatidium B, with the weaker discharge, is reduced by 11/sec.

An interesting release from inhibition was achieved as follows: records were taken from two ommatidia, A and B, one millimetre apart. When both were illuminated together their responses were less than when illuminated alone, as in Fig. 654. Additional ommatidia were now illuminated; they were chosen so that they were too far away to affect ommatidium A, hence this additional illumination only inhibited B. The effect of this inhibition of B was to increase the discharge from A because the inhibitory effect of B on A depended on the frequency of discharge in B. When the inhibitory effects of two receptors were studied, separately and together, on a third receptor, it was found that their effects would summate, but not necessarily in a simple additive manner.

If they were so far apart as to have no effect on each other, then the effects were simply additive; if they were close enough to interact, then their combined effects were less than the sum of their individual effects. It was possible, however, to describe the interaction between receptors by a series of simple linear equations.

MECHANORECEPTORS

A number of further studies, notably those of Katz and of Kuffler & Eyzaguirre on stretch receptors, and those of Alvarez-Buylla & de Arellano, Gray & Sato, and Loewenstein on the Pacinian corpuscle, have strengthened the concept of the generator potential.

The Vertebrate Spindle

Structure. The spindle is essentially a group of specialized muscle fibres, provided with a sensory innervation that makes the system sensitive to stretch—it is a stretch receptor.* Fig. 655 is a diagrammatic illustration of its innervation. Three ordinary—*extrafusal*—muscle fibres are shown; between them are two

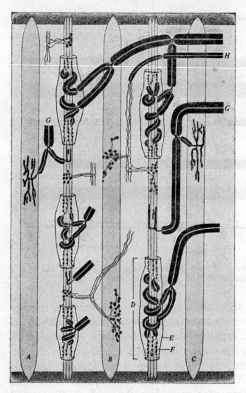

Fig. 655. Schematic representation of the innervation of frog muscle. Two spindle systems are shown between three extrafusal fibres, A, B and C. A and C are "twitch" fibres and B a "tonic" one. Each of the two types has its characteristic end-plate. Both types of end-plate also occur on different fibres of the intrafusal bundle between the sensory endings. (E. G. Gray. *Proc. Roy. Soc.*)

---

* The spindle lies in parallel with the main mass of muscle fibres and so is stretched when the muscle is stretched, and released from tension when the muscle contracts. By contrast, the *tendon organ of Golgi*, lying in the tendon principally near musculo-tendinous junctions, is in series with the contractile fibres; contraction of the fibres causes an increased tension that leads to a sensory discharge. During muscular contraction, therefore, the discharge from the tendon organ is augmented, that from the spindle is diminished. Hunt (1952) has reviewed the function of the spindle in reflex activity. Motor activity in intrafusal fibres has been recorded by intracellular electrodes by Koketsu & Nishi (1957) who observed both local and spike potentials—the resting potential recorded was exceptionally low (35–45 mV).

spindle systems. The middle fibre is a "slow" fibre (p. 1175), innervated by small motor nerve fibres with grape-like endings; the outer, extrafusal, fibres, A and C, are twitch fibres with typical end-plate innervation from large myelinated motor fibres. The sensory regions of the spindles—D—are encapsulated and are innervated by large sensory fibres which coil around the intrafusal fibres, and finally end in fine varicose processes (F). Although the spindle subserves an essentially sensory role, it nevertheless has a motor innervation; in the spindle of the frog, as Fig. 655 shows, this is achieved by both small and large motor nerve innervation and, as indicated, the same axons, by branching, may innervate both intra- and extrafusal fibres.* Stretching of the muscle causes a discharge in the sensory fibres from a spindle; this is the basis of the well known stretch-reflex. When the muscle contracts, by contrast, any tonic discharge that may have been taking place is inhibited. The function of the motor innervation is of some interest, and has been investigated most thoroughly in the mammal by Kuffler, Hunt & Quilliam; in this order, the motor innervation of the intrafusal fibres is exclusively by fine (3–8 μ diameter) fibres; stimulation of these fibres caused an increased sensory discharge in the afferent fibres; in other words, the intrafusal fibres were made more sensitive to the degree of stretch prevailing in the muscle. The motor innervation therefore permits of an adjustment of the sensitivity of the receptor to meet different states of contraction or relaxation.

Electron-Microscopy

The capsule, which gives the spindle its shape, is a continuation of the endothelial sheath of the sensory axons; these, after entering the capsule, break up into numerous non-myelinated terminal branches, which run alongside the intrafusal muscle fibres and are characterized by varicose swellings to create the appearance of beaded chains. In the electron-microscope the same appearance of a chain of beads, or microspindles linked together by thin tubes, is observed; within the bulbous expansions there are often numerous small mitochondria (Fig. 656, Pl. LVII). In contrast to the motor fibres to a muscle, these sensory fibres are not closely invested with Schwann cells. The relationship between the sensory fibre and the intrafusal muscle fibre is of fundamental importance for the understanding of the transducer process, whereby a stretch of the muscle excites the nerve fibres. As Fig. 657 shows, the swellings lie in depressions of the muscle fibre; the approach of the two cell membranes is closer than at the end-plate, the gap being of the order of 100–200A; at points the gap is crossed by filaments that seem to connect the two cells and possibly serve to transmit to the axon the effects of alteration of muscle length. Examination of the intrafusal muscle fibres, within the capsule, reveals alternating discontinuities in their structure; in some "compact" regions, the sensory contacts were made with a fibre that had the usual striations revealing contractile power (p. 1347) but in other regions, in series with the compact regions, there were very obvious differences of structure; in longitudinal section the fibre seemed fenestrated, being split up into a complicated framework with interstitial spaces filled with a dense network of fine connective-tissue fibrils that are derived from the external diffuse layer of the muscle's surface membrane complex, i.e., what was called the sarcolemma and what Katz has named the ectolemma. Thus, in these reticular zones the muscle fibre is embedded in a meshwork of fibrils (Fig. 658, Pl. LVIII). The muscle fibre loses some 85 per cent. of its contractile myofilaments, so that this region will not partake of a contraction taking place in the rest of the muscle fibre; in fact it will be stretched, and so the nervous connections here will mediate the response to contraction of the intrafusal fibres in response to its motor innervation. By contrast, the compact regions will only be stretched when the muscle fibre as a whole is lengthened, i.e., in response to a pull on the fibre as when the stretch-reflex is elicited.

* The mammalian (cat and rabbit) spindle has been exhaustively described by Barker (1948) and Boyd (1962); E. G. Gray (1957) has drawn attention to points of similarity with, and difference from, the frog spindle.

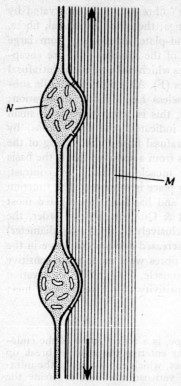

The response to a stretch of the muscle has two components, a dynamic, or phasic, component depending on the velocity of stretch, and a static, or tonic, component depending on the actual extension. In the compact zone we have the system of interdigitating myofibrils which, when the muscle contracts, slide over each other; and it is likely, on mechanical grounds, that the nerve fibres making contact with this zone would subserve the response to sustained stretch. The nerve fibre in the reticular zone, with its much lower presumptive viscosity, would respond much more readily to phasic changes in length, *i.e.*, the velocity component.

Generator Potential. Katz recorded the potential changes occurring in single fibres from a frog's spindle. A very definite electrotonic potential preceded the discharge of action potentials, the greater the potential the greater the frequency of discharge; the potential was not abolished by concentrations of local anæsthetics that were sufficient to prevent the development of action potentials.

This "spindle potential" belongs to the general class of *generator potentials*. It presumably represents, like the end-plate potential, a region of depolarization that behaves as a sink for current and activates the sensory neurone. Like the propagated spike, it is sensitive to Na^+-lack; Fig. 659 from Calma (1965) shows the progressive changes in generator potential and spike in

FIG. 657. Illustrating the bulbous swellings on the nerve fibre, N, and the indentations in the intrafusal muscle fibre, M. (Katz. *Phil. Trans.*)

response to weak stretches, as the preparation is kept in a Na^+-free medium. In general, it was found by both Calma and by Ottoson (1964) that some generator activity persists in the Na^+-free medium.

Invertebrate Stretch-Receptor

An analogue to the spindle of vertebrate muscle was described by Alexandrowitz in the lobster, and its physiological properties were investigated by Wiersma and by Kuffler. As Fig. 660 illustrates, it consists of just two specialized muscle fibres, described as bundles of myofibrils, containing specialized regions in which the dendrites of sensory neurones ramify. The muscle fibres are supplied by motor fibres which ramify to form numerous junctions, as with ordinary invertebrate fibres. On stimulation of the motor nerve they contract; one of the fibres gives a slow response, by contrast with that of the other which is a rapid twitch. The two fibres are described therefore as "slow" and "fast" respectively. Contraction of the muscle, or a mechanical stretch, gives rise to an afferent discharge, as with the frog spindle; the fast neurone shows rapid adaptation, whilst the slow neurone maintains a high rate of discharge during a sustained stretch.

Intracellular Records

This invertebrate system has been studied in some detail by Eyzaguirre & Kuffler with the aid of intracellular electrodes which may be inserted into the

PLATE LVII

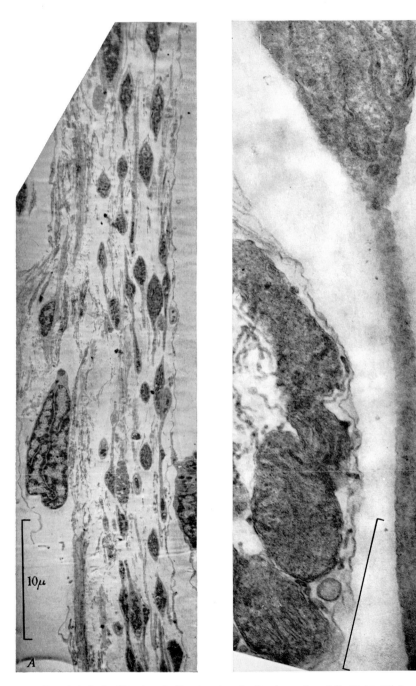

FIG. 656. Left: Muscle spindles showing the "varicose threads". *Right*: Higher magnification of the same structures. (Katz. *Phil. Trans.*)

PLATE LVIII

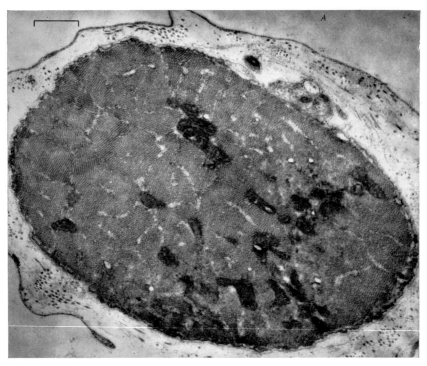

FIG. 658. Cross-sections of the same intrafusal fibre. *Above,* at the entry into the capsular region. *Below,* in the reticular zone. (Katz. *Phil. Trans.*)

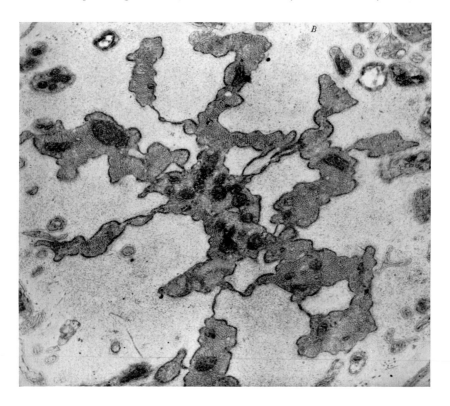

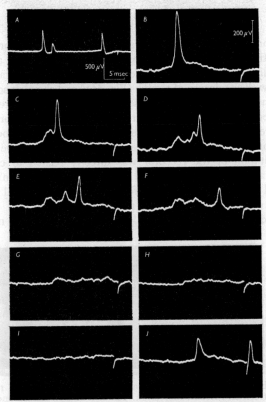

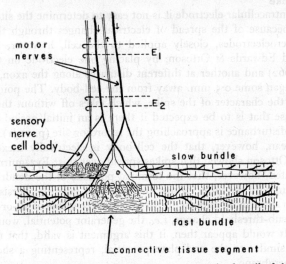

FIG. 659. Effects of sodium depletion on spindle potentials. *A*: Potentials generated by brief stretch in normal Ringer's solution. *B*: 5 min; *C*, *D*, *E*, *F*: 8–12 min; *G*, *H*: 15 min; *I*: 17 min after changing the external Ringer's solution to sodium-free Ringer's solution. *J*: 2 min after return of muscle into normal Ringer's solution. Downward deflexion at the end of the sweep is due to end of pulse energizing electro-magnetic puller. Voltage calibration constant from *B* to *J*. Time calibration in *A*. (Calma. *J. Physiol.*)

FIG. 660. Simplified scheme of stretch receptor-organ in the tail of the lobster. The two fine muscle strands are divided by a segment of connective tissue in which are embedded terminals of a sensory fibre with its cell body nearby. In addition to the sensory neurone, several other fibres innervate each muscle strand. Only two motor fibres are represented here. Positions of stimulating and recording electrodes (E_1 and E_2) on the common nerve trunk are indicated. (Kuffler. *J. Neurophysiol.*)

sensory neurones by virtue of their accessibility. The neurones exhibited a resting potential of 70–80 mV; stretching a muscle caused a depolarization of several millivolts, the degree of depolarization—the generator potential— being graded with the degree of stretch (Fig. 661) and showing some adaptation, in the sense that its magnitude decreased during a sustained stretch. Increasing the rate or degree of stretch led to greater depolarization, from which action potentials were fired. By blocking discharge in the nerve with local anæsthetics, large generator potentials could be recorded, and it could be shown clearly that the adaptation of the receptor was directly related to changes in this potential. An analysis of the potential changes taking place in the cell body, dendrites, and axon suggested that deformation of the terminal portions of the dendrites set up the true generator potential, which spread electrotonically to the cell body, or soma, to give a "pre-potential", *i.e.*, the measured effect

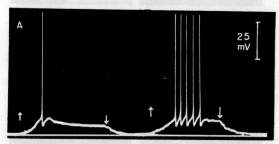

FIG. 661. Intracellular recording from a slow sensory cell. The receptor has been stretched twice in succession. The first stretch sets up one conducted soma impulse, whilst the second stretch causes five discharges after an initial depolarization of 12 mV. (Eyzaguirre & Kuffler. *J. gen. Physiol.*)

of the generator potential on the soma. At a critical level of this pre-potential (8–12 mV in slow fibres; 17–22 mV in fast fibres), an impulse fired off.

Origin of Spike

With an intracellular electrode it is not easy to determine the site of initiation of a spike because of the spread of electrical changes through the cell. With external microelectrodes, closely applied to the cell, however, this becomes possible and Edwards & Ottoson, by placing one electrode on the cell-body (E in Fig. 662) and another at different distances along the axon, showed that the spike began some 0·5 mm. away from the cell-body. The point of origin is revealed by the character of the spike, which takes off without the preliminary positive phase that is to be expected if there is an initial period in which the propagated disturbance is approaching the recording site (p. 1051). This finding does not mean, however, that the cell-body is electrically inexcitable, since Edwards & Ottoson showed that spikes spread over this; Eyzaguirre & Kuffler's study indicated, however, that the terminal portions of the dendrites were not invaded, mainly on the grounds that the generator potential persisted after the passage of an antidromic impulse; if they had been capable of all-or-none type of activity any sub-threshold change, *i.e.*, the generator potential, would have been wiped out. It would appear then, if this argument is valid, that the generator potential is similar to the end-plate potential, representing a short-circuiting of the cell membrane.

Frequency and Generator Potential. The relationship between the magnitude of the generator potential and the degree of stretch was determined by Terzuolo and Washizu who prevented the development of spikes, which would

have obscured the measurement of the generator potential, by passing an inward
current adjusted so as just to prevent the firing of an impulse at a given degree
of stretch. From the measured resistance and current-strength, the voltage-drop
required to prevent firing was calculated and this was taken as the magnitude of
the excess of the generator potential above the firing level, assuming that the
threshold for firing remained constant. The results of two experiments are shown
in Fig. 663, where it is seen (a) that the generator potential, computed in this
way, is a linear function of the muscle-length; since (b) the impulse-frequency
is also a linear function of stretch, it follows that the frequency of discharge
is linearly related to the magnitude of the generator potential.

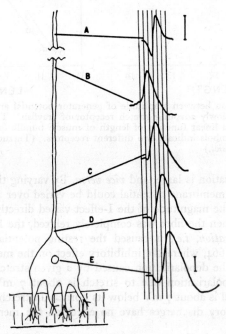

Fig. 662. Site of initiation of impulse in lobster receptor cell during excitation by
stretch. One electrode was kept fixed on the cell at E, while the other was
moved to different locations. The earliest response, not preceded by a
positive deflexion, is at B. This is the site of impulse initiation, about 0·5
mm. from the cell body. Conduction spreads from B towards A and also E.
Vertical lines give 0·1 msec. intervals. (Edwards & Ottoson. *J. Physiol.*)

Equilibrium Potential. If the generator potential is a depolarization resulting from an indiscriminate increase in membrane permeability, we may expect
an equilibrium, or reversal, potential in the region of zero; in fact, by extrapolation Terzuolo & Washizu found this to be the case.

Inhibition. This invertebrate system is of special interest since not only
does it permit intracellular recording from a sensory cell close to its dendrites,
but also the activity of the cell is subject to inhibitory influences from another
axon. Stimulation of this axon causes complete cessation of the steady discharge
in the slow afferent neurone resulting from a maintained stretch. Essentially,
the effect of the inhibitory impulse (I-impulse) is to stop the depolarization
of the generator potential, *i.e.*, it represents a tendency to repolarize the
membrane. The degree of repolarization produced by an inhibitory impulse
depends on the degree of depolarization that has already occurred, being large

40*

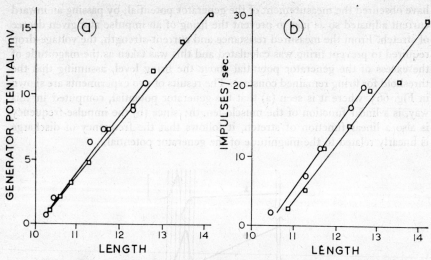

FIG. 663. Relation between amplitude of generator potential and impulse frequency in slowly adapting stretch receptor of crayfish. These two parameters are a linear function of length of muscle bundle (abscissæ). The
different symbols indicate two different receptors. (Terzuolo & Washizu.
J. Neurophysiol.)

when the depolarization is large, and *vice versa*. By varying the stretch applied
to a fast fibre, the membrane potential could be varied over a wide range, and
it was found that the magnitude of the I-effect varied directly with the degree
of polarization. When the fibre was completely relaxed, the I-impulse actually
caused a *depolarization, i.e.,* it caused the resting potential to fall. This is
illustrated by Fig. 664, where the inhibitory effect on the membrane potential
is plotted against the depolarization caused by a given stretch; it will be seen
that when the depolarization due to stretch is about 7 mV, *i.e.,* when the
membrane potential is about 7 mV below its normal unstretched value of about
77 mV the inhibitory discharges have no effect on the membrane potential,

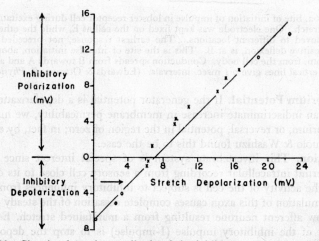

FIG. 664. Showing how the amplitude of the inhibitory potential varies with
the state of polarization of the receptor cell. Abscissa: Different amounts
of depolarization produced by varying amount of stretch. Ordinate:
Amplitude of inhibitory potential. (Kuffler & Eyzaguirre. *J. gen. Physiol.*)

whilst when the potential is above this reversal point the inhibitory discharge causes a hyperpolarization, or repolarization as Kuffler prefers to call it. Once again, then, we find that the direction of the inhibitory potential-change depends on the state of polarization of the post-synaptic cell; if this is high, then it depolarizes; if it is low it repolarizes, and thus drives the membrane potential to the so-called equilibrium, or reversal, potential. The effect of this is, of course, to prevent the depolarization that is the initial step in the firing off of discharges.

Increased Permeability to K^+

The inhibitory discharge is associated with an increase in membrane conductance, as revealed by the increased rate of decay of the potential induced in the soma by an antidromic impulse. According to Edwards & Hagiwara it is because of an increased permeability to K^+ that the conductance increases; this presumably tends to establish the resting potential at the Nernst K^+-potential determined by the concentrations of this ion across the membrane. Thus, during the establishment of the generator potential there is presumably a generalized increase in permeability so that the inward penetration of Na^+ causes depolarization; a specific increase in permeability to K^+ will tend to re-establish the Nernst K^+-potential.*

Permeability to Cl^-

By inserting microelectrodes and passing current across the cell membrane, Hagiwara, Kusano & Saito were able to vary the membrane potential in both directions and study the effects of an inhibitory discharge; their results agreed with those of Eyzaguirre & Kuffler in showing that the reversal potential was close to the resting potential; by allowing chloride to diffuse from the electrode into the cell, however, the reversal potential was reduced, indicating an influence on chloride as well as on potassium conductance by the inhibitory discharge.

Effects of Tetrodotoxin

Loewenstein, Terzuolo & Washizu (1963) found that a concentration of 1 to 5.10^{-6} g./ml. would block spike activity in the crayfish stretch receptor, without affecting the steady component of the generator potential. In the related stretch-receptor of the lobster, Albuquerque & Grampp (1968) were able to obtain reversible block of impulse activity in concentrations as low as 4.10^{-8} g./ml.; there was no effect on the steady level of the generator potential, but the hyperpolarization that follows this potential was reduced to 65 per cent. Other changes were an increase in resting potential by 4·8 mV and a 47 per cent. reduction in membrane resistance. The changed resting potential and membrane resistance might be explained on the basis of the Goldman equation using Brinley's (1965) figures for relative permeability and ionic concentration for the giant axon of the lobster; thus a reduced permeability to Na^+ associated with an increased permeability to K^+ could raise resting potential and also membrane conductance. The fact that the generator potential was unchanged in spite of the increased conductance means that tetrodotoxin increased the generator current, perhaps by activating the transport of an ion concerned in the process.†

Effects of Lithium

Li^+ will replace Na^+ in the genesis and conduction of the spike (see, for example, Hodgkin & Katz, 1949) but, as Obara & Grundfest (1968) have

* Edwards, Terzuolo & Washizu (1963) found that both action potential and generator potential were abolished by changing to Na^+- and Cl^--free Ringer by sucrose substitution; with 25 per cent. Na^+ and Cl^-, the generator potential occurred, but not the action potential.

† Nakajima (1964) compared the effects of tetrodotoxin on the electrical responses of fast and slowly adapting receptors; they found no difference in the generator potential with maintained stretch that would account for a difference in adaptation; in both, the generator potential declined rapidly from an initial peak.

emphasized, this is not true of the depolarizing electrogenesis of certain types of electrically inexcitable membrane, such as crayfish muscle fibres (Ozeki & Grundfest, 1967). The generator potential of the stretch-receptor could still be evoked when all Na^+ was replaced by Li^+, being only slightly smaller than normal, but the cell underwent a slow depolarization, probably due to the blocking of an electrogenic component in the Na^+-extrusion mechanism since, as we have seen (p. 634), Li^+ is unable to utilize the carrier-mechanism for Na^+-extrusion in several systems. The finding that the non-spike generator potential was not affected by Na^+-substitution indicates that the membrane responsible is not selective with respect to Li^+ and Na^+.

OTHER CATIONS

When the effects of other cation substitutions were examined, Obara (1968) found that tris(hydroxymethyl)aminomethane (Tris), trimethylammonium (TMA), choline, and triethylammonium (TEA) were all effective in allowing the development of a generator potential in response to stretch, the order of decreasing effectiveness of cations was: $Na \doteq K > Li > Tris > TMA > Choline$ or TEA.

The Pacinian Corpuscle

Structure. The Pacinian corpuscle is an ellipsoidal body, with a long axis some 2 mm. long, embedded in either muscle or connective tissue, *e.g.*, the cat's mesentery. As seen in the light-microscope, it consists of a number of

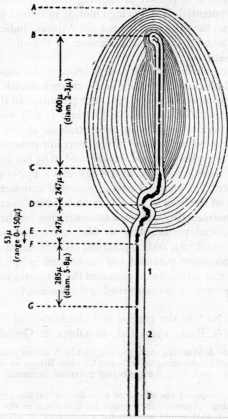

FIG. 665. Diagram of a typical Pacinian corpuscle. (Quilliam & Sato. *J. Physiol*).

concentric lamellæ, with the nerve embedded in its centre (Fig. 665); in the electron-microscope its appearance is rather more complex. The terminal nerve fibre in the centre (Figs. 666a and b) is surrounded by an inner core consisting of some 60 closely packed concentric laminæ, bilaterally arranged so that there are two opposing groups, one on either side of the nerve fibre, separated by longitudinally arranged clefts; in section, therefore, the laminæ appear as half-rings. These lamellæ contain mitochondria and are to be regarded as protoplasmic extensions of cell bodies situated in the intermediate, or growth, zone, the processes passing down the clefts separating the two opposed groups of lamellæ. The outer zone consists of some 30 concentric lamellæ that com-

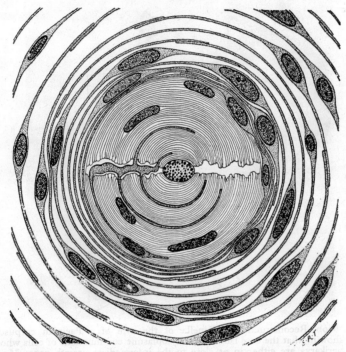

FIG. 666a. Schematic section through central region of Pacinian corpuscle, illustrating relationship of cells of intermediate growth zone to the lamellæ of the inner core. Note that the core region is bilaterally organized. (Pease & Quilliam. *J. biophys. biochem. Cytol.*)

pletely envelop the inner zones, *i.e.*, they are not made up of two halves; these are the lamellæ normally recognized by the light-microscopist. As shown by Hubbard, the spacing between successive lamellæ decreases from outside inwards in a regular logarithmic progression. The lamellæ are built up of flattened cells—0·2 μ thick on the average—overlapping each other so that each one constitutes a protoplasmic continuum like the capillary endothelium. Collagenous fibrils are conspicuous in the interlamellar spaces, forming a lacework that presumably gives the corpuscle its structural rigidity. The corpuscle is fluid-filled and under some turgor; it is essentially incompressible, therefore, so that a mechanical stimulus, if it is to affect the nerve fibre within, must be a displacement of one part in relation to another.

Nerve and Lamellæ

The terminal portion of the nerve fibre, as it enters the core, loses its myelin sheath completely, and then, within a few micra, its Schwann sheath, so that ultimately the core lamellæ make contact with the bare axon; in this terminal region the fibre becomes oval in cross-section, and the number of mitochondria within it is striking.

It will be recalled that the functional membrane enclosing a nerve fasciculus is probably the perineurium (p. 1035); this is a squamous epithelial structure, some five cell layers thick in the sciatic nerve; as the fasciculus divides and the number of contained fibres becomes smaller, the membrane thins until it is one cell thick. According to Shanthaveerappa & Bourne (1963), the Pacinian corpuscle is to be regarded as a bag of squamous stratified epithelium, each lamella as it is peeled off

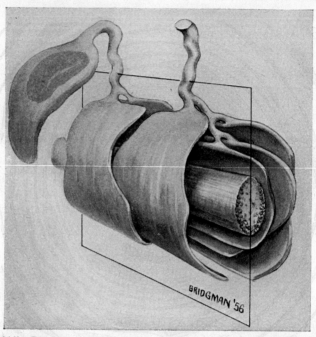

FIG. 666b. Reconstruction of lamellæ of the core of a Pacinian corpuscle showing that the core lamellæ are cytoplasmic continuations of cells whose perikarya are either in or close to the intermediate growth zone. Major cytoplasmic arms from these perikarya extend into the clefts of the core and from there branch laterally to form lamellæ which interdigitate with those of other cells. (Courtesy T. A. Quilliam.)

being revealed as a single sheet of this epithelium. The most superficial layers when peeled back could be seen to be continuous with the outermost layer of the perineurial epithelium of the sensory axon, whilst the deeper layers were not, but seemed to be derived from the superficial layers. On this view, then, the laminæ of the corpuscles are modified perineurium and not modified fibroblasts.

Generator Potential. Alvarez-Buylla & de Arellano recorded monophasic action potentials from a nerve-corpuscle preparation. With subliminal mechanical stimuli they obtained a subthreshold local response in the fibre; as the stimulus-intensity increased, this potential grew until, at a critical point, an action potential was fired off. With a maintained gentle pressure on the corpuscle, a repetitive response was obtained which showed marked adaptation, in the sense that the frequency of discharge rapidly diminished, although the mechanical stimulus was sustained. With a short stimulus—50 msec.—only two spikes were obtained, one at "on" and the other at "off".

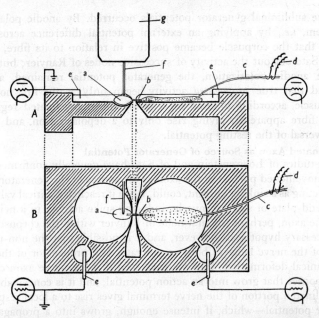

FIG. 667. Diagram illustrating method of recording potentials from axon leading from an isolated Pacinian corpuscle. A, transverse section. B, plan. a, Pacinian corpuscle; b, first node of Ranvier; c, mesenteric nerve; d, electrodes for antidromic stimulation; e, recording electrodes; f, rod from mechanical stimulator; g, mechanical stimulator. (Gray & Sato. *J. Physiol.*)

Gray & Sato studied the potentials in rather more detail; their recording system is illustrated in Fig. 667 which shows two pools of Ringer separated by an air-gap; the corpuscle was in the left-hand pool and the nerve fibre passed through the gap into the right-hand pool. Recording leads ran from the two pools, so that the potential across the gap was measured. The changes of potential following a mechanical stimulus consisted of three phases (Fig. 668):

FIG. 668. Records of mechanical stimulus to and potential from isolated Pacinian corpuscle. Downward deflection of the upper record indicates onset and magnitude of mechanical stimulus. Note slowly-rising generator potential leading to a diphasic spike. (Gray & Sato. *J. Physiol.*)

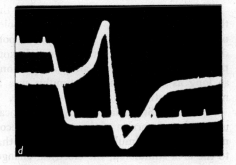

a slow initial phase of negativity at the left-hand electrode—the generator potential—followed by a diphasic change due to spike activity at the successive nodes of the myelinated fibre (Quilliam & Sato). This diphasic activity could be blocked by procaine, leaving the slow generator-potential nearly unaffected. By increasing the intensity of the mechanical stimulus, the size of the generator potential could be increased up to a maximum which may have been equal to the resting potential of the nerve fibre. Summation of successive subliminal generator potentials occurred. By anodic polarization of the system, *i.e.*, by applying an external potential difference across the two pools so that the corpuscle became positive in relation to its fibre, Diamond, Gray & Sato cut out the activity of successive nodes of Ranvier; but, however high the anodic polarization, the generator potential remained, and it was concluded that true *propagated* activity began only at the first node (within the corpuscle, according to Quilliam & Sato), the unmyelinated region of the terminal fibre apparently giving rise only to a depolarization, and not to an active reversal of the resting potential.

Unmyelinated Axon as Source of Generator Potential

Later studies of Loewenstein and of Sato have generally confirmed the view of the unmyelinated portion of the axon as the source of the generator potential; this, by acting as a sink for current, could, when it reached a critical value, behave like the end-plate of skeletal muscle, inducing spike activity in a neighbouring part of the axon, perhaps the first node of Ranvier within the corpuscle. This is not a necessary hypothesis, however, and it is possible that the non-myelinated portion of the nerve behaves both as generator and propagator of the response to mechanical deformation; thus we have seen that the nerve axon can initiate local responses that grow into an action potential, and it is conceivable that the non-myelinated portion of the nerve terminal gives rise to a local response—the generator potential—which, if intense enough, grows into a propagated spike.

Decapsulated Corpuscles. Loewenstein & Rathkamp dissected away the outer core and practically all of the inner core and yet obtained identical responses to mechanical stimulation, so that it is difficult to escape the conclusion that it is not the corpuscle, but its contained axon, that is responsible for all the electrical responses to mechanical stimulation. In decapsulated preparations, the first node is exposed whilst the terminal part of the axon can also be seen by phase-contrast microscopy; this permits the compression of different regions in an attempt to determine the origin of the all-or-none potential. Compression of the first node caused the disappearance of every sign of all-or-none activity, only the generator potential remaining, whilst compression of the terminal myelinated portion of the axon, as close as 10μ to the first node, had no effect on the action potential.

This strongly supports the view that the first node is the site of origin of the propagated activity. Crushing the whole length of the unmyelinated portion of the axon abolished the generator potential. By selectively compressing different portions of the unmyelinated axon, as in Fig. 669, it could be shown that, so long as there was an intact portion next to the myelinated region, and so long as the mechanical stimulus was able to deform this intact portion, mechano-electric conversion was possible.

Localized Stimulation. By highly localized mechanical stimulation of the unmyelinated portion of the axon, and recording the generator potential at and near the site of stimulation, it was found that the electric response was confined to the area stimulated, the electrical changes that occurred at a distance being apparently accounted for by electrotonic spread. Thus, according to this study, the generator potential is of the non-propagated type similar to the post-synaptic potentials discussed earlier. In a sense, therefore, we may consider the whole non-myelinated terminal as made up of a number of independent receptor sites, not, of course, anatomically discrete but functionally so. Since the area of excited membrane increased with increasing mechanical stimulus, Loewenstein considered that this represented a recruitment, by summation, of individual receptor sites. Certainly the experimental curve obtained by plotting generator against

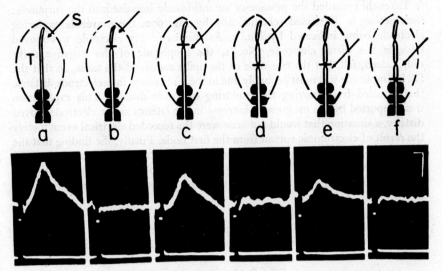

FIG. 669. Mechano-electric conversion after partial compression of the non-myelinated ending. A subthreshold mechanical stimulus of constant strength is
applied to the inner core of a decapsulated corpuscle (*a* to *f*) ; only in *a, c,*
and *e* is a generator response detected. Lower beam signals relative magnitude and duration of mechanical stimuli ; upper beam, the electric activity
of the ending led off the myelinated axon. The arrows of the diagram
indicate the zone of application of mechanical stimuli. The horizontal lines
across the nerve ending (T) indicate the central boundary of compressed area
of ending. At *a*, generator response of the intact ending; at *b*, after compression of a distal portion of ending; at *c*, after moving stimulus application
point beyond the compressed zone; at *d*, after compressing a zone located
centrally with respect to stimulus application point; and at *f*, after compression of entire length of ending. (Loewenstein. *Ann. N. Y. Acad. Sci.*)

stimulus-strength had the shape predicted on the basis of a summation of
localized "current-sinks" (Loewenstein, 1959; 1961).

Propagation of the Spike. It would appear from the work of Gray, Diamond,
and Loewenstein that the unmyelinated portion of the axon is unable to *conduct*
a propagated impulse, *i.e.*, that its electrical changes are not of the regenerative
all-or-none type. Later work has cast doubt on this view and suggests that the
non-myelinated portion of the nerve may give rise to local non-propagated
responses—the generator potential—and also to all-or-none spike activity. Hunt

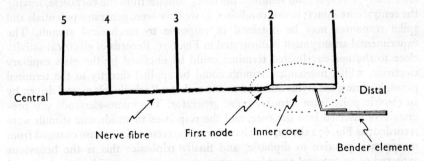

FIG. 670. Schematic view of recording and stimulating arrangement. 1, stationary recording electrode; 2, roving recording electrode; 3, earthed electrode;
4 and 5, stimulating electrodes. Interrupted line indicates approximate
original outline of corpuscle. (Hunt & Takeuchi. *J. Physiol.*)

& Takeuchi recorded the passage of an antidromic impulse into the corpuscle; their set-up is illustrated schematically by Fig. 670, the moveable exploring electrode being indicated by No. 2. As this exploring electrode was moved towards the distal electrode, No. 1, the amplitude of the negative spike did, indeed, fall, but the rise-time of the spike remained the same, so that the fall in amplitude was most probably due to the diminished interelectrode distance that resulted from moving the exploring electrode distally. This explanation was supported by the progressive increase in the latency as the electrode moved distally, a situation that would not arise were the recorded electrical event merely the result of electrotonic spread from the first node. Finally, the finding that the negative spike was followed by a positive wave strongly suggested that the impulse passed beyond the exploring electrode into the unmyelinated region, thereby acting as a sink and making the exploring electrode a positive source of current (p. 1051).

Non-Myelinated Conduction

Later studies of Sato & Ozeki (1963) and Ozeki & Sato (1964) have left little doubt that the non-myelinated region *can* conduct propagated spike activity. In

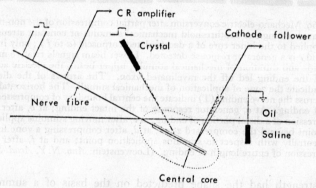

FIG. 671. Schematic drawing of stimulating and recording arrangement. Interrupted line indicates approximate original outline of corpuscle. (Ozeki & Sato. *J. Physiol.*)

the earlier study the corpuscle was held at an oil–water interface in such a way that the primary recorded electrical activity must have arisen from regions closer to the terminal than the first node of Ranvier. Both graded responses and all-or-none spike activity were recorded in response to mechanical stimuli. In the later study, Ozeki & Sato removed the outer lamellæ from the corpuscle, leaving the central core intact; in this condition, as we have seen, generator potentials and spike responses may be obtained in response to mechanical stimuli. The experimental arrangement is illustrated in Fig. 671. Records of electrical activity close to the non-myelinated terminal could be obtained by the glass capillary electrode, whilst mechanical stimuli could be applied directly to the terminal through a fine glass stylus attached to one end of a Rochelle salt crystal driven by an electric pulse from a square-pulse generator. The micro-electrode was progressively inserted into the core, and the responses to antidromic stimuli were recorded; as Fig. 672 shows, the character of the recorded response changed from monophasic-positive to diphasic, and finally triphasic; this is the behaviour expected of an external recording system as the active electrode gets closer and closer to the conducting axon (Murakami, Watanabe & Tomita, 1961) so that it is very likely that propagated activity did, indeed, extend to the non-myelinated region. Mechanical stimulation likewise gave rise to all-or-none spike activity in the unmyelinated region, the amplitude being some 80 per cent. of that of the antidromic spike, presumably because the mechanical stimulus, by stretching the axonal membrane, alters its electrical characteristics.

Refractory Period and Summation. The methods of stimulation and recording employed by Ozeki & Sato allowed the study of a number of features of the receptor. Thus, immediately after antidromic stimulation, a mechanical stimulus fails to evoke a response; as with electrical stimulation of nerve, this period of refractoriness can be resolved into an absolutely refractory period of 2 msec., and a relatively refractory period, during which the threshold to mechanical

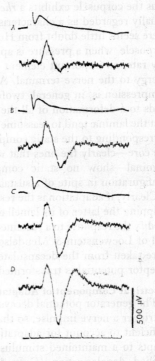

Fig. 672. Responses of the nerve terminal to antidromic stimuli, recorded as the micro-electrode was being advanced to the terminal. Upper traces show the antidromic impulse in the axon. Time marker, 1 msec. (Ozeki & Sato. *J. Physiol.*)

stimuli is higher than normal, of some 10 msec. When two subliminal stimuli were applied to the unmyelinated axon, their receptor potentials could summate and give rise to a spike; this could happen if the two stimuli were applied to the same, or spatially separate, portions of the terminal.

Effects of Sodium Deficiency. Exchanges of Na$^+$ between the corpuscle and the medium are slow so that in order to reduce the Na$^+$-content to 5 per cent. of its original value it would have to be soaked in a Na$^+$-free medium for some 5 hr. (Gray & Sato, 1955); hence failure to block the generator potential in a Na$^+$-free medium might well be due to this factor. When the corpuscle was perfused with a Na$^+$-free medium by way of the mesenteric artery (Diamond, Gray & Inman, 1958), the generator potential fell, and the action potential was blocked within 30–105 sec.; the generator potential was never completely abolished, suggesting that current was carried by other ions as well as Na$^+$. We have seen that tetrodo-toxin specifically blocks the propagated spike activity that depends on activated Na$^+$-permeability, leaving other types of activity relatively unaffected. Experiments with the neurotoxin have given conflicting results; Loewenstein, Terzuolo & Washizu (1963) found that 1.10^{-5} g/ml. blocked the spike activity in the nerve from the Pacinian corpuscle but left the generator potential virtually unchanged, whereas Nishi & Sato (1966) found that 5.10^{-7} g./ml. blocked the propagated response leaving only the generator potential; as time progressed, the generator potential diminished to about 60 per cent. of its control value. Both results agree, however, in showing that, as with procaine, the propagated spike activity is more susceptible to block than the non-propagated activity.

Adaptation. The striking feature of the response of the Pacinian corpuscle is its rapid adaptation. By contrast with a stretch receptor, which settles down to a steady discharge lasting as long as the stretch is maintained, the corpuscle ceases to discharge quite soon after a constant pressure is established. This decline in spike activity is reflected in a decline in generator potential, so that within a few milliseconds of initiating a constant pressure both electrical changes have ceased. As one might expect, a slowly increasing pressure stimulus may fail to excite when a compression of equal magnitude, applied rapidly, does; thus the corpuscle exhibits a *rheobase* (Nishi & Sato, 1968). Adaptation has been usually regarded as a characteristic of the nerve rather than of the receptor, but there seems little doubt from Hubbard's study of the changes taking place in the corpuscle, when a pressure is applied and maintained, that, with this receptor at any rate, an important element in adaptation is the mechanical transmission of energy to the nerve terminal. According to Hubbard, the effect of a sustained compression is, in general, twofold; a dynamic process, occurring immediately, leads to a deformation of all the laminæ; this deformation is not maintained, so that the laminæ tend to reassume their original positions, the residual deformation corresponding to the static component of the stimulus. The laminæ belonging to the core—clearly the ones that will transmit the mechanical energy to the nerve terminal—show no static component, so that they reassume their normal configuration in spite of a sustained pressure.

Clearly, if adaptation is the result of the lamellar arrangement of the corpuscle, stripping the latter of its lamellæ, leaving only the basic core, should profoundly modify the response to a sustained compression. The work of Ozeki & Sato (1965) and of Loewenstein & Mendelson (1965), in which electrophysiological records were taken from the decapsulated corpuscle, confirm this view in so far as the receptor potential is transformed from a transient to a sustained depolarization.

Electrical Component of Adaptation

The generator potential decays rapidly, and for 2 or 3 msec. at most is sufficient to trigger a nerve impulse, so that this circumstance, itself, might be considered sufficient to account for adaptation. In fact, however, the neurone itself rapidly adapts to a maintained stimulus, whether this is in the form of an electrically applied depolarization (Gray & Matthews, 1951) or a sustained generator potential in the decapsulated corpuscle (Loewenstein & Mendelson, 1965). The question naturally arises as to why the nerve terminal of the Pacinian corpuscle is enclosed in the elaborate series of laminæ if the adaptation that they are ideal to subserve is an act of supererogation. Loewenstein & Mendelson have discussed this with some acumen without, however, reaching a conclusion.

Fast and Slow Adapters

It must be appreciated, however, that the excitation parameters of the nerve fibre supplying a receptor are not sufficient, in themselves, to characterize the behaviour of the receptor in response to its natural stimulus. Vallbo (1964), for example, has studied accommodation of single sensory fibres, measured by the current-gradient, dI/dt, required to excite (p. 1090), and compared the accommodation measured in this way in fibres from rapidly and slowly adapting mechano-receptors of the toad skin. In fact, he found no difference in fibres from rapidly or slowly adapting receptors.* With a slowly adapting receptor, maximum mechanical stimulation gave some ten times the number of action potentials obtained by a step-current of long duration, so that it appears as if the nerve terminals in the receptor were

---

* Accommodation was slightly faster in motor than in sensory fibres, thus confirming earlier work.

accommobating much less than the axon, if we assume that under these conditions a generator potential was in fact being sustained at a steady level. With fast adapting receptors, the same number of action potentials could be obtained with mechanical and electrical stimulation. Another difference between electrical and mechanical stimulation was the failure to obtain low frequencies of discharge by electrical stimulation, although these were regularly obtained by stretch of the skin.

The Off-Effect. As indicated earlier, Gray & Sato observed a discharge on cessation of a mechanical stimulus; in the absence of nerve action potentials this can be recognized as a transient hyperpolarization, so that the action potential is the result of depolarization of the hyperpolarized membrane: it is a "post-hyperpolarization response" (Fig. 673). This OFF-hyperpolarization has been examined by Ilyinsky (1965) and Nishi & Sato (1968). Ilyinsky observed that, if the corpuscle was rotated through 90° about its longitudinal axis, the hyper-polarization was converted to a depolarization, whilst the ON-depolarization was converted into an ON-hyperpolarization, and she explained this on the basis of the elliptical cross-section of the nerve terminal; thus a pressure applied along the short axis of the ellipse would increase the ratio a/b, where a and b are the long

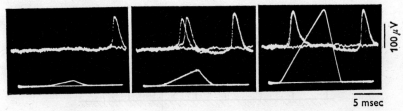

FIG. 673. ON- and OFF-responses of a decapsulated terminal produced by compressions of varying rate of increase. In each trace, compressions of constant rate of increase were applied twice successively. Upper trace: Response. Lower trace: Potential change applied to crystal. (Nishi & Sato. *J. Physiol.*)

and short axes respectively, and thus increase the area of the membrane. Pressure applied along the long axis would have the opposite effect, so that, if increasing its area depolarized the membrane, the effects would be explained.

Viscous-Elastic System. Experimentally, Hubbard showed that, whereas the immediate effects of a compression were transmitted to the centre of the corpuscle, the resulting deformation was not sustained, in spite of the maintained deformation of the outer lamellæ. Loewenstein & Skalak (1966) have developed a theoretical mechanical model and analysed its responses to compression in an attempt to relate the changes occurring in the core to the electrophysiological events. Essentially the model presupposes that the initial effects of pressure are transmitted by viscous flow of fluid as the deformation, imposed by the compression, leads to a redistribution of the fluid contents. Clearly, this will cease as the compression is sustained, whereas the force of compression is resisted by the elastic forces developed in the outside laminæ. Since Hubbard found that, during sustained compression, there was a static component of deformation in the inner lamellæ, transmission of this must be brought about by some non-viscous element, presumably the occasional attachments between lamellæ seen by Pease & Quilliam. Fig. 674 illustrates the model, in which dash-pots and weak springs, S, serve to transmit forces from one lamina to the next, whilst the more powerful springs, M, represent the elastic compliances of the individual laminæ. From the known dimensions of the system, and employing Young's modulus for arterial walls to estimate the compliance of M,[*] the authors were able to predict static displacements of the individual laminæ,

* The compliance of the weaker springs, S, was estimated by making a series of computational runs with arbitrary values of the appropriate coefficient until fit was found with Hubbard's displacement figures for the corpuscle.

in response to a sustained compression, that agreed with those measured by Hubbard reasonably well. The dynamic responses are essentially what would be expected intuitively; a rapid alteration of pressure would result in a great deal of transmitted pressure at the centre, whereas slow application would result in little or none. On this basis, too, the off-effect is seen as a mirror image of the on-effect, both being due to a deformation of the axon terminal.

Response to Vibration. Because of its rapid adaptation, the Pacinian corpuscle may be expected to subserve responses to rapidly repeated mechanical stimuli, and Hunt & McIntyre and Hunt have shown that fibres in the interosseous nerve in the leg of the cat respond to very slight vibrations by bursts of discharges, the receptors responsible being most probably the Pacinian corpuscles. Sato examined the responses of isolated corpuscles to sinusoidal mechanical stimuli and showed that they could, indeed, respond with action potentials to stimuli of frequencies ranging from some 20/sec to 500/sec, or, if the corpuscles were examined in situ at body temperature, up to 1,000/sec. At low frequencies the corpuscle responded to one cycle of vibration with one or two impulses; at frequencies of vibration greater than about 200/sec it responded to two or more cycles with only one spike potential. As we should expect from Loewenstein & Cohen's study, the threshold was strongly influenced by frequency of stimulation, so that at 20/sec, or greater than 500/sec, the isolated corpuscle failed to respond with a spike potential although in vivo responses at 1,000/sec could be obtained,

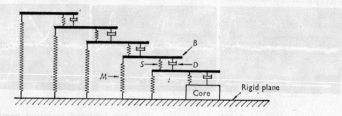

FIG. 674. Mechanical analogue of the Pacinian corpuscle. Lamella compliance is represented by springs M; radial spring compliance, by springs S; and the fluid resistance, by dashpots D. (Loewenstein & Skalak. *J. Physiol.*)

presumably because of the higher temperature that increases amplitude and rate of rise of the generator potential (Inman & Peruzzi; Ishiko & Loewenstein). When the amplitude of the individual generator potentials was plotted against frequency of stimulation, it passed through a maximum at about 200 cycles per sec, and this corresponded fairly closely with the optimum frequency for lowering the threshold, so that according to Sato the effects of varying frequency of stimulation are largely, if not completely, accounted for by the effects on the generator potential. The rise in threshold and fall in amplitude of the generator potential at very low frequencies conform with the earlier discovery that the amplitude of the generator potential depended on how rapidly the mechanical deformation was brought about; the effect of high frequencies likewise conforms with the observed depressing effect of one stimulus on a succeeding one if it falls less than 10 msec after. As we have seen, Loewenstein & Cohen's studies suggest a further complication in that potentiation may also occur, whilst the variation in the threshold of the response of the first node to a generator potential further confuses the issue.

Acetylcholine and the Sensory Response

We may ask, finally, whether the initiation of impulses at the terminals of sensory nerves is brought about by the liberation of acetylcholine. According to Brown & Gray, a cutaneous injection of this substance causes a discharge in the sensory nerve, an effect that may be blocked by large doses of the same substance or nicotine. The important point, however, is that this blocking action does not extend to an inhibition of the effects of mechanical stimulation, and for this reason Brown & Gray answered the question, posed above, in the negative. They pointed to the different effects of acetylcholine on the neurones of a ganglion; the postganglionic cell, on the "receiving end", is sensitive to the drug, whereas the preganglionic cell, on the "transmitting end", is not,* but it does *liberate acetylcholine*. The sensory nerve is a "receiver" and in conformity with this viewpoint, should be sensitive to acetylcholine, but should not liberate it. Subsequent studies of the pharmacology of the sensory receptor, in particular the pressure receptors in the blood vessels, have reinforced this opinion. The stimulating action of acetylcholine on the sensory receptors of the skin seems to be an action *sui generis* on the nerve terminals, and not secondary to any action on the muscle fibres (Douglas & Gray); moreover, as seen in the carotid sinus (Diamond) and frog skin (Jarett), the action of acetylcholine summates with the normal stimulus. We may assume, then, that acetylcholine depolarizes the nerve terminals, in the same way that a mechanical stimulus would appear to do; the fact that a receptor, blocked to the action of acetylcholine—for example, by excess acetylcholine, curarine, or hexamethonium—responds to a mechanical stimulus simply means, as Thesleff has shown for the neuromuscular junction, that this type of block is not a depolarization block.† Of some interest, in this connexion, is the finding of large amounts of a cholinesterase in the nerve ending of the Pacinian corpuscle, amounts as large as those found in an end-plate (Loewenstein & Molins). The esterase differs from the typical cholinesterase in a number of particulars, however.

* Perfusion of the stellate ganglion with a solution containing acetylcholine initiates discharges only in the postganglionic fibres.
† Gray & Diamond (1957) have summarized the pharmacological aspects of sensory receptors. Adrenaline potentiates the effects of mechanical stimulation of the Pacinian corpuscle (Loewenstein, 1956). According to Wiersma, Furshpan & Florey (1953), acetylcholine increases the sensory discharge from the invertebrate receptor; this is apparently *not* due to an action on the intrafusal fibres. In the vertebrate spindle, on the other hand, the increased discharge due to acetylcholine probably does result from contraction of the intrafusal fibres (Hunt, 1952).

REFERENCES

ALBUQUERQUE, E. X. & GRAMPP. W. (1968). Effects of tetrodotoxin on the slowly adapting stretch receptor neurone of lobster. *J. Physiol.*, **195**, 141–156.
ALEXANDROWITZ, J. S. (1951). "Muscle Receptor Organs in the Abdomen of *Homarus vulgaris* and *Palinurus vulgaris*." *Quart. J. micr. Sci.*, **92**, 163–199.
ALVAREZ-BUYLLA, R., & DE ARELLANO, J. R. (1953). "Local Responses in Pacinian Corpuscles." *Amer. J. Physiol.*, **172**, 237–244.
BARKER, D. (1948). "The Innervation of the Muscle-spindle." *Quart. J. micr. Sci.*, **89**, 143–186.
BARRON, D. H., & MATTHEWS, B. H. C. (1938). "The Interpretation of Potential Changes in the Spinal Cord." *J. Physiol.*, **92**, 276.
BERNHARD, C. G. (1942). "Isolation of Retinal and Optic Ganglion Response in the Eye of *Dytiscus*." *J. Neurophysiol.*, **5**, 32.
BOYD, I. A. (1962). The structure and innervation of the nuclear bag muscle fibre system and the nuclear chain muscle fibre system in mammalian muscle systems. *Phil. Trans.*, **245**, 81–136.
BRINLEY, F. J. (1965). Sodium, potassium, and chloride concentrations and fluxes in the isolated giant axon of *Homarus*. *J. Neurophysiol.*, **28**, 742–772.
BROWN, G. L., & GRAY, J. A. B. (1948). "Some Effects of Nicotine-like Substances and their Relations to Sensory Nerve Endings." *J. Physiol.*, **107**, 306.
CALMA, I. (1965). Ions and the receptor potential in the muscle spindle of the frog. *J. Physiol.*, **177**, 31–41.
DIAMOND, J. (1955). "Observation on the Excitation by Acetylcholine and by Pressure of Sensory Receptors in the Cat's Carotid Sinus." *J. Physiol.*, **130**, 513–532.
DIAMOND, J., GRAY, J. A. B. & INMAN, D. R. (1958). The relation between receptor potentials and the concentration of sodium ions. *J. Physiol.*, **142**, 382–394.
DIAMOND, J., GRAY, J. A. B. & SATO, M. (1956). The site of initiation of impulses in Pacinian corpuscles. *J. Physiol.*, **133**, 54–67.

DOUGLAS, W. W., & GRAY, J. A. B. (1953). "The Excitant Action of Acetylcholine and other Substances on Cutaneous Sensory Pathways and its Prevention by Hexamethonium and d-Tubocurarine." *J. Physiol.*, **119**, 118–128.

DOWLING, J. E. (1968). Discrete potentials in the dark-adapted eye of the crab *Limulus*. *Nature*, **217**, 28–31.

EDWARDS, C. & HAGIWARA, S. (1959). Potassium ions and the inhibitory process in the crayfish stretch receptor. *J. gen. Physiol.*, **43**, 315–321.

EDWARDS, C., & OTTOSON, D. (1958). The site of impulse initiation in a nerve cell of a crustacean stretch receptor. *J. Physiol.*, **143**, 138–148.

EDWARDS, C., TERZUOLO, C. A. & WASHIZU, Y. (1963). The effect of changes of the ionic environment upon an isolated crustacean sensory neuron. *J. Neurophysiol.*, **26**, 948–957.

EYZAGUIRRE, C. & KUFFLER, S. W. (1955). Processes of excitation in the dendrites and in the soma of single isolated sensory cells of the lobster and crayfish. *J. gen. Physiol.*, **39**, 87–119.

FRANKENHAEUSER, B. & VALLBO, A. B. (1965). Accommodation in myelinated nerve fibres of *Xenopus laevis* as computed on the basis of voltage clamp data. *Acta physiol. scand.*, **63**, 1–20.

GRAY, E. G. (1957). "The Spindle and Extrafusal Innervation of a Frog Muscle." *Proc. Roy. Soc., B*, **146**, 416–430.

GRAY, J. A. B. & DIAMOND, J. (1957). Pharmacological properties of sensory receptors and their relation to those of the autonomic nervous system. *Brit. med. Bull.*, **13**, 185–188.

GRAY, J. A. B. & MATTHEWS, P. B. C. (1951). A comparison of the adaptation of the Pacinian corpuscle with the accommodation of its own axon. *J. Physiol.*, **114**, 454–464.

GRAY, J. A. B., & SATO, M. (1953). "Properties of the Receptor Potential in Pacinian Corpuscles." *J. Physiol.*, **122**, 610–636.

GRAY, J. A. B. & SATO, M. (1955). The movement of sodium and other ions in Pacinian corpuscles. *J. Physiol.*, **129**, 594–607.

HAGIWARA, S., KUSANO, K., & SAITO, S. (1960). Membrane changes in crayfish stretch receptor neuron during synaptic inhibition and under action of gamma-aminobutyric acid. *J. Neurophysiol.*, **23**, 505–515.

HARTLINE, H. K., & RATLIFF, F. (1957). "Inhibitory Interaction of Receptor Units in the Eye of *Limulus*." *J. gen. Physiol.*, **40**, 357–376.

HARTLINE, H. K., & RATLIFF, F. (1958). "Spatial Summation of Inhibitory Influences in the Eye of *Limulus*, and the Mutual Interaction of Receptor Units." *J. gen. Physiol.*, **41**, 1049–1066.

HARTLINE, H. K., WAGNER, H. G., & RATLIFF, F. (1956). "Inhibition in the Eye of *Limulus*." *J. gen. Physiol.*, **39**, 651–673.

HODGKIN, A. L. & KATZ, B. (1949). Effect of Na^+ on the electrical activity of the giant axon of the squid. *J. Physiol.*, **108**, 37.

HUBBARD, S. J. (1958). "A Study of Rapid Mechanical Events in a Mechanoreceptor." *J. Physiol.*, **141**, 198–218.

HUNT, C. C. (1952). "Muscle Stretch Receptors; Peripheral Mechanisms and Reflex Function." *Cold Spr. Harb. Symp. quant. Biol.*, **17**, 113–123.

HUNT, C. C. (1961). On the nature of vibration receptors in the hind limb of the cat. *J. Physiol.*, **155**, 175–186.

HUNT, C. C., & McINTYRE, A. K. (1960). Characteristics of responses from receptors from the flexor longus digitorum muscle and the adjoining interosseus region of the cat. *J. Physiol.*, **153**, 74–87.

HUNT, C. C., & TAKEUCHI, A. (1962). Responses of the nerve terminal of the Pacinian corpuscle. *J. Physiol.*, **160**, 1–21.

HUXLEY, A. F. (1959). Ionic movements during nerve activity. *Ann. N.Y. Acad. Sci.*, **81**, 221–246.

ILYINSKY, O. B. (1965). Processes of excitation and inhibition in single mechanoreceptors (Pacinian corpuscles). *Nature*, **208**, 351–353.

INMAN, D. R., & PERUZZI, P. (1961). The effects of temperature on the responses of Pacinian corpuscles. *J. Physiol.*, **155**, 280–301.

ISHIKO, N., & LOEWENSTEIN, W. R. (1961). Effects of temperature on the generator and action potentials of a sense organ. *J. gen. Physiol.*, **45**, 105–124.

JARETT, A. S. (1956). "The Effect of Acetylcholine on Touch Receptors in Frog's Skin." *J. Physiol.*, **133**, 243–254.

KATZ, B. (1950). "Depolarization of Sensory Terminals and the Initiation of Impulses in the Muscle Spindle." *J. Physiol.*, **111**, 26.

KATZ, B. (1961). The terminations of the afferent nerve fibre in the muscle spindle of the frog. *Phil. Trans. B*, **243**, 221–240.

KOKETSU, K., & NISHI, S. (1957). "Action Potentials of Single Intrafusal Muscle Fibres of Frogs," *J. Physiol.*, **137**, 193–209.

KUFFLER, S. W. (1954). "Mechanism of Activation and Motor Control of Stretch Receptors in Lobster and Crayfish." *J. Neurophysiol.*, **17**, 558–574.

KUFFLER, S. W., & EYZAGUIRRE, C. (1955). "Synaptic Inhibition in an Isolated Nerve Cell." *J. gen. Physiol.*, **39**, 155–184.

KUFFLER, S. W., HUNT, C. C., & QUILLIAM, J. P. (1951). "Function of Medullated Small-nerve Fibres in Mammalian Ventral Roots: Efferent Muscle Spindle Innervation." *J. Neurophysiol.*, **14**, 29–54.

LASANSKY, A. (1967). Cell junctions in ommatidia of *Limulus. J. Cell Biol.*, **33**, 365–383.

LOEWENSTEIN, W. R. (1956). "Excitation and Changes in Adaptation by Stretch of Mechanoreceptors." *J. Physiol.*, **133**, 588–602.

LOEWENSTEIN, W. R. (1959). The generation of electric activity in a nerve ending. *Ann. N.Y. Acad. Sci.*, **81**, 367–387.

LOEWENSTEIN, W. R. (1961). Excitation and inactivation in a receptor membrane. *Ann N.Y. Acad. Sci.*, **94**, 510–534.

LOEWENSTEIN, W. R., & COHEN, S. (1959). After-effects of repetitive activity in a nerve ending. *J. gen. Physiol.*, **43**, 335–345.

LOEWENSTEIN, W. R. & MENDELSON, M. (1965). Components of receptor adaptation in a Pacinian corpuscle. *J. Physiol.*, **177**, 377–397.

LOEWENSTEIN, W. R., & MOLINS, D. (1958). Cholinesterase in a receptor. *Science*, **128**, 1284.

LOEWENSTEIN, W. R., & RATHKAMP, R. (1958). The sites for mechano-electric conversion in a Pacinian corpuscle. *J. gen. Physiol.*, **41**, 1245–1265.

LOEWENSTEIN, W. R. & SKALAK, R. (1966). Mechanical transmission in a Pacinian corpuscle. An analysis and a theory. *J. Physiol.*, **182**, 346–378.

LOEWENSTEIN, W. R., TERZUOLO, C. A. & WASHIZU, Y. (1963). Separation of transducer and impulse-generating processes in sensory receptors. *Science*, **142**, 1180–1181.

MacNICHOL, E. F., WAGNER, H. H., & HARTLINE, H. K. (1953). "Electrical Activity recorded within Single Ommatidia of the Eye of *Limulus.*" *XIX Int. Congr. Physiol.* Abstracts of Communications, pp. 582–583.

MOODY, M. F. & ROBERTSON, J. D. (1960). The fine structure of some retinal photoreceptors. *J. biophys. biochem. Cytol.*, **7**, 87–92.

MURAKAMI, M., WATANABE, K. & TOMITA, T. (1961). Effect of impalement with a micropipette on the local cell membrane. Study by simultaneous intra- and extracellular recording from the muscle fiber and giant axon. *Jap. J. Physiol.*, **11**, 80–88.

NAKAJIMA, S. (1964). Adaptation in stretch receptor neurons of crayfish. *Science*, **146**, 1168–1170.

NISHI, K. & SATO, M. (1966). Blocking of the impulse and depression of the receptor potential by tetrodotoxin in non-myelinated nerve terminals in Pacinian corpuscles. *J. Physiol.*, **184**, 376–386.

NISHI, K. & SATO, M. (1968). Depolarizing and hyperpolarizing receptor potentials in the non-myelinated nerve terminal in Pacinian corpuscles. *J. Physiol.*, **199**, 383–396.

OBARA, S (1968). Some effects of organic cations on generator potential of crayfish stretch receptor. *J. gen. Physiol.*, **52**, 363–386.

OBARA, S. & GRUNDFEST, H. (1968). Effects of lithium on different membrane components of crayfish stretch receptor neurons. *J. gen. Physiol.*, **51**, 635–654.

OTTOSON, D. (1964). The effect of sodium deficiency on the response of the isolated muscle spindle. *J. Physiol.*, **171**, 109–118.

OZEKI, M. & GRUNDFEST, H. (1967). Crayfish muscle fiber: ionic requirements for depolarizing synaptic electrogenesis. *Science*, **155**, 478–481.

OZEKI, M. & SATO, M. (1964). Inititiation of impulses at the non-myelinated nerve terminal in Pacinian corpuscles. *J. Physiol.*, **170**, 167–185.

OZEKI, M. & SATO, M. (1965). Changes in the membrane potential and the membrane conductance associated with a sustained compression of the unmyelinated nerve terminal in Pacinian corpuscles. *J. Physiol.*, **180**, 186–208.

PEASE, D. C., & QUILLIAM, T. A. (1957). "Electron Microscopy of the Pacinian Corpuscle." *J. biophys. biochem. Cytol.*, **3**, 331–342.

QUILLIAM, T. A., & SATO, M. (1955). "The Distribution of Myelin on Nerve Fibres from Pacinian Corpuscles." *J. Physiol.*, **129**, 167–176.

SATO, M. (1961). Response of Pacinian corpuscles to sinusoidal vibration. *J. Physiol.*, **159**, 391–409.

SATO, M. & OZEKI, M. (1963). Response of the non-myelinated nerve terminal in Pacinian corpuscles to mechanical and antidromic stimulation and the effect of procaine, choline and cooling. *Jap. J. Physiol.*, **13**, 564–582.

SHANTHAVEERAPPA, T. R. & BOURNE, G. H. (1963). New observations on the structure of the Pacinian corpuscle and its relation to the perineural epithelium of peripheral nerves. *Amer. J. Anat.*, **112**, 97–109.

STÄMPFLI, R., & NISHIE, K. (1956). Effects of calcium-free solutions on membrane-potential of myelinated fibres of the Brazilian frog, *Leptodactylus ocellatus*. *Helv. physiol. Acta*, **14**, 93–104.

TERZUOLO, C. A., & WASHIZU, Y. (1962). Relation between stimulus strength, generator potential and impulse frequency in stretch receptor of crustacea. *J. Neurophysiol.*, **25**, 56–66.

TOMITA, T., & FUNAISHI, A. (1952). "Studies on Intraretinal Action Potential with Low-resistance Microelectrode." *J. Neurophysiol.*, **15**, 75–84.

VALLBO, A. B. (1964). Accommodation of single myelinated nerve fibres from *Xenopus laevis* related to type of end organ. *Acta physiol., scand.*, **61**, 413–428.

VALLBO, A. B. (1964). Accommodation related to inactivation of the sodium permeability in single nerve fibres from *Xenopus laevis*. *Acta physiol. scand.*, **61**, 429–444.

WATERMAN, T. H., & WIERSMA, C. A. G. (1954). "The Functional Relation between Retinal Cells and Optic Nerve in *Limulus*." *J. exp. Zool.*, **126**, 59–85.

WIERSMA, C. A. G., FURSHPAN, E., & FLOREY, E. (1953). "Physiological and Pharmacological Observations on Muscle Receptor Organs of the Crayfish *Cambarus Clarkii*, Girard." *J. exp. Biol.*, **30**, 136–150.

CHAPTER XVIII

EXCITABILITY OF CARDIAC MUSCLE

PHENOMENA associated with the excitability of cardiac muscle are of some interest in so far as they present an example of spontaneous activity that may be modified by both inhibitory and excitatory nervous impulses.

The Heart

The mammalian heart consists, essentially, of two pairs of muscular chambers, the auricles and ventricles; they are separated by a connective-tissue ring, so that the propagation of electrical activity from one pair to the other is achieved by a specialized conducting system, the *bundle of His or Kent*, which is a set of muscle fibres, originating on the auricle at the AV-node and ramifying in the muscle of the ventricles (Fig. 675); these fibres are described as *Purkinje fibres*. Electrical activity normally is initiated at a specialized region on the right auricle—

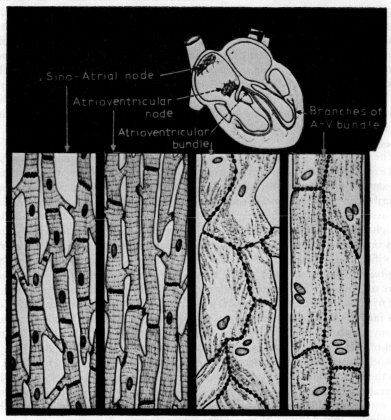

Fig. 675. A schematic representation of the gross anatomical, histological, and to some extent the ultrastructural organization of the conducting system of the ox heart. The four areas represent *camera lucida* drawings at a magnification of about 130 times. The size of the intercalated disks and the desmosomes is somewhat exaggerated in order to facilitate a comparison of their ultrastructure. (Rhodin, del Missier & Reid. *Circulation.*)

the *sino-auricular* or *SA-node*—whence it spreads over the auricles to the *auriculo-ventricular*, or *AV-node*; from here the activity is conducted by way of the Purkinje fibres of the bundle of His to the ventricles. Anatomically, therefore, we may expect to distinguish three specialized zones, the two nodes and the Purkinje fibres (Fig. 675).

Cardiac Muscle. Heart muscle or *myocardium* is similar, in microscopic appearance, to striated voluntary muscle, exhibiting the characteristic cross-banding that divides it into sarcomeres (p. 1348) and the less marked longitudinal striations due to the presence of myofibrils, which may be resolved into thinner myofilaments under the electron-microscope. Functionally, the heart muscle behaves as a syncytium, in the sense that an impulse initiated at a point spreads over the whole. With few exceptions, classical histologists ascribed this functional continuity between cells to the presence of protoplasmic bridges between them, but recent studies with the electron-microscope have demonstrated that muscle cells, whilst organized into fibres by end-to-end aggregation, retain their individuality, in the sense that they are everywhere bounded by cell membranes. Classical histologists recognized the presence of the so-called *intercalated discs*, dense regions transecting the muscle fibres, but most considered that the myofibrils were continuous across these and thus ran the length of the muscle. In the electron-microscope, however, van Breemen, and later Sjöstrand & Andersson, recognized that these intercalated discs were boundaries between adjacent cells, usually end-to-end, and represented regions where the continuity of the myofibrils was interrupted by the adjacent plasma membranes; they are, in fact, regions where the myofibrils are attached to the plasma membranes of their respective cells. One is illustrated in Fig. 676, Pl. LIX, from the work of Muir; at high resolution it can be seen to consist of interdigitating cell membranes of adjacent cells, the two membranes maintaining a separation of some 300A.

Specialized Contacts

The myocardium and its conducting tissue behave as a syncytium, in the sense that excitation at one point spreads from cell to cell; for this reason it is of interest to examine the contacts between adjacent cells to ascertain whether junctions, described as *zonula* or *macula occludens*, are present, since these would provide low-resistance pathways between the cells. Examination of the intercalated disc at high resolution reveals that it is not only a region where membranes of adjacent cells interdigitate, but it contains regions of membrane fusion called, when first identified in this tissue by Sjöstrand, Andersson-Cedergren & Dewey (1958), *longitudinal connexions*, and which have subsequently been called *nexuses* or, more usually, *tight junctions* or *zonulae occludentes*. As illustrated schematically in Fig. 677, then, the intercalated disc is a composite structure which may be divided into *intermyofibrillar* and *intersarcoplasmic regions*. In the interfibrillar portion, the myofibrils insert into the plasma membranes of adjacent cells at the Z-band level of each myofibril; the branching of the filaments of the I-band to form a network causes an increased density near the membrane extending as much as 1 μ into the I-band; and it is this density that doubtless permits the resolution of the intercalated disc in the light-microscope. In the intersarcoplasmic region we find the zonula occludens, running for a considerable distance parallel with the myofibrils; in addition, there is the characteristic desmosome or macula adhaerens.* Fig. 678, Pl. LX is an electron micrograph of a portion of an

---

* Dewey (1969) points out that the term intercalated disc has been used variously; thus it has been used to describe single regions, such as the desmosomes and intermyofibrillar insertions; they recommend it be used to cover the whole composite of structure, whose true significance only becomes clear when the complete geometry of the cell is considered.

PLATE LIX

IC   **Z**

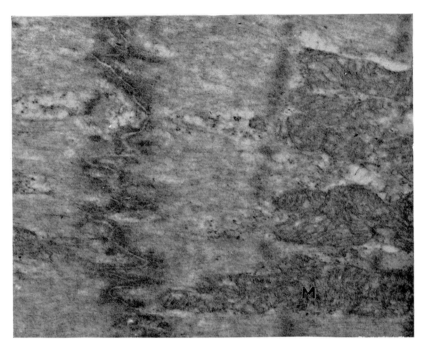

FIG. 676. Longitudinal section through rabbit cardiac muscle showing fine structure of the intercalated discs (IC). The myofibrils, formed of myofilaments, can be seen separated from each other by sarcoplasm containing mitochondria. At the disc the myofilaments end in the dense material adjacent to the cell membrane. Z, Z-membrane. Osmium fixation. × 25,000. (Muir. *J. biophys. biochem. Cytol.*)

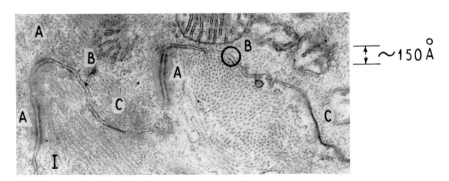

FIG. 679. Junction between two Purkinje cells showing intercalated disc (A), ordinary contact (B), and longitudinal connexion (C). × 25,000. (Courtesy, Dr. J. Rhodin. New York University School of Medicine.)

[To face p. 1274.

PLATE LX

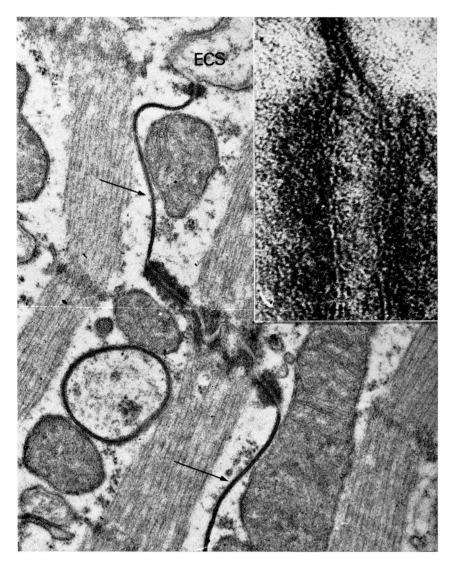

Fig. 678. Illustrating a limited region of intercalated disc between two cardiac
muscle cells at a branch point of a muscle fibre. The arrows indicate the
region of nexus, occurring only in the regions of sarcoplasmic columns
(intersarcoplasmic region of the disc) as the disc passes from one sarco-
meric level to another. Note that the adjacent plasma membranes form a
nexus immediately upon turning in at the disc (near region at ECS).
Where the disc crosses a sarcomere (interfibrillar region of intercalated
disc), regions of attachment of myofilaments occur. Desmosomes occur in
the intersarcoplasmic regions of the disc. × 31,000. (Dewey & Barr.
J. Cell Biol.)

Inset: Portion of intersarcoplasmic region of intercalated disc. Cell
membranes adjacent to the desmosome fuse to form a nexus. × 320,000.
(Barr, Berger & Dewey. *J. gen. Physiol.*)

intercalated disc; two regions of membrane fusion are obvious, the upper one continuous with an opening into the extracellular space; desmosomes are on each side of the interfibrillar region. The inset to Fig. 678 shows higher magnification of a portion of desmosome opening into a zonula occludens.

Purkinje and Sub-Mammalian Fibres. On geometrical considerations we may expect to see the intercalated disc in fibres that are essentially cylindrical with the myofibrils terminating in the portion of plasma membrane that covers each end. As Dewey (1969) points out, if the fibres were fusiform, then the myofibrils would insert into the sarcolemma all along the cell surface; again, there might be regions in the muscle where the fibres made contact along their lateral surfaces.

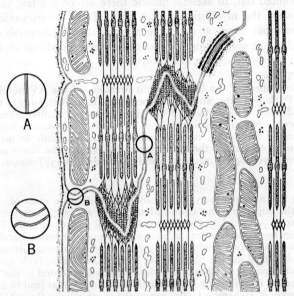

FIG. 677. Schematic drawing of portion of intercalated disc of mammalian cardiac muscle. A shows nexus along intersarcoplasmic portion of the intercalated disc and B shows the region of gap in continuity with the extracellular space along the lateral margin of the fibre. In the upper right a desmosome is illustrated in an intersarcoplasmic portion of the disc. (Dewey. *Comparative Physiology of the Heart: Current Trends.* Birkhaüser Verlag.)

In both cases, well defined discs would not be apparent, and this accounts for the absence of the disc in all non-mammalian vertebrates. In mammalian hearts certain portions may lack intercalated discs for the same reason. Thus the Purkinje fibres, *i.e.*, the cells constituting the terminal fibres of the conducting system of the *ungulate* heart, are composed of cylindrical to fusiform cells, which make contact with their neighbours cells over large areas. As Rhodin, del Missier & Reid (1961) showed, the essential feature of the contact in the Purkinje fibre is the desmosome; in addition, and associated with it, is the longitudinal connexion or zonula occludens (Fig. 679, Pl. LIX).*

* The Purkinje fibre, as correctly defined, is a constituent of the conducting system of the *ungulate* heart; in other mammalian hearts there are conducting fibres, similar in having a large diameter with few myofibrils and containing large quantities of glycogen, but are dissimilar in having intercalated discs. Dewey recommends confining the use of the term Purkinje fibre to the ungulate heart. In embryonic and fœtal life intercalated discs are not seen in the light-microscope, and for this reason it was considered that the adult disc had little physiological significance as a boundary between cells. Muir has shown that the specialized contacts in the common myocardium are of the multiple-desmosome type

Frequency of Occlusion

The almost invariable association of the light-microscopically visible zonula adhaerens with the zonula occludens permits the establishment, from light-microscopical observation, of the frequency with which the space between a given pair of cells has been occluded; Fig. 680 illustrates the situation as ascertained by Johnson & Sommer for a fine, 30–80 μ diameter, strand of cardiac ventricular muscle consisting of some 13 fibres of about 10 μ in diameter.

Mitochondria

As we should expect of a tissue carrying out continuous and ceaseless mechanical work, cardiac tissue is well supplied with mitochondria; thus, as Kisch has pointed out, in skeletal muscle there are only a few sarcosomes to be seen between the myofibrils in each field of the electron-microscope; in heart, on the other hand, at the same magnification, innumerable masses of these organelles are seen filling the spaces between individual myofibrils, and also between the sarcolemma and the myofibrils.

Specialized Tissue

The terminal fibres of the left and right branches of the AV-bundle are often described as Purkinje fibres; and in birds and some mammals, such as the pig, they are large and easily differentiated from the nearby ventricular fibres. In other species, however, they are smaller and not so easily differentiated, namely in man and cat, whilst in the rabbit, rat and guinea pig there is little or no histological differentiation, at any rate in the light-microscope, so that the fibres are identified by their electrophysiological features (Nandy & Bourne, 1963; Johnson & Sommer, 1968).*

SA-NODE

In general the SA-node may be identified as a discrete mass of fibres at the junction of the superior vena cava and the right auricle; these are in obvious contrast to the large fascicles of atrial muscle. Small strands of nodal fibres radiate outwards from the node, proper, to course between contiguous bundles of atrial muscle; these small fibres gradually enlarge and, after varying distances, become continuous with atrial muscle fibre. As we shall see, the sinus node acts as a pace-maker, so that normally rhythmic electrical activity is initiated in the nodal cells; these may thus be identified by the pacemaker potentials that lead to a propagated spike; Trautwein & Uchizono (1965) examined these cells in the rabbit after they had been identified electrophysiologically; in the electron microscope they had a large nucleus and sparse myofibrils by comparison with the myocardial cells; they were fusiform, packed together, and surrounded by a basement membrane; infrequently contact areas of the desmosome type were observed. When a pace-maker cell made contact with an ordinary myocardial cell, a desmosome type junction was formed, with the myofibrils of each cell apparently attached to the cytoplasmic surfaces of the respective cell membranes at the region of contact.†

AV-NODE

The AV-node in the dog is some 2 mm by 2 mm. in area and 0·5–1·0 in maximum thickness; it is situated at the junction between auricle and ventricle, and its anterior margin is continuous with the AV-bundle whilst posteriorly and superiorly it merges with the atrial myocardial fibres; the fibres within the node are finer than the adjacent myocardial fibres with fewer myofibrils; nerves may be seen in the node and these continue into the AV-bundle (James, 1964).

Electrical Activity

Electrogram. The heart passes through a cycle of contraction—*systole*—and relaxation—*diastole*; these muscular events, as with voluntary muscle, are

---

up to birth and only later does the intercalated disc develop. Thus the Purkinje fibre is to be regarded as a primitive rather than a specialized type.

* In birds there is no discrete SA-node (Truex & Smythe, 1965).

† James (1962) has described the sinus node in the dog; here the predominant fibre is a miniature Purkinje fibre, occurring in interlacing bundles. Within the centre are what he called syncytial cells which were continuous with the node fibres and related to the nerve fibres.

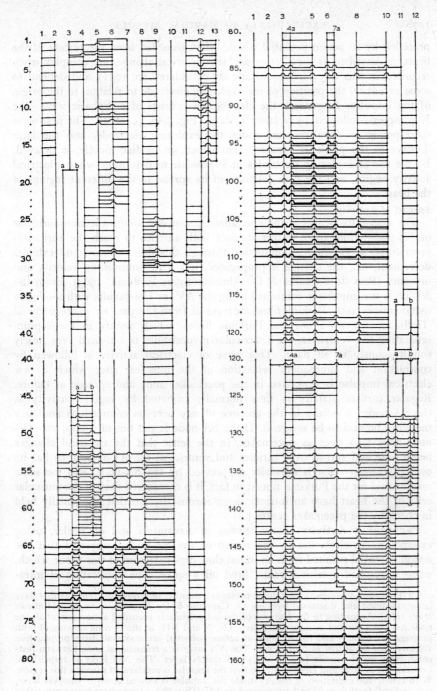

FIG. 680. Distribution of probable low resistance connections between fibres in a
segment of a strand of ventricular muscle, determined from serial transverse sections. The numbers running horizontally denote the thirteen fibres,
represented by the vertical lines, that were identified in the first section.
The numbers running vertically denote the section (∼1 μ thick). A horizontal
line joining two (or more) vertical lines (beginning and ending with a solid
dot) signifies that the two (or more) fibres are judged occluded with one
another in that section. The heavy horizontal lines represent two (or more)
horizontal lines. When a vertical line ends prematurely, this signifies that
the fibre became attenuated and finally vanished. When a vertical line
splits, e.g. No. 3 in section No. 18 splits into 3a and 3b, this indicates that
the fibre split into two components (and *vice versa*). (Johnson & Sommer.
J. Cell Biol.)

preceded by an action potential which, when recorded from the surface of the heart, is described as the *electrogram*; the electrical changes are complex when recorded in this way for reasons already considered (p. 1051), and depend to some extent on the position of the exploring electrode in relation to the origin of the impulse. In general, the electrogram consists of a diphasic wave—the R complex—followed by a later T-wave (Fig. 686, p. 1285). The first wave corresponds to depolarization under the exploring electrode, and the later T-wave to the phase of repolarization which, in cardiac muscle, is delayed. By the use of exploring electrodes, it is possible to determine where electrical activity begins, and the time-relations of its spread to the different regions of the heart.

Spread of Activity

The negativity begins at the specialized pacemaker zone—the sino-auricular or SA-node; from there it spreads over both auricles to reach the auricular-ventricular, or AV-node. Conduction through this tissue is slow and exhibits decrement, *i.e.*, the change is propagated, but with diminishing velocity and intensity (Paes de Carvalho & De Almeida 1960; Hoffman, 1965). From the AV-node the impulse is continued along the AV- or His-bundle with a velocity that increases, as the Purkinje fibres are reached, from 1 to 3 m./sec. in the mammal (Hoffman, 1965).* From the Purkinje fibres, which ramify in the septum separating the ventricles, the ventricular myocardium is activated very nearly simultaneously in all parts. This wave of electrical activity is followed by contraction and subsequent relaxation of the muscles, after which a new electrical impulse is generated in the pacemaker zone and spreads as before. Regular cardiac activity is thus normally initiated by regular activity at the SA-node; it occurs in the absence of any nervous connections and may therefore be said to be *myogenic*. If the SA-node is put out of action, the AV-node may take over as pacemaker, in the sense that the electrical changes begin here and spread over auricles and ventricles; the AV-node thus has its own rhythmicity that is normally suppressed by the SA-node, and the same may be said for the Purkinje fibres; in fact, it is considered that all the muscular cells of the heart have an intrinsic spontaneous rhythm that is normally held in check by the pacemaker centre.†

Pacemaker Activity. Early studies of Arvanitaki and of Bozler, using external micro-electrodes in the region of the SA-node, suggested that the action potential resulted from an initial slow depolarization of this tissue which, when it reached a critical value, fired off a spike. The application of intra-

---

* Propagation through the AV-node presents a number of interesting features that have been examined and discussed by Paes De Carvalho & De Almeida, 1960, and Hoffman *et al.* 1963). According to the features of the intracellularly recorded action potential, the node has been divided into three regions, AN, N and NH; in the N zone slowing of both propagation velocity and rising phase of action potential are maximal; here propagation velocity may be as low as 0·02 m/sec. The AN layer is a transitional zone between fast-conducting atrial muscle (0·7 m./sec.) and the N layer. The NH layer is transitional between the N layer and His bundle, so that the impulse is accelerated through this region. With retrograde excitation of the AV-node the impulse is slowed down in the NH layer, reduced still further in N and accelerated in AN. Thus the intermediate layers can either slow down or accelerate propagation and are thus not regions of decremental conduction, by contrast with the N zone. As Hoffman has emphasized, delay and decrement in the N zone may be the cause of the so-called *Wenckebach periods*, *i.e.*, gradual decrease in responsiveness to repetitive stimuli with eventual failure of conduction. These would be due to sequential changes in the extent to which decrement took place in the N region and the resulting sequential changes in the amplitude of the local response in the NH-fibre. If the action potential met a cell that was only partially repolarized after its delayed action potential, this new action potential would be blocked.

† This discussion is confined to the myogenic heart-beat; the arthropods (insects, crustaceans) have hearts whose beats originate in ganglia on the surface.

cellular electrodes has fully confirmed this picture. For instance, Draper &
Weidmann (1951) inserted an electrode into single Purkinje fibres of the dog's
heart; they found an average resting potential of 90 mV and an action potential,
which could occur either spontaneously or as a result of an electrical stimulus,
showing an overshoot of some 31 mV (Fig. 681). The time-scale of the action
potential was altogether slower than that of nerve or striated muscle, the
whole cycle lasting 300–500 msec. The slowness is almost exclusively due to
the lengthy recovery phase since the upstroke of the action-potential lasts only
about 1 msec in the cat ventricle.

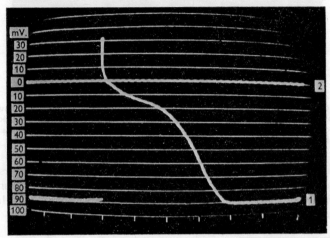

FIG. 681. Monophasic action potential from single Purkinje fibre of dog's heart.
Trace 1 obtained with electrode inside fibre; trace 2 with tip in contact
with bathing solution. Time marks at intervals of 100 msec. (Draper &
Weidmann. *J. Physiol.*)

Nodal Records

Usually the spontaneous action potentials rose suddenly from the base-line
of the resting potential, but with some insertions of the electrode, presumably
at the origin of pacemaker activity in the fibres, the spike was preceded by a
slow depolarization (Fig. 682) which led to the spike. A better preparation for
studying pacemaker activity is obviously the intact heart with an electrode in
the SA-node*; Fig. 683 shows a series of records taken by Hutter & Trautwein
from the sinus venosus (A and B) and the auricle of the frog. The auricular
action potential (C) rises from the resting potential with no preliminary
depolarization; it is simply an action potential propagated from the pacemaker
region. The record A, from the sinus venosus, shows an initial depolarization
—which, in different preparations, varied between 3 and 15 mV—leading to a
spike. In record B the "pacemaker potential" is not so high; it is taken from
the same sinus but a different fibre has been punctured; presumably this fibre
is at some distance from the true pacemaker centre; its rate of depolarization is
slower, so that the action potential, spreading from the pacemaker zone, puts
an end to this before it can reach the level of depolarization necessary to fire
the spike.

A similar pacemaker activity was recorded by Paes De Carvalho, Mello &
Hoffman (1959) in the rabbit; the initial pacemaker depolarization was, on average,

---

* The pacemaker region of the frog's heart is the *sinus venosus*, formed by the junction
of the three venæ cavæ.

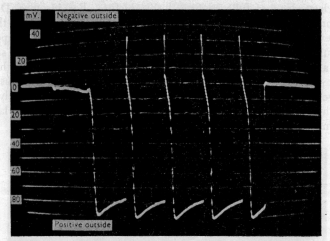

FIG. 682. Rhythmical action potentials from single Purkinje fibre showing pacemaker activity. Note preliminary phase of slow depolarization leading to the spike. The interval between successive action potentials was 1·4 sec. (Draper & Weidmann. *J. Physiol.*)

18 mV, and there was a smooth transition from this to the spike; fibres nearby gave rather similar action potentials, but the initial depolarization was smaller—4 mV—and the transition to the spike was abrupt, indicating that the cell had been fired by an adjacent, pacemaker, cell.

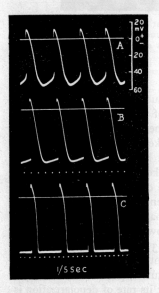

FIG. 683. Action potentials from the frog's heart. A and B are records from different fibres in the sinus venosus. C, is a record from an auricular fibre. In each case a line is drawn through zero potential. Note absence of preliminary depolarization in auricular record; also that the depolarization is smaller in record B than in A. (Hutter & Trautwein. *J. gen. Physiol.*)

Control of Frequency

This difference between the two regions—namely the slower rise of pacemaker potential—demonstrates the probable mechanism by which a given zone can act as pacemaker to others. Clearly, the zone with the most rapid rise of pacemaker potential will initiate a spike first; this spike will propagate rapidly to the other zones and render useless any pacemaker activity that may be present. When the pacemaker is inactivated, then the zone with the next most rapid rate of development of the pacemaker potential will take over; normally, this would be the AV-node. In general, we may expect, then, that the rate of spontaneous depolarization at the SA-node, and the critical degree of polarization necessary to initiate a spike, will determine the frequency of the heart-beat. In fact, a lowered temperature has been shown by Coraboeuf & Weidmann to slow both frequency of heart-beat and rate of depolarization, whilst raising the concentration of Ca^{++}, which increases the critical depolarization necessary to excite, slows the rate of beating (Weidmann, 1955).*

Resting Potential. It is reasonable to assume that the resting potential of cardiac muscle is similar in origin and magnitude to those in voluntary muscle and nerve. Analysis of the intracellular concentrations of K^+ and Na^+, summarized by Trautwein, indicates a high internal concentration of K^+ (130–160 mM) and a low concentration of Na^+ (7–13 mM) in mammalian auricles and ventricles, so that a resting potential of the order of 90 mV may be expected. Some values, taken from a collation by Trautwein, are summarized in Table LXXX together with those for the action potentials. Values for the mammal are of the same order of magnitude as for skeletal muscle and suggest that the potential

TABLE LXXX

Values of the Resting and Action Potentials in Different Cardiac Tissues.
(After Trautwein, 1961.)

Species	Tissue	Resting Potential mV	Action Potential mV
Frog . .	Ventricle	62–85	81–103
	Auricle	67–77	82–104
Rabbit . .	Ventricle	80	102
	Auricle	77–78	92
Dog . .	Papillary	85	105
	Auricle	85	100
	Purkinje	89–90	121

is largely determined by the distribution of K^+. As Table LXXX shows, there seem to be no significant differences between auricle, Purkinje fibre and ventricle so far as the magnitude of the resting potential is concerned. In the SA-node, however, it would seem from the studies of Hutter & Trautwein that the resting potential is significantly lower than that in the auricle, but, in a tissue in which spontaneous activity is occurring, the "resting potential' is only an ideal, since what is measured is the maximum membrane potential during diastole, and it may well be that, because of the development of the pacemaker potential—*i.e.*, a depolarization—the membrane never achieves a real resting potential, so that it is

* Van Mierop (1967) has studied the pacemaker zone in the chick's embryonic heart; by 4–6 days the configuration of the action potential was the same as in the adult; at 31 hr. (10 somites) an action potential could be recorded from the SA-zone followed by one in the ventricle, and even at 28 hr., or 8 somites, action potentials could be recorded from the SA- and bulbo-ventricular zones, the SA-potential always preceding the bulbo-ventricular by about 100 msec. Thus electrical activity occurs before there is any discernible muscular contraction, and the SA-zone is the pacemaker zone *from the first*. This contrasts with the hitherto accepted view (Patten, 1950) that during development there was a succession of pacemaking zones, shifting from the ventricular to the atrial regions as fusion of the two sides of the primordial heart proceeded; this view was based on the observation that the bulbo-ventricular zone started beating first. We may note that, although the internal concentration of Na^+ is reported as being very high in early embryonic heart muscle, the overshoot depends on external Na^+ as though the internal concentrations were comparable to that of more mature tissue (Yeh & Hoffman, 1968).

usual to refer to the *maximum diastolic potential* when speaking of pacemaker tissue.*

Effect of External K⁺. The effects of varying the external concentration of K⁺ on the resting potential have been examined by several workers; they agree in showing a fair concordance with the requirements of a Nernst K-potential at high values of external K⁺, but in the region of the normal blood value there is a considerable flattening of the curve suggesting the importance of other ions in this region (Fig. 684). By assuming that the permeability to Na⁺ is a significant factor, Vaughan Williams modified the Nernst equation along the lines already indicated (p. 571) and obtained a better fit of the experimental points. Kanno & Matsuda (1968), studying the AV-nodal cells of the perfused toad heart, also found a serious deviation of E_m, the membrane potential, from E_K, the Nernst equilibrium potential for K⁺, when the external concentration of K⁺ was low.

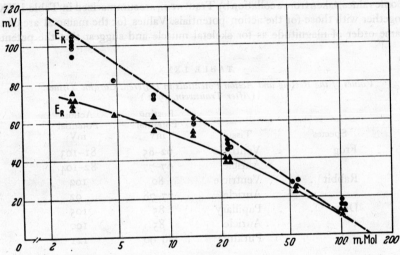

Fig. 684. Membrane potential of dog auricular fibres as a function of the external concentration of K⁺ as abscissa. The dashed line represents the theoretical Nernst potentials. Circles represent potentials obtained when K⁺-permeability was increased by treatment with acetylcholine. (Trautwein & Dudel. *Pflüg. Arch.*)

They obtained a fit for the Goldman equation with values of $P_K:P_{Na}:P_{Cl}$ employed by Hodgkin & Katz for *Loligo*, namely 1:0·04:0·45 for the lower range of $[K]_{out}$ and 1:0·025:0·3 for the higher range.

Specialized Regions. It is interesting that de Mello & Hoffman have observed that the specialized pacemaker fibres from the SA and AV nodes are influenced less strongly by raised external K⁺-concentration than the unspecialized auricular fibres; thus, at an external concentration of 13·5 mM, propagated activity in auricular fibres failed, whilst even in 21·6 mM, AV and SA fibres continued to show spontaneous activity, their resting and action potentials being only moderately reduced. The normal resting potential of auricular fibres was found to be some 20 mV less than the calculated Nernst potential; in the

* Marshall (1957) compared pacemaker with non-pacemaker regions of the auricle; the resting potential of the non-pacemaker region was 73 mV compared with the maximum diastolic potential of 61 mV for the pacemaker region; the amplitude of the action potential was 89 mV in the former instance and 66 mV in the latter. Danielson (1964) has made an elaborate study of the concentrations of Na⁺, K⁺ and Cl⁻ in sinus venosus, auricle and ventricle of the toad heart, comparing these with those in the sartorius muscle.

specialized fibres of the SA node this discrepancy was much greater, namely 37 mV. These findings would be consistent with a relatively high permeability of the pacemaker fibres to Na^+ or, what seems more probable in the light of Noble's studies, with a relatively low permeability to K^+ so that the membrane potential is less than the K^+ equilibrium potential.

Chloride

The distribution of chloride between auricular muscle cell and plasma has been studied by Lamb; the intracellular concentration, based on an extracellular space of 25 per cent., was 19·5 mM whilst the extracellular concentration was 108 mM. This would correspond to a chloride-potential of 46 mV, considerably less than the measured membrane potential. A similar discrepancy was found by Verdonck, De Clercq & Carmeliet (1965) in the frog ventricle, the internal concentration being given by the regression:—

$$[Cl]_i = 4·1 + 0·3\,[Cl]_o$$

and considerably higher than required of a passive distribution.

Extracellular Space

As Danielson (1964) and Page (1965) have emphasized, the assessment of the internal concentrations of ions depends on the value taken for the extracellular space; and considerable ambiguity arises from the finding that the mannitol-space is very considerably larger than the inulin-space (Fig. 685). As we have seen, the inulin-space may well give a low estimate of the extracellular space, because it may be excluded from certain regions. On the basis of the two spaces the internal concentrations were (in mEq./litre):—

	K^+	Na^+	Cl^-
Inulin	162 ± 3	43 ± 5	46 ± 2
Mannitol	208 ± 6	5 ± 2	17 ± 2

On the basis of diffusion through a sheet of tissue, Page & Bernstein (1964) calculated that there would be diffusion channels in the tissue occupying approximately the volume of the inulin-space.*

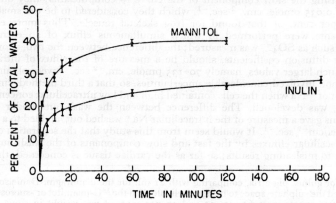

FIG. 685. Showing difference in mannitol- and inulin-spaces of papillary muscle. (Page. *Ann. N.Y. Acad. Sci.*)

---

* Page (1962) ascertained that both sucrose and mannitol remained extracellular by their effects on the cellular volume. They attribute the greater uptake of mannitol than of inulin to entry into the sarcotubular system (p. 1349) by mannitol. Goodford & Lullmann (1962) have shown that ethanesulphonate slowly penetrates heart muscle cells so that it can only be used as an extracellular marker by extrapolation; they computed a space of 26 ml./100 g.

Apart from the ambiguity arising from different volumes of distribution of inulin and mannitol, there is the difficulty due to the degree of blotting and state of contraction of the muscle when being prepared for analysis; vigorous contraction will extrude extracellular fluid, whilst blotting allows accumulation of Na^+ by damaged cells. Keenan & Niedergerke (1967) have made allowance for these errors and conclude that the intracellular concentrations of Na^+ and K^+ are 8 and 163 mM/kg. cell H_2O; thus the intracellular Na^+ is remarkably low.

Ionic Fluxes

The potassium-fluxes have been estimated by several workers and have usually been described as complex because of the apparent presence of at least two compartments; e.g., Haas & Glitsch (1962). However, Burrows & Lamb (1962) and Goerke & Page (1965) found a simple behaviour, at least 90 per cent. of the K^+ being exchangeable and behaving as though in a single compartment. The fluxes calculated by Burrows & Lamb are shown in Table LXXXI where other values have been collected by the authors.

TABLE LXXXI

Ion Contents and Fluxes of Different Excitable Tissues. (Burrows & Lamb, 1962.)

	Ion Concentration (mM)		Flux* ($p$mole/cm²/sec.)	
	Na	K	Na	K
Squid axon . . .	50	400	30	16
Frog muscle . . .	15	137	4	5
Rat diaphragm . .	18	149–158	28	21
Guinea-pig *Tænia coli* .	56–85	119–98	200–300	2–4
Rat auricle . . .	—	143	—	11–17
Chick cultured cells . .	16	186	14·2	10·7
Erythrocytes . . .	16–20	150	0·05	0·03

As Goerke & Page point out, however, the difficulty in calculating the area of the muscle fibre makes the estimate for cardiac muscle approximate.

Sodium Fluxes

Keenan & Niedergerke (1967), have estimated Na^+-efflux by wash-out experiments, using the slow component of the curve to compute this; they obtained a value of 0·035 pmole. cm.$^{-2}$ sec.$^{-1}$, which they considered to be too small, being only 1 per cent. of that found for frog skeletal muscle. Thus more elaborate experiments were performed in which simultaneous efflux of an extracellular marker, such as $SO_4^=$, was measured; the difference between the two, allowing for different diffusion coefficients, should be a measure of the efflux of intracellular Na^+. Much larger values, namely 70–135 pmole. cm.$^{-2}$ sec.$^{-1}$ were obtained, but the method involved considerable uncertainties, so that a third method, employing wash-out of ^{24}Na when the cells contained high and low intracellular concentrations of ^{23}Na, was developed. The difference between the wash-outs under the two conditions gave a measure of the intracellular Na^+ washed out, and led to a value of 50 pmole./cm^{-2}/sec.$^{-1}$. It would seem from this study that the separation of extra- and intracellular effluxes by the fast and slow components of the wash-out curves can lead to misleading results; so far as the cardiac tissue is concerned, the intra-

wet wt. for guinea pig atria, comparing with 21 ml. for the diaphragm. Danielson (1964) considered the sulphate-space to be more reliable than the ^{14}C-mannitol or -sucrose spaces; the sulphate spaces were 63, 49, 22 and 17 per cent. of wet weight in sinus venosus, auricle, ventricle and sartorius muscle respectively of the toad.

* The circumstance that the Na^+-flux is greater than the K^+-flux does not mean a greater Na^+-permeability; thus the passive inward flux of Na^+ occurs under a large gradient of concentration as well as of potential, whilst that of K^+ is against a gradient of concentration. Thus the relative permeabilities are given by:

$$P_{Na}/P_K = \frac{Na \; Influx}{K \; Influx} \cdot \frac{K_{Out}}{Na_{Out}}$$

cellular efflux is apparently included in the fast component along with the extra-cellular efflux, whilst the slow component is of unknown significance.[*]

Sodium Pump

Page & Storm (1965) allowed papillary muscle to cool and accumulate Na^+ and lose K^+; on rewarming, accumulation of K^+ and extrusion of Na^+ occurred, so that in 30 minutes the normal ionic contents were reached. The maximal uptake of K^+, and extrusion of Na^+, were about the same. Interestingly, the membrane potential returned to its original value much more rapidly than the ionic concentrations, so that the membrane potential, E_m, was now greater than E_K, the equilibrium potential, the reverse of the usual situation. As with skeletal muscle (p. 635), this may be the result of the operation of an electrogenic pump.[†]

Action Potential. The action potentials of the different regions of the heart have some characteristic differences as Fig. 686 shows; thus the ventricular muscle fibres are characterized by a long-lasting overshoot giving a plateau;

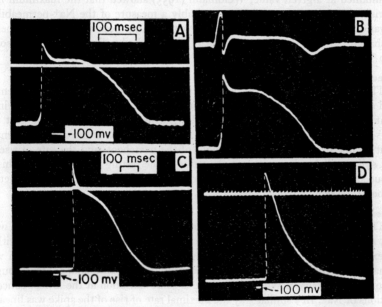

Fig. 686. Intracellular records of action potentials at different sites of the heart. A. Ventricle. B. Ventricle showing simultaneous electrogram. C. Purkinje fibre. D. Auricle (time-scale shows 10 and 50 msec. intervals). (Brooks, Hoffman, Suckling & Orias. *Excitability of the Heart.* Grune & Stratton.)

in the Purkinje fibre, the plateau belongs to the repolarization phase. The auricular action potential is usually altogether sharper, the plateau being absent; its shape[‡] varies, however, according to the degree of inhibition present.

---

[*] Haas, Glitsch & Kern (1964) have shown that the efflux of ^{24}Na is decreased by replacing external Na^+ by Li^+ or sucrose, presumably indicating exchange-diffusion. From the measured flux of 5 mM/kg./min., equivalent to 483 coulombs/kg./min., and the measured value of the membrane potential, E_m, of -68 mV, and the estimated value of the Na^+ equilibrium potential, E_{Na}, of $+30$ mV, they computed an energy requirement of $483 \times 98 = 47,500$ coulombs/mV/kg./min. $= 11 \cdot 3$ cal./kg./min.

[†] As the authors point out, the membrane potential was necessarily measured in surface cells whilst the ionic concentrations were mean values for the whole tissue; and it might be argued that the surface cells had recovered their normal ionic contents. However, Page & Storm consider that the fluxes would have had to be too high for this to happen.

[‡] By increasing the sweep-speed, Wright & Ogata have shown that the frog's auricular action potential is more complex, exhibiting a "spike and dip" before the plateau; thus the repolarization phase has a notch. They suggested that there were two phases of depolarization, a rapid one giving the initial spike, and a slower one continuing after the spike.

The upstrokes of the spikes are all very short, so that the characteristic feature of the cardiac action potential is the delay in repolarization, which may be due to a lower active permeability to K+ than that in nerve (p. 1078).

In general, when nodal cells are compared with the conducting system, we find in the former a lower resting potential, slower rising time of the spike, and lower conduction velocity (Hecht, 1965).

Sodium Permeability

It is reasonable to assume that the rising phase is due to a sudden increase in permeability to Na+; Draper & Weidmann showed that the magnitude of the overshoot* was directly proportional to the external concentration of Na+, and by applying a voltage-clamp technique, whereby the membrane potential of a spontaneously active Purkinje fibre could be suddenly established and maintained at a given value, Weidmann (1955) showed that the maximum rate of rise of the action potential—presumably a measure of the Na+-permeability —varied in a characteristic manner with the clamp-potential, the curve being S-shaped, so that the maximal rate of rise reached a limiting value of some 500 V/sec. at a voltage of 90 mV.

Effects of Resting Potential Changes

The effects of varying resting potential on rate of rise of action potential and overshoot are shown in Fig. 687 from Weidmann. These curves are reminiscent of the Na+-activation curves of Hodgkin & Huxley and indicate that the Na+-permeability mechanism is maximally activated when the resting potential is at 90 mV or above. Thus the height of the overshoot, and the rate of rise of the spike, will be dependent on the value of the resting potential; when this is reduced, as in anoxia or with high external K+, we may expect these parameters to be decreased, as indeed they are.

External Sodium

The effects of altered external concentration of Na+ are quite consistent with the sodium hypothesis, the overshoot being linearly related to the logarithm of external concentration and becoming zero at about 20 mM; the system was behaving approximately as a Na-electrode (Draper & Weidmann, 1951) although, according to Brady & Woodbury, there is some deviation at the lowest concentrations of Na+. In a similar way the maximal rate of rise of the spike was linearly related to the external Na+ concentration (Délèze; Brady & Woodbury).†

In the frog's ventricle the decline in height of the spike was very much less than the change in Na+-potential, namely 17 mV for a tenfold reduction in concentration; the rate of rise of the spike, however, was a linear function of $[Na^+]_{out}$, so that it is in the region of overshoot that deviations occur, and Niedergerke & Orkand consider that the ratio P_{Na}/P_K may alter, being determined effectively by the concentration of Ca^{++} in the medium. Thus increasing $[Ca^{++}]_{out}$ increased the overshoot; moreover, if $[Na^+]_{out}$ was varied, keeping the ratio $[Ca^{++}]/[Na^+]^2$

---

* Cranefield, Eyster & Gilson also showed that the overshoot of the turtle heart was reduced by reducing the sodium concentration in the perfusion fluid. It is interesting that Eyster, Meek, Goldberg & Gilson (1938) using external electrodes, recognized that the action potential was more than a depolarization, i.e., that it was a reversal of polarization, thus anticipating its more unequivocal demonstration with intracellular electrodes.

† The effects of varying $[Na^+]_{out}$ on the various parameters of the heart cycle—maximal diastolic potential, spike overshoot, depolarization and repolarization times—are probably different according to the particular heart tissue; they have been examined recently by Toda (1968) in the pacemaker cells of the SA-node. This author draws attention to the effect of reduced $[Na^+]_{out}$ on the uptake of noradrenaline by sympathetic nerve terminals observed by Iversen & Kravitz (1966); this might well be a confusing factor in the situation.

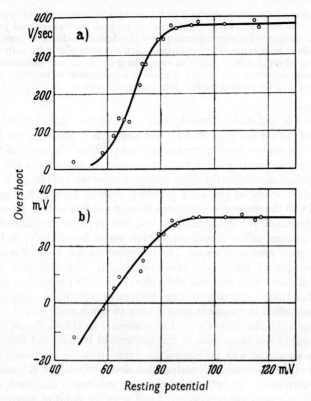

FIG. 687. Effect of resting potential (abscissa) on (a) the rate of rise of spike and
(b) the magnitude of the overshoot of sheep Purkinje fibre. (Trautwein,
after Weidmann. *Ergebn. d. Physiol.*)

constant, the variation of overshoot was now in accordance with the theoretical
Na^+-potential. The fact that the ratio $[Ca^{++}]/[Na^+]^2$ was important suggests
that the effect of Ca^{++} is on the cell membrane rather than as a carrier of current,
since, as Wilbrandt & Koller (1948) have argued, the effective concentrations of
Ca^{++} and Na^+ at the membrane are related to the bulk concentrations in the
medium in this way. Thus the Na^+-hypothesis accounts qualitatively for the
effects of altered resting potential and altered external concentration of Na^+ on
the spike of the action potential; since the resting potential can be affected by
numerous influences—excess CO_2, high external K^+, low temperature and
digitalis—the primary effect may well be through the Na^+-activity mechanisms.

Very Low External Na^+-Concentration

Brady & Woodbury and Casteels (1962) found that a minimum of 10 per cent. of
the normal concentration of Na^+ was necessary to permit a conducted spike, but
there have been reports of maintained activity in much lower concentrations; as Brady
& Tan (1966) pointed out, however, this might well be due to the presence of
extracellular Na^+ that had not been completely washed out.* They employed in
their study the frog's atrial trabeculæ containing less than 100 fibres, so that
equilibration with the medium occurred within a few seconds. They found that,
when the medium had less than 5–10 per cent. of normal Na^+, in the presence of
normal concentrations of K^+ and Ca^{++}, the heart was incapable of propagating a

---

* Sommer & Johnson (1968) emphasize the very tight packing of the Purkinje fibres in
large hearts, a factor that, by reducing the intercellular spaces to mere 200A clefts, might
well make it difficult to reduce the extracellular concentration of Na^+ by mere soaking of
the tissue in Na^+-free medium.

spike; if, however, the concentrations of K^+ and Ca^{++} were reduced in proportion to the reduction in Na^+, even 2 per cent. of normal Na^+ permitted propagated activity. This seems to be just one example of the delicate balance between divalent and monovalent ions required to maintain normal excitability. As with the squid axon, Li^+ could replace Na^+ so far as supporting the spike was concerned; unlike the squid axon, however, the effects of choline substitution were small, indicating that current could be carried by this ion during the spike.

Flux Changes

As with nerve and skeletal muscle, each cycle of the action potential should be associated with a net gain of Na^+ and loss of K^+, although the increased efflux of K^+ should not be so marked as in nerve (Noble). The study of Rayner & Weatherall (1959) on active auricles has demonstrated an increased efflux of K^+, and in ventricular fibrillation the effect was much more marked (Briggs & Holland, 1961). In their study of the cat's papillary muscle, Goerke & Page (1965) emphasized that the tissue concentrations were in a steady state as far as K^+ was concerned, regaining K^+ as fast as it was lost, and therefore they considered that to observe an extra efflux during contraction would be unlikely; in fact they, themselves, were unable to do so. The phenomenon has been examined with considerable care by Lamb & McGuignan (1968), who have also measured changes in efflux of other ions and molecules; with their preparation they could measure alterations occurring during successive 100 msec. intervals. The K^+-influx and efflux in quiescent muscle were 16 $pmole.cm^{-2}/sec.^{-1}$, computed on the basis of a cell diameter of $3\cdot5\,\mu$. Contraction *reduced* both fluxes, the reduction occurring at the same time as the mechanical twitch; and this could be shown to be associated with the contraction rather than the electrical event, and may be due to the mechanical distortion that slows diffusion of K^+ and permits greater reabsorption by the cells. With the other substances examined, *e.g.*, Na^+ Ca^{++}, $SO_4^=$, the peak effluxes occurred just after the point of maximal rate of contraction.

Membrane Conductance

Weidmann, by inserting two micro-electrodes into Purkinje fibres, one for recording and one for passing a pulse of current, showed that the membrane resistance, calculated from the electrotonic response to pulses, fell rapidly during the rising phase, as indicated in Fig. 688; the plateau, representing delayed repolarization, was associated with a return of the membrane resistance, and the final down-stroke, representing repolarization of the membrane, corresponded with another fall. It could well be, then, that the characteristic shape of the action potential depends on a delayed permeability to K^+.

Thus the clue to the behaviour of heart muscle, both in respect to its pacemaker potentialities and its long drawn-out action potential, may well reside in the elucidation of the K^+-permeability mechanism. The anion-conductance of heart muscle is low, in contrast with that of skeletal muscle, so that in sodium-deficient media the membrane conductance is essentially a measure of the partial K^+-conductance. Hall, Hutter & Noble showed that the dependence of the K^+-conductance on the membrane potential was complex, and qualitatively different from that of skeletal muscle and of nerve.

RECTIFICATION

We have already seen that the conductance of an ion across a membrane sustaining a difference of potential will differ according to the direction of the current-flow in a way that is predictable from the constant field equation; that is, in spite of a constant permeability, P_K, the membrane will show rectification. This normal rectification will be manifest as a greater conductance when positive current moves out of the fibre, *i.e.*, for a depolarizing potential—than for the reverse direction of current-flow. The Purkinje fibre also shows *anomalous rectification*, in that an outward electrochemical gradient, causing an outward movement of positive current, causes a decrease in *permeability* to K$^+$ and therefore a decrease in conductance; the reverse occurs for inward currents. This is similar to the behaviour of skeletal muscle described by Katz and by Hodgkin & Horowicz and Adrian & Freygang (p. 575). When the depolarizing current is strong enough to displace the membrane potential about 30 mV, a new effect is manifest in that depolarization now causes an increased conductance similar to the *delayed rectification* described by Hodgkin & Huxley for squid nerve, *i.e.*, the increased K$^+$-permeability that serves to repolarize the fibre after the Na$^+$-spike. The magnitude of the increase in K$^+$-conductance is less

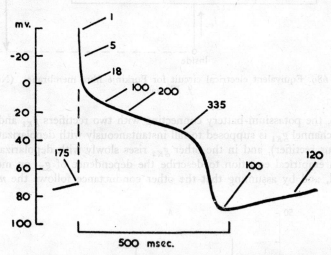

FIG. 688. Change in electrical resistance across membrane of Purkinje fibre during the action potential. The figures attached to the curve indicate membrane resistance in relative units. (Weidmann. *Ann. N.Y. Acad. Sci.*)

than that for the squid axon, so that it does not bring the total K$^+$-conductance back to its resting value, whilst the speed with which the conductance increases when a depolarization is initiated is altogether slower. Expressed mathematically, the K$^+$-conductance is described by the equation:—

$$g_K = (g_K')F_1(E_m - E_K) + F_2(E_m, t)$$

where g_K' is the conductance at the resting potential and the function F_1 is less than unity when the argument is positive and greater than unity when negative. The function F_2 is similar to the Hodgkin-Huxley K$^+$-current equations describing their delayed rectification.*

Calculated Action Potential. Noble has applied the general concepts implied in the Hodgkin-Huxley equations to the situation given by the Purkinje fibre where the K$^+$-conductance behaves as indicated above. Thus the equivalent circuit of the squid fibre (Fig. 555, p. 1082) was modified as illustrated by

* Johnson & Tille (1961) have claimed that the current-voltage relationships in ventricular muscle fibres show no anomalous rectification of the type described by Hall *et al.* In a recent paper, however, Noble (1962) has shown that the arrangement of the recording and current-passing electrodes employed by Johnson & Tille was such as to make the system remarkably insensitive to changes in membrane resistance, and would fail to detect even large non-linearities in the current-voltage relationship.

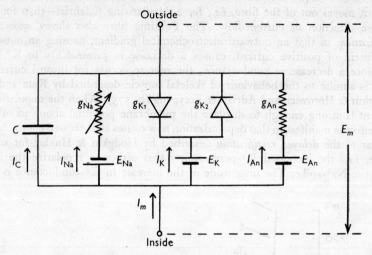

FIG. 689. Equivalent electrical circuit for Purkinje fibre membrane. (Noble. *J. Physiol.*)

Fig. 689, the potassium-battery connecting with two rectifiers g_{K1} and g_{K2}; in the one channel g_{K1} is supposed to fall instantaneously with depolarization (the anomalous rectifier), and in the other g_{K2} rises slowly with depolarization. By using an empirical equation to describe the dependence of g_{K1} on membrane potential, and by assuming that the other conductance follows the *n*-type of

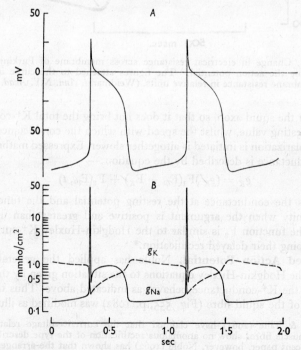

FIG. 690. A, computed action and pacemaker potentials. B, time-course of conductance changes on a logarithmic scale. Continuous curve, g_{Na}; interrupted curve, g_K. (Noble. *J. Physiol.*)

equations employed by Hodgkin & Huxley for the K^+-conductance of the squid axon, a theoretical voltage-current curve could be calculated that agreed reasonably well with that determined by Hall *et al.* As we have seen, Weidmann's voltage-clamp studies indicated that the Na^+-current might well be described by the same equations as in the squid axon; hence Noble set up equations defining m, h, α_m and β_m, and by employing a computer for the solution of the set of differential equations he deduced the action potentials shown in Fig. 690. It will be seen that the equations predict not only the shape of the action potential but the existence of pacemaker activity, so that the repolarization following the spike and plateau is not complete, a new phase of depolarization leading to a new spike.

Changes in Conductance

On the same figure may be seen the changes in the sodium- and potassium-conductance during the cycle. During the spike, g_{Na} rises to a very high value because m rises much faster than h, the inactivation term, falls, but within milliseconds the fall in h reduces g_{Na} again; because inactivation is not complete g_{Na} settles down to a fairly constant value about eight times larger than the smallest value in diastole. The relative constancy of the sodium-conductance over the period corresponding to the plateau of action potential is due to the fact that m and h change in opposite directions so that m^3h remains almost constant. At the end of the plateau the rise in h no longer fully counteracts the fall in m, so that g_{Na} falls to its diastolic value and the membrane repolarizes rapidly. During the same cycle, g_K falls rapidly—the anomalous rectifier effect—but during the plateau it rises slowly because of the delayed rectifier effect, g_{K2}. At the end of the action potential there is a further increase in potassium conductance when g_{K1} increases as a result of the repolarization of the membrane. Since g_{K2} takes some time to fall again, the total potassium conductance exceeds its end-diastolic value for several hundred milliseconds during the pacemaker potential.

Ionic Currents

The calculated Na^+- and K^+-currents are indicated in Fig. 691 where they have been given the same sign so that their relative magnitudes are seen more easily. The feature that determines the plateau is the approximate equality of the two currents; this leads to a somewhat extravagant loss of K^+ and gain of Na^+ per cycle of change, by contrast with the squid axon.

Pacemaker Activity

As Trautwein & Kassebaum emphasized, the probable cause of the pacemaker activity is essentially the finite sodium conductance that holds the membrane potential away from the K^+-, or resting, potential; thus at the height of repolarization the membrane potential, E_{max}, is $-$ 65 mV compared with a considerably higher value of the K^+-potential, or the measured resting potential in the absence of all pacemaker activity. Put semi-quantitatively, the condition is that the sodium-current, I_{Na}, should be greater than the potassium-current, I_K, which depends on $(E - E_K) \times g_K$. As $(E - E_K)$ increases, *i.e.*, as depolarization proceeds, the potassium-current should rise and put a stop to pacemaker activity were it not for the decrease in potassium conductance that takes place during depolarization, and the small increase in g_{Na}. Both these points are brought out by Figs. 690 and 691, whilst experimental support was provided by Trautwein & Kassebaum who showed that pacemaker activity could be reduced by lowering the external sodium-concentration. Again, Gross-Schulte & Trautwein found that the resting potential of the rabbit sinus, *i.e.*, its maximum diastolic potential, rises to 89 mV from 84 mV when the external concentration of Na^+

is reduced by 90 per cent., whilst papillary muscle, which has no pacemaker activity, was unaffected. If the resting potential is held below the K⁺-potential by virtue of a measurable Na⁺-permeability, then reducing the external concentration of Na⁺ should bring the resting potential closer to the K⁺-potential. However, a high permeability to Na⁺ is not adequate to cause pacemaker activity as such, and it must be the associated decrease in permeability to K⁺ during diastole, revealed by the increase in membrane resistance as depolarization increases, that is the additional factor. Thus any factor that decreased potassium conductance (g_K) might be expected, other things being equal, to promote pacemaker activity; Vassalle (1965) found that raising the external concentration

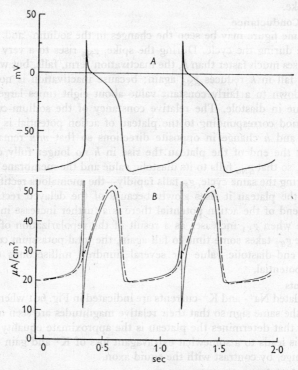

Fig. 691. A, computed action and pacemaker potentials as in Fig. 690. B, computed ionic currents; continuous curve, I_{Na}; interrupted curve, I_K. (Noble. *J. Physiol.*)

of K⁺ from 2·7 mM to 5·4 mM suppressed spontaneous activity, and this was associated with an increased g_K. In a later study, employing a voltage-clamped Purkinje fibre, Vassalle (1966) confirmed that during diastole the membrane resistance steadily increased, although the membrane was clamped at its maximal diastolic value. His voltage-clamp studies indicated that pacemaker activity was due to a time-dependent drop in g_K together with a voltage-dependent drop in g_K and, finally, a voltage-dependent rise in g_{Na} as the threshold is approached.

Anodal Pulses

We have already seen that an anodal pulse applied on the plateau of the TEA-treated axon may cause a sudden repolarization of the membrane—a *regenerative repolarization*. Weidmann (1951) demonstrated a similar all-or-nothing repolarization of the Purkinje fibre; and an analysis of the effects of current pulses, as predicted by Noble's equations, showed that, when the applied polarizing potential was beyond a certain size, the effect was to repolarize the membrane completely; below this size, the passive repolarization due to the current pulse was followed by a return to the plateau of depolarization.

Repetitive Stimulation

Noble's equations account for the shortening of the duration of the action potential caused by an increase in the frequency of stimulation, described by Hoffman & Suckling and others. The cause is the slow decay of g_{K_2}, the delayed rectifier effect, after the decay of the action potential. Thus the second of two action potentials in rapid succession is shorter because g_{K_2} starts off at a higher value and so takes less time to rise to the value required to initiate repolarization of the membrane.

Permeability Changes

An increase in anion permeability, or the "leak conductance", may be brought about by replacing the Cl^- of the Ringer's solution by NO_3^- (Hutter & Noble; Carmeliet); theoretically this should about halve the duration of the plateau, an effect that was actually found. An increase in K^+-conductance should work in the same direction, reducing the duration of the plateau; as we shall see, vagus stimulation does precisely this, an effect that is probably due to increased permeability to K^+ induced by the liberated acetylcholine.

Anion Substitutions

The effects of anion substitutions, as determined experimentally by Hutter & Noble, agree remarkably well with the predictions from the equations when the anion conductance is altered. Hutter & Noble's study indicated that the order of anion permeability was $I > NO_3 > Br > Cl$ (the reverse of that found in skeletal muscle, Hutter & Padsha) so that the anion conductance in NO_3^--Ringer should be greater than in Cl^--Ringer. The equations predict a greatly increased frequency of pacemaker activity, whilst if anion conductance exceeds a certain value the fibre should be arrested. In fact the frequency of activity increases in NO_3^- and is abolished in I^-.*

Delayed Currents

It is agreed that the initial inward current responsible for the ascending phase of the action potential is carried by Na^+, as in the squid axon, but there has been disagreement as to the carrier of the repolarizing outward current; according to Noble (1962) and Vassalle (1966), this is determined almost exclusively by K^+, and may be resolved into two components in accordance with their dependence on membrane potential and time to develop. Voltage-clamp experiments of Deck & Trautwein (1964) appeared to contradict this view, but this may have been because the experimental arrangement did not permit the direct measurement of the initial Na^+-current so that this had to be obtained by subtraction of the current in Na^+-free medium from the total. If the membrane properties in a Na^+-free medium were different from those in a normal one,† then this approach would be unsound and could account, at least in part, for the discrepancy between the work of Deck & Trautwein and later studies of McAllister & Noble (1967) and

* Brady & Woodbury (1960), in addition to examining the effects of altered external Na^+- and K^+-concentrations on the action potential of the frog ventricle, formulated equations, analogous with the Hodgkin-Huxley equations, to describe the action potential. They assumed a decrease in K^+-conductance with depolarization whilst the Na^+ inactivation and -activation mechanisms were assumed to have fast and slow components. With these modifications, the Hodgkin-Huxley equations accounted reasonably well for the shape of the action potential.

† In a later study, Dudel et al. (1967b) compared the currents in the presence and absence of tetrodotoxin, and concluded that a Na^+-free medium does, indeed, alter the permeability relationships, causing either an increased positive current or decreased negative current. Tetrodotoxin abolished the large negative current due to activated Na^+-permeability but failed to influence the steady-state negative current that persists during a sustained depolarization and that had been attributed by Deck & Trautwein to a persisting Na^+-permeability.

Noble & Tsien (1968), in which it was shown that the reversal potential for the late current was about 100 mV positive, *i.e.*, about equal to the K^+-potential, so that these authors concluded that the declining phase of the action potential was, indeed, determined by the appearance and growth of phases of high K^+-conductance.

Slowly Developing K^+-Conductance

Noble's more recent analysis of the slowly developing phase, described as g_{K_2}, has shown that it differs in several respects from that defined in his earlier study. McAllister & Noble showed that the slow outward current would develop at membrane depolarizations that were too low to activate the Na^+-permeability and so could be examined in isolation, whilst the fact that they occurred at small degrees of depolarization emphasized their importance in the development of the

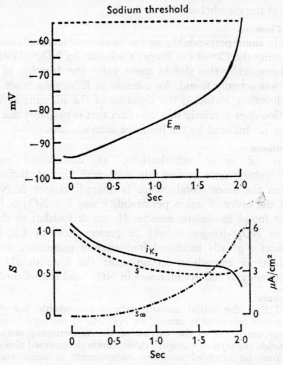

FIG. 692. Mechanism of pacemaker potential based on new model for K-current. *Top:* Variation in membrane potential during pacemaker activity, replotted from Vassalle. *Bottom:* s_∞ (t) relation. Note that, although s does not fall below a certain value and actually increases towards the end of the pacemaker potential, i_{K_2} falls continuously. This is a consequence of the negative slope in the rectifier function. (Noble & Tsien. *J. Physiol.*)

pacemaker potential. An additional feature is the dual type of rectification; the initial time-independent g_{K_1} shows only inward, or anomalous, rectification, in the sense that more current is required to hyperpolarize the membrane than to depolarize it, consequently g_{K_1} is low at the plateau of the action potential. g_{K_2} also shows inward rectification in respect to fast changes of potential, but it slowly rectifies in the opposite direction, so that repolarization of the membrane is favoured, and according to Noble & Tsien it is this phase of potassium conductance that is mainly responsible for repolarization of the membrane. As with the Na^+-system, the time-constant for development of the K^+-current and the steady-state fraction of activation of the K^+-permeability are functions of membrane potential, without, however, any indication of an inactivation process. The time-scale of the events, moreover, is some two orders of magnitude longer, and is comparable with the K^+-inactivation variable, k, in myelinated cells (Frankenhaeuser, 1962) and in skeletal muscle (Adrian, Chandler & Hodgkin, 1968). Employing the experimentally determined parameters of the slow conductance change, Noble & Tsien (1968) altered their description of the pacemaker activity; it now seems that it is the decline of g_{K_2} at moderate depolarizations—because of the inward rectification—that triggers off the spike rather than activation of the Na^+-current. This is illustrated by Fig. 692 which shows, above, an experimentally determined pacemaker potential and, below, the probable course of the slowly developing K^+-current; thus towards the end of the pacemaker depolarization, the K^+-current falls sharply although the activation of the K^+-mechanism, indicated by s,* is actually increasing as a result of the depolarization, and this is because of the negative slope conductance of the rectifier function for i_{K_2} observed experimentally at membrane potentials beyond about 25 mV positive to E_K. With regard to the repolarization phase of the action potential, Noble & Tsien's analysis indicates that, because of the inward going rectification of g_{K_2}, the amount of repolarizing current supplied by K^+-ions will be smaller, so that repolarization will be a slow process.

Effect of Tetrodotoxin

Dudel *et al.* (1967, *b*) found that, at concentrations greater than 10^{-6} g./ml., the rate of rise of the spike was decreased, whilst the spike-height was depressed. At 2.10^{-5} g./ml. the fibres became inexcitable. Voltage-clamp experiments showed

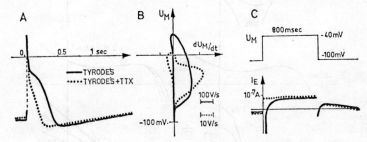

FIG. 693. Effects of tetrodotoxin on intracellularly recorded action potentials (A), rate of rise of these potentials (B) and on currents during clamping of voltage from -180 mV to -40 mV and, after 800 msec., back to -100 mV (C). Continuous curves represent controls in Tyrode's solution, dashed curves measured in presence of 10^{-5} g/ml tetrodotoxin. Ordinates in A and B represent membrane potential, in C clamp current; abscissae in A and C represent time, in B rate of change of potential. Note that in B the curves are plotted at different calibrations. (Dudel, Peper, Rüdel & Trautwein. *Pflüg. Arch.*)

that the initial strong inward current was abolished, whilst the sustained outward current was unaffected, and this presumably accounts for the more rapid repolarization observed (Fig. 693, A and C). As Fig. 693 shows, the slow diastolic depolarization was unaffected, suggesting that this is due to a non-Na^+ current. Thus Deck & Trautwein had concluded that the late steady-state current was largely due to Na^+, *i.e.*, that the active Na^+-permeability had not decayed as rapidly as happens in the squid axon; the effects of tetrodotoxin indicate, however, that this current, so important for the repolarization of the membrane, is carried either by K^+, as suggested by Noble, or by both K^+ and Cl^-.

Chloride Current

The effects of anion substitutions described earlier suggest that the chloride-permeability might well play a role in the repolarization phase, since substitution by an impermeant anion prolongs the declining phase of the action potential. Thus a passage of Cl^- down its gradient of electrochemical potential when the fibre is depolarized would contribute towards repolarization. Dudel *et al.* (1967a) found that the delayed outward current observed in the voltage-clamped Purkinje fibre

---

* Noble (1962) originally described the time- and voltage-dependent development of g_{K_2} as a function of n^4 in accordance with Hodgkin-Huxley notation; it seems, however, from McAllister & Noble's (1966, 1967) results that it is best described by n^1, and they have changed the nomenclature, describing g_{K_1} as a function of s, which, like n, may be described as a function of time:—

$$ds/dt = a_s (1-s) - \beta_s s$$

was strongly affected by substitution of chloride by maleate or acetylglycinate; this maleate-sensitive current became large at a membrane potential of −20 mV and remained at this high value at further depolarization up to high positive values of the potential. It would thus be of great significance for repolarization. The altered membrane-permeability responsible for the high Cl^- currents was time-dependent, being inactivated with a time-constant of about 50 msec.

CONTRIBUTIONS OF Cl^- AND K^+

If, then, a considerable proportion of the current flowing during strong depolarization is carried by Cl^-, it remains to determine the contribution of K^+ during those phases that are not dominated by Na^+-permeability. According to Dudel et al. (1967c), the K^+-currents are relatively simple, exhibiting little dependence on time. The results of Dudel et al.'s analyses are indicated schematically in Fig. 694, in which the currents due to the separate ions are plotted against potential.

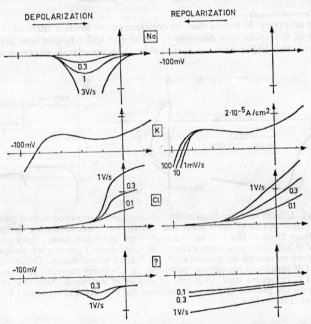

FIG. 694. Schematic survey of the hypothetical components of the membrane current that flows when a Purkinje fibre is de- and repolarized with different constant speeds. The diagrams labelled by a question mark show the negative non-sodium current. Ordinate: Ionic current per unit area. The calibration of the ordinates is equal for all diagrams. Abscissa: Membrane potential. (Dudel, Peper, Rüdel & Trautwein. *Pflüg. Arch.*)

Because of inactivation, the Na^+- and Cl^--currents will depend on how rapidly the potentials are established; thus, as Fig. 694 shows, the inward Na^+-current, conventionally described as negative, is very small if depolarization is brought about slowly; at 3 V/sec. it becomes very high and is responsible for the rising phase of the spike; when the membrane potential becomes positive, *i.e.*, at the peak of the spike, Na^+-current is about zero; under these conditions the process of repolarization is due to a large Cl^--current and a smaller K^+-current. In the region of the resting potential, K^+-conductance is important; and variations in this are probably largely responsible for the pacemaker potential.

General Conclusion. To summarize a rather intricate situation, all workers seem now agreed that the rising phase of the spike is determined by a rapidly inactivated Na^+-permeability system, whilst the repolarization phase and pacemaker depolarization are determined by other ions; so far as repolarization is concerned, the results from Trautwein's laboratory emphasize the combined effects of Cl^-- and K^+-permeabilities, whereas Noble invokes a more complex K^+-system as the prime determinant. Pacemaker activity seems to be largely determined by K^+-permeability, since it is unaffected by tetrodotoxin, whilst the Cl^--permeability is too low at the relatively high values of membrane potential at which pacemake activity is manifest.

Electrotonic Interaction. The heart tissue behaves as though its individual cells were electrotonically coupled; morphological evidence for this connection is strong, in the sense that numerous zonulae and maculae occludentes may be identified (Fig. 680, p. 1277). Electrophysiological evidence of low-resistance pathways between adjacent cells is contradictory, however. Thus Fänge *et al.* (1956), in their study of trypsin-dispersed heart cells, found that, if they inserted microelectrodes into two cells of a formation up to 1 mm. apart, electrotonic coupling could be observed. However, the studies of Tarr & Sperelakis (1964), who inserted double microelectrodes, of variable separation, into heart tissues, indicated that, if the separation was greater than 60 μ, there was no interaction; if the distance was less, interaction could be observed, presumably because the electrodes were in the same cell. Tille (1966) likewise found that many pairs of cells failed to show interaction, and he was able to differentiate, on the surface of the ventricle, *P-cells*, which showed strong electrotonic coupling at distances less than 1 mm., from the great majority of *V-cells*, which showed no interaction even at very small distances. Electrophysiologically the P-cells could be distinguished by their high resistance, long time-constant and low threshold; and histologically by their tendency to form loops and interconnections. Tarr & Sperelakis (1967) found many pairs of cells in which an intermediate degree of coupling was observed; this was hard to explain since if, as they thought, coupling was only observed when the electrodes were in the same cell, the amounts should only be zero and 100 per cent. They concluded that intermediate degrees were observed when the electrodes were in adjacent cells that had been damaged by the insertion of microelectrodes. Thus, when the resting potentials were greater than 50 mV no large degree of coupling could be observed; they assumed that the damage was manifest in the reduction of the resistance through the specialized contact regions, which permitted a variable degree of coupling. Tarr & Sperelakis estimated that the resistance between two cells less than 500 μ apart was almost twice that between a cell and its extracellular fluid, and concluded that there was no low-resistance pathway between cardiac cells. More direct experiments of Trautwein (1961) and of Weidmann (1966), however, tend to confirm the existence of low-resistance pathways between cells; to quote the latter work, Weidmann separated the two halves of a bundle of ventricular fibres by passing them through a tightly fitting aperture, and he measured the distribution of concentrations of ^{42}K along the length when one side was exposed to this isotope and the other to a Ringer's solution without it. The distribution was compared with the calculated distribution on the basis of diffusion through a series of intercalated discs; the value of the "space constant", λ, so derived, namely the distance over which the concentration fell to $1/e$ of its value, was 1·5 mm.; the time-constant, τ, for efflux was 48 minutes, and since the apparent diffusion coefficient is given by $\lambda^2 = \tau$, the computed coefficient was $7\cdot9.10^{-6}$ cm²/sec., only three times lower than the self-diffusion coefficient of K^+ in 150 mM KCl. Clearly, if diffusion occurred intracellularly, there were no significant diffusion barriers from cell to cell.*

---

* Kriebel (1968) studied the transmission of current pulses across the tunicate heart, *e.g.*, *Ciona intestinalis*, which is one cell-layer thick, in the form of a tube; transmission across this sheet indicated the presence of tight junctions between cells, and this was

Cultured Cells. Fänge, Persson & Thesleff (1956) were the first to describe the preparation and electrophysiology of cultured chick embryonic heart cells, disintegrated from the tissue by trypsin treatment. More elaborate studies have been carried out by Lehmkuhl & Sperelakis (1963) and De Haan & Gottlieb (1968); the cells reassembled in groups, some in pairs and some making rosettes and other formations. These usually contracted synchronously at 30–130 beats/min., one cell acting as a pacemaker so that, on occasions, transmission-block could be observed, in the sense that a driven cell would contract only once for every two beats of the pacemaker cell.* Intracellular recording revealed the development of pacemaker potentials from which the spike arose, as in the pacemaker region of the intact heart; sometimes a spike would arise abruptly from the pacemaker potential, indicating that the cell was driven by a neighbouring cell. Separate cells were spontaneously active so that it is not necessary for one to have contact with a pacemaker cell to develop an action potential. Transmitter drugs, such as acetylcholine, epinephrine, etc., had no effect on frequency of firing of either pacemaker or driven cells; tetrodotoxin, which abolishes Na^+-dependent spike activity in other excitable tissues, surprisingly had no effect. The average membrane potential was $59 \pm 1 \cdot 2$ mV and the spike-height was $71 \cdot 2 \pm 1 \cdot 5$ mV; the membrane parameters were $R_m = 480$ ohm cm^2; $C_m = 20$ $\mu F/cm^2$; time-constant $\tau_m = 9 \cdot 3$ msec. (Sperelakis & Lehmkuhl, 1964). When external K^+-concentration was raised, the membrane potential varied in accordance with the Nernst equation, when the concentration was high, but at low values, as with the intact heart, the variation could only be explained in terms of the Goldman equation on the basis of permeability ratios: $P_{Na}/P_K = 0 \cdot 05$ and $P_{Cl}/P_K = 0 \cdot 25$.

Inhibition and Acceleration

Modification of pacemaker activity will produce slowing—*negative*—or acceleration—*positive chronotropic action*. In general, a changed frequency of beat can be brought about in three ways, as illustrated by Fig. 695; by altering the threshold depolarization at which the spike fires (A), by changing the slope of diastolic depolarization (B) or by changing the early diastolic potential after repolarization, *i.e.*, a change of E_{Max} (C).

Innervation. Nervous control seems to be exerted largely through the pacemaker and conducting zones; thus Trautwein & Uchizono (1963) in their, study of the SA-node, found nerve fibres in 50 out of 500 photographs of the muscular layer of the pacemaker zone, whereas in 200 photographs of the atrial myocardium there were only three. These authors found axons filled with vesicles making close relation with muscle cells, suggesting a synaptic arrangement, but no characteristic synaptic structures, such as those seen in the sympathetic ganglion or central nervous system, were observed. Johnson & Sommer (1967), in their study of a ventricular strand that consisted, apparently, of Purkinje fibres, described the presence of nerve fibres with numerous boutons along the course of their axons; these contained vesicles, some of which had the dense core that is associated with catechol amine-containing nerve endings (p. 1195).

Vagal Action. Stimulation of the vagus nerve causes the heart to slow and, if the stimulation is intense enough, to stop. The electrical changes taking place have been studied by a number of workers, and the results fit in with those

---

confirmed in the electron microscope. Electrotonic coupling between cells in the longitudinal direction was revealed by the "sucrose-gap technique". With this technique a local region is treated with sucrose solution to raise its external resistance; appreciable passage of current across the gap is thus only possible if there is a low-resistance pathway within the cells.

* Forcing the cells into contact did not impose synchrony on the separate cells, however (Sperelakis & Lehmkuhl, 1964). We may note that Ba^{++} rapidly depolarized the cells, but also converted non-pacemaker into pacemaker cells, possibly by its effect on the K^+-conductance, decreasing this and thus creating the condition for pacemaker activity; at any rate the input resistance increased (Sperelakis & Lehmkuhl, 1966). De Haan & Gottlieb (1968) have discussed a number of features of the action potential in the cultured cell.

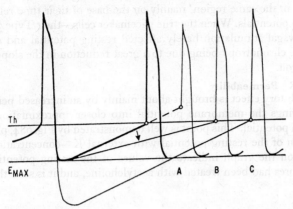

FIG. 695. Illustrating the mechanisms of slowing of the heart. The normal
sequence of activation is seen in the two heavily outlined complexes. Slowing
will occur (A) by increasing threshold (Th); (B) by decreasing slope of
diastolic depolarization; or (C) by increasing the early diastolic potential
after repolarization (change of E_{MAX}). (Hecht. Ann. N.Y. Acad. Sci.)

obtained on other systems exhibiting the phenomenon of inhibition, for example
the spinal cord motor neurones. Hutter & Trautwein, and Del Castillo & Katz,
recorded from the pacemaker region of the frog and turtle hearts; they showed
that repetitive stimulation of the vagus, which stopped the heart, caused a
progressive rise in the resting potential, a hyperpolarization as high as 33 mV
being attained. As Hutter & Trautwein pointed out, the resting potential of the
frog's sinus venosus is considerably lower than that of the auricle (55 mV
compared with 70 mV), so that inhibition of the sinus caused it to behave like
the auricle.

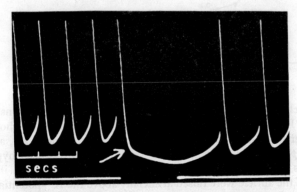

FIG. 696. Vagal inhibition of pacemaker function. Vagal stimulation of spontaneously beating frog sinus venosus, as indicated by break in reference
line. Increase in resting potential indicated by arrow. Note also narrowing
of action potentials after resumption of spontaneity. (Hecht. Ann. N.Y. Acad.
Sci.)

Fig. 696 illustrates the effect of vagal stimulation on the spontaneously beating
frog sinus venosus; block was associated with a rise in resting potential (arrow)
and when stimulation had ceased the next spike arose off a much slower phase of
depolarization, whilst the spikes were narrowed. More recently Toda & West
(1967) have studied the effects of vagal stimulation on pacemaker cells of the
SA-node of the rabbit; it was important to distinguish these cells from three

other types in the same region, mainly on the base of their time relations to the atrial action potentials. When the true pacemaker cells—their Type 3 cells—were examined, vagal stimulation barely affected resting potential and spike-height, the negative chronotropy being due to a great reduction in the slope of diastolic depolarization.

Increased K⁺-Permeability

The inhibitory effect is brought about mainly by an increased permeability to K^+ that brings the membrane potential into closer approximation to the K^+ equilibrium potential. This point is well demonstrated by Fig. 684, p. 1282, where the variation of the resting potential with external K^+-concentration is shown; the circles on the graph indicate the values of the resting potential when the auricular fibres had been treated with acetylcholine, and it is seen that the fibres

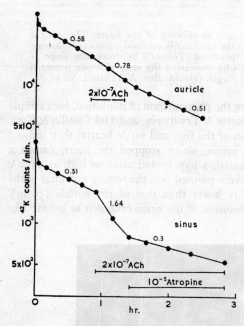

FIG. 697. Effect of acetylcholine on the exchange of ^{42}K between inside and outside of sinus venosus (lower graph) and right auricle (upper graph) of a tortoise heart. The tissues were loaded with the isotope and its rate of loss measured. The figures close to the graphs give the rate-constants, in hr. $^{-1}$, at different stages of the experiment. (Harris & Hutter from Hutter. Brit. med. Bull.)

behave much more closely as K^+-electrodes under these conditions (Trautwein & Dudel). Furthermore, the hyperpolarizing action of acetylcholine on the auricle reverses its direction at the K^+-equilibrium potential, i.e., acetylcholine causes a *depolarization* when the membrane potential has been hyperpolarized beyond this equilibrium potential (Trautwein & Dudel). Again, if the resting potential of the auricle were raised artificially above the K^+-potential, by anodic polarization, then an increased K^+-permeability should accelerate the return to normal, i.e., it should *depolarize*, and this was observed when acetylcholine was applied (Trautwein & Dudel).

The inhibitory action of a nerve impulse may well be completely accounted for by this specific increase in permeability to K^+; thus, moderate stimulation causes a slowing of the heart, associated with a reduction in the rate of rise of the pacemaker potential. An increased permeability to K^+ would oppose the pacemaker depolarization, especially if this were due to a specific increase in sodium permeability; Draper & Weidmann's finding that the rate of rise of this pacemaker potential may be decreased by perfusing with Na^+-poor solutions certainly suggests, moreover, that an increase of Na^+-permeability is at the basis of this change. Stronger stimulation causes heart-block; under these conditions, stimulation of the sinus electrically has no effect, the tissue being completely refractory; presumably the permeability to K^+ is so great that any increase in Na^+-permeability is inadequate to depolarize, influx of Na^+ being balanced by an equally rapid outflux of K^+. Similar results were obtained on the action potential of the auricular fibres as a result of vagal stimulation; the resting potential remained constant, whilst the action potential either failed, or was reduced in size and shortened in duration. Corresponding with these changes we should expect to find a decrease in membrane resistance, and an increased rate of exchange of ^{42}K, as a result of vagal stimulation or treatment with acetylcholine; both these predictions are verified (Trautwein, Kuffler & Edwards, 1956; Harris & Hutter, 1956; Danielson, 1964); Fig. 697 shows the remarkable increase in the rate of escape of ^{42}K from the sinus venosus on treatment with acetylcholine.

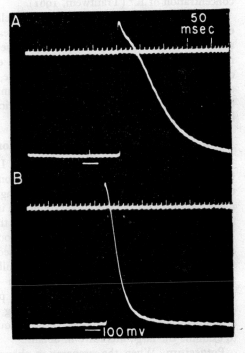

FIG. 698. A. Control action potential recorded from a single fibre of the isolated dog auricle. B. Same auricle in presence of 5.5×10^{-4} M acetylcholine. Note absence of any decrease in amplitude of resting or action potentials. (Hoffman & Suckling. *Amer. J. Physiol.*)

With the auricle, as Hoffman & Suckling showed, because the membrane potential, E_m, is close to E_K, the effect of inhibitory impulses is largely on the rate of repolarization, which increases, the resting potential and spike-height being largely unaffected (Fig. 698). The slowing here must therefore be due to a slowing of the diastolic depolarization phase, which depends on a decrease in K^+-permeability (Fig. 695, B). Mammalian ventricular tissue is unaffected by acetylcholine; Ware & Graham (1967) showed that the frog's ventricular action potential was shortened and made more like an atrial action potential when treated with acetylcholine; and this was associated with a *negative inotropic* effect, in the sense that the force of contraction was reduced.

Acceleration. The heart-rate may be increased by stimulation of the sympathetic nerve supply, or by perfusing with adrenaline; under these conditions, the pacemaker depolarizes more rapidly, so that the threshold for firing the spike is reached sooner; associated with this, there is an increase in spike-height, *i.e.*, the size of the overshoot, and also in the rate of rise of the spike. With the quiescent sinus, or Purkinje fibres, adrenaline may cause excitation, which begins as an oscillating depolarization that eventually exceeds threshold (Otsuka, 1958); furthermore, abnormally small action potentials may be converted to their normal form (Trautwein, 1957; Otsuka, 1958; Trautwein & Schmidt, 1960). Thus an explanation of the action of the sympathetic or adrenaline must take into account the increase in steepness of the diastolic depolarization (pacemaker), the hyperpolarization of the resting membrane, and the increase in the spike-overshoot and rate of rise.*

Efficiency of Na+-Pump

Explanations based on increased K+- or Na+-permeability fail to cover all these points, and it has been suggested that the effect is rather to improve the efficiency of the active-transport mechanisms maintaining the normal high concentration of K+ (Trautwein, 1961).

This view is partly supported by Waddell's measurements of the influx and efflux of ^{42}K of control and adrenaline-treated auricles; influx was increased appreciably from about 4 to 7 $p$mole/cm^2 sec whilst efflux was not increased so much; this difference between influx and efflux accounts for the increase in concentration of K+ caused by adrenaline, and is presumably due to more efficient working of the Na+-pump. The small increase in efflux that occurs in both quiescent (left) and active (right) auricles may, however, indicate a small increase in passive permeability to K+.

The Effects of Altered Ionic Composition

The classical experiments of Ringer and Locke showed that an excised heart would only continue to beat if its perfusion medium contained Na+, K+ and Ca++ in appropriate proportions. In analysing the effects of altered ionic environment we must clearly take into account the influence on the various parameters governing Na+- and K+-permeabilities; thus we have seen that a raised K+-concentration will decrease the overshoot because of the increased degree of inactivation of the Na+-mechanism caused by the reduced resting potential; the lowered level of the resting potential will also affect the other phases of the action potential—plateau and repolarization—and also the rate of depolarization during the development of the pacemaker potential. Clearly, then, the effects of variations in ionic environment are going to be complex, and in the absence of adequate experimental studies on the various parameters it becomes premature to examine the large literature in detail.

Potassium. When the preparation shows pacemaker activity, lowered K+ causes an increase in the frequency, owing to both the lowered resting potential and an increased rate of development of the pacemaker potential. Increased K+ can also do this, although the usual effect is a loss of spontaneous activity due to depolarization. The effects on the action potential, namely, decreased rate of rise and amount of overshoot, caused by both high and low K+, are also probably related to the depolarization, since, as we have seen, a maintained depolarization inactivates the Na+-permeability system.†

* According to Reuter (1965), adrenaline increases the rates of uptake and release of ^{45}Ca by the guinea pig's auricle, so that the increased *force* of contraction—positive inotropic action—may be due to this, rather than to any effect on excitability. When contracture is induced in the depolarized muscle, *i.e.*, when the contractile mechanism is activated independently of the action-potential, adrenaline may decrease the force of contraction due, presumably to the diminished entry of Ca++ (Graham & Lamb, 1968).

† The effects of Rb+ on the Purkinje fibre are not simple; as with other excitable tissues it depresses the resting potential and, according to Müller (1965), it is less effective than

Calcium. The effects of K$^+$ cannot be considered in isolation from those Ca^{++}, since the depolarizing action of a low K$^+$-concentration may be mitigated by reducing the Ca^{++}-concentration, whilst the effects of excess K$^+$ can be counteracted by raising the concentration of Ca^{++} too. For example, with a concentration of Ca^{++} of o·27 mM, raising the K$^+$-concentration to 8·0 mM lowered the resting potential from 90 to 70 mV, but with a Ca^{++}-concentration of 8·1 mM the resting potential was normal, in spite of the raised K$^+$-concentration. These results emphasize that the resting potential is not simply a potassium potential, otherwise it would be uninfluenced by the concentration of Ca^{++}.

Threshold

In general, the effects of Ca^{++} are not easy to summarize, or to explain, since they vary according to the type of heart tissue examined. We have seen that Ca^{++}-lack increases spontaneous activity, be it in isolated axon, neuro-muscular junction, or ganglionic synapse. The effects are predictable in accordance with the Hodgkin-Huxley equations on the basis of an increase in the velocity constants determining Na$^+$- and K$^+$-permeabilities as a function of membrane potential. It seems likely that a similar mechanism will explain some of the effects on the heart. Weidmann (1955) showed, by using two intracellular electrodes—one to stimulate and one to record—that a raised Ca^{++}-concentration increased the threshold, in the sense that a greater depolarization was required to initiate a spike; the reverse effect was obtained with lowered Ca^{++}. Since changes of Ca^{++}-concentration produced no change in the rate of spontaneous depolarization in pacemaker regions, the increased rate of heart activity that occurred with lowered Ca^{++} was due to the lowered threshold, a smaller amount of spontaneous depolarization being necessary to set off a spike. So far as the Na$^+$-activation mechanisms were concerned, the effects of high and low Ca^{++}-concentrations were similar to those on the squid.

Ca^{++}- Na$^+$-Antagonism. Ca^{++} and Na$^+$ frequently exhibit an antagonism in their actions, as though they both compete for the same membrane- or carrier-site; thus when the force of contraction of ventricular muscle is studied, this is found to be closely related to the ratio: $[Ca^{++}]/[Na^+]^2$ suggesting competition for the same anionic site; Stanley & Reiter (1965) found that the duration of the action potential was proportional to the same ratio, suggesting antagonistic effects of the two ions on the membrane, or competition for a carrier in a current-carrying mechanism. Again Schaer (1968) found mutual competition between

TABLE LXXXII

Negative Chronotropic Effects of Mg^{++} with Varying Concentrations of Ca^{++} and Na$^+$. (Schaer, 1968.)

[Na$^+$] (Per cent. Normal Ringer)	[Ca^{++}] (mM)	Interval Lengthening/mM Mg^{++} (msec./mM)
100	2·5	23·6
100	0·62	37·4
50	0·62	77·2
50	0·16	136·1
25	0·16	293

K$^+$, indicating a smaller permeability. Rb$^+$ increased the efflux of ^{42}K from the fibre and thus shortened the action potential; on the other hand Rb$^+$ decreased the influx of ^{42}K more effectively than K$^+$. The results are consistent with a carrier-mediated permeability to K$^+$, with Rb$^+$ having a higher affinity for the carrier than K$^+$, provided several *ad hoc* hypotheses are made.

Ca^{++}, Na^+ and Mg^{++} when he studied the rate of beat of the guinea pig heart; as Table LXXXII shows, the negative chronotropic effects of Mg^{++} are greater the smaller the concentrations of Ca^{++} and of Na^+.

Calcium Current. Niedergerke & Orkand (1966) have examined the effects of Ca^{++} on the spike in the frog's heart, working from the hypothesis that some of the inward current might be carried by Ca^{++}; such an event would be consistent with the observed influx of ^{45}Ca with each beat (Niedergerke, 1963). They found that the size of the spike-overshoot and of the resting potential varied with $[Ca^{++}]$ (Fig. 699), effects that could be explained on the basis of a participation of Ca^{++}-ions in the spike at a time when the Na^+- and K^+-permeabilities were diminishing. However, when repetitive stimulation was used, the increase in overshoot caused by high $[Ca^{++}]$ was inhibited, whilst in low $[Ca^{++}]$, repetitive stimulation was nearly without effect. Evidence against the Ca^{++}-carrying hypothesis is provided by Dudel, Peper & Trautwein (1966), who measured the currents in voltage-clamped Purkinje fibres in a Na^+-free solution; they considered that the current

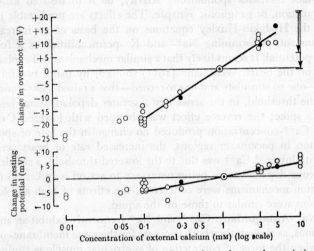

FIG. 699. Changes in resting potential and overshoot due to the variation of the external calcium concentration. Effects are plotted as the potential difference with respect to the control potentials (in 1 mM-Ca Ringer). The two arrows represent results of two experiments with 10 mM-Ca-Ringer, length of arrows indicates range of values, their direction the change of the overshoot with time. (Niedergerke & Orkand. *J. Physiol.*)

measured under these conditions was the delayed K^+-current, so that if Ca^{++} could carry an appreciable inward current, increasing its concentration should neutralize the outwardly directed K^+-current. In fact, Ca^{++} had no effect on the current, although there was a marked change in the current-voltage relation when all Ca^{++} was removed, but this was shown not to be due to withdrawal of inward Ca^{++}-current but rather to a change in membrane conductance.*

Calcium and Contraction. The movements of Ca^{++} into and out of contractile tissue are of significance mainly because the concentration of ionized Ca^{++} within the contractile cell is related to the activation of the contractile process; the matter will be discussed in detail in Chapter XXI, and here it is

* Reuter (1966) found that the concentration of Ca^{++} affected the current-voltage relation in Purkinje fibres, but decided that there was no effect on g_K since Ba^{++}, which is said to reduce g_K, had a qualitatively different effect from that of Ca^{++}. In his later (1967, 1968) studies he concluded in favour of an inward Ca^{++}-current, significant presumably for excitation-contraction coupling (p. 1397). The fluxes into atrial mammalian and ventricular tissue have been measured recently by Reuter & Seitz (1968); once again the ratio of $[Ca^{++}]/[Na^+]^2$ assumes significance presumably because of the competition for a carrier that has affinities for both ions.

sufficient to emphasize that cardiac muscle does not differ fundamentally from skeletal muscle in respect to the process of excitation-contraction coupling, *i.e.*, the process by which the spread of an action potential can lead to the activation of the contractile machinery of the muscle fibre. Thus the extra influx of Ca^{++} during the action potential, described by Winegrad & Shanes (1962), and Niedergerke (1963) is probably the essential step in the coupling; alternatively, or additionally, the fundamental process may be the release of internal Ca^{++} from its binding on membranous sites within the cell; this release would increase its exchangeability with isotopic Ca^{++} and thus account for the shortening of the half-life of exchange that occurs with activity, described, for example, by Hoditz & Lüllman (1964).*

Anions. As indicated earlier, the effects of anionic substitutions are predictable in terms of their effects on the "leak" or anion-conductance, and thus exert a critical action on the phases where the inward and outward currents are nearly balanced, *i.e.*, on the plateau and the pacemaker potential.

Nitrate

Thus nitrate is considered to have a higher permeability than that of Cl^-; if the anion-equilibrium potential is lower than the K^+-equilibrium potential, then an increased anion permeability will have a depolarizing effect, *i.e.*, it will increase the inward current. This should increase the frequency of beat, as found. When anion permeability becomes very high, on the other hand, the increased permeability will tend to stabilize the resting potential at the anion-potential and so abolish pacemaker activity (Noble, 1962). When the tissue is exposed for some time to nitrate-Ringer the concentration of Na^+ in the fibres rises, and so the magnitude of the Na^+-potential, and thus the degree of overshoot, will decrease on this count (Petersen & Feigen).

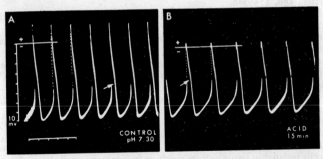

FIG. 700. Showing effects of acid pH on threshold and slope-change of pacemaker fibres. Excess hydrogen ions of the bathing solution raises threshold and decreases the slope of diastolic depolarization thereby slowing the rate or decreasing automaticity. The less negative threshold value is responsible for the obvious change in rise-time and the decrease of overshoot. (Hecht. *Ann. N.Y. Acad. Sci.*)

pH. Hecht (1965) has discussed the effects of pH on cardiac muscle; as Fig. 700 shows, an increased concentration of H^+-ions suppresses spontaneous activity, the effect being to raise the threshold and to decrease the slope of diastolic depolarization, thereby slowing the rate, or decreasing automaticity.

* Imai & Takeda (1967) compared the behaviour of cardiac and smooth muscle in media with very low concentrations of Ca^{++}; at a concentration of less than 10^{-7} M development of tension in cardiac muscle was completely abolished and they concluded that cardiac muscle relied exclusively on influx of Ca^{++} from outside for the development of tension; with smooth muscle it was impossible to abolish development of tension entirely, so that release from stores was considered to be a factor in addition to influx from outside.

Antifibrillary Drugs. Fibrillation in an auricle represents a condition of hyperexcitability, so that the cycle of events in the development of the action potential from the pacemaker potential takes place more rapidly. Quinidine is an effective agent in restoring the beat to normal, and its effects seem to be due to the slowing of the rate of rise of the spike, associated with a raised threshold and prolongation of the effective refractory period. Thus the drug has no effect on the resting potential nor yet on the intracellular concentration of K^+ (Goodford & Vaughan Williams); it shifts the Na^+-activation curves of Fig. 687 to the right, so that at a given resting potential the inactivation of the Na^+-mechanism is greater than normal; thus the amount of Na^+-activation necessary to start a spike takes longer to develop, *i.e.*, the phase of pacemaker depolarization lasts longer (Szekeres & Vaughan Williams).

REFERENCES

ADRIAN, R. H., CHANDLER, W. K. & HODGKIN, A. L. (1968). Voltage clamp experiments in striated muscle fibers. *J. gen. Physiol.*, **51**, 188–192 s.

ARVANITAKI, A. (1938). Étude expérimentale sur le myocarde d'hélix. *Actualités sci. industr.*, **762**.

BARR, L., DEWEY, M. M. & BERGER, W. (1968). Propagation of action potentials and the structure of the nexus in cardiac muscle. *J. gen. Physiol.*, **48**, 797–823.

BOZLER, E. (1954). The initiation of impulses in cardiac muscle. *Amer. J. Physiol.*, **138**, 273–282.

BRADY, A. J. & TAN, S. T. (1966). The ionic dependence of cardiac excitability and contractility. *J. gen. Physiol.*, **49**, 781–791.

BRADY, A. J., & WOODBURY, J. W. (1960). The sodium-potassium hypothesis as the basis of electrical action in frog ventricle. *J. Physiol.*, **154**, 385–407.

BRIGGS, A. H., & HOLLAND, W. C. (1961). Transmembrane fluxes in ventricular fibrillation. *Amer. J. Physiol.*, **200**, 122–124.

BURROWS, R., & LAMB, J. F. (1962). Sodium and potassium fluxes in cells cultured from chick embryo heart muscle. *J. Physiol.*, **162**, 510–531.

CARMELIET, E. E. (1961). Chloride ions and the membrane potential of Purkinje fibres. *J. Physiol.*, **156**, 375–388.

CASTEELS, R. G. (1962). Effect of sodium-deficiency on the membrane activity of the frog's heart. *Arch. int. Physiol.*, **70**, 599–610.

CORABOEUF, E., & WEIDMANN, S. (1954). "Temperature Effects on the Electrical Activity of Purkinje Fibres." *Helv. physiol. acta*, **12**, 32–41.

CRANEFIELD, P. F., EYSTER, J. A. E., & GILSON, W. E. (1951). "Effects of Reduction of External Sodium Chloride on the Injury Potentials of Cardiac Muscle." *Amer. J. Physiol.*, **166**, 269–272.

DANIELSON, B. G. (1964). The distribution of some electrolytes in the heart. *Acta physiol. scand.*, **62**, Suppl. 236.

DECK, K. A. & TRAUTWEIN, W. (1964). Ionic currents in cardiac excitation. *Pflüg. Arch. ges. Physiol.*, **280**, 63–80.

DEHAAN, R. L. & GOTTLIEB, S. H. (1968). The electrical activity of embryonic chick heart cells isolated in tissue culture singly or in interconnected sheets. *J. gen. Physiol.*, **52**, 643–665.

DEL CASTILLO, J., & KATZ, B. (1955). "Production of Membrane Potential Changes in the Frog's Heart by Inhibitory Nerve Impulses." *Nature*, **175**, 1035.

DÉLÈZE, J. (1959). Effects of K-rich and Na-deficient solutions on transmembrane potentials. *Circulation Res.*, **7**, 461–465.

DEWEY, M. M. (1969). The structure and function of the intercalated disc in vertebrate cardiac muscle. *Comparative Physiology, the Heart; Current Trends. Experientia*, Suppl. 15, pp. 10–28.

DEWEY, M. M. & BARR, L. (1964). A study of the structure and distribution of the nexus. *J. Cell Biol.*, **23**, 553–585.

DRAPER, M. H., & WEIDMANN, S. (1951). "Cardiac Resting and Action Potentials recorded with an Intracellular Electrode." *J. Physiol.* **115**, 74–94

DUDEL, J., PEPER, K. & TRAUTWEIN, W. (1966). The contribution of Ca^{++} ions to the current voltage relation in cardiac muscle (Purkinje fibres). *Pflüg. Arch. ges. Physiol.*, **288**, 262–281.

DUDEL, J., PEPER, K., RÜDEL, R. & TRAUTWEIN, W. (1967, a). The dynamic chloride component of membrane current in Purkinje fibers. *Pflüg. Arch. ges. Physiol.*, **295**, 197–212.

DUDEL, J., PEPER, K., RÜDEL, R. & TRAUTWEIN, W. (1967, b). The effect of tetrodotoxin on the membrane current in cardiac muscle. (Purkinje fibers). *Pflüg. Arch. ges. Physiol.*, **295**, 213–226.

DUDEL, J., PEPER, K., RÜDEL, R. & TRAUTWEIN, W. (1967, c). The potassium component of membrane current in Purkinje fibers. *Pflüg. Arch.* **296**, 308–327.

EYSTER, J. A. E., MEEK, W. J., GOLDBERG, H., & GILSON, W. E. (1938). "Potential Changes in an Injured Region of Cardiac Muscle." *Amer. J. Physiol.*, **124**, 717–728.

FÄNGE, R., PERSSON, H. & THESLEFF, S. (1956). Electrophysiologic and pharmacological observations on trypsin-disintegrated embryonic chick hearts cultured *in vitro*. *Acta physiol. scand.*, **38**, 173–183.

FRANKENHAEUSER, B. (1962). Delayed currents in myelinated nerve fibres of *Xenopus lævis* investigated with voltage clamp technique .*J. Physiol.*, **160**, 40–45.

GOERKE, J. & PAGE, E. (1965). Cat heart muscle *in vitro*. VI. Potassium exchange in papillary muscles. *J. gen. Physiol.*, **48**, 933–948.

GOODFORD, P. J. & LÜLLMANN, H. (1962). The uptake of ethanesulphonate -35S ions by muscular tissue. *J. Physiol.*, **161**, 54–61.

GOODFORD, P. J., & VAUGHAN WILLIAMS, E. M. (1962). Intracellular Na- and K- concentrations of rabbit atria, in relation to the action of quinidine. *J. Physiol.*, **160**, 483–493.

GRAHAM, J. A. & LAMB, J. F. (1968). The effect of adrenaline on the tension developed in contractures and twitches of the ventricle of the frog. *J. Physiol.*, **197**, 479–509.

GROSS-SCHULTE, E., & TRAUTWEIN, W. (1960). Der Einfluss der extracellulären Natriumkonzentration auf das Membranpotential der spontätigen Vorhoffaser des Herzens. *Pflüg. Arch.*, **272**, 39.

HAAS, H. G., & GLITSCH, H. G. (1962). Kalium-Fluxe am Vorhof des Froschherzens. *Pflüg. Arch.*, **275**, 358–375.

HAAS, H. G., GLITSCH, H. G. & KERN, R. (1964). Zum Problem der gegenseitigen Beeinflussung der Ionenfluxe am Myokard. *Pflüg. Arch. ges. Physiol.*, **281**, 282–299.

HALL, A. E., HUTTER, O. F., & NOBLE, D. (1963). Current-voltage relations of Purkinje fibres in sodium deficient solutions. *J. Physiol.*, **166**, 225–240.

HARRIS, E. J., & HUTTER, O. F. (1956). "The Action of Acetylcholine on the Movements of Potassium Ions in the Sinus Venosus of the Heart." *J. Physiol.*, **133**, 58–59 P.

HECHT, H. H. (1965). Comparative physiological and morphological aspects of pacemaker tissues. *Ann. N.Y. Acad. Sci.*, **127**, 49–83.

HODITZ, H. & LÜLLMANN, H. (1964). Die Calcium—Umsatzgeschwindigkeit ruhender und kontraktierender Vorhofmuskulatur *in vitro*. *Pflüg. Arch. ges. Physiol.*, **280**, 22–29.

HOFFMAN, B. F. (1965). Atrioventricular conduction in mammalian hearts. *Ann. N.Y. Acad. Sci.*, **127**, 105–112.

HOFFMAN, B. F., MOORE, E. N., STUCKEY, J. H. & CRANEFIELD, P. F. (1963). Functional properties of the atrioventricular conduction system. *Circ. Res.*, **13**, 308–328.

HOFFMAN, B. F., & SUCKLING, E. E. (1953). "Cardiac Cellular Potentials: Effect of Vagal Stimulation and Acetylcholine." *Amer. J. Physiol.*, **173**, 312–320.

HOFFMAN, B. F., & SUCKLING, E. E. (1954). "Effect of Heart Rate on Cardiac Membrane Potentials and the Unipolar Electrogram." *Amer. J. Physiol.*, **179**, 123–130.

HUTTER, O. F. (1957). Mode of action of autonomic transmitters on the heart. *Brit. med. Bull.*, **13**, 176–180.

HUTTER, O. F., & NOBLE, D. (1961). Anion conductance of cardiac muscle. *J. Physiol.*, **157**, 335–350.

HUTTER, O. F., & PADSHA, S. M. (1959). Effects of nitrate and other anions on the membrane resistance of frog skeletal muscle. *J. Physiol.*, **146**, 117–132.

HUTTER, O. F., & TRAUTWEIN, W. (1956). "Vagal and Sympathetic Effects on the Pacemaker Fibres in the Sinus Venosus of the Heart." *J. gen. Physiol.*, **39**, 715–733.

IMAI, S. & TAKEDA, K. (1967). Calcium and contraction of heart and smooth muscle. *Nature*, **213**, 1044–1045.

IVERSEN, L. L. & KRAVITZ, E. A. (1966). Sodium dependence of transmitter uptake at adrenergic nerve terminals. *Molec. Pharmacol.*, **2**, 360–362.

JAMES, T. N. (1962). Anatomy of the sinus node of the dog. *Anat. Rec.*, **143**, 251–265.

JAMES, T. N. (1964). Anatomy of the A-V node of the dog. *Anat. Rec.*, **148**, 15–18.

JOHNSON, E. A. & SOMMER, J. R. (1967). A strand of cardiac muscle. Its ultrastructure and the electrophysiological implications of its geometry. *J. Cell Biol.*, **33**, 103–129.

JOHNSON, E. A., & TILLE, J. (1961). Investigations of the electrical properties of cardiac muscle fibres with the aid of intracellular double-barrelled electrodes. *J. gen. Physiol.*, **44**, 443–467.

KANNO, T. & MATSUDA, K. (1968). The effects of external sodium and potassium concentration on the membrane potential of atrioventricular fibers of the toad. *J. gen. Physiol.*, **50**, 243–253.

KEENAN, M. J. & NIEDERGERKE, R. (1967). Intracellular sodium concentration and resting sodium fluxes of the frog heart ventricle. *J. Physiol.*, **188**, 235–260.

KISCH, B. (1956). "The Sarcosomes of the Heart." *J. biophys. biochem. Cytol.*, **2**, Suppl., 361–362.

KRIEBEL, M. E. (1968). Electrical characteristics of tunicate heart cell membranes and nexuses. *J. gen. Physiol.*, **52**, 46–59.

LAMB, J. F. (1961). The chloride content of rat auricle. *J. Physiol.*, **157**, 415–425.

LAMB, J. F. & McGUIGAN, J. A. S. (1968). The efflux of potassium, sodium, chloride, calcium and sulphate ions and of sorbitol and glycerol during the cardiac cycle in frog's ventricle. *J. Physiol.*, **195**, 283–315.

LEHMKUHL, D. & SPERELAKIS, N. (1963). Transmembrane potentials of trypsin-dispersed chick heart cells cultured *in vitro*. *Amer. J. Physiol.*, **205**, 1213–1220.

MARSHALL, J. M. (1957). Effects of low temperatures on transmembrane potentials of single fibres of the rabbit atrium. *Circulation Res.*, **5**, 664–669.

McALLISTER, R. E. & NOBLE, D. (1966). The time and voltage dependence of the slow outward current in cardiac Purkinje fibres. *J. Physiol.*, **186**, 632–662.

McALLISTER, R. E. & NOBLE, D. (1967). The effect of subthreshold potentials on the membrane current in cardiac Purkinje fibres. *J. Physiol.*, **190**, 381–387.

DE MELLO, W. C., & HOFFMAN, B. F. (1960). Potassium ions and electrical activity of specialized cardiac fibres. *Amer. J. Physiol.*, **199**, 1125–1130.

MUIR, A. R. (1957). "An Electron Microscope Study of the Embryology of the Intercalated Disc in the Heart of the Rabbit." *J. biophys. biochem. Cytol.*, **3**, 193–202.

MUIR, A. R. (1957). "Observations on the Fine Structure of the Purkinje Fibres in the Ventricles of the Sheep's Heart." *J. Anat.*, **91**, 251–258.

MÜLLER, P. (1965). Potassium and rubidium exchange across the surface membrane of cardiac Purkinje fibres. *J. Physiol.*, **177**, 453–462.

NANDY, K. & BOURNE, G. H. (1963). A study of the morphology of the conducting tissue in mammalian hearts. *Acta anat.*, **53**, 217–226.

NIEDERGERKE, R. (1963). Movements of Ca in frog heart ventricles at rest and during contractures. *J. Physiol.*, **167**, 515–550.

NIEDERGERKE, R. (1963). Movements of Ca in beating ventricles of the frog heart. *J. Physiol.*, **167**, 551–580.

NIEDERGERKE, R. & ORKAND, R. K. (1966). The dual effect of calcium on the action potential of the frog's heart. *J. Physiol.*, **184**, 291–311.

NOBLE, D. (1962). A modification of the Hodgkin-Huxley equations applicable to Purkinje fibre action and pace-maker potentials. *J. Physiol.*, **160**, 317–352.

NOBLE, D. (1962). The voltage dependence of the cardiac membrane conductance. *Biophys. J.*, **2**, 381–393.

NOBLE, D. & TSIEN, R. W. (1968). The kinetics and rectifier properties of the slow potassium current in cardiac Purkinje fibres. *J. Physiol.*, **195**, 185–214.

ORSUKA, M. (1958). Die Wirkung von Adrenalin auf Purkinje-Fasern von Säugetieren. *Pflüg. Arch. ges. Physiol.*, **266**, 512–517.

PAES DE CARVALHO, A. & DE ALMEIDA, D. F. (1960). Spread of activity through the atrioventricular node. *Circulation Res.*, **8**, 801–809.

PAES DE CARVALHO, A., DE MELLO, W. C. & HOFFMAN, B. F. (1959). Electrophysiological evidence for specialized fiber types in rabbit atrium. *Amer. J. Physiol.*, **196**, 483–488.

PAGE, E. (1962). Cat heart muscle *in vitro*. II. The steady state resting potential in quiescent papillary muscles. *J. gen. Physiol.*, **46**, 189–199.

PAGE, E. (1962). Cat heart muscle *in vitro*. III. The extracellular space. *J. gen. Physiol.*, **46**, 201–213.

PAGE, E. (1965). Ion movement in heart muscle: tissue compartments and the experimental definition of driving forces. *Ann. N.Y. Acad. Sci.*, **127**, 34–48.

PAGE, E. & BERNSTEIN, R. S. (1964). Cat heart muscle *in vitro*. V. Diffusion through a sheet of right ventricle. *J. gen. Physiol.*, **47**, 1129–1140.

PAGE, E. & STORM, S. R. (1965). Cat heart muscle *in vitro*. VIII. Active transport of sodium in papillary muscles. *J. gen. Physiol.*, **48**, 957–972.

PATTEN, B. M. (1950). *Early Embryology of the Chick*. London: K. K. Lewis.

PETERSEN, N. S., & FEIGEN, G. A. (1962). Effect of [NO₃] on atrial action potentials and contraction as modified by [Na] and [Ca]. *Amer. J. Physiol.*, **202**, 950–956.

RAYNER, B., & WEATHERALL, M. (1959). Acetylcholine and potassium movements in rabbit auricles. *J. Physiol.*, **146**, 392–409.

REUTER, H. (1965). Uber die Wirkung von Adrenalin auf den cellulären Ca-Umsatz des Meerschweinchenvorhofs. *Arch. exp. Path. Pharmak.*, **251**, 401–412.

REUTER, H. (1966). Strom-Spannungsbeziehungen von Purkinje-Fasern bei verschiedenen extracellularen Calcium-Konzentrationen und unter Adrenalinwirkung. *Pflüg. Arch. ges. Physiol.*, **287**, 357–367.

REUTER, H. (1967). The dependence of slow inward current in Purkinje fibres on the extracellular calcium-concentration. *J. Physiol.*, **192**, 479–492.

REUTER, H. & SCHOLZ, H. (1968). Uber den Einfluss der extracellulären Ca-Konzentration auf Membranpotential und Kontraktion isolierter Herzpräparate bei graduierter Depolarisation. *Pflüg. Arch. ges. Physiol.*, **300**, 87–107.

REUTER, H. & SEITZ, N. (1968). The dependence of calcium efflux from cardiac muscle on temperature and external ion composition. *J. Physiol.*, **195**, 451–470.

RHODIN, J. A. G., DEL MISSIER, P., & REID, L. C. (1961). The structure of the specialized impulse-conducting system of the steer heart. *Circulation*, **24**, 349–367.

SCHAER, H. (1968). Antagonische Wirkungen von Magnesium-, Calcium- und Natriumionen auf die Impulsbildung im Sinusknoten des Meerschweinchenherzens. *Pflüg. Arch. ges. Physiol.*, **298**, 359–371.

SJÖSTRAND, F. S. & ANDERSSON, E. (1954). Electron microscopy of the intercalated discs of cardiac muscle tissue. *Experientia*, **10**, 369–370.

SJÖSTRAND, F. S., ANDERSSON-CEDERGREN, E., & DEWEY, M. M. (1958). The ultra-structure of the intercalated discs of frog, mouse and guinea pig cardiac muscle. *J. Ultrastr. Res.*, **1**, 271–287.

SOMMER, J. R. & JOHNSON, E. A. (1968). Cardiac muscle. A comparative study of Purkinje fibers and ventricular fibers. *J. Cell Biol.*, **36**, 497–526.

SPERELAKIS, N. & LEHMKUHL, D. (1964). Effect of current on transmembrane potentials in cultured chick heart cells. *J. gen. Physiol.*, **47**, 895–927.

SPERELAKIS, N. & LEHMKUHL, D. (1966). Ionic interconversion of pacemaker and non-pacemaker cultured chick heart cells. *J. gen. Physiol.*, **49**, 867–895.

STANLEY, E. J. & REITER, M. (1965). The antagonistic effects of sodium and calcium on the action potential of guinea pig papillary muscle. *Arch. exp. Path. Pharmak.*, **252**, 159–172.

SZEKERES, L., & VAUGHAN WILLIAMS, E. M. (1962). Antifibrillatory action. *J. Physiol.*, **160**, 470–482.

TARR, M. & SPERELAKIS, N. (1964). Weak electrotonic interaction between contiguous cardiac cells. *Amer. J. Physiol.*, **207**, 691–700.

TARR, M. & SPERELAKIS, N. (1967). Decreased intercellular resistance during spontaneous depolarization in myocardium. *Amer. J. Physiol.*, **212**, 1503–1511.

TILLE, J. (1966). Electrotonic interaction between muscle fibers in the rabbit ventricle. *J. gen. Physiol.*, **50**, 189–202.

TODA, N. (1968). Influence of sodium ions on the membrane potential of the sino-atrial node in response to sympathetic nerve stimulation. *J. Physiol.*, **196**, 677–691.

TODA, N. & WEST, T. C. (1967). Interactions of K, Na, and vagal stimulation in the S-A node of the rabbit. *Amer. J. Physiol.*, **212**, 416–423.

TRAUTWEIN, W. (1957). In *Rhythmustörungen des Herzens*. Ed. K. Spang. Stuttgart: Thieme.

TRAUTWEIN, W. (1961). Elektrophysiologie der Herzmuskelfaser. *Ergebn. Physiol.*, **51**, 131–198.

TRAUTWEIN, W., & DUDEL, J. (1958). Zum Mechanismus der Membranwirkung der Acetylcholin an der Herzmuskelfaser. *Pflüg. Arch.*, **266**, 324–334.

TRAUTWEIN, W., & DUDEL, J. (1958). Hemmende und "erregende" Wirkungen des Acetylcholin am Warmblüterherzen. Zur Frage der spontanen Erregungsbild. *Pflüg. Arch.*, **266**, 653–664.

TRAUTWEIN, W., & KASSEBAUM, D. G. (1961). On the mechanism of spontaneous impulse generation in the pacemaker of the heart. *J. gen. Physiol.*, **45**, 317–330.

TRAUTWEIN, W., KUFFLER, S. W., & EDWARDS, C. (1956). "Changes in Membrane Characteristics of Heart Muscle during Inhibition." *J. gen. Physiol.*, **40**, 135–145.

TRAUTWEIN, W., & SCHMIDT, R. F. (1960). Zur Membranwirkung des Adrenalins an der Herzmuskelfaser. *Pflüg. Arch.*, **271**, 715–726.

TRAUTWEIN, W. & UCHIZONO, K. (1963). Electron microscopic and electrophysiologic study of the pacemaker in the sino-atrial node of the rabbit heart. *Z. Zellforsch.*, **61**, 96–109.

TRUEX, R. C. & SMYTHE, M. Q. (1965). Comparative morphology of the cardiac conduction tissue in animals. *Ann. N.Y. Acad. Sci.*, **127**, 19–33.

VAN BREEMEN, V. L. (1953). Intercalated discs in heart muscle studied with the electron microscope. *Anat. Rec.*, **117**, 49–56.

VAN MIEROP, L. H. S. (1967). Location of pacemaker in chick embryo heart at the time of initiation of heartbeat. *Amer. J. Physiol.*, **212**, 407–415.

VASSALLE, M. (1965). Cardiac pacemaker potentials at different extra- and intracellular K concentrations. *Amer. J. Physiol.*, **208**, 770–775.

VASSALLE, M. (1966). Analysis of cardiac pacemaker potential using a " voltage clamp " technique. *Amer. J. Physiol.*, **210**, 1335–1341.

VAUGHAN WILLIAMS, E. M. (1959). The effect of changes in extracellular potassium concentration on the intracellular potentials of isolated rabbit atria. *J. Physiol.*, **146**, 411–427.

VERDONCK, F., DE CLERCQ, D. & CARMELIET, E. (1965). Intracellular Cl concentration in frog ventricle as a function of the intracellular Na and Cl concentration. *Arch. int. Physiol.*, **73**, 381–382.

WADDELL, A. W. (1961). Adrenaline, noradrenaline and potassium fluxes in rabbit auricles. *J. Physiol.*, **155**, 209–220.

WARE, F. & GRAHAM, G. D. (1967). Effects of acetylcholine on transmembrane potentials in frog ventricle. *Amer. J. Physiol.*, **212**, 451–455.

WEIDMANN, S. (1951). Effect of current flow on the membrane potential of cardiac muscle. *J. Physiol.*, **115**, 227–236.

WEIDMANN, S. (1955). "Effects of Calcium Ions and Local Anæsthetics on Electrical Properties of Purkinje Fibres." *J. Physiol.*, **129**, 568–582.

WEIDMANN, S. (1956). *Elektrophysiologie der Herzmuskelfaser.* Huber. Bern.

WEIDMANN, S. (1957). "Resting and Action Potentials of Cardiac Muscle." *Ann. N.Y. Acad. Sci.*, **65**, 663–678.

WEIDMANN, S. (1966). The diffusion of radiopotassium across intercalated disks of mammalian cardiac muscle. *J. Physiol.*, **187**, 323–342.

WILBRANDT, W. & KOLLER, H. (1948). Die Calciumwirkung am Froschherzen als Funktion des Ionengleichgewichts Zwischen Zellmembran und Umgebung. *Helv. Physiol. acta*, **6**, 208–221.

WINEGRAD, S., & SHANES, A. M. (1962). Calcium flux and contractility in guinea-pig atria. *J. gen. Physiol.*, **45**, 371–394.

WRIGHT, E. B., & OGATA, M. (1961). Action potential of amphibian single auricular muscle fibre: a dual response. *Amer. J. Physiol.*, **201**, 1101–1108.

YEH, B. K. & HOFFMAN, B. F. (1968). The ionic basis of electrical activity in embryonic cardiac muscle. *J. gen. Physiol.*, **52**, 666–681.

CHAPTER XIX

ELECTRICAL ACTIVITY IN SMOOTH MUSCLE

SMOOTH, or plain, muscle, in contrast to striated or voluntary muscle, is typified by the sheets of contractile tissue surrounding the viscera and blood vessels. Anatomically, the muscles are characterized by the extreme shortness of their fibres—in the arterial wall, for example, they are only 0·02 mm. long (the "average length" for mammalian tissues is given as 50–150 μ and the breadth 5–50 μ in the central region)—and, physiologically, by their slow, sustained, and spontaneous contractions which contrast with the rapid twitches of most voluntary fibres—twitches that normally only take place in response to a nerve stimulus. Moreover, smooth muscle is characterized by a "tonus" that permits the maintenance of a sustained and powerful contraction over long periods of time with a very small expenditure of energy. The fact that the smooth muscle of the gut or ureter shows spontaneous activity, independently of nervous stimuli, does not mean, however, that *in situ* this activity is not well co-ordinated; in fact, the gut is very well supplied with nerves which are both excitatory and inhibitory, and the strong susceptibility of the smooth muscle of many organs to drugs such as acetylcholine, and blocking agents such as nicotine, indicates that nervous activity is important in initiating and controlling contractions.

Classification of Smooth Muscle

A great deal of confusion, or at any rate of conflict, has been caused by the separation, on a gross morphological basis, of contractile tissue into striated and non-striated muscles; thus Eccles has supported his thesis of the essential similarity in electrical characteristics in the two types by citing a smooth muscle found in the invertebrate *Beroe* which, but for its absence of striations, has indeed many, if not all, of the electrical characteristics of mammalian skeletal muscle. Again, his own work was largely carried out on a sheet of smooth muscle fibres, the nictitating membrane of the cat, an effector which is by no means typical of smooth muscle generally. Bozler has suggested, on the basis of his pioneering studies, that, from a functional point of view, this classification of muscles into the striated and smooth varieties is unsound, and only tends to obscure the issue; each muscle must be considered on its merits and its function ascertained in terms of (*a*) the nerve supply to the fibres, (*b*) the nature of the chemical transmitter substance, *i.e.*, acetylcholine or adrenaline, and (*c*) the electrical characteristics of the tissue. Thus, there seems little doubt that the nictitating membrane is either electrically inexcitable or very nearly so, but one cannot conclude from this that all smooth muscle is electrically inexcitable.

Single- and Multi-Unit Muscles

According to Bozler, a more reasonable classification is on the basis of electrical behaviour; certain smooth muscles behave as "multi-unit" tissues like striated voluntary muscle, in the sense that their electrical and mechanical responses to stimulation can be analysed into changes taking place in functionally separate fibres, responding in an all-or-none fashion to stimulation.* Other

* Bozler includes the musculature of the blood vessels in the multi-unit type but points out that the spontaneously beating non-innervated veins of the bat must be included in the visceral group (Mislin, 1948). That the muscles in the small arteries of frogs may be syncytial is suggested by the work of Fulton & Lutz (1942) who showed that stimulation of a minute nerve caused a contraction of the arteries limited to a small

muscles, on the other hand, behave as though they consist of a single unit, or at any rate multiples of large units, a *syncytium*, the individual fibres being in functional, if not actual, protoplasmic continuity with each other. This description of the electrical aspects of smooth muscle may therefore be profitably divided into two aspects, (*a*) the multi-unit muscle, typified by some lamellibranch muscles, in which the evidence for a syncytium is doubtful, and where activity is largely governed by discharges in true motor nerves, and (*b*) what Bozler calls the visceral type, exemplified by the muscles of the intestine, the pregnant uterus, and ureter—muscles that are spontaneously active by virtue of their own inherent rhythmicity; the nerves, with which they are well supplied, mainly modifying excitability (according to Bozler). Amongst the visceral muscles there is a wide spectrum of behaviour, from the vas deferens, with features of excitability very similar to those of skeletal muscle, to intestinal muscle, in which behaviour may be characterized as essentially syncytial.

In the present chapter it is convenient to concentrate attention on vertebrate smooth muscle; some interesting features of invertebrate smooth muscle, exhibiting the "catch mechanism", will be described in Chapter XXII.

THE VISCERAL TYPE OF MUSCLE

Structure. Forming as they do a large part of the structure of the wall of the gastro-intestinal tract, the uterus, and the ureter, the smooth muscles of these tissues are responsible for their powers of contraction which takes the form, characteristically, of peristaltic movements, as a result of which the contents of the hollow organ are driven in a preferred direction. In the electron-microscope, the smooth muscle fibres are seen as spindle-shaped cells with ellipsoidal nuclei (Fig. 701, Pl. LXI); the cytoplasm is filled with *myofilaments*, contractile elements probably consisting of actomyosin (p. 1353) some 100–200A in diameter. They contrast with the myofilaments of voluntary muscle by the absence of cross-striations and their rather irregular arrangement. The well ordered hexagonal packing, to give myofibrils, seen in the voluntary muscles is absent, so that the cell consists really of a single bundle of filaments, in the clefts of which are contained the nucleus, mitochondria and endoplasmic reticulum. In the smooth muscle of the bladder, described by the pioneering electron microscopical study of Cæsar, Edwards & Ruska (1957), the cells were found to be closely apposed to each other in many regions, whilst in others they were separated by collagen fibres (Fig. 702, Pl. LXII).

Cell Contacts

As we shall see, the visceral types of smooth muscle, whose electrophysiological behaviour we shall be discussing, may be regarded physiologically as being made up of syncytial units, *i.e.*, units composed of groups of cells exhibiting at least a functional cytoplasmic continuity. The extent of this syncytial behaviour passes through the extremes of the vas deferens with largely single-unit behaviour comparable with that of skeletal muscle, to the smooth muscle of the gut, *e.g.*, the tænia coli, where the units of activity must be very large. Since functional cytoplasmic continuity can probably be achieved by the tight junction, or zonula occludens of Farquhar & Palade (p. 492), we may expect to find differences in the packing of the muscle cells and the numbers of these specialized contacts according as we examine, say, the nictitating membrane and vas deferens on the one hand, and intestinal muscle on the other. In smooth muscle the specialized

number of vessels; treatment with cocaine, which abolished nervous activity, and stimulation of one of these small arteries caused a contraction that spread to the same extent as before.

PLATE LXI

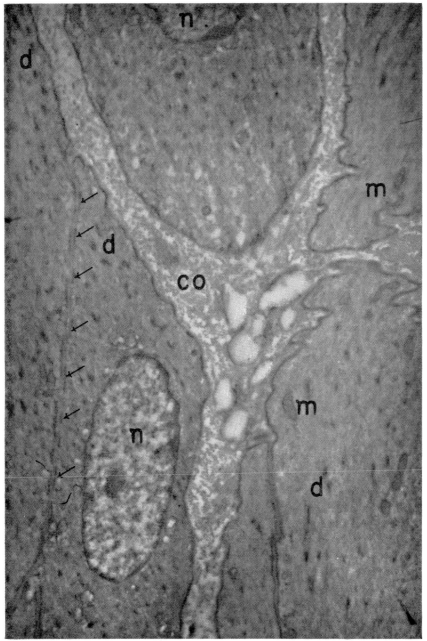

FIG. 701. Smooth muscle cells of mouse urinary bladder showing closely apposed and more widely separated cells. Irregular thickness and density of plasma membranes are seen (arrows), especially where the cells are in close contact. Collagen fibrils (*co*) run essentially transversely to the long axes of the cells. The cytoplasm shows myofilaments with no fibril formation, scattered mitochondria (*m*) and dark spots of unknown nature (*d*). Nuclei (*n*) are visible in upper centre and lower left. Osmium fixation × 12,600. (Caesar, Edwards & Ruska. *J. biophys. biochem. Cytol.*)

[*To face p.* 1312.

PLATE LXII

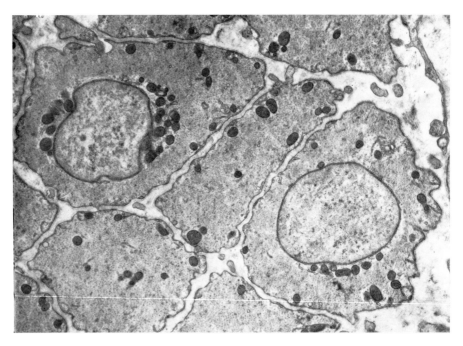

Fig. 702. Longitudinal intestinal muscle in cross section. The cells are loosely arranged and touch each other over broad planar areas. × 7000 *ca*. (Lane & Rhodin. *J. Ultrastr. Res.*)

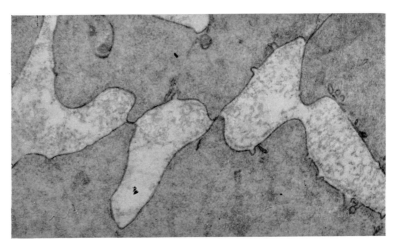

Fig. 703. Cross-section of three smooth-muscle cells showing three nexuses. The less dense regions are the spaces between adjacent cells. × 15,000. (Dewey & Barr. *Science.*)

contact was first described by Dewey & Barr (1962, 1964) and called by them the *nexus*; here the outer dense lines of the apposing membranes apparently fuse to give what is now described as the characteristic five-layered appearance. Three such nexuses are shown in Fig. 703, Pl. LXII, whilst Fig. 704, Pl. LXIII is a more recent electron micrograph at high magnification (Oosaki & Ishii, 1964).

Variations in Cell Relation

There have been several studies in which smooth muscles have been compared with respect to their cellular relations; for example Evans & Evans (1964) compared the chick amnion with the nictitating membrane; the amnion cells were very irregular in outline, exhibiting frequent zonulæ occludentes, whereas the nictitating membrane cells were regularly arranged, surrounded by basement

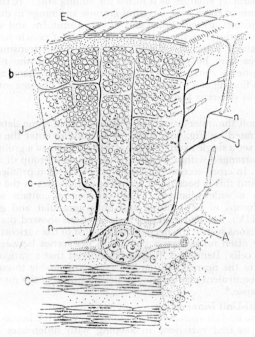

FIG. 705. Simplified diagram of a section through the taenia coli and underlying tissue. *A*, Auerbach's plexus; *b*, smooth muscle cell bundle; *C*, circular smooth muscle layer; *c*, connective tissue; *E*, serosal epithelium; *G*, ganglion cell; *J*, junction of muscle bundles; *n*, nerve bundle. (Bennett & Rogers. *J. Cell Biol.*)

membrane and separated from each other by collagen fibrils, and with no hint of tight junctions. Again, Lane & Rhodin (1964) compared the vas deferens and intestinal smooth muscle; the vas deferens cells showed a peg and socket relationship, in which an evagination of one cell fitted into an invagination of the adjacent cell; the contact points of intestinal cells were large planar areas, but because of their smaller frequency the total areas of close intercellular apposition were about the same; the contact regions of intestinal cells differed, however, in that the cell membranes fused to form the nexus of Dewey & Barr, whereas in the regions of contact between vas deferens cells the membranes remained some 100A apart with no evidence of fusion (see, also, Merrillees, 1968). A similar inter-digitation of cellular protrusions and invaginations was described in the tænia coli by Bennett & Rogers, and it was here, also, that tight junctions were found.

Bennett & Rogers emphasized the arrangement of the cells in bundles, as illustrated schematically in Fig. 705. The smooth dilator muscle of the iris exhibited interlocking of adjacent cells, and often large areas of fusion of adjacent membranes could be demonstrated (Richardson, 1964).

Innervation. The nature of the relation between the smooth muscle fibre and its nerve is a subject of considerable interest and has been examined especially by Richardson (1962, 1964), Taxi (1964), Thæmert (1963, 1966), Lane & Rhodin (1964), Bennett & Rogers (1967) and Merrillees (1968).

Vas Deferens

As described by Richardson (1962), the nerve fibres of the vas deferens finally lose their Schwann covering and run naked between cells to come into close apposition with one, forming a depression on its surface leaving a separation of some 180–250A. Its diameter enlarges as it forms the ending and here the axon is packed with vesicles; some of these have a central granule and range in diameter from 300 to 900A, and the rest are agranular and of more uniform and smaller diameter (450–600A). Where the contents of the granular type of vesicle have been studied, they have contained catechol amines, presumably the transmitter, whilst the agranular vesicles may be classed with those found at the nerve-muscle and numerous nerve-nerve synapses of the central nervous system, and may well be cholinergic. The finding of a mixture of vesicle types is of some interest, and is not peculiar to the vas deferens (Lane & Rhodin, 1964).

Tænia Coli

In the tænia coli the innervation has been studied in some detail by Bennett & Rogers; as illustrated by Fig. 705, small nerve bundles enter the muscle bundles; since it is rare to see a single axon, it appears that all fibres of a group leave together; these axons are arranged within a Schwann sheath as a group of 2–30 fibres or as individual fibres. In cross-section the diameters of the axon profiles vary from 0·08 to less than 4 μ, and this is because of recurring varicosities in the axon which give its cross-section a wider diameter. The varicosities contain widely dispersed neurotubules, clumps of neurofibrils and both agranular and granular vesicles (Fig. 706, Pl. LXIV). Longitudinal and serial sections showed that, because of the interweaving of axons within the nerve bundle, most of the varicosities occur on the periphery where often no Schwann cytoplasm intervenes between them and the adjacent muscle cells; Bennett & Rogers considered that a varicosity could be as close as 1,000A to the nearest cell, and it seems reasonable to conclude that this varicosity is the equivalent of the synaptic nerve ending of the muscle end-plate or nerve-nerve synapse.[*]

Single- and Multi-Unit Innervation

Just as with the muscle-muscle cell relations, so with the nerve-muscle relations we may expect to find variations in keeping with differences of physiological behaviour, those muscles that behave as though their fibres were organized into large units having the less discrete types of innervation. Taxi (1964) showed that the ciliary muscle and vas deferens had the discrete type of innervation with probably an individual "synapse" on each muscle fibre, whilst intestinal smooth muscle is supplied with bundles which apparently never reduce to single fibres; serial sections showed that many muscle fibres never made close relations with a nerve fibre, *i.e.*, less than 500A separation. The nictitating membrane was considered by Taxi to have an intermediate form of innervation; thus the formation of gutters in the muscle cell was not general as in the vas deferens, but the contacts were within 80–100A. The number of nerve fibres was not striking so that it was doubted whether there could be individual innervation of the muscle fibres.[†]

Smooth Muscle as Functional Syncytium

Present-day views on the electrical events associated with spontaneous and induced activity in visceral smooth muscle are essentially those reached by Bozler on the basis of a series of studies that may justly be described as

---

[*] Merrillees (1968) emphasizes that there are no close contacts between axon and muscle cell in guinea pig intestinal muscle by contrast with the situation in the vas deferens.

[†] Thaemert (1966) has described some interesting three-dimensional reconstructions of nerve-smooth muscle relations.

PLATE LXIII

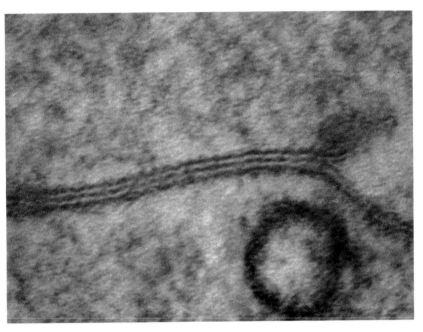

FIG. 704. Tight junction between two cells of smooth muscle of the rat small intestine. At the right of the Figure, the two outer layers of the cell membranes fuse to form the intermediate line of the junction. × 400,000 ca. (Oosaki & Ishii. *J. Ultrastr. Res.*)

PLATE LXIV

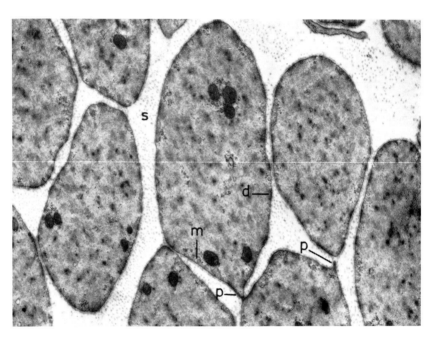

FIG. 706. Transverse section through basally situated layers of tænia coli. Note wide intercellular spaces (*s*) and finger-like projections (*p*) from the surfaces of the muscle cells. *m*, microvesicles; *d*, dark area. × 10,500 *ca*. (Bennett & Rogers. *J. Cell Biol.*)

classical. As mentioned earlier, the visceral type of muscle, exemplified by the wall of the gut, undergoes spontaneous and rhythmic contractions; Bozler's studies strongly suggest that the behaviour is what one might expect of a syncytium, the whole muscle behaving as a "giant fibre", complicated, however, by very frequent changes in the excitability of different parts. Thus, if the muscle were a single unit of uniform excitability, we should expect a contraction, initiated at one point, to travel uniformly over the whole of the muscle; under ideal conditions such behaviour may be seen—at least in strips of excised tissue —but, in general, the propagated disturbance fades out, or changes intensity, as it proceeds—a phenomenon that is interpreted by Bozler as due to changes in excitability encountered during the passage of the wave of action potential. Thus, the uterus, during anœstrus, exhibits low excitability, and electrical stimuli are not propagated; during œstrus, on the other hand, electrical excitability is high and the response to a stimulus is a well co-ordinated wave of contraction passing through the whole structure.* As we shall see, a better representation is given by picturing smooth muscle as made up of large units, namely bundles of single cells with electrotonic connexions; these bundles are, however, not discretely insulated from each other but exhibit functional connexions which presumably have as their morphological basis the anastomoses between bundles. We may now briefly review some of Bozler's studies.

Electrical Events. A single stimulus applied to a strip of ureter, for example, caused a propagated action potential, moving in both directions, associated with

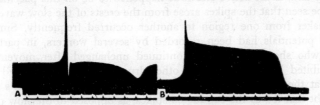

FIG. 707. A. Diphasic recording, with "differential electrodes," of action potential of rat's ureter. B. Monophasic action potential obtained by integration. Time-scale, 1/5 sec. (Bozler. *Experientia.*)

a contraction, both events apparently beginning at the cathode; increasing the strength of the stimulus had very little effect on the response, so that the "all-or-none" law apparently applied with this tissue. The action potential, recorded "diphasically", was typical of that recorded from the heart, showing R, S and T waves (Fig. 707), a finding that strongly suggests that we are dealing with a syncytium since, as Bozler emphasized, recording with external electrodes from a system composed of numerous small units, each activated independently, would not give this type of response. By the use of "differential electrodes"—*i.e.*, very fine electrodes so close together that the difference of potential across them was a measure of the rate of change of potential under the active electrode—Bozler deduced that the monophasic record, *i.e.*, the course of the potential under the active electrode, was that shown in Fig. 707, B, consisting of a brief spike followed by a long plateau of negativity, similar to that obtained from heart. This monophasic action potential depended for its validity on a number of assumptions, namely that conduction between the leads was uniform with respect to speed and intensity, and that the distance between the leads was small compared with the length of the active region.

* Melton (1956) found that the uterus of the castrated female was electrically inex-citable, a condition that could be reversed by treatment with œstrogen.

By a more direct approach, Ichikawa & Bozler obtained monophasic records that agreed remarkably well with those deduced from the differential recording.*

Slow Potentials and Spikes

Spontaneous activity, as for example during peristalsis of the intestine, was associated always with bursts of impulses whose frequency generally ran parallel with the intensity of the muscular contraction. In examining the ureter, Bozler observed that activity began at a definite pacemaker region close to the renal end, in the sense that, when two electrodes were placed on the ureter, negativity always began at the electrode closest to this region. This pacemaker region showed "slow potentials" which rose to a few tenths of a millivolt; these slow potentials usually led to the firing off of spikes, which were conducted down the ureter. The slow potentials seemed to be similar to those recorded from the pacemaker region of the heart which, provided that they reached a certain value—depending on the excitability of the tissue—led to the firing off of a propagated impulse. In both cases, the slow potential is a local change being propagated only by electrotonic spread. In the small intestine of the guinea pig the study of the initiation of spontaneous discharges was harder for technical reasons, but in this tissue too it was possible to show that slow potentials occurred with a frequency of approximately one to two per sec. In addition, bursts of impulses, accompanying waves of contraction (peristalsis) were observed. Usually the peristaltic waves started out from the same region repeatedly and, if the active electrode happened to be on this pacemaker zone, it could be seen that the spikes arose from the crests of the slow waves. A shift of pacemaker from one region to another occurred frequently. Similar slow rhythmic potentials had been recorded by several workers, in particular by Berkson who showed that they continued unchanged after movements had been inhibited by adrenaline and other drugs; as Berkson pointed out, this meant that the slow potentials were not action potentials. Bozler's work left little doubt that it was the conducted spikes, as opposed to the slow potentials, that were immediately associated with the contraction of smooth muscle.

Inhibition

Stimulation of the sympathetic nervous system, or the application of adrenaline, inhibits the activity of most visceral smooth muscle, and Bozler found that this was associated with a marked decrease in electrical excitability, often to the extent that muscular conduction was temporarily abolished, although the spontaneously occurring negativity at a pacemaker region persisted.

Co-ordinated Contractions

Segmenting movements are essentially localized contractions, and it is interesting that, when these occur in intestinal muscle, the discharges are confined to a region of a centimetre or two; presumably the excitability is restricted to a small region so that the impulses fade out when they get beyond this. This localized excitability may also be a factor in the co-ordinated peristaltic movement which drives a bolus forwards along the gut; Bozler found that each slow wave of peristalsis, travelling, in the dog, at 1–2 cm./min.,

---

* The recording of true monophasic action potentials, by placing an electrode on an injured region and another on a normal region, demands that the distance between the two electrodes should not be too great (Bishop *et al.*, 1926); the limiting distance will necessarily be much smaller in smooth muscle so that it is not easy to avoid spread of damage to the region under the active electrode. This difficulty is overcome to some extent by sucking a highly localized region of the muscle into a micropipette, which acts as the inactive electrode; the tissue sucked in becomes anoxic and so behaves as an injured region.

was accompanied by a discharge, the greatest discharge being on the oral side of the bolus; beyond this the activity faded out. If the distention of the gut was responsible for a decreased excitability, in the region of greatest distention the impulse would necessarily fade out. We must remember, however, that this form of co-ordinated contraction is dependent on nervous activity too, since painting the tissue with nicotine inhibits it, the nicotine blocking synaptic transmission in the ganglia in intimate association with the muscle.

Myogenic Contraction

Bozler's studies on the ureter have been largely confirmed and extended by Prosser and his collaborators; they showed that the rate of conduction of impulses was of the order of 3–4 cm./sec. Propagation could be blocked by such agents as nicotine, hexamethonium and procaine, but only at such high concentrations that it was unlikely that their effect was on nerve or ganglia, the more so as histological study indicated that ganglia were lacking.* It seems safe to conclude, therefore, that conduction along the muscle is purely a muscular phenomenon, so far as the ureter is concerned at least, in spite of the apparently co-ordinated nature of the wave of contraction as it passes down the ureter, forcing fluid to flow in one direction. This view is strengthened by the observation of Evans & Schild that the chick amnion, which consists only of a sheet of epithelium apposed to a sheet of smooth muscle, both sheets being only a single layer of cells thick and entirely devoid of nerve fibres, shows both local and co-ordinated activity. Prosser & Rafferty showed that electrical activity, initiated at a point on the amnion, was propagated over the muscle at about 3 cm./sec. At about 10-days' old, rhythmic activity in the amnion ceased, and this may well be due to the circumstance that the muscle cells no longer overlap at this age but become separated by connective tissue; if propagation depends on close contact between cells, i.e., if it is dependent on the presence of "nexuses", this becomes intelligible. In the vas deferens, on the other hand, the fact that every single cell responds to nerve stimulation with a spike suggests that the co-ordination of the wave of contraction is determined mainly by neural activity, and it is very likely that the contribution of nervous activity to co-ordination varies with the tissue and its function.

Uterus

In the uterus, during spontaneous activity, the same general phenomena have been observed by Melton and Jung. Thus, Melton described two types of potential fluctuation; a slow wave of less than 1 mV lasting about 0·5 sec, and a fast spike of 1–2 mV lasting 100 msec. The slow waves usually were maintained indefinitely, by contrast with the spikes which only occurred in bursts and were associated with mechanical contraction. By recording from different regions simultaneously, it could be shown that the slow waves occurred independently, whereas if one region showed a burst of spikes, these occurred at all the other electrodes after a conduction time corresponding to a velocity of propagation of some 0·26–6·0 cm./sec. In a similar way, Jung showed that the bursts of spikes in the uterus, associated with a contraction-wave, were often preceded by a series of slow oscillations of potential, corresponding to the slow waves of Bozler. These oscillations continued between the waves of contraction, so that Jung concluded that the interval between contractions was a period of reduced excitability rather than one of reduced local electrical activity, i.e., the pacemaker activity continues but the threshold for firing of spikes is raised.

* At high amplification small "pre-spikes", in advance of the ordinary action potentials, were recorded; they travelled some 10 times faster, and were blocked by tetracaine; presumably they represented nerve impulses with some function other than activation of the muscles.

It must be emphasized that these results of Bozler and others were obtained with extracellular recording; in so far as the muscle cells are very small, the records usually represented the integrated responses in many fibres, so that the interpretation must be equivocal. One cannot be sure, therefore, whether the long-lasting monophasic action potentials described by Bozler are analogous with the action potentials recorded, for example, from a single Purkinje fibre by an intracellular electrode. We may pass, therefore, to some studies of intra-cellular recording, but first we may summarize Bozler's main conclusions:—

(1) Contraction is preceded by depolarization which takes the form of a burst of spikes; in general, these spikes are preceded by a slow pace-maker depolarization which is not propagated, and which may occur independently of any muscular contraction should it not give rise to spike activity.

(2) The response to electrical stimulation is all-or-none, although this characteristic may be masked by the fluctuations in excitability, both temporal and spatial, that take place.

(3) The electrical activity can best be explained on the basis that the tissue is functionally a syncytium.

(4) The speed of propagation and its extent are limited by the excitability of the tissue.

(5) Some autonomous co-ordination, such as that exhibited in the wave of peristalsis passing down the ureter, is possible, and is probably the result of the effect of stretch on electrical activity.

Intracellular Recording

The Resting Potential. The average length of a visceral smooth muscle fibre is $50-150\,\mu$, whilst the average width in the central region, where it is greatest, is $5-10\,\mu$; intracellular recording presents formidable difficulties, not only because of the difficulty of insertion in such a small cell but because of the damage that the cell must sustain when a relatively large portion of its area is punctured. Greven (1953) made use of the circumstance that the smooth muscle cells of the salamander's stomach are large—$1\cdot1$ mm. long and some $20\,\mu$ broad. On inserting a $0\cdot5\,\mu$ KCl-filled micropipette electrode, resting potentials up to 70 mV, which sank fairly rapidly to some 20 mV, were recorded. The fall was probably the result of damage but, as with the heart, in a spon-taneously active tissue it is difficult to speak of a true "resting potential", and it may be that some of the variability in magnitude, observed by Greven and subsequent workers, was due to different phases of activity. Woodbury & McIntyre profited by the circumstance that the cells of the uterus are very much enlarged during pregnancy, increasing from 5 to $9-14\,\mu$ in diameter. They found potentials in the range of 21-31 mV for man, 35-52 mV for the rabbit, and 27-60 mV for the guinea pig. They argued that the low values recorded could not have been due to a low value of internal potassium con-centration, since analyses indicated a theoretical potential of 81 mV; they suggested that the main reason was the spontaneous activity of the cells; at any rate, cooling the tissue, which reduced spontaneous activity, caused a rise in resting potential. Bülbring & Hooton measured rather larger resting potentials in both the sphincter pupillaris of the iris (chosen because it shows no spontaneous activity), and the isolated tænia coli muscle of the guinea pig,*

---

* The tænia coli is a longitudinal muscle on the cæcum of some mammals; in the guinea pig it is less than 1 mm. wide, only 0·5 mm. thick.

the value for the former being 30–80 mV and for the latter 20–80 mV. In a later study on the tænia coli, Holman reported a mean value of 51 mV. In the circular intestinal muscle Sperelakis & Prosser found a range of 20 to 60 mV. In the guinea pig ureter, Kuriyama, Osa & Toida (1967) found values of 50–65 mV.

Resting Potential. The low values of resting potential by comparison with skeletal muscle may well be due to the lower internal concentration of K^+ and higher concentration of Na^+; thus Goodford & Hermansen computed an intracellular concentration of some 119 mM for K^+ and 56 mM for Na^+ although the uncertainties as to the true extracellular space make the exact estimate of

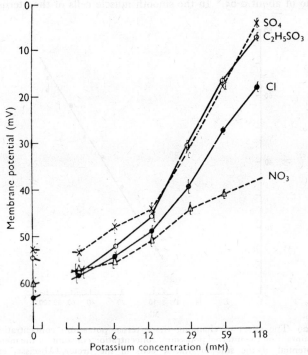

FIG. 708. The relation between membrane potential of tænia coli and the logarithm of the external potassium concentration in the presence of different anions. Cl^- ●, NO_3^- △, SO_4^{2-} ×, and $C_2H_5SO_3^-$ O. (Kuriyama. *J. Physiol.*)

internal concentration difficult.* The calculated K^+-potential came out at some 78 mV, so that the resting potential is not that of a simple K^+-electrode, presumably because of some Na^+-permeability.

Effects of External Potassium. The effects of varying the extracellular concentration of K^+ on the membrane potential of the guinea pig tænia coli were studied by Kuriyama (1963); the results confirmed earlier work of Holman and Burnstock & Straub in showing that, although the potential decreases with increasing external concentration, the relationship is not that predictable from the Nernst relationship. Thus, at high concentrations there is, indeed, a linear relationship between potential and logarithm of concentration, but the decrease in potential per tenfold rise in concentration is only 38 mV compared with the theoretical 58 mV. Below 30 mM the relationship is not linear (Fig. 708).

* Goodford (1964) has summarized the chemically determined concentrations of Na^+, K^+ and Cl^- in the guinea pig tænia coli.

A serious error in this type of measurement, however, is the failure to determine whether the internal concentration of K^+ has changed when the outside concentration has been altered experimentally (p. 570), and Bennett (1966c) has examined the published measurements in the light of Boyle & Conway's equations describing the Donnan equilibrium between the muscle cell and its environment. As a result of his analysis it would seem that Kuriyama's failure to obtain the 58 mV change in potential with tenfold change of external K^+-concentration was, indeed, due to ignoring the change in internal concentration. At low concentrations of external K^+ the deviation from linearity could be accounted for on the basis of the Goldman equation on the assumption of a Na^+/K^+ permeability ratio of about 0·05.* In the smooth muscle cells of the uterus, Casteels

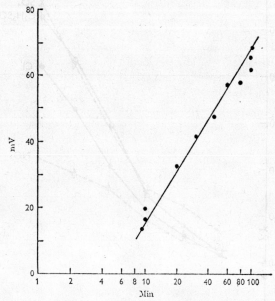

Fig. 709. The effect of changing the external potassium concentration and maintaining the internal potassium concentration constant on the membrane potential of the smooth muscle cells of the ureter. Abscissa, external potassium concentration, $\log_{10}$ scale. Ordinate, depolarization of the membrane potential. (Bennett & Burnstock. *J. Physiol.*)

& Kuriyama (1965) demonstrated a reasonable concordance between the predicted behaviour on the basis of a K^+-electrode and that experimentally observed; these authors estimated the internal concentrations of Na^+, Cl^- and K^+ when the external solution was varied; only at low concentrations of K^+ was there deviation from linearity between membrane potential and $\log [K^+]_{Out}$.† Finally, Bennett & Burnstock (1966) have applied the sucrose-gap technique

* Casteels & Kuriyama (1966) have re-investigated the effects of external K^+-concentration on membrane potential in the tænia coli, taking into account the altered internal concentrations of K^+ and Cl^- when the external solution was varied. Their results suggest some contribution of Cl^- to the membrane potential; furthermore, Cl^- was not distributed between cell and medium in accordance with the Gibbs-Donnan equilibrium.

† The main uncertainty in estimating the internal concentrations depended on the choice of a value for the extracellular space; on the basis of the distribution of inulin, this was 37·5 per cent., giving an estimated $[K^+]_{In}$ of 113 meq./litre; on the basis of the distribution of ethanesulphate, the space was 56·9 per cent. giving an estimated $[K^+]_{In}$ of 163 meq./litre. The changes in membrane potential during pregnancy could be accounted for by altered permeability to K^+ and Na^+ since there were no significant changes of internal ionic concentration.

(p. 1050) to the estimation of the resting potential of the ureter smooth muscle; when $[K^+]_{out}$ was varied over a range of 10 to 100 meq./litre, keeping the product $[K^+] \times [Cl^-]$ (and thus the internal concentration of K^+) constant a linear relation between potential and log $[K^+]_{out}$ was obtained (Fig. 709) with a slope of 53 mV per tenfold change in concentration.

The Sodium Pump

In isolated uterine segments the internal concentration of K^+ was found by Daniel & Robinson to be 139 mM, whilst that of Na^+ was 8·5 mM; this corresponded with a K^+-potential of -91 mV and a Na^+-potential of 75 mV. Here, too, the resting potential (50 mV according to Daniel & Singh) is less than the K^+-potential. Just as with striated muscle, cooling the uterine segments in Ringer's solution for 24 hours caused a loss of K^+ and uptake of Na^+; and a large proportion of these changes could be reversed by re-warming, provided that K^+ was in the medium. Thus the smooth muscle seems to have the same type of K^+-linked Na^+-pump as that in skeletal muscle and nerve. A study of the effects of anaerobiosis and metabolic poisons suggested that the main energy source was glycolytic, as with the mammalian erythrocyte, since anaerobiosis and cyanide had little effect, whilst iodoacetate and fluoride inhibited extrusion of Na^+. However, the fact that DNP was effective as an inhibitor does suggest that oxidative phosphorylation is important; in view of the apparently high permeability of smooth muscle cells to Na^+, an extremely efficient and highly active pump mechanism would be required to maintain a low internal concentration of Na^+, and it would be surprising if the relatively inefficient glycolytic source of energy were exploited.

Chloride. The chloride content of tænia coli has been measured by Goodford (1964); this rises from 55–76 meq./kg. immediately after killing to some 96 meq./kg. after incubation for an hour in Krebs' medium. Exchanges with $^{36}Cl^-$ were rapid and could be resolved into a slow component, presumably intracellular, and two rapid components, presumably extracellular.

The Action Potential. Woodbury & McIntyre observed small spontaneously occurring action-potentials of some 2 mV with their intracellular recordings from uterine cells; much larger depolarizations occurred by adding oxytocin —a hormone causing strong uterine contraction—to the bathing medium, but an actual reversal of potential, or "overshoot", was rarely observed. In a later study (1956) they employed a flexibly mounted electrode which remained in position inside an impaled cell in spite of movements, and so could be used in studies of the uterus *in situ*. With this preparation, their largest resting potential was 38 mV, and this was associated with an action potential of 48 mV on treatment with oxytocin, *i.e.*, there was a reversal of 10 mV. In general, however, reversal was unusual, the mean resting potential being 32·6 mV and the mean action potential 21.9 mV. When the animal's temperature fell the resting potential rose.

Hormonal Influences

Subsequent work, notably that of West & Landau, Marshall & Csapó, and Kuriyama & Csapó, has emphasized the importance of the hormonal state of the animal, at the time of removal of the uterus, on the degree of spontaneous activity, and on the responses to drugs such as acetylcholine, which is strongly excitatory (Schmidt & Huber). When the uterus was "œstrogen dominated" by being removed post-partum, spontaneous activity was greater than when it was removed in late pregnancy. Oxytocin provoked large burst of spikes whilst progesterone inhibited most of the spontaneous activity. Membrane potentials of single cells averaged 45–50 mV in the spontaneously active post-partum uterus, whilst they were higher, at 55–60 mV, in the late-pregnancy organ, thus indicating that spontaneous activity is associated with depolarization of the cell; the difference is reminiscent of the discrepancy between pacemaker and non-pacemaker sites on the auricle.

Slow Waves and Spikes. Bülbring, working on the tænia coli, observed relatively rapid "spikes" of about 150 msec. duration, consisting of a depolarization of only 1–8 mV, occurring singly, in pairs, or in bursts, and usually

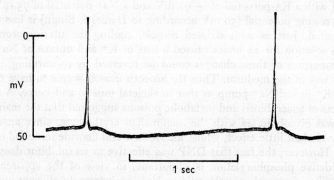

FIG. 710. Typical records of spikes taken during normal spontaneous activity in the guinea pig's tænia coli. Holman. *J. Physiol.*)

superimposed on a slow fluctuation in the resting potential. Later studies with more refined technique (Holman, 1958) showed that the depolarization involved in a typical spike was considerably greater than 1–8 mV, an actual overshoot of some 12 mV being recorded; *i.e.*, the normal spike-height was of the order of 50–75 mV in amplitude (Fig. 710); the duration, moreover, was only 15 msec. Bülbring had noted that the spike discharges were usually superimposed on a slow fluctuation in the resting potential—the *slow wave*—which was probably similar to the pacemaker activity of Bozler. The relation

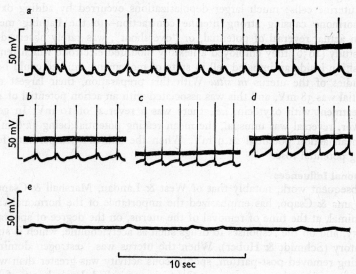

FIG. 711. Patterns of spontaneous activity recorded intracellularly from the guinea pig's tænia coli. (Bülbring, Burnstock & Holman. *J. Physiol.*)

of the slow wave—which had a duration of about o·1 to 1 sec.—to the spikes was studied in some detail by Bülbring, Burnstock & Holman. They showed that, although the spike could, in general, be regarded as the product of the pacemaker activity represented by the slow potential, nevertheless, in any given preparation, the two types of potential change were not rigidly linked. Fig. 711 illustrates a series of records taken from different preparations. It will be seen that in records *b*, *c*, and *d* the frequency of the spikes was the same as that of the slow waves; in record *a* this was not true, although the spikes bore an obvious relation to the slow waves; in record *e* there was no spike activity.

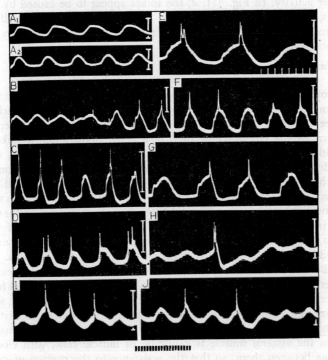

FIG. 712. Spontaneous slow waves and spikes in the cat's longitudinal intestinal muscle. Vertical bar at right of each record indicates 20 mV. Time bar at bottom gives 1-sec intervals for all but *E*. A_1: steep rising phase and slow falling phase; A_2: sinusoidal waves. *B*: summation of electrically driven slow waves with spikes on last two waves; small vertical deflections are stimulus artefacts. *C*, *E*, *G* (2nd wave) show interference of spike with down-phase of slow wave; *D*, *F*: spikes occurring early in slow waves fail to hasten down-phase. *G*: steep rise may indicate slow wave driven from adjacent cell. *H*, *I*, *J* (and spike): reduction of succeeding slow wave by undershoot (hyperpolarization after spike). Double spikes in *D*, *E*, *H*. (Tamai & Prosser. *Amer. J. Physiol.*)

An essentially similar dependence-independence relationship was described by Tamai & Prosser (1966) in the longitudinal intestinal muscle of the cat. Fig. 712 is a typical illustration of the slow potential leading to a spike, in which case the slow potential may be argued to be a pacemaker; however, the variation in the pattern of the relations of slow wave to spikes, and the observation that an altered ionic environment, *e.g.*, doubling external K^+-concentration, could modify the spike but not the "pacemaker", indicated that the two phenomena were not *necessarily* related as cause and effect.

Pacemaker Activity. Bulbring *et al.* considered that the results were best interpreted on the assumption that the slow potential is a pacemaker activity which, in any given muscle cell, will determine a spike, provided that a threshold depolarization—which is probably variable in time and from cell to cell—is reached. The change of potential recorded in any given cell will depend not only on how close the cell is to a region of pacemaker activity, but also on the conducted effects of spikes from nearby cells. Thus, the spikes in Fig. 711, *a* were partly locally initiated and partly conducted, the argument being that a conducted spike would bear no definite relation to the pacemaker activity. In Fig. 711, *d*, the spikes appear to be caused by the slow potentials, wiping these out, whereas in *b* and *c* the spikes may well be conducted from nearby cells, similar to the spikes in the auricle which are initiated in the pacemaker at some distance.*

Spatial Variations

When two neighbouring cells were impaled, and recordings of their spontaneous activities were taken simultaneously, it was found that the spikes generally had the same frequency but were not necessarily synchronous; they were therefore generated independently, but some mutual dependence was shown by the circumstance that, in a given preparation, a spike in one cell would show up as a graded potential in the other which might, or might not, produce a spike; in the latter case, this spike appeared in addition to the regular rhythmic spikes of this cell. In general, the results showed that the activity in one cell was reflected in that of another cell if the latter was not more than a few cell-lengths away. The fluctuating phase-difference of the electrical changes in a pair of impaled cells, and the varying magnitude of the pre-potentials, suggested that the influence of local pacemakers was not fixed, but varied both temporally and spatially with respect to the two cells.

Fibre Bundles as Units. Later studies with intracellular recording carried out by Prosser, Tomita, and Bennett and their colleagues have contributed further to our understanding of the origin of spike activity in any given cell, and have largely confirmed Bülbring's view of the relation of slow-wave activity to the spike. The concept emerging from these studies is that the individual unit of activity in the visceral muscle is a bundle of muscle cells, with electrotonic connexions between them, so that the resistance to the passage of current is smaller when current passes through these connexions than across the remaining portion of the cell membrane (Nagai & Prosser, 1963). These bundles, which in intestinal muscle are some $100\,\mu$ in diameter, are made up of some 200–300 cells; anastomoses between bundles give them an indefinite length and may ultimately lead to spread of activity over wide areas.

Pacemaker and Driven Cells

Thus intracellular records from the guinea pig tænia coli suggest that spontaneous spike activity is of two kinds, the *pacemaker type*, in which the spikes quite clearly arise from a slowly increasing depolarization of the cell, and the *driven type*, where the repolarizing phase is slow and often continuous with the next action potential; the absence of a slow depolarization before the spike, and the initiation of a spike at a high membrane potential, suggested that the cells

---

* With external recording the shape of the slow-wave may vary greatly, according to the method employed. Bortoff (1967) has examined the problem from the general aspect of recording activity at a distance from its origin; he has shown that the records are a combination of the actual changes of potential at the membrane and of the "field potential" as created by the membrane currents, and related to the membrane potential through its second derivative, *i.e.*, d^2V/dt^2. Thus little can be deduced from the shape of the slow potential wave.

were driven by some form of pacemaker activity in their vicinity. The two types are illustrated in Fig. 713. In both types of cell the spike was genuinely regenerative —*i.e.*, not simply electrotonically transmitted; this is revealed by the effects of polarizing currents passed through the cell (Kuriyama & Tomita, 1965). The same currents had remarkably little effect on the slow pre-potentials, however, suggesting that in the cell that had been impaled most of this activity was electrotonically transmitted from other pacemaker cells of the postulated unit. Thus spontaneous activity is, indeed, determined by pacemaker activity in certain cells and leads to spikes which can set off, by electrotonic transmission, spikes in adjacent cells and thus lead to propagation beyond the limits of the individual cell. Since pacemaker activity in a given cell is not always affected by

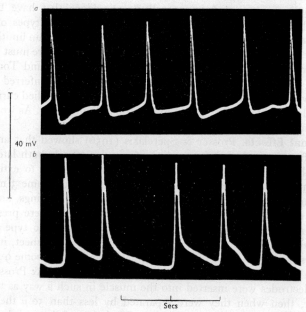

FIG. 713. Two contrasting types of activity recorded in the smooth muscle cells of the taenia coli. (*a*) Action potentials preceded by slow depolarizations ("pace-maker-like" activity). (*b*) Action potentials with slow repolarization phases ("driven" activity). (Bennett, Burnstock & Holman. *J. Physiol.*)

applied currents, or by spike activity, we may postulate, with Bennett, that genuine pacemaker activity occurs in relatively few cells, and that this is transmitted electrotonically to others nearby.

Transmembrane Stimulation

The importance of the unit, as opposed to the individual cell, in determining electrical behaviour is revealed by the almost invariable failure to induce an action potential in an impaled tænia coli cell by transmembrane stimulation (Kuriyama & Tomita, 1965); the situation is similar to that described by Rushton (1938) when too small an area of axon was stimulated.* Essentially, then, the

* Intracellular pulses applied to the smooth muscle cells of the guinea pig vas deferens often led to local spike-like responses that were confined to the cell stimulated, *i.e.*, the absence of contraction indicated lack of propagation; the spikes were graded up to a maximum amplitude of 75 mV, but in some cells self-regenerative spikes occurred if they were depolarized by greater than 35 mV. By contrast with the graded spikes following electrical stimulation, those following nerve stimulation were always all-or-none (Bennett

propagated action potential occupies an area larger than that of a single cell. It is certain that there is a limiting thickness of tissue required for conduction; thus Burnstock & Prosser reduced the width of a strip of smooth muscle and found the conduction-distance to remain at 25–40 mm. until the diameter was reduced to 100 μ when none occurred; again, Nagai & Prosser showed that, when using stimulating electrodes of less than 100 μ diameter, conducted spikes could not be obtained. Further evidence indicating this coupling between cells, giving rise to both pacemaker and driven activity, is given by studying the effects of nervous stimulation (p. 1336).

Mechanism of Conduction

The transmission of electrical activity, and thus of contraction, depends, then, on electrotonic connexions between cells,* connexions that have been well established by the electron microscopical studies of several types of visceral muscle. We must assume that, for activity to be initiated, a certain limiting group of cells must be activated together, and that for propagation there must be spread of activity from one unit to another. Thus, as Prosser, Barr and Tomita have emphasized, the cable properties of the muscle as a whole, inferred from the conduction velocity, strength-duration curve, and spread of applied currents, will be necessarily greatly different from those of the individual cell. As we shall see, these are, indeed, different by an order of magnitude.

Directional Effects. Prosser & Sperelakis (1956) showed that an impulse could pass from a single point on a ring of circular muscle both laterally and longitudinally; the lateral transmission decremented rapidly, to extinguish at about 3 mm., with the result that a ribbon of active tissue, some 3 mm. wide, spiralled round the ring. When a ring was cut to form two rings, an impulse, initiated in one ring, could spread to the other if the two were pressed close together, a mode of transmission reminiscent of the ephaptic type (p. 1204). When an impulse was found to travel diagonally over a flat sheet, its velocity could always be compounded of a longitudinal component of some 9·4 cm./sec. and a much slower transverse velocity of 10 mm./sec. (Nagai & Prosser, 1963). If double electrodes were inserted into the muscle in such a way as to be both intracellular, then when they were separated by less than 50 μ there was no difference in the latencies of responses to stimulation of the muscle, recorded from the two electrodes; if the separation was greater than 50 μ, then the difference of latency corresponded to a velocity of 10 cm./sec.; this indicates that the delay at each junction is constant.

Tænia Coli. Essentially similar results were obtained by Bülbring, Burnstock & Holman (1958) on the tænia coli; they recorded intracellularly the effects of exciting the muscle locally, either with a fine external electrode or by trans-membrane stimulation with an internal electrode. The results with external

---

& Merrillees, 1966; Hashimoto, Holman & Tille, 1966). Tomita (1966) has summarized the different responses to intra- and extracellular stimulation, in the guinea pig tænia coli; with intracellular stimulation there is no modification of the frequency of spontaneous spike discharge, no electrotonic spread from one cell to the next, and no spike response. With extracellular stimulation, spike frequency is modified, there is electrotonic spread, and the induced spike has a slow component. His statement that intracellular stimulation leads to no recordable change in neighbouring cells is in conflict with the results of Bulbring et al. (1958).

* Barr, Berger & Dewey (1968) found that treatment of the tænia coli with hypertonic sucrose blocked propagation; since their earlier study on the heart (Barr et al. 1965) had shown that hypertonicity broke the nexus and thus the coupling between cells, they considered that blockage in the case of smooth muscle was due to this, and their electron-microscopical studies confirmed this, the rupture being reversible. Tomita (1966) was able to obtain propagated spikes in hypertonic media, however, but this may be because the nexuses were not ruptured in the depth of the tissue.

stimuli, in which a large number of cells were necessarily stimulated, were similar to those obtained by others. With a preparation that was not spontaneously active, a single impulse gave a single response with a latency—*i.e.*, conduction time— proportional to the distance from the point of stimulation, and giving a conduction rate of 6·7–8·8 cm./sec. at 38°C with no decrement. If the tissue was spontaneously active, then by stimulating at frequencies within plus or minus 25 per cent. of the spontaneous rhythm, the system could be "driven", in the sense that the spontaneous rhythm was replaced by the rhythm of stimulation. Excitability and conduction were normal when nervous participation was excluded by treatment of the muscle with atropine, or when ganglion-free strips were employed. With intracellular stimulation spike activity was induced but its propagation, as measured by activity in nearby cells, was very limited, so that spread beyond some 0·7 mm. in the longitudinal direction was not observed.*

Electrical Parameters. The electrical parameters determining excitability and propagation include the membrane capacity and resistance; these have been measured by the now classical technique of passing current across the tissue and determining its temporal and spatial decay constants. From the studies of Nagai & Prosser, and of Tomita, it emerged that there were large discrepancies between the estimated values, according as current was passed across the individual cell membrane with an intracellular electrode, or across the tissue with extracellular electrodes. Thus the time-constant, τ, with intracellular current application is only 2–4 msec. for the guinea pig tænia coli, whereas with external application it is 60–100 msec. (Tomita, 1966).† The space-constant, λ, for the tænia coli with external current application is some 1·6 mm.; in the cat's intestinal muscle Nagai & Prosser (1963) found that λ varied according to the direction of measurement, being 1·03 ± 0·16 mm. in the longitudinal direction, *i.e.*, some 10–20 cell lengths, and 0·27 mm. transversely, *i.e.*, some 50–90 cell widths. With intracellular polarization it is difficult to detect electrotonic spread beyond 100 μ, so that the space-constant, *i.e.*, the distance at which the size of the potential change is $1/e$ (37 per cent.) that at the point of application, must be much smaller than this. As Tomita (1967) has pointed out, with a cable, the area of a membrane increases linearly with distance from point of application of the current; with a three-dimensional syncytium, it will increase probably as the cube of distance and thus account for the sharp decay.

Low-Resistance Bridges

These results with external polarization strongly indicate that the cells are connected by relatively low-resistance bridges; in fact, using microelectrodes, Nagai & Prosser found that the resistance between adjacent cells was about a quarter of that between a given cell and its extracellular space; this means that the specific membrane resistance in the region of the supposed tight junctions is low, since the area of this junction must be small compared with the total cell area. When microelectrodes were in adjacent cells, the current-voltage relation was quite linear during depolarizations and hyperpolarizations; this was in contrast with the situation when one electrode was in the extracellular space, in which case rectification was present, as with most excitable membranes. This indicates a difference between the membranes at the site of junctions and elsewhere.

SUCROSE-GAP TECHNIQUE

Barr, Berger & Dewey (1968) employed the sucrose-gap technique to demonstrate the existence of an intercellular current pathway. Thus, by causing isotonic sucrose to flow over a limited region, current movements on the outside of the tissue in the node are virtually abolished because of the high resistance of the sucrose-bathed tissue. A spike will not be propagated beyond the gap because of this restriction on

---

* Later studies on the tænia coli and cat's intestinal muscle have shown that transmembrane pulses through an intracellular electrode usually fail to induce a propagated spike (p. 1325).

† Nagai & Prosser found a time-constant of 31 ± 9 msec. for intracellular polarization and 133 ± 0·24 msec. for external polarization in the cat's intestinal muscle, *i.e.*, a much smaller discrepancy than that found by Kuriyama & Tomita (1965) on the guinea pig tænia coli. In the guinea pig's vas deferens, the time-constants (time to reach 84 per cent. or erf 1 of the peak value) were 80–100 msec. and 3·5–5 msec. for external and internal polarization respectively (Tomita, 1967).

current flow, and a monophasic action potential will be recorded from electrodes on either side. It was possible to allow external currents to flow by simply shunting the gap with a resistance, and in this case the spike became diphasic and the gap no longer acted as a block. To obtain transmission, currents must have moved internally from cell to cell.

Membrane Capacity and Resistance

As Tomita (1966, 1967) has pointed out, the fact that the cells are interconnected means that, when current is applied through a single cell, it spreads electrotonically in three directions so that the spatial and temporal decay of current will be very much sharper than in a cable-like membrane. In consequence, the calculated resistance and capacity of the membrane, derived from these decay functions, may well be in error, the capacity too high and the resistance too small. Thus Nagai & Prosser estimated a membrane capacity of 40–45 μF/cm.^{2} and Kuriyama & Tomita (1965) of 10 μF/cm.^{2} for cat's intestine and guinea pig tænia coli respectively. These are very large values, and on morphological grounds the smooth muscle cell should have a low capacity by comparison with skeletal muscle, since a large contribution to the latter is made by the sarcoplasmic tubular system (Falk & Fatt, 1964; Adrian & Peachey, 1965) which is lacking in smooth muscle. The specific resistance, calculated by multiplying the input resistance with intracellular polarization by the cell surface area, is low, 560 and 320 ohm.cm.^{2} in cat intestine and tænia coli respectively, and Tomita considered that the low values were due to the uneven spread of current, so that simple cable-theory was inapplicable; and models based on interconnected capacities and resistances showed that large discrepancies between results with external and internal current application could be obtained. Tomita therefore calculated the membrane capacity from different experimental parameters; thus Tasaki & Hagiwara (1957) had shown that the membrane capacity could be calculated from the propagation velocity and time-course of rise of the foot of the action potential.* Using experimental values, namely a time-constant for the foot of the tænia propagated spike of 6 msec., a propagation velocity of 7 cm./sec., and membrane time- and space-constants of 80 msec. and 1·6 mm. respectively, they obtained a value of 2·8 μF/cm.^{2}, much smaller than the 10 μF/cm.^{2} calculated from intracellularly applied currents, and of the order of magnitude expected from experiments on frog muscle fibres by Falk & Fatt (1964) and Adrian & Peachey (1965). The membrane resistance can be calculated from the membrane capacity and the time-constant ($\tau = C \times R$), and comes out at 2–3.10^4 ohm.cm.^{2}, and is thus of the same order of magnitude as that of frog slow muscle fibre, but more than 30 times that calculated from intracellular current application,† namely 320 ohm.cm.^{2}.

Propagated Spike. Tomita (1966, 1967) has examined the propagated spike recorded intracellularly, in tænia coli when stimulated with large extracellular electrodes; the preparation was in a hypertonic medium which abolished spontaneous activity, possibly by hyperpolarizing the cells; and the passive spread from the stimulating electrodes was prevented by an insulating partition between stimulating and recording sites. Under these conditions, applied current pulses

* The equation employed was:

$$C_m = \frac{a}{2v^2 R_1 T}$$

a is the radius of the fibre, v the velocity of spike propagation, R_1 the specific internal resistance, and T the time-constant for the exponential rise of the foot of the spike.

† Kobayashi, Prosser & Nagai (1967), in the course of an elaborate experimental analysis of the differences between intra- and extracellular recording, have calculated the space-constant, λ, from the equation: $\lambda = \sqrt{(R_m/R_i) \times r/2}$ where R_m and R_i are the membrane and internal resistances, and r the radius of a hypothetical conducting bundle. Using their low value of 1,050 ohm.cm^2 for R_m, they obtained a value of 1·1 mm. for λ compared with an experimental one of 1·1 mm. Again Abe & Tomita (1968) have made new measurements of spatial and temporal decay of applied currents in tænia coli and deduced the space constant and other parameters. The membrane capacity is probably 2–3 μF/cm.^{2} and resistance 30–50 kΩcm.^{2}. Finally, Kurijama, Osa & Toida (1967) have compared the membrane parameters of the ureter with published values for the tænia coli. They have also compared electrical activity and membrane parameters in different intestinal muscles; they conclude that, whilst the membrane parameters of the individual cells are similar, the time-constants, measured with transmembrane currents, the chronaxies, and the conduction velocities differ, presumably because of the differing numbers and arrangements of cells in the unitary bundles.

were conducted electrotonically to the intracellular recording electrode, and the conventional cable properties of the system could be measured.

When the duration of the applied external stimulus was shortened, Tomita (1967) often obtained graded spike activity, and propagation was decremental. There seemed to be a critical amplitude of spike required that it conduct normally; when larger than this critical value, it increased in amplitude as it propagated to reach its full size; when the induced spike was below this critical size it decremented and finally became only a passive electrotonic potential. On the basis of his studies Tomita postulated the existence of three types of membrane on the smooth muscle cell; *A*, at the nexus, conducting externally applied current from cell to cell; *B*, an area immediately surrounding this, giving rise to the slow potential; *C*, the remainder, giving rise to spikes when area *B* builds up to a sufficient degree of depolarization.

Importance of Cell Contacts. By comparing the electron-microscopy of different visceral smooth muscles with their electrical characteristics, in particular their conduction velocities, the importance of cell contacts was emphasized. In Table LXXXIII some of these measurements have been tabulated, and it is seen that there is a good correlation between conduction velocity and cell length, the

TABLE LXXXIII

Relation of Certain Structural Features of Smooth Muscle with Conduction Velocity.
(Prosser, Burnstock & Kahn, 1960.)

Muscle	Conduction Velocity (cm./sec.)	Cell Length (μ)	Cell Diameter (μ)	Extracellular Space (per cent.)
Pig œsophagus	15	220	6	4·4
Guinea-pig				
tænia coli	7	150	6	12·0
Cat small intestine				
circular	4	120	6	9·0
longitudinal . . .	3	120	6	12·5
Dog retractor				
penis	1	90	6	18·2
Pig carotid artery . . .	nil	30	3	39·0

greater the cell length the greater the velocity. If the regions of specialized contact between cells are the only regions where transmission from cell to cell is possible, and if there is some delay here, then clearly the fewer junctions that have to be crossed the greater will be the conduction velocity. The larger the cells the fewer will be the number of junctions necessary to be crossed in any length of muscle. Again Table LXXXIII shows an inverse correlation between conduction velocity and extracellular space; the pig œsophagus, with an extracellular space of 4·4 per cent., has the greatest velocity, whilst there is no conduction of impulses in the muscle of the pig carotid artery with a space of 39 per cent. In this last muscle there are no specialized areas of contact between cells. Thus the more compact the muscle the greater will be the number of contacts and the greater the conduction velocity, other things being equal.

Effect of Stretch on Spike Discharge. Bülbring noted that the size of the resting potential depended on the degree of stretch imposed on the muscle; thus, when held at its *in situ* length, the average potential was 60 mV whilst, when stretched so tight that no movement could be detected, the potential was only 43 mV. On the other hand, the frequency of the spontaneous spike discharge increased with increasing stretch. In the longitudinal muscle of the colon, Gillespie observed a similar effect of stretch, the resting potential being reduced by some 5 mV and the spike discharge becoming continuous when the muscle was tightly stretched. Stimulation of the sympathetic nervous supply, which is inhibitory to this muscle, caused the resting potential to return to its unstretched value whilst it abolished spike activity and contraction.

Tension and Spike Activity

By measuring the tension in the muscle simultaneously with its electrical activity, Bülbring (1955) showed that it was essentially the change in tension, rather than in length, that determined the effects of stretch on resting potential and frequency of spike discharge, an increase in tension being associated with depolarization of the cell membrane and increased frequency of discharge. As Bülbring pointed out, therefore, the smooth muscle fibre differs radically from the striated fibre in being sensitive to stretch and, in so far as this causes depolarization with spike discharge, it resembles the stretch-receptor of striated muscle.

In a later study Bülbring & Kuriyama (1963) used as a criterion of the degree of stretch the ratio:—

$$\frac{\text{Weight (mg.)}}{\text{Length (mm.)}},$$

since this is a measure of the cross-sectional area; using this criterion, smooth curves could be plotted showing tension, spike-frequency, membrane potential, and so on, as a function of stretch. Except at the very highest degree of stretch, spike frequency and tension ran parallel. Thus, stretching of muscle, in so far as it develops tension, acts as a stimulus for contraction since there is a good correlation between frequency of discharge and force of contraction. We have here, then, in electrophysiological terms, an explanation for the vigorous activity developed in such organs as the intestine or ureter when they are distended artificially.

Effects of Na$^+$ and Ca^{++}. The effects of ions and drugs on spike activity of smooth muscle are more conveniently studied by extra-cellular electrodes, and this has been made possible by Burnstock & Straub's application of the sucrose-gap technique of Stämpfli (p. 1050) to this muscle. The spike potentials are remarkably similar to those obtained by intracellular recording, with the exception that the amplitude is reduced. It must be appreciated, of course, that the picture may be confused by asynchrony of spontaneously occurring spikes since the technique must necessarily involve recording of activity in many cells.

Na$^+$-Deficient Solutions

The fact that the action potential of the guinea pig tænia coli shows a reversal, or overshoot, would suggest that the mechanism is essentially the same as that involved in skeletal muscle and nerve, being determined by altered permeability of the excitable membrane to Na$^+$ and K$^+$. Studies on the effects of reducing the concentration of Na$^+$ in the outside medium, however, are perplexing. Holman showed that spike-height could be normal when the bathing solution contained as little as 20 mM Na$^+$; below this concentration the spike-height was reduced, and at 2 mM electrical activity was abolished. Even more striking is the observation of Singh & Acharya that, after soaking stomach muscle for hours in a non-electrolyte medium, both contraction and action potentials could be recorded.*

In a systematic study of the effects of Na$^+$-concentration on the size and rate of rise of the action potential, Bülbring & Kuriyama have confirmed that the

---

* Essentially similar results of Na$^+$-deficient solutions have been described by Burnstock & Straub (1958), Axelsson (1961), Daniel & Singh (1958), Bozler (1960) and Kolodny & Van der Kloot (1961).

size of the spike and overshoot are not dependent on the external concentration of Na^+ between 0 and 137 mM; on the other hand, the rates of rise and fall were dependent. Thus on placing a muscle in sodium-free solution (Tris chloride), all spontaneous activity ceased after 2–5 minutes, but then, after a pause of 5–10 minutes, spontaneous activity returned. The membrane potential rose during the first few minutes and then slowly fell over a period of 30 minutes to give a maximum depolarization of about 15 mV; the size of the action potentials became regular and an overshoot was recorded for at least 20–30 minutes in the absence of Na^+; the rates of rise and fall of the spike decreased whilst the local potential was abolished, the membrane potential being stable between spikes.*

Ca^{++} CURRENT

In the tænia coli, then, it appears that the slow-wave activity depends on an increased permeability to Na^+, but the current during spike activity can clearly be carried by some other ion than Na^+, and it is interesting that Nonomura, Hotta & Ohashi (1966) found no effect of concentrations of tetrodotoxin as high as 5.10^{-6} g./ml. on the resting potential or on the amplitude of spontaneous spikes. 0·5 mM $MnCl_2$ abolished spontaneous discharge without causing hyperpolarization, thus differentiating it from adrenaline (p. 1339). These two findings suggest that Ca^{++} carries the inward current in Na^+-free solutions since Hagiwara & Nakajima showed that Mn^{++} inhibited the Ca^{++}-spike in the barnacle muscle fibre (p. 1112) whilst tetrodotoxin did not. In the vas deferens, which gives a graded spike activity in response to intracellular stimulation, Na^+-deficient solutions had no effect; when the muscle was stimulated with large external electrodes there was little or no effect on the propagated spike-threshold, overshoot or maximum rate of rise; on this muscle, too, tetrodotoxin had no effect. According to Bozler (1960) and Axelsson (1961), Ca^{++} is, indeed, necessary for the development of spikes in Na^+-deficient media.†

Calcium

The effects of Ca^{++} on smooth muscle electrical activity are confusing and have been summarized by Bülbring & Kuriyama, who have themselves examined the effects of excess and deficiency on the behaviour of the tænia coli muscle. The effects are consistent with the view that Ca^{++} influences the Na^+-permeability activating mechanism, as with the squid axon. It will be recalled that the effects of an increase of concentration were equivalent to those of a hyperpolarization, in that the steady-state inactivation of the Na^+-mechanism was reduced whilst the degree of depolarization required to activate the Na^+-mechanism was correspondingly raised.

As we have seen, the effects on activity in the squid axon are complex and can only be predicted by application of the Hodgkin-Huxley equations to the system. In the squid axon, myelinated nerve and skeletal muscle, excess of Ca^{++} decreases the rate of rise of the spike by virtue of its reducing Na^+-permeability, i.e., by antagonizing the effects of depolarization during the spike. By contrast, in smooth muscle an excess of Ca^{++} actually increases the rate of rise of the spike, but this is probably because the membrane potential in this tissue is normally low; if this means that the Na^+-mechanism is normally inactivated, then excess Ca^{++} manifests its effects predominantly on the steady-state condition of

* Exchange of Na^+ between smooth muscle and medium is rapid (Goodford, 1962), so that if the internal concentration of Na^+ falls in proportion to the fall in external Na^+, the equilibrium Na^+-potential, which presumably determines the height of the spike, may not be strongly affected (Goodford, 1962).

† According to Marshall (1963) uterine muscle shows some dependence on Na^+ for spike activity; the spike-height remains normal down to 50 per cent. normal Na^+ but after that falls and at about 10 per cent. spike activity fails.

Na^+-activation. When the membrane potential of nerve is held at a low value, then excess Ca^{++} does increase the rate of rise of the spike.

EFFECTS ON MEMBRANE POTENTIAL

According to Kuriyama (1964) the resting membrane potential of the vas deferens smooth muscle increases from -50 mV in 0.25 mM Ca^{++} to -90 mV in 25 mM Ca^{++}, due perhaps to a decrease in Na^+-conductance; at any rate the influence was not on Cl^- since the same effect was obtained when the slowly penetrating ethane sulphonate ion was substituted for Cl^-. A complicating factor in the effects of Ca^{++} may well be that it acts as a significant carrier of current, even when the medium is not Na^+-deficient (Bennett, 1967), whilst its function in the coupling between excitation and mechanical contraction (Hurwitz, 1965; Daniel, 1965) may well be another obscuring factor.

Unstable Membrane Potential. In general, then, we may regard the resting membrane potential of such spontaneously active muscles as the guinea-pig tænia coli, uterus, etc. as being essentially unstable because of a high permeability to Na^+; the level of this permeability may well be determined by the concentration of Ca^{++} in the medium, and this will affect both the "resting" level and the degree of stability of this level. The spike may be assumed to be associated with an increased permeability to Na^+ and K^+ so timed that the increased permeability to Na^+ leads to a small reversal of the polarity of the membrane potential. In electrolyte-deficient solutions it would seem that, unless local, residual concentrations of ions in the extracellular space are capable of causing current-flow, the spike is due to a breakdown of permeability leading perhaps to an escape of internal organic anions; if these escaped faster than K^+, the membrane would depolarize and the potential would reverse its sign (Goodford, 1962). In addition, alkali-earth cations may act as charge carriers.

The Ureter. The external recordings of Bozler suggest that the ureter's electrical activity consists in the development of a plateau of depolarization from which spikes are fired; intracellular recordings essentially confirm this, although, as Fig. 714 illustrates, the behaviour in response to stimulation is variable. The spikes showed an overshoot of up to 10 mV. As with other smooth muscles, tetrodotoxin had no effect on spike activity, whereas $MnCl_2$ blocked this whether it was spontaneous or in response to an external stimulus. The effects of varying the external concentration of Ca^{++} suggest that this ion may carry current during the spike (Bennett *et al.* 1962). Thus in a Na^+-free medium, spike activity was only possible if the concentration of Ca^{++} was raised from 2.5 to 5.0 mM (Kobayashi, 1965).

Arterial Smooth Muscle. According to Keatinge (1968), the smooth muscle of the sheep's carotid artery does not exhibit spontaneous electrical activity, but when kept in a Ca^{++}-free medium spontaneous spike activity develops; this is abolished by reducing the external concentration of Na^+, which presumably carries the inward current; at any rate there is an increased influx of ^{24}Na when activity is brought about by placing the tissue in a Ca^{++}-free medium; the extra influx per spike was 1.2 pmole.cm.$^{-2}$; if, as seemed likely, not all the cells were active, the figure may compare well with that of 15 pmol.cm.$^{-2}$ for the cephalod giant axon (Hodgkin & Keynes, 1955).* In spite of this dependence on Na^+ for carrying the inward current, tetrodotoxin did not affect spike activity (Keatinge, 1968).†

---

* The extracellular space derived from the distribution of ^{14}C-sucrose was approximately 50 per cent.; to identify a slowly exchanging intracellular fraction, by contrast with a rapidly exchanging extracellular fraction, was difficult since it became clear that a large part of the slowly exchanging fraction was extracellular, and probably bound to collagen etc. Thus the total slow-phase Na^+ was 5.9μ mol./g. of which 3.95 were bound and 1.96 intracellular, corresponding to an intracellular concentration of 7.3 m-mole/kg. water.

† Bozler, in his classification of smooth muscle, described those of blood vessels as belonging to the multiple-unit type, but this is not generally true; thus Bolton (1968), for example, finds that the longitudinal muscle of the mesenteric artery of the domestic fowl behaves very similarly to visceral muscle; this analogy extends to the actions of noradrenaline and adrenaline, which cause inhibition and reduced frequency of action potentials.

Spikes of guinea-pig ureter

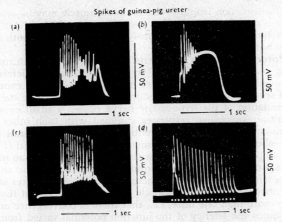

FIG. 714. Intracellular records of different patterns of action potentials of guinea-pig ureter elicited by extracellular stimulation. *a–d* are recorded from different cells, and the distances between stimulating and recording electrodes are not constant. (*a*) Complicated plateau phase and spikes of various size; (*b*) large and rapidly rising plateau and spikes changing to damped oscillation; (*c*) small plateau and spikes of consistent size; (*d*) spikes triggered by repetitive stimulation (6 c/s). (Kuriyama, Osa & Toida. *J. Physiol.*)

Nervous Transmission

Although the phenomena of excitation and conduction so far described are apparently truly myogenic, the extraordinarily rich innervation of such visceral smooth muscles as those in the wall of the gut leaves little doubt that the excitability of the cells may be modified by nervous activity, whilst direct studies on stimulation of these nerves show also that spike activity can be initiated in quiescent muscle.

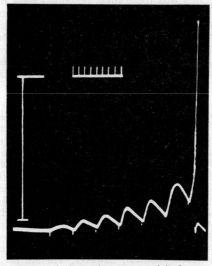

FIG. 715. Intracellular record of membrane potential of a smooth-muscle cell of the vas deferens during repetitive stimulation of the hypogastric nerve. Note how the depolarizations build up to spike-height. Stimulus duration, 1 msec. Time-marker, 100 msec. Vertical calibration, 50 mV. (Burnstock & Holman. *J. Physiol.*)

The smooth muscle of portal vein is also of the single-unit type, records of action potentials being obtained by the sucrose-gap technique; noradrenaline and sympathetic stimulation are excitatory, however, by contrast with the tænia coli (Holman *et al.* 1968).

The mode of transmission from nerve to smooth muscle may be studied in a normally quiescent muscle such as the vas deferens, which is activated by the sympathetic system, or in the spontaneously active intestinal smooth muscle, which is activated by both.

Junction Potentials of Vas Deferens. Burnstock & Holman measured intracellular resting potentials of 57–80 mV in the cells of the vas deferens; these are higher than those in spontaneously active muscles, as we should expect, since here the absence of spontaneous activity presumably betokens a more stable membrane potential. Stimulation of the hypogastric nerve caused a small depolarization—a junction potential; successive responses summed with each other so that with repetitive stimulation the junction potential would build up until it exceeded a critical value—about 37 mV—when a spike would be discharged (Fig. 715). The spikes had an amplitude of 57–90 mV with overshoots of up to 20 mV. Every cell impaled gave a junction potential on stimulation of the nerve, and since the size of the junction potential increased with increasing numbers of fibres stimulated, we may conclude, not only that every muscle cell is innervated, but that they receive impulses from more than one fibre, *i.e.*, that there is convergence on the smooth muscle cells. Since the latency of the junction potential varied from 20 msec. to as high as 60 msec., it may be that the potential can be initiated, not only by localized liberation at a nerve-muscle contact, but also by diffusion of liberated transmitter from neighbouring cells, as postulated by Rosenblueth to account for the special features of excitation in the nictitating membrane. The decay of the junction potential was slow, and could be best accounted for by delay in destruction of the transmitter, which in adrenergic systems is apparently slow (Brown, Davies & Gillespie).*

Miniature Potentials

Burnstock & Holman described spontaneous miniature potentials varying in size from less than 1 mV to 12 mV with occasional "giants" of up to 22 mV; the spontaneous discharge never led to action potentials, nor yet to muscular contraction, so that it is unlikely that these miniature potentials may be regarded as the basis for spontaneous muscular activity. The effects of partial denervation and of treatment with reserpine—which depletes the nerves of their stores of transmitter —both strongly indicated that these small potentials were neural in origin, and might well have resulted from liberation of packets of transmitter. Thus, treatment with reserpine reduced both the frequency of the spontaneous potentials and the amplitude of the junctional potentials in response to stimulation; the reduction amounted to 90 per cent., which corresponded precisely with the reduction in amount of catechol amines in the nerves. It seems likely from this, and other findings, that the junction potential is equivalent to a mobilization of small units of activity, just as in the vertebrate end-plate.

Effects of Ca++

Lowering the concentration of Ca^{++} decreased the frequency of spontaneous junction potentials whilst raising the concentration initially increased the frequency although it fell subsequently. Raising the concentration of Ca^{++} increased the amplitude of the junctional potential in response to nerve stimulation and raised the threshold for triggering a spike; to a large extent, as with skeletal muscle, Ca^{++} and Mg^{++} were antagonistic in their effects on the junction (Kuriyama, 1964). Bouillin (1967) has shown that the release of noradrenaline from the sympathetic nerve terminals in the intestine is increased by Ca^{++}. The vas deferens thus shows great similarities to the skeletal muscle fibre, at any rate in so far as the extensive influence of the nerve supply is concerned; and it is interesting that in the electron miscroscope Merrillees, Burnstock & Holman (1963) observed very few regions of close contact between muscle cells, which were usually some 500–800A apart. It seemed doubtful, however, whether each muscle cell could

---

* Bennett & Merrillees (1966) examined the excitatory junction potentials of the vas deferens in response to hypogastric nerve stimulation. They estimated an equilibrium potential of −60 mV by extrapolation. They attributed the long time-course of the junction potential to temporal dispersion in the release of transmitter, due to the low velocity of transmission in nerve terminals, and to the absence of any enzymatic mechanism for destroying liberated transmitter, the situation in smooth muscle being analogous with that of eserinized skeletal muscle. Tomita (1967) has discussed the decay of the junction potential in some detail.

receive a close type of contact with a single nerve fibre, and they concluded that activation occurred, not only by transmitter release, but also by electrotonic spread of activity between cells.

Nervous Influences on Intestine. The pioneering studies on intestinal muscle were carried out by Gillespie (1962); the parasympathetic (pelvic) nerves increase muscular activity, and in muscle showing spontaneous spikes it was found that these could be "driven" so as to synchronize with the nerve stimuli up to a maximum of 2/sec.; as the frequency increased, the responses became irregular and the baseline rose, indicating depolarization. Finally, there were no spikes, and the membrane remained in a stable state of depolarization with contraction maintained at its maximal value. Single stimuli caused a slow prolonged depolarization after a latency of some 400 msec. and lasting some 600 msec.; if it was large enough it led to a spike potential.

Fast-rising junction potentials merged without discontinuity into a spike, so that in this tissue a single stimulus could cause a spike by contrast with the vas deferens which requires the summation of several. In the instances where the spike rose straight off the junction potential it is likely that the latter initiated the spike; at the other extreme, Gillespie & Mack (1964) found cells in which the spike was so delayed as to appear on the declining phase of the junction potential, and it is likely that this appearance was due to conduction from an adjacent pacemaker cell. This work has been extended very considerably by the studies of Bennett and his colleagues. In the tænia coli, sympathetic innervation is by way of perivascular nerves and is inhibitory, whilst the intramural innervation, derived from Auerbach's plexus (Fig. 705, p. 1313) is mainly inhibitory, but not entirely. The excitatory action is cholinergic and blocked by atropine, so that stimulation of the intramural nerve supply, in the presence of atropine, gives purely inhibitory activity, as revealed by relaxation of the muscle (Burnstock, Campbell & Rand, 1966).

Inhibitory Activity

Repetitive stimulation of the perivascular nerves caused a decrease in the frequency of spontaneous spikes, an effect that was blocked by bretyllium and guanethidine, indicating the adrenergic character of the nerves. A single stimulus had no recordable effect, whilst repetitive stimulation built up a hyperpolarization of the muscle cell that modified the character of the spike, if it occurred, or blocked it completely. Thus the "driven" type of activity (p. 1325) could be blocked by stimulating at 60/sec., and this only resumed several seconds after cessation of the stimulation; the degree of depression was clearly related to the size of the hyperpolarization, and this increased with the frequency of stimulation. When spike activity was blocked, a small membrane depolarization could often be observed at the time when a spike would have been expected; its time-relations were similar to those of the foot of the action potential, so that it may be that here one was observing an electrotonically transmitted influence of spike activity in a neighbouring cell.

The hyperpolarization following stimulation was smaller in the pacemaker type of cell, and the effect on the spikes was to reduce their frequency and to increase the level of membrane potential at which the spike began, and to increase the potential to which the membrane repolarized; inhibition thus caused an increased spike-height. There was also an increase in rate of rise and fall, effects that may have been due to the hyperpolarization increasing the Na^+-activation mechanism.

With the intramural nerves, in the presence of atropine to block excitatory activity, a single pulse produced a hyperpolarization, or inhibitory junction potential, of up to 25 mV; its time-course for decay was long (500 msec. compared with 3 msec. for decay of an applied hyperpolarization), so that it may be that transmitter action is prolonged. This inhibitory action summated up to frequencies of stimulation of 10/sec. but at higher frequencies the hyperpolarization decreased; it was unaffected by adrenergic blocking agents, and since it occurs in the presence of atropine, we may postulate some, at present, unidentified transmitter.

In many cells the amplitude of the inhibitory junction potential was unaffected by altering the membrane potential; this is contrary to the usual behaviour at junctions where, as we have seen, hyperpolarization increases, and depolarization decreases, the junction potential, so that at a certain potential—the reversal potential—the junction potential is zero. This applies also to sympathetic inhibitory activity, so that it would seem that the measured junction potentials in a given impaled cell are usually due to events taking place in other cells and transmitted electrotonically (Bennett & Rogers, 1967).

Excitatory Response

When the intramural nerves are stimulated in the absence of atropine, both excitatory and inhibitory responses may be obtained (Bennett, 1966a), and since it is likely that cells may receive both types of innervation, the response of a given impaled cell is likely to represent the balance of opposing influences. Fig. 716 illustrates both an excitatory and inhibitory response in cells less than 1 mm. apart. In general, some 10 out of 80 cells gave excitatory junction potentials and 70 out of 80 inhibitory potentials. With a quiescent cell the spike could be seen to rise directly off the junction potential, whilst with spontaneously active cells the rising phase was more complete, the spike only starting after the junction potential had reached its full height, so that it is possible that the spike was initiated by the junction potential in quiescent cells but in spontaneous cells the spike might have

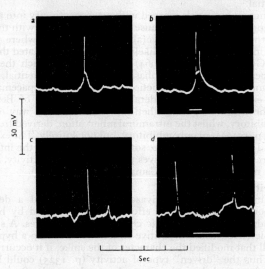

FIG. 716. Effects of intramural nerve stimulation on two quiescent cells from the same preparation less than 1 mm apart. Stimulation with single pulses in *a* and *c*. Repetitive stimulation at 20 and 10 c/s in *b* and *d* respectively, during the period given by the horizontal line. Note the rebound action potential at the end of the I.J.P. in *c* and *d*. (Bennett. *J. Physiol.*)

been initiated elsewhere and have invaded the impaled cell during the junction potential. With repetitive stimulation there was never any summation of excitatory junction potentials, and if the stimulation was prolonged, the spikes were abolished, perhaps because inhibitory influences were predominating.

Rebound Activity

A striking feature of the response to inhibitory stimulation is the rebound, in the form of increased spike activity. In a normally quiescent cell, the response to inhibitory stimulation is an inhibitory junction potential, and this is followed by a spike (Fig. 717). When a spontaneously active cell is stimulated, firing ceases but is resumed at the end of stimulation at a higher frequency than before; with repetitive stimulation, the greater the summated junction potential the greater the duration and frequency of the rebound discharge. Thus with a low frequency of stimulation, a single rebound spike follows each stimulus, whereas, when the frequency is increased, a sustained inhibitory junction potential results, giving rise to a burst of activity on cessation. Since spike activity has its correlate in muscular contraction, there are some interesting consequences; thus stimulation of an inhibitory nerve can give rise to a powerful after-contraction or, if the frequency is low, it may lead to contraction during stimulation (Bennett, 1966b; Campbell, 1966).

Effects of Muscle Stimulation

When intestinal muscle is stimulated by external electrodes, the responses can represent the direct activation of the muscle cells by the depolarizing current, or they may be indirect effects of stimulation of the intramural nerves. There seems little doubt from Bülbring & Tomita's (1967) study on the tænia coli that the hyper-polarizing responses described by Bennett and his colleagues are, indeed, neuro-genic. Kuriyama, Osa & Toida (1967) have made use of chemicals such as tetrodo-toxin, atropine, Ba^{++} etc., to differentiate muscular from neural responses to electrical stimulation; they concluded that the spike occurring in response to a short pulse is a direct muscle response, whilst the slow depolarization, delayed hyperpolarization (inhibitory junction potential) and post-inhibitory rebound are all neurogenic; thus tetrodotoxin blocked all these activities except the spike.

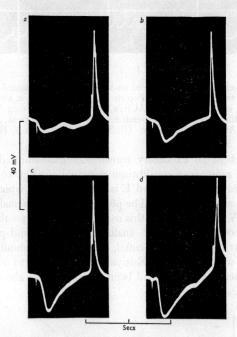

FIG. 717. The effects of increasing the strength of stimulation across the taenia coli on the I.J.P. The pulse strengths increased from *a* to *d*. Note the inflexion on the recovery phases of the I.J.P.'s in *b* and *d*. (Bennett, Burn-stock & Holman. *J. Physiol.*)

Effects of Acetylcholine and Adrenaline

Acetylcholine. So far as the smooth muscle of the intestine is concerned, acetylcholine increases the force of its contraction, whilst adrenaline has an inhibitory action, and we may assume that these drugs are mimicking the actions of excitatory and inhibitory nerves, producing excitatory and inhibitory junction potentials (p. 1335). Bülbring found, in fact, that acetylcholine did lower the resting potential of the tænia coli, increased the spike frequency, and prolonged the duration of the spike. The primary effect of the drug may well be its depolarizing activity, the increased spike frequency being the consequence of the lowered resting potential that would favour the development of pacemaker activity; at any rate Bülbring found that cathodal polarization would mimic the action of acetylcholine. More elaborate studies by Burnstock (1958) and Bülbring & Kuriyama have generally confirmed these effects of acetylcholine (Fig. 718), the depolarization being associated with increased excitability to direct stimulation and with an increased conduction velocity.

Effects of Membrane Potential

Bennett (1966) has analysed the effects of alterations of membrane potential on this transmitter action, and shown that $V - E_m$ is a linear function of E_m, as

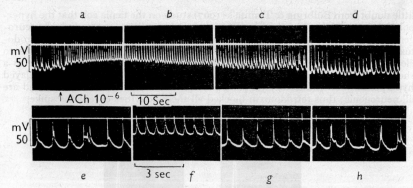

FIG. 718. The effect of acetylcholine on the membrane potential and electrical activity of tænia coli muscle. Top row: slow film speed, *a*, start of exposure to ACh; *b*, end of exposure to ACh; *c*, 70 sec. after removing ACh; *d*, 2 min. later. Bottom row: fast film speed, *e*, before ACh, *f*, at peak, *g*, 80 sec. after removing ACh, *h*, 2 min. later, (Bülbring & Kuriyama. *J. Physiol.*)

one would expect of an excitatory transmitter that shunted the membrane potential through a circuit of equilibrium potential E_e (Fig. 598, p. 1145). Here E_m is the initial resting potential and V is the potential to which the membrane is polarized by an applied current. The plot indicated a reversal, or equilibrium, potential of 26 mV (Fig. 719), and this may represent the result of an increased permeability to both Na^+ and K^+ analogous with the end-plate potential of skeletal muscle. At the reversal potential, the transmitter should have no effect, and this explains why Bülbring & Kuriyama were unable to obtain any further depolarization when the muscle had been depolarized to about -14 mV with high external K^+.

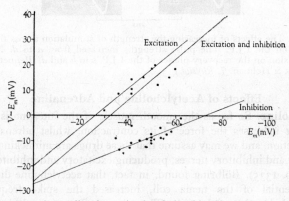

FIG. 719. Experimentally determined relationship between the membrane potential change, $V - E_m$, caused by the application of a transmitter substance to the smooth muscle cells of the taenia coli, and the initial resting potential, E_m, of these cells. Closed squares, application of acetylcholine. Closed circles, application of adrenaline. (Bennett. *Nature.*)

Increase in Permeability

That the effect is a non-selective increase in permeability is suggested by several studies; for example, Bülbring & Kuriyama found that the depolarization occurred in Na^+-free solutions, whilst Durbin & Jenkinson found increased inward and outward fluxes of K^+, Cl^- and Br^-; the evidence suggested, too, that permeability to Na^+ and Ca^{++} was also increased, although the interpretation of flux studies with these ions was complicated by the existence of inexchangeable fractions.

Action on Depolarized Muscle

Evans, Schild & Thesleff (1958) showed that carbachol, a stable analogue of acetylcholine, and acetylcholine increase the tension developed by smooth muscle, even though it is depolarized in KCl or K_2SO_4 (Evans, Schild & Thesleff, 1958); since the penetration of Ca^{++} may well be the link between the action potential and the mechanical contraction (Shanes, 1958), the increased tension caused by carbachol and acetylcholine may be due to an increased permeability to Ca^{++}; certainly the increase in tension is prevented if there is no Ca^{++} in the medium (Durbin & Jenkinson, 1961); moreover, Ca^{++} itself will cause contraction in a depolarized muscle (Edman & Schild, 1962).

Adrenaline. This drug causes a hyperpolarization of the tænia coli and a diminution in spontaneous spikes; its effects may be matched in this respect by anodic polarization. The effect depended on the membrane potential, so that if this was above 65–70 mV there was no hyperpolarization, but spike discharge was blocked. This suggests that the reversal potential for the inhibitory transmitter action of adrenaline is about 65–70 mV, and Bennett (1966) has shown that, when $V - E_m$ is plotted against E_m, a straight line is obtained, as would be expected on the basis of the simple inhibitor circuit of Fig. 631 (p. 1207), giving a reversal potential of 73 mV (Fig. 719). This is in the region of 80 mV predicted on the assumption that adrenaline increases permeability to K^+ to the point that the membrane potential is the K^+-potential. When the external concentration of K^+ is high, and the membrane partially depolarized, the membrane potential is very close to the K^+-potential, so that we should not expect adrenaline to have a hyperpolarizing effect; and this was indeed found by Bülbring & Kurijama.

Combined Excitation and Inhibition

By the simple application of Ohm's Law to the equivalent circuits represented by the membrane electromotive force with parallel inhibitory and excitatory e.m.f.'s, E_i and E_e, the combined effects of the two transmitters, acting simultaneously, can be predicted to give the line of Fig. 719; the effect is to bring the resting potential down to -30 to -40 mV.

Increased K+-Permeability

Direct proof of the action of adrenaline in increasing K^+-permeability was provided by Jenkinson & Morton (1967); earlier studies of Born & Bülbring (1956) and Huter, Bauer & Goodford (1963) had, indeed, shown an increased uptake of ^{42}K but no comparable change in efflux. Jenkinson & Morton pointed out that the blocking of spike activity would reduce efflux of K^+ and so mask any effect due to adrenaline itself. By studying the muscle in isotonic K_2SO_4, under which condition spike activity is blocked, they showed that noradrenaline increased the rate of efflux of ^{42}K from preloaded muscles; as Fig. 720 shows, there was no significant effect on loss of ^{36}Cl. Phentolamine, which blocks the action of noradrenaline, blocked the increase in ^{42}K-efflux, whilst a β-inhibitor (p. 1340), pronethalol, had no effect. Thus the inhibitory action of noradrenaline in blocking spike activity through membrane hyperpolarization represents a reaction with α-receptors (Jenkinson & Morton, 1967b). Exchanges of Na^+ are so rapid that comparable experiments with this ion were difficult to carry out, but by artificially increasing the resting Na^+-content it was possible to establish whether significant differences in uptake of ^{24}Na occurred in the presence and absence of noradrenaline. In fact, no difference was observed, although it was possible to establish an increased uptake with the cholinergic drug, carbachol.

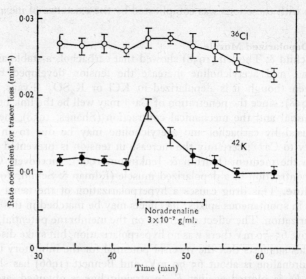

FIG. 720. The effects of noradrenaline on the rate of loss of ^{42}K (●) and ^{36}Cl (○), from the guinea pig taenia coli. (Jenkinson & Morton. *J. Physiol.*)

Relation to Metabolism

That the energy supply and metabolism of tissues are related to the action of adrenaline has been clear for a long time, and Bülbring and her colleagues have been concerned to relate the specific metabolic events with the hyperpolarization caused by adrenaline, along with its blockage of spike activity and mechanical relaxation. Adrenaline was shown by Axelsson *et al.* to depend for its inhibitory action to a large extent on the possibilities of metabolism; thus, if this was reduced by glucose-depletion or by iodoacetate, the inhibitory action of the hormone was prevented. It was considered that adrenaline acted by stimulating the phosphorylase responsible for breakdown of glycogen, but later studies ruled this out. There is little doubt that adrenaline increases the concentration of organic phosphates in the tænia coli, the time-course of its action in this respect being similar to that of blockage of the spike discharge (Bueding *et al.* 1967); since ATP is a potent inhibitor of smooth muscle contraction (Gillespie, 1934) there is some substance for this view relating metabolism to activity. When the muscle is held in a glucose-free medium under anærobic conditions, adrenaline now fails to increase the concentration of phosphorylated compounds, and it also fails to block spike discharge; on readmission of O_2, with or without readmission of glucose, the biochemical and physiological effects of adrenaline are restored.*

Alpha- and Beta-Actions

Adrenaline has two types of action on tissues, suggesting that there are two types of receptor with which it may react; in accordance with Ahlquist's (1948, 1962) nomenclature, those actions, largely excitatory, that are blocked by such "classical" adrenaline antagonists as ergotoxin, dibenamine and phentolamine, are the result of reaction with the α-receptor; reaction with the β-receptor, largely inhibitory, is inhibited by dichloroisoprenaline, and pronethalol. Noradrenaline, the adrenergic transmitter, has a mainly α-action, whilst isoprenaline (isopropylnoradrenaline) has mainly β-action. The inhibitory action of adrenaline on intestinal smooth muscle is unusual since it is excitatory to most others, such as the nictitating membrane, arterial smooth muscle, and so on; furthermore, the inhibitory action is due to activation of both α- and β-receptors, since it requires both α- and β-blocking agents to prevent the action of adrenaline completely (Ahlquist & Levy, 1959). The

* Adrenaline increases the concentration of cyclic AMP in smooth muscle (Bueding *et al.* 1966); Bartelstone, Nasyth & Telford (1967) have established correlations between activities that are likely to increase intracellular concentration of cyclic AMP and those that cause contraction of blood vessels *in vivo*.

β-action of adrenaline is revealed as an inhibition of the contraction caused by adding Ca^{++} to completely depolarized muscle (Jenkinson & Morton, 1967), *i.e.*, it seems likely that the action is on the inside of the muscle rather than on its membrane properties, and it is tempting to speculate that it is this β-activity that is the basis of the effects described by Bueding *et al.* (1967). The hyperpolarization and blockage of spike activity may well be exclusively the result of α-activity, since Jenkins & Morton (1967b) showed that the same effects could not be obtained by isoprenaline, whilst the action could be blocked by phentolamine but not by pronethalol. Jenkinson & Morton suggest, therefore, that the β-mediated activity, shown under these artificial conditions of depolarized muscle, occurs normally and that it causes relaxation by a mechanism that does not involve a change in K^+-permeability, but depends on intracellular changes such as those described by Bueding *et al.*; certainly intestinal smooth muscle relaxes more rapidly following activation by α- as compared to β-receptors (Van Rossum & Mujić, 1965).

MUSCLES OF *PHASCOLOSOMA* AND *THYONE*

In some smooth muscles, notably the long retractors of the wall of *Thyone* and the proboscis retractors of the worm *Phascolosoma*, electrical stimulation leads to propagation but, according to Prosser's studies, this is entirely due to, and limited by, nervous propagation, so that in this case, although the muscle appears to behave like the syncytial type, its physiology is really determined by the unitary behaviour of the individual fibres. An initial study of the proboscis retractor of *Phascolosoma* by Prosser, Curtis & Travis suggested the typical syncytial behaviour; according to conditions of stimulation, the response to an electrical stimulus was a fast-conducting spike, propagating at 1·6 m.p.s., or a slow wave of negativity, conducting at 0·34 m.p.s. The slow potential, like the action potential of *Mytilus*, showed the phenomenon of summation with repetitive stimulation, so that at 60/sec. a smooth plateau of negativity was obtained. Contraction of the muscle correlated well with the potential changes, the spike being associated with a brief twitch, whilst the slow wave was associated with a sustained contraction. The authors concluded from the general features of the response to electrical stimulation that the muscle, which is made up of small fibres some 1–2 mm. long and 3–6 μ in diameter, was a syncytium, but there was no histological evidence for the presence of two types of fibre, corresponding to the twitch and tonus types of contraction. However, Prosser & Melton, by analysis of the nerve supply to the muscle, by studying the effects of nerve degeneration, and by the use of micro-recording electrodes, showed that the conduction was by way of the nerves, the muscle being apparently incapable of a propagated response. In this muscle, then, the individual fibres may indeed be treated as single functional units; the appearance of slow and fast responses is due to the different innervations, but whether this means that different muscle fibres are involved in the responses, or that the same fibres respond to different nerves, remains to be seen. (It should be mentioned that the electrical responses were responses of the muscle, the action potentials in the nerve would have been too small to detect with the external recording set-up.) Why such a system of discrete units should show action potentials with external electrodes has been discussed by Prosser & Melton, who have argued that, if the individual small fibres were depolarized asymmetrically in series, an action potential could be recorded, by contrast with the situation where each fibre was symmetrically and completely depolarized. A study of the effects of inter-electrode distance, and other factors, on the shape of the action potential agreed with this interpretation. A similar situation appears to be present in the long retractors of the body wall of *Thyone*; Prosser *et al.* failed to record propagated action potentials in response to an electrical stimulus at a distance greater than 2–5 mm. from the cathode, and contraction of the muscle was largely confined to the space between the stimulating electrodes. Examination of the nervous distribution (Prosser, 1954) showed that propagation was entirely confined to the field of distribution of the small branches of radial nerves.

REFERENCES

ABE, Y. & TOMITA, T. (1968). Cable properties of smooth muscles. *J. Physiol.*, **196**, 87–100

ADRIAN, R. H. & PEACHEY, L. D. (1965). The membrane capacity of frog twitch and slow muscle fibres. *J. Physiol.*, **181**, 324–336.

AHLQUIST, R. P. (1948). A study of the adrenotropic receptors. *Amer. J. Physiol.*, **153**, 586–600.

AHLQUIST, R. P. (1962). The adrenotropic receptor. *Arch. int. Pharmacodyn.*, **139**, 38–41.

AHLQUIST, R. P. & LEVY, B. (1959). Adrenergic receptive mechanism of canine ileum. *J. Pharmacol.*, **127**, 146–149.

AXELSSON, J. (1961). Dissociation of electrical and mechanical activity in smooth muscle. *J. Physiol.*, **158**, 381–398.

AXELSSON, J., BUEDING, E., & BÜLBRING, E. (1961). The inhibitory action of adrenaline on intestinal smooth muscle in relation to its action on phosphorylation activity. *J. Physiol.*, **156**, 357–374.

BARR, L. (1963). Propagation in vertebrate visceral smooth muscle. *J. theoret. Biol.*, **4**, 73–85.

BARR, L., BERGER, W. & DEWEY, M. M. (1968). Electrical transmission at the nexus between smooth muscle cells. *J. gen. Physiol.*, **51**, 347–368.

BARR, L., DEWEY, M. M. & EVANS, H. (1965). The role of the nexus in the propagation of action potentials of cardiac and smooth muscle. *Fed. Proc.*, **24**, 142.

BARTELSTONE, H. J., NASMYTH, P. A. & TELFORD, J. M. (1967). The significance of adenosine cyclic 3′, 5′-monophosphate for the contraction of smooth muscle. *J. Physiol.*, **188**, 159–176.

BENNETT, M. R. (1966, a). Transmission from intramural excitatory nerves to the smooth muscle cells of the guinea-pig taenia coli. *J. Physiol.*, **185**, 132–147.

BENNETT, M. R. (1966, b). Rebound excitation of the smooth muscle cells of the guinea-pig taenia coli after stimulation of intramural inhibitory nerves. *J. Physiol.*, **185**, 124–131.

BENNETT, M. R. (1966, c). Model of the membrane of smooth muscle cells of the guinea-pig taenia coli muscle during transmission from inhibitory and excitatory nerves. *Nature*, **211**, 1149–1154.

BENNETT, M. R. (1967). The effect of cations on the electrical properties of the smooth muscle cells of the guinea-pig vas deferens. *J. Physiol.*, **190**, 465–479.

BENNETT, M. R. & BURNSTOCK, G. (1966). Application of the sucrose-gap technique to determine the ionic basis of the membrane potential of smooth muscle. *J. Physiol.*, **183**, 637–648.

BENNETT, M. R., BURNSTOCK, G. & HOLMAN, M. E. (1966). Transmission from perivascular inhibitory nerves to the smooth muscle of the guinea-pig taenia coli. *J. Physiol.*, **182**, 527–540.

BENNETT, M. R., BURNSTOCK, G. & HOLMAN, M. E. (1966). Transmission from intramural inhibitory nerves to the smooth muscle of the guinea-pig taenia coli. *J. Physiol.*, **182**, 541–558.

BENNETT, M. R., BURNSTOCK, G., HOLMAN, M. E. & WALKER, J. W. (1962). The effect of Ca^{2+} on plateau-type action potentials in smooth muscle. *J. Physiol.*, **161**, 47–48 P.

BENNETT, M. R. & MERRILLEES, N. C. R. (1966). An analysis of the transmission of excitation from autonomic nerves to smooth muscle. *J. Physiol.*, **185**, 520–535.

BENNETT, M. R. & ROGERS, D. C. (1967). A study of the innervation of the taenia coli. *J. Cell Biol.*, **33**, 573–596.

BERKSON, J. (1933). "An Enquiry into the Origin of the Potential Variations of Rhythmic Contraction in the Intestine; Evidence in Disfavour of Muscle Action Currents." *Amer. J. Physiol.*, **104**, 67–72.

BISHOP, G. H., ERLANGER, J., & GASSER, H. S. (1926). "Distortion of Action Potentials as recorded from the Nerve Surface." *Amer. J. Physiol.*, **78**, 592–609.

BOLTON, T. B. (1968). Electrical and mechanical activity of the longitudinal muscle of the anterior mesenteric artery of the domestic fowl. *J. Physiol.*, **196**, 283–292.

BOLTON, T. B. (1968). Studies on the longitudinal muscle of the anterior mesenteric artery of the domestic fowl. *J. Physiol.*, **196**, 273–281.

BORN, G. V. R., & BÜLBRING, E. (1956). "The Movement of Potassium between Smooth Muscle and Surrounding Fluid." *J. Physiol.*, **131**, 690–703.

BORTOFF, A. (1967). Configuration of intestinal slow waves obtained by monopolar recording techniques. *Amer. J. Physiol.*, **213**, 157–162.

BOUILLIN, D. J. (1967). The action of extracellular cations on the release of the sympathetic transmitter from peripheral nerves. *J. Physiol.*, **189**, 85–99.

BOZLER, E. (1938). "Electric Stimulation and Conduction of Excitation in Smooth Muscle." *Amer. J. Physiol.*, **122**, 614–623.

BOZLER, E. (1939). "Electrophysiological Studies on the Motility of the Gastrointestinal Tract." *Amer. J. Physiol.*, **127**, 301–307.

BOZLER, E. (1942). "The Activity of the Pacemaker Previous to the Discharge of a Muscular Contraction." *Amer. J. Physiol.*, **136**, 543–552.

BOZLER, E. (1948). "Conduction, Automaticity, and Tonus of Visceral Muscles." *Experientia*, **4**, 213–218.

BOZLER, E. (1960). Contractility of muscle in solutions of low electrolyte concentration. *Amer. J. Physiol.*, **199**, 299–300.

BROWN, G. L., DAVIES, B. N., & GILLESPIE, J. S. (1958). The release of chemical transmitter from the sympathetic nerves of the intestine of the cat. *J. Physiol.*, **143**, 41–54.

BUEDING, E., BÜLBRING, E., GERCKEN, G., HAWKINS, J. T. & KURIYAMA, H. (1967). The effect of adrenaline on the adenosine triphosphate and creatine phosphate content of intestinal smooth muscle. *J. Physiol.*, **193**, 187–212.

BUEDING, E., BUTCHER, R. W., HAWKINS, J., TIMMS, A. R. & SUTHERLAND, E. W. (1966). Effect of epinephrine on cyclic adenosine 3′, 5′-phosphate and hexose phosphates in intestinal smooth muscle. *Biochim. biophys. acta*, **115**, 173–178.

BÜLBRING, E. (1954). "Membrane Potentials of Smooth Muscle Fibres of the Tænia Coli of the Guinea Pig." *J. Physiol.*, **125**, 302–315.

BÜLBRING, E. (1955). "Correlation between Membrane Potential, Spike Discharge and Tension in Smooth Muscle." *J. Physiol.*, **128**, 200–221.

BÜLBRING, E., BURNSTOCK, G., & HOLMAN, M. (1958). "Excitation and Conduction in the Smooth Muscle of the Isolated Tænia Coli of the Guinea Pig." *J. Physiol.*, **142**, 420–437.

BÜLBRING, E., & HOOTON, I. N. (1954). "Membrane Potentials of Smooth Muscle Fibres in the Rabbit's Sphincter Pupillæ." *J. Physiol.*, **125**, 292–301.

BÜLBRING, E., & KURIYAMA, H. (1963). Effects of changes in the external sodium and calcium concentrations on spontaneous electrical activity in smooth muscle of guinea-pig tænia coli. *J. Physiol.*, **166**, 29–58.

BÜLBRING, E., & KURIYAMA, H. (1963). Effects of changes in ionic environment on the action of acetylcholine and adrenaline on the smooth muscle cells of guinea-pig tænia coli. *J. Physiol.*, **166**, 59–74.

BÜLBRING, E. & KURIYAMA, H. (1963). The effect of adrenaline on the smooth muscle of guinea-pig tænia coli in relation to the degree of stretch. *J. Physiol.*, **169**, 198–212.

BÜLBRING, E. & TOMITA, T. (1967). Properties of the inhibitory potential of smooth muscle as observed in the response to field stimulation of the guinea-pig tænia coli. *J. Physiol.*, **189**, 299–315.

BURNSTOCK, G. (1958). The effects of acetylcholine on membrane potential, spike frequency, conduction velocity and excitability in the tænia coli of the guinea-pig. *J. Physiol.*, **143**, 165–182.

BURNSTOCK, G. (1958). The action of adrenaline on excitability and membrane potential in the tænia coli of the guinea-pig and the effect of DNP on this action and on the action of acetylcholine. *J. Physiol.*, **143**, 183–194.

BURNSTOCK, G., CAMPBELL, G. & RAND, M. J. (1966). The inhibitory innervation of the taenia of the guinea-pig caecum. *J. Physiol.*, **182**, 504–526.

BURNSTOCK, G., & HOLMAN, M. E. (1961). The transmission of excitation from autonomic nerve to smooth muscle. *J. Physiol.*, **155**, 115–133.

BURNSTOCK, G., & HOLMAN, M. E. (1962). Spontaneous potentials at sympathetic nerve endings in smooth muscle. *J. Physiol.*, **160**, 446–460.

BURNSTOCK, G., & HOLMAN, M. E. (1962). Effect of denervation and of reserpine treatment on transmission at sympathetic nerve endings. *J. Physiol.*, **160**, 461–469.

BURNSTOCK, G. & PROSSER, C. L. (1960). Conduction in smooth muscle: comparative electrical properties. *Amer. J. Physiol.*, **199**, 553–559.

BURNSTOCK, G., & STRAUB, R. W. (1958). A method for studying the effects of ions and drugs on the resting and action potentials in smooth muscle with external electrodes. *J. Physiol.*, **140**, 156–167.

CAESAR, R., EDWARDS, G. A., & RUSKA, H. (1957). "Architecture and Nerve Supply of Mammalian Smooth Muscle Tissue." *J. biophys. biochem. Cytol.*, **3**, 867–878.

CAMPBELL, G. (1966). Nerve-mediated excitation of the taenia of the guinea-pig caecum. *J. Physiol.*, **185**, 148–159.

CASTEELS, R. & KURIYAMA, H. (1965). Membrane potential and ionic content in pregnant and non-pregnant rat myometrium. *J. Physiol.*, **177**, 263–287.

CASTEELS, R. & KURIYAMA, H. (1966). Membrane potential and ion content in the smooth muscle of the guinea-pig's taenia coli at different external potassium concentrations. *J. Physiol.*, **184**, 120–130.

DANIEL, E. E. (1965). Attempted synthesis of data regarding divalent cations in muscle function. In *Muscle*. Ed. W. M. Paul *et al.* Oxford: Pergamon. pp. 295–313.

DANIEL, E. E., & ROBINSON, K. (1960). The secretion of sodium and uptake of potassium by isolated uterine segments made sodium-rich. *J. Physiol.*, **154**, 421–444.

DANIEL, E. E., & ROBINSON, K. (1960). The relation of sodium secretion to metabolism in isolated sodium-rich uterine segments. *J. Physiol.*, **154**, 445–460.

DANIEL, E. E., & SINGH, H. (1958). The electrical properties of the smooth muscle membrane. *Canad. J. Biochem. Physiol.*, **36**, 959–975.

DEWEY, M. M., & BARR, L. (1962). Intercellular connections between smooth muscle cells: the nexus. *Science*, **137**, 670–672.

DEWEY, M. M. & BARR, L. (1964). A study of the structure and distribution of the nexus. *J. Cell Biol.*, **23**, 553–585.

DURBIN, R. P., & JENKINSON, D. H. (1961). The calcium dependence of tension development in depolarized smooth muscle. *J. Physiol.*, **157**, 90–96.

DURBIN, R. P., & JENKINSON, D. H. (1961). The effect of carbachol on the permeability of depolarized smooth muscle to inorganic ions. *J. Physiol.*, **157**, 74–89.

ECCLES, J. C. (1936). "Synaptic and Neuromuscular Transmission." *Ergebn. Physiol.*, **38**, 339–444.

EDMAN, K. A. P. & SCHILD, H. O. (1962). The need for calcium in the contractile responses induced by acetylcholine and potassium in the rat uterus. *J. Physiol.*, **161**, 424–441.

EVANS, D. H. L. & EVANS, E. M. (1964). The membrane relationships of smooth muscles: an electron microscope study. *J. Anat.*, **98**, 37–46.

EVANS, D. H. L., & SCHILD, H. O. (1956). "Reactions of the Chick Amnion to Stretch and Electrical Stimulation." *J. Physiol.*, **132**, 31 P.

EVANS, D. H. L., SCHILD, H. O. & THESLEFF, S. (1958). Effects of drugs on depolarized plain muscle. *J. Physiol.*, **143**, 474–485.

FALK, G. & FATT, P. (1964). Linear electrical properties of striated muscle fibres observed with intracellular electrodes. *Proc. Roy. Soc., B*, **160**, 69–123.

FULTON, G. F., & LUTZ, B. R. (1942). "Smooth Muscle Motor-units in Small Blood Vessels." *Amer. J. Physiol.*, **135**, 531–534.

GILLESPIE, J. H. (1934). The biological significance of the linkages in adenosine triphosphoric acid. *J. Physiol.*, **80**, 345–359.

GILLESPIE, J. S. (1962). The electrical and mechanical responses of intestinal smooth muscle cells to stimulation of their extrinsic parasympathetic nerves. *J. Physiol.*, **162**, 76–92.

GILLESPIE, J. S. (1962). Spontaneous mechanical and electrical activity of stretched and unstretched intestinal smooth muscle cells and their response to sympathetic nerve stimulation. *J. Physiol.*, **162**, 54–75.

GILLESPIE, J. S. & MACK, A. J. (1964). The electrical response of intestinal smooth muscle to stimulation of the extrinsic or intrinsic motor nerves. *J. Physiol.*, **170**, 19–20 P.

GOODFORD, P. J. (1962). The sodium content of the smooth muscle of the guinea-pig taenia coli. *J. Physiol.*, **163**, 411–422.

GOODFORD, P. J. (1964). Chloride content and ^{36}Cl uptake in the smooth muscle of the guinea-pig taenia coli. *J. Physiol.*, **170**, 227–237.

GOODFORD, P. J., & HERMANSEN, K. (1961). Sodium and potassium movements in the unstriated muscle of the guinea-pig taenia coli. *J. Physiol.*, **158**, 426–448.

GOODFORD, P. J. & LÜLLMANN, H. (1962). The uptake of ethanesulphonate -^{35}S ions by muscular tissue. *J. Physiol.*, **161**, 54–61.

GREVEN, K. (1953). "Über Ruhe- und Aktionspotentiale der glatten Muskulatur nach Untersuchungen mit Glaskapillarelektroden." *Z. Biol.*, **106**, 1–15.

HASHIMOTO, Y., HOLMAN, M. E. & TILLE, J. (1966). Electrical properties of the smooth muscle membrane of the guinea-pig vas deferens. *J. Physiol.*, **186**, 27–41.

HOLMAN, M. E. (1958). "Membrane Potentials recorded with High-resistance Micro-electrode; and the Effects of Changes in Ionic Environment on the Electrical and Mechanical Activity of the Smooth Muscle of the Tænia Coli of the Guinea Pig." *J. Physiol.*, **141**, 464–488.

HOLMAN, M. E., CASBY, C. B., SUTHERS, M. B. & WILSON, J. A. F. (1968). Some properties of the smooth muscle of rabbit portal vein. *J. Physiol.*, **196**, 111–132.

HURWITZ, L. (1965). Calcium and its interrelations with cocaine and other drugs in contraction of intestinal muscle. In *Muscle*. Ed. W. M. Paul. *et al.* N.Y.: Pergamon.

HÜTER, J., BAUER, H. & GOODFORD, P. J. (1963). Die Wirkung von Adrenalin auf Kalium-Austausch und Kalium-Konzentration im glatten Muskel. *Arch. exp. Path. Pharmak.*, **246**, 75–76.

ICHIKAWA, S. & BOZLER, E. (1955). Monophasic and diphasic action potentials of the stomach. *Amer. J. Physiol.*, **182**, 92–96.

JENKINSON, D. H. & MORTON, I. K. M. (1967, a). The effect of noradrenaline on the permeability of depolarized intestinal smooth muscle to inorganic ions. *J. Physiol.*, **188**, 373–386.

JENKINSON, D. H. & MORTON, I. K. M. (1967, b). The role of α- and β-adrenergic receptors in some actions of catecholamines on intestinal smooth muscle. *J. Physiol.*, **188**, 387–402.

JUNG, H. (1956). "Erregungsleitung und Erregungsbildung am Uterus." *Z. Geburts. Gynäkol.*, **147**, 51–71.

KEATINGE, W. R. (1968). Ionic requirements for arterial action potential. *J. Physiol.*, **194**, 169–182.

KEATINGE, W. R. (1968). Sodium flux and electrical activity of arterial smooth muscle. *J. Physiol.*, **194**, 183–200.

KOBAYASHI, M. (1965). Effects of Na and Ca on the generation and conduction of ex-citation in the ureter. *Amer. J. Physiol.*, **208**, 715–719.

KOBAYASHI, M., PROSSER, C. L. & NAGAI, T. (1967). Electrical properties of intestinal muscle as measured intracellularly and extracellularly. *Amer. J. Physiol.*, **213**, 275–286.

KOLODNY, R. L., & VAN DER KLOOT, W. G. (1961). Contraction of smooth muscle in non-ionic solutions. *Nature*, **190**, 786–788.

KURIYAMA, H. (1963). The influence of potassium, sodium and chloride on the membrane potential of the smooth muscle of tænia coli. *J. Physiol.*, **166**, 15–28.

KURIYAMA, H. (1964). Effect of calcium and magnesium on neuromuscular transmission in the hypogastric nerve-vas deferens preparation of the guinea-pig. *J. Physiol.*, **175**, 211–230.

KURIYAMA, H., & CSAPÓ, A. (1961). A study of the parturient uterus with the micro-electrode technique. *Endocrinology*, **68**, 1010–1025.

KURIYAMA, H., OSA, T. & TOIDA, N. (1967). Membrane properties of the smooth muscle of guinea-pig ureter. *J. Physiol.*, **191**, 225–238.

KURIYAMA, H., OSA, T. & TOIDA, N. (1967). Electrophysiological study of the intestinal smooth muscle of the guinea-pig. *J. Physiol.*, **191**, 239–255.

KURIYAMA, H., OSA, T. & TOIDA, N. (1967). Nervous factors influencing the membrane activity of intestinal smooth muscle. *J. Physiol.*, **191**, 257–270.

KURIYAMA, H. & TOMITA, T. (1965). The responses of single smooth muscle cells of guinea-pig taenia coli to intracellularly applied currents, and their effect on the spontaneous electrical activity. *J. Physiol.*, **178**, 270–289.

LANE, B. P. & RHODIN, J. A. G. (1964). Cellular interrelationships and electrical activity in two types of smooth muscle. *J. Ultrastr. Res.*, **10**, 470–488.

MARSHALL, J. M. (1963). Behaviour of uterine muscle in Na-deficient solutions: effects of oxytoxin. *Amer. J. Physiol.*, **204**, 732–738.

MARSHALL, J. M., & CSAPÓ, A. I. (1961). Hormonal and ionic influences on the membrane activity of uterine smooth muscle cells. *Endocrinology*. **68**, 1026–1035.

MELTON, C. E. (1956). "Electrical Activity in the Uterus of the Rat." *Endocrinology* **68**, 139–149.

MERRILLEES, N. C. R. (1968). The nervous environment of individual smooth muscle cells of the guinea-pig vas deferens. *J. Cell Biol.*, **37**, 794–817.

MERRILLEES, N. C. R., BURNSTOCK, G. & HOLMAN, M. E. (1963). Correlation of fine structure and physiology of the innervation of smooth muscle in the guinea-pig vas deferens. *J. Cell Biol.*, **19**, 529–550.

MISLIN, H. (1948). "Das Elektrovenenogram der isolierten Flughautvene (*Chiroptera*)." *Experientia*, **4**, 28.

NAGAI, T. & PROSSER, C. L. (1963). Patterns of conduction in smooth muscle. *Amer. J. Physiol.*, **204**, 910–914.

NAGAI, T. & PROSSER, C. L. (1963). Electrical parameters of smooth muscle cells. *Amer. J. Physiol.*, **204**, 915–924.

NONOMURA, Y., HOTTA, Y. & OHASHI, H. (1966). Tetrodotoxin and manganese ions: effects on electrical activity and tension in taenia coli of guinea pig. *Science*, **152**, 97–99.

OOSAKI, T. & ISHII, S. (1964). The ultrastructure of the regions of junction between smooth muscle cells in the rat small intestine. *J. Ultrastr. Res.*, **10**, 567–577.

PROSSER, C. L. (1954). "Activation of a Non-propagating Muscle in Thyone." *J. cell. comp. Physiol.*, **44**, 247–253.

PROSSER, C. L., BURNSTOCK, G., & KAHN, J. (1960). Conduction in smooth muscle: comparative structural properties. *Amer. J. Physiol.*, **199**, 545–552.

PROSSER, C. L., CURTIS, H. J., & TRAVIS, D. M. (1951). "Action Potentials from Some Invertebrate Non-striated Muscles." *J. cell. comp. Physiol.*, **38**, 299–319.

PROSSER, C. L., & MELTON, C. E. (1954). "Nervous Conduction in Smooth Muscle of *Phascolosoma* Proboscis Retractors." *J. cell. comp. Physiol.*, **44**, 255–275.

PROSSER, C. L., & RAFFERTY, N. S. (1956) "Electrical Activity in Chick Amnion.' *Amer J. Physiol.*, **187**, 546–548.

PROSSER, C. L. & SPERELAKIS, N. (1956). Transmission in ganglion-free circular muscle from cat intestine. *Amer. J. Physiol.*, **187**, 536–545.

RICHARDSON, K. C. (1962). The fine structure of autonomic nerve endings in smooth muscle of the rat vas deferens. *J. Anat.*, **96**, 427–442.

RICHARDSON, K. C. (1964). The fine structure of the albino rabbit iris with special reference to the identification of adrenergic and cholinergic nerves and nerve endings in its intrinsic muscles. *Amer. J. Anat.*, **114**, 173–205.

RUSHTON, W. A. H. (1938). Initiation of the propagated disturbance. *Proc. Roy. Soc.*, B, **124**, 210–243.

SCHMIDT, R. F., & HUBER, U. (1960). Der Einfluss von Acetylcholin und Orasthin auf das Ruhe- und Aktionspotential des Ureters. *Pflüg Arch. ges. Physiol.*, **270**, 308–318

SHANES, A. M. (1958). Electrochemical aspects of physiological and pharmacological action in excitable cells. *Pharmacol. Rev.*, **10**, 59–273.

SINGH, I., & ACHARYA, A. K. (1957). "Excitation of Unstriated Muscle without any Ionic Gradient across the Membrane." *Indian J. Physiol. Pharmacol.*, **1**, 265–269.

SPERELAKIS, N., & PROSSER, C. L. (1959). Mechanical and electrical activity in intestinal smooth muscle. *Amer. J. Physiol.*, **196**, 850–856.

TAMAI, T. & PROSSER, C. L. (1966). Differentiation of slow potentials and spikes in longitudinal muscle of cat intestine. *Amer. J. Physiol.*, **210**, 452–458.

TASAKI, I. & HAGIWARA, S. (1957). Capacity of muscle fiber membrane. *Amer. J. Physiol.*, **188**, 423–429.

TAXI, J. (1964). Étude, au microscope électronique, de l'innervation du muscle lisse intestinal, comparée à celle de quelques autres muscles lisses de mammifères. *Arch. de Biol.*, **75**, 301–328.

THAEMERT, J. C. (1963). The ultrastructure and disposition of vesiculated nerve processes in smooth muscle. *J. Cell Biol.*, **16**, 361–377.

THAEMERT, J. C. (1966). Ultrastructural interrelationships of nerve processes and smooth muscle cells in three dimensions. *J. Cell Biol.*, **28**, 37–49.

TOMITA, T. (1966). Electrical responses of smooth muscle to external stimulation in hypertonic solution. *J. Physiol.*, **183**, 450–468.

TOMITA, T. (1966). Membrane capacity and resistance of mammalian smooth muscle. *J. theoret. Biol.*, **12**, 216–227.

TOMITA, T. (1967). Spike propagation in the smooth muscle of the guinea-pig taenia coli. *J. Physiol.*, **191**, 517–527.

TOMITA, T. (1967). Current spread in the smooth muscle of the guinea-pig vas deferens. *J. Physiol.*, **189**, 163–176.

VAN ROSSUM, J. M. & MUJIĆ, M. (1965). Classification of sympathomimetic drugs on the rabbit intestine. *Arch. int. Pharmacodyn.*, **155**, 418–431.

WEST, T. C., & LANDA, J. (1956). "Transmembrane Potentials and Contractility in the Pregnant Rat Uterus." *Amer. J. Physiol.*, **187**, 333–337.

WOODBURY, J. W., & MCINTYRE, D. M. (1954). "Electrical Activity in Single Muscle Cells of Pregnant Uteri studied with Intracellular Ultramicroelectrodes." *Amer. J. Physiol.*, **177**, 355–360.

WOODBURY, J. W., & MCINTYRE, D. M. (1956). "Transmembrane Action Potentials from Pregnant Uterus." *Amer. J. Physiol.*, **187**, 338–340.

5. *The Mechanism of Contraction of Muscle*

CHAPTER XX

THE STRUCTURE OF MUSCLE

MECHANICAL work of the organism is carried out by muscular, or contractile, tissue which in higher forms is made up of cells highly differentiated to permit of a rapid and reversible shortening and relaxation. As with the conduction of an impulse, it seems very likely that contractility, as such, is a widely spread characteristic of cells, for example, the streaming of the slime mould, *Phycomyces*, has been interpreted on the basis of an inherent contractility of parts of its protoplasm; and, again, the ciliary action of many unicellular organisms, including some bacteria, may be included in any classification of muscular activity. In the higher forms we may differentiate two main types of specialized contractile tissue, *voluntary* or *skeletal* muscle, controlling the movements of the limbs and other movable structures, and *smooth muscle*, generally arranged in sheets around hollow organs, such as the intestine and bladder.

The unit of the voluntary muscle is the *muscle fibre*, some 10–100 μ thick, and with a length that varies considerably with the muscle, from a few millimetres to several centimetres. The fibre results from the fusion of many cells, *myoblasts*, during development and contains many nuclei (in fibres several centimetres long there may be several hundred nuclei). Bundles of fibres are held together with elastic connective tissue. Most muscles are continued at one or both ends into dense tendinous bands which make attachments to the bone or other structure that they operate. The fibres of a given muscle need not necessarily, in fact they most frequently do not, run the whole length of the muscle; in the frog sartorius they probably do but in other muscles they may begin and end in connective tissue.

General Features of the Striated Muscle Fibre

A- and I-Discs

An individual muscle fibre is illustrated in Fig. 721; it is striated in both longitudinal and transverse directions, the former being due to the fact that the fibre is built up of numerous *myofibrils* some 1 to 2 μ thick, or less, embedded in a structureless medium, the *sarcoplasm*. The cross striations, which give the name to this type of muscle, are due to alternations in the optical and staining properties of the fibrils along their axes, so that the whole fibre appears to be made up of a series of discs: the A- or *anisotropic* discs,* which are highly refractile and appear dark under the microscope; and the I-, or *isotropic* discs, which are less highly refractile and appear light. As their names imply, the A-discs are doubly refracting, whereas the I-discs are only very weakly so, their double refraction amounting to only about 10 per cent. of that of the A-discs. These optical properties are accompanied by a difference

---

* Also called Q discs.

in staining characteristics; thus, with the Heidenhain or methylene blue stains, the A-bands become dark and the I-bands remain light, whilst with silver staining the banding is reversed, the I-bands preferentially taking up the stain. The I-band is crossed by a thin, doubly-refracting *Z-line* (Krause's membrane), which appears as a membrane dividing the whole muscle fibre transversely, since it is apparently not confined to the myofibrils but crosses the sarcoplasm too. When muscle fibres are macerated they tend to break at the Z-line, hence the unit, or *sarcomere*, of the fibre is defined as the region enclosed between two successive Z-lines. The light-microscope has revealed other discontinuities, some of which will be considered later; for example, the H-band in the middle of the A-band.

FIG. 721. Part of a rabbit's striated muscle fibre. × 600 approx. (Microphotograph by F. J. Pittock, F.R.P.S., from preparation by K. C. Richardson.)

Myofibrils

With the application of thin sectioning technique by Bennett & Porter (1953) the relationship of the myofibrils to the sarcoplasm became clearer. According to these workers, the fibrils are arranged in groups of four to twenty or more, lying closely together to form "muscle columns", separated from each other by a thick (0·2–0·5 μ) layer of sarcoplasm. Much thinner (400–800A) sheets of sarcoplasm separate the individual myofibrils. Most of the nuclei lie immediately underneath the limiting membrane (sarcolemma), although some of them are deeper in the fibril. Bennett & Porter noted that, in many preparations, individual myofibrils had contracted during fixation to different extents; as a result, the banding of successive myofibrils was out of phase, and examination of the Z-lines and myofilaments revealed no distortions such as would have been expected had adjacent myofibrils been connected laterally across the intervening sarcopasm by a definite Z-membrane, as had frequently been suggested. The myofibrils were seen to be made up of bundles of myo-

PLATE LXV

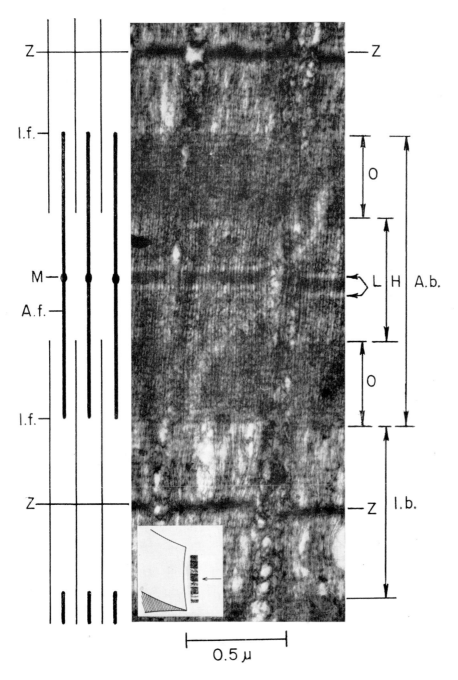

FIG. 722. Electron-micrograph of part of muscle fibre to show the different structural components in a longitudinal section of resting myofibrils at equilibrium length. The sarcomere length was the same as before fixation. The different regions are indicated at the side. The direction of the cutting knife is indicated in the inset. (Carlsen, Knappeis & Buchtal. *J. biophys. biochem. Cytol.*)

PLATE LXVI

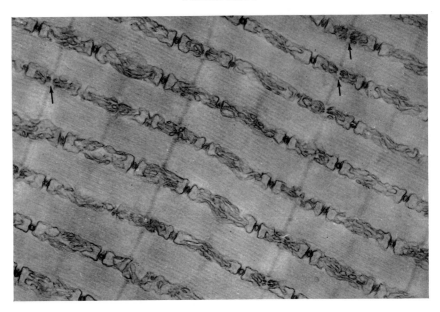

FIG. 723. Illustrating the sarcoplasmic reticulum. A small area of a muscle fibre has been cut longitudinally; the triads consist of a slender intermediate tube flanked by two larger lateral channels; these run across the broad face of the myofibrils in planes perpendicular to the page. The longitudinally orientated tributaries of the triads form tight networks parallel to the myofibrils. The branches of the two triads in the same sarcomere are continuous in the region of the H-band, but often appear to be discontinuous at the level of the Z-line (arrows). × 22,500. (Fawcett & Revel. *J. biophys. biochem. Cytol.*)

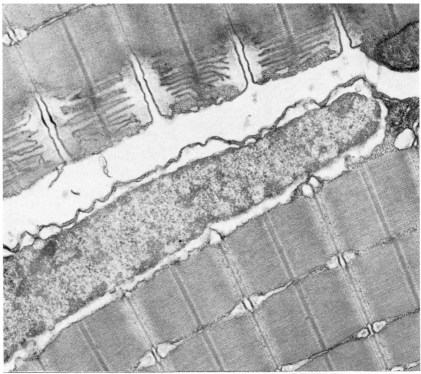

FIG. 725. Longitudinal section through the periphery of two adjacent muscle fibres; the lower one shows a nucleus. In the adjacent fibre, above, the T system appears in wide communication with the outside. × 25,000. (Frazini-Armstong & Porter. *J .Cell Biol.*)

filaments. Fig. 722, Pl. LXV is an electron micrograph of a portion of a muscle fibre showing the orderly arrangement of the myofilaments within the individual myofibrils. The characteristic cross-striations are seen to correspond with regions of different electron-density.

Mitochondria. Wedged in between the myofibrils were numerous large particles that were apparently analogous with mitochondria of other cells, particles that had been described by light-microscopists as the *interstitial granules* of Kolliker, or the *sarcosomes* of Retzius. In the particular muscle described by Bennett & Porter (breast muscle of the chicken) their number was not great, and they were not associated with any particular band. Muscle fibres are of two main types; in the one, characteristic of muscles that are used continuously, such as the flight muscles of the pigeon and of many insects, or cardiac muscle, the sarcoplasm is abundant, and the numerous large mito-chondria are arranged along the myofibrils in a characteristic manner, usually in association with the A-band, although Weinstein has recently emphasized their close relationship with the Z-line. The muscle in which this type of fibre predominates is red, by comparison with the whitish muscle in which fibres of the second type—namely fibres with few mitochondria—predominate. These mitochondria have been isolated by a number of workers, for example by Chappell & Perry, and have been shown to possess all the oxidative activity associated with the citric-acid cycle (p. 287), whereas the glycolytic activity of the fibre seems to be associated with the non-particulate matter of the sarco-plasm.

Sarcotubular System. We may recall that Palade & Porter described a series of interconnected tubules in the cytoplasm of all the cells they examined —the so-called *endoplasmic reticulum*. Within the muscle fibre Porter and Porter & Palade described an exceptionally well ordered arrangement of tubular structures, organized in relation to the individual myofibrils, to which they gave the name of *sarcoplasmic reticulum*. Tangential sections close to the myofibril often revealed a plexus of smooth-surfaced tubules closely applied to the surface, and examination of many sections in the salamander, *Ambystoma*, muscle led to a picture of a series of tubules orientated both longitudinally and transversely so as to cover the sarcomere of a given fibril with a network of interconnected tubules.

Triads

Andersson-Cedergren was the first to demonstrate the dual character of the system, which could be divided into a *transverse T-system*, close to the level of the A-I junction, or in other species at the Z-band, and an L-system whose organization was largely longitudinal. The two systems were probably not continuous with each other but at certain regions they came into close contact, so that a longitudinal section revealed characteristic *triads* (Fig. 723, Pl. LXVI), a central flattened profile of a transverse tubule sandwiched between two larger profiles, which could be shown to be sections of the L-system and were described as *terminal cisterns* since they were, in fact, the dilated ends of flat cisterns belonging to the L-system.

Transverse T-System

Later studies have led to the picture of the sarcotubular system shown in Fig. 724. The transverse tubules apparently represent invaginations of the sarco-lemma (Fig. 725, Pl. LXVI) and presumably provide a rapid conduction pathway for the action potential into the depths of the fibre (Franzini-Armstrong & Porter, 1964); continuity with the outside medium is demonstrated by the appearance of ferritin particles (H. E. Huxley, 1964), ThO_2 (Birks, 1965), fluorescent dyes (Endo, 1966) or horseradish peroxidase (Eisenberg & Eisenberg, 1968) within the central elements of the triad after exposure of the muscle fibre to these markers, whilst serial sections, of some types of muscle fibre at any rate, indicate an unobstructed opening of the T-system to the surface.* As Franzini-Armstrong & Porter (1964) describe it, the T-system at any one Z-level may be regarded as a net composed of tortuous tubules and nodal points, through the openings in which run the fibrils; this net interrupts the continuity of the sarcoplasmic reticulum, or L-system, at every Z-level.

L-System

The longitudinal system is described as the *sarcoplasmic reticulum* proper and, being unconnected with the T-system, has no direct connexion with the extracellular space; there are specialized contact regions, at the triads, and these may be analogous with tight junctions. As indicated by Fig. 724 and Fig. 726, Pl. LXVII, the L-system is a hollow three-dimensional lacework located in the space between the myofibrils, and surrounding each one. At the centre of the A-bands it is in the form of a collar perforated by a number of holes similar to the pores of the nuclear membrane (with which the sarcoplasmic reticulum is continuous) in that their circumference consists of the sealed membranes of the cisterns. This fenestrated collar connects with the terminal cisterns at the Z-lines, and the form of the triad results from the section of these terminal cisterns and the transverse tubules.†

Ontogenetic Development

The studies of Ezerman & Ishikawa (1967) and Ishikawa (1968) on the developing explanted muscle fibre, *in vitro*, show unequivocally that the T-tubules originate as invaginations of the sarcolemma—caviolæ—which may appear as a single pocket of 810A diameter communicating with the surface through a tunnel-like channel of 300–500A diameter, or they may be multiple, extending by way of channels from the original caveola. The sarcoplasmic reticulum, as opposed to the transverse T-system, seems to arise by growth of the rough-surfaced endoplasmic reticulum; a vesicle of this system forms tubular projections which eventually become a honeycombed network in relation to the myofibrils. Ezerman & Ishikawa describe a variety of stages in the development of the triads, from casual (dyadic) contacts between the invaginated T-tubule and a vesicle of the sarcoplasmic reticulum to the fully developed triad as seen in the adult vertebrate fibre.

Fast-Acting Muscles

If, as Huxley & Taylor's studies on local stimulation suggest, the initiation of contraction depends on transmission into the depth of the fibre along elements of the endoplasmic reticulum, then in exceptionally fast muscles we might expect to find this structure more highly developed than in slow ones. Although this is certainly not true of insect muscles, it may well be so of vertebrate striated muscles, since Fawcett & Revel found a highly developed system in the swim-bladder

---

* Rayns, Simpson & Bertaud (1968) have emphasized the paucity of direct morphological evidence for continuity between extracellular space and the T-system; their own study, involving both freeze-etching and classical thin-sectioning, has shown that the T-tubules open into subsarcolemmal caveolæ, which are open to the extracellular space; thus the communication of the T-system with the exterior exists, but is more complex than originally thought.

† Peachey & Schild (1968) have described the distribution of the T-system along the sarcomeres of frog and toad muscle as revealed by the accumulation of ferritin particles; the particles are highly concentrated at the Z-line and fall off rapidly to the A-I junction. Hill (1964) observed accumulation of ^{131}I-albumin at the Z-line and at the A-I boundary, suggesting the presence of the T-system at this boundary too, but Peachey & Schild were unable to confirm this with respect to ferritin. The volume of the muscle accessible to ^{131}I-albumin at the Z-line was 0·1 per cent., whilst the estimated geometric volume of the T-system was 0·2 per cent.

PLATE LXVII

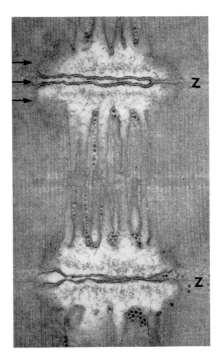

Fig. 726. Longitudinal section of muscle fibre that includes an expanse of sarcoplasm, and the surface of the underlying fibril includes a layer of sarcoplasmic reticulum. It can be seen that, from one Z line (Z) to the next, the SR is a continuous structure with transversely (terminal sacs at I band level) as well as longitudinally orientated elements (along the A band). At the Z line the transverse (T) system intrudes between the two terminal sacs in the form of a continuous cistern. The three constitute a triad, indicated by three arrows. × 30,000. (Franzini-Armstrong & Porter. *J. Cell Biol.*)

Fig. 737. Synthetic myosin filaments with bare central shaft and projections all the way along the rest of the length of the filament. × 145,000. (Huxley. *J. mol. Biol.*)

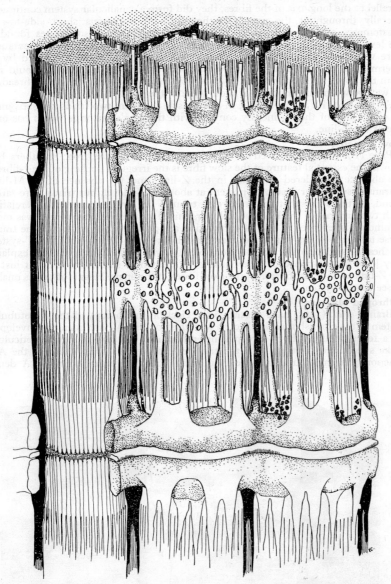

FIG. 724. Three-dimensional reconstruction of the sarcoplasmic reticulum associated with several myofibrils. Note continuity of the transverse tubules, the presence of intermediate cisternæ, and fenestrations of the collar at the centre of the A-band. (Peachey. *J. Cell Biol.*)

muscle of the toadfish, whilst Revel described a similar system in the cricothyroid muscle of the bat. It is this muscle that is responsible for emission of the high-frequency signals that guide the bat during flight, and Revel computed that this muscle must contract within a few msec after arrival of a nerve impulse, reach maximal contraction in at most 4 msec. and relax in 1 msec. As with the other muscles studied, the reticulum consisted of transverse elements forming a neck-lace round the myofibril in the region of the junction of the A- and I-bands; and these transverse elements were connected with a longitudinal system which, in this muscle, was far more complex, forming several layers of longitudinal tubules at the level of the I-band. Although these channels were mainly orientated

43*

parallel to the long axis of the fibres, they did form a canalicular system continuous laterally throughout the muscle, either through numerous short side-to-side anastomoses or by way of more direct transverse channels sometimes found at the level of the Z-lines. The triads formed the principal transverse elements and were well developed; these consisted of two terminal cisternæ separated by an intermediate element that was shown to be a continuous tubule that could run from the space surrounding one myofibril to that surrounding the next, threading its way between mitochondria if these were in the way.

As Fig. 723, Pl. LXVI shows, in this muscle there are two triads per sarcomere on either side of the H-line, by contrast with the frog's sartorius which has only one at the Z-line.

Slow Muscles

Page (1965) has compared slow and twitch fibres in the frog; as others had shown, the essential feature of the slow fibre is the irregular packing of the I-fibrils, the absence of ordered structure in the Z-line, and the absence of an M-line. A transverse tubular system is present at about the Z-level, but triads are rarely found, in the sense that the transverse tubules did not come into close relation with two cisterns of the longitudinally organized system; instead a *dyad* was more common. Differences in rate of contraction are probably not related to the transverse tubular systems but to reaction rates between the filaments. The T-system reaches the surface of the fibre at the A band as well as the Z-line, and this explains why local stimulation causes response over the whole sarcomere and not just at the Z-line as with fast fibres. In the slow fibres of the snake Hess (1965) was unable to see any T-system.

Arthropod Muscles

Brandt *et al.* (1965) have described the electron microscopy of the sarcotubular system in the crayfish muscle fibre; as with the frog, the A-band region is enveloped by a fenestrated girdle of sarcoplasmic reticulum; sacculations of the reticulum make specialized contacts with the transverse system to form dyads at the A-I junctions; at these junctions the membranes are 150A apart with a 35A dense

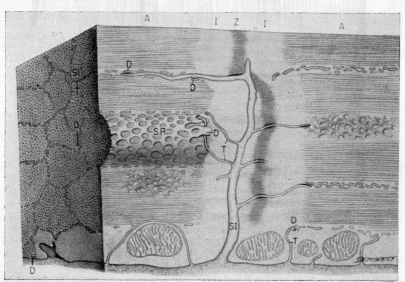

FIG. 727. Composite representation of the fine structure of a crayfish muscle fibre. A fenestrated envelope of sarcoplasmic reticulum (*SR*) is shown surrounding one myofibril at the junction of the longitudinal and cross-sectional views. Two types of invagination of the plasma membrane are distinguished by the presence of the sarcolemmal coat in the sarcolemmal invaginations and their large diameter (*SI*), and the absence of the sarcolemmal coat in the finer radial tubules (*T*). The sarcolemmal invaginations and their branches also give rise to many tubules. Diads (*D*) may be formed between the *SR* and the tubules, the sarcolemmal invaginations, or the plasma membrane. (Brandt, Reuben, Girardier & Grundfest. *J. Cell Biol.*)

plate in the gap. As Fig. 727 illustrates, the transverse system results from an invagination of the sarcolemma which branches into a series of tubules.

The insect system has been described by Hagopian & Spiro (1967) and Smith (1966); the sarcoplasmic reticulum proper is well developed and runs parallel with the myofibrils, and continues both longitudinally and transversely; it comes into relation with the T-system to form dyads; dyads are also formed when the sarcoplasmic reticulum forms close contacts with the plasma membrane proper. As with the other systems studied, ferritin only appears in the T-system (Smith, 1966).

Sarcolemma. Localized injury to a muscle fibre may cause a so-called "retraction clot" to appear, whereby the internal structures of the fibre contract locally to leave a clear translucent membrane bridging the gap. Bowman, who described this membrane, gave it the name *"sarcolemma"*; when he examined it by modern methods of phase-contrast microscopy, Barer was unable to detect any structure. Electron-microscopical studies on thin sections, *e.g.*, those of Porter & Palade, Robertson, Andersson-Cedergren, and several others, agree in describing the sarcolemma as a complex, consisting first of the plasma membrane of the muscle fibre, made up of two dark zones separated by a light zone— Robertson's "unit membrane". Above this is a 200–300A thick less dense zone, presumably consisting of basement-membrane material; and finally there is a thin feltwork of fine fibrils; some of these could be identified by their characteristic 700A repeat as collagen. The inner dense layer of the surface membrane complex appeared to be connected to the Z-bands of the underlying fibrils, producing a scalloped effect that had been observed before in studies of the "sarcolemma". That the "basement-membrane layer" is not an artefact, but a stable structure separate from the plasma membrane, was shown by Birks, Katz & Miledi who found that when the muscle fibre atrophied the plasma membrane shrank away from the basement-membrane layer, which retained the shape of the normal unatrophied fibre.

Discontinuity of Fibrils

The question whether the myofibrils are continuous, through holes in the sarcolemma, with the tendon fibrils has been a subject of controversy for many years; Long's work has demonstrated the absence of such a continuity, the myofibrillæ ending within the sarcolemma. The muscle fibre exerts its pull on the tendon fibre, therefore, through the agency of reticular fibrils, which are continuous with the fibrils of the tendon, and continue over the surface of the muscle fibre as the sarcolemmal and endomysial reticulum.

The Actomyosin System

We have seen that, under the light-microscope, the individual muscle fibre appeared to contain numerous myofibrils which, with the electron-microscope, could be resolved into myofilaments; by analogy with the fibrous structures considered earlier we may expect these myofilaments to be built up of bundles of long-chain protein molecules, the so-called crystallites. Early studies by Edsall and v. Muralt of the protein extracted from muscle revealed the presence of a high concentration of a globulin-type protein—*myosin*—extracted with 1·2 M KCl at pH 8 and precipitation by dilution of the extract to 0·1–0·05 M KCl. It showed marked double-refraction of flow and formed thixotropic gels in concentrations of greater than 2 per cent., both properties indicating a marked asymmetry of its particles. On squirting a solution into distilled water, threads were formed which were used to study its X-ray diffraction spectrum.

X-Ray Diffraction

When a pool of myosin sol is left to dry, a film remains, which may be studied by X-rays. On passing the X-rays perpendicular to the surface of the film a simple ring photo is obtained, similar to that given by disorientated keratin in its natural state; whilst, if the beam of X-rays is parallel to the surface, the picture is one of fibres lying roughly parallel to the surface of the film, but themselves randomly orientated in its plane. On stretching this film, however, the fibres tend to orientate themselves with their long axes in the direction of stretch, in the same way as the long molecules of rubber orientate themselves on stretching. Under these conditions the wide-angle X-ray photo becomes almost indistinguishable from that given by keratin fibres, and similar, moreover, to the X-ray photo of the intact muscle fibre. Myosin is thus an α-type of fibrous protein, and it was considered by Astbury that contraction might consist of a further folding of the α-chains, as observed in super-contraction of keratin, whilst passive extension, due to stretch, would be associated with a change from α- to β-configuration. Attempts to demonstrate changes in the wide-angle axial repeat, as a result either of contraction or of extension, were uniformly unsuccessful; moreover, Elliott (1952) showed that the change from α- to β- could be recognized by the infra-red spectrum, and he found no changes in this spectrum during changes of length of muscle. Employing a similar technique, Malcolm was able to dispose of the suggestion of Pauling & Corey that contraction involved a transformation of β- to α-myosin; thus, he showed that the infra-red spectrum of relaxed muscle indicated the presence of very little β-material.

Actin and Myosin

A great advance in knowledge followed the discovery that the "myosin", extracted by the classical procedures, was really a mixture of at least two proteins, *actin* and *myosin*, which were capable of interaction to give a complex protein, *actomyosin*. Thus, Schramm & Weber in 1942 showed that "myosin" could be fractionated into a rapidly sedimenting, high-molecular weight, compound—*S-myosin*—with a high and anomalous viscosity; and a less rapidly sedimenting fraction, which they called L-myosin.* It is the high-molecular weight substance, S-myosin, that is now called *actomyosin*, as a result of Szent-Györgyi's studies; it results from the combination of *actin* with what we may now call true myosin. Actin itself was isolated by Straub and shown to exist in two forms, globular, or *G-actin*, with a molecular weight of 76,000; and a polymerized fibrous form, *F-actin*; according to Jakus & Hall and Perry & Reed, F-actin is formed from G-actin by the linear aggregation of corpuscular particles. Only F-actin combines with myosin to give actomyosin. Myosin and F-actin are thus two fibrous components of the muscle fibre, and a variety of studies on artificial actomyosin threads, and of macerated and glycerol-extracted muscle fibrils from which the remaining proteins have been removed, have shown that the contractility of muscle may be largely reproduced by these so-called "models". We may defer until later a detailed account of the model experiments carried out by various workers; it is sufficient for our present purpose to appreciate that the structural components responsible for the contractile and tensile properties of muscle consist of two proteins, actin and myosin, capable of reversible combination under the influence of ATP; during the process of contraction in these model systems ATP is broken down to ADP and inorganic phosphate, whilst the two proteins appear to combine to give a shortened fibril. Relaxation is associated with a reversal of this combination.

---

* Actomyosin, directly extracted from muscle, is frequently described as myosin B, by contrast with myosin A, the true myosin.

The A- and I-Bands

With this knowledge, we may now approach the detailed examination of the structure of the muscle fibril, and of the changes that may be observed to take place during contraction and passive extension. The most striking anatomical feature of the skeletal muscle fibre is the cross-banding of the individual fibrils, revealed as alternate regions of high refractive index—the A-bands—and low refractive index—the I-bands. The A-band is anisotropic, exhibiting the phenomenon of birefringence; the anisotropy is due to two causes*; first, the submicroscopic units are rod-shaped and lie with their axes more or less parallel with the axis of the fibre (form double refraction), and secondly, these units are themselves built up of molecules in an orderly or crystalline fashion (micellar, or *eigen*, double refraction). Both types of birefringence are positive, *i.e.*, they are due to particles arranged with their long axes parallel to the axis of the fibre, so that a fundamental structure, composed of long rod-shaped bundles of polypeptide chains is consistent with the birefringence of the A-bands. The I-bands have only 10 per cent. of the birefringence of the A-bands, and are generally described as isotropic; this comparative isotropy could be due, either to a lack of order in the arrangement of the submicroscopic units in this region, or to the presence of some strongly negatively birefrigent material which cancelled the positive birefringence of the contractile units.†

A-Substance

In the electron-microscope this periodicity is revealed, in osmium stained preparations, as an alternation in electron-density, suggesting that the

Fig. 728. Schematic illustration of two sarcomeres at about resting length.
(After Huxley. *Brit. med. Bull.*)

characteristic difference between the two bands rests on the presence of extra material—*A-substance*—in the A-band. This is evident from Fig. 722, Pl. LXV, which brings out also the characteristic Z-lines which bisect the I-bands, and divide the myofibril into a series of sarcomeres. Within the A-band there is a region of lesser density—the H-band of the classical microscopists. Thus, schematically, the muscle fibril may be indicated by Fig. 728. Continuity within the sarcomere is apparently provided by the "protofibrils" that appear, in the usual sections, to run the whole length of the segment; in view of the frequency with which, in macerated tissue, breaks in the fibrils occur at the Z-lines it may well be that the sarcomere is, indeed, the true contractile unit, continuity along individual myofibrils being maintained by the Z-substance.

Myosin as A-Substance. The high electron-density, birefringence and refractive index of the A-band were shown, independently by Hasselbach and Hanson & H. E. Huxley, to be due to myosin, thus confirming the earlier surmises made by Weber and Amberson *et al.* Myosin may be extracted

---

* The micellar birefringence is some 35 per cent. of the total. (Fischer, 1944.)
† We may note, also, that stretching a striated muscle has little (Fischer, 1944) or no (Buchtal *et al.*, 1936) effect on its birefringence; if the I-regions were regions of disorder we should expect stretching the fibre to reduce this disorder and so increase the birefringence of the whole fibre appreciably.

from muscle, leaving the actin behind; when this is done, the fibrils retain their filamentous appearance, but in the electron-microscope Hasselbach found that the A- and I-bands were now no longer distinguishable, both containing fine fibrils that apparently ran the length of the sarcomere. The Z-bands were still present, a fact suggesting that they were composed of something different from myosin. Hasselbach concluded from this that the muscle fibril contained actin filaments running the length of the sarcomere; between these filaments the spaces, in the A-band, were filled with myosin.*

Double Array of Filaments

X-Ray Diffraction. A more detailed picture of the ultra-structure of the muscle fibril has been provided by the studies of H. E. Huxley & Hanson. Huxley examined the X-ray diffraction pattern of muscle; the wide-angle pattern revealed only the 5·1A axial repeat of myosin, as Astbury had found. Narrow-angle studies were handicapped by the necessity to use dried muscle, but the experimental difficulties in the way of using fresh muscle were overcome by H. E. Huxley. He found that the transverse pattern revealed the presence of long molecules, parallel to the fibre-axis, some 450A apart. On stretching the muscle, the spacing decreased inversely as the square root of the length, as we should expect. When ATP was removed, i.e., when the muscle went into rigor, or when the ATP was extracted with glycerol, there was a reversal in the intensities of the reflections due to the 450A spacing and its second order reflection, namely 225A; the latter now became the more intense. This indicated that regions of high electron-density were appearing in the contracted muscle at specific places between the long molecules, and Huxley suggested that these regions of high density corresponded to chemical linkages between a main array of myosin fibrils and a secondary array lying between them. Thus, in the relaxed state, the main array would be well orientated and impose their arrangement on the X-ray diffraction pattern; the secondary array, being free to move and thus not well orientated, would not contribute appreciably to the pattern. If, during contraction, the secondary array became linked to the primary array at specific points, this might cause them to become well orientated and so introduce a new reflection.

Electron-Microscopy. Striking confirmation of this view was provided by the electron-microscopical study of cross-sections (H. E. Huxley, 1953) and longitudinal sections (H. E. Huxley, 1957) of the muscle fibre. Thus, cross-sections of muscle fibrils in the A-segments revealed a double array of filaments. A primary array of thick filaments, some 100A thick, packed hexagonally some 300–350A apart in the fixed and dried specimen. Between these, in trigonal positions (i.e., such that each was associated with three thick filaments) were thin filaments, some 40A in diameter. Huxley observed, however, that this double array was confined to the A-segment, the I-segment having only thin filaments, which were presumably continuous with those in the A-segment. He suggested, then, as a working hypothesis, the structure illustrated

---

* Finck, Holtzer & Marshall (1956) prepared an antibody to myosin and labelled it with fluorescein; on applying this to a glycerol-extracted muscle it was found firmly attached to the A-band; the H-bands, within the A-bands, were only weakly stained. Hanson & H. E. Huxley came to the same conclusion on the basis of extracting myosin from glycerol-extracted fibrils, and they later proved, by interference-microscopy, that at least 90 per cent., and probably all, of the myosin was indeed concentrated in the A-band. It should be noted that A. G. Szent-Györgyi, Mazia & A. Szent-Györgyi (1955) have argued that under Hanson & Huxley's conditions of extraction of myosin only about 70 per cent. of the protein extracted was myosin, the remainder being unidentified material (the soluble sarcoplasmic proteins are washed out before myosin extraction).

schematically in Fig. 729. According to this, the thin actin filaments run from the Z-line through the I-segment to end in the H-region of the A-band where they are connected, by way of hypothetical S-filaments, to actin filaments running from the next Z-line. The primary array of myosin thick filaments are confined to the A-segment.

In the electron-micrograph of Fig. 722, the arrays of filaments may be resolved, whilst the bands resulting from variations in electron-density of the myofilaments, and from overlap of the two types, are interpreted schematically at the sides. Thus the H-band is less dense because of the absence of overlap, and so on.

FIG. 729. Schematic diagram illustrating the arrangement of filaments in striated muscle. The thick bars represent myosin in the A-band, whilst the thinner bands, extending from the vertical Z-bands and joined to each other in the middle of the A-band, are the actin filaments. (H. E. Huxley. *Brit. med. Bull.*)

Sliding Filament Hypothesis. On the basis of this structure, the shortening of the muscle fibre might result from the sliding of the thin I-filaments along the thick A-filaments, presumably as a result of the alternate formation and breaking of chemical bridges between the actin and myosin molecules, mediated by ATP. This model of the muscle fibril and its contraction was suggested independently by H. E. Huxley & Hanson and by A. F. Huxley on the basis of evidence derived from a number of sources.

H. E. Huxley pointed out that, if the filaments in the A-band were arranged in a hexagonal array to give a cross-section, as in Fig. 730a, then, in order to obtain a longitudinal section revealing one thin filament between two thick ones, as in Fig. 730b, the section must be only some 250A thick and must be cut in the direction indicated in Fig. 730a. If the section is thicker, it will include the thick

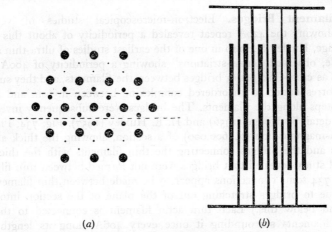

(a)   (b)

FIG. 730. (a) Diagram showing end-on view of a double hexagonal array of filaments; the two dotted lines indicate the outline of a longitudinal section, about 250A thick, such that the appearance will be that indicated in (b). (H. E. Huxley. *J. biophys. biochem. Cytol.*)

filaments lying above or below the thin filaments, and these will then be seen instead of the thin filaments. Thus, if the section is thicker than about 250A only thick filaments will be seen. Again, a section only 150A thick, and cut in the direction indicated in Fig. 731a, would give the appearance of two thin filaments between each pair of thick filaments, but once again a thicker section would include the thick filaments lying above and below the two thin filaments, and these would hide the presence of the latter and give rise to the appearance of a simple array of thick filaments. Huxley pointed out that the sections used in earlier studies were probably considerably thicker than 150–250A, and so could not have revealed the double array; by cutting exceptionally thin sections he obtained electron-micrographs giving both the pictures indicated schematically in Figs. 730 and 731. It will be seen from Huxley's electron-micrographs (Figs. 732 and 733, Pl. LXVIII) that the thin filaments apparently terminate in the H-zone of the A-band.

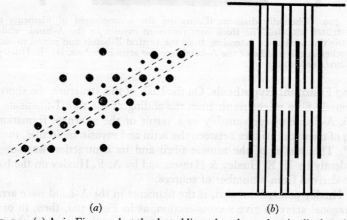

(a)   (b)

FIG. 731. (a) As in Fig. 730, but the dotted lines show how a longitudinal section, some 150A thick, may be cut so as to give the appearance shown in (b), with two secondary filaments between each pair of primary filaments. (H. E. Huxley. *J. biophys. biochem. Cytol.*)

Interfilament Bridges. Electron-microscopical studies of vertebrate muscle showing the 420A repeat revealed a periodicity of about this length; thus Hodge, Huxley & Spiro in one of the earliest studies of ultra-thin sections of muscle, observed "cross-striations" showing a periodicity of 400A, which appeared as cross-linkages or bridges between the filaments, and they suggested that it corresponded to an ordered complexing between myosin and actin in regular steps along the filaments. The bridges were subsequently investigated in more detail by Hodge (1956) and H. E. Huxley (1957). Fig. 734, Pl. LXIX is a high-magnification (× 600,000) of a section, showing the thick and thin filaments and the bridges connecting the thin filaments with the thick ones. A careful study revealed that bridges were not formed between thin filaments. (In Fig. 734 some connections appear to be made between thin filaments, but this is due to bridges stretching out of the plane of the section into planes above and below this.) Each thin actin filament is connected to the three primary filaments surrounding it once every 400A along its length; these links are fairly evenly spaced along the filament to give a periodicity of about 133A. According to Huxley, this arrangement is consistent with a helical structure for both the primary and secondary filaments, the bridges occurring

PLATE LXVIII

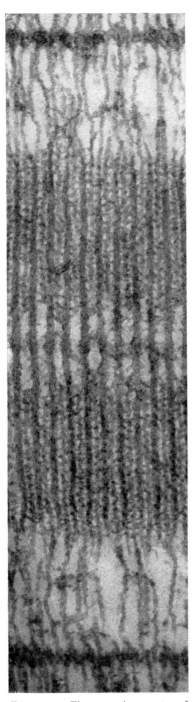 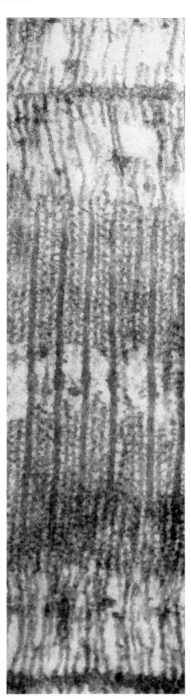

FIG. 732. Electron-micrograph of thin section through myofibril sectioned as indicated in Fig. 730, such that primary filaments alternate with secondary filaments. Note termination of the secondary filaments at the edges of the H-zone. Glycerinated psoas muscle. Osmium fixation. × 150,000 approx. (H. E. Huxley. *J. biophys. biochem. Cytol.*)

FIG. 733. Electron-micrograph of thin section through myofibril cut as indicated in Fig. 731, such that two secondary filaments lie between each pair of primary filaments. Glycerinated psoas muscle. Osmium fixation. × 150,000 approx. (H. E. Huxley. *J. biophys. biochem. Cytol.*)

PLATE LXIX

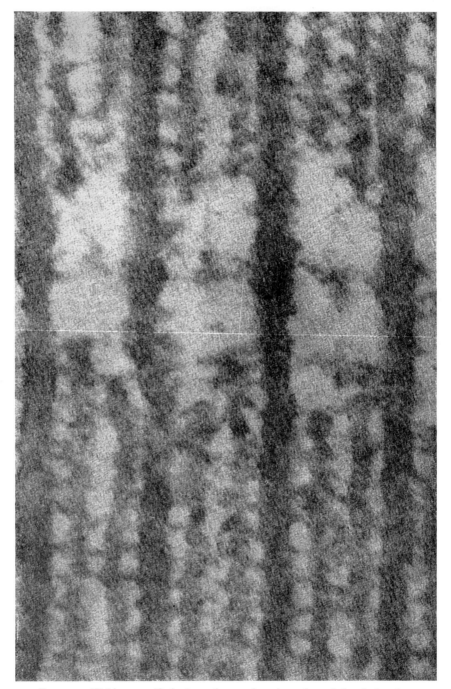

FIG. 734. Highly magnified view of central region of an A-band showing primary and secondary filaments and the bridges between them. The appearance of bridges between pairs of secondary filaments is probably due to those bridges that extend out of the plane of the section to the primary filaments in the layers above and below. Glycerinated psoas muscle. Osmium fixation. × 600,000. (H. E. Huxley. *J. biophys. biochem. Cytol.*)

at definite nodes on the helices. Thus, at resting length there would be some 54 bridges in the region of overlap.

Isolated Filaments. By mechanically disrupting glycerinated fibres in a medium favourable to dissociation of the actomyosin complex (a relaxing medium, p. 1404), H. E. Huxley showed that the individual filaments could be separated and examined by negative staining in the electron-microscope. Two types of filament were found, corresponding closely in dimensions with those of the thin and thick filaments. The lengths of the thin filaments, and of the isolated Z-lines with their thin filaments attached, were not dependent on sarcomere length, thus strengthening the view that during changes of muscle

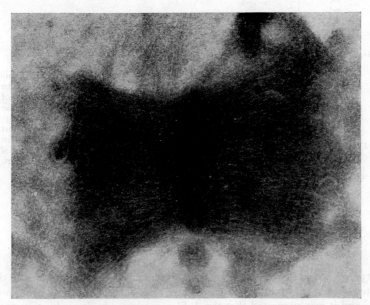

FIG. 735. Isolated I-segment showing array of thin filaments attached to Z-line. (× 45,000.) (H. E. Huxley. *Circulation.*)

length the constituent fibrils do not change in length, but rather slide alongside each other. The thick filaments had a diameter of 100A and were 1·5 μ long, with projections from the ends probably corresponding to the interfilament bridges. The thin filaments were 1 μ long; groups were found attached to the Z-disc to give an isolated I-segment (Fig. 735).

I-Filaments

Hanson & Lowy (1963) prepared a suspension of thin filaments from the glycerol-extracted muscle, and they compared this with F-actin prepared in the classical way; they found no difference so that F-actin, prepared, for example, by the aggregation of globular G-actin units, is essentially a suspension of I-filaments. When negatively stained, they were shown to consist of double-stranded helices, crossing over at regular intervals; the helices were built up of globular subunits of some 50A diameter. The pitch of the helical path pursued by each chain is approximately 700A, there being 13 monomers per turn with a spacing of about 55A (Fig. 736). Hanson & Lowy suggested that tropomyosin, which constitutes quite a considerable proportion of the I-band protein, is arranged as filaments in the groove of the double helix. According to H. E. Huxley (1962) the polarity of the helix of the I-filament is reversed on opposite sides of the Z-line; this would be expected were the Z-line to be the reference point for shortening, the filaments on either side having to slide in opposite directions to cause shortening of the sarcomeres.

A-Filaments

H. E. Huxley (1963) caused aggregation of myosin molecules by reducing the ionic strength of a solution of the protein; filaments were formed which were remarkably similar in appearance to the A-filaments of muscle, not only in respect to length but also in respect to shape, having a bare central shaft and lateral projections from the rest of the length similar to the interfilament bridges

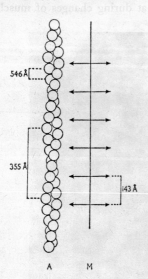

546 Å

355 Å

143 Å

A   M

FIG. 736. Illustrating relative longitudinal positions of cross-bridges and actin monomers in myosin and actin filaments. Orientation of myosin cross-bridges not shown. Note very close approximation to repeat of longitudinal alignment after ∼ 710 Å intervals, if actin subunit repeat is 54·6 Å along either chain, if actin chains are helices of pitch 2 × 355 Å, and if cross-bridge repeat is 143 Å. (Huxley & Brown. *J. mol. Biol.*)

(Fig. 737, Pl. LXVII). Examination of the individual molecules showed that they consisted of a straight rod with a globular end, the total length being some 1500A, and it was suggested that the A-filaments were built up by a staggered arrangement, as illustrated in Fig. 738. In this way we can account for the central bare region and the opposite polarity of the molecules on both sides of the centre of the filament, an arrangement that would permit the A-filament to pull I-filaments in opposite directions, so that both were pulled towards the centre of the A-filament and caused shortening of the sarcomere.

ANTIBODY STAINING

A more elaborate model of the myosin filament is one put forward by Pepe (1967), on the basis of his studies with anti-myosin treatment of the glycerinated fibre; after fixation and staining with uranyl acetate and lead citrate, regions of uptake of the antibody appeared as two bands in the middle part of each half of the A-band; each of these bands contained seven equally spaced dense lines. The model is illustrated in Fig. 739, and is based on the well-established asymmetry of the myosin molecule, consisting of a bulky head and thin tail; the thin tails are packed centrally leaving the bulky heads more superficially to form the cross-bridges. The arrangement gives rise to tapering at each end.

Actin + Myosin Filaments

Huxley found that, if the actin filaments were treated with myosin,* this attached to the actin giving filaments of 200–300A diameter, considerably thicker

---

* Actually *H-meromyosin*, the heavier particle derived from treatment of myosin with trypsin; the lighter particle, *L-meromyosin*, does not combine with actin.

FIG. 738. Arrangement of myosin molecules to produce filaments with a straight shaft and globular projection on each side. The polarity of the myosin molecules is reversed on either side of the centre, but all the molecules on the same side have the same polarity. (H. E. Huxley. *J. mol. Biol.*)

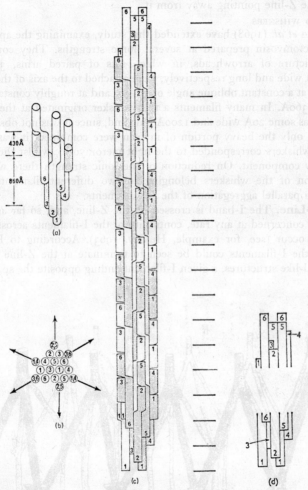

FIG. 739. Diagrammatic representation of the packing of the myosin molecules in the filament. (a) Representation of six molecules viewed from the surface of the filament. Molecules 1 and 3 are centrally located. The cross-bridges come off at the top of the overlap region of the molecules. (b) A cross-section of the filament for reference. (c) An entire one-half of the filament. The group of six molecules shown in (a) have been alternately shaded for clarity. The top is the tapered end of the filament. The bottom is in the pseudo-H-zone. Cross-bridges come off at the top of the overlap regions. (d) Representation of the two tapered ends of the filament showing the asymmetry. (Pepe. *J. mol. Biol.*)

than those of actin; they had a very strongly defined axial period of 366 ± 15A close to that of the actin double helix. The composite filaments had a structural polarity, seen in the tendency of their rather complicated internal structure to give an appearance of arrowheads all pointing in the same sense along whole lengths of any given filament. This structural polarity is a property of the actin filament since, when gaps in the arrowhead arrangement occurred as a result of adding too little myosin, the arrowheads on either side of the gap still pointed in the same way. When myosin was added to I-filaments, still attached to the Z-line, as in Fig. 735, a similar polarity was observed, the arrowheads on either side of the Z-line pointing away from it.

ARMS AND WHISKERS

Ikemoto *et al.* (1968) have extended this study, examining the appearance of natural actomyosin prepared at several ionic strengths. They confirmed the typical picture of arrowheads, in which sets of paired arms, 30–60A and 100–300A wide and long respectively, were attached to the axis of the composite filaments at a constant oblique angle of 30–40° and at roughly constant intervals of about 360A. In many filaments a thin whisker originated at the end of the arm; it was some 20A wide and 1200A long and, since it was not observed when actin and only the heavy portion of myosin were combined, it was concluded that the whiskers corresponded to the light meromyosin, whilst the arms were the heavy component. On reduction of the ionic strength there was a lateral aggregation of the whiskers belonging to two different filaments, and this initiated a parallel aggregation of the two filaments.

The Z-Line. The I-band is crossed by the Z-line, and, so far as vertebrate muscle is concerned at any rate, continuity of the I-filaments across the Z-line does not occur (see, for example, Huxley, 1963). According to Knappeis & Carlsen, the I-filaments could be seen to terminate at the Z-line where they joined rod-like structures, a given I-filament ending opposite the space between

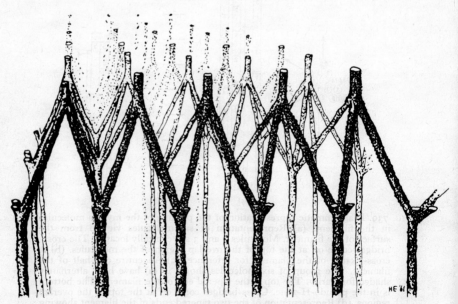

FIG. 740. Illustrating the attachment of each I-filament to four Z-filaments. The I-filaments have been drawn with twice their actual distance as compared with their diameter. The thickness of the I-filaments corresponds to a magnification of about 400,000. (Knappeis & Carlsen. *J. Cell Biol.*)

two filaments on the opposite side. By studying the changes in appearance when small changes in the plane of section were made, Knappeis & Carlsen concluded that the rod-like body to which the I-filament attached really consisted of four filaments, so that the actual arrangement could be illustrated schematically by Fig. 740, the four filaments in the Z-line actually making a pyramid.

Tropomyosin Filament

Huxley examined negatively stained crystals of tropomyosin and demonstrated a remarkable lattice-work of fine filaments, resembling closely the lattice-work seen in cross-sections of the Z-line; and it has been suggested that tropomyosin filaments intertwine with I-filaments at the Z-line, providing a fibrillar

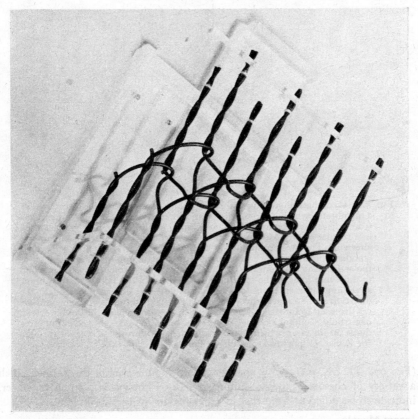

FIG. 741. View of a model portraying two tiers of I-band filaments in which Z-band structure is represented as the interlocked loops of actin strands (dark wires). These loops arise as the uncoiled ends of I-band actin double helices. (Kelly. *J. Cell Biol.*)

continuity across this region. A postulated arrangement is that suggested by Kelly (1967) illustrated in Fig. 741; this author has drawn an analogy between the looping behaviour of actin filaments as they approach the Z-band and the looping of intracellular filaments in epithelial cells at the desmosome (p. 492). On this basis, each actin filament provides two strands which wind round strands from the next sarcomere and return to fold with an adjacent strand. Grossly, such filaments appear to give off four strands, but only two belong to it and the other two are the approaching and departing legs of a looped strand from the opposite side.*

Membranous Nature of Z-Line

Franzini-Armstrong & Porter (1964) emphasized, on the basis of their electron microscopical study of a variety of species, the membranous character of the material in the Z-disc and suggested the model of Fig. 742; they pointed out that the myofibril always begins at the cell surface, or the myotendon junction, with the plasma membrane representing the first Z-line; again, in cardiac muscle at the intercalated disc, the region where one fibre terminates and the next begins is always in phase with a Z-line. On this basis the dense material beneath the plasma membrane at the desmosome part of the intercalated disc

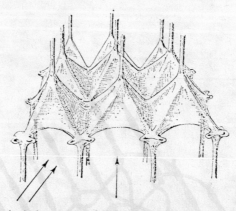

FIG. 742. Hypothetical structure of Z-disc which is represented as composed of a membrane, which is stretched in opposite directions by the actin filaments from adjacent sarcomeres. The latter, upon joining the Z disc, are disposed in a regular square pattern in which the elements from the two sarcomeres alternate. The stretch applied to the membrane causes it to form ridges (condensations) which run between the alternating points of application of the actin filaments from the two sarcomeres. These thickenings correspond to the dense lines of the electron micrographs. In the latter the membrane stretching between the ridges is not visible.—An observer looking in the direction of the single arrow would see that the thickenings of the membrane are disposed to form a zigzag. Looking in the direction of the double arrow, one would see how a change in orientation of the plane of sectioning produces an image in which the actin filaments seem to pass straight through the Z disc. (Franzini-Armstrong & Porter. *Z. Zellforsch.*)

(Fig. 678, Pl. LX, p. 1274) is analogous with the material of the Z-line, which a variety of chemical studies, summarized by Franzini-Armstrong & Porter, indicate to be different from that in other regions of the fibril.

Insect Muscle

In the flight muscle of the belostomatid water bug, *Lethocerus*, the structure of the Z-line seems simpler, consisting of an interdigitation of I-filaments within amorphous material, without any interconnecting filaments, as illustrated in Fig. 743; this arrangement, on cross-section, would give a hexagonal array of sectioned I-filaments (Ashurst, 1967).

The M-Line. In the electron microscope the M-line appears as a 750A wide line of high electron-density; it is bordered by two 250A wide lines of very low electron density (Fig. 722, Pl. LXV); this line is in the middle of the H-zone, *i.e.*

* Reedy, in the discussion of the paper by Hanson & Lowy (1964), has proposed a model for the I-filament–Z-filament relationship, the Z-filaments being regarded as unwound regions of a four-stranded cable of I-filament.

the zone free from overlap of the A- and I-filaments, and thus represents a feature belonging to the A-filaments; and it would seem that the high electron-density is due to arrays of cross-bridges connecting each A-filament with its six neighbours. Knappeis & Carlsen (1968) showed that the high electron-density was associated with both the A-filaments and the spaces between them, and they were able to

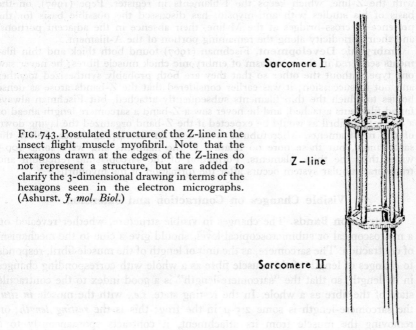

FIG. 743. Postulated structure of the Z-line in the insect flight muscle myofibril. Note that the hexagons drawn at the edges of the Z-lines do not represent a structure, but are added to clarify the 3-dimensional drawing in terms of the hexagons seen in the electron micrographs. (Ashurst. *J. mol. Biol.*)

Sarcomere I

Z-line

Sarcomere II

resolve a series of so-called *M-filaments*, running parallel with the A-filaments in the M-line; in addition, they showed the presence of M-bridges connecting the A-filaments transversely; their model is illustrated in Fig. 744. The low electron-density of the borders of the M-line was due to the absence of the M-bridges and also of the projections that arise from the A-filaments outside the M-zone.* It

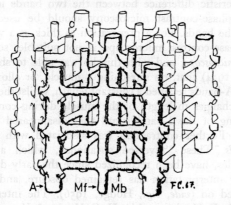

FIG. 744. Arrangement of M bridges (*Mb*) and M filaments (*Mf*) between the A filaments (*A*). On the reconstruction, the spacing between A filaments is 30% larger than in electron micrographs. (Knappeis & Carlsen. *J. Cell Biol.*)

* The region containing the M-line and the adjacent light regions is described as the *pseudo-H-zone*, and is the region where there are no cross-bridges observable on the A-filaments; it is interrupted, however, by the M-line where the cross-bridges make contact with adjacent A-filaments (Huxley, 1964/65; Pepe, 1967).

would seem that the main function of the M-line is to keep the A-filaments in position in the longitudinal, as well as the transverse, direction. The "fence" of M-bridges and filaments keeps each sliding I-filament in its prismatic tube where the I-filaments are no longer guided by the projections from the A-filaments when they enter the M-region during shortening; the M-line is thus analogous with the Z-line, which keeps the I-filaments in register. Pepe (1967), on the basis of his studies with anti-myosin, has discussed the possible basis for the presence of cross-bridges at the M-line, their absence in the adjacent portions, and their periodicity along the remaining portion of the A-filament.

Embryonic Development. Fischman (1967) found both thick and thin filaments scattered in the cytoplasm of embryonic chick muscle fibres; he never saw one type without the other so that they are both probably synthesized together and not in succession; it was earlier considered that the Z-bands arose as dense bodies to which the thin filaments subsequently attached, but Fischman always found filaments attached, and he never saw a Z-band a sarcomere length ahead of a developing fibril as would be expected if the Z-band organized the laying down of the myofilaments. Microtubules were seen in developing fibres, below the sarcolemma, but these bore no relation to the filaments and were some 100A wider than the thick filaments. Formation of the sarcoplasmic reticulum and transverse tubular system occurs after the formation of the myofibrils.

Visible Changes on Contraction and Extension

Contraction Bands. The changes in visible structure, whether revealed on a microscopical or submicroscopical level, should give a clue to the mechanism of contraction. The sarcomere, as the unit of length of the muscle fibril, responds to changes of length of the muscle fibre as a whole with corresponding changes in its length, so that the "sarcomere-length" is a good index to the contractile state of the fibre as a whole. In the resting state, *i.e.*, with the muscle *in situ*, the sarcomere-length is some 2·5 μ in the frog; this is the *resting length*; on removing the muscle from its attachment, it contracts spontaneously to a sarcomere-length of 2·0 μ, the so-called *equilibrium length*. The changes in which we are interested are in the relative lengths of the A- and I-bands. Early studies of Buchtal, Knappeis & Lindhard, who prepared light-micrographs of the muscle during isometric contraction, indicated that it was the A-band that changed most during contraction, but, as A. F. Huxley pointed out, the characteristic difference between the two bands is one of refractive index, so that phase-contrast microscopy should be used for the accurate delimitation of the bands. For a preparation as thick as a whole muscle fibre, the ordinary phase-contrast microscope is not suitable, so Huxley designed an *interference microscope*, and with this he was able to show (A. F. Huxley & Niedergerke, 1954) that, on stretching a muscle, or allowing it to contract moderately, the A-bands retained a constant length, whilst the I-bands shortened or lengthened; changes in length of the muscle during contraction were thus due to shortening of the I-bands until they disappeared at 65 per cent. of resting length.* Further contraction was associated with the appearance of *contraction bands*. These "reversed striations", which appear in strongly contracted muscles, have been recognized since the early days of light-microscopy, but their interpretation has remained obscure, and even now is not completely agreed on (*vide*, *e.g.*, Hodge, 1956). The interpretation of these bands given by A. F. Huxley, and by H. E. Huxley & Hanson, is based on their observation that it is the I-band that gets shorter during contraction, so that

---

* De Villafranca (1957) has confirmed the relative constancy of the A-band during stretch, and the considerable increase in the H-zone; this author observed glycerol-extracted fibrils under the phase-contrast microscope. Knappeis & Carlsen (1956), studying muscles fixed at different degrees of stretch in the electron-microscope, found the A-band constant for contractions between 5 and 35 per cent.

when the I-band has completely disappeared the Z-line comes in contact with the A-band to give an exceptionally dense region (Fig. 745). Further contraction of the fibre must necessarily be accompanied by a shrinkage of the A-segment; if this were completely homogeneous, the fibril would not appear segmented at all, except for the extra-dense regions at the ends of the sarcomeres, due to the presence of Z-material. However, the A-band is normally not homogeneous, as we have seen, owing to the presence of the less dense H-region in the centre. As the muscle shrinks, this H-region becomes less distinct, so that at 90 per cent. of resting length it is indistinguishable from the rest. At about 80 per cent. of resting length, three lines of exceptionally high density appear in the A-band, one at the centre and one at either end. These two become included in the contraction bands at 65 per cent. of resting length, whilst the dark central line—M-line—is replaced by two lines which are eventually incorporated into the contraction bands at 30 per cent. of resting length.

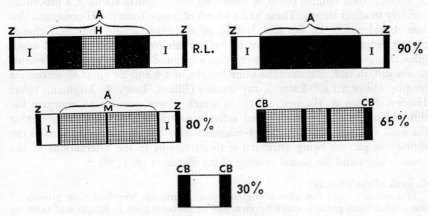

FIG. 745. Illustrating the changes taking place in the striation of a muscle on contraction, leading to the formation of contraction bands.

Electron-Microscopy of Shortening and Lengthening. The complete explanation of the light-microscopical appearance of a muscle fibre at all stages of shortening is difficult, but the changes leading to the disappearance of the I-bands may be explained on the basis of the sliding of the I-filaments into the A-band. This was demonstrated by H. E. Huxley & Hanson's study, in which they extracted myosin from glycerol-extracted fibres at various stages of contraction. Fibres extracted at normal length showed a gap at the H-region, suggesting that the actin filaments did not extend right across the A-band; when the fibre was extracted at 90 per cent. of resting length, the gaps no longer appeared, suggesting that at this length the actin filaments had retracted into the H-region.

Carlsen, Knappeis & Buchtal examined normal muscle fibres in the electron-microscope when they had been fixed under known conditions of tension and alteration of length; they took care to examine, in some detail, the effects of preparative procedure on the fibre diameter and sarcomere length.* Their results, in general, were consistent with the sliding-filament hypothesis and were therefore in agreement with the studies of Huxley & Hanson and A. F. Huxley

---

* For example, the direction of the fibrils in relation to the sectioning knife should be specified; if the knife is parallel to the fibril length the diameter is smaller than if it is at right-angles to this.

on glycerinated fibres. Thus two separate types of filament, thick and thin, were identified, the thin filaments lying between the thick A-filaments; the zone of overlap—O-band of Fig. 722—decreased with passive stretch, whilst it increased with shortening. The diameter of the A-filaments barely altered during shortening although Sjöstrand & Andersson-Cedergren had claimed that the diameter of these increased, and on this account were inclined to reject the sliding-filament theory. Carlsen *et al.* emphasized that sliding of I-filaments into the A-band could account for a shortening to a sarcomere-length of less than 1.6μ, and they presumed that further shortening was associated with coiling of filaments.

Constant Filament Length

An essential feature of the sliding filament theory is that both types of myo-filament should retain the same length, and Page (1964) has examined this point with great care, taking into account the effects of various preparative procedures on the apparent length as seen in the electron microscope. She used, as her standard of comparison, the lengths of I-filament that were still attached to Z-discs; their lengths could be measured after negative staining, a procedure unlikely to affect length. These had a length of 2.05μ from the H-zone, and this was considered to be the true length, so any other length indicated change during preparation. It turned out that glutaraldehyde fixation produced negligible changes in length so that, using this, she found that the A- and I-filaments *in situ* did, in fact, maintain the same lengths of 1.6 and 2.05μ at all sarcomere lengths above 2.1μ.* Later X-ray studies (Elliott, Lowy & Millman, 1965; Huxley, Brown & Holmes, 1965), in which it was possible to compare the diffraction patterns of resting and actively contracting muscles, revealed that the axial spacings of the A- and I-filaments were unchanged, alterations in the diffraction pattern being attributable to alterations in the orientations of the cross-bridges and the lateral spacing of the filaments (p. 1376).

Growth of the Muscle

Goldspink (1968) has shown that increase in muscle length during growth of the animal takes place initially by virtue of an increase both in length and number of sarcomeres; at a certain age, however, further increase in length takes place exclusively as a result of the addition of new sarcomeres to the fibre. In the mouse the sarcomere length in the newborn was 2.3μ, increasing to 2.8μ in the adult, and this was due entirely to a reduction in overlap of the A- and I-filaments, indicating that these filaments are of a constant size during development.

Constant Volume of Filament Lattice

Huxley (1957) noted that, with passive stretch of the muscle, the filament lattice of the A- and I-bands maintained a constant volume at all sarcomere lengths from the point at which I-filaments abut in the H-zone to the point at which all the I-filaments leave the A-band myosin array. This means that the actin-myosin centre-to-centre distance varies between 190 and 260A, in frog muscle, and 220 and 280 in mammalian muscle. The sum of the radii of the actin and myosin filaments in the living state is probably 150A, so that the remainder varies between 40 and 110A, in the first case, and 70 and 130A in the second; *i.e.* variations of greater than 100 per cent. can occur in the surface-to-surface distances of the filaments; and this raises an interesting point as to how the cross-linkages, that presumably determine the interfilament forces during contraction, are able to stretch across (Elliott, 1964).

During active contraction too, the surface-to-surface actin-to-myosin distance

---

* Page & Huxley (1963) measured the periodicity of the cross striations in the I-filaments under conditions where length was unaffected by fixation; this was 406A; thus the length of any I-filament should be obtained by multiplying the number of periods by 406A.

varies between 50A at 3.5μ sarcomere length and 130A at 1.8μ sarcomere length (Elliott, 1967).

Extreme Shortening

The electron-microscopy of extreme shortening, leading to the large contraction bands, has not been completely elucidated; Gilev has induced localized contraction in a single fibre in such a way that all stages of contraction of sarcomeres could be observed in a single field. This permitted him to identify a number of successive stages that were consistent with the view that up to a shortening of 79 per cent. of resting length there is a simple sliding of I-filaments into the A-band; beyond this, the appearances suggested a spiralization of the I-filaments in the region of the Z-line, and spiralization of the A-filaments too. It seems definite that in vertebrate muscles the I-filaments do not penetrate the opposite Z-band during extreme shortening; in the striated scutal depressor muscle of the barnacle, *Balanus nubilis*, however, Hoyle *et al.* (1965) noted that contractions down to 30 per cent. of resting length could be achieved by nerve stimulation; under these conditions the Z-lines apparently broke up, allowing the I-filaments to interdigitate with the A-filaments of the next sarcomere.

Effects of Stretch on Bands. The effects of stretching the passive muscle fibre are consistent with the view of H. E. Huxley & Hanson on the muscle structure. Thus, A. F. Huxley & Niedergerke (1954) showed, by interference-microscopy of isolated fibres, that stretch was associated with an increase in

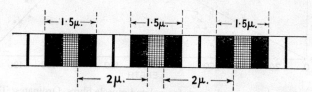

FIG. 746. Diagram showing the dimensions that remain constant as a striated muscle fibre is stretched.

length of the I-segment, that of the A-segment being virtually unchanged. Huxley & Niedergerke made the important discovery that the distance between the edges of the H-regions remained constant too, so that a lengthening of the intervening I-band must have been accompanied by a corresponding expansion of the H-region (Fig. 746). If the H-zone represents a region where there is no overlapping of actin and myosin filaments, the extension of the sarcomere, by pulling the actin filaments out of the A-band, will increase the length of the H-zone exactly in proportion to the increase in length of the sarcomere, *i.e.*, of the I-zone. It was this finding that led A. F. Huxley & Niedergerke to formulate a theory of muscle structure that was essentially the same as that put forward independently by H. E. Huxley & Hanson, who also obtained the same results on studying isolated glycerol-extracted myofibrils with the phase-contrast microscope. If myosin was extracted from these fibrils at various stages of stretch, it was found that the gap corresponding to the H-zone increased with increasing degree of stretch.

Maximum Length for Contraction. On the basis of the sliding-filament theory, the tension developed in an isometric twitch should decrease as the muscle is stretched, since stretching means reducing the degree of overlap of the I- and A-filaments. If the tension developed is proportional to the number of chemical links established between the two filaments, then the greater the overlap the greater the tension developed, and *vice versa*. Qualitatively this is true, and is reflected in the well-established length-tension diagram of Fig. 747, which shows that tension decreases on either side of a maximum which corresponds approximately with the resting length. The point at which zero tension is developed should be the point at which the sarcomere length has been increased to the degree that there is no overlap; with the frog muscle this corresponds to a sarcomere length of $3 \cdot 5 \mu$, the sum of the lengths of the A- and I-filaments ($1 \cdot 5 + 2 \cdot 0 \mu$). The ordinary length-tension diagram is complicated, however, by the circumstance that all the sarcomeres do not stretch to the same extent, those near the tendons stretching much less than those in the middle; hence it could well happen that the sarcomeres at the ends of the muscle would still have considerable overlap, whilst those in the middle had been stretched well beyond the point of overlap, giving an actual gap between the I- and A-filaments. The muscle fibres as a whole would be stretched beyond the

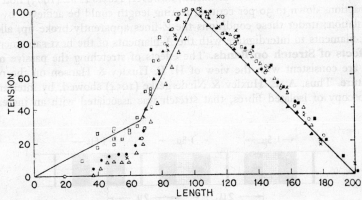

FIG. 747. Length-tension diagram for isolated muscle fibres. Ordinates: Tension developed (total minus resting) in per cent. of maximum developed. Abscissæ: Length, in per cent. of resting length. (Ramsey. *Medical Physics*. Chicago. Year Book Publishers.)

theoretical limit, but the less stretched sarcomeres at the end would develop tension.

Variation of Sarcomere Length

In fact, A. F. Huxley & Peachey showed that the sarcomere length of a single fibre fell off considerably at the ends. Thus, in a given stretched fibre the middle sarcomeres had a length of some 4μ whilst the lengths at the ends could be as short as 2μ. When the fibre so stretched was caused to contract isometrically, tension developed, and microscopical examination of the sarcomeres at the ends of the fibre showed that their lengths were, indeed, less than the critical value of about $3 \cdot 5 \mu$.

Tension at Constant Length

In a later study on single muscle fibres (Gordon, Huxley & Julian, 1966), the length of a selected part could be held constant by a feed-back servo; the length of this was measured continuously by a photo-electronic spot-follower which fed back to a moving-coil apparatus, which pulled on the tendon at one end as soon as shortening occurred. Stimulation under these conditions caused the development of tension, which was recorded at the other tendon. Thus the length of a selected part could be held constant while tension was developed in it. Before stimulation, the fibre was examined microscopically and the sarcomere spacings of different marked regions measured accurately; the length-servo was started, and finally a tetanus was established. When the fibre was stretched so that the length of the sarcomere under observation was greater than that expected to permit the development of tension, then the actual tension developed was much less than if the tendons had been held at both ends, showing that a considerable part of the tension, developed under the latter condition, is due to shortening of some sarcomeres at the expense of others.

EXTRAPOLATED MINIMUM TENSION

Even at high degrees of stretch there was always some development of tension which increased with time, presumably because there was some irregularity in striation-spacing so that overlap of thick and thin filaments existed locally; during stimulation this would increase with time, so that the minimum tension should be derived by extrapolation to zero time. As Fig. 748 shows, when the

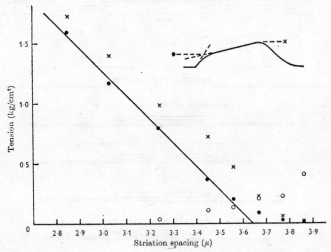

FIG. 748. Graph of tension versus striation-spacing for muscle fibre. Crosses are maximum tension developed in servo-controlled tetani. Filled circles tension to same tetani extrapolated back to start of tetanus. Inset shows method of extrapolation. Open circles show resting tension at each length. (Gordon, Huxley & Julian. *J. Physiol.*)

extrapolated tension was plotted against sarcomere length, the tension extrapolated to zero at a sarcomere length of 3·65 μ. There is some residual tension at this length, but the greatest deviation from the ideal result was only 3–5 per cent. of the tension developed at optimum length.*

Stripped Fibres

Podolsky (1964) pointed out that the demonstration of very low tension at large sarcomere lengths was only a confirmation of the sliding filament theory if it could be shown that the limiting factor was not the spread of excitation at the greater sarcomere length. He employed fibres whose sarcolemmas had been stripped off by Natori's (1954) technique, as this permits rapid and complete activation of the fibre by application of a drop of high-Ca^{++} Ringer (p. 1400). The sarcomere-length at which no shortening occurred was 3·65 μ, rather greater than the sum of the filament lengths, namely 3·5 μ. As Podolsky pointed out, this sarcomere length is that at which there is no observable shortening, and not necessarily the same as that at which no tension develops, λ_0, since there could be

* Proof that the slow rise was due to irregular spacing was provided by the effects of different treatments of the servo-controlled fibre between tetani; those fibres that were stretched between tetani gave a smaller rise than those allowed to relax or shorten between tetani.

an alteration in compliance during stimulation such that a small tension would just be balanced, leading to no measurable shortening. Podolsky showed, from simple calculations, however, that the value of λ_0 would lie between 3·65 and 3·75 μ.*

Length-Tension Diagram. As indicated earlier, the tension that a muscle fibre may develop if it is held at a fixed length and stimulated tetanically will, if the sliding filament theory is correct, be maximal at the point of complete overlap of I-filaments with A-filaments, and zero at the point of no overlap. Intermediate stages are revealed by the length-tension diagram of Fig. 747.

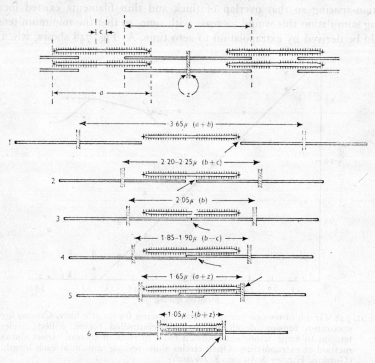

FIG. 749. *Below:* Illustrating the critical stages in the increase of overlap between thick and thin filaments as a sarcomere shortens.

Above: Schematic diagram of filaments, indicating nomenclature for the relevant dimensions. (Gordon, Huxley & Julian. *J. Physiol.*)

Here the extra tension developed during stimulation is plotted as a function of percentage of resting length. Gordon, Huxley & Julian (1966) have used their servo-controlled fibres to examine the length-tension diagram in terms of the measured striation spacing, which in the frog is made up of thick filaments of length 1·6 μ, thin filaments of 2·05 μ, a region lacking cross-bridges of 0·15–0·20 μ, and the Z-line of 0·05 μ. Six possible stages are illustrated schematically in Fig. 749 with the appropriate sarcomere lengths; the corresponding situations are indicated in the experimental length-tension diagram measured at the same time (Fig. 750). Corners B and D clearly coincide well with stages 2 and 5 respectively of the development of overlap; corner C may be related to stage 3

* The contribution of the sarcolemma to the passive tension developed on stretching a fibre is negligible at lengths up to 1·4 times L_0, the resting length; as length increases, the sarcolemma contributes, so that the elastic modulus of the stripped fibre is some fifth of the normal fibre (Natori, 1954).

where the A-filaments abut, but corner E does not fit satisfactorily with any of the landmarks of the shortening process, whilst any effect of stage 4, overlap of the bare regions of the A-filaments, is not reflected in the length-tension diagram. The corner D, where the slope of the length-tension line increases as shortening gets greater, corresponds almost exactly with the striation-spacing corresponding to the collision of the ends of thick filaments with Z-lines; this would increase resistance, whilst the crumpling or folding of thick filaments would reduce the number of bridges capable of generating tension.*

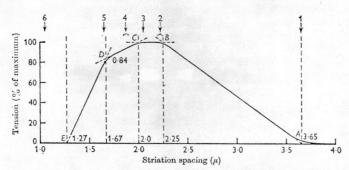

FIG. 750. Experimental curve showing tension as a function of striation spacing. The numbers refer to the conditions illustrated by Fig. 749. (Gordon, Huxley & Julian. *J. Physiol.*)

Crayfish Fibre

The length-tension diagram for single fibres of the crayfish muscle was determined by Zachar & Zacharová, using the tension developed in complete depolarization by KCl as a measure of maximum tension at any given length. Maximum tension occurred at 1·25 resting length, and this corresponded to a sarcomere length of 10·5 μ; tension vanished at a length of 16·3 μ, *i.e.* some five times greater than in the frog. On the basis of the sliding filament theory the maximum tensions developed should be in relation to the degrees of overlap in the two systems; the size of the A-filaments, which determines the overlap, is 3·95 μ in the crayfish compared with 1·6 μ in the frog, and the forces developed were 8·2 and 3·6 kg./cm². respectively.

Further X-Ray Studies

Narrow-Angle Diffraction. The necessity of fixing muscle for study in the electron microscope restricts the amount of useful information that can be derived with respect to the exact ordering of the protein molecules in both resting and contracted states, so that H. E. Huxley (Huxley & Brown, 1967; Huxley, 1967) has returned to the X-ray analysis of muscle in resting and contracted states, using improved instruments that increase resolution and reduce time of exposure. The low-angle diffraction diagram illustrated by Fig. 751 shows a regular arrangement of reflexions on layer-lines, with characteristic variations of intensity along these—a diffraction pattern characteristic of a helical arrangement of scattering matter. As Elliott (1964) and Worthington (1959) had shown, this low-angle pattern belongs to myosin, the main, if not exclusive, constituent of the A-filaments; the separation of the layer-lines

---

* As Edman (1966) has pointed out, the tension developed on the basis of the sliding filament hypothesis, and on the assumption that this tension depends only on the number of bridges formed between the filaments, will be determined entirely by the state of the filaments in respect to each other and not by the previous history of this relationship. His experiments fully support this view, the actual tension ultimately developed by a fibre being determined by its sarcomere length irrespective of the load it had to carry, or the degree of shortening that was permitted while on its way to this length.

corresponds to a helical repeat of 429A; and this is a multiple of a true repeat of 143A. The distribution of intensities along the layer lines is identified with that expected of a 6/2 helix, *i.e.* one in which there are *pairs* of cross-bridges, one on either side of the filament, repeating at the 143A spacing with successive

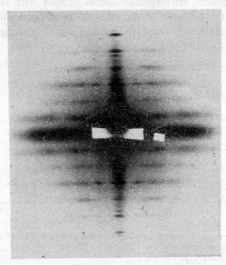

FIG. 751. Low-angle X-ray diffraction diagram from frog sartorius muscle showing system of layer-line reflections arising from helical arrangement of cross-bridges on myosin filaments. (Huxley. *J. gen. Physiol.*)

repeats rotated with respect to each other by 60°, so that the whole structure repeats at 3 × 143A to give the 429A repeat. (Fig. 752).

Wide-Angle Diffraction. The wide-angle diagram is the well-known *k-m-e-f* picture first described by Astbury; it belongs to the actin filament which

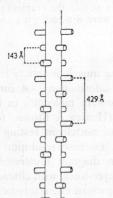

FIG. 752. Schematic diagram showing arrangement of cross-bridges on 6/2 helix. Helical repeat is 429A, but true meridional repeat (*i.e.* periodicity of the variation in density of structure projected on to long axis of filaments) is 143A. (Huxley & Brown. *J. mol. Biol.*)

consists of a double helix of globular subunits as suggested by Selby & Bear in 1956; the narrow-angle spacings indicate a helix in which the pitch is 355A × 2 to 370A × 2, with 13–13½ subunits per turn in each of the chains.* The double

* By a pitch of 2 × 360A is meant that each of the two chains of actin subunits has a pitch of 720A, but the actual repeat of the whole chain is 360A because of the half-period stagger of the component helices.

helix is illustrated in Fig. 736, p. 1360, which also shows how the myosin cross-bridges would be placed on the neighbouring A-filament. It will be seen that there is an approximate repeat period of their alignment every 710A.

Longitudinal Spacings. In addition to these prominent features, Huxley has shown that there are others that are harder to interpret, such as strong meridional reflexions indicating longitudinal spacings of 385 and 442A, which cannot be fitted into the periodicities of either actin or myosin units and so presumably arise from some other material that is attached to these. In the electron microscope the periodicity of 388A can be shown to correspond with that observed as a cross-striation in the I-band, whilst the larger spacing of 442A can be seen in the A-band.

Tropomyosin

Crystals of the protein, tropomyosin, have a periodicity of some 380–390A (Cohen & Longley, 1966); and it is interesting that the fluorescent-antibody studies of Endo et al. (1966) and of Pepe (1966) indicate that this protein is spread along the I-filaments.

Tropomyosin, as we shall see, as ordinarily prepared may be resolved into a "native tropomyosin" and tropomyosin B; and the native tropomyosin was shown by Ebashi & Kodama (1965) to be a combination of tropomyosin B with a new protein, *troponin*. Hanson (1967) found that aggregates of actin showed the 400A periodicity which disappeared when the actin was purified, and reappeared when tropomyosin was added; the tropomyosin may have contained troponin, since Ohtsuki et al. (1967) were able to show a period distribution of ferritin-labelled anti-troponin in the I-filaments; with 24 periods on a 1μ I-filament, this would correspond to a 400A periodicity.

Different Periodicities. It is probably an important feature of the actin and myosin filaments that their periodicities, not only helical but also of their subunits, are different; this ensures that all cross-bridges do not occupy the same relative positions with respect to sites on the actin filaments, an arrangement that would allow a relatively steady tension to be developed in a given thin filament by an unsynchronized activity of a number of cross-linkages attaching to it. The additional periodicity of 444A in the A-filaments has not so far been associated with a protein.*

Changes in Rigor. When the muscle passes into rigor, *i.e.*, when permanent attachments of the cross-bridges may be assumed to have taken place, then there is little change in the wide-angle repeat of the actin picture indicating that there has been very little change in the packing of the actin monomers when the cross-bridges attach to them. By contrast, the pattern of the myosin filaments changes considerably, indicating a movement of the cross-bridges, so that their positions correspond to repeats on the pattern of the actin filaments. The fundamental meridional 149A repeat, representing the long-spacing of the cross-bridges, remains unchanged. In general, it appears that the myosin helix changes its shape when the muscle passes into rigor. When the highly-stretched muscle is put into rigor the changes are not now so great, and this corresponds with the circumstance that overlap between the filaments is confined to only a small region; nevertheless, the X-ray diagram indicates that considerable disorganization of the cross-bridge pattern takes place, so that the actin filaments presumably stabilize the cross-link material in a stable pattern.

---

* When myosin was extracted from the glycerinated muscle all the reflexions between 60 and 430A disappeared, with the exception of a faint diffuse reflexion at 370A; the 59A actin reflexion remained, but it was considerably less well ordered, indicating that the I-filaments need the A-filaments to maintain their exact alignment.

Filament Spacing in Rigor

Examination of the equatorial reflexions during rigor revealed significant changes suggesting that the I-filaments, occupying the trigonal positions in the hexagonal array of I- and A-filaments, had increased their density at the expense of the A-filaments. In a re-examination of these reflexions, Huxley (1968) computed that, in the region of overlap of the A- and I-filaments the relative densities of the (A-plus I-material): I-material changed from 3 : 1 to about 1·8 : 1. In normal resting muscle this 3 : 1 proportion indicates that the mass per unit length of A-filament is some four times that of the I-filament, since there are twice as many I-filaments as A-filaments. The change in this ratio in rigor indicates a transfer of some 0·3 of the material of the A-filament to each I-filament in the region of overlap. If the molecular weight of a myosin molecule is 540,000, then this would be compatible with the shift of the globular portion of the heavy meromyosin molecule (p. 1377) from the A-filament to the I-filament. Thus, if may be that the globular portion of the heavy meromyosin molecule is attached to the backbone of the A-filament by way of the linear portion of the molecule; flexible regions at each end of this linear portion would enable it to function as a link which could accommodate closely to substantial variations in interfilament spacing between actin and myosin. The force of contraction would be developed by a change in the effective orientation of the globular portion of the heavy meromyosin molecule in relation to sites on the actin filament, whilst the tension would be transmitted by way of the flexible region.*

Normal Contraction. When muscle is made to contract as a result of electrical stimulation,† it is found that the repeating periodicities along the lengths of the actin and myosin filaments, and hence their overall lengths, are unchanged,‡ whilst the intensities of the off-meridional part of the layer-lines become weaker, indicating a substantial change in the helical pattern of cross-bridges; a likely reason for the weakening of the pattern is the continual movement of the cross-bridges so that at any given moment only a fraction of the whole occupy an ordered position. The change in the helical pattern probably consists, too, in a movement of the cross-bridges not only longitudinally, consistent with their carrying out unsynchronized longitudinal movement of attachment and detachment, but also moving radially and circumferentially while carrying out searching movements for an actin site at an appropriate level and facing in the correct direction.

THE MUSCLE PROTEINS

It will be of value, at this point, to describe in greater detail, the structural proteins of muscle. Weber, in 1950, described the muscle proteins as consisting of non-fibrous albumin (myogen A and myogen B) and globulin X, representing some 20 per cent. each of the total proteins, whilst the remainder was made up of actin, myosin, and tropomyosin. The albumin and globulin presumably contribute nothing to the structure, and may be described as sarcoplasmic proteins consisting mainly of the larger number of enzymes that are concentrated here.

* The behaviour of the interfilament spacings of glycerol-extracted muscle under various conditions has been examined by Rome (1967); alterations in the electrolyte content of the medium, or of the pH, have large effects, suggesting that electrostatic forces operate to maintain the spacing.

† In order to obtain exposures of contracting muscle for periods long enough to give an adequate diffraction pattern the muscle is repeatedly stimulated and only photographed during its contracted state.

‡ Actually there is a measurable change in the 143A meridional reflexion, namely an increase to 144·6A, indicating an increase in the distance between cross-bridges in the axial direction.

Myosin*

Myosin belongs to the *k-m-e-f* group of fibrous proteins, showing the characteristic 5·1A axial repeat in the unstretched condition; according to Pauling & Corey, therefore, it has a helical structure.

Molecular Weight. The molecular weight and shape of myosin are vexed questions that have been discussed recently by Dreizen *et al.* (1967). Kielley & Harrington's (1960) picture of a 420,000 weight molecule, some 1650A long, is considered to be incorrect, a value of 500,000 being favoured although this is not unequivocally proved since there is some dependence on the method of preparation, ammonium sulphate precipitation giving a product with an estimated molecular weight of 600,000 and the conventional Szent-Gyorgyi technique giving one of 500,000. According to Dreizen *et al.* the 600,000 product is not homodisperse and when they extrapolated the results of sedimentation equilibrium they concluded that a molecular weight of 500,000 with an uncertainty of 3–4 per cent. was the most likely value; this agrees well with Mueller's (1964) estimate, using the Archibald sedimentation technique, namely 524,100. The electron microscopical picture of the myosin molecule was presented by Rice (1961) who employed a mica-replica technique; the rods visualized in this way had a mean length of 1,100A, with a globular enlargement at one end. Zobel & Carlson (1963) found a mean length of 1590 ± 165A; the rod part had a diameter of 20 ± 5A; the head was 400 ± 70A long and about 35A in diameter. There were a large number of particles of 1,150A length that may well have been molecules that had lost their heads. As indicated earlier, subsequent studies of Huxley have confirmed and elaborated on these features of the myosin molecule.

Subunits. By treating myosin with a hydrogen-bond rupturing agent, such as 5M guanidine hydrochloride, they demonstrated the break-up of the molecule into three apparently identical subunits of molecular weight $2·19.10^5$; after allowing for the incorporation of some guanidine hydrochloride and N-ethyl maleimide† on the particles, the corrected molecular weight of the subunit was 206,000, corresponding to one-third of the original molecule. This might well have represented a single strand of a triple helix. From the amino-acid analysis, it was computed that each chain would have 1,747 residues which, on the basis of an α-helix, would give a chain some 2,620A long. If this was wound as a supercoil, with an axial period of 62A, the minimum length of 2,240A was deduced; this is long by comparison with the estimated length of 1,650A based on viscosity and other measurements. Kielley & Harrington suggested, therefore, that the coil was folded back on itself, so that one end would be thicker than the other.

In view of the present estimate of the molecular weight as 500,000 the concept of a triple helix breaking down to three subunits is probably incorrect, and the results may be interpreted in terms of a two-chain helix.

Meromyosins

A. G. Szent-Gyorgyi found that when myosin was subjected to a short digestion with trypsin it was split into two well defined units, which he called L- and H-meromyosin, L and H indicating "light" and "heavy" respectively, since the molecular weight of L-meromyosin was 96,000 and that of H-meromyosin was 232,000; both molecules were highly asymmetrical rods, L-meromyosin being some 549 × 16·6A and H-meromyosin 435 × 29A.

It was originally thought that the molecules occurred in the proportion of two L to one H, but subsequent work indicates a one-to-one ratio (Lowey & Holtzer, 1959). In the electron microscope H-meromyosin preparations contained many tadpole-shaped particles, the tails being shorter and narrower than myosin rods; the very small diameter of 5–10A suggests a single chain. L-meromyosin preparations contained thin rods of diameter 20A and length around 900A (Rice, 1964). The ATPase activity and power to unite with actin reside in the heavy meromyosin. Subsequent studies on breaking off subunits have left a confused picture

---

* Unfortunately there is still a confusing tendency to refer to myosin proper as *myosin A*, and its combination with actin, as directly extracted from muscle, as *myosin B*, rather than actomyosin. As we shall see, this "natural actomyosin" contains additional proteins and is thus different from "artificial actomyosin" formed from pure preparations of actin and myosin.

† N-ethylmaleimide was added to the solutions to prevent the formation of S–S linkages between the SH-groups of any of the liberated subunits.

which has been summarized by Dreizen *et al.*; Fig. 753 illustrates their interpretation. According to this, the long tail is made up of a double helix, and light meromyosin represents most of the tail. The *f*-subunits are split off in 5 M guanidine from the heavier subunits resulting from the splitting off of small g-subunits in alkaline solution. Thus the myosin molecule has an axial helical core composed of 2 *f*-subunits of weight 215,000 that extends into a globular head region which contains 3 g-subunits of average weight 20,000.

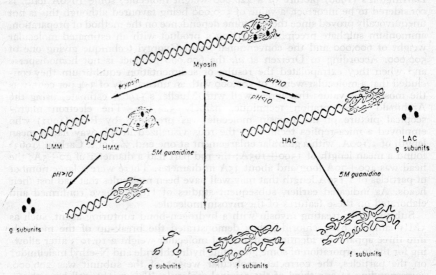

FIG. 753. A schematic diagram, for the subunit structure and interactions of myosin. LMM, light meromyosin; HMM, heavy meromyosin; LAC, light alkali component; HAC, heavy alkali component. (Dreizen *et al.* *J. gen. Physiol.*)

Actin

G-F Transformation. Actin accounts for some 20–25 per cent. of the residue remaining from extraction of the sarcoplasmic proteins; the extract obtained in the usual way has a low viscosity, by contrast with the highly viscous solutions that are obtained on addition of salt, or making the medium more acid than pH 6. This change in viscosity was assumed to be the result of a change from a globular *G-actin* to a fibrous *F-actin*, by linear aggregation of the globular G-actin molecules; and this view was supported by electron-microscopical studies of films of actin sols at different stages in the G-F conversion. Thus Rozsa, Szent-Gyorgyi & Wyckoff (1949) induced the growth of very long filaments, some 100A in diameter, which appeared to have been formed by the approximately lengthwise aggregation of ellipsoidal rodlets, some 300A long by 100A wide.

As we have seen (p. 1359), Hanson & Lowy have visualized the individual actin filaments under high resolution with negative staining, and have shown that they are composed of two strands wound round each other; each strand consists of a single series of regularly spaced nodules, of diameter about 55A and probably equivalent to the monomer of G-actin of molecular weight 60,000 to 70,000.

ATP. Straub & Feuer and Laki, Bowen & Clark reported that actin contained ATP as a prosthetic group, which was converted to ADP on polymerization to F-actin; subsequent work has generally confirmed this finding, and Mommaerts (1952) has shown that one molecule of inorganic phosphate is liberated per 56,000 g. of actin; this suggests that G-actin and ATP react in equimolecular proportions:

$$G + ATP \rightleftharpoons G.ATP \qquad \qquad \text{(1)}$$

In the presence of salt, G.ATP would polymerize as follows:

$$G.ATP \rightarrow F.ADP + P_i \qquad \qquad \text{(2)}$$

the F-actin containing ADP. Studies with ^{14}C-labelled ATP have indicated that reaction (1) is indeed reversible, since the labelled ATP exchanges rapidly and completely with the unlabelled ATP of G-actin. By contrast, there is no exchange of labelled ATP or ADP with the bound ADP of F-actin. When the F-actin is dialysed against distilled water to remove salt, it becomes depolymerized:

$$F.ADP \rightarrow G.ADP$$
$$G.ADP + ATP \rightarrow G.ATP + ADP$$

ATP being required to convert G.ADP to G.ATP (Martonosi, Gouvea & Gergely).

When G.ATP is placed in an ATP-free solution it gradually loses its power to polymerize; this is presumably because the complex with ATP dissociates in accordance with reaction (1), the free G-actin slowly changing to an inactive form; excess ATP, by driving the reaction to the right, favours the formation of G.ATP, which is polymerizable (Asakura).

Calcium. In view of the importance of Ca^{++} in the activation of muscle, its relation to actin is of some interest. Chramback, Bárány & Finkelman showed that Ca^{++}, like ATP, reacts with G-actin stoichiometrically to form a complex, containing equimolecular proportions. The Ca^{++} so bound is completely exchangeable with ^{45}Ca in the medium. It may be removed by several procedures that likewise remove its bound ATP; when the Ca^{++} is removed all the characteristic properties, namely, its ability to polymerize, its ability to form actomyosin, and the ability to activate the ATPase of L-myosin in presence of Mg^{++}, are lost too. All these characteristic properties of G-actin were shown to be a linear function of the Ca^{++} content until maximum activity was attained at an actin: Ca^{++} ratio of $1 : 1$. When the G-actin polymerized to F-actin, the Ca^{++} became inexchangeable with ^{45}Ca in the medium. It would seem, therefore, that Ca^{++} forms a bridge between the nucleotide and protein moieties of the complex, as suggested originally by Straub & Feuer (Barany, Finkelman & Therattil-Antony).

Physiological Significance of G-F Change. It seems unlikely that G-F transformation has any physiological significance; Szent-Györgyi considered that the conversion of G-actin to F-actin was a necessary element in the contractile process but, as Perry (1956) has pointed out, the conditions within the cell would demand that actin be in the F-form; moreover, if the secondary filaments described by H. E. Huxley are, indeed, actin, they must be F-actin. Finally Martonosi, Gouvea & Gergely have exploited the exchangeability of ATP bound to G-actin to demonstrate that this form of actin does not occur in either resting or active muscle; by injection of ^{32}P-labelled phosphate into an animal, the ATP became labelled but none appeared in the ADP of actin, in spite of muscular exercise; if the F-actin had been transformed to G-actin its ATP would have become exchangeable.

α-Actinin. Ebashi & Ebashi (1965) found that crude extracts of muscle, used for preparing "native tropomyosin" (p. 1380), contained another protein factor that promoted the superprecipitation of synthetic actomyosin, and they called the isolated factor α-actinin. Thus an actinin-free preparation of actin reacted only slowly, and to a small extent, with myosin; addition of α-actinin caused a high degree of superprecipitation. It seems that α-actinin strengthens the interaction between actin and myosin under conditions of high ATP concentration that would otherwise favour dissociation. α-actinin reacts with actin, so that the actin prepared in the conventional manner is really a mixture of the two. A highly purified preparation of α-actinin contained three components in the ultracentrifuge, with sedimentation coefficients of 6·2, 10, and 25·6; all these accelerated the superprecipitation of actin. The 6·2 S and 25·6 S components were globular with molecular weights, determined by light-scattering, of 156,000 and 3,360,000 respectively (Nonomura, 1967).

β-Actinin. Maruyama (1965) isolated a protein from KI-extracted actin that inhibited the tendency of the actin filaments to form a network of fibrils; and it may be that this protein is the one invoked by Huxley (p. 1381) to account for the uniform lengths of actin filaments.

Tropomyosin

The protein tropomyosin, first described by Bailey, is derived from muscle by treatment of the minced tissue with alcohol and subsequent extraction with molar KCl. It is an α-fibrous protein with a coiled-coil structure producing a rod-shaped

490A $\times$ 20A molecule of weight 74,000 (Holtzer, Clark & Lowey, 1965). As
Cohen & Longley (1966) have shown, it tends to form paracrystals in the presence
of Ca^{++} which, on cross-section, have a remarkable similarity to cross-sections
of the Z-band of striated muscle. The most characteristic feature of this globulin-
type protein is its amino-acid composition, since it is completely lacking in some,
such as tryptophan and proline, and has unusually large amounts of others,
namely, the dicarboxylic acids.*

Relation to Paramyosin. Tropomyosin acquired a renewed interest when it
was discovered by Bailey that if invertebrate smooth muscles, such as those of
Pinna or *Octopus*, were extracted according to the technique required for extracting
tropomyosin from vertebrate striated muscle, large quantities of a globulin-type
protein with essentially similar amino-acid composition to that of vertebrate
tropomyosin could be extracted.

Troponin and Native Tropomyosin. Perry & Grey noticed that synthetic
actinomyosin, prepared from the purified proteins, was not sensitive to the
relaxing action of EDTA, unlike natural actinomyosin (myosin B), so that it
appeared that the synthetic product was not the complete model required to
produce a Ca^{++}-sensitive system. Ebashi & Ebashi (1964) showed that, during
preparation of the actin and myosin, a protein, resembling Bailey's tropomyosin A,
had been lost; they called this protein, because of its resemblance to tropomyosin,
native tropomyosin. The protein component, native tropomyosin, could be removed
from natural actomyosin by trypsin digestion, and the actinomyosin now became
insensitive to the action of Ca^{++}. Addition of native tropomyosin restored the
sensitivity, so that native tropomyosin is a necessary component of the superprecipitation of actinomyosin by Ca^{++}. Although in many ways native tropomyosin resembled Bailey's tropomyosin, the latter did not have the power of
making actinomyosin sensitive to Ca^{++}, and its molecular weight was less.
Tropomyosin, as ordinarily prepared, reacts with actin stoichiometrically; Ebashi
& Kodama (1965) separated from native tropomyosin a new protein which they
called *troponin*; this separation from native tropomyosin caused a fall in viscosity,
and when added back the viscosity returned. It seems that troponin combines
with tropomyosin to give native tropomyosin, and the latter is the constituent of
muscle that promotes the Ca^{++}-sensitive actin-myosin interaction. Thus we may
envisage a 3-protein interaction—troponin + tropomyosin F-actin (Ebashi &
Kodama, 1966). When tropomyosin and actin combine together the combination
is loose, so that when a pellet of the combination is re-suspended in an aqueous
medium, a great deal of the tropomyosin dissociates; if troponin is added, the
degree of this reversibility is greatly decreased (Drabikowski, Kominz & Maruyama,
1968b). Profiting by the great sensitivity of cardiac actinomyosin system to Sr^{++}
compared with the skeletal muscle system, Ebashi, Ebashi & Kodama (1967)
were able to show that this sensitivity was a property of the troponin in the
system, so that skeletal muscle troponin plus cardiac actomyosin + tropomyosin
did not have high sensitivity. They concluded, in consequence, that it was troponin
that was responsible for Ca^{++}-sensitivity of the complete system. In support of
this contention, Wakabayashi & Ebashi (1968) showed that Ca^{++} induced a
reversible change in the sedimentation pattern of troponin; thus at 10^{-5} M the
pattern showed a single peak, but on reducing the concentration to the point where
full relaxation of an actomyosin system would be observed, the peak became
broader and frequently split into two or more peaks. The results show that, at
low free Ca^{++}-concentration, troponin molecules have a tendency to aggregate.

Localization. The localization of tropomyosin and these other components,
actinin and troponin, on the muscle filaments has not been completely elucidated;
we have seen that both filaments have long axial spacings that seem not to belong
to the actin or myosin helices and so have been attributed to the presence of
additional material. Pepe (1966) employed an antibody staining technique and
showed that the anti-tropomyosin antibody stained the I-band; none was taken
up by the Z-line, however, a region where it has been considered that tropomyosin

* Bailey & Rüegg (1960) have described two forms of tropomyosin in the smooth
muscle of lemellibranch adductors; the globulin type responsible for the paramyosin
properties of the fibre is called tropomyosin A; tropomyosin B is considered to be a
component in the actomyosin system and, according to Kominz, Saad & Laki (1957), it
may be similar to the tropomyosin of vertebrates. These authors have examined the
amino-acid composition of tropomyosins from a wide variety of classes of the vertebrate
phylum and in different invertebrate phyla.

contributes to structure. Endo *et al.* (1966) removed tropomyosin and troponin from glycerinated muscle by trypsin digestion; preparations of these proteins were labelled with fluorescein and added to the treated muscle. The whole of the thin filaments and the H-zone fluoresced with labelled tropomyosin; a similar picture was obtained with labelled troponin, but the Z-line contained no label. If the muscle was treated for a long time with trypsin, troponin was no longer taken up, presumably because now all the tropomyosin, to which troponin must attach itself, was removed.* When the muscle was treated with ferritin-labelled anti-troponin, a periodic distribution was observed along the I-band and overlap region of the A-band; the period was some 400A and may well correspond to that described in X-ray and electron-microscopical studies (Worthington, 1959; see p. 1375).†

At any rate, a study of the localization of troponin in paracrystals of tropomyosin indicated that its periodicity could be explained on the assumption that a troponin molecule attached itself to a specific site on the tropomyosin molecule; thus the periodicity of 370 to 390A exhibited by troponin reveals, essentially, the underlying periodicity of the tropomyosin molecules (Nonomura, Drabikowski & Ebashi, 1968).

M-Protein. The M-line is a region of density on the myosin filaments (Fig. 722, Pl. LXV); studies with fluorescent antibodies indicated that the M-line bound to it antibodies contained in antisera to the structural proteins; and Masaki, Takaiti & Ebashi (1968) were able to extract from chicken skeletal muscle a protein, free from myosin, the antibodies to which attached exclusively to the M-line. In the electron microscope it could be shown that the "M-protein" markedly accelerated the lateral aggregation of myosin, an effect that was exerted on L-meromyosin but not on H-meromyosin.

Δ **Protein.** During the course of his studies on the effects of protein extractants on the histological appearance of the muscle fibre, Amberson analysed electrophoretically the proteins obtained by an extraction procedure that removed the myosin from muscle but not the actin. He found three fractions; the first two were sarcoplasmic proteins and myosin, whilst the third was complex, consisting of a fibrous protein, called Δ protein, together with a complex formed by it with myosin, and possibly other materials. According to Kominz (1966) Amberson's Δ protein is probably Ebashi & Ebashi's native tropomyosin; it certainly combines with actin (Amberson, 1967).

Significance of Additional Proteins. The extra proteins, tropomyosin, troponin, actinin, and possibly an as yet undiscovered one on the thick filaments giving rise to the 444A repeat, are presumably concerned in the interaction between actin and myosin and possibly, also, in the assembly process of the subunits into helices, determining the lengths of these. To consider the first possibility, it is essential, on any sliding filament hypothesis, that the nature of the reaction between I-filaments and cross-bridges should change as their mutual positions change, otherwise it is difficult to see how a complete cycle of chemical events can proceed in an orderly fashion leading from attachment to subsequent release; the presence of different proteins along the helix could alter the reactivity of the helix with position. With regard to the assembly of subunits to molecules of fixed length, Huxley (1963) suggested that this could be achieved if two proteins of different periodicities polymerized together, so that they came into register after a certain number of repeats and so gave rise to filaments that were particularly stable at a certain length. So far as the actin filaments are concerned, it is possible that Maruyama's (1965) protein preparation that restricts the polymerization of Straub's G-actin is of significance in this respect.

* The contributions of these digested proteins to the strength of the filament is indicated by the break-up of the trypsin-digested filaments when treated with ATP (Ebashi, 1966).

† The interaction of tropomyosin, actin and actinin has been studied by Drabikowski, Nonomura & Maruyama (1968a); it appears that tropomyosin prevents the precipitation of actin by actinin, and so presumably weakens the interaction between these two proteins.

REFERENCES

AMBERSON, W. R. (1967) Complex formation between delta protein and F-actin. *J. cell. Physiol.*, **70**, 91–103.

AMBERSON, W. R., SMITH, R. D., CHINN, B., HIMMELFARB, S. & METCALF, J. (1949). On the source of birefringence within the striated muscle fibre. *Biol. Bull.*, **97**, 231.

AMBERSON, W. R., WHITE, J. I., BENSUSAN, H. B., HIMMELFARB, S. & BLANKENHORN, B. E. (1957). △-Protein, a new fibrous protein of skeletal muscle: preparation. *Amer. J. Physiol.*, **188**, 205–211.

ANDERSSON-CEDERGREN, E. (1959). Ultrastructure of motor end plate and sarcoplasmic components of mouse skeletal muscle fibre as revealed by three-dimensional reconstructions from serial sections. *J. Ultrastr. Res.*, Suppl., **1**, pp. 1–191.

ASAKURA, S. (1961). The interaction between G-actin and ATP. *Arch. Biochem. Biophys.*, **92**, 140–149.

ASHURST, D. E. (1967). Z-line of the flight muscle of belostomatid water bugs. *J. mol. Biol.*, **27**, 385–389.

BAILEY, K. (1957). Invertebrate tropomyosin. *Biochim. biophys. Acta.*, **24**, 612–618.

BAILEY, K. & RUEGG, J. C. (1960). Further chemical studies on the tropomyosins of lamellibranch muscle with special reference to *Pecten maximus*. *Biochim. biophys. Acta.*, **38**, 239–245.

BÁRÁNY, M., FINKELMAN, F. & THERATTIL-ANTONY, T. (1962). Studies on the bound calcium of actin. *Arch. Biochem. Biophys.* **98**, 28–45.

BENNETT, H. S., & PORTER, K. R. (1953). "An Electron Microscope Study of Sectioned Breast Muscle of the Domestic Fowl." *Amer. J. Anat.*, **93**, 61–105.

BIRKS, R., KATZ, B., & MILEDI, R. (1959). Dissociation of the "surface membrane complex" in atrophic muscle fibres. *Nature*, **184**, 1507–1508.

BIRKS, R. I. (1965). The sarcoplasmic reticulum of twitch fibres in the frog sartorius muscle. In *Muscle*, Ed. W. M. Paul, E. E. Daniel, C. M. Kay & G. Monckton. London: Pergamon. pp. 199–216.

BRANDT, P. W., REUBEN, J. P., GIRARDIER, L. & GRUNDFEST, H. (1965). Correlated morphological and physiological studies on isolated single muscle fibers. I. Fine structure of the crayfish muscle fiber. *J. Cell Biol.*, **25**, (Pt. 2 of No. 3), 233–260.

BUCHTAL, F., KNAPPEIS, G. G. & LINDHARD, J. (1936). Die Struktur der quergestreiften lebenden Muskelfaser des Frosches in Ruhe und während der Kontraktion. *Skan. Arch. Physiol.*, **73**, 163.

CARLSEN, F., KNAPPEIS, G. G., & BUCHTAL, F. (1961). Ultrastructure of the resting and contracted striated muscle fibre at different degrees of stretch. *J. biophys. biochem. Cytol.*, **11**, 95–117.

CHAPPELL, J. B. & PERRY, S. V. (1953). The respiratory and adenosinetriphosphatase activities of skeletal-muscle mitochondria. *Biochem. J.*, **55**, 586–595.

CHRAMBACH, A., BÁRÁNY, M. & FINKELMAN, F. (1961). The bound calcium of actin. *Arch. Biochem. Biophys.* **93**, 456–457.

COHEN, C. & LONGLEY, W. (1966). Tropomyosin paracrystals formed by divalent cations. *Science*, **152**, 794–796.

DRABIKOWSKI, W., NONOMURA, Y. & MARUYAMA, K. (1968, a). Effect of tropomyosin on the interaction between F-actin and the 6S component of α-actinin. *J. Biochem.* (Tokyo), **63**, 761–765.

DRABIKOWSKI, W., KOMINZ, D. R. & MARUYAMA, K. (1968, b). Effect of troponin on the reversibility of tropomyosin binding to F-actin. *J. Biochem.* (Tokyo), **63**, 802–804.

DREIZEN, P., GERSHMAN, L. C. TROTTA, P. P. & STRACHER, A. (1967). Myosin. Subunits and their interactions. *J. gen. Physiol.*, **50**, Pt. 2, No. 6, 85–113.

EBASHI, S. (1966). Structural proteins controlling the interaction between actin and myosin. In *Symposion über progressive Muskeldystrophie*. Ed. Erich Kuhn, Berlin: Springer. pp. 506–513.

EBASHI, S. & EBASHI, F. (1964). A new protein component participating in the superprecipitation of myosin B. *J. Biochem.* (Tokyo), **55**, 604–613.

EBASHI, S. & EBASHI, F. (1965). α-Actinin, a new structural protein from striated muscle. I. Preparation and action on actomyosin-ATP interaction. *J. Biochem.* (Tokyo), **58**, 7–12.

EBASHI, S., EBASHI, F. & KODAMA, A. (1967). Troponin as the Ca^{++}-receptive protein in the contractile system. *J. Biochem.* (Tokyo), **62**, 137–138.

EBASHI, S. & KODAMA, A. (1965). A new protein factor promoting aggregation of tropomyosin. *J. Biochem.* (Tokyo), **58**, 107–108.

EBASHI, S. & KODAMA, A. (1966). Interaction of troponin with F-actin in the presence of tropomyosin. *J. Biochem.* (Tokyo), **59**, 425–426.

EDMAN, K. A. P. (1966). The relation between sarcomere length and active tension in isolated semitendinosus fibres of the frog. *J. Physiol.*, **183**, 407–417.

EISENBERG, R. S. & EISENBERG, B. (1968). The extent of the disruption of the transverse tubular system in glycerol treated skeletal muscle. *Fed. Proc.*, **27**, 247.

ELLIOTT, A. (1952). Infra-red spectra of muscle. *Proc. Roy. Soc.*, B, **139**, 526–527.

ELLIOTT, G. F. (1964). X-ray diffraction studies on striated and smooth muscles. *Proc. Roy. Soc. B.*, **160**, 467–472.

ELLIOTT, G. F. (1967). Variations of the contractile apparatus in smooth and striated muscles. X-ray diffraction studies at rest and in contraction. *J. gen. Physiol.*, **50**, No. 6, Pt. 2, 171–184.

ELLIOTT, G. F., LOWY, J. & MILLMAN, B. M. (1965). X-ray diffraction from living striated muscle during contraction. *Nature*, **206**, 1357–1358.

ENDO, M. (1966). Entry of fluorescent dyes into the sarcotubular system of the frog muscle. *J. Physiol.*, **185**, 224–238.

ENDO, M., NONOMURA, Y., MASAKI, T., OHTSUKI, I. & EBASHI, S. (1966). Localization of native tropomyosin in relation to striation pattern. *J. Biochem.* (Tokyo), **60**, 605–608.

EZERMAN, E. B. & ISHIKAWA, H. (1967). Differentiation of the sarcoplasmic reticulum and T system in developing chick skeletal muscle *in vitro*. *J. Cell Biol.*, **35**, 405–420.

FAWCETT, D. W., & REVEL, J. P. (1961). The sarcoplasmic reticulum of a fast-acting muscle. *J. biophys. biochem. Cytol.*, **10**, Suppl. 89–111.

FINCK, H., HOLTZER, H., & MARSHALL, J. M. (1956). "Immunochemical Study of the Distribution of Myosin in Glycerol Extracted Muscle." *J. biophys. biochem. Cytol.*, **2**, Suppl., 175–177.

FISCHER, E. (1944). "Birefringence of Striated and Smooth Mammalian Muscles." *J. cell. comp. Physiol.*, **23**, 113.

FISCHMAN, D. A. (1967). An electron microscope study of myofibril formation in embryonic chick skeletal muscle. *J. Cell Biol.*, **48**, 557–575.

FRANZINI-ARMSTRONG, C. & PORTER, K. R. (1964). Sarcolemmal invaginations constituting the T system in fish muscle fibres. *J. Cell Biol.*, **22**, 675–696.

FRANZINI-ARMSTRONG, C. & PORTER, K. R. (1964). The Z disc of skeletal muscle fibrils. *Z. Zellforsch.*, **61**, 661–672.

GILEV, V. P. (1962). A study of myofibril sarcomere structure during contraction. *J. Cell Biol.*, **12**, 135–147.

GOLDSPINK, G. (1968). Sarcomere length during post-natal growth of mammalian fibres. *J. Cell Sci.*, **3**, 539–548.

GORDON, A. M., HUXLEY, A. F. & JULIAN, F. J. (1966). The variation in isometric tension with sarcomere length in vertebrate muscle fibres. *J. Physiol.*, **184**, 170–192.

GORDON, A. M., HUXLEY, A. F. & JULIAN, F. J. (1966). Tension development in highly stretched vertebrate muscle fibres. *J. Physiol.*, **184**, 143–169.

HAGOPIAN, M. & SPIRO, D. (1967). The sarcoplasmic reticulum and its association with the T system in an insect. *J. Cell Biol.*, **32**, 535–545.

HANSON, J. (1967). Axial period of actin filaments. *Nature*, **213**, 353–356.

HANSON, J., & HUXLEY, H. E. (1953). "Structural Basis of the Cross-Striations in Muscle." *Nature*, **172**, 530–532.

HANSON, J., & HUXLEY, H. E. (1955). "The Structural Basis of Contraction in Striated Muscle." *Symp. Soc. Exp. Biol.*, **9**, 228–264.

HANSON, J. & LOWY, J. (1963). The structure of F-actin and of actin-filaments isolated from muscle. *J. mol. Biol.*, **6**, 46–60.

HANSON, J. & LOWY, J. (1964). The structure of actin filaments and the origin of the axial periodicity in the I-substance of vertebrate striate muscle. *Proc. Roy. Soc. B.*, **160**, 449–458.

HASSELBACH, W. (1953). "Elektronenmikroskopische Untersuchungen an Muskelfibrillen bei totaler und partieller Extraktion des L-Myosins." *Z. Naturf.*, **8**b, 449–454.

HESS, A. (1965). The sarcoplasmic reticulum, the T-system, and the motor terminals of slow and twitch muscle fibers in the garter snake. *J. Cell Biol.*, **26**, 467–476.

HILL, D. K. (1964). The space accessible to albumin within the striated muscle fibre of the toad. *J. Physiol.*, **175**, 275–294.

HODGE, A. J. (1956). "The Fine Structure of Striated Muscle." *J. biophys. biochem. Cytol.*, **2**, Suppl. 131–142

HODGE, A. J., HUXLEY, H. E. & SPIRO, D. (1954). Electron microscope studies of ultrathin sections of muscle. *J. exp. Med.*, **99**, 201–206.

HOLTZER, A., CLARK, R. & LOWEY, S. (1965). The conformation of native and denatured tropomyosin B. *Biochem.*, **4**, 2401–2411.

HOYLE, G., McALEAR, J. H. & SELVERSTON, A. (1965). Mechanism of supercontraction in a striated muscle. *J. Cell Biol.*, **26**, 621–640.

HUXLEY, A. F. (1956). "Interpretation of Muscle Striation: Evidence from Visible Light Microscopy." *Brit. med. Bull.*, **12**, 167–170.

HUXLEY, A. F. (1964). Evidence for continuity between the central elements of the triads and extracellular space in frog muscle. *Nature*, **202**, 1067–1071.

HUXLEY, A. F., & NIEDERGERKE, R. (1954) "Measurement of Muscle Striations in Stretch and Contraction." *J. Physiol.*, **124**, 46–47 P.

HUXLEY, A. F., & PEACHEY, L. D. (1961). The maximum length for contraction in vertebrate striated muscle. *J. Physiol.*, **156**, 150–165.

HUXLEY, A. F. & TAYLOR, R. E. (1955). Activation of a single sarcomere. *J. Physiol.*, **130**, 40–50.

HUXLEY, H. E. (1953). "X-ray Analysis and the Problem of Muscle." *Proc. Roy. Soc*, B, **141**, 59–66.

HUXLEY, H. E. (1953). "Electron Microscope Studies of the Organization of the Filaments in Striated Muscle." *Biochim. biophys. Acta*, **12**, 387–394.

HUXLEY, H. E. (1956). The ultra-structure of striated muscle. *Brit. Med. Bull.*, **12**, 171–173.

HUXLEY, H. E. (1957). "The Double Array of Filaments in cross-striated Muscle." *J. biophys. biochem. Cytol.*, **3**, 631–648.

HUXLEY, H. E. (1961). The contractile structure of cardiac and skeletal muscle. *Circulation*, **24**, 328–335.

HUXLEY, H. E. (1962). Studies on the structure of natural and synthetic protein filaments from muscle. *Proc. Vth Int. Congr. Electron Microscopy.* Vol. **2**, O-1.

HUXLEY, H. E. (1963). Electron microscope studies on the structure of natural and synthetic protein filaments from striated muscle. *J. mol. Biol.*, **7**, 281–308.

HUXLEY, H. E. (1964/65). The fine structure of striated muscle and its functional significance. *Harvey Lectures*, **60**, 85–118.

HUXLEY, H. E. (1967). Recent X-ray diffraction and electron microscope studies of striated muscle. *J. gen. Physiol.*, **50**, 71–81.

HUXLEY, H. E. (1968). Structural difference between resting and rigor muscle; evidence from intensity changes in the low-angle equatorial X-ray diagram. *J. mol. Biol.*, **37**, 507–520.

HUXLEY, H. E. & BROWN, W. (1967). The low-angle X-ray diagram of vertebrate striated muscle and its behaviour during contraction and rigor. *J. mol. Biol.*, **30**, 383–434.

HUXLEY, H. E., BROWN, W. & HOLMES, K. C. (1965). Constancy of axial spacings in frog sartorius muscle during contraction. *Nature*, **206**, 1358.

HUXLEY, H. E., & HANSON, J. (1954). "Changes in the Cross-striations of Muscle during Contraction and Stretch and their Structural Interpretation." *Nature*, **173**, 973–976.

HUXLEY, H. E., & HANSON, J. (1957). "Quantitative Studies on the Structure of Cross-striated Myofibrils. I. Investigation by Interference-microscopy. II. Investigations by Biochemical Techniques." *Biochim. biophys. Acta*, **23**, 229–249; 250–260.

IKEMOTO, N., KITAGAWA, S., NAKAMURA, A. & GERGELY, J. (1968). Electron microscopic investigations of actomyosin as a function of ionic strength. *J. Cell Biol.*, **39**, 620–629.

ISHIKAWA, H. (1968). Formation of elaborate networks of T-system tubules in cultured skeletal muscle with special reference to the T-system formation. *J. Cell Biol.*, **38**, 51–66.

JAKUS, M. A., & HALL, C. E. (1947). "Studies of Actin and Myosin." *J. biol. Chem.*, **167**, 705.

KELLY, D. F. (1967). Models of muscle Z-band fine structure based on a looping filament configuration. *J. Cell Biol.*, **34**, 827–840.

KIELLEY, W. W. & HARRINGTON, W. F. (1960). A model for the myosin molecule. *Biochim. biophys. Acta*, **41**, 401–421.

KNAPPEIS, G. G., & CARLSEN, F. (1956). "Electron Microscopical Study of Skeletal Muscle during Isotonic (Afterload) and Isometric Contraction." *J. biophys. biochem. Cytol.*, **2**, 201–211.

KNAPPEIS, G. G., & CARLSEN, F. (1962). The ultrastructure of the Z disc in skeletal muscle. *J. Cell Biol.*, **13**, 323–335.

KNAPPEIS, G. G. & CARLSEN, F. (1968). The ultrastructure of the M line in skeletal muscle. *J. Cell Biol.*, **38**, 202–211.

KOMINZ, D. R. (1966). Interactions of calcium and native tropomyosin with myosin and heavy meromyosin. *Arch. Biochem. Biophys.*, **115**, 583–592.

KOMINZ, D. R. & MARUYAMA, K. (1967). Does native tropomyosin bind to myosin. *J. Biochem.* (Tokyo), **61**, 269–271.

KOMINZ, D. R., SAAD, F. & LAKI, K. (1957). Vertebrate and invertebrate tropomyosins. *Nature*, **179**, 206–207.

LAKI, K., BOWEN, W. J. & CLARK, A. (1950). Adenosine triphosphate and the polymerization of actin. *J. gen. Physiol.*, **33**, 437–443.

LONG, M. E. (1947). "Development of the Muscle-Tendon Attachment in the Rat." *Amer. J. Anat.*, **81**, 159.

LOWEY, S. & HOLTZER, A. (1959). The homogeneity and molecular weights of the meromyosins and their relative proportions in myosin. *Biochim. biophys. Acta.*, **34**, 470–484.

MALCOLM, B. R. (1955). "Some Observations on the Infra-red Spectrum of Muscle." *Symp. Soc. Exp. Biol.*, **9**, 265–270.

MARTONOSI, A., GOUVEA, M. A. & GERGELY, J. (1960). The interaction of ^{14}C-labelled adenine nucleotides with actin. *J. biol. Chem.*, **235**, 1700–1703.

MARTONOSI, A., GOUVEA, M. A. & GERGELY, J. (1960). G-F transformation of actin and molecular contraction (experiments *in vivo*). *J. biol. Chem.*, **235**, 1707–1710.

MARUYAMA, K. (1965). A new protein-factor hindering network formation of F-actin in solution. *Biochim. biophys. Acta.*, **94**, 208–225.

MASAKI, T., TAKAITI, O. & EBASHI, S. (1968). "M-Substance", a new protein constituting the M-line of myofibrils. *J. Biochem.*, (Tokyo), **64**, 909–910.

MOMMAERTS, W. F. H. M. (1952). The molecular transformations of actin. I. Globular actin. II. The polymerization process. III. The participation of nucleotides. *J. biol. Chem.*, **198**, 445–457; 459–467; 469–475.

MUELLER, H. (1964). Molecular weight of myosin and meromyosins by Archibald experiments performed with increasing speed of rotations. *J. biol. Chem.*, **239**, 797–804.

NATORI, R. (1954). The property and contraction process of isolated myofibrils. *Jikeikai Med. J.*, **1**, 119–126.

NONOMURA, Y. (1967). A study on the physico-chemical properties of α-actinin. *J. Biochem.* (Tokyo), **61**, 796–802.

NONOMURA, Y., DRABIKOWSKI, W. & EBASHI, S. (1968). The localization of troponin in tropomyosin paracrystals. *J. Biochem.* (Tokyo), **64**, 419–422.

OHTSUKI, I., MASAKI, T., NONOMURA, Y. & EBASHI, S. (1967). Periodic distribution of troponin along the thin filament. *J. Biochem.* (Tokyo), **61**, 817–819.

PAGE, S. (1964). Filament lengths in resting and excited muscles. *Proc. Roy. Soc. B.*, **160**, 460–466.

PAGE, S. G. (1965). A comparison of the fine structures of frog slow and twitch muscle fibres. *J. Cell Biol.*, **26**, 477–497.

PAGE, S. G. & HUXLEY, H. E. (1963). Filament lengths in striated muscle. *J. Cell Biol.*, **19**, 369–390.

PAULING, L., & COREY, R. B. (1951). "The Structure of Hair, Muscle and Related Proteins." *Proc. Nat. Acad. Soc. Wash.*, 37, 261–271.

PEACHEY, L. D. (1965). The sarcoplasmic reticulum and transverse tubules of the frog's sartorius. *J. Cell Biol.*, **25**, 209–231.

PEACHEY, L. D. & SCHILD, R. F. (1968). The distribution of the T-system along the sarcomeres of frog and toad sartorius muscles. *J. Physiol.*, **194**, 249–258.

PEPE, F. A. (1966). Some aspects of the structural organization of the myofibril as revealed by antibody-staining methods. *J. Cell Biol.*, **28**, 505–525.

PEPE, F. A. (1967). The myosin filament. I. Structural organization from antibody staining observed in electron microscopy. II. Interaction between myosin and actin filaments observed using antibody staining in fluorescent and electron microscopy. *J. mol. Biol.*, **27**, 203–225; 227–236.

PERRY, S. V. (1956). Relation between chemical and contractile function and structure of the skeletal muscle cell. *Physiol. Rev.*, **36**, 1–76.

PERRY, S. V. & GREY, T. C. (1956). Ehtylenediaminetetra-acetate and the adenosinetriphosphatase activity of acomyosin systems. *Biochem. J.*, **64**, 5 P.

PERRY, S. V., & REED, R. (1947). "An Electron Microscope and X-ray Study of Actin. I. Electron Microscope." *Biochim. biophys. Acta*, **1**, 379.

PODOLSKY, R. J. (1964). The maximum sarcomere length for contraction of isolated myofibrils. *J. Physiol.*, **170**, 110–123.

PORTER, K. R. (1961). The sarcoplasmic reticulum. Its recent history and present status. *J. biophys. biochem. Cytol.*, **10**, Suppl. 219–226.

PORTER, K. R., & PALADE, G. E. (1957). Studies on the endoplasmic reticulum. III. *J. biophys. biochem. Cytol.*, **3**, 269–299.

RAMSEY, R. W. (1944). Muscle physics. In Medical Physics. Ed. Glasser. Chicago: Year Book Publishers.

RAYNS, D. G., SIMPSON, F. O. & BERTAUD, W. S. (1968). Surface features of striated muscle. I. and II. *J. Cell. Sci.*, **3**, 467–474; 475–482.

REVEL, J. P. (1962). The sarcoplasmic reticulum of the bat cricothyroid muscle. *J. biophys. biochem. Cytol.*, **12**, 571–588.

RICE, R. V. (1961). Conformation of individual macromolecular particles from myosin solutions. *Biochim. biophys. Acta.*, **52**, 602–604.

RICE, R. V. (1964). Electron microscopy of macromolecules from myosin solutions. In *Biochemistry of Muscular Contraction.* Ed. J. Gergely. Boston: Little-Brown. pp. 41–58.

ROBERTSON, J. D. (1956). "Some Features of the Ultrastructure of Reptilian Skeletal Muscle." *J. biophys. biochem. Cytol.*, **2**, 369–380.

ROME, E. (1967). Light and X-ray diffraction studies of the filament lattice of glycerol-extracted rabbit psoas muscle. *J. mol. Biol.*, **27**, 591–602.

ROZSA, G., SZENT-GYÖRGYI, A. & WYCKOFF, R. W. (1949). The electron microscopy of F-actin. *Biochim. biophys. Acta.*, **3**, 561.

SCHRAMM, G., & WEBER, H. H. (1942). "Uber Monodisperse Myosinlösungen." *Koll. Z.*, **100**, 242.

SELBY, C. C., & BEAR, R. S. (1956). "The Structure of Actin-rich Filaments of Muscles according to X-ray Diffraction." *J. biophys. biochem. Cytol.*, **2**, 71–85.

SJÖSTRAND, F. S., & ANDERSSON-CEDERGREN, E. (1957). The ultrastructure of the skeletal muscle myofilaments at various states of shortening. *J. Ultrastr. Res.*, **1**, 74–108.

SMITH, D. S. (1966). The organization of flight muscle fibers in the odonata. *J. Cell Biol.*, **28**, 109–126.

SMITH, D. S. (1966). The organization and function of the sarcoplasmic reticulum and T-system of muscle cells. *Progr. Biophys.*, **16**, 109–142.

STRAUB, F. B. & FEUER, G. (1950). Adenosinetriphosphate the functional group of actin. *Biochim. biophys. Acta.*, **4**, 455–470.

SZENT-GYÖRGYI, A. G. (1953). Meromyosins, the subunits of myosin. **42**, 305–320.

SZENT-GYÖRGYI, A. G. (1955). "Structural and Functional Aspects of Myosin." *Adv. Enzymol.*, **16**, 313–360.

SZENT-GYÖRGYI, A. G., MAZIA, D., & SZENT-GYÖRGYI, A. (1955). "On the Nature of the Cross-striation of Body Muscle." *Biochim. biophys. Acta*, **16**, 339–342.

DE VILLAFRANCA, G. W. (1957). "Observations on the Anisotropic Band of Cross-striated Muscle." *Exp. Cell Res.*, **12**, 410–413.

WAKABAYASHI, T. & EBASHI, S. (1968). Reversible change in physical state of troponin induced by calcium ions. *J. Biochem.* (Tokyo), **64**, 731–732.

WEINSTEIN, H. J. (1954). "An Electron Microscope Study of Cardiac Muscle." *Exp. Cell Res.*, **7**, 130–146.

WORTHINGTON, C. R. (1959). Large axial spacings in striated muscle. *J. mol. Biol.*, **1**, 398–401.

ZACHAR, J. & ZACHAROVÁ, D. (1966). Potassium contractures in single muscle fibres of the crayfish. *J. Physiol.*, **186**, 596–618.

ZOBEL, C. R. & CARLSON, F. D. (1963). An electron microscopic investigation of myosin and some of its aggregates. *J. mol. Biol.*, **7**, 78–89.

CHAPTER XXI

MECHANICAL AND THERMAL ASPECTS OF MUSCULAR CONTRACTION

MECHANICAL ASPECTS

The Latent Period

Stimulation of muscle, as we have seen, is followed by the passage of a wave of action potential extending over its whole length; immediately after the electrical event the muscle contracts. Contraction has been studied under a variety of mechanical conditions, and with apparatus varying greatly in complexity. A generalized experimental set-up is illustrated in Fig. 754; the muscle, most commonly the frog's sartorius, is clamped rigidly at its pelvic end and attached at the tibial end by a wire to a light lever, *b*. Shortening of the muscle

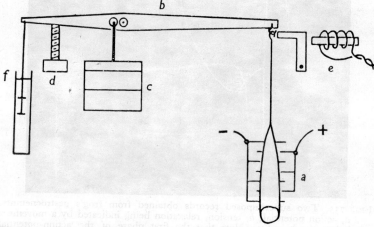

FIG. 754. Generalized experimental set-up for the study of the contraction of muscle. (Wilkie. *J. Physiol.*)

may be recorded by a photoelectric device (not shown) which indicates movement of the lever. The muscle may be caused to lift a load, *c* (which is placed near the fulcrum of the lever to reduce inertia), in which case the contraction is called *isotonic* to distinguish it from the case where shortening is prevented by connecting the muscle directly to a tension-measuring device, for example, the transducer, *f*, consisting of a small valve whose anode can be moved from the outside, a change of tension being converted to a change of voltage, whilst the actual amount of movement of the anode is negligible. This type of contraction is called *isometric*. The stop, *d*, permits the initial length of the muscle to be adjusted, whilst the electromagnetic stop, *e*, allows the experimenter to release the muscle at various moments after a stimulus.

Latency-relaxation. The interval between the direct electrical stimulation of the muscle and the beginning of the upstroke of shortening is the *latent period*; its duration depends so strongly on the inertia of the ordinary recording apparatus that Roos, who reduced this inertia to a minimum, considered that there was no real latent period. Subsequent work—for example that of Sandow and Abbott & Ritchie—has shown that there is, indeed, a measurable period of time between an electrical stimulus—applied simultaneously over the whole length of the muscle to reduce the effect of conduction-time—and the onset of shortening. Moreover, it has revealed that the development of tension is preceded by relaxation—the so-called *latency-relaxation*, so that the time-course of events consisted of (*a*), a quiescent period of about 1·5 msec. from the moment of stimulation to the beginning, (*b*), of the so-called *latency-relaxation* which lasts for some 1·5 msec. and passes abruptly into (*c*), the phase of rising tension, the first 0·5 msec. of which is required to bring the tension back to its initial value. The latency, as ordinarily recorded, is thus of the order of 3·5 msec. at 23° C; if, however, we take as a measure of the

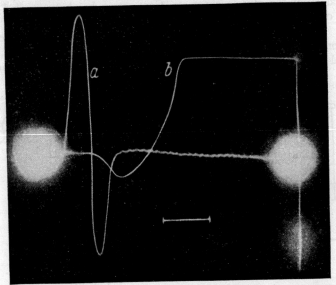

FIG. 755. Two superimposed records obtained from frog's gastrocnemius: *a*, action potential; *b*, tension, relaxation being indicated by a movement below the base-line. Note that the first phase of the action-potential precedes the onset of latency-relaxation. Time-mark, 10 msec. (Schaefer & Göpfert. *Pflüg. Arch.*)

mechanical latent period the time between the stimulus and the moment when the tension begins to rise from its minimum value, the period is 3·0 msec. Fig. 755 is a simultaneous recording of tension and action potential taken from the work of Schaefer & Göpfert, whilst Fig. 756 illustrates diagrammatically the various stages in development of tension. At 0° C the whole process of contraction is much slower; in this event Abbott & Ritchie found the latency to the first drop in tension to be 6·7–7·9 msec. The time to re-crossing the base-line, *i.e.*, the ordinarily measured latent period, was 14–17 msec. These workers found that the amount of latency-relaxation decreased with the initial length of the muscle, so that if a muscle had a normal resting length of 40 mm., and it was allowed to shorten to 30 mm. before stimulation, there was no latency-relaxation, the latent period being now much shorter, and uncomplicated by an initial fall in tension. As Sandow has emphasized, latency-relaxation may be one of the earliest signs of molecular changes occurring in the contractile apparatus leading to the development of tension.

Repulsion Between Filaments

Thus the resting tension of muscle, revealed by its resistance to passive stretch and its tendency to shorten on release from its attachments, could be due to a mutual electrostatic repulsion between filaments; since the lattice of filaments retains a constant volume during passive stretch this means that they must move closer to each other, so that the force necessary to move them closer, against the electrostatic repulsion, will appear as the resting tension. Activation of the contractile machinery may well be associated with an initial reduction in this electrostatic repulsion leading to relaxation (Huxley & Brown, 1967).

Contraction Wave. Studies with whole muscle, and more recently on isolated fibres by Ramsey & Street and by Buchtal, have shown that contraction does not begin instantaneously along the whole length of the muscle fibre but progresses as a wave from the point of stimulation, so that in the absence of a load the more proximal part of the fibre contracts strongly against the more

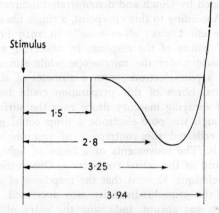

Fig. 756. Diagrammatic representation of latency-relaxation. Numbers indicate times, in msec. after stimulation, at which the various phases of the tension take place for a muscle loaded with 8g. (Sandow. *J. cell. comp. Physiol.*)

distal, relaxed, part. By comparing the effects of stimulating a muscle first at one end and then simultaneously at many points along its length, Abbott & Ritchie calculated the velocity of propagation of muscular activity and showed that this agreed well with the measured velocity of propagation of the action potential (56 cm./sec. at 0° C.). An interesting feature brought out by this work was the very rapid development of the maximum velocity of shortening; thus, the latent period under these conditions was 24 msec., whilst 27 msec. after the stimulus the full speed of shortening had been reached, and this lasted for some 70 per cent. of the whole duration of shortening.

Summation and Tetanus

When the muscle is stimulated during the period of contraction a process of summation is observed, the second stimulus causing the tension to develop to a considerably higher value; with repetitive stimulation at a high rate, the summated effects of successive stimuli lead to the development of a tension which may be nearly ten times as great as that developed in a single twitch, the muscle being said to be in *tetanus*. The frequency of stimulation necessary to obtain a complete tetanus will clearly depend on the contraction-time of the muscle, *i.e.*, the time from the beginning of the action potential to the point of maximum tension, a very rapid muscle requiring a high frequency. A muscle with a long refractory period, extending into the relaxation period, cannot develop a complete tetanus. The important feature of summation is that the tension developed is higher than in a single twitch; this is important both theoretically and practically; theoretically, because, as we shall see, it gives a clue to the time-course of the fundamental change taking place in the contractile machinery as a result of a single stimulus; and practically, because it permits of a graded response to stimulation.

The All-or-None Law

Just as with a bundle of nerve fibres, increasing the strength of stimulus above threshold produces increasing muscular responses until the maximal stimulus is reached, when further increase has no effect. That this increase in response was due to the successive entry of new muscle fibres into the total response was suggested by Gotch and demonstrated indirectly by Keith Lucas in 1905 and 1909. According to this viewpoint, a single fibre always responded maximally, *i.e.*, the effect was "all-or-none". In 1919 Pratt confirmed the so-called "quantal" nature of the response by observing single fibres of the frog's sartorius muscle under the microscope while stimulating the surface of the muscle with a pore-electrode, with a diameter of about $7\,\mu$. The contraction of individual fibres of this preparation could be recorded by the ingenious device of spraying mercury drops over the surface of the muscle; on stimulating through the pore-electrode a drop could generally be found whose movements reflected the contraction of just one or two fibres immediately beneath it. The movements of a beam of light, reflected off this droplet, gave a record of the actual contraction. Observations carried out by Pratt, using this technique, showed that the responses of a single fibre were all-or-none; when the stimulus-intensity was increased continuously, any increase in response was abrupt, indicating the entry of a new, maximally contracting, unit.

Local Responses. Extension of this technique to the very thin sheet of muscle constituting the retrolingual membrane of the frog—where the isolation of activity in single fibres was more precise—and the development of techniques for the study of single fibres dissected out completely from a muscle, brought to light many apparent contradictions to the all-or-none law. Thus stimulation of a single fibre with increasing strength of stimulus produced contractions of increasing strength; these graded responses differed in an important respect, however, from the maximal response, in that they were not associated with a wave of action potential; in keeping with this finding they were generally localized to a limited region in the neighbourhood of the stimulating electrode. In some cases, however, the whole fibre went into a prolonged contraction, as opposed to a twitch, but this differed once again from a normal tetanus in the absence of action potentials. This type of contraction is not confined to single fibres but was observed long before in whole muscles and is described as a *contracture*, generally produced by injurious agents (*e.g.*, NH_4Cl); it also develops on prolonged application of high hydrostatic pressures (Brown). It was considered by Gelfan that the local response of the fibre was a consequence of the highly localized conditions of stimulation, the electrotonic currents being so concentrated as to fail to activate the conduction mechanism (*i.e.*, to depolarize a sufficient area of membrane to induce a propagated all-or-none disturbance), yet was sufficient to activate the contractile mechanism. This viewpoint has since been confirmed by the more precise studies of Huxley & Taylor on single fibres; these authors, as we shall see, confined the stimulus to just a portion of a sarcomere, showing that the highly localized application of a current-pulse failed to cause spread of electrical activity. Because of this, contraction was confined to the region stimulated. Thus the all-or-none behaviour of a single muscle fibre is a consequence of the all-or-none nature of the propagated disturbance in the normal fibre. When propagation fails, the activation of the contractile machinery remains localized too, and the contraction is confined to this activated region.

THE ACTIVE STATE

Viscous-Elastic Properties of Muscle

If we come now to consider the mechanical work done when a muscle shortens against a load, we find that this varies very considerably with the load to be lifted; experiments on the isolated frog muscle, and on the human forearm, by Hill showed that the work done decreased rapidly as the speed of shortening increased. It was found that the work done could be approximately expressed by the simple equation:—

$$W = W_o(1 - k/t)$$

where W is the work actually done, t is the time of shortening, and W_o and k are constants. Since a small load is lifted more rapidly than a large one, the work done decreases with decreasing load. As Hill pointed out, the conditions of muscular contraction appeared to be similar in many respects to the behaviour of a viscous-elastic system; as a result of the contractile process, the viscosity of the muscle increased so that contraction was associated with a considerable loss of energy in overcoming internal friction. On this view, W_o is the ideal work which would be obtained when no energy was lost by viscous resistance, *i.e.*, the maximum work calculated from the length-tension diagram. Thus, if the muscle were a simple elastic system, like a wire spring, the work obtainable from it would be entirely determined by the tension it developed at different lengths, and could be calculated from a length-tension diagram.

Effect of Quick Stretches

The fundamental difference between an elastic and a viscous-elastic system is well brought out by further experiments of Gasser & Hill; these authors studied the effects of applying sudden stretches to a tetanized muscle, or, alternatively, of allowing a tetanized muscle to contract suddenly to a shorter length, and recording the changes of tension in consequence. Fig. 757 illustrates the general results of stretching. The horizontal lines indicate the static

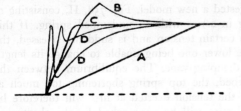

STATIC ISOMETRIC TENSION
AT LONGER LENGTH

STATIC ISOMETRIC TENSION
AT SHORTER LENGTH

ZERO TENSION

FIG. 757. Illustrating the effects of stretching a frog's sartorius at varying speeds during a maximal tetanus. A, very slow stretch; D, very rapid; B, fairly rapid; C, intermediate between B and D. (Gasser & Hill. *Proc. Roy. Soc.*)

isometric tensions developed by the muscle at two lengths. Curve D shows the effect of a sudden stretch; the tension rises rapidly, as we might expect, but then falls rapidly to rise again slowly to a value characteristic of the greater length. If the stretch is not quite so rapid, the tension rises above the static isometric value and approaches the latter from above (Curve B), whilst a very slow stretch (Curve A) gives a continuous rise in tension to the new level. The reverse phenomena were also described; a sudden release of a tetanized muscle to a shorter length caused the tension to fall almost to zero and then to rise to the value characteristic of its new length; only if the shortening was very slow did the tension approach its new value along a straight line. In a passive (unstimulated) muscle, no such phenomena were observable, the muscle adopting a new tension for each length immediately. These effects were considered to fit well with the concept of a damped spring as in Fig. 758, I, or more simply still, a thin india-rubber balloon filled with a very viscous material.

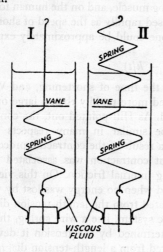

FIG. 758. Viscous - elastic models to illustrate the behaviour of muscle. System I. Simple damped spring. System II. Damped spring (contractile elements) in series with undamped spring (elastic elements). (Levin & Wyman. *Proc. Roy. Soc.*)

Levin-Wyman Model. Later Levin & Wyman drew attention to the inadequacy of this simple model; they pointed out that a quick release of the hook attached to the damped spring must result in a *complete loss of tension,* the whole being taken up at first by the viscous resistance of the vane; only as the spring adopted its new length would the tension rise to the value characteristic of the new length. Now although Hill's results on quick releases indicated a drop in tension below the normal isometric value for the new length, this drop was not to zero unless the shortening was greater than 15 per cent. Levin & Wyman suggested a new model, Fig. 758, II, consisting of a damped viscous-elastic spring in series with an undamped spring. If this model is at equilibrium under a certain tension and is suddenly released, the top spring will shorten first, the lower one being unable to change its length immediately on account of the damping vane. The equilibrium between the two springs will thus be disturbed, the top spring shortening too much at first, the bottom one too little; the tension exerted at first will therefore be too small, but not zero. Equilibrium will be approached by the shortening of the lower spring and an extension of the upper one. To transpose this mechanical model into terms of muscular contraction is another matter. Tentatively, we may regard the contractile process as a sliding of the myofilaments; this we may look upon as the shortening of the damped spring, *i.e.*, the rearrangement of myofilaments is by no means frictionless but, on the contrary, is damped, owing to the high viscosity, or molecular friction. The shortening of the fibrils is associated with the stretching of connective tissue, a process similar to the stretching of a metal spring, taking place with little or no frictional loss, so that the elastic energy may be stored as potential energy during shortening. These elastic elements are thus the undamped spring. According to Gasser & Hill's measurements, if a muscle in isometric tetanus is suddenly released to 85 per cent. of its length, the tension falls to zero, whereas with smaller releases some tension remains; Levin & Wyman concluded, therefore, that in an isometric contraction the undamped elastic elements were stretched by an amount equalling 15 per cent. of the muscle's resting length. These free elastic elements may be regarded as buffers, protecting the muscle against too sudden changes in tension; moreover, since they take up the tension developed in the true contractile elements, a very sudden shortening is not associated with a temporary complete loss of tension in the muscle— a very important factor in the smooth performance of muscular tasks.

Effects of Stretches Applied During Contraction. The work of Gasser & Hill on the effect of stretching a contracting muscle has provided some

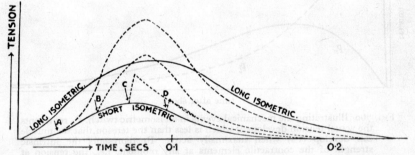

FIG. 759. Illustrating the effect, on the tension developed, of stretching a muscle at various moments during a twitch. (After Gasser & Hill. *Proc. Roy. Soc.*)

information regarding the true time-course of the fundamental contractile process. To appreciate the full value of these experiments we must form a clear picture of the events taking place during a twitch. The stimulus sets off the fundamental contractile process which involves, essentially, a shortening of myofibrils; the development of tension in the muscle requires time, the elastic elements having to be stretched; clearly, the tension could be developed more rapidly if the muscle were artificially stretched during the contractile process, the actual tension developed depending on the vigour with which the damped elements were contracting. Some estimate of the force with which these elements could pull, at different times during a twitch, should therefore be obtained by applying stretches at different moments. The results of an experiment of this nature are shown in Fig. 759. The curves in full lines represent the observed tension during isometric twitches at two lengths. With the muscle fixed at the shorter length, it was stretched suddenly to the greater length at different moments after the stimulus; it will be seen that the tension developed by stretching at A was very considerably greater than that developed by stimulating the muscle at the greater length. Clearly we are tapping here practically all the available force in the contractile process; as the stretches are applied later and later, the tension developed becomes smaller until, finally, the tension developed during stretch is no greater than that in passive muscle. The results show that the energy of the twitch is made available sooner than is indicated by the curve of isometric tension, and that in a single twitch the maximal tension of which the muscle is capable, for a given length, is not attained.

Rapid Onset of Active State

Subsequent studies of Hill (1949, 1950), in which the apparatus was improved so that stretches could be applied very rapidly and at precisely determined points during the development of tension, showed that the onset of the activated state of the contractile machinery was very rapid indeed. Thus, the application of a stretch of 15 per cent., just after the end of the latent period, led to the development of a tension that was actually greater than the maximal tetanic tension for that length; quite soon, however, the tension fell to this value. In general, the results were consistent with the view that, on stimulation, the contractile elements developed their maximal power virtually instantaneously, this maximal power being that observed in an isometric

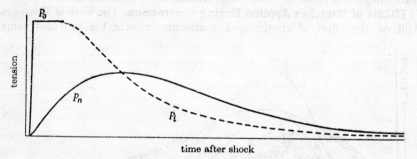

Fig. 760. Illustrating the mechanical events in an isometric twitch. P_n represents the actual tension developed, which is less than the tension that the muscle can develop (P_o) when maximally active. P_i represents the intrinsic strength of the contractile elements at any moment, *i.e.*, the tension at which they neither lengthen nor shorten. (Modified from Hill. *Proc. Roy. Soc.*)

tetanus at the given length. This was maintained for a short time and subsequently subsided. Schematically the events taking place in an isometric twitch may be represented by Fig. 760. P_o represents the maximal tension that the muscle can develop; as the figure indicates, this maximally "active state" is established almost instantaneously, and is maintained for a "plateau period", after which it wanes along the broken curve indicated by P_i, P_i being the tension of which the muscle is capable at a given moment. The contractile elements pull against the elastic elements and stretch them; as a result, a tension, P_n, is developed in the tendon which, during the rising phase of development, is necessarily less than P_i, the tension that *could* be developed were the elastic elements fully stretched. At a certain point, P_n and P_i become equal, and this will be at the peak of the tension-time curve. Beyond this point, the process will be reversed, the elastic elements stretching the shortened contractile elements.

Twitch- and Tetanus-Tension. The reason why the tetanus-tension is considerably greater than the twitch-tension is now clear. After a single stimulus the contractile elements develop their maximal activity at once, but maintain this for only a short time. In order that the muscle may develop tension, these contractile elements must stretch the series elastic elements; and this takes time—the heavier the load the longer is taken—so that by the time these elastic elements have stretched appreciably the activity of the contractile elements has begun to wane. Prolongation of the active state should, therefore, increase the tension developed in a twitch. This prolongation may be achieved by various means; thus caffeine, some anions, and adrenaline all do this and thereby increase the twitch-tension. Another means of prolonging the active state is to stimulate the muscle a second time, the interval between stimuli being so short that the active state has not had time to decay. Two such stimuli may therefore be expected to cause a more powerful—*summated*—twitch. If a series of successive stimuli fall while the contractile elements are in their maximal state of activity, then this maximally active state will be prolonged until fatigue sets in, and the *tetanus-tension* developed will be the maximal activity of which the muscle is capable at that length.

Duration of Active State

It will be clear from the schematic curves of Fig. 760 that, for a given intensity of the active state, the maximal tension developed in a twitch will depend on the duration of the active state, and the rate at which the elastic elements can shorten; a rise in temperature affects these two parameters differently, its influence being predominantly to increase the rate of decay of the active state; at 20° C, therefore, the ratio of twitch-tension/tetanus-tension is 0·22, whilst at 0° C it is 0·68 (Hill, 1951). Again, we may note that isometric conditions are particularly unfavourable for the development of twitch-tension. The force-velocity relationship tells us that the greater the load the slower will be the rate of shortening; at infinite load, *i.e.*, in an isometric contraction, external shortening is of course zero but the contractile elements do contract against the elastic elements. Their rate of contraction will be minimal so that, for a given duration of the active state, the tension developed will be very low.

Contracture

The phenomenon of contracture, in which a muscle exhibits a sustained, reversible contraction with the liberation of heat and formation of lactic acid, but shows no propagated spike potentials, as in the sustained contraction of a tetanus, has been cited as an example of the independence of the contractile and electrical phenomena in contraction. However, this is not necessarily true since the phenomenon is most often due to a sustained depolarization.

Veratrine Contracture. A typical agent for producing contractures is veratrine; on stimulating a veratrine-poisoned muscle it exhibits a twitch followed by a sustained contraction, or contracture; if only half the muscle is poisoned only half is involved in the contracture. Strong direct currents, heat, acids, narcotics, potassium, citrate, caffeine, acetylcholine, nicotine, and many other drugs are capable of producing contractures. The most instructive study of the phenomenon is that of Kuffler (1946) on a single fibre preparation; a drop of veratrine was applied locally to a fibre, and a subsequent stimulus gave rise to a series of propagated spikes (as many as 300/sec.) which rapidly gave way to a pronounced negativity of the electrode on the poisoned region; at first the negativity showed oscillations—abortive spikes—but finally became smooth. The fibre exhibited a local contracture as long as the negative potential lasted. A number of other facts all pointed to the contracture as being essentially a non-propagated contraction, of which the graded responses described earlier are merely transient examples; the sustained nature is apparently due to the prolonged depolarization caused by the agent, be it a maintained constant current (the contracture is at the cathode), acetylcholine, potassium, and so on. Thus the condition is similar to that of nerve treated by certain blocking agents, in the sense that the conducting mechanism is out of action through depolarization;* if, however, the contractile mechanism is triggered by a depolarization of the muscle membrane, we may understand why it is that the fibre remains contracted over the poisoned region. We have seen that anodal polarization restores conduction to a nerve, blocked with a narcotic, and it is interesting that the same procedure will cause relaxation of a contracture.

Depolarization as Trigger

Thus, as Katz (1950) has suggested in a review of the subject, the phenomena of contracture, or local contractions, strongly indicate that the trigger for activating the contractile mechanism is the depolarization of the muscle membrane by the wave of action potential, or its negative after-potential; if the action potential is blocked at any point in its path along the muscle fibre, the contractile process is likewise blocked, and the contraction remains local. Repolarization of the membrane brings the contraction to an end, so that any substance interfering with this, such as a narcotic or veratrine, may cause a sustained contraction or contracture. As we shall see, however, relaxation will occur in a fast skeletal muscle in the presence of complete depolarization, so that we must postulate an active relaxation process too.

Potentiation

In lower doses the typical contracture agents, such as veratrine, caffeine, quinine, may be described as potentiators of contraction, in the sense that they increase the tension developed in a single twitch;† associated with this, the duration of the period of contraction is prolonged. As Etzensperger (1962) emphasized, it is largely because these agents prolong the action potential that the duration of the contractile process is lengthened; we have seen that in a twitch, as opposed to a tetanus, an increased duration of the "active state" of the contractile process allows the muscle to develop a greater maximal tension. The fact that the period during which the action potential is above the threshold depolarization for triggering the contraction is only some 2 msec., whilst the active state of contraction lasts much longer, means that the depolarization triggers off an intermediate process that continues to activate the contractile process; it is this intermediate process, and its subsequent reversal to cause relaxation, that constitute the *coupling between excitation and contraction*, the theme of a great deal of modern work inspired by the early work of Sandow. The importance of the relaxation process is revealed by the observation that a muscle, depolarized completely in isotonic KCl, will relax spontaneously within a minute after its contracture. That this is not due to any failure of the contraction machinery *per se*, induced by the high K^+ or prolonged depolarization, is shown by the fact that a contracture agent, such as caffeine, added to the medium, will cause development of a prolonged contraction.

Fast and Slow Muscles. We may note here that the fast (phasic) and slow (tonic) muscles, typified by the sartorius and rectus abdominis of the frog, differ with respect to the contractures induced by immersion in isotonic KCl; it is the phasic muscle that spontaneously relaxes, whereas the tonic contracture is maintained for long periods (Sandow, 1955). In this respect, as Pauschinger & Brecht have emphasized, the fast sartorius muscle is able to break the link

---

* According to Feng, for example, a wave of action potential cannot cross a region of contracture.

† Because the tension developed in a tetanus will be unaffected by changes in the duration of the active state of muscle (p. 1438) the potentiating effect is best indicated by the ratio of the peak twitch and tetanus tensions. For a discussion of some objections to the procedure, Sandow & Isaacson (1966) should be consulted.

between excitation and contraction, since it remains depolarized in the high-K^+ solution, yet fails to sustain its contraction.

The Coupling of Excitation with Contraction

In normal muscle, a contraction is initiated by the spread of the action potential over its surface. A great deal of evidence has shown that mere depolarization of the membrane need not trigger off the contractile mechanism, however, so that an intermediate step, between membrane depolarization and activation of contraction, has been postulated. Under appropriate experimental conditions, therefore, we may activate the contractile mechanism independently of the condition of the membrane potential.

Depolarization and Tension. Some of the phenomena bearing on the relation of membrane potential to contraction that have led to this viewpoint

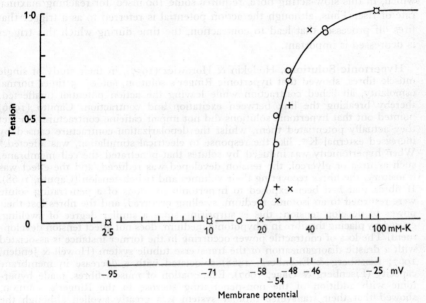

FIG. 761. Relation between peak tension and potassium concentration or membrane potential. Circles and plus-signs (O, $+$) refer to a fibre where tension only was measured, whilst crosses ($\times$) refer to a fibre in which both tension and membrane potential were measured. (Hodgkin & Horowicz. *J. Physiol.*)

may now be briefly mentioned. First, we may note that depolarization is closely related, causally, to the development of tension; we may show this by avoiding the development of the all-or-none type of depolarization, which gives us only one very high degree of depolarization amounting to membrane reversal. Thus, if we place a muscle in a Ringer's solution containing higher and higher concentrations of K^+, the membrane becomes progressively depolarized; at a critical point, which corresponds to a depolarization to a membrane potential of 50–55 mV, tension develops; as the depolarization is increased, the tension increases, so that there is no all-or-none effect so far as development of tension is concerned, the all-or-none phenomenon being essentially the result of the all-or-none nature of the spike potential, which is of sufficient height normally to cause maximum activation of the contractile machinery. The steep relationship between membrane potential and tension is indicated by Fig. 761, and it is seen that maximal tension was developed in about 100 mM KCl, corresponding to a depolarization to 25 mV. It is interesting, moreover, that the critical level of depolarization for development of tension corresponds approximately with the critical level for development of the action potential.

Using a voltage-clamp technique that permitted the measurement of membrane currents during step-wise depolarizations of the muscle membrane, Constantin (1968) showed that the threshold depolarization required to induce a microscopically visible local contraction coincided quite closely with the threshold for onset of increased K^+-conductance, i.e. in the region of a membrane potential of -48 to -40 mV. As Dudel, Morad & Rüdel (1968) have shown in their study of the voltage-clamped crustacean muscle fibre, it is the degree of depolarization, and not the flow of current, that determines the force of contraction; however, the duration of the depolarization is important not only for development of maximal tension but also for the rate of rise of tension which, in this slow-acting fibre, requires some 160 msec. for reaching maximal rate of rise. Thus, although the action potential is referred to as a trigger that fires off processes that lead to contraction, the time during which this trigger is depressed is important.

Hypertonic Solutions. Hodgkin & Horowicz (1957), in their study of single muscle fibres, showed that hypertonic Ringer's solution, some 2·5 times normal osmolality, abolished contraction while leaving the action potential unaffected, thereby breaking the link between excitation and contraction. Caputo (1966) pointed out that hypertonic solutions did not impair caffeine contractures, in fact they actually potentiated them, whilst the depolarization-contracture caused by increased external K^+, like the response to electrical stimulation, was affected.[*] When hypertonicity was induced by solutes that penetrated the cell membrane, such as urea or glyercol, the tension developed was reduced, and the effect was transitory, the fibres recovering their volumes and twitch-tension (Caputo, 1968). If fibres that had been exposed to hypertonic solutions of a penetrating solute were returned to an isotonic medium, swelling occurred, and the fibres lost their power to develop tension; this is surprising since a similar degree of swelling, caused by placing the fibre in a hypotonic medium, does not affect tension development. The loss of contractile power occurring in the former instance is associated with a drastic disorganization of the transverse tubule system (Howell & Jenden, 1967; Eisenberg & Eisenberg, 1968) associated with a decrease in membrane capacity (Eisenberg & Gage, 1967). Examination of muscle fibres, made hypertonic with addition of the non-penetrating sucrose to the Ringer's solution, showed that their transverse tubular system was greatly swollen although the remainder of the fibre was shrunken (Huxley, Page & Peachey, 1963).[†]

Potentiators of Contraction. Certain anions, such as NO_3^-, and CNS^-, prolong the twitch and increase the tension developed, because they prolong the active state (p. 1415). When we repeat the depolarization experiments using KNO_3 instead of KCl, the relationship between membrane potential and tension changes, so that a given degree of depolarization becomes more effective in developing tension.

Unlike the potentiation of veratrine and caffeine, that caused by anions is unaccompanied by any significant effect on the action potential, so that we must appreciate that potentiation of twitch tension may be achieved in two ways: influence on the action potential, and by a more direct action on the contractile machinery or the link between this and the electrical event.

Function of Calcium. The experiments discussed above and many others,

---

[*] Fujino & Fujino (1964) removed the inhibitory effect of hypertonic solutions by simply soaking the fibre in isotonic KCl or K_2SO_4 before exposing to the hypertonic solution; the treatment did not affect the osmotic shrinking that occurs in the hypertonic medium.

[†] See appendix to paper by Dydynska & Wilkie (1963).

demonstrating as they do the possibility of independent activation of the contractile mechanism, demand a link between this and membrane depolarization.
The importance of Ca^{++} in the external medium in the development of tension,
as revealed particularly in studies on heart muscle such as those of Wildbrandt &
Koller and Luttgau & Niedergerke, has led to the suggestion that activation of
the contractile mechanism is brought about by an increase in the concentration
of ionized Ca^{++} in the fibre, perhaps in a highly localized region. Thus depolarization might cause an increased influx of Ca^{++} from the outside medium, or it
might release unionized Ca^{++} from binding sites on the membrane, or in the
fibre; the bulk of the evidence indicates that it is the release of bound, or
"sequestrated", Ca^{++} within the fibre that is the coupling process. Thus, we
may consider three processes: (1) The depolarization of the membrane that in
some way causes the release of this bound Ca^{++} so that the internal concentration of ionized Ca^{++} is raised above a threshold value. (2) The catalysis of the
contractile process by Ca^{++}. (3) The removal of the liberated Ca^{++} that
permits relaxation—the *relaxing factor*.

Local Activation of Muscle

First, however, let us look at some interesting experiments on the local
activation of muscle carried out by A. F. Huxley & Taylor. As Hill emphasized
in 1949, diffusion of a molecule or ion, liberated at the surface, throughout the
fibre would be too slow to permit the rapid onset of the active state, so that the
effects of the electrical change on the membrane must be transmitted deep into
the cell if activation of all the elements within it is to be rapid enough. A. F.
Huxley's studies on local activation of single muscle fibres have shed some light
on the problem. By passing a current through a pipette, with an area of only a
few micra, applied to the muscle surface, Huxley & Taylor confined the potential
changes to the region of contact of the pipette. The area of depolarization was
thus too small to cause a propagated action potential, and only local contractions
were obtained, which were not all-or-none but increased with increasing
current-strength.

Z-Line

With frog muscle, the local contraction never occurred if the pipette was
applied to an A-band; by contrast, when it was on the I-band there was a contraction which consisted of a movement of the A-bands, flanking the I-band,
towards one another; it was thus symmetrical about the Z-line, and examination
of the two halves of an I-band showed that they always contracted to equal
extents. It appeared, therefore, that the site of activation for the I-band was the
Z-line. In the crab and lizard muscles, on the other hand, the point of symmetry
for contraction was found to be the region of contact of the A- and I-bands.
When a local contraction was produced, the Z-line was pulled over to the side
of the I-band to which the pipette was applied until it came in contact with the
A-band. The other half of the I-band never took part in the contraction.

Importance of Triads

The difference between the frog, on the one hand, and the crab and lizard
on the other, may well reside in the different organizations of the sarcoplasmic
reticulum in the two cases; in the frog the triads (p. 1349) are situated at the
Z-line whereas in the lizard these are arranged, as in *Ambystoma* and fish
muscles, at the junction of the A- and I-bands. It may be, as Huxley & Taylor
suggested, that the effects of the membrane depolarization are conducted,
perhaps as electrotonic spread, along the central tubule of the triad which, as
we have seen (p. 1349), is continuous transversely across the width of the fibre.

Although direct openings of the T-tubules on to the surface of the fibre have not always been observed, the experiments with ferritin and other particulate matter, showing ready penetration into the T-system (p. 1349), leave little doubt as to the functional continuity between extracellular space and the T-tubules. As a working hypothesis, then, we may assume that the electrical effects of the action potential spread, either actively or electrotonically, into the T-system and there cause release of Ca^{++}; since the volume of the T-system is quite inadequate to hold the stores of Ca^{++} within the fibre (Huxley, 1964; Peachey, 1965), we may assume that the release is from the terminal cisterns, induced by some membrane interaction at the triad.*

Crayfish Fibre

In the crayfish† muscle fibre, Sugi & Ochi (1967) observed graded local contractions, which were quite strictly confined to the sarcomere under the micropipette electrode used for stimulation; as the area or intensity of stimulation increased, the transverse spread became greater, so that ultimately contraction spread right across the fibre. When brief high-voltage pulses were passed through the electrode, Sugi & Ochi observed a circumferential spread of contraction (Fig. 762, B), so that the central myofibrils were relaxed. In the crayfish fibre the response to a nervous stimulus is normally graded, in contrast to the all-or-none response of a

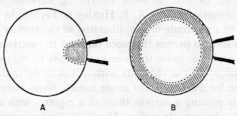

FIG. 762. Illustrating two different types of transverse spread of contraction initiated by local activation of crayfish muscle fibres. Each circle represents cross-section of the fibre. Pipette for local activation is shown at the right-hand side of each circle. Shaded area represents contracted region. (Sugi & Ochi. *J. gen. Physiol.*)

vertebrate twitch fibre; Sugi & Ochi's studies on the recruitment of more and more myofibrils as the stimulus intensity or area increased indicate that the basis of graded activity is essentially the recruitment of active myofibrils, rather than the variation of contractile strength of all the myofilaments activated together.‡

Intracellular Injection

The first to emphasize the possible significance of Ca^{++} were Heilbrunn & Wiercinski, who injected various ions into single muscle fibres; a Ringer's solution containing Na^+, K^+ and Ca^{++} caused contraction, whilst if Ca^{++} was omitted, or replaced by Mg^{++}, there was no contraction. Ba^{++} also caused contraction. Later Niedergerke repeated this experiment, controlling the amount of Ca^{++} injected by iontophoresis; under these conditions local reversible

---

* Peachey (1965) draws attention to the serrate appearance of the junctions at the triad, reminiscent of the septate junctions examined by Loewenstein (p. 494).

† The sarcotubular system of the crayfish has been described by Grundfest and his colleagues (Brandt *et al.*, 1965, 1968) in conjunction with a puzzling swelling and shrinkage of the T-system according to the direction of passage of an applied current across the muscle fibre. Their results suggested that the membrane of the T-tubule was selectively anion-permeable, and they argued that transmission of electrical changes into the interior of the fibre was across this anion-permeable membrane.

‡ We may note that Sugi & Ochi found circumferential spread in frog fibres as well, the spread occurring at a rate of some 0·8 to 6 cm./sec., so that it appears that the transverse tubular system can propagate some sort of change confined, however, to the more superficial fibrils. Natori (1965) showed that skinned fibres would give a propagated contraction after an electrical stimulus; presumably propagation occurred through the sarcotubular system.

contractions could be demonstrated. Employing isolated myofibrils, Podolsky applied Ca++ directly through a micropipette to different regions. When the pipette was opposite a Z-line, shortening occurred in both halves of the I-band to which this belonged.*

Caldwell & Walster (1963) were able to inject modified Ringer's solution into a cannulated crab muscle fibre; Fig. 763 shows the effects of increasing concentrations of Ca^{++}; as more Ca^{++} is injected, the duration of the contraction increases, and relaxation becomes slower as though the limiting factor in relaxation were the removal of the injected ion. Ba^{++} and Sr^{++} were effective in about the same concentrations.

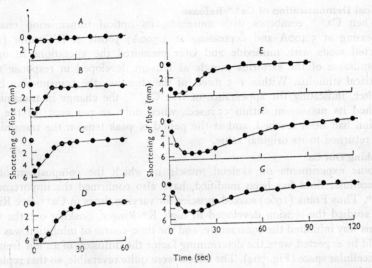

FIG. 763. The effects of the injection of solutions containing various concentrations of $CaCl_2$ and of distilled water on a *Maia* muscle fibre bathed in crab saline. The shortening produced at various times after the injection of each solution over an 8 mm length of the fibre is shown. Injections: *A*, distilled water; *B*, 1 mM-$CaCl_2$; *C*, 5 mM-$CaCl_2$; *D*, 10 mM-$CaCl_2$; *E*, 20 mM-$CaCl_2$; *F*, 59 mM-$CaCl_2$; *G*, 100 mM-$CaCl_2$. (Caldwell & Walster. *J. Physiol.*)

Chelation of Internal Ca++

Injection into the crab fibre of the Ca^{++}-chelator, EGTA, which leaves most of the Mg^{++} in an ionized form, abolished the response to electrical stimulation, to depolarization by high K^+, or to treatment with 2–5 mM caffeine, provided the internal concentration of EGTA was raised above 2–3 mM; the internal concentration of Ca^{++} is of this order, so that it appears that an amount equivalent to the total fibre-Ca^{++} must be injected in order to suppress contraction completely; with lower concentrations weak activity could be evoked (Ashley et al., 1965). The authors considered that EGTA was free in the sarcoplasm until the Ca^{++}, with which it complexed, had been released, and they concluded that, under the conditions of their experiments, all the sequestered Ca^{++} was released so that the concentration left over, after binding with all the EGTA, was adequate for threshold contraction.

* When the micropipette was displaced from the Z-line there was an asymmetry in the contraction, the half of the I-band nearer the pipette contracting more than the other half. This indicates that the regions of overlap between A- and I-bands may be separately excited.

Threshold Concentration of Ca++

The actual concentration of ionized Ca^{++} in a buffered solution of EGTA-Ca complex was probably little affected by dilution within the fibre after injection, *i.e.*, the Ca^{++}-concentration was effectively buffered. If this was true, then the threshold concentration in the sarcoplasm, necessary to induce contraction, was $0.3–1.5\,\mu M$, similar to the estimate based on model studies, namely greater than $0.3\,\mu M$ (Portzehl, Caldwell & Rüegg, 1964). These authors computed, from their study of injection into the crab muscle fibre, that Ca^{++} would have to be removed at a rate of 3 μmoles muscle/min. to account for the relaxation time; estimates of uptake by rabbit isolated particles are in the region of 30 μmoles/min.

Optical Demonstration of Ca++-Release

When Ca^{++} combines with murexide, its optical transmission changes, increasing at 5,400A and decreasing at 4,700A; Jöbsis & O'Connor (1966) injected toads with murexide and later measured the alterations in optical transmission of the sartorius muscle as tension developed in response to an electrical stimulus. Within 1–5 msec. of the stimulus, the transmission began to alter, indicating the appearance of free Ca^{++}; the change in transmission reached its maximum within 55 msec., when only 20 per cent. of the peak tension had been reached, and at the period of peak tension the transmission had returned to its original value.

Washing Out Ca++

Some experiments on skeletal muscle in which the composition of the extracellular fluid has been modified, have also confirmed the importance of Ca^{++}. Thus Frank (1960) soaked muscles for varying times in Ca^{++}-free Ringer and studied the tension developed in high K^{+}-Ringer. Soaking out the Ca^{++} in this way inhibited the contracture, and the time-course of inhibition was what would be expected were the determining factor the diffusion of an ion from the extracellular space (Fig. 764). The effects were quite reversible, so that replacing Ca^{++} in the medium restored the contracture. Interestingly, the caffeine contracture was not affected appreciably by soaking out the Ca^{++}; this demonstrates that the contractile mechanism, *per se*, was uninfluenced by soaking the muscle

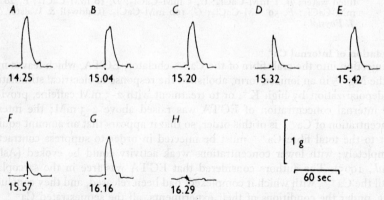

FIG. 764. Inhibition of contractures produced by isotonic potassium chloride in the frog's toe muscle by soaking the muscle in a calcium-free solution. A, C, E and G control responses produced by a solution containing only 123 mM potassium chloride and 1·08 mM calcium chloride. Time (min.) that muscle was kept in a calcium-free solution before testing with a 123 mM potassium chloride solution, B, O; D, 1; F, 3; and H, 6. Dots immediately below each record indicate point at which potassium chloride solutions put in bath. Time of day recorded below each test record. (Frank. *J. Physiol.*)

in Ca^{++}-free solutions, but later studies have shown that, in a Ca^{++}-free medium, the number of responses that may be evoked by caffeine is strictly limited.

Calcium Fluxes

The simplest hypothesis, and one presented by Shanes, is that excitation leads to an increased influx of Ca^{++} from the outside medium; in this respect, muscle would behave like the squid giant axon (p. 1045). Studies with the radioactive isotope of Ca^{++}, ^{45}Ca, carried out by Harris and Bianchi & Shanes, among others, leave little doubt that the influx of this ion increases greatly during an electrical stimulus. However, later studies have shown that stimulation of muscle increases the efflux from muscles previously loaded with ^{45}Ca, so that excitation and contraction are not accompanied by any significant *net* changes of Ca^{++}-content. Furthermore, a simple computation shows that the amount of Ca^{++} entering during an action potential is quite inadequate to cause activation of more than 0·4 per cent. of the total number of myosin molecules (Winegrad, 1961). Since most of the Ca^{++} in muscle seems to be in a bound or sequestered form, not exchanging readily with ^{45}Ca in the medium (Harris, 1957; Curtis, 1966), it seems safe to conclude, at any rate with twitch-type skeletal muscle, that the action potential in some way releases the sequestered Ca^{++} into the sarcoplasm. With heart muscle the situation may well be different since Winegrad has shown that the influx of ^{45}Ca into the isolated auricle increases when this is made to beat; the faster the beat the greater the uptake-per-beat. Since the tension increases in a parallel manner we have here an excellent correlation between Ca^{++}-influx and tension developed. This point will be discussed further in Chapter XXII.

Calcium Substitutions

The importance of the release of bound calcium is revealed by some interesting experiments of Frank (1962), who studied the ability of other ions to substitute for Ca^{++} in restoring contractility to muscles soaked in a Ca^{++}-free medium. A number of divalent cations, such as Zn^{++}, Cu^{++} and Cr^{++}, were effective in giving large contractures when the muscle, previously soaked in a Ca^{++}-free medium, was placed in high-K^+ Ringer's solution. These ions did not substitute for Ca^{++} in all respects, however. Thus, if Ca^{++} was restored, the muscle responded to a raised K^+-concentration repeatedly, so that alternating between Ca^{++}-Ringer with normal K^+- and Ca^{++}-Ringer with high K^+, gave a series of contractile responses. When Cr^{++} or another of these divalent cations was substituted for Ca^{++}, the muscle responded for the first few exposures approximately as well as if the divalent ion had been Ca^{++}, but after repeated exposures the response deteriorated. Rather similar effects were obtained using caffeine to substitute for Ca^{++}; thus we have seen that caffeine can induce a contracture in a Ca^{++}-free medium; however, if the muscle was kept in a Ca^{++}-free medium and tested for a caffeine contracture every 10–20 minutes, the caffeine-induced contractures diminished rapidly and eventually disappeared. Replacing the muscle in a Ca^{++}-containing solution restored its ability to give a caffeine-contracture.

Mobilization of Bound Ca^{++}

It would seem, therefore, that the effects of these Ca^{++}-substitutions, be they divalent ions—such as Cr^{++} or Zn^{++}—or caffeine, depend for their effectiveness on the presence of a store of bound calcium in the fibre or at its surface. Repeated testing of a muscle in a Ca^{++}-free medium with caffeine or divalent cations accelerates the decline of its responsiveness, as though each test involved the mobilization of some of the store of bound calcium.

This view was supported by Caldwell & Walster's microinjection experiments when they showed that Ba^{++} and Sr^{++} were just as effective as Ca^{++} in inducing contraction; in addition, they showed that injected caffeine[*] was just as effective, mole for mole, and this equivalence suggests that it is the release of bound Ca^{++} by the injected substances that is the essential feature in common.[†]

Relaxation

The contracture of a twitch muscle, induced by a sustained depolarization, is not maintained; after a minute, relaxation is complete, and according to Hodgkin & Horowicz' study, the process can be accounted for on the basis of an exponential inactivation process with a rate-constant of 30 sec^{-1} corresponding to a half-life of 23 msec. This inactivation might well be the removal of the liberated Ca^{++}; and the results of many studies concur in attributing a major role to the sarcoplasmic reticulum in this accumulation. A great deal of the evidence in favour of this was derived from experiments on model systems based on the interaction between the muscle proteins, actin and myosin, induced by Mg^{++}, ATP and Ca^{++}. Some of this work will be discussed in greater detail later, and for the moment it suffices to know that these model systems could be made to simulate the contractile process remarkably well, but subsequent relaxation was only achieved if certain "relaxing factors" were present; these were derived from muscle homogenates, and one such factor could be separated by ultra-centrifugation at 35,000 g and was thus associated with the microsome fraction.

RELAXING FACTOR

Nagai, Makinose & Hasselbach examined the particulate fraction in the electron microscope and found it to consist of a number of membrane-covered vesicles presumably derived from the sarcoplasmic reticulum; and in a more elaborate study, in which the activities of different preparations were compared with their electron-microscopical appearance, Muscatello, Andersson-Cedergren & Azzone were able to show that this activity was confined to the vesicular fraction obviously derived from the sarcoplasmic reticulum; the mitochondria were definitely excluded and it was shown that the Kielley-Meyerhof "soluble ATPase", which had been shown to have relaxing activity, was virtually identical with this microsomal preparation. Treatment of the microsomes with agents, such as detergents, that broke up membranous structures abolished the relaxing activity (Nagai et al., 1960).[‡]

Uptake of Ca^{++} by Sarcoplasmic Microsomal Fraction

An obvious mechanism by which the microsomes could cause relaxation would be through the accumulation of Ca^{++} within their membrane-bound vesicles by the operation of an ATP-activated Ca^{++}-pump, as suggested by Annemarie Weber. In this way the sarcotubular system could act as storage site for sequestered Ca^{++} at the same time as it caused relaxation. Such an accumulation was demonstrated by Ebashi & Lipman (1962) and Hasselbach &

---

[*] Axelsson & Thesleff (1958) had failed to observe a contracture with intracellularly injected caffeine; since caffeine penetrates the cell rapidly (Bianchi, 1962), this was puzzling.

[†] To a large extent Sr^{++} can replace Ca^{++} so far as maintaining resting potential and excitation-contraction coupling; Edwards, Lorkovic & Weber (1966) have shown that this extends to the myofibrillar ATPase activity; they emphasize the difference between Sr^{++} and the other divalent cations studied by Frank, which do not substitute completely for Ca^{++}. These authors consider that these other cations, such as Ni^{++}, exert their effects on Ca^{++} release without entering the cell, i.e. by a membrane action.

[‡] Relaxing factors apparently share in common the power to chelate or sequester Ca^{++}; thus "excess ATP" causes relaxation in models; some of the "supernatant" factors from muscle homogenates may also consist of substances that complex Ca^{++}.

Makinose (1961). According to Hasselbach & Makinose, the amount taken up was far too great to be accounted for by binding to protein, and they concluded that an active calcium-pump accumulated the Ca^{++} within the vesicles of the microsomal fraction. Weber, Herz & Reiss showed that the relaxing factor microsomes were able to remove the exchangeable Ca^{++}-bound to the myofibrils, the amount removed correlating well with the reduction in ATPase activity of the myofibrils, and the degree to which they were superprecipitated, *i.e.*, contracted. The relaxing factor—microsomes—was able to reduce the concentration of ionized Ca^{++} in the medium to below 0·01 μM, a concentration that had been shown to be low enough to prevent actomyosin ATPase activity.

If, as suggested by Annemarie Weber, actomyosin depends for its ATPase activity on the formation of a Ca-actomyosin complex, then removal of Ca^{++} from this complex by relaxing factor would account for the latter's effect.

OXALATE AND ATP

Figure 765 illustrates the uptake of Ca^{++} by sarcoplasmic vesicles as a function of concentration in the medium; when oxalate was present, accumulation was more rapid and continued to a greater extent, since precipitation within the cell

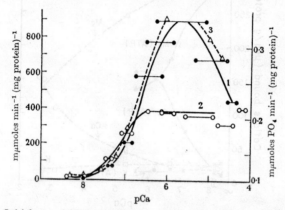

FIG. 765. Initial rate of Ca accumulation (curves 1 and 2, left ordinate) and ATP hydrolysis (curve 3, right ordinate) by sarcoplasmic reticulum as a function of pCa. Curve 1 shows accumulation in presence of 5 mM oxalate; curves 2 and 3 in absence of oxalate. Note that although accumulation of Ca in absence of oxalate (Curve 2) flattens off at high concentrations of Ca (low pCa), the rate of splitting of ATP continues to rise. (Weber *et al. Proc. Roy. Soc.*)

prevented efflux; in the absence of oxalate the rate flattened off at high concentrations of Ca^{++} owing to the greater effects of efflux, but it is interesting that the splitting of ATP, which is presumably a necessary step in the influx, continued to increase in parallel with the influx when oxalate was present.*

Hasselbach & Makinose (1963) showed that accumulation would take place in the presence of ATP until the concentration of Ca^{++} in the medium was less than 10^{-6} M, although the concentration inside remained the same owing to precipitation of oxalate; at this point the internal concentration was some 500 times the external. A simple model of uptake would consist of a carrier capable of complexing with Ca^{++}; on phosphorylation by ATP on the surface, its affinity

* According to Martonosi & Feretos (1964) the effect of oxalate may be mimicked by other substances that precipitate Ca^{++}, *e.g.* F^-, pyrophosphate, etc. This may explain the "supernatant co-factor" in relaxation described by Briggs, Kaldor & Gergely (1959). Hasselbach & Makinose (1963) found that uptakes of Ca^{++} and oxalate were equivalent. The inhibitory action of oxalate on the ATPase activity of relaxing factor is due to its favouring uptake of Ca^{++}, which proceeds to the point where the concentration is below that required for ATPase. Excess Ca^{++} also inhibits, and the optimum concentration in the presence of 4 mM $MgCl_2$ and 4 mM ATP is 10 μM. These facts go a long way to reconciling many contradictions in the literature regarding the action of Ca^{++}.

with Ca^{++} would increase by a factor of 500 over the affinity in the unphosphory-lated condition; thus dephosphorylation on the inside would cause accumulation.* As to the condition of the Ca^{++} within the vesicles, it is unlikely that it is all bound in a non-ionic form; thus it could be argued that accumulation was the result of binding so that energy is provided in increasing the affinity for the binding sites; Martonosi & Feretos argued, however, that the affinity of the binding sites would have to correspond to a dissociation constant of 10^{-6} M to produce the concentration gradients actually found, whereas that of oxalate is some 1,000 times higher and yet this favours uptake. Nevertheless, as Carvalho & Leo (1967) have shown, there is considerable binding of Ca^{++} within the vesicles, since they estimate that the internal concentration of ionized Ca^{++} is only some 1 per cent. of the total; this binding, in their view, is increased by ATP whilst that of Mg^{++} is reduced by an amount corresponding to the extra Ca^{++} bound; this is illustrated by Fig. 766, where it is seen that, at a given P_{Ca} (the negative

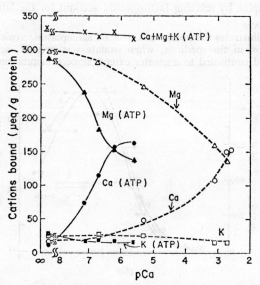

FIG. 766. Illustrating the effects of ATP on the binding of cations to microsomes of rabbit skeletal muscle. Broken lines: Binding in absence of ATP. Full lines: Binding in presence of ATP. Note that for a given pCa, ATP causes an increased binding of Ca^{++} but a corresponding decrease in binding of Mg^{++}. (Carvalho & Leo. *J. gen. Physiol.*)

logarithm of the concentration of Ca^{++}) the presence of ATP increased the binding of Ca^{++} at the expense of Mg^{++}. Thus these authors speak of an "active" and "passive" binding of Ca^{++} by the sarcoplasmic reticulum. Furthermore, the exchangeability of the bound Ca^{++}, whether this is "intrinsic" or induced by ATP, depends on the presence of ATP in the medium (Carvalho, 1968).

QUANTITATIVE ASPECTS

Weber & Herz (1963) estimated that if 1 mole of Ca^{++} was liberated for 1 mole of myosin, to cause maximal contraction, then the amount liberated would be 0·15 μmole per ml. of fibre; the relaxing granules from the same amount of fibre can remove 0·05 to 0·5 μmoles of Ca^{++} in 100–400 msec., so that the system is probably adequate to carry out this function during the period of relaxation of muscle (Hasselbach, 1964).

T-FRACTION

A. Weber *et al.* (1964) found that a heavier fraction of the microsomes, sedi-menting at 4,000–8,000 g would also accumulate Ca^{++} and cause relaxation in

---

* Hasselbach & Makinose point out that, during the process of accumulation, trans-phosphorylation of ADP and ATP takes place, as revealed by the transfer of ^{32}P from $AD^{32}P$ to ATP.

models; they considered at first that these were mitochondria which, as we have seen, also accumulate Ca^{++}; however, the absence of inhibition by dicumarol or CN^- plus oligomycin rules the mitochrondia out, so that the heavier fraction might well be the fragmented T-system, which would produce heavier vesicles.*

Localization of Ca^{++}

Constantin, Franzini-Armstrong & Podolsky (1965) applied oxalate to fibres, whose sarcolemma had been stripped, and examined the localization of the precipitated material in the electron microscope. The material was found in the terminal sacs adjacent to the I-band, with none elsewhere including the transverse T-system; if the fibre was treated with Ca^{++} first and then with oxalate, there was a considerable increase in the number of dense spots. If this study gives a true picture of Ca^{++} accumulation, then it would appear that it is the sarcoplasmic reticulum proper that is responsible for accumulation, and, presumably, release. Since the effect of the electrical stimulus is probably transmitted along the transverse T-system, it seems likely that the coupling between this and the Ca^{++}-release mechanism takes place across the triads, i.e., the regions where the two systems come into close apposition.

TERMINAL CISTERNS

Using a radioautographic technique with high resolution in the electron microscope, Winegrad (1965, 1968) showed that, after soaking the muscle in ^{45}Ca Ringer, the activity was confined to the terminal cisterns of the sarcoplasmic reticulum; if the muscles had been stimulated tetanically while exposed to ^{45}Ca, and the localization was observed at different periods after stimulation, then there were three regions at which accumulation occurred, namely the terminal cisterns, the intermediate cisterns and longitudinal tubules, and the A-band portion of the myofibrils. The terminal cisterns were labelled more rapidly than the fibrils, and both processes were accentuated by stimulation; as recovery from stimulation proceeded, the amount of ^{45}Ca in the terminal cisterns increased at the expense of that in the longitudinal cisterns and tubules.

Hypothetical Scheme

The simplest interpretation is that stimulation releases Ca^{++} from the terminal cisterns, whilst relaxation occurs as a result of the binding of this by the longitudinal cisterns and tubules; final recovery depends on the transport of this bound calcium back to the terminal cisterns. In the resting condition, exchange of ^{45}Ca would occur between the transverse tubules and the terminal cisterns. Thus, since the first part of the sarcoplasmic reticulum to show an increase in ^{45}Ca uptake after a tetanus is the longitudinal cisterns and tubules, it is likely that they contain the Ca^{++}-sequestering system at the basis of relaxation (Winegrad, 1968). These cisterns are the site of ATPase activity with optimum activity at pH $7\cdot2$, by contrast with the myofibrillar ATPase that is active at pH $9\cdot2$ (Gauthier & Padykula, 1965).†

* Seraydarian & Mommaerts (1965) have made a systematic study of the fractionation of muscle microsomes in a density gradient; two main fractions may be separated at 15,000–41,000 and 441,000–150,000 g, the heavier of these having high powers of Ca^{++}-accumulation and being clearly derived from the sarcoplasmic reticulum, whilst the lighter fraction might well have had a different origin, at least in part. Briggs & Fleishman (1965) have prepared particle-free supernatants from muscle homogenates that will bind Ca^{++}; the degree of binding was measured by allowing competition between the preparations and a standard exchange-resin, chelex-100, and an appropriate "stability constant", corresponding to a dissociation constant, was measured; the constants were of the same order as that of EGTA, so that the material would be effective as a relaxing agent.

† Elison et al. (1965) showed that glycerol-extracted muscle fibres would take up Ca^{++} from the medium in the presence of ATP; incubation with a bile salt, which presumably destroyed the membrane system of the muscle, abolished this. When the

Isolated Terminal Sacs

Ikemoto et al. (1968), with the aid of tryptic digestion, isolated vesicles, with a globular head and tail, from the microsome fraction of muscle; these vesicles may well have represented the terminal sacs of the sarcoplasmic reticulum, and it is interesting that they had both high ATPase activity and also high accumulating activity for Ca^{++}.

Binding of Ca^{++}

Winegrad argued that the Ca^{++} taken up by the longitudinal cisterns and tubules was bound on the surface, and an estimate of the amount immediately after a tetanus, namely 0·2 μmole/g. muscle, indicated what it was adequate to remove that necessary to produce maximal contraction (0 1–0 2 μmole); consequently, relaxation need not necessarily have to wait on an active Ca^{++}-pump, the latter's activity being required to remove this bound Ca^{++} back to the terminal cisterns. In this event the ATP splitting of relaxation would be used after relaxation, and Kushmerick* has shown that there is a significant amount of splitting in the first few seconds after recovery from a tetanus.

Localization of ATPase

The accumulation of Ca^{++} apparently requires ATP, so that it is reasonable to seek a Ca^{++}-sensitive ATPase in the sarcoplasmic reticulum, but as Tice & Engel (1966) and Sommer & Hasselbach (1967) have emphasized, the mere localization of phosphate-splitting is no indication of the presence of a Ca^{++}-sensitive ATPase specifically involved in this accumulation. According to Sommer & Hasselbach the "basic ATPase" that is not stimulated by Ca^{++} is much less sensitive to inactivation by fixation procedures and so is likely to be the one identified in cytochemical studies. A difficulty in identifying a Ca^{++}-sensitive ATPase is the fact that EDTA chelates the Pb^{++} used to precipitate the liberated phosphate (Tice & Engel). There is certainly ATPase activity in the terminal sacs as well as a uniform distribution throughout the rest of the sarcoplasmic reticulum; and Tice & Engel have suggested that it is the activity in the terminal sacs that is Ca^{++}-activated, the uniformly distributed activity being the "basic ATPase".

Post-tetanic Potentiation

The duration of the active state of muscle following a stimulus is increased if the muscle is previously stimulated tetanically; Winegrad (1968) has pointed out that, since the rate of uptake by isolated sarcoplasmic reticulum decreases with increasing uptake, it will be unable to remove the sarcoplasmic Ca^{++} so completely after a tetanus as after a single twitch, so that the Ca^{++} liberated after a stimulus now has longer to act, i.e., the active state is prolonged.

Electrical Stimulation of Microsomes

Lee et al. (1965, 1966) passed square pulses of some 2 mV and 10 msec. duration at a rate of 60/min. through a suspension of microsomes, and found that this reduced accumulation of Ca^{++}, or, if this had occurred maximally, caused release. If the stimulation occurred when the microsomes were mixed with a model contractile system—e.g., a myofibril suspension—then this was caused to contract, presumably by inhibition of Ca^{++}-uptake by the microsomes.

Skinned Muscle Fibre

Constantin & Podolsky (1967) employed a technique, developed by Natori (1954), by which a portion of the sarcolemma is peeled off, and stimuli are applied through electrodes attached to the unskinned ends, with the preparation under oil; the composition of the sarcoplasm could be altered by local application of a drop of solution. A pulse of d.c.-current caused contraction of the skinned region;† this was localized to a variable number of sarcomeres, with relaxed sarcomeres between; at a given point, several sarcomeres contracted together. Constantin & Podolsky postulated that the T-system of transverse tubules became, in some way, sealed off after removal of the sarcolemma; and if the active transport processes

---

accumulation occurred in the presence of oxalate, Pease, Jenden & Howell (1965) observed accumulations in the whole of the L-system; it was impossible to say whether accumulation had also occurred in the T-system since this did not survive the treatment.

* Quoted by Winegrad in his 1968 paper. The interested reader should consult this paper for a most interesting discussion of many of the features of the Ca^{++} uptake and release processes.

† The unskinned regions did not contract, presumably because they had become depolarized as a result of damage.

operated across the sarcolemma, the concentrations of K^+ in the tubules would be low and those of Na^+ and Cl^- high, and thus permit the establishment of a resting potential. Thus the contractile machinery would be ready for activity, induced by a depolarization of the T-membrane. On this basis, bathing the skinned fibre with high Cl^- should depolarize, and act as an effective stimulus; in fact it activated the fibre, whilst treatment with an ATPase inhibitor, strophanthidin, rendered the fibre electrically inexcitable. These results, striking as they are in demonstrating electrical excitability of the T-system, pose a number of problems, which have been discussed by Constantin & Podolsky. Thus, contraction is not maximal at the anode, and the spread, as the potential is increased, should be towards the cathode, but in fact spread occurs in both directions from the originally responsive sarcomeres. Again, the computed potentials at points remote from the electrode are much smaller than would be expected to excite, unless some longitudinal continuity between sarcomeres is postulated; this might be through the sarcoplasmic reticulum at the specialized contacts at the triads. The graded nature of the response is consistent with Huxley & Taylor's hypothesis that spread along the T-system is electrotonic; on the other hand, the occasional observation of an extensive response, propagated for up to 200 μ, suggests that the membrane is capable of regenerative activity; and, if this were so, one contradiction would be resolved, namely the fact that, if Ca^{++} is released by the action potential, the currents flowing passively during the passage of this are inadequate to supply one charged activator per molecule of contractile protein.

Effects of External Calcium

The effects of modifying the concentration of Ca^{++}, in the medium surrounding a muscle, are necessarily complex, since this ion can influence several parameters that determine the force of contraction and the duration of the active state; thus the effect on ionic permeability will modify the resting potential and the action potential; the action potential is accompanied by an increased permeability to Ca^{++}, so that the amount entering the fibre will presumably be influenced by the external concentration; modifications of the sarcoplasmic concentration brought about in this way must influence the contractile process, as we have seen.

RESTORATION OF CONTRACTION

It seems very likely that the decreased depolarization-contracture tension, obtained by placing a muscle in high K^+-Ringer, that follows soaking in Ca^{++}-free medium is due to the progressive fall in resting potential. This depolarization, if sustained for a short time, breaks the link between membrane potential and the contractile machinery, so that the muscle remains in a relaxed state, although the membrane is sufficiently depolarized that, if this degree of depolarization had been induced at first, considerable tension would have developed. In this connexion we may note that the amount of tension that can be developed, when the partially depolarized muscle is completely depolarized, depends on the degree of this partial depolarization; and we may plot a Restoration-Potential curve, as in Fig. 767. Thus depolarization to, say, -39 mV means that when the muscle is transferred to 160 mM K^+ to produce complete depolarization, the tension developed will be some 40 per cent. of that obtained if the muscle had been maintained at its normal resting potential before the complete depolarization. It is clear from this curve that, if soaking a muscle in a Ca^{++}-free medium reduced its membrane potential to -30 mV, no contracture-tension would develop on placing it in 160 mM K, an effect that might have been attributed to the "leaching out" of Ca^{++} from the muscle, but which is undoubtedly due to the effect of depolarization, *per se*, in breaking the link between tension and membrane potential.*

* Parallel studies on membrane potential, electrical excitability of the muscle fibre, and development of contractile strength, carried out by Edman & Grieve (1966), have shown that, in general, the failure of muscle fibres in a Ca^{++}-free medium to contract

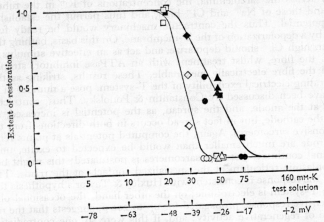

Fig. 767. Restoration-Potential Curves. When a muscle is depolarized in high-K⁺ solutions, to the extent indicated on the abscissa, the tension developed on complete depolarization depends on the initial degree of depolarization to which it was submitted. As ordinate is plotted, the tension developed by the muscle when completely depolarized, as a fraction of the tension it would have developed if not previously partially depolarized. Thus, with a previous depolarization to −39 mV, the tension developed with complete depolarization is only some 0.3–0.4 of that which would have been developed had the muscle been directly depolarized. The Figure shows the shifting of the curve due to the presence of caffeine (filled symbols). (Lüttgau & Oetliker. *J. Physiol.*)

SHIFT OF POTENTIAL CURVES

For a given degree of depolarization, however, Tension-Potential and Restoration-Potential curves are affected by external Ca⁺⁺-concentration. Thus the Restoration curve is shifted to the right at high Ca⁺⁺-concentrations, indicating that the muscle can be depolarized to a greater degree and retain the same residual contracting power; the Tension-Potential curve is also shifted to the right, so that a greater degree of depolarization is required to develop a given tension at high concentrations of Ca⁺⁺ (Jenden & Reger, 1963; Lüttgau, 1963; Lüttgau & Oetliker, 1968). Again, soaking in low-Ca⁺⁺ medium shifts the membrane Potential-Tension curve to the left, thus potentiating contraction (Caputo & Gimenez, 1967). Thus, whereas the most prominent effects of external Ca⁺⁺ seem to be exerted through the membrane potential, there are other effects that are exerted on a later stage of the excitation-contraction coupling system; these could be effected through actual gains or losses of Ca⁺⁺ by the fibre, or they could represent an action of altered membrane characteristics on the release and storage mechanisms for this divalent cation.

Rate of Relaxation

The rate of relaxation from a contracture is presumably determined by rate of removal of Ca⁺⁺; Lüttgau's work suggested that the relaxation phase was dependent on the ionic content of the medium; and the effects of changes occurred sufficiently rapidly to suggest that they were mediated through the cell membrane, as though the rate were voltage-dependent. Foulks & Perry (1966) have developed this viewpoint by studying the effects of sudden changes of the medium on tension; thus, when peak tension had been developed in a high-K⁺ medium, a sudden reduction in external K⁺-concentration retarded the relaxation that would have occurred with sustained exposure to the high-K⁺ medium. Usually the relaxation

in response to electrical stimulation is due to failure of the action potential; the resting potential declined progressively along with the decline in spike-height and twitch-tension.

could be resolved into a rapid and slow component, the latter corresponding to that initially developed at the same external K^+-concentration. This suggests that the sudden repolarization, caused by reducing external K^+-concentration, exerted its effect by altering the rate of sequestration of Ca^{++} to a new value appropriate to the new degree of depolarization, perhaps by altering the affinity of the carrier-mechanism for the ion. The effects of varying external Ca^{++}-concentration were essentially similar to Lüttgau's, and could be interpreted in terms of an alteration in the position of the curve relating relaxation-rate to membrane potential; this is illustrated by Fig. 768, which shows both Tension-Potential and Relaxation Rate-Potential curves for normal (1·08 mM) and high (5·4 mM) Ca^{++}-concentrations; high Ca^{++} shifts both curves to the right, showing that a higher degree of depolarization (higher K^+-concentration) is required to cause a given degree of tension or rate of relaxation.[*]

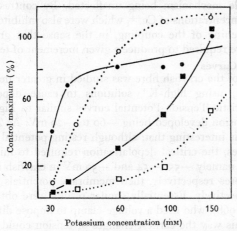

FIG. 768. The effect of increased $[Ca]_0$ on the relation between $\log [K]_0$ and (1) contracture tension (●, ○), and (2) the rate of relaxation (■, □), in frog toe muscle, both being expressed as per cent of the maximum value observed during control contractures in media containing 1·08 mM calcium. At normal (1·08 mM) $[Ca]_0$ (continuous lines), the curves for both contracture tension (●) and relaxation rate (■) rise steeply at lower $[K]_0$ (less depolarization) than corresponding curves (○, □) in media containing an increased (5·4 mM) $[Ca]_0$ (interrupted lines). Exposure to increased $[Ca]_0$ began 1–2 min before the onset of each contracture and continued until relaxation was complete. Note that, at high values of $[Ca]_0$, a higher degree of depolarization is required to cause a given degree of tension or rate of relaxation. (Foulks & Perry. *J. Physiol.*)

Invertebrate Muscle. In muscle fibre the response to an electrical stimulus is not the all-or-none propagated action potential, so that it is possible to reduce the membrane potential in a graded manner and study the changes in tension corresponding to each degree of depolarization. Orkand, using the crayfish muscle fibre, showed that, when the depolarization increased beyond a threshold value of 20 mV (*i.e.*, when the resting potential had fallen from 80 to 60 mV), tension developed in the fibre and increased steeply with increasing depolarization.[†] The significant factor was not so much the depolarization but the

---

[*] Frankenhaeuser & Lännergren (1967) have generally confirmed Lüttgau's findings; they emphasize the rapidity with which an altered external concentration of Ca^{++} can affect the contracture tension; it is not so rapid, however, that it could not have been mediated by diffusion along the T-system of tubules.

[†] It is interesting that the *duration* of the depolarization is important, so that the tension developed by a brief strong depolarization can be matched by a long supra-threshold depolarization; this emphasizes, once again, that the activation of the contractile machinery is not an all-or-none process. Besides the tension developed, the latent period depends on the degree of depolarization; thus Goodall was able to reduce this by

absolute value of the membrane potential, so that if the membrane had been hyperpolarized, by passing a constant current, then a greater degree of depolarization was required to produce a given increment of tension. Furthermore, experiments showed that the effective stimulus was the fall of potential, and not the passage of current *per se*; thus the current could be altered by changing the membrane resistance, but the tension corresponded to the altered membrane potential. This was illustrated by the action of GABA, which reduced the membrane resistance; in consequence, an increased membrane current was required to produce a given degree of depolarization, and the inhibitory action of this substance on the contraction of the fibre could be attributed entirely to this influence on membrane resistance, the coupling between membrane potential and the contractile mechanism being unaffected. By contrast, the effects of Mn^{++} and high concentrations of Ca^{++}, which were also inhibitory, seemed to be related to a weakening of the coupling, in the sense that greater degrees of depolarization were required to produce a given increment of tension.

Tension-Potential Curves

The behaviour of the crayfish fibre was studied in greater detail by Zachar & Zacharová (1966); using high-K^+ solutions to cause depolarization, they obtained characteristic Tension-Potential curves similar to Fig. 761, the limits between which tension developed being -60 to -20 mV. As Zachar & Zacharová point out, it is interesting that, although resting potential is very different in different muscles, the critical depolarization required to initiate contraction is about the same, namely -55, -54 and -50 mV in crayfish fibres and phasic and tonic frog fibres respectively, their membrane potentials being -79, -92 and -63 mV respectively. Essentially similar results were obtained by Dudel, Morad & Rudel (1968), who used a voltage-clamp to impose different potentials on the fibre; in this way the earliest changes of tension could be followed. As Orkand had found, the duration of the depolarization was important, so that with depolarization to -30 mV, a pulse of 2–20 msec gave a tension of only a few mg, whilst pulses lasting a second or more gave steeply rising tensions to reach 200 mg. If the stimulus lasted for 10–20 sec, the force declined—the usual relaxation during continued depolarization.

Effects of Calcium

Studies on the effects of Ca^{++} on these invertebrate fibres reveal some differences, suggesting that the influx of this ion may be a more important link between excitation and contraction than in vertebrate twitch muscle; thus Gainer (1968) was unable to correlate the reduction in KCl contracture of lobster muscle, brought about by soaking in Ca^{++}-free medium, with any alteration in resting potential; and the importance of influx of Ca^{++} during KCl contracture was shown by the much larger responses of muscles, previously soaked in a Ca^{++}-free medium, if the medium in which they were stimulated contained a high concentration of Ca^{++}. This does not mean, however, that the internal stores of Ca^{++} are without significance, since the caffeine-induced contracture did not show a similarly strong dependence on external Ca^{++}. Unlike the situation described by Frank (p. 1402), a variety of divalent cations, such as Ba^{++}, Sr^{++}, could not substitute for Ca^{++} in promoting KCl-contractures in a Ca^{++}-free medium.

INFLUX OF CA^{++}

Measurements of ^{45}Ca-influx revealed, not only a much larger resting influx than in frog muscle, but also a massive increase during KCl-contracture,

---

stimulating with transverse fields so that the depolarization was greater than that occurring during the action potential.

namely 3,590 μmoles/kg compared with 13 μmoles/kg in the frog, observed by Bianchi & Shanes (1959). It would seem, then, that in invertebrate skeletal muscle the influx of Ca^{++} contributes greatly to excitation-contraction coupling; the deep clefts formed by invagination of the sarcolemma into the substance of the fibre, corresponding to the T-system of the vertebrate fibre, would promote passage of Ca^{++} into the fibre, whilst the absence of a triad system would be consistent with the lesser importance of release of sequestrated Ca^{++}.

Caffeine and Other Potentiators. In low concentrations (*ca.* 1 mM), caffeine increases the tension developed in response to a single stimulus—an effect that may be attributed to a prolongation of the active state of the contractile machinery (Ritchie, 1954)—and increases the rate of rise of tension (Sandow *et al.*, 1964).

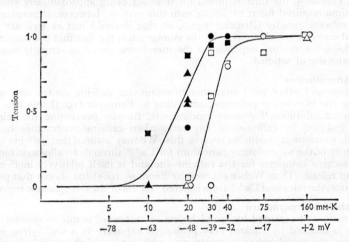

FIG. 769. Effects of caffeine on the contracture caused by depolarization with high-K^+. The ordinate shows the contracture-tension developed by the degree of depolarization (or concentration of K^+) indicated on the abscissa. Filled symbols correspond to the presence of 1·5 mM caffeine. (Lüttgau & Oetliker. *J. Physiol.*)

At high concentrations it causes a contracture which, unlike the depolarization-contracture of high K^+, lasts for a long time with only a slow development of relaxation, which may never reach completion. The time-course of relaxation depends on the membrane potential, so that if the membrane is completely depolarized the contracture, induced by caffeine (after the depolarization contracture has relaxed), remains steady. Expressed formally, the effect of caffeine is to shift the curve of tension versus membrane potential to the left (Fig. 769).

Effect on Accumulation of Ca^{++}

If the relaxation of the depolarized muscle is due to the re-accumulation of Ca^{++} by the sarcoplasmic reticulum, we may ask whether caffeine acts by preventing this re-accumulation. In this way it could lengthen the duration of the active state of the contractile machinery during a twitch, and also prevent complete relaxation in a depolarized muscle. In fact 8–10 mM caffeine does inhibit the uptake of Ca^{++} by microsome preparations of muscle, whilst, if added after maximal accumulation, it causes some 20–40 per cent. of the sequestered Ca^{++} to be released. With a pCa of about 7, a concentration that permits relaxation of actomyosin models, caffeine released enough Ca^{++} to account for contraction (Herz & Weber, 1965).* Furthermore, there was an excellent parallelism between

* Carvalho & Leo (1967) and Carvalho (1968) found no influence of caffeine on the amount of Ca^{++} bound by their preparations of sarcoplasmic reticulum; however, when maximal uptake was induced, caffeine did release Ca^{++}, as indicated by the action on an actomyosin system. In this respect we must note that there are apparently differences

conditions that favoured release of Ca^{++} from sarcoplasmic fragments and those that favoured contracture, both being inhibited by procaine, for example, whilst the speeds of action of the drug on the two processes, contracture and Ca^{++}-release, were similar (Weber & Herz, 1968). Caffeine also increases the efflux of Ca^{++} from muscle (Bianchi, 1961), but this may be secondary to the release of sequestered Ca^{++} since caffeine causes a contracture in muscles treated with EDTA to remove all external calcium. The effects of lowered temperature are consistent with this view; Sakai (1965) found that sensitivity to caffeine was increased by cooling, and this could be due to inhibition of the Ca^{++}-pump leading to a reduction in the amount of Ca^{++} restored to the sarcoplasmic reticulum (Lüttgau & Oetliker, 1968). The fact that caffeine, in concentrations too low to cause a contracture, will permit high K^+ to cause a contracture in a Ca^{++}-free medium suggests that depolarization and caffeine act in the same way to release sequestered calcium.* Lüttgau & Oetliker suggested that caffeine acts on the T-system, the time required for it to act being approximately equivalent to the time required for it to diffuse into this system; however, it penetrates the cell membrane rapidly (Bianchi, 1962), so that it could just as well act on the terminal cisterns, which seem to be the storage sites; the fact that the amplitude of the caffeine contracture depends on the membrane potential strongly suggests a membrane site of action.†

Local Anaesthetics

Procaine and other local anaesthetics antagonize caffeine contracture without affecting the membrane potential; according to Feinstein (1963) this is a simple competitive inhibition for some binding site, thereby preventing the release of Ca^{++}, induced by caffeine; at any rate, when caffeine contracture has been induced, procaine is unable to reverse this. We may assume that caffeine reduces the affinity of the sarcoplasmic reticulum for Ca^{++} through an allosteric modification; procaine competes for the caffeine-site and is itself relatively ineffective in inducing release. Thus Weber (quoted by Sandow, 1965) has shown that procaine antagonizes the release of Ca^{++} from isolated vesicles of the sarcoplasmic reticulum.‡

Acetylcholine-Contracture

The contracture caused by acetylcholine or carbachol is due to depolarization, since it is blocked by curare; very prolonged exposure to a Ca^{++}-free solution eliminates the acetylcholine contracture, which returns if the muscle is replaced in Ca^{++}-Ringer and then tested in Ca^{++}-free Ringer. According to Frank (1963), acetylcholine-contracture is due to release of sequestered Ca^{++} by virtue of the depolarizing action of the transmitter. However, as Jenkinson & Nicholls (1961) have shown, acetylcholine causes contracture of depolarized, chronically denervated rat diaphragm if Ca^{++} is in the bathing medium, so that it would appear that an increased influx of Ca^{++}, caused by acetylcholine, is the prime event here.§

---

between frog and mammalian muscle with regard to susceptibility to caffeine action (see, for example, Weber & Herz, 1968). Furthermore, if accumulation is measured in the presence of oxalate, caffeine has no action because the internal concentration of Ca^{++}-ions is held low by the oxalate.

* Isaacson & Sandow (1967) showed that the increased influx and efflux of ^{45}Ca, caused by caffeine, would occur in Ca^{++}-free medium.

† Gebert (1968) has pointed to a dual action of caffeine; at concentrations up to 4 mM the effect is single, consisting of the development of tension with subsequent relaxation; at higher concentrations a second stage becomes manifest after 1–5 minutes, during which tension again develops, this time irreversibly, in the sense that washing out the caffeine fails to induce relaxation. This late development of tension is unaffected by procaine or depletion of Ca^{++}, and the muscle in this condition is irresponsive to electrical stimulation.

‡ Benoit, Carpeni & Przybyslawski (1964) have described the effects of quinine; the contracture develops slowly and relaxes slowly; it persists in Ca^{++}-free medium even after repeated testing with caffeine; unlike caffeine it will not cause contracture of a depolarized muscle. Its action is attributed to a reduction in the threshold depolarization necessary for development of tension. According to Carvalho (1968) quinine causes release of passively bound Ca^{++} from sarcoplasmic reticulum.

§ It will be recalled that the contraction of depolarized smooth muscle, induced by acetylcholine, may be prevented by removing Ca^{++} from the medium, so that with this type of muscle there is little doubt that an influx of Ca^{++} can be important.

Anions and Heavy Metals

Anions of the lyotropic series —NO_3^-, Br^-, CNS^-— and heavy metal cations, such as Zn^{++}, Cd^{++}, UO_2^{++}, are potentiators of contraction, in so far as they lengthen the active state of muscle; their mechanisms of action are different, since the anions lower the mechanical threshold without affecting the spike potential, so that at a given degree of depolarization the mechanical tension is greater. The heavy metals, on the other hand, leave the threshold unchanged but prolong the falling phase of the spike potential (Sandow & Isaacson, 1966); this action is sufficiently rapid to suggest a surface membrane effect; and this is confirmed by the finding that CaEDTA reverses the effects of Zn^{++} although it is unable to penetrate the cell. The effects of the anions occur rapidly, too, so that a surface action must be kept in mind. Carvalho (1968) has shown that both types of potentiator tend to inhibit binding of Ca^{++} by sarcoplasmic reticulum, in the sense that they displace Ca^{++}, but because of the unlikelihood of their acting intracellularly it must be assumed that their effects are exerted on cell or tubular membranes, perhaps decreasing permeability ions to and thus prolonging the falling phase of the action potential. Certainly ions of the lyotropic series are able to inhibit carrier-mediated permeability (Davson, 1940), whilst Zn^{++} reduces the amplitude of the late after-potential of Freygang, Goldstein & Hellam (1964), a phenomenon considered to be due to the increased efflux of K^+ taking place during the repolarizing phase of the action potential.[*]

Electrical Mechanism of Coupling. The means by which depolarization of the fibre membrane can lead to liberation of Ca^{++} into the sarcoplasm is, of course, the crucial problem in excitation-contraction coupling. Studies on the electrical parameters of the muscle membrane, by Katz, Adrian & Freygang and Falk & Fatt, have brought to light anomalies that indicate that the simple equivalent circuit of capacity and resistance in parallel is inadequate to characterize the responses of the membrane to applied currents.

Current Pathways

In the first place, the capacity of 5–$8 \, \mu F/cm^2$ is considerably higher than that encountered in the majority of cells, suggesting that the effective area of membrane is larger than that indicated by the superficial geometry of the fibre; this difficulty could be resolved if the transverse tubular system were included

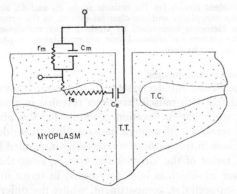

FIG. 770. The equivalent circuit for muscle membrane proposed by Falk and Fatt superimposed on a schematic diagram of the ultrastructure of a muscle fibre in order to suggest a possible relation between them. The location of r_e is tentative. *T.T.* refers to the transverse tubular system and *T.C.* refers to the terminal cisternæ of the sarcoplasmic reticulum. (Freygang, Rapoport & Peachey. *J. gen. Physiol.*)

* Kao & Stanfield (1968) have examined the effects of the lyotropic series of anions on both electrical and mechanical parameters of the voltage-clamped twitch muscle fibre; the strongest correlation was between the thresholds for delayed rectification and for mechanical contraction. The authors hesitate, however, to attribute causality to the relation between the two.

in the effective surface. Many features of the anomalous electrical characteristics can be reconciled if the transverse tubular system interacts with the sarcoplasmic reticulum proper, so that the current pathways become complex and an equivalent network, containing an additional resistance, r_e, and capacitance, C_e, are invoked, as illustrated by Fig. 770, r_e being the resistance of the terminal sac membrane and C_e the capacity of the T-membrane. At present it is considered that the depolarization of the surface membrane is transmitted electrotonically along the transverse system,* so that we may expect a flow of positive current from the sarcoplasm into the tubules. According to Freygang's hypothesis, this flow of positive current consists of a flux of K^+ along the tubules of the sarcoplasmic reticulum, proper, to the terminal cisterns, causing here, by replacement at adsorption sites, the release of Ca^{++}. However, estimates of the amount of current that could be carried show that it would be inadequate to release the amounts of Ca^{++} required for activation of the contractile machinery, so some amplification is necessary.

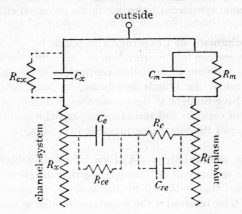

FIG. 771. Equivalent circuit for the muscle fibre. R_e and C_e act as a coupling between the myoplasm and the channel-system in the cross-section of the fibre. The elements connected by dashed lines are those for which no quantitative evidence is adduced, but are postulated on physical grounds. (Falk & Fatt. *J. Physiol.*)

K^+ and Cl^--Selective Membranes

The importance of the tubular system in the interpretation of many of the phenomena associated with muscle has been revealed in a number of studies; we have already seen that its location accounts for the phenomena of local stimulation. The responses of a single muscle fibre to sudden changes in the concentrations of ions in the medium surrounding it, studied by Freygang, may be interpreted in terms of the sarcotubular system, since they have two components, the slower of which is best interpreted in terms of diffusion of ions within a second, intracellular, compartment, whilst the differences in response to K^+ and Cl^- suggest that, whereas the membrane potential across the fibre is responsive mainly to Cl^-, that across the internal system is responsive mainly to K^+. Estimates of the volume of the K^+-sensitive region amounted to 0·2–0·5

---

* Falk & Fatt (1964) summarized the conflicting estimates of the capacity of muscle fibres; they pointed out that differences could arise from an inadequate electrical model for the fibre and developed the equivalent circuit of Fig. 771. They calculated the potential changes that would occur across C_e when the action potential occurred, and showed that they would consist of a distorted picture of the changes across the plasma membrane; the maximum depolarization would be some 20 mV greater than the mechanical threshold (p. 1397), and thus capable of maximal activity.

per cent. of the volume of the fibre, which corresponds reasonably well with Peachey's estimate of the volume of the transverse tubular system, namely some 0·2 per cent.

Late After-Potential

A train of impulses in a muscle fibre leads to a prolongation of the action potential, in the sense that the phase of repolarization is slowed to give a "late after-potential"; according to Freygang et al. (1967), this can be accounted for by an accumulation of K^+ in the transverse tubular space as a result of the action potential. In a hypertonic medium the action potential was prolonged some five-fold, and this could be accounted for if the volume of this space were enlarged, a simple electrical analysis indicating that the effect would be only to expand the time-scale of the spike; in fact, the measured swelling of the T-system was some 4·8-fold, compared with an estimated 5·2-fold from the electrical measurements. As to why the T-system should swell when the medium is made hypertonic with sucrose is not completely clear; if diffusion and flow outwards were restricted, an osmotic influx from sarcoplasm to T-system would cause a temporary swelling.*

Destruction of the T-System

Treatment of a muscle fibre with a Ringer's solution made hypertonic with a penetrating molecule, such as glycerol, urea, or ethylene glycol, causes a shrinkage and slow return to normal volume as the non-electrolyte penetrates. On subsequent transfer to normal isotonic Ringer's solution the muscle under-goes a temporary swelling, since it is now hypertonic to its environment. Such a treatment, for a reason difficult to understand, causes destruction of the transverse tubular system (Howell & Jenden, 1967; Eisenberg & Eisenberg, 1968). The electrical properties of the muscle fibre should now be simpler, and the capacity smaller; in fact Eisenberg & Gage (1967) found a capacity of 2·25 $\mu F/cm^2$, compared with a normal value of 6·7, whilst the action potentials revealed no after-depolarization; in fact, an after-hyperpolarization was usually observed (Gage & Eisenberg, 1967). As we should expect, the destruction of the transverse T-system broke the link between excitation and contraction so that, although action potentials were evoked, there was no twitch.† In a com-bined physiological and electron microscopical study, Nakajima, Nakajima & Peachey (1969) showed that the altered physiological aspect (i.e., more rapid repolarization) was associated with the failure of ferritin to enter the T-system.

FURTHER MECHANICAL AND THERMAL ASPECTS OF CONTRACTION

Fenn Effect. The rather simple scheme of an undamped and a damped elastic system in series, illustrated by Fig. 758, p. 1392, accounted for many of the mechanical properties of muscle, and was widely accepted for a long time;

---

* Swelling of the T-system in hypertonic sucrose solution should increase the capacity, C_e, of Falk & Fatt's equivalent circuit, since this is a function of membrane area; Freygang, Rapoport & Peachey (1967) found that the measured effect corresponded with that predicted on the basis of the electron-microscopical changes; however, the resistance, R_e, was not decreased, as would be expected. Sperelakis & Schneider (1968) found that membrane resistance decreased in hypertonic Cl^--free solutions, whilst there was a small increase in Cl^--containing hypertonic media. Dydynska & Wilkie (1963) found that the osmotic behaviour of muscle in hypertonic solutions suggested the presence of some sucrose within the muscle fibres, the sucrose-space being so much greater than the histologically observed extracellular space; Huxley, Page & Wilkie, in an appendix to their paper, described the swelling of the T-system, and shrinking of the sarcoplasmic reticulum, observed in the electron microscope.

† Gage & Eisenberg were able to record end-plate potentials under these conditions.

however, Fenn pointed out the true implications of the model and showed that they were inconsistent with some of his findings. Thus the model envisaged contraction as a primary shortening of the damped elements (the lower spring in Fig. 758, II) which, pulling against the elastic elements, stored their contractile energy as potential energy, and it was these elastic elements that were considered to do the mechanical work by subsequently shortening. On this basis, therefore, the muscle, on stimulation, developed a given amount of heat, and a given amount of potential energy in the elastic elements, both these quantities varying with the length of the fibres of the muscle. The amount of this elastic energy that could be recovered as mechanical work would depend on the art of the experimenter with no relation to the energy liberated. Fenn, in 1923, showed that the amount of actual energy liberated by a muscle (*i.e.*, heat plus external work) was not constant for any given initial condition of the muscle, but varied with the load, a large load resulting in the liberation of more energy than a small load. Thus, in some way, the muscle can adjust its output of energy to the task that is to be performed, a property quite foreign to the Levin-Wyman model in its simple form. As Fenn aptly remarked in 1924, we may think of the muscle lifting a weight either by (*a*) doing work on a spring and then allowing the latter to shorten and lift the weight, or (*b*) by raising the weight by a chain and windlass; each link, as it is wound up, requiring extra energy *at the moment of winding*; the *Fenn effect* disposes of the first alternative. In 1935 Fenn & Marsh re-emphasized the inadequacy of the spring model and showed that, under conditions of isotonic contraction, the rate of shortening should vary linearly with the load if the muscle did indeed behave in the simple manner demanded by the model. For example, in an isotonic contraction, the undamped element would be stretched until it just began to lift the weight, and after this it would retain a constant length and constant force to the end of contraction. The speed of contraction would thus be determined, for this particular load, by the viscous elements. The same would be true of other loads so that, if the model is correct, we must expect the speed of contraction to vary with the load in the manner predicted of a simple damped system, *i.e.*, it should follow the equation: $P = P_o - kv$, where P is the load, or tension developed, v is the velocity of contraction and P_o the isometric tension; Fenn & Marsh showed that the relationship was not linear, but could be described by the equation:—

$$P = P_o e^{-av} - kv$$

Force-Velocity Relationship

With small loads the velocity of contraction is high; as the load increases, the velocity falls off, but not so rapidly as would be expected for the Levin-Wyman model; as Fenn points out, such a relationship would be given by a system in which the energy made available for contraction increased as the load increased. This point will be considered further in the light of studies on heat production during contraction; for the present, we may note that Levin & Wyman's picture of contractile elements in series with elastic elements is the one that is currently accepted, the essential difference between the system envisaged by Levin & Wyman and that accepted today being that the damped spring, representing the contractile elements, which would give a linear relationship between force and velocity, is replaced by contractile elements which, in this respect, follow a "characteristic equation" rather than a linear one, *i.e.*, they follow the exponential relationship between force and velocity given by Fenn, or the rather simpler hyperbolic relationship brought out by Hill's studies (p. 1428).

Thermal Aspects of Contraction

Since the transformations of energy are associated with the liberation of heat, the course of heat production during contraction and relaxation can provide some idea of the timing of chemical events during the process. The hope has frequently been expressed that such thermodynamic studies will also provide useful clues as to the intimate mechanisms involved; it must be emphasized, however, that unless the exact nature of the chemical and physical processes is already understood, the true interpretation of the time-course of events is impossible. Muscular contraction requires the expenditure of more energy than the normal basal requirements, and this therefore requires the liberation of energy bound up in chemical substances, *e.g.*, adenosine triphosphate, phosphagen, etc., capable of reacting to produce energy in the required forms; this liberation will be associated with the production of heat, which is to be regarded as a necessary waste; but unless the efficiencies of the various metabolic steps are well understood, *i.e.*, the amounts of waste heat in proportion to the total amount of energy made available, it is impossible, from purely thermal measurements, to estimate the exact course of events during contraction and relaxation. Thus contraction is probably the result of the sliding of myofilaments, together with the stretching of the elastic components of the muscle—sarcolemma, elastic connective tissue, etc. These may be regarded as purely mechanical events, which may be associated either with the absorption of heat, *i.e.*, cooling, as for example when rubber contracts, or the liberation of heat, as in a frictional process. The actual rearrangement of the myosin linkages, associated with sliding, may require a preliminary supply of chemical energy, and further supplies may be required while contraction is maintained, to hold the myofilaments in their new state; if this is true, the development of tension in the muscle will be preceded and accompanied by the liberation of waste heat associated with the chemical transformations. Relaxation will undoubtedly involve further frictional and elastic recoils which will be associated with the liberation of heat, although some element of this purely mechanical aspect may be accompanied by cooling.* Relaxation presumably involves an active removal of Ca^{++} by the sarcoplasmic reticulum, a process that will be associated with evolution of heat; the removal of the Ca^{++} from a hypothetical link in the myosin-actin reaction may involve liberation or absorption of heat.

The importance of the purely mechanical aspect of heat production is well brought out by the heat production of muscle (*a*) when it is allowed to relax with the weight which it has lifted, and (*b*) when it is allowed to relax without a weight. In the former case Hill has shown that the heat production is greater by an amount equal to the potential energy of the lifted weight; it follows, therefore, that the potential energy of the weight is transformed completely into heat. The importance of the metabolic aspect, on the other hand, is illustrated by the effect of stretching the actively contracting muscle; this is associated with a diminished heat production which is best explained by a diminution in the rate of liberation of chemical energy, although the tension developed may be the same as in the shorter muscle. The thermodynamic facts, therefore,

* The thermo-elastic aspects of stretch have been discussed by Hill & Howarth (1959). When a wire, for example, is stretched, it cools; and this is due to the increased potential energy of the atoms in this state, and is equivalent to the cooling of water as its molecules are released from their lattice in the ice crystal, *i.e.*, as they pass into a state of higher potential energy. Relaxed muscle has a rubber-like thermo-elasticity, warming on stretch; this is converted to the metal wire-like elasticity when it is activated, so that the thermo-elastic effect of stretching an actively contracting muscle is to cool it.

reveal only gross changes not easily susceptible of interpretation; in fact their interpretation must rely on the particular model of the contractile process that one finally adopts. Nevertheless, the rapidity and accuracy with which they may be recorded, thanks to the experimental skill shown by A. V. Hill and Hartree in particular, have permitted certain interesting, if tentative, conclusions to be drawn.*

Heat Production During a Tetanus

Initial and Delayed Heats. The contraction of muscle, be it a twitch or a sustained tetanus, is accompanied by the liberation of excess heat, occurring, under aerobic conditions, in two principal phases, the *initial* and *delayed heats*.

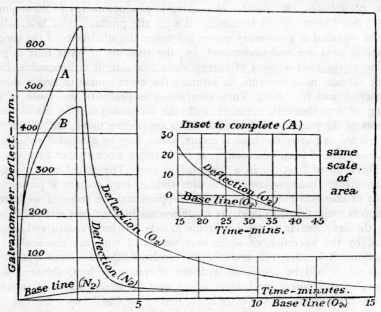

FIG. 772. Galvanometer records of heat-production of muscle in a series of twitches. A, in oxygen; B, in nitrogen. The total heat is given by the area of each curve above its appropriate base-line. *Inset:* The later part of curve A, reduced in scale five times horizontally, increased five times vertically. The muscle gave 51 twitches during the first 2 mins. 40 secs. of the recording. (Hill. *Proc. Roy. Soc.*)

The resting heat of frog muscle at $15°$ C has been given by A. V. Hill as $6·3.10^{-5}$ cal./g.sec. With tetanic stimulation, the heat production rises rapidly, as shown in Fig. 772; according to Hartree, the heat production during the maintenance of tension is in the region of 2.10^{-1} cal./g.sec. With cessation of stimulation the heat production falls abruptly to a value in the region of only $0·7$ per cent. of the initial rate, and soon falls much lower; nevertheless this

* We may mention here that a passive (unstimulated) muscle responds to a stretch by an increased O_2-consumption (see, for example, Feng, 1932). The associated heat production has been measured by Clinch (1968), who has summarized the recent literature on this subject and described some interesting observations; the effect in frog sartorius begins at $1·2$ L_o, and rises steeply with increasing stretch; potentiators of contraction, such as caffeine, nitrate, etc., increase the evolution of heat for a given degree of stretch. The beginning of the response at $1·2$ L_o corresponds to the point at which the sarcolemma takes up tension; and we may assume that this tension activates the excitation-contraction coupling mechanism; at any rate, changes, such as increased CO_2-tension, that reduce the stretch-effect accelerate relaxation.

recovery heat is so prolonged that the *total* recovery heat is of the same order of magnitude as the *initial* heat. According to Hill's later measurements, the ratio of the total energy liberated to the initial energy set free is equal to 2 under a variety of conditions of stimulation, both isometric and isotonic. Under isometric conditions, where all the initial energy appears as heat, this means that the recovery heat and initial heat are equal. When the muscle is caused to contract in N_2 the initial heat is altered very little (Fig. 772), whilst the recovery heat becomes almost zero. An analysis of the events under these conditions indicates that there is a certain quantity of delayed anaerobic heat, amounting to about 8 per cent. of the initial heat; it is liberated within about 20 sec. of cessation of a short tetanus, so that delayed, or recovery, heat of a muscle in N_2 is soon over.

Recovery Heat

The small reduction in the initial heat, on stimulation in N_2, is due entirely to the fact that the recovery heat is abolished, since during a tetanus there must be an element of recovery heat; this fact, together with the observation that a muscle may develop just as high a tension under anaerobic conditions as when O_2 is present, indicates that the energy required for the contraction of muscle is derived from non-oxidative reactions. The restoration of this energy, on the other hand, relies exclusively on oxidative processes. It will be seen later that the recovery heat is greater when the muscle does mechanical work during its contraction than when it does none, in fact the heat is greater by an amount equal to the thermal equivalent of this work. The recovery heat is the outward manifestation of a metabolic process that restores energy to the muscle; the heat itself is so much lost energy, so that the *recovery energy required* must be at least equal to this extra heat production plus the energy involved in the work done; since the heat production involved in doing work is found to be equal to the work done, the extra recovery energy, necessary when the muscle does work, is at least twice the extra recovery heat. Consequently we may tentatively assume that, when a certain amount of *heat* is liberated by muscle in the recovery process, the actual amount of *energy* utilized is twice this; thus the rate of recovery heat is of the order of 2·8 to 6 times the resting heat, so that the rate of O_2-*consumption* must, in these circumstances, be some 5·6 to 12 times the normal resting value; calculations made by Hill indicate that such a high rate is possible under the conditions of his experiments.

Chemical Events

The energy for contraction is *ultimately* derived from the oxidation of carbohydrate—glycogen—to CO_2 and H_2O, but the fact that a muscle can contract vigorously for a time in the complete absence of O_2 indicates that the oxidative reaction is concerned with the restitution of energy derived from non-oxidative sources; this is especially apparent in the greatly increased O_2-consumption of the organism *after* vigorous muscular work has been performed, a phenomenon illustrating the concept of "oxygen-debt", elaborated by A. V. Hill. The currently accepted view of the chain of reactions is as follows:

(1) ATP————————————→ADP   + H_3PO_4
  (Adenosine triphosphate)   (Adenosine diphosphate)

(2) Restitution Process (*a*)   ADP + Phosphoryl creatine ——→ Creatine + ATP

(3) Restitution Process (*b*)   Rephosphorylation of creatine.

It is important to appreciate that, so far as is known, the breakdown of phosphoryl creatine is obligatorily connected with the resynthesis of ATP from ADP, the enzyme concerned being *creatine phosphoryl transferase.* Thus the experimental demonstration of splitting of phosphoryl creatine during muscular activity has been used as *a priori* evidence for a previous splitting of ATP to ADP.

Breakdown of ATP. The breakdown of ATP has been considered to be the event most closely connected with contraction, and it has been customary to equate the free-energy change associated with the *hydrolysis* of ATP with that made available during contraction; the amount of this has been stated to be equal to some 10 kcal/mole.* It must be appreciated that ATP presumably enters into the contractile processes as a substrate, and during this process it may well be converted to ADP, but this is almost certainly not a hydrolytic process, and in fact, as we shall see, in normal muscle no conversion may be demonstrated, so that to equate the energy required for muscular contraction with the assumed free energy of the hydrolysis of ATP is hardly justifiable.

Anaerobic Restitution

Restitution process (*b*) is endothermic and requires energy which is ultimately derived from the breakdown of glycogen. Under *anaerobic conditions* glycogen (or glucose) is "glycolysed" to lactic acid through a number of stages involving the phosphorylation of organic intermediate compounds, as we have already seen (p. 288).

During glycolysis, a part of the energy set free is restored to the creatine by reconversion to creatine phosphate. We can thus understand how, even under strictly anaerobic conditions, the muscle is able to function normally; lactic acid, however, accumulates and causes an early onset of fatigue; and complete recovery from anaerobic conditions involves a further restitution process, namely the oxidation of part of the lactic acid to CO_2 and H_2O, and the reconversion of the remainder to glycogen. Restitution process (*b*) may be described as:—

$$C_6H_{10}O_5 + H_2O \rightarrow 2\ C_3H_6O_3$$
$$\text{Glycogen unit} \qquad\qquad\qquad \text{Lactic acid}$$

during the course of which creatine is rephosphorylated; whilst subsequent restitution consists in:—

$$1/5 \left\{ 2\ C_3H_6O_3 + 6\ O_2 \rightarrow 6\ CO_2 + 6\ H_2O \right\}$$

$$4/5 \left\{ 2\ C_3H_6O_3 \rightarrow C_6H_{10}O_5 + H_2O \right\}$$

Some of the energy of oxidation of lactic acid appears as heat, the remainder being used in the endothermic re-synthesis of glycogen. The cycle is now complete, the heat and mechanical work performed being equivalent to the heat of combustion of the glycogen that has disappeared.

Aerobic Restitution

Normally, however, muscular contraction is carried out under adequately aerobic conditions, in which case lactic acid is not formed; the process of glycolysis takes place only up to the stage of formation of pyruvate but then this compound, instead of being dehydrogenated by lactic dehydrogenase, is oxidized by the series of reactions known as the citric acid-cycle, a part of the

---

* Early estimates of the amount of energy available ranged between 10 and 12 kcal per mole of ATP; Podolsky & Morales (1956) have measured the total heat (enthalpy) and found a much lower value, namely, 4·7 kcal/mole, so that the available free energy would be rather less than this.

energy made available by these oxidative reactions being utilized in the synthesis of ATP in a similar but more generous fashion to that occurring in glycolysis. In this connection the high concentration of mitochondria in the type of muscle that is involved in sustained postural activity acquires significance. As Harman and Edwards & Ruska have pointed out, this sustained activity will require a highly efficient supply of energy, one that could be provided by the oxidative breakdown of carbohydrate, by contrast with the less efficient anaerobic glycolysis. This oxidative activity is mediated by the enzymes contained in the mitochondria which, in this type of muscle, are arranged in close relationship with the myofibrils. In the pale type of muscle, on the other hand, it may well be that the less efficient glycolytic system will suffice; the enzymes belonging to this system are apparently distributed throughout the sarcoplasm, so that no special accumulation of particulate matter is necessary for this supply of energy.

Quantitative Studies on Chemical Reactions. The development of analytical techniques, whereby a muscle may be rapidly frozen immediately after contraction and its content in various phosphate fractions—ATP, ADP, phosphoryl creatine, etc.—determined and compared with control muscles, has permitted the comparison of the work done with the chemical changes taking place at the time of, or immediately after, contraction. Carlson & Siger, for example, employed a muscle poisoned with a low concentration of iodoacetic acid and caused to contract in an atmosphere of nitrogen. Under these conditions they showed that the creatine phosphoryl transferase system was not poisoned, so that any breakdown of ATP would be associated with its resynthesis, accompanied by the breakdown of phosphoryl creatine to creatine and phosphate. Replacement of this lost phosphoryl creatine would be prevented by the iodoacetate, which inhibited anaerobic glycolysis; similarly, the anaerobic conditions would prevent resynthesis of ATP more directly. Carlson & Siger measured the changes in phosphoryl creatine content with increasing numbers of twitches, and plotted these changes against the number of twitches; by extrapolation they obtained a value of 0·286 μmole of phosphoryl creatine split per twitch (isometric). This splitting is compatible with the activation heat derived from thermal measurements, in the sense that the free-energy change would be adequate to provide this amount of heat. The concentration of ATP did not alter, but this could have been due to the resynthesis from phosphoryl creatine; thus, when the level of this fell to 25 per cent. of its resting value, the ATP level did begin to fall, an effect predictable from the experimentally determined equilibrium constant for the transphosphorylation reaction. Later studies (Carlson, Hardy & Wilkie, 1963, 1967; Maréchal & Mommaerts, 1963; Sandberg & Carlson, 1966) have amply confirmed the relation between liberated heat and the splitting of phosphoryl creatine, so that an equivalence of some 10–11 kcal of measured thermal and mechanical energy and 1 mole of phosphoryl creatine has been established (Fig. 773). If the only net chemical change under these conditions, during a complete cycle of contraction and relaxation, is the conversion of phosphoryl creatine to creatine and phosphate, this correspondence, whilst it does not prove the involvement of a hydrolytic reaction in the process, does at least show that splitting of phosphoryl creatine may be used as an index of the work done, and heat evolved, in muscular contraction.

Length-Tension Diagram

When splitting of phosphoryl creatine was used as an index of energy liberation, an equivalent length-energy diagram could be determined (Sandberg & Carlson, 1966), the maximum occurring at rather shorter lengths than that for

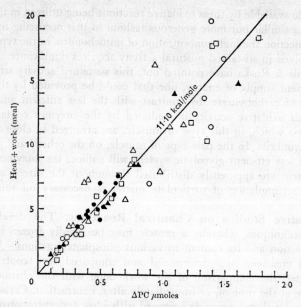

FIG. 773. The Heat + Work (Enthalpy) of muscle plotted against the splitting of phosphoryl creatine (△ PC). The different symbols refer to different modes of contraction. (Wilkie. *J. Physiol.*)

tension, and the energies falling off less rapidly at shorter and greater lengths; from an analysis of the curve, Sandberg & Carlson deduced that a certain amount of splitting of phosphoryl creatine occurred independently of the development of tension, so that at zero tension some 27 per cent. of the total, measured at optimum length, still occurred. This remained constant at different lengths, so that when this tension-independent amount was subtracted at each length, a corrected length-splitting curve was obtained that coincided reasonably well with the length-tension curve (Fig. 774).

The Immediate Source of Energy. These studies of Carlson, Mommaerts

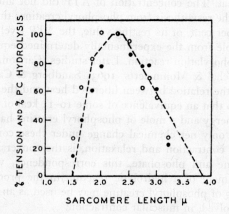

FIG. 774. Showing that the tension-sarcomere length curve may also be represented as a phosphoryl creatine split-sarcomere length curve. Solid circles indicate hydrolysis of PC as percent of that at resting length. Open circles, tension as percent of that at L_0, taken from Gordon *et al.* (1966). (Sandberg & Carlson. *Biochem. Z.*)

and Wilkie have obviously opened the way to biochemical studies that should provide the link between the thermal and mechanical events, studied by Hill, and the chemical changes taking place in the muscle. Earlier attempts to demonstrate changes in ATP content of muscle during contraction failed, however; this could have been due to the too rapid reconversion of ADP to ATP, or because ATP is *not* converted to ADP during contraction or relaxation in a metabolically normal muscle.*

FDNB-POISONED MUSCLE

When synthesis of ATP in muscle is blocked by treatment with fluorodinitro-benzene (FDNB),† which specifically inhibits the creatine-ATP phosphoryl transferase (Infante & Davies, 1965), it would seem that the only metabolic source of energy for muscle during the first three seconds of stimulation is derived directly from ATP, since synthesis of ATP through glycolysis does not become significant until then. Under these conditions, stimulation of muscle caused a loss of ATP; at slow rates of stimulation this was equivalent to that expected if ATP were involved in the sarcoplasmic reticulum Ca^{++}-pump, but at faster rates this was less per impulse. The activation heat of muscle is much greater than the thermal equivalent of the ATP used, on the assumption that the enthalpy of hydrolysis is 10 kcal/mole; the heat associated with shortening was likewise much greater than the thermal equivalent of the extra ATP split during this process, so that Davies, Kushmerick & Larson (1967) concluded that shortening heat was not the degraded free-energy of ATP.

ATP AND RELAXATION

Mommaerts & Wallner (1967) have examined the claim that some ATP is used up during relaxation as well as contraction, comparing FDNB-poisoned muscles, frozen at the height of contraction, in a tetanus and after relaxation; they found no difference in the ATP-contents of the two groups of muscles, so that we must conclude that, in a tetanus, the utilization of ATP in re-accumulation of Ca^{++} by the sarcoplasmic reticulum takes place very rapidly after the development of the active state following each stimulus.

ATP-CONTENT AND TENSION

Murphy (1966) used FDNB poisoning to control the ATP content of the muscle, stimulating it for variable times and thus depleting the muscle to different extents. He estimated the contractile power of the muscle by measuring its maximal tetanic isometric tension, and found this to be proportional to the logarithm of the ATP content; the degree of association of the contractile elements, measured by the tension developed when the muscle was quickly stretched passively, was inversely proportional to the logarithm of ATP content. This conforms with expectation, since model experiments have shown that ATP not only promotes the interaction of myosin and actin that leads to development of tension, but it also, in appropriate concentration, prevents a

* Carlson & Siger and Mommaerts never found any significant change in ATP concentration as a result of brief stimulation; studies with ^{32}P-labelled ATP failed to reveal any redistribution of the label with ADP (Fleckenstein, Janke & Davies, 1956; Dixon & Sacks, 1958). We may note that there is evidence of a structurally bound phosphate in rectus abdominis muscle; if labelled with ^{32}P there is a loss of label on contracture (Cheesman & Hilton, 1966); this phosphate is probably different from the ethanol-insoluble ADP fraction of muscle described by Janke (1968); it is probably structurally bound, but shows little exchange with labelled phosphate, by contrast with the rapid labelling of Cheesman & Hilton's fraction.

† The changes in organic phosphate content of muscles treated with FDNB have been examined in some detail by Maréchal & Beckers-Bleukx (1966); because of the myokinase activity, any ADP formed is converted to ATP and AMP, and the latter is deaminated to IMP. There would appear to be a reduction in the ATP-cost during a maintained tetanus in the poisoned muscle.

firm linkage between the two, such as that found in rigor mortis when the ATP content of muscle has fallen to a low value (p. 1440).

LOCALIZATION OF ORGANIC PHOSPHATES

By injecting a frog with tritium-labelled creatine and causing the creatine phosphate to be precipitated *in situ* by treating the sectioned fibre with lanthanum, D. K. Hill showed that some two-thirds of the creatine phosphate, as revealed by radioautography, was concentrated in narrow bands on either side of the Z-line, the remainder being evenly distributed over the A-band. Labelled adenine nucleotides, precipitated in the same way, were concentrated in a narrow band inside the I-band near the A-I boundary.

Delayed Heat. As we have indicated earlier, delayed heat is undoubtedly connected with restitution processes, but since all the ATP in muscle would probably be exhausted within o·5 sec. of stimulation if it were not replaced, we must conclude that some of the restitution processes take place in the initial phases of heat production. We may note that D. K. Hill has developed a method for the rapid determination of the course of oxygen consumption by stimulated muscle; he has found that the oxygen consumption occurs entirely after activity, and its time-course follows closely that of the oxidative delayed heat, *i.e.*, the part of the delayed heat that is missing in the absence of oxygen.

Further Studies on the Thermal Events

Analysis of Initial Heat. The development of rapid methods of heat measurement has permitted a more exact analysis of the time-course of the thermal changes during the period of evolution of initial heat. In general, the

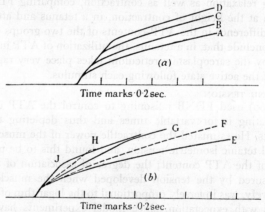

FIG. 775. Heat production during isotonic shortening from the start. Tetanus at 0° C; time, 0·2 sec. (*a*) Shortening different distances under constant load of 1·9 g. A, isometric; B, 3·4 mm.; C, 6·5 mm.; D, 9·6 mm. (*b*) Shortening constant distance of 6·5 mm. under different loads. E, isometric; F, 31·9 g.; G, 23·7 g.; H, 12·8 g.; J, 1·9 g. (Hill. *Proc. Roy. Soc.*)

period has been divided into a *shortening heat*, a *maintenance heat*, and a *relaxation heat*. Shortening and maintenance heats are well illustrated by Fig. 775; with a constant load on the muscle, the magnitude of the steep rise (the shortening heat) increases with the degree of shortening permitted by the experimental set-up. The maintenance heat is represented by the constant-slope portion of the curve; since the weight lifted was the same in all cases the curves run approximately parallel (and parallel also to the isometric curve, A). It will be noted that even in isometric contraction there is a significant "shortening heat"; this may be said to represent the shortening of the contractile component of the muscle, stretching the elastic elements, the tendon and the lever itself. Thus, even if no shortening at all were permitted, *i.e.*, if the contraction were truely isometric, the contractile elements could shorten against inert elastic elements; only in this way, moreover, can we explain the fact that tension does not develop immediately, in isometric contraction, to its full value. The maintenance heat, represented by the slope of the curve after the initial steep rise, remains remarkably constant, and is not greatly influenced by the length of the muscle.

Activation Heat

Analysis of the thermal events in a single twitch, as opposed to a tetanus, showed an initial outburst, before shortening had begun, followed by a constant evolution of heat running parallel with shortening (Fig. 776); the initial outburst may be called the *activation heat*, being the manifestation of the chemical events corresponding to the development of the active state; the later phase has been described as the shortening heat, as before. In a tetanus the summated activation heats represent the maintenance heat.

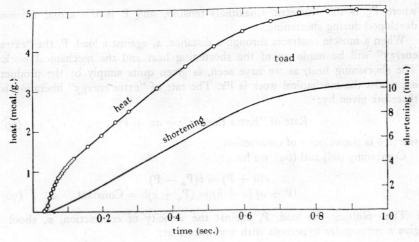

FIG. 776. Heat and shortening, simultaneously recorded in the isotonic twitch of a pair of toad's semi-membranosus muscles at 0° C. under 6 g. load. (Hill. *Proc. Roy. Soc.*)

Relaxation Heat

The magnitude of the relaxation heat depends greatly on whether the muscle is permitted to lower its load; if so, and the load is great, the curve shows a rapid up-stroke; if the muscle has been maintained in the isometric state, on the other hand, only a small hump appears which is probably the mechanical effect of the release of the elastic element.* The relaxation heat may therefore be classed primarily as the degradation of potential energy in the extensile system of the muscle; metabolic events are, of course, taking place at the same time, but in this region of the curve they appear to be neither exothermic nor endothermic.

* This hump on the heat curve during relaxation from an isometric contraction has been examined by Hill (1961b), and is largely explainable on the basis of the thermo-elastic effect, *i.e.*, the absorption of heat when an elastic body is stretched, and the evolution of heat when it is relaxed. That the thermo-elastic effect may result in measurable heat absorption during the development of tension has been demonstrated very elegantly in a series of experiments by Woledge (1961).

Shortening Heat

Hill's studies on the tetanus, and later on the single twitch, indicated that the shortening heat was a simple linear function of the shortening:

$$\text{Shortening Heat} = a.x$$

where x is the shortening. The constant, a, moreover, is related to the maximum tension the muscle can develop (the isometric tetanic tension, P_o), so that the ratio:

$$a/P_o = \text{constant.}$$

Extra-Energy Liberation. The relative constancy of the maintenance heat permits the introduction of the concept of *extra-energy liberation* associated with shortening; thus during an isometric tetanus no work is done; the heat produced is maintenance heat, and the "extra-energy" is zero. Hill found experimentally that, under a variety of conditions of shortening, the rate of liberation of "extra-energy" was a linear function of the tension developed, *i.e.*,

$$\text{Rate of "Extra Energy"} = b(P_o - P) \quad . \quad . \quad . \quad . \quad (68)$$

where P_o is the isometric (maximal) tension, and P is the actual tension developed during shortening.

When a muscle contracts through a distance, x, against a load, P, the "extra energy" will be made up of the shortening heat and the mechanical work. The shortening heat, as we have seen, is given quite simply by the product ax, whilst the mechanical work is Px. The rate of "extra energy" liberation is therefore given by:

$$\text{Rate of "Extra Energy"} = av + Pv \quad . \quad . \quad . \quad . \quad . \quad (69)$$

where v is the velocity of contraction.

Combining (68) and (69), we have:

$$v(a + P) = b(P_o - P)$$

or,
$$(P + a)(v + b) = (P_o + a)b = \text{Constant} . \quad . \quad . \quad (70)$$

Thus plotting the load, P, against the velocity of contraction, v, should give a rectangular hyperbola with asymptotes at:

$$P = -a, \text{ and } v = -b$$

Hence a, a thermal constant of the muscle, may be obtained from purely mechanical measurements (Fig. 778, p. 1431).

Force-Velocity Relationship. This relationship between tension developed and velocity of contraction is known as the *Force-Velocity Relationship*. Essentially it is a statement of the fact that the more rapidly a muscle is allowed to shorten the less is the force developed. Since the force developed must depend on some energy-giving process, this means that the rate of movement of the myofilaments in relation to each other in some way influences the supply of contractile energy. This point was emphasized by Podolsky, who expressed the experimental basis of the Fenn effect by Fig. 777. Thus, if the total heat of shortening is independent of contraction speed, the rate of heat production during shortening must increase linearly with speed. Hence on plotting the energy-flux represented by this shortening heat against velocity of shortening we have the straight line. If the muscle is doing work we must add this to obtain the total energy-flux. The rate of work production depends too, on speed, as shown by the force-velocity equation; thus it may be calculated and added to the heat to give the curved line; we may note that at zero velocity and maximum velocity the work term is zero. The curve shows that the energy liberated increases with the velocity of shortening; this means that the chemical reaction linked to the contractile mechanism must increase with speed in this way too.*

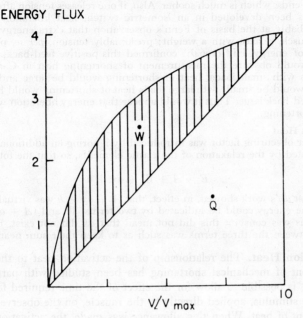

FIG. 777. Relation between energy-flux and velocity in living muscle. Open region, rate of heat production; shaded region, rate of work production. The unit for the ordinate is the rate of heat production during isometric contraction (the maintenance heat). (Podolsky. "Biophysics of Physiological and Pharmacological Actions." *Amer. Ass. Adv. Sci.*)

Non-Linear Relation of Heat to Shortening. According to Hill's formulation, the total energy liberated in a contraction, be it a twitch or a tetanus, is given by:

$$E = A + W + ax$$

where A is the activation or maintenance heat, W is the work, and x the shortening. Carlson *et al.*, in their study of the thermal and chemical events during contraction (p. 1423), were able to formulate the liberated energy entirely in terms of A and W, and concluded that Hill's shortening heat was non-existent. They pointed out, moreover, that on Hill's formulation, the maximal energy output would occur with very small loads (where shortening was large), whereas on the basis of the simpler formulation the maximum energy would be with medium-sized loads, as originally found by Fenn, and as they found when energy liberation was equated with the splitting of phosphoryl creatine. Hill's equation, however, applies to the contractile phase, whereas the studies of Carlson *et al.* involved the complete cycle of contraction and relaxation. Hill (1964a) showed that shortening heat did, indeed, exist, but in these more precise experiments he found that the relation was not linear with shortening, so that the "constant", which must now be indicated by α to distinguish it from the mechanical parameter, a, was a function of tension and given by:

$$\alpha/P_o = 0.16 + 0.18\, P/P_o.$$

* The optimum load for efficiency, *i.e.*, for performance of maximal work with minimal energy consumption, is given by $0.45\, P_o$; the optimum load for development of power is smaller, namely $0.3\, P_o$ (Hill, 1964d).

Positive Tension Feed-Back

The cause of the discrepancy between Carlson *et al.*'s and Hill's work probably resides in the effect of tension on prolonging the activation heat, a form of positive feed-back (Hill, 1964b). Thus it has been known for a long time that, during an isometric twitch, the heat liberation continues until tension has disappeared, whereas with an isotonic twitch, with small load, heat liberation ends when shortening ends, which is much sooner. Also, if one releases tension after maximum tension has been developed in an isometric twitch, heat liberation ceases, and this is probably at the basis of Fenn's observation that extra energy is liberated when a muscle relaxes with a weight; presumably tension, *per se*, promotes the liberation of energy. Hill (1964b) confirmed this positive feed-back, and showed that this would obscure the measurement of shortening heat in Carlson *et al.*'s work; thus with small loads, heat of shortening would be large and so positive feed-back would be small; with large loads heat of shortening would be small and positive feed-back large. Thus it would appear that energy liberation was independent of shortening.

Relaxation Heat

A further obscuring factor was introduced by ignoring an additional heat term, h, represented by the relaxation of the elastic elements, so that the total energy is given by:

$$E = A + W + ax + h$$

Carlson *et al.*'s work showed, in effect, that $A + ax + h$ was virtually constant, *i.e.*, that the energy could be indicated by two terms, W and $(A + ax + h)$, but because this was constant this did not mean that ax did not exist, but that the relation between the three terms was such as to keep their sum nearly constant.

Activation Heat. The relationship of the activation heat to the course of development of mechanical shortening has been studied with particular care by Hill. It is possible to allow for the effect of the time required for propagation of the stimulus, applied directly to the muscle, on the observed course of development of heat. When this allowance was made, the activation heat of a muscle, theoretically stimulated at all points simultaneously, was shown to begin some 10 msec. after the stimulus (in the frog at $0°$ C; 25 msec. in the case of the toad); moreover, the rate of liberation of activation heat apparently starts at its maximum, and then declines to zero as the shortening heat becomes manifest. The delay in development of heat may possibly be related to the time required for the action potential to reach its maximum, which would be of the order of 7 msec. (Eccles, Katz & Kuffler, 1941); thus, as Hill expresses it, the stimulus has a trigger action on some chemical process which develops its maximal rate at the beginning; the heat developed is the sign of this chemical process, which puts the muscle in a state of activity that may be regarded as a readiness to shorten, to exert a mechanical force, or to do mechanical work. The actual proof that the liberation of heat preceded the first signs of mechanical activity is not easy in normal muscle because of the instrumental delays in recording heat by comparison with tension; by exploiting Howarth's discovery that treatment of a muscle with hypertonic Ringer's solution delays the onset of contraction, Hill (1958) was able to prove that, under these conditions at any rate, the evolution of heat began definitely before any signs of altered tension.*

The All-or-None Principle. The effects of both tetanic and single-shock stimulation have shown that the response, in terms of energy liberation, is not

---

* Gibbs, Ricchiuti & Mommaerts (1966) have developed a simple method of determining activation heat, based on the extra heat liberated by a second stimulus; thus, if we take as the base-line the heat produced by the first twitch, a second stimulus falling within the fusion interval will produce an increment of the heat which will increase with the interval between stimuli to reach a plateau as a more and more complete occurrence of the full activation cycle is allowed. Under appropriate conditions the extra heat at the plateau is equal to the activation heat. Nitrate increased this activation heat.

all-or-none. If shortening is kept constant, however, we may expect a reasonably constant liberation of energy. The following figures, derived from experiments on the special ergometer which allowed variations in work with the same amount of shortening, reveal the remarkable constancy of the heat under these conditions:—

Work/Heat (per cent.).	6	12	16	19	22	25	27	27	30	32	35
Heat (per cent.) . .	100	99	99	100	99	93	98	103	100	95	102

where the upper row indicates the percentage of the total energy appearing as mechanical work, and the lower row represents the heat liberated expressed as a percentage of that with the smallest load. So far as the all-or-none principle is concerned, therefore, it is only the activation heat that may be regarded as an event independent of the mechanical conditions of contraction, but even here, as we have seen, the positive feed-back ensures that activation heat increases with prolongation of the tension, and thus with the magnitude of the load.

Shape of Force-Velocity Curve. The fact that the same parameters can be derived from heat and mechanical measurements means that the shape of the

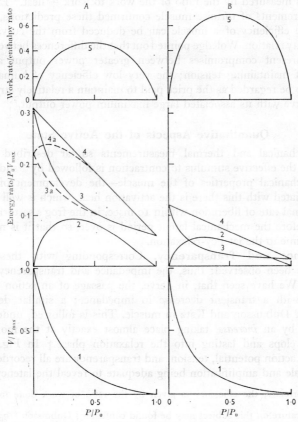

Fig. 778. Diagrammatic comparison of the properties of frog muscle (A) with those predicted for tortoise muscle (B). The lines show how velocity (1), work rate (2), heat rate (3), work rate + heat rate (4), and work rate/(work rate + heat rate) (5), vary with the load on the muscle.

A. Frog muscle: summary of the results of Hill (1938: continuous lines) and of Hill (1964: interrupted lines).

B. Tortoise muscle: Line 1 from the results of Katz (1939). Lines 2, 3, 4 and 5 are predictions, on the basis of Katz' results. (Woledge. *J. Physiol.*)

force-velocity curve, from which these parameters may be derived, is closely related to the heat production of a given muscle, and Woledge (1968) has put this thesis to an interesting experimental test by comparing the mechanical and thermal features of contraction in frog and tortoise muscle. Fig. 778 illustrates the force-velocity curves of the two muscles (A1 and B1), that of the tortoise being steeper. Curves A2, 3 and 4 represent the experimentally measured variations of heat rate (2), work rate (3) and work rate + heat rate (4) as functions of tension development whilst curve 5 is essentially a measure of the efficiency of the muscle, being the ratio of the rate of work and the rate of energy liberation. The corresponding curves for the tortoise muscle have been computed on the basis of the measured force-velocity curve (1) and on the assumption that the heat production follows the simple relation described by Hill (1938).

Predicted Efficiency

The most interesting feature of this analysis is that heat production—both "shortening" and "maintenance"—should be much less than in frog muscle whilst the efficiency, indicated by curve 5, should be much greater if the efficiency is measured by the ratio of the work to work + heat.* Experimental heat measurements of tortoise muscle confirmed these predictions, so that, in general, the efficiency of a muscle can be deduced from the curvature of its force-velocity relation. Woledge pointed out that the differences between different species represent compromises between greater power output and greater economy of maintaining tension; the very low efficiency of human and frog muscle is to be regarded as the price paid to maintain a relatively straight force-velocity curve with its associated large maximum power output.

Quantitative Aspects of the Active State

The mechanical and thermal measurements so far described have suggested that the effective stimulus to contraction is followed by a sudden change in the mechanical properties of the muscle—the development of the *active state*; associated with this there is the activation heat, which is well on the way to its maximal rate of liberation within 10 msec. in the frog's sartorius at 0° C, *i.e.*, well before the mechanical latent period is over, so that it is more closely related in time to the latency-relaxation.

Impedance and Transparency. Corresponding with these changes, others have been observed; thus, the impedance and transparency of muscle both alter. We have seen that, in nerve, the passage of an action potential is associated with a transient decrease in impedance; a similar decrease was observed by Dubuisson and Katz in muscle. This is followed, under isometric conditions, by an *increase*, taking place almost exactly at the moment when tension develops and lasting into the relaxation phase.† In Fig. 779, from D. K. Hill, action potential, tension, and transparency are all recorded together, the time scale and amplification being adequate to reveal the latency-relaxation

---

* More correctly, the efficiency is given by the ratio of work done to free-energy liberated.

† The literature on this subject may be found confusing; Dubuisson (1935) originally described two falls, an initial rapid one presumably associated with the action potential, and a slower one associated with contraction; Bozler described only an increase in impedance and later Dubuisson (1937) showed that, under strictly isometric conditions, the initial fall is, indeed, followed by a rise. Katz observed a fall in the end-plate region associated with the end-plate and action potentials; in his review of the subject Dubuisson (1950) describes an initial fall—the a-wave—followed by two rises, b_1 and b_2. There seems no doubt that the initial fall in impedance is associated with passage of the excitation wave.

of Sandow; corresponding with the latter there is an increase in transparency which passes through a maximum as the tension begins to develop. A study of the characteristics of the early increase in transparency, and the later decrease, suggests that they have different origins, the former being probably due to a decreased absorption of light whilst the latter is a scattering phenomenon.

Pressure. The effects of hydrostatic pressure on muscle also reveal the occurrence of some change. Exposing a muscle to a uniform hydrostatic pressure of 272 atmospheres for a fraction of a second causes a typical twitch, differing from that obtained by the electrical stimulation of muscle or its nerve by the complete absence of any wave of action potential. Moreover, it was

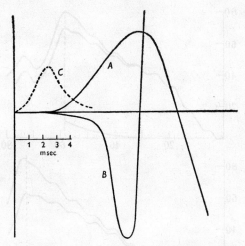

Fig. 779. Changes in opacity (A), and mechanical response (B), on stimulation of muscle. (C) shows course of action potential, not recorded, but as given by Katz. (D. K. Hill. *J. Physiol.*)

found that a pressure-stimulus was capable of summating its effect with that of an electrical stimulus. The pressure-stimulus was unaffected by the refractory state of the muscle following electrical stimulation, so that the effects of summation could be studied with extremely short intervals between the electrical and pressure-stimuli. Brown showed that a maximal effect was obtained with an interval of about 3·5 msec., and the limiting period for augmentation was 5 msec. A rise in hydrostatic pressure prolongs the active state of muscle, and by measuring the effects of pressure-stimuli, following an original electrical stimulus at different intervals, Brown was, in effect, measuring the duration of the active, or *alpha* state as he called it.

Duration of Active State

Onset. Whatever the actual state of the contractile machinery during activity may be, it is of interest to delimit its time-course with precision, and some experiments of several investigators, devoted to this point, may now be mentioned. Hill's precise studies of the effects of quick stretches indicated that the active state was fully developed during the latent period of frog or toad muscle, but even at 0° C these muscles were too rapid to permit measuring the actual course of development of inextensibility. By studying tortoise muscle, which has a speed only one tenth to one twentieth that of frog muscle, this difficulty could be overcome. Unfortunately, stretching the unstimulated tortoise muscle produces a considerable rise in tension that only slowly falls, by contrast with frog and toad muscle where negligible changes in tension are caused by the small stretches necessary for these studies. It was therefore necessary to compare the effects of stretches applied to stimulated and un-stimulated muscles by superimposing the records obtained in the two cases. Fig. 780 illustrates a pair of superimposed records obtained (a) by stretching a muscle immediately after application of a shock and (b) without any previous shock. The curves of tension begin to deviate some 30 msec. after the shock; this means that before 30 msec. the shock had had no measurable influence on the elastic properties of the muscle, the stretch-response being the same.

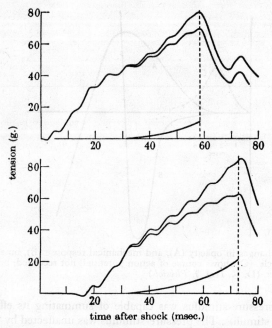

Fig. 780. Effect of stretching tortoise muscle. Lower curve of each pair shows the tension developed when the unstimulated muscle is stretched, whilst the upper curves show tension developed when the stretch follows a maximal stimulus occurring at zero time. The small curves at the bottom indicate the difference in tension due to the shock. (Hill. *Proc. Roy. Soc.*)

Thus, the beginning of the active state occurs some 30 msec. after the stimulus, which is about half way through the latent period of this muscle. A study of heat production in tortoise muscle (Hill, 1950) showed that this, too, began after the first 60 per cent. of the latent period. As Hill pointed out, the latency-relaxation and change of transparency also begin at this point.

   Plateau. So much for the beginning; what about the duration of the plateau of maximal activity, and the time-course of its decay? The duration of the plateau has been measured by several methods. For example, if the curves showing the development of tension with time for (a) an isometric twitch, and (b) an isometric tetanus are superimposed, the curve for the twitch will deviate from that for the tetanus at the moment when the maximally active state begins to decay. With this method, Macpherson & Wilkie found a duration for the fully active state in frog sartorius at o° C of 44 ± 5 msec. It follows from the above consideration that the minimum frequency of stimulation, necessary to achieve a smooth tetanus, will be determined by the duration of the maximally active state; Ritchie (1954) found this to be 40/sec., giving a period of 25 msec. for the duration of maximal activity; this is considerably shorter than the earlier estimate, but is probably nearer to being correct since it was found that, on cessation of a tetanus, the time elapsing till a fall in tension could be recorded was 35 msec. In a tetanus the elastic elements are fully stretched; on cessation of the stimuli, the active state following the last stimulus will be pulling against fully stretched elements; as soon as maximal activity falls off this will be reflected as a change in tension. The time of 35 msec. included the latent period between the electrical stimulus and the earliest moment at which tension develops; this was found to be 10 msec. under conditions where the record was not obscured by latency relaxation, i.e., at short initial length (p. 1388). Thus the duration of maximal activity, according to this estimate, is 25 msec.

Importance of Refractory Period

In mammals, at their body temperature, the minimal separation of two shocks is much less, namely 2–4 msec (Buller & Lewis, 1965; Wilander, 1966), but Desmedt & Hainaut (1968) have argued, from their own experiments on human muscles in which they measured both spike response and tension development, that the limiting factor may well be the duration of the refractory period of the muscle's excitable membrane. Thus the second stimulus, delivered before the critical interval, fails to increase the rate of tension development beyond that in a single twitch, not because the active state generated by the previous stimulus was still maximal (i.e., impossible to increase) but simply because the second stimulus was ignored by the still irresponsive muscle membrane. When they plotted the interval between the electrical responses to two stimuli, and the corresponding interval between the mechanical responses, against the interval between stimuli, Desmedt & Hainaut found complete overlap of the curves. As these authors argue, if the development of the active state represents the liberation of Ca^{++} within the fibre, and if the intensity of the actin-myosin reaction depends on the concentration of Ca^{++}, then it is unlikely that the active state is maximal for a single twitch and that it achieves this condition instantaneously.*

Decay of Active State. Activity of the contractile machinery does not fall off abruptly; this was revealed by Gasser & Hill's quick-stretch studies. The actual time-course of decay was measured by Ritchie (1954), who utilized the principle that the degree to which the contractile elements were active was measured by the point in a twitch where the elastic elements were neither shortening nor lengthening, i.e., at the point of maximum tension of the isometric twitch. This will be clear from Fig. 760 (p. 1394), the value of P_i being equal to P_n where the curves cross. If the maximum of the tension-time curve could be made to occur at different times, clearly it would be possible

---

* The earlier studies of Buller & Lewis (1965) and Wilander (1966) have also emphasized the inadequacy of the two-stimulus estimate of the plateau of the active state; thus on this basis the minimum durations for fast and slow muscles were 2·75 and 3·3 msec. respectively, but these limits could well have been imposed by the refractory period. The minimum fusion frequencies for fast and slow muscles were about 125 and 50 pulses per sec. respectively, and at these frequencies maximal tension was developed; these figures would indicate much longer plateaux, but Buller & Lewis found that the rate of rise of tension continued to increase up to frequencies of 600 and 310 pulses per sec., and on this basis the plateaux would have durations of some 2 to 3 msec. Wilander was forced to draw a distinction between the active states for shortening and for development of tension; thus with isometric contraction, the optimum frequency was 200/sec. whilst with isotonic contraction against a small load the maximum rate of shortening was obtained with a frequency of 425/sec.

to derive a series of values of P_i, indicating the variation with time of the tension of which the muscle is capable. Ritchie varied the time for attainment of maximum tension by varying the moment, after a stimulus, at which the muscle was released, so that it shortened to a new length and then developed its isometric tension. The later this release occurred, the smaller was the tension developed, and the more slowly it happened. A series of records is illustrated in Fig. 781, and when the maximum tension was plotted against time after the stimulus, a curve of active state against time could be obtained. It was found that at 0° C in the frog's sartorius the active state lasted for some 400msec.

Series Elastic Element. So much for the contractile elements; how about the series elastic element? Does this display constant characteristics throughout the cycle of contraction and relaxation? Wilkie has devised a simple method of analysing the behaviour of this component. A muscle is arranged as in Fig. 754 (p. 1387), so that on stimulation it pulls on a lever which is prevented from

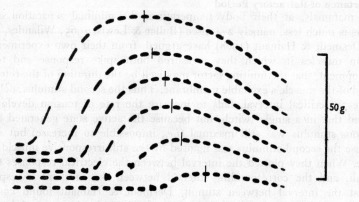

FIG. 781. Tension-time curves of frog sartorius released at increasing intervals after a single stimulus. Top record is the ordinary isometric twitch; subsequent records have been preceded by quick releases of about 3 mm. to the standard length timed at about 10, 40, 60, 100 and 175 msec. after the stimulus. The vertical bars mark the peaks of the contraction curves. (Ritchie. *J. Physiol.*)

moving by a stop. At any given moment, the stop is released by an electro-magnet, so that the muscle immediately shortens. Shortening occurs in two stages—an initial very rapid process, representing the release of the stretched elastic elements; this will continue until the muscle has acquired the length corresponding to the isotonic load, *i.e.*, until it starts to lift the load. Further shortening will now consist of shortening of the contractile elements, which is a slow process, so that the onset of this will be indicated by a break in the record. The rapid shortening thus corresponds to the adjustment of the elastic elements—initially in a highly stretched state—to the load on the muscle; if the load is large, very little adjustment will be necessary, so that the rapid shortening will be small; if the load is small, the elastic elements are much too extended; they must relax considerably before they come into equilibrium with the load, *i.e.*, the rapid shortening will be large in extent. We may thus compute the amount of extension of the elastic elements that is necessary to lift any given load, and a curve of extension against load, which is described as the stress-strain curve for the series elastic elements, may be plotted. Wilkie found that this curve was the same whether the muscle was released 100 or 600 msec. after the stimulus, so that it would seem that the series elastic element is independent of the state of the contractile mechanism, *i.e.*, it is probably not a part of the actomyosin system.

Relaxation. When a muscle contracts, lifting a load, and subsequently relaxes so that the load is dropped to its original position, Jewell & Wilkie showed that, although the muscle had returned to its original length, it had not lost all the tension it developed in this twitch. For example, if a muscle lifted a load of 9 g it returned to its original length by about half a second after the beginning of contraction; the tension, however, did not fall to zero at the end of this time but declined exponentially over a further period of half a second. This is shown by Fig. 782; the upper tracings represent the shortening–time curves when the muscle lifts 3, 5 and 9 g. respectively. The lower traces represent the tension developed under the three conditions. As the muscle returns to its original length the tension falls abruptly, and then more slowly along an exponential curve. The interesting feature revealed by these curves, and other studies, was that the initial rapid fall of tension was correlated with the increase of length during relaxation;

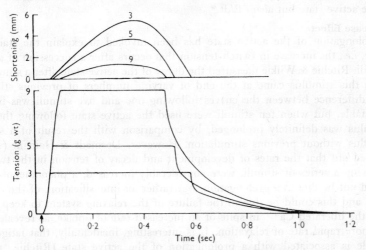

FIG. 782. Tracings of simultaneous records of the length and tension changes in a frog sartorius (30 mm, 5 mg, 0°C) during after-loaded isotonic twitches against loads of 3, 5 and 9 g wt. (Jewell & Wilkie. *J. Physiol.*)

thus, if the muscle was made to lengthen more during relaxation, by pulling on it, the initial drop was greater, whilst the slower decline followed the same exponential course, but from a lower level. Consequently, the rate of the slow decline of tension seemed to be independent of length whilst the rapid phase varied directly with the amount of lengthening. The explanation for this finding, like that of the length–tension diagram, is presumably related to the bridges formed between the contractile filaments; sudden lengthening of the muscle reduces the degree of overlap, and so reduces the tension.

Effects of Drugs and Ions. The myogram of a *tetanus*, *i.e.*, the time-course of the development of tension with repetitive stimuli, will be determined by the stress-strain relationship of the series elastic component, and by the characteristic relation between force and velocity. A change in the myogram of a *twitch* could be determined by either of these factors, but also by changes in the duration and intensity of the active state.

Effects on Intensity of Active State

In the past it has been customary to treat the intensity of the active state as something that attains a fixed height almost instantaneously and then declines, so that increases in the tension developed in a twitch have been ascribed to a prolongation of the active state, and some plausibility for this view is given by the frequent finding that, whereas twitch-tension may be increased by a drug, the tetanus-tension is unaltered. Thus the augmentation of twitch-tension by adrenaline (Goffart & Ritchie, 1952) and by tetraethylammonium (TEA), tetramethylammonium (TMA) and choline have been explained in terms of increased duration of the active state, and this may well be due to their actions in prolonging the action potential (Edwards, Ritchie & Wilkie, 1956). These effects, and those of anions such as NO_3^- and Br^-, have been discussed earlier in relation to their probable influence on the release mechanism of Ca^{++} in excitation-contraction coupling, so that we may equate prolongation of the active state with increased release of Ca^{++}, or delayed sequestration by the sarcoplasmic reticulum. It is difficult to believe, on this basis, that various potentiating factors will not also influence the *intensity* of the active state, which would be revealed by an increase in rate of rise of tension in the twitch (dP/dt); in fact, Sandow & Brust (1966) observed that caffeine not only increased duration of the active state but also dP/dt.*

Staircase Effect

Prolongation of the active state has been invoked to explain the staircase effect, *i.e.*, the increase in twitch-tension that occurs after a succession of single stimuli. Ritchie & Wilkie measured the decay of the active state after a stimulus, when this stimulus came at the end of varying numbers of previous stimuli. The difference between the curves following one and five stimuli was barely detectable, but when ten stimuli were used the active state following the last stimulus was definitely prolonged, by comparison with the result of a single stimulus without previous stimulation. However, Desmedt & Hainaut (1968) pointed out that the rates of development and decay of tension in the twitch, following a series of stimuli, were considerably increased, a phenomenon that would not be due to a prolongation, but rather an intensification, of the active state, and this could be due to the failure of the relaxing system to keep pace with the liberated Ca^{++} in spite of an increased rate of uptake, as revealed by the more rapid rate of relaxation. It is interesting, incidentally, that fatigue in muscle is associated with a prolongation of the active state (Ritchie, 1954; Gabel *et al.*, 1968).

Short-Range Elastic Component

D. K. Hill has recently described some phenomena of slow small stretches of muscle that may well have a bearing on many aspects of filament interaction. Fig. 783 shows the tension developed as a muscle is passively stretched; it will be seen that tension develops rapidly up to a plateau and then remains nearly the same while stretching continues; this second phase has the characteristics of a frictional resistance, being more or less independent of velocity. On allowing the muscle to relax passively the process reverses itself. The phenomenon is very much more marked in hypertonic solution, and thus permits accurate study under these conditions. In general, Hill concluded that the short-range elastic component represented an interaction between filaments similar to that taking place during contraction, the interaction lasting sufficiently long to produce a frictional resistance during the observed stretches. By contrast, during contraction, the life of an interfilament link must be very short since a muscle can change its length within a few msec. By an ingenious approach the effects of stimulation on the

---

* Gabel, Carson & Vance (1968) have argued that the duration of the plateau of the active state is given by the time between stimulation and the point when the second derivative of the tension, d^2P/dt^2, begins to decrease; using a tension-recording device that gave this second derivative directly, they showed a good agreement between plateau times derived on this basis and on the conventional twin-pulse method. The authors argue, further, that the third derivative will be a measure of the *intensity* of the active state; conventionally the tetanus tension is used to indicate this, and they found that tetanus tension and d^3P/dt^3 varied in the same way with temperature.

short-range elastic component could be indirectly measured; it did, in fact, become vanishingly small. On the basis of an interfilament action during the relaxed state, revealed by these experiments, we may explain latency-relaxation as an inhibition of this interaction.

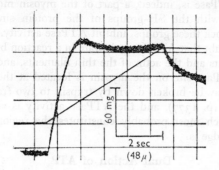

FIG. 783. The effect of a slowly applied stretch (thin line) on the development of tension (thicker oscillograph record). (D. K. Hill. *J. Physiol.*)

THE THEORY OF THE CONTRACTILE PROCESS

Morphological studies of muscle during the relaxed and contracted states have provided a convincing picture of the basic mechanical process, namely that of the sliding of the A- and I-filaments in relation to each other; it is now time to try to present some coherent picture of the possible chemical and physical events that bring about this interaction between the filaments. The theory that has found widest acceptance is that put forward by A. F. Huxley, whilst the metabolic implications of this theory, and the possible molecular basis for it, have been developed by Davies. Before presenting these specula-tions, it would be profitable to describe some experiments in which the inter-actions of the protein skeleton of muscle, actin and myosin, have been investi-gated. As long ago as 1932, Boehm & Weber prepared threads of "myosin" by extracting muscle with KCl solution; these were, in reality, actomyosin, and Boehm & Weber showed that, from the points of view of X-ray diffraction, birefringence, and mechanical properties, they had some analogy with muscle.

Actin and Myosin as ATPases

Perhaps the most important contribution to our knowledge, and one that was the stimulus to a most fruitful series of investigations, was the discovery by the Russian workers, Engelhardt & Ljubimova in 1939, that these myosin threads had ATPase activity, *i.e.*, that when ATP was added to their bathing medium, its breakdown to ADP and inorganic phosphate was catalysed by the threads. With the discovery of Szent-Györgyi that myosin, as ordinarily prepared, was a mixture of myosin and actomyosin, the problem arose as to which part of the complex contained ATPase activity; it emerged that the activity belonged to myosin, actin being free of activity; when myosin was combined with actin as actomyosin it retained its activity, but in a modified form. Thus myosin ATPase is activated by low concentrations of Ca^{++} and inhibited by Mg^{++}, whereas actomyosin ATPase is activated by Mg^{++}. Hence the degree to which the enzyme is affected by the Mg^{++} and Ca^{++} concentration will depend on the degree to which it exists as free myosin or as actomyosin; for example, Hasselbach has shown that, at high ionic strengths or under other conditions favourable for dissociation of the actomyosin complex, the enzymatic activity is characteristic of myosin and is therefore activated by Ca^{++} and inhibited by Mg^{++}. Although it was considered that the ATPase was an attached "impurity", derived from the sarcoplasmic proteins, it is now generally agreed that the ATPase is, indeed, a part of the myosin molecule, an activity that is associated with the SH-groups of the protein since compounds as Salyrgan, which block these groups, inhibit ATPase activity.

There is little doubt that ATP is a substrate in a reaction between the myosin of the thick filaments and the actin of the thin filaments, and we may presume that the active ATPase site on the myosin is located at the cross-bridges; at any rate myosin may be broken down by trypsin to two fractions, heavy and light meromyosins (p. 1377), and the ATPase activity is associated with the heavy fraction which most probably constitutes the region of the molecule where the cross-bridge is.

Dual Action of ATP

Another fundamental discovery, which led to the development of various models, was that of A. Szent-Györgyi who showed that, under appropriate conditions, ATP could cause contraction of actomyosin threads. If an acto-myosin sol was studied, ATP would cause the phenomenon of "superpre-cipitation", or coagulation, consisting of a heavy flocculation of the protein; when added to actomyosin gels a vigorous contraction, or syneresis, was caused, the water being driven out of the gel structure with an associated coalescence of the fine actomyosin filaments that held the gel in the solid state. On the basis of these phenomena, Szent-Györgyi considered that the super-precipitation of an actomyosin sol, the syneresis of an actomyosin gel, and the shortening of an actomyosin fibre could be used as models of the process of contraction in the intact muscle fibre. It was objected, however, that acto-myosin fibres do not contract anisodiametrically like muscle fibres, *i.e.*, they do not fatten as they shorten, but merely shrink as a result of the expulsion of water. Secondly, Buchtal *et al.* (1947) observed that, whereas an unloaded actomyosin fibre shrank on treatment with ATP, a loaded fibre actually elongated. Both of these objections have been answered by Szent-Györgyi; he points out that in a thread the actomyosin molecules are quite unorientated, so that even if they did contract with an associated increase in thickness, this increase would not be observable; if the thread is first stretched to bring the molecules parallel with the fibre-axis, ATP then causes shortening with widening. The behaviour of the loaded thread is explained by the action of ATP on the linkages between actomyosin molecules; ATP loosens these, so that, if the contractile force generated by ATP is insufficient to allow the load to be lifted, the thread will extend because of this loosening.

Rigor. This apparent contradiction in the behaviour of ATP, namely its contractile action, which is presumably associated with the formation of chemical bonds between actin and myosin, and its relaxing action, associated with the loosening of the actomyosin link, must now be discussed, since it is fundamental to the understanding of the behaviour of models. The function of ATP as a relaxer, or plasticizer of the actomyosin complex, is revealed most strikingly by studies on the extractability of the muscle proteins in rigor. As is well known, a certain time after death the muscle goes into rigor, in which condition it is stiff and may no longer be extended passively with small forces. Deuticke observed as long ago as 1930 that the amount of protein extractable from a muscle in rigor was decreased considerably below that obtained from freshly excised muscle; under his conditions of extraction, this would have meant that the amounts of actin and myosin bound as an actomyosin complex were large in rigor. Associated with this decrease in extractability of protein there was a decrease in the ability of the muscle to form organic phosphates; since ATP is involved in the phosphorylation of organic compounds, Deuticke's results may be explained on the grounds that the concentration of ATP decreased in rigor and so left larger proportions of the muscle proteins in the difficultly soluble actomyosin form. This view of the nature of rigor—namely a loss of plasticity due to the destruction of ATP, was confirmed by the later studies of Erdos and Bate-Smith & Bendall, who showed that the post-mortem development of inelasticity of muscle, which culminates in rigor, was correlated with the disappearance of ATP from muscle.

ATP as Plasticizer. Under some conditions at any rate, ATP acts as a plasticizer, and that this is to be correlated with its ability to prevent the association of actin and myosin in a complex or, if this is already formed, to dissociate it—*i.e.*, to relax the muscle—is indicated by a variety of findings. For instance, Dainty *et al.* found that the addition of ATP to "myosin" (actomyosin) solutions caused a rapid decrease in the double refraction of flow (by 48 per cent.), and a smaller decrease in viscosity; again, Schramm & Weber found that the addition of ATP to the "heavy component of myosin" (actomyosin) caused a reversible fall in the sedimentation constant to the value typical of their "light myosin". Both these findings indicated that ATP dissociated the actomyosin complex, and in a more recent study A. Weber (1956) has shown by ultracentrifugal analysis that ATP does, indeed, split off myosin from purified actomyosin, the myosin being identified by its sedimentation constant and ATPase activity. Tentatively it might be assumed that the following reactions took place:

$$(1) \quad M.ATP \rightarrow M \sim P + ADP$$
$$(2) \quad M \sim P + A \rightarrow AM + P$$
$$(3) \quad AM + ATP \rightarrow M.ATP + A$$

In this way ATP may both initiate the active state and also, in accordance with reaction (3), act as a plasticizer (Needham).

Bearing these fundamental points in mind, namely that the contractile proteins are actin and myosin capable of reversible combination with each other; that the myosin, either alone or in combination with actin, acts as an ATPase; and that ATP is able, according to the prevailing conditions, to cause either association of actin and myosin or dissociation, we may now proceed to discuss the behaviour of some models in more detail.

Models

Actomyosin Threads. The models that are currently studied are the orientated actomyosin threads of Portzehl & Weber, prepared by extracting actomyosin from a muscle brei with 0·6 M KCl, separating the actomyosin from other proteins by repeated precipitation and dissolution, and extruding a concentrated solution of the protein through an orifice to give threads in which the long molecules are orientated in the direction of the fibre-axis. With an optimal concentration of protein of some 6–10 per cent., these fibres will contract anisodiametrically on addition of ATP and Mg^{++}, to develop a maximal tension amounting to some 200–300 g./cm.^{2} (Portzehl, 1951). This is smaller than the tension developed by muscle, or other models (4 kg./cm.^{2}), presumably because of the relatively low concentration of actomyosin.

Glycerol-Extracted Fibres. A model originally developed by Szent-Györgyi is the glycerol-extracted bundle of muscle fibres; thus the psoas muscle of the rabbit may be thoroughly extracted with aqueous glycerol, as a result of which most of the sarcoplasmic material, including the ATP, is removed, whilst the contractile material, actomyosin, remains in the fibrils which retain their characteristic banded appearance. On placing this "muscle" in Ringer solution containing 0·001 M Mg^{++} and 0·02 per cent. ATP, the muscle shortened and developed as much tension as a normal muscle on electrical stimulation. The objection to large bundles of muscle fibres is that ATP cannot diffuse readily to the central regions, so that contraction induced by this substance tends to be asymmetrical. To avoid this difficulty, single isolated fibres may be extracted with glycerol in the same way (Weber, 1951), in which case they may develop tensions of the order of 4 kg./cm.², *i.e.*, comparable with those developed by normal muscle. Finally, the myofibrils of a muscle may be separated, either by using a collagenase preparation with which to remove connective tissue, or simply by mechanical agitation (Perry, 1952). This last preparation cannot be used to study the development of tension, but the changes in volume that the fibrils undergo parallel the contraction and relaxation that take place in the macroscopical models; moreover, as we have seen, they are excellent objects for phase-contrast study of the changes in banding that occur during contraction and passive extension.

Behaviour of Models

Plasticizing and Contractile Actions of ATP. These models will all simulate muscle in showing a contraction, with the development of tension, in the presence of ATP and Mg^{++}. Associated with the contraction there is a splitting of the ATP, the actomyosin behaving as an ATPase. In one important respect, however, the models differ from real muscle; contraction in the models is not followed by relaxation, in fact, when the ATP is used up, the extensibility of the model is characteristic of muscle in rigor, so that the force required to extend it is some ten times that required to extend normal muscle. It would seem, then, that ATP exerts a dual action on the model, as well as on the muscle, inducing both contraction and the plasticity necessary for relaxation; when the ATP has been used up, therefore, the model remains shortened and with a modulus of elasticity similar to that of muscle in rigor; when ATP is still present the model is shortened but may be extended by stretching, with a modulus between those characteristic of relaxed and rigor muscles (Portzehl, 1952). To demonstrate the dual action conclusively, however, it is necessary to dissociate the contractile and plasticizing effects, and this was done by Portzehl, either by removing Mg^{++} from the system or, better still, by inhibiting the ATPase activity of the actomyosin with an SH-reagent, such as Salyrgan. If this is done, the model fails to develop tension with ATP but remains plastic; its state under these conditions is analogous with normal muscle that has been stimulated to contract without a load; this remains shortened, but may be extended by plastic stretch to its original length. This plasticizing effect is not peculiar to ATP and seems to depend on the presence of a polyphosphate chain in the molecule since ADP, sodium triphosphate ($Na_3P_3O_{10}$) or pyrophosphate ($Na_4P_2O_7$) are also effective, in marked contrast to the contractile effect which is peculiar to ATP.* Since pyrophosphate will

* ITP (inosine triphosphate) is also effective in inducing contraction. The shortening of glycerol-extracted muscle induced by ADP is apparently due to the presence of myokinase, which catalyzes the reaction:—

$$2 ADP \rightleftharpoons ATP + AMP$$

also, like ATP, dissociate the actomyosin complex, it would seem that plasticizing activity, and thus relaxation, consists essentially in reversal of the combination of actin and myosin that seems to be the feature of contraction. The contractile action of ATP may be considerably reduced by increasing the ATP concentration above a critical value determined by the ionic strength of the medium, temperature, etc. The plasticizing action is not decreased under these conditions so that, as Bozler showed, adding excess ATP to a fibre-model, contracting under the influence of ATP, may cause relaxation.

Relaxing Factor. Thus, so far as models are concerned, when ATP is present, and can be broken down by the actomyosin ATPase, e.g., in the presence of Mg^{++}, the model goes into contraction; if the ATP is washed out, tension and length remain largely unchanged and the model is inelastic; if ATP is present but cannot be split, as when Salyrgan is added, then the model remains at its equilibrium length and is plastic, showing a low modulus of elasticity. We may ask, now, what is it in normal unstimulated muscle that prevents the ATP, normally present in it, from exerting its contractile effect, assuming, of course, that the analogy between model and muscle can be carried thus far. The answer is given by what has been called the relaxing factor, and which has turned out to be the Ca^{++}-accumulating system of the sarcoplasmic reticulum. The reaction between actin and myosin requires Ca^{++}, and if this is held below a critical concentration the muscle will remain in its relaxed state (p. 1401).

Energy of Contraction. Addition of ATP to the model system causes it to contract and enables it to perform work; during the process the ATP is broken down to ADP and inorganic phosphate, whilst the myosin, one of the reactants in the process, behaves as an ATPase presumably favouring transphosphorylation reactions that lead, ultimately, to the formation of ADP. As indicated earlier, it is customary to speak of the energy involved in contraction as being in some way bound in the "high-energy phosphate compound", ATP; and this energy has been equated to the free-energy of hydrolysis of ATP. Whilst it is doubtless incorrect to speak in this way, since the energy released results from mutual interaction and cannot be said to belong to any of the reactants individually, we may nevertheless seek to associate ATPase activity with ability to contract. Such a correlation has been demonstrated most convincingly by Weber & Portzehl. For example, all inhibitors of ATPase block contraction in glycerol-extracted fibres; lowering of temperature causes parallel decreases in ATPase activity and contractile force; the reduced contraction due to raised ATP concentration is associated with decreased ATPase activity, and so on. When the effects of Ca^{++} on activation of contraction of models prepared from a variety of muscles are compared with the effects on ATPase activity of the same preparations there is likewise a good correlation (Schädler, 1967). The correlation, although strong, is not perfect, as Mommaerts and his colleagues have pointed out; thus, Bowen & Kerwin showed that, by varying the medium in which glycerol-washed muscle was suspended, variations in contractile force were obtained that were not paralleled by variations in ATPase activity. Because of this lack of correlation, Mommaerts and Morales & Botts incline to the view that it is essentially the act of binding of ATP to the actomyosin system that leads to contraction, not its breakdown to ADP; the latter process would therefore take place during relaxation. It should be emphasized, however, that

and thus makes a supply of ATP available (A. Weber, 1951). Mg^{++} seems to be necessary for the plasticizing action of ATP and of pyrophosphate (Bozler, 1952, 1954).

we do not have to envisage the hydrolysis of ATP as the exergonic reaction that is linked to an endergonic contractile process; under certain highly artificial conditions hydrolysis occurs, a process, incidentally, that is associated with a remarkably small free-energy change (Podolsky & Morales, 1956), but this could be accidental; it is better to regard the ATPase activity as a reflexion of a more complex enzymatic activity through which reactive groupings on actin and myosin are enabled to interact.

A. F. Huxley's Theoretical Model

Experimental Basis. The main experimental findings regarding the contraction of muscle are as follows:

(1) Shortening is apparently the result of a sliding of the thin filaments between the thick filaments.

(2) The maximum tension developed in a tetanus depends on the length of the muscle, so that when the muscle is stretched beyond a certain point this becomes zero.

(3) The force developed decreases with increasing velocity of shortening; thus the rate of work production decreases with speed of contraction.

(4) The energy made available for contraction increases with the work done —the Fenn effect.

(5) The heat of shortening is independent of contraction speed, so that the rate of liberation of heat increases linearly with rate of shortening.

Control of Chemical Reactions

If we plot the total energy liberated against speed of contraction we obtain the curve of Fig. 777 (p. 1429), which is made up of a linear increase in heat production plus a variable amount of work, ranging between zero at zero rate of

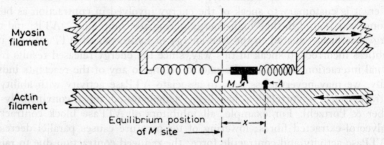

FIG. 784. Diagram illustrating the mechanism by which it is assumed that tension is generated. The part of a fibril that is shown is in the right-hand half of an A-band, so that the actin filament is attached to a Z-line which is out of the picture to the right. The arrows give the direction of the relative motion between the filaments when the muscle shortens. (A. F. Huxley. *Progress in Biophysics.*)

shortening—isometric contraction—and zero again at the maximum rate of shortening (without a load). This curve emphasizes the fact that the energy made available depends on the speed of shortening, and thus indicates that the chemical reactions are in some way controlled in their velocity by this shortening process. Such a control could be exerted if chemical links between the thin and thick filaments were being formed and broken during the shortening process, provided that the number of links possible depended on the length of the muscle; the sliding filament theory renders this a necessary consequence, if the actin filaments, say, have groups that will react with groups on the myosin filaments.

Interfilament Links. The model put forward by A. F. Huxley is illustrated by Fig. 784. The thick and thin filaments lie parallel, and the figure shows the region of overlap, the Z-line, belonging to the thin filament, lying on the right. Shortening of the sarcomere will thus represent a movement of the thin filament to the left. On the thick filament there is a reactive site, M, capable of linking with A by virtue of a chemical affinity. This reaction is assumed to occur spontaneously:

$$A + M \rightarrow AM \text{ (Rate-constant: } f) \quad . \quad . \quad . \quad . \quad . \quad (1)$$

The probability that the reaction occurs will be given by f, and this will vary according to the position of M in relation to its mean position, O. Thus the farther M moves over to the right, $i.e.$, the greater the positive value of x, the greater will be the chance of a link being formed between M and a group A on the thin filament. When the link is formed, the tension in the spring will pull the thin filament towards O, $i.e.$, it will favour shortening. The group M is supposed to oscillate naturally about its mean position, but the chance of forming a link when it is on the left of O is zero. This imparts a one-way character to the pull caused by a given link.

Breaking the Links

In order that measurable shortening may occur, links must not only be formed, but also broken, so that M may form a new link with another A farther to the right. The breaking of the link may be considered to follow from the reaction:

$$AM + XP \rightarrow AXP + M \text{ (Rate constant: } g) \quad . \quad . \quad . \quad . \quad (2)$$

where XP is a phosphate compound uniting with a site near A. This reaction is an energy-releasing reaction. The chance that this breaking of the link will occur is given by g; it will be small when M is to the right of the mean position, and will increase to a high value as soon as it has crossed to the left of this. Thus, during an oscillation to the right, the chances of forming a link are high, and the chances of its being broken are low; to the left there is no chance of making a link and such links that have been made will tend to oppose shortening; the high probability of breaking of the link on this side, however, keeps this opposing force small unless shortening is very rapid, in which case the number of links on the left-hand side becomes significant.

Effect of Velocity

A simple mathematical treatment enables the computation of the relative numbers of links on each side of the mean position to be computed at any instant during shortening at different speeds. Four different conditions are illustrated in Fig. 785; it is seen that when $V = O$, $i.e.$, in an isometric contraction, there are no links on the left, whilst the number on the right is the maximum compatible with the assumed probabilities of formation and breaking of the links. At the other extreme, where the muscle contracts without a load, the numbers of links on both sides approximate each other, and the tension developed is zero.

Force-Velocity Relationship

The tension developed is clearly determined by the relative numbers of links formed at the different points in the range of oscillation of all the M groups capable of forming a link with an A group. The energy liberated will represent the sum of the mechanical work (which will be a simple function of the distance moved by a given link and the elastic constant of the series-elastic elements) and the heat produced, which will depend on the rate of breaking of the links. Huxley showed that a mathematical formulation on this basis allowed the derivation of a force-velocity relationship similar to Hill's,

Blocking of Sites

Before indicating several other qualitative inferences that could be drawn we must first introduce a third chemical reaction, in addition to (1) and (2), in order to account for the absence of reaction between sites before stimulation. Reaction (2) forms AXP and stops A from reacting, and it can only be reactivated by a third reaction:

$$AXP \rightarrow A + X + PO_4 \quad . \quad . \quad . \quad . \quad . \quad . \quad (3)$$

Thus, at rest, reaction (3) is inhibited, so that the A sites are blocked; activation may be considered to be a rapid conversion of AXP to A which can now react with M.

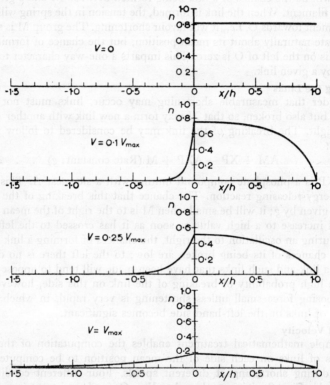

FIG. 785. Variation of n (proportion of sites at which links between actin and myosin are in existence) with x (position of A-site relative to equilibrium position of M-site), for the steady state in isometric contraction (top) and in shortening at three different speeds. (A. F. Huxley. *Progress in Biophysics.*)

If a muscle is tetanized and allowed to shorten under a small load until it reaches a stop, we may then put a load on it about equal to the isometric tension that it would develop at this shorter length. It is found that the muscle does not hold this load, there is an initial slip as the muscle lengthens, and then it shortens to the stop position. The explanation is that during the rapid shortening the number of links formed is small, as Fig. 785 shows; if the weight is applied before sufficient links can be formed, then it pulls the filaments apart. This stretch increases the length of the muscle and thus the number of links that can be formed, and this in turn increases the tension developed in them and so brings the lengthening to an end; finally the muscle re-shortens.

Two Stages in Relaxation

The activation process, as defined by Huxley's treatment, consists of two steps: the removal of the phosphate groups from A, and the reaction of A with M; in a similar way, relaxation would require two steps, and on the basis of any theory we must postulate that the development and decay of the active state will take measurable times, determined by the probabilities of the formation (f) and the breakage (g) of the AM links. Huxley pointed out that there should be some evidence for two stages in contraction and relaxation; thus in relaxation the cessation of shortening, which depends on the making of links as well as breaking, should not necessarily lead to a cessation of tension, since the loss of this depends on the breaking of links only. This prediction is verified by Jewell & Wilkie's study described earlier; here it was shown that the tension during relaxation did, indeed, outlive the lengthening; moreover, the initial rapid loss of tension could be increased by increasing the length during relaxation, a process that would, of course, favour breakage of links.*

Quick-Release

Again, the phenomena revealed by quick-release can be interpreted. Thus, when a muscle is only released to the extent that the series elastic component can relax, the tension will fall to zero but the links will not break because shortening has been so small. In this case the redevelopment of tension will be very rapid— more rapid than at the beginning of a tetanus when links must form. Again, if release is greater than that required to relax the series elastic component, linkages will now be broken and so the tension developed by a quick stretch immediately after this release will be smaller than that developed by a similar stretch during a sustained tetanus at the same length.

Molecular Model

Davies (1963) has put forward a model of the contractile process that takes into account many of the biochemical features of muscular contraction. According to this, the H-meromyosin, which contains the ATPase reactive group of the myosin molecule and is presumed to occupy the cross-bridge portion, consists of poly-peptide chains that can adopt the fully extended β- or the helical α-configuration according to the electrostatic repulsions that operate at a given moment. If attach-ment is made when the chain is extended, then contraction of the muscle will be induced by a reversion to the α-helical configuration, the shortening of the poly-peptide chains leading to a sliding of the actin filament over the myosin filament.

The postulated mechanism involves a number of *ad hoc* hypotheses; thus, as indicated in Fig. 786, the H-meromyosin cross-bridge is assumed to contain a molecule of ATP at its end, being held there electrostatically by three positive charges that are presumably Mg^{++}-ions, forming electrostatic links with negative groupings on the cross-bridge. The extra negative charge of the ionized ATP is supposed to be repelled by a fixed negative charge on the cross-bridge, and this is considered adequate to force the H-meromyosin into its extended β-configura-tion, which is thus the resting state of the cross-bridge. Activation of the muscle releases Ca^{++}, which forms a link between the bound ADP of actin; this electro-static bond between meromyosin and actin reduces the repulsion between the

* Podolsky (1961) has very elegantly exhibited the essential principles underlying Huxley's treatment of contraction; he has discussed Huxley's model in relation to one of his own where the structural basis of shortening is a coiling of the I-filaments, which would thus shorten and pull the Z-line towards the A-band. More recently Woledge (1968) has discussed the factors determining efficiency of muscle in terms of the sliding-filament theory; he considers that the conversion of chemical free energy into work can be highly efficient, and that inefficiencies arise through failure of the system to make use of the liberated free energy, e.g., bridges might persist after passing into a configuration such that they exerted an opposing force; they might break prematurely, before performing the maximum possible work; and so on. In an interesting discussion he shows how a reduction in these inefficiencies would increase the curvature of the force-velocity relation.

46*

ATP and the postulated fixed negative charge, and thus permits adoption of the α-helical form, which is held in this stable state by hydrogen and hydrophobic bonds, as in the elastic recovery of stretched keratin. The development of tension causes sliding of the filaments.

It is next postulated that the contraction brings the ATP, involved in the cross-linkage, into contact with the ATPase-grouping of the H-meromyosin; this hydrolyses the ATP and thereby breaks the link between actin and myosin. The ADP bound to the cross-bridge is now rephosphorylated by cytoplasmic ATP, and this re-establishes the negative charge that leads to re-extension of the polypeptide chain. The whole process repeats itself while Ca^{++} is available in the sarcoplasm, so that the duration of the active state is the time required for the sarcoplasmic reticulum to remove most of the ionized Ca^{++}.

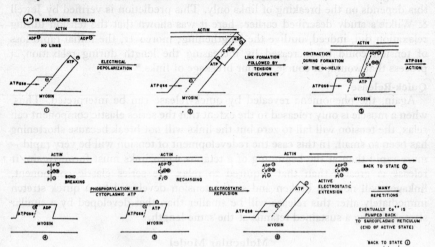

FIG. 786. Illustrating Davies' model for the Ca^{++} -dependent contraction of muscle involving α-helix formation and ATP -dependent extension of the cross-bridges. (Davies. *Nature*.)

The cycle is indicated in Fig. 786. On this basis the energy for the contraction comes from the formation of the numerous hydrogen and hydrophobic bonds that come into play when the Ca^{++}-link permits the adoption of the α-helical conformation; the activation heat might well be the heat evolved when Ca^{++} is released from its binding on the sarcoplasmic reticulum, from which must be subtracted an absorption of heat as it forms the postulated link. Immediately on activation, and before shortening or development of tension begins, the response to stretch will have changed because of the formation of the Ca^{++}-links.

As Davies has shown, not only can many qualitative facts be made to fit the theory reasonably well, but the theory permits a number of quantitative estimates of such factors as the heat changes during shortening and stretching of active fibres, the force developed and work done per ATP molecule, the length of the latent period, and so on.

REFERENCES

ABBOTT, B. C., & RITCHIE, J. M. (1951). "The Onset of Shortening in Striated Muscle." *J. Physiol.*, **113**, 336–345.

ABBOTT, B. C., & RITCHIE, J. M. (1951). "Early Tension Relaxation during a Muscle Twitch." *J. Physiol.*, **113**, 330–335.

ASHLEY, C. C., CALDWELL, P. C., LOWE, A. G., RICHARDS, C. D. & SCHIRMER, H. (1965). The amount of injected EGTA needed to suppress the contractile responses of single *Maia* muscle fibres and its relation to the amount of calcium released during contraction. *J. Physiol.*, **179**, 32–33 P.

AXELSSON, J. & THESLEFF, S. (1958). Activation of the contractile mechanism in striated muscle. *Acta physiol. scand.*, **44**, 55–66.

BATE-SMITH, E. C. & BENDALL, J. R. (1947). Rigor mortis and ATP. *J. Physiol.*, **106**, 177.

BIANCHI, C. P. (1961). Calcium movements in muscle. *Circulation*, **24**, 518–522.

BIANCHI, C. P. (1962). Kinetics of radiocaffeine uptake and release in frog sartorius. *J. Pharmacol.*, **138**, 41–47.

BIANCHI, C. P., & SHANES, A. M. (1959). Calcium influx in skeletal muscle at rest, during activity, and during potassium contracture. *J. gen. Physiol.*, **42**, 803–815.

BIANCHI, C. P., & SHANES, A. M. (1960). The effect of ionic milieu on the emergence of radiocalcium from tendon and from sartorius muscle. *J. cell comp. Physiol.*, **56**, 67–76.

BENOIT, P. H., CARPENI, N. & PRZYBYSLAWSKI, J. (1964). Sur la contracture provoquée par la quinine chez le muscle strié de grenouille. *J. de Physiol.* (Paris), **56**, 289–290.

BOWEN, W. J. & KERWIN, T. D. (1955). The rate of dephosphorylation of adenosine triphosphate and the shortening of glycerol-washed muscle fibres. *Biochim. biophys. Acta.*, **18**, 83–86.

BOZLER, E. (1935). Change of a.c. impedance of muscle produced by contraction. *J. cell. comp. Physiol.*, **6**, 217.

BOZLER, E. (1952). Evidence for an ATP-actomyosin complex in relaxed muscle and its response to calcium ions. *Amer. J. Physiol.*, **168**, 760–765.

BOZLER, E. (1954). Relaxation in extracted muscle fibres. *J. gen. Physiol.*, **38**, 149–159.

BRANDT, P. W., REUBEN, J. P., GIRARDIER, L. & GRUNDFEST, H. (1965). Correlated morphological and physiological studies on isolated single muscle fibers. I. *J. Cell Biol.*, **25**, Pt. 2, No. 3, 233–260.

BRANDT, P. W., REUBEN, J. P. & GRUNDFEST, H. (1968). Correlated morphological and physiological studies on isolated single muscle fibers. II. *J. Cell Biol.*, **38**, 115–129.

BRIGGS, F. N. & FLEISHMAN, M. (1965) Calcium binding by particle-free supernatants of homogenates of skeletal muscle. *J. gen. Physiol.*, **49**, 131–149.

BRIGGS, F. N., KALDOR, G. & GERGELY, J. (1959). Participation of a dialysable cofactor in the relaxing factor system of muscle. I. *Biochim. biophys. Acta*, **34**, 211–218.

BROWN, D. E. S. (1941). "Regulation of Energy Exchange in Contracting Muscle." *Biol. Symp.*, **3**, 161.

BUCHTAL, F., DEUTSCH, A. & MUNCH-PEDERSEN, A. (1947). Effect of ATP on myosin threads. *Acta. physiol. scand.*, **13**, 167.

BULLER, A. J. & LEWIS, D. M. (1965). Rate of tension development in isometric tetanic contractions of mammalian fast and slow skeletal muscle. *J. Physiol.*, **176**, 337–354.

CALDWELL, P. C. & WALSTER, G. (1963). Studies on the micro-injection of various substances into crab muscle fibres. *J. Physiol.*, **169**, 353–372.

CAPUTO, C. (1966). Caffeine- and potassium-induced contractures of frog striated muscle fibers in hypertonic solutions. *J. gen. Physiol.*, **50**, 129–139.

CAPUTO, C. (1968). Volume and twitch tension changes in single muscle fibers in hypertonic solutions. *J. gen. Physiol.*, **52**, 793–809.

CAPUTO, C. & GIMENEZ, M. (1967). Effects of external calcium deprivation on single muscle fibers. *J. gen. Physiol.*, **50**, 2177–2195.

CARLSON, F. D., HARDY, D. J. & WILKIE, D. R. (1963). Total energy production and phosphocreatine hydrolysis in the isotonic twitch. *J. gen. Physiol.*, **46**, 851–882.

CARLSON, F. D., HARDY, D. & WILKIE, D. R. (1967). The relation between heat produced and phosphorylcreatine split during isometric contraction of frog's muscle. *J. Physiol.*, **189**, 209–235.

CARLSON, F. D., & SIGER, A. (1960). The mechanochemistry of muscular contraction. I. The isometric twitch. *J. gen. Physiol.*, **43**, 33–60.

CARVALHO, A. P. (1968). Calcium-binding properties of sarcoplasmic reticulum as influenced by ATP, caffeine, quinine, and local anaesthetics. *J. gen. Physiol.*, **52**, 622–642.

CARVALHO, A. P. (1968). Effects of potentiators of muscular contraction on binding of cations by sarcoplasmic reticulum. *J. gen. Physiol.*, **51**, 427–442.

CARVALHO, A. P. & LEO, B. (1967). Effects of ATP on the interaction of Ca^{++}, Mg^{++}, and K^+ with fragmented sarcoplasmic reticulum isolated from rabbit skeletal muscle. *J. gen. Physiol.*, **50**, 1327–1352.

CHEESMAN, D. F. & HILTON, E. (1966). Exchange of structurally bound phosphate in muscular activity. *J. Physiol.*, **183**, 675–682.

CLINCH, N. F. (1968). On the increase in rate of heat production caused by stretch in frog's skeletal muscle. *J. Physiol.*, **196**, 397–414.

CONSTANTIN, L. L. (1968). The effect of calcium on contraction and conductance thresholds in frog skeletal muscle. *J. Physiol.*, **195**, 119–132.

CONSTANTIN, L. L. & PODOLSKY, R. J. (1967). Depolarization of the internal membrane system in the activation of frog skeletal muscle. *J. gen. Physiol.*, **50**, 1101–1124.

CONSTANTIN, L. L., FRANZINI-ARMSTRONG, C. & PODOLSKY, R. J. (1965). Localization of calcium-accumulating structures in striated muscle fibers. *Science*, **147**, 158–160.

CURTIS, B. A. (1966). Ca fluxes in single twitch muscle fibers. *J. gen. Physiol.*, **50**, 255–267.

DAINTY, M., KLEINZELLER, A., LAWRENCE, A., MIALL, M., NEEDHAM, J., NEEDHAM, D. M. & SHEN, S. C. (1944). Changes in anomalous viscosity and flow-birefringence of myosin solutions in relation to ATP and muscular contration. *J. gen. Physiol.*, **27**, 355.

DAVIES, R. E. (1963). A molecular theory of muscle contraction. *Nature*, **199**, 1068–1074.

DAVIES, R. E., KUSHMERICK, M. J. & LARSON, R. E. (1967). ATP, activation, and the heat of shortening of muscle. *Nature*, **214**, 148–151.

DAVSON, H. (1940). "The Influence of the Lyotropic Series of Anions on Cation Permeability." *Biochem. J.*, **34**, 917–925.

DESMEDT, J. E. & HAINAUT, K. (1968). Kinetics of myofilament activation in potentiated contraction: staircase phenomenon in human skeletal muscle. *Nature*, **217**, 529–532.

DIXON, G. J., & SACKS, J. (1958). Occurrence of phosphate exchanges in muscular contraction. *Amer. J. Physiol.*, **193**, 129–134.

DUBUISSON, M. (1935). "L'ionogramme de la contraction musculaire étudié à l'oscillographe cathodique." *Arch. int. Physiol.*, **41**, 177.

DUBUISSON, M. (1937). "Impedance Changes in Muscle during Contraction and their possible Relation to Chemical Processes." *J. Physiol.*, **89**, 132.

DUBUISSON, M. (1950). "Discussion on Muscle." *Proc. Roy. Soc., B*, **137**, 63.

DUDEL, J., MORAD, M. & RÜDEL, R. (1968). Contractions of single crayfish muscle fibers induced by controlled changes of membrane potential. *Pflüg. Arch. ges. Physiol.*, **299**, 38–51.

DYDYNSKA, M. & WILKIE, D. R. (1963). The osmotic properties of striated muscle fibres in hypertonic solutions. *J. Physiol.*, **169**, 312–329.

EBASHI, S. & LIPMANN, F. (1962). Adenosine triphosphate-linked concentration of calcium ions in a particulate fraction of rabbit muscle. *J. biophys. biochem. Cytol.*, **14**, 389–400.

ECCLES, J. C., KATZ, B. & KUFFLER, S. W. (1941). Nature of the end-plate potential in curarized muscle. *J. Neurophysiol.*, **5**, 211.

EDMAN, K. A. P. & GRIEVE, D. W. (1964). On the role of calcium in the excitation-contraction process of frog sartorius muscle. *J. Physiol.*, **170**, 138–152.

EDWARDS, C., LORKOVIĆ, H. & WEBER, A. (1966). The effect of the replacement of calcium by strontium on excitation-contraction coupling in frog skeletal muscle. *J. Physiol.*, **186**, 295–306.

EDWARDS, C., RITCHIE, J. M., & WILKIE, D. R. (1956). "The Effect of some Cations on the Active State of Muscle." *J. Physiol.*, **133**, 412–419.

EDWARDS, G. A., & RUSKA, H. (1955). "The Function and Metabolism of certain Insect Muscles in relation to their Structure." *Quart. J. micr. Sci.*, **96**, pt. 2, 151–159.

EISENBERG, R. S. & EISENBERG, B. (1968). The extent of the disruption of the transverse tubular system in glycerol treated skeletal muscle. *Fed. Proc.*, **27**, 247.

EISENBERG, R. S. & GAGE, P. W. (1967). Frog skeletal muscle fibers: changes in electrical properties after disruption of transverse tubular system. *Science*, **158**, 1700–1701.

ELISON, C., FAIRHURST, S., HOWELL, J. N. & JENDEN, D. J. (1965). Calcium uptake in glycerol-extracted rabbit psoas muscle fibers. I. *J. cell. comp. Physiol.*, **65**, 133–140.

ENGELHARDT, W. A. & LJUBIMOWA, M. N. (1939). Myosine and ATPase. *Nature*, **144**, 668.

ETZENSPERGER, J. (1962). Études des réponses électriques et méchaniques de la fibre musculare striée intoxiquée par la vératrine. *C. r. Soc. Biol.*, **156**, 1125–1131.

FALK, G. & FATT, P. (1964). Linear electrical properties of striated muscle fibres observed with intracellular electrodes. *Proc. Roy. Soc. B.*, **160**, 69–123.

FEINSTEIN, M. B. (1963). Inhibition of caffeine rigor and radiocalcium movements by local anaesthetics in frog sartorius muscle. *J. gen. Physiol.*, **47**, 151–172.

FENG, T. P. (1932). "The Thermo-Elastic Properties of Muscle." *J. Physiol.*, **74**, 455.

FENN, W. O. (1923). "Quantitative Comparison between the Energy liberated and the Work performed by the Isolated Sartorius Muscle of the Frog." *J. Physiol.*, **58**, 175.

FENN, W. O. (1924). "Relation between Work Performed and Energy Liberated in Muscular Contraction." *J. Physiol.*, **58**, 373.

FENN, W. O., & MARSH, B. S. (1935). "Muscular Force at Different Speeds of Shortening." *J. Physiol.*, **85**, 277.

FLECKENSTEIN, A., JANKE, J., & DAVIES, R. E. (1956). Der Austausch von radioaktivem Phosphat mit dem α-, β-, und γ- Phosphor von ATP und mit Kreatinphosphat bei der Konkraktur des Froschrectus durch Acetylcholin, Nicotin und Succinylbischolin. *Arch. exp. Path. Pharmakol.*, **228**, 596–614.

FOULKS, J. G. & PERRY, F. A. (1966). The relation between external potassium concentration and the relaxation rate of potassium-induced contractures in frog skeletal muscle. *J. Physiol.*, **186**, 243–260.

FRANK, G. B. (1960). Effects of changes in extracellular calcium concentration on the potassium-induced contracture of frog's skeletal muscle. *J. Physiol.*, **151**, 518–538.

FRANK, G. B. (1962). Utilization of bound calcium in the action of caffeine and certain multivalent cations on skeletal muscle. *J. Physiol.*, **163**, 254–268.

FRANK, G. B. (1963). Utilization of bound calcium in the acetylcholine contracture of frog skeletal muscle. *J. Pharmacol.*, **139**, 261–268.

FRANKENHAEUSER, B. & LÄNNERGREN, J. (1967). Effect of calcium on mechanical response of single twitch muscle fibres of *Xenopus lævis*. *Acta. physiol. scand.*, **69**, 242–254.

FREYGANG, W. H., GOLDSTEIN, D. A. & HELLAM, D. C. (1964). The after-potential that follows trains of impulses in frog muscle fibers. *J. gen. Physiol.*, **47**, 929–952.

FREYGANG, W. H., RAPOPORT, S. I. & PEACHEY, L. D. (1967). Some relations between changes in the linear electrical properties of striated muscle fibers and changes in ultrastructure. *J. gen. Physiol.*, **50**, 2437–2458.

FUJINO, S. & FUJINO, M. (1964). Removal of the inhibitory effect of hypertonic solutions on the contractability in muscle cells and the excitation-contraction link. *Nature*, **201**, 1331–1333.

GABEL, L. P., CARSON, C. & VANCE, E. (1968). Active state of muscle and the second and third derivatives of twitch tension. *Amer. J. Physiol.*, **214**, 1025–1030.

GAGE, P. W. & EISENBERG, R. S. (1967). Action potentials without contraction in frog skeletal muscle fibers with disrupted transverse tubules. *Science*, **158**, 1702–1703.

GAINER, H. (1968). The role of calcium in excitation-contraction coupling of lobster muscle. *J. gen. Physiol.*, **52**, 88–110.

GASSER, H. S., & HILL. A. V. (1924). "Dynamics of Muscular Contraction." *Proc. Roy. Soc., B*, **96**, 398.

GAUTHIER, G. F. & PADYKULA, H. A. (1965). Cytochemical studies of adenosine triphosphatase activity in the sarcoplasmic reticulum. *J. Cell Biol.*, **27**, 252–260.

GEBERT, G. (1968). Caffeine contracture of frog skeletal muscle and of single muscle fibres. *Amer. J. Physiol.*, **215**, 296–298.

GELFAN, S. (1930). "Studies of Single Muscle Fibres. I. The All-or-None Principle." *Amer. J. Physiol.*. **93**, 1.

GELFAN, S., & GERARD, R. W. (1930). "Studies of Single Muscle Fibres. II. Further Analysis of the Grading Mechanism." *Amer. J. Physiol.*, **95**, 412.

GIBBS, C. L., RICCHIUTI, N. V. & MOMMAERTS, W. H. F. M. (1966). Activation heat in frog sartorius muscle. *J. gen. Physiol.*, **49**, 517–535.

GOFFART, M., & RITCHIE, J. M. (1952). "The Effect of Adrenaline on the Contraction of Mammalian Muscle." *J. Physiol.*, **116**, 357–371.

GOODALL, M. C. (1958). Dependence of latent period in muscle on strength of stimulus. *Nature*, **182**, 1736–1737.

HARMAN, J. W. (1956). The cytochondria of cardiac and skeletal muscle. *Int. Rev. Cytol.*, **5**, 89–146.

HARRIS, E. J. (1957). Output of ^{45}Ca from frog muscle. *Biochim. biophys. Acta*, **23**, 80–87.

HARTREE, W. (1931). Analysis of the intial heat production of muscle. *J. Physiol.*, **72**, 1.

HARTREE, W. (1932). Analysis of delayed heat production of muscle. *J. Physiol.*, **75**, 273.

HASSELBACH, W. (1964). Relaxation and the sarcotubular calcium pump. *Fed. Proc.*, **23**, 909–912.

HASSELBACH, W. & MAKINOSE, M. (1961). Die Calciumpumpe der "Erschlaffungsgrana" des Muskels und ihre Abhängigkeit von der ATP-Spaltung. *Biochem. Z.*, **333**, 518–528.

HASSELBACH, W. & MAKINOSE, M. (1963). Über den Mechanismus des Calcium transportes durch die Membranen des Sarkoplasmatischen Reticulums. *Biochem. Z.*, **339**, 94–111.

HEILBRUNN, L. V., & WIERCINSKI, F. J. (1947). "The Action of various Cations on Muscle Protoplasm." *J. cell. comp. Physiol.*, **29**, 15–32.

HERZ, R. & WEBER, A. (1965). Caffeine inhibition of Ca uptake by muscle reticulum. *Fed. Proc.*, **24**, 208.

HILL, A. V. (1938). The heat of shortening and the dynamic constants of muscle. *Proc. Roy. Soc., B*, **126**, 136–195.

HILL, A. V. (1949). "Heat of Activation and Heat of Shortening in a Muscle Twitch." *Proc. Roy. Soc., B*, **136**, 195.

HILL, A. V. (1949). "Energetics of Relaxation in a Muscle Twitch." *Proc. Roy. Soc., B*, **136**, 211.

HILL, A. V. (1949). Work and heat in a muscle twitch. *Proc. Roy. Soc., B*, **136**, 220.

HILL, A. V. (1949). "The Onset of Contraction." *Proc. Roy. Soc.*, B, **136**, 242.

HILL, A. V. (1949). "The Abrupt Transition from Rest to Activity in Muscle." *Proc. Roy. Soc.*, B, **136**, 399.

HILL, A. V. (1950). "Development of the Active State of Muscle during the Latent Period." *Proc. Roy. Soc.*, B, **137**, 320.

HILL, A. V. (1950). "A Note on the Heat of Activation in a Muscle Twitch." *Proc. Roy. Soc.*, B, **137**, 330.

HILL, A. V. (1951). "The Influence of Temperature on the Tension developed in an Isometric Twitch." *Proc. Roy. Soc.*, B, **138**, 349-354.

HILL, A. V. (1953). "The Mechanics of Active Muscle." *Proc. Roy. Soc.*, B, **141**, 104-117.

HILL, A. V. (1953). "The 'Plateau' of Full Activity during a Muscle Twitch." *Proc. Roy. Soc.*, B, **141**, 498-503.

HILL, A. V. (1953). "A Reinvestigation of Two Critical Points in the Energetics of Muscular Contraction." *Proc. Roy. Soc.*, B, **141**, 503-510.

HILL, A. V. (1958). The priority of the heat production in a muscle twitch. *Proc. Roy. Soc.*, B, **148**, 397-402.

HILL, A. V. (1961). The heat produced by a muscle after the last shock of a tetanus. *J. Physiol.*, **159**, 518-545.

HILL, A. V. (1964, a). The effect of load on the heat of shortening of muscle. *Proc. Roy. Soc. B.*, **159**, 297-318.

HILL, A. V. (1964, b). The effect of tension in prolonging the active state in a twitch. *Proc. Roy. Soc.*, B, **159**, 589-595.

HILL, A. V. (1964, c). The variation of total heat production in a twitch with velocity of shortening. *Proc. Roy. Soc.*, B, **159**, 596-605.

HILL, A. V. (1964, d). The efficiency of mechanical power development during muscular shortening and its relation to load. *Proc. Roy. Soc. B.*, **159**, 319-324.

HILL, A. V., & HOWARTH, J. V. (1959). The reversal of chemical reactions in contracting muscle during an applied stretch. *Proc. Roy. Soc.*, B, **151**, 169-193.

HILL, D. K. (1940). "Time Course of the O_2-Consumption of Stimulated Frog's Muscle." *J. Physiol.*, **98**, 207.

HILL, D. K. (1940). "Time Course of Evolution of Oxidation Recovery Heat of Frog's Muscle." *J. Physiol.*, **98**, 454.

HILL, D. K. (1949). "Changes in Transparency of a Muscle during a Twitch." *J. Physiol.*, **108**, 292.

HILL, D. K. (1962). The location of creatine phosphate in frog's striated muscle. *J. Physiol.*, **164**, 31-50.

HILL, D. K. (1968). Tension due to interaction between the sliding filaments in resting striated muscle. The effect of stimulation. *J. Physiol.*, **199**, 637-684.

HODGKIN, A. L., & HOROWICZ, P. (1957). The differential action of hypertonic solutions on the twitch and action potential of a muscle fibre. *J. Physiol.*, **136**, 17-18P.

HODGKIN, A. L., & HOROWICZ, P. (1960). Potassium contractures in single muscle fibres. *J. Physiol.*, **153**, 386-403.

HODGKIN, A. L., & HOROWICZ, P. (1960). The effect of nitrate and other anions on the mechanical response of single muscle fibres. *J. Physiol.*, **153**, 404-412.

HOWELL, J. N. & JENDEN, D. J. (1967). T-tubules of skeletal muscle: morphological alterations which interrupt excitation-contraction coupling. *Fed. Proc.*, **26**, 553.

HUXLEY, A. F. (1957). Muscle structure and theories of contraction. *Progr. Biophys.*, **7**, 255-318.

HUXLEY, A. F., & TAYLOR, R. E. (1958). Local activation of striated muscle fibres. *J. Physiol.*, **144**, 426-441.

HUXLEY, H. E. (1964). Evidence for the continuity between the central elements of the triads and extracellular space in frog sartorius muscle. *Nature*, **202**, 1067-1071.

HUXLEY, H. E. & BROWN, W. (1967). The low-angle X-ray diagram of vertebrate striated muscle and its behaviour during contraction and rigor. *J. mol. Biol.*, **30**, 383-434.

IKEMOTO, N., SRETER, F. A., NAKAMURA, A. & GERGELY, J. (1968). Tryptic digestion and localization of calcium uptake and ATPase activity in fragments of sarcoplasmic reticulum. *J. Ultrastr. Res.*, **23**, 216-232.

INFANTE, A. A. & DAVIES, R. E. (1965). The effect of 2, 4 -dinitrofluoro-benzene on the activity of striated muscle. *J. biol. Chem.*, **240**, 3996-4001.

ISAACSON, A. & SANDOW, A. (1967). Quinine and caffeine effects on ^{45}Ca movements in frog sartorius muscle. *J. gen. Physiol.*, **50**, 2109-2128.

JANKE, J. (1968). Die Aufteilung des intracellularen ADP, ATP und Orthophosphate in der ruhenden und tetanisch gereizten Froschmuskulatur mittels einer fraktionierten Extraktion. *Pflüg. Arch. ges. Physiol.*, **300**, 1–22.

JENDEN, D. J. & REGER, J. F. (1962). Calcium deprivation and contractile failure in frog sartorius muscles. *J. Physiol.*, **164**, 22–23 P.

JENKINSON, D. H. & NICHOLLS, J. G. (1961). Contractures and permeability change produced by acetylcholine in depolarized denervated muscle. *J. Physiol.*, **159**, 111–127.

JEWELL, B. R., & WILKIE, D. R. (1960). The mechanical properties of relaxing muscle. *J. Physiol.*, **152**, 30–47.

JÖBSIS, F. F. & O'CONNOR, M. J. (1966). Calcium release and reabsorption in the sartorius muscle of the toad. *Biochem. Biophys. Res. Comm.*, **25**, 246–252.

KAO, C. Y. & STANFIELD, P. R. (1968). Actions of some anions on electrical properties and mechanical threshold of frog twitch muscle. *J. Physiol.*, **198**, 291–309.

KATZ, B. (1939). The relation between force and speed in muscular contraction. *J. Physiol.*, **96**, 45–64.

KATZ, B. (1942). "Impedance Changes in Frog Muscle associated with Electrotonic and 'Endplate' Potentials." *J. Neurophysiol.*, **5**, 169.

KATZ, B. (1950). Discussion on muscluar contraction. *Proc. Roy. Soc. B*, **137**, 40.

KIELLEY, W. W. & MEYERHOF, O. (1948). A new Mg-activated ATPase in muscle. *J. biol. Chem.*, **176**, 591.

KUFFLER, S. W. (1946). "Relation of Electric Potential Changes to Contracture in Skeletal Muscle." *J. Neurophysiol.*, **9**, 367.

LEE, K. S., LADINSKY, H., CHOI, S. J. & KASUYA, Y. (1966). Studies on the *in vitro* interaction of electrical stimulation and Ca^{++} movement in sarcoplasmic reticulum. *J. gen. Physiol.*, **49**, 689–715.

LEE, K. S., TANAKA, K. & YU, D. H. (1965). Studies on the adenosine triphosphate, calcium uptake and relaxing activity of the microsomal granules from skeletal muscle. *J. Physiol.*, **179**, 456–478.

LEVIN, A., & WYMAN, J. (1927). "The Viscous Elastic Properties of Muscle." *Proc. Roy. Soc., B*, **101**, 218.

LÜTTGAU, H. C. (1963). The action of calcium ions on potassium contractures of single muscle fibres. *J. Physiol.*, **168**, 679–697.

LÜTTGAU, H. C., & NIEDERGERKE, R. (1958). The antagonism between Ca and Na ions on the frog's heart. *J. Physiol.*, **143**, 486–505.

LÜTTGAU, H. C. & OETLIKER, H. (1968). The action of caffeine on the activation of the contractile mechanism in striated muscle fibres. *J. Physiol.*, **194**, 51–74.

MACPHERSON, L., & WILKIE, D. R. (1954). "The Duration of the Active State in a Muscle Twitch. *J. Physiol.*, **124**, 292–299.

MARÉCHAL, G. & BECKERS-BLEUKX, G. (1966). Adenosine triphosphate and phosphorylcreatine breakdown in resting and stimulated muscles after treatment with 1-fluro-2, 4 -dinitrobenzene. *Biochem. Z.*, **345**, 286–299.

MARÉCHAL, G. & MOMMAERTS, W. F. H. M. (1963). The metabolism of phosphocreatine during an isometric tetanus in the frog muscle. *Biochim. biophys. Acta.*, **70**, 53–67.

MARTONOSI, A. & FERETOS, R. (1964). The uptake of Ca^{++} by sarcoplasmic reticulum fragments. *J. biol. Chem.*, **239**, 648–657.

MARTONOSI, A. & FERETOS, R. (1964). Sarcoplasmic reticulum II. Correlation between adenosine triphosphate activity and Ca^{++} uptake. *J. biol. Chem.*, **239**, 659–668.

MOMMAERTS, W. F. H. M. (1954). "Is Adenosine Triphosphate broken down during a single Muscle Twitch." *Nature*, **174**, 1083–1084.

MOMMAERTS, W. F. H. M. (1954). The biochemistry of muscle. *Ann. Rev. Biochem.*, **23**, 381–404.

MOMMAERTS, W. F. H. M., OLMSTED, M., SERAYDARIAN, K., & WALLNER, A. (1962). Contraction with and without demonstrable splitting of energy-rich phosphate in turtle muscle. *Biochim. biophys. Acta*, **63**, 82–92.

MOMMAERTS, W. F. H. M., SERAYDARIAN, K., & MARÉCHAL, G. (1962). Work and chemical change in isotonic muscular contractions. *Biochim. biophys. Acta*, **57**, 1–12.

MOMMAERTS, W. F. H. M., SERAYDARIAN, K., & WALLNER, A. (1962). Demonstration of phosphocreatine splitting as an early reaction in contracting frog sartorius muscle. *Biochim. biophys. Acta*, **63**, 75–81.

MOMMAERTS, W. F. H. M. & WALLNER, A. (1967). The break-down of adenosine triphosphate in the contraction cycle of the frog sartorius muscle. *J. Physiol.*, **193**, 343–357.

MORALES, M. F. & BOTTS, J. (1952). A model of the elementary process in muscle action. *Arch. Biochem.*, **37**, 283–300.

MURPHY, R. A. (1966). Correlations of ATP content with mechanical properties of metabolically inhibited muscle. *Amer. J. Physiol.*, **211**, 1082–1088.

MUSCATELLO, U., ANDERSSON-CEDERGREN, E. & AZZONE, G. F. (1962). The mechanism of muscle-fibre relaxation adenosine triphosphatase and relaxing activity of the sarcotubular system. *Biochim. biophys. Acta.*, **63**, 55–74.

NAGAI, T. MAKINOSE, M. & HASSELBACH, W. (1960). Der physiologische Erschlaffungs-faktor und die Muskelgrana. *Biochim. biophys. Acta.*, **43**, 233–238.

NAKAJIMA, S., NAKAJIMA, Y. & PEACHEY, L. D. (1969). Speed of repolarization and mor-phology of glycerol-treated muscle fibres. *J. Physiol.*, **200**, 115–116P.

NATORI, R. (1954). The property and contraction process of isolated myofibrils. *Jikeikai Med. J.*, **1**, 119–126.

NATORI, R. (1965). Propagated contractions in isolated sarcolemma-free bundles of myofibrils. *Jikeikai Med. J.*, **12**, 214.

NEEDHAM, D. M. (1960). Biochemistry of muscular action. In *Structure and Function of Muscle*. Ed. Bourne. New York: Academic Press, pp. 55–104.

NIEDERGERKE, R. (1955). "Local Muscular Shortening by Intracellularly Applied Calcium." *J. Physiol.*, **128**, 12–13 P.

ORKAND, R. K. (1962). The relation between membrane potential and contraction in single crayfish muscle fibres. *J. Physiol.*, **161**, 143–159.

ORKAND, R. K. (1962). Chemical inhibition of contraction in directly stimulated crayfish muscle fibres. *J. Physiol.*, **164**, 103–115.

ORKAND, R. K. (1968). Facilitation of heart muscle contraction and its dependence on external calcium and sodium. *J. Physiol.*, **196**, 311–325.

PAUSCHINGER, P., & BRECHT, K. (1959). Uber die Beziehung von Erregung und Kon-traktion an der quergestreiften Muskelfaser. *Naturwiss.*, **46**, 267–268.

PAUSCHINGER, P., & BRECHT, K. (1961). Influence of calcium on the potassium-contracture of "slow" and "fast" skeletal muscle fibres of the frog. *Nature*, **189**, 583–584.

PAUSCHINGER, P., & BRECHT, K. (1961). Der Einfluss von Änderungen der extracellulären Calciumkonzentration auf die Ermüdungskontraktur von Skeletmuskeln. *Pflüg. Arch.*, **272**, 254–261.

PEACHEY, L. D. (1965). The sarcoplasmic reticulum and transverse tubules of the frog's sartorius. *J. Cell Biol.*, **25**, Pt. 2, No. 3, 209–231.

PEACHEY, L. D. (1966). The role of transverse tubules in excitation contraction coupling in striated muscles. *Ann. N.Y. Acad. Sci.*, **137**, 1025–1037.

PEASE, D. C., JENDEN, D. J. & HOWELL, J. N. (1965). Calcium uptake in glycerol-extracted rabbit psoas muscle fibers. II. *J. cell. comp. Physiol.*, **65**, 141–153.

PERRY, S. V. (1952). The bound nucleotide of the isolated myofibril. *Biochem. J.*, **51**, 495–499.

PODOLSKY, R. J. (1961). The nature of the contractile mechanism in muscle. In *Biophysics of Physiological and Pharmacological Actions*. Ed. Shanes. Washington: Amer. Ass. Adv. Sci., pp. 461–482.

PODOLSKY, R. J. (1962). The structural changes in isolated myofibrils during calcium-activated contraction. *J. gen. Physiol.*, **45**, 613–614A.

PODOLSKY, R. J. & MORALES, M. F. (1956). The enthalpy change of adenosine triphosphate hydrolysis. *J. biol. Chem.*, **218**, 945–959.

PORTZEHL, H. (1951). Muskelkontraktion und Modellkontraktion. II. *Z. Naturf.*, **6b**, 355–361.

PORTZEHL, H. (1952). Der Arbeitzyklus geordneter Aktomyosinsysteme. *Z. Naturf.*, **7b**, 1–10.

PORTZEHL, H., CALDWELL, P. C. & RÜEGG, J. C. (1964). The dependence of contraction and relaxation of muscle fibres from the crab *Maia squinado* on the internal concentration of free calcium ions. *Biochim. biophys. Acta.*, **79**, 681–591.

PORTZEHL, H. & WEBER, H. H. (1950). Zur Thermodynamik der ATP Kontraktion des Aktomyosinfadens. *Z. Naturf.*, **5b**, 123.

PRATT, F. H., & EISENBERGER, J. P. (1919). "Quantal Phenomena in Muscle." *Amer. J. Physiol.*, **49**, 1.

RAMSEY, R. W. (1944). Muscle Physics. *Medical Physics.* Ed. Glasser. Year Book Pub-lishers, Chicago.

RAMSEY, R. W. & STREET, S. F. (1940). The isometric length-tension diagram of isolated skeletal muscle fibres of the frog. *J. cell. comp. Physiol.*, **15**, 11.

RITCHIE, J. M. (1954). "The Duration of the Plateau of Full Activity in Frog Muscle." *J. Physiol.*, **124**, 605–612.

RITCHIE, J. M. & WILKIE, D. R. (1955). The effect of previous stimulation on the active state of muscle. *J. Physiol.*, **130**, 488–496.

ROOS, J. (1932). The latent period of skeletal muscle. *J. Physiol.*, **74**, 17.

SAKAI, T. (1965). The effects of temperature and caffeine on activation of the contractile mechanism in the striated muscle fibres. *Jikeikai Med. J.*, **12**, 88–102.

SANDBERG, J. A. & CARLSON, F. D. (1966). The length dependence of phosphorylcreatine hydrolysis during an isometric tetanus. *Biochem. Z.*, **345**, 212–231.

SANDOW, A. (1944). "General Properties of Latency-relaxation." *J. cell. comp. Physiol.*, **24**, 221.

SANDOW, A. (1955). "Contracture Responses of Skeletal Muscle." *Amer. J. phys. Med.*, **34**, 145–160.

SANDOW, A. (1965). Excitation-contraction coupling in skeletal muscle. *Pharmacol. Rev.*, **17**, 265–320.

SANDOW, A. & IASACSON, A. (1966). Topochemical factors in potentiation of contraction by heavy metal cations. *J. gen. Physiol.*, **49**, 937–961.

SANDOW, A., TAYLOR, S. R., ISAACSON, A. & SEGUIN, J. J. (1964). Electrochemical coupling in potentiation of muscular contraction. *Science*, **143**, 577–579.

SCHÄDLER, M. (1967). Proportionale Aktivierung von ATPase-Aktivität und Kontraktions-spannung durch Calciumionen in isolierten contractilen Strukturen verschiedener Muskelarten. *Pflüg. Arch. ges. Physiol.*, **296**, 70–90.

SCHAEFER, H., & GÖPFERT, H. (1937). "Aktionstrom und optisches Verhalten des Froschmuskels in ihrer zeitlichen Beziehung zur Zuckung." *Pflüg. Arch.*, **238**, 684.

SCHRAMM, G. & WEBER, H. H. (1942). Uber monodisperse Myosinlosungen. *Koll. Z.*, **100**, 242.

SERAYDARIAN, K. & MOMMAERTS, W. F. H. M. (1965). Density gradient separation of sarcotubular vesicles and other particulate constituents of rabbit muscle. *J. Cell Biol.*, **26**, 641–656.

SOMMER, J. R. & HASSELBACH, W. (1967). The effect of glutaraldehyde and formaldehyde on the calcium pump of the sarcoplasmic reticulum. *J. Cell Biol.*, **34**, 902–905.

SPERELAKIS, N. & SCHNEIDER, M. F. (1968). Membrane ion conductances of frog sartorius fibers as a function of tonicity. *Amer. J. Physiol.*, **215**, 723–729.

SUGI, H. & OCHI, R. (1967). The mode of transverse spread of contraction initiated by local activation in single frog muscle fibers. *J. gen. Physiol.*, **50**, 2167–2176.

SZENT-GYORGYI, A. (1947). *Chemistry of Muscular Contraction.* New York: Academic Press.

TICE, L. W. & ENGEL, A. G. (1966). Cytochemistry of phosphatases of the sarcoplasmic reticulum II. *In situ* localization of the Mg-dependent enzyme. *J. Cell Biol.*, **31**, 489–499.

WEBER, A. (1951). Muskelkontraktion und Modellkontraktion. *Biochem. biophys. Acta*, **7**, 214–224.

WEBER, A. (1956). The ultracentrifugal separation of L-myosin and actin in an actomyosin sol under the influence of ATP. *Biochim. biophys. Acta*, **19**, 345–351.

WEBER, A. & HERZ, R. (1963). The binding of calcium to actomyosin systems in relation to their biological activity. *J. biol. Chem.*, **238**, 599–605.

WEBER, A. & HERZ, R. (1968). The relationship between caffeine contracture of intact muscle and the effect of caffeine on reticulum. *J. gen. Physiol.*, **52**, 750–759.

WEBER, A., HERZ, R. & REISS, I. (1963). On the mechanism of the relaxing effect of fragmented sarcoplasmic reticulum. *J. gen. Physiol.*, **46**, 679–702.

WEBER, A., HERZ, R. & REISS, I. (1964). The regulation of myofibrillar activity by calcium. *Proc. Roy. Soc. B.*, **160**, 489–501.

WEBER, H. H. & PORTZEHL, H. (1954). The transference of the muscle energy in the contraction cycle. *Progr. Biophys.*, **4**, 60–111.

WILANDER, B. (1966). Active state durations of rat gastrocnemius muscle. *Acta. physiol. scand.*, **68**, 1–17.

WILBRANDT, W. & KOLLER, H. (1948). Die Calcium-Wirkung am Froschherzen als Funktion des Ionengleichgewichts zwischen Zellmembran und Umgebung. *Helv. physiol. acta*, **6**, 208–221.

WILKIE, D. R. (1949). "Relation between Force and Velocity in Human Muscle." *J. Physiol.*, **110**, 249.

WILKIE, D. R. (1956). "Measurement of the Series Elastic Component at Various Times during a Single Muscle Twitch." *J. Physiol.*, **134**, 527–530.

WINEGRAD, S. (1961). The possible role of calcium in excitation-contraction coupling of heart muscle. *Circulation*, **24**, 523–529.

WINEGRAD, S. (1965). Autoradiographic studies of intracellular calcium in frog muscle. *J. gen. Physiol.*, **48**, 455–479.

WINEGRAD, S. (1968). Intracellular calcium movements of frog skeletal muscle during recovery from tetanus. *J. gen. Physiol.*, **51**, 65–83.

WOLEDGE, R. C. (1961). The thermoelastic effect of change of tension in active muscle *J. Physiol.*, **155**, 187–208.

WOLEDGE, R. C. (1968). The energetics of tortoise muscle. *J. Physiol.*, **197**, 685–707.

ZACHAR, J. & ZACHAROVÁ, D. (1966). Potassium contractures in single muscle fibres of the crayfish. *J. Physiol.*, **186**, 596–618.

CHAPTER XXII

VARIATIONS IN THE STRUCTURE AND PERFORMANCE OF THE CONTRACTILE MACHINERY

Slow and Fast Fibres

The great bulk of experimental work on muscular contraction has been carried out on vertebrate skeletal muscle, and of this type of muscle it is the "twitch", as opposed to the "slow", muscle fibre that has received most attention; this is because rapidity of action has been developed to its pitch in this type of muscle fibre. Thus striated skeletal muscle may be made up of twitch or slow fibres, or mixtures of these, exhibiting different contractural features that are not merely the consequence of differing innervation, the slow fibres having a less ordered myofibrillar system (*Felderstruktur*) and a dyad rather than a triad form of relationship between the T- and L-sarcotubular systems suggesting a different, and perhaps less highly developed, excitation-contraction coupling mechanism (see, *e.g.*, Hess, 1965).[*]

Mechanical Features. In general, the mechanical feature that distinguishes the slow fibre is the slower maximal velocity of shortening, whilst the maximum tetanic tension, P_o, is not different. As we should expect, the fusion-frequency for a tetanus is smaller in the slow muscle; for example, Buller & Lewis (1965) found 50 and 125 pulse/sec. for mammalian soleus (slow) and flexor hallucis longus (fast) respectively, and this has been ascribed to differing durations of the active state.

Contracture

The duration of the contracture induced by high external K^+ is greatly different according as a twitch muscle or a slow muscle is examined; in the case of the twitch muscle the response is brief whilst in a slow muscle, such as the rectus abdominis, the contracture is sustained. Since in both cases the membrane is depolarized, this means, essentially, that the twitch muscle is able to uncouple its excitation-contraction mechanism rapidly whilst the slow muscle is not. Pauschinger & Brecht showed that if the external concentration of Ca^{++} was increased tenfold, the contractures of both types of muscle were prolonged so that the twitch muscle's response approached that of slow muscle in normal Ringer. In a Ca^{++}-free medium, on the other hand, both muscles now behaved like twitch muscles, giving only brief responses.[†] Again, slow muscles after repeated stimulation go into a "fatigue-contracture" by contrast with fast

---

[*] The absence of a significant T-system, with access to the extracellular space, probably explains the lower computed membrane capacity of slow muscle; thus the high membrane capacity of twitch muscle is probably due to the use of the geometrical area of the fibre membrane whereas the correct value should be this plus the area of the T-system (Adrian & Peachey, 1965).

[†] Kutscha (1961) showed that the twitch of sartorius induced by acetylcholine could be prevented by washing out the Ca^{++}; Ba^{++} and Sr^{++} could replace Ca^{++} but Mg^{++} antagonized its effect. When the denervated diaphragm in a high-K^+ Ringer's solution is treated with acetylcholine it goes into contracture, although the membrane potential remains at approximately zero. This seems to be due to an increased influx of Ca^{++}, since the effect is abolished by absence of this ion from the medium, whilst studies with [45]Ca indicate an increased influx (Jenkinson & Nicholls, 1961). We may note that smooth muscle, completely depolarized in KCl, contracts and then relaxes; it contracts in response to acetylcholine (Evans, Schild & Thesleff, 1958) but not in the absence of Ca^{++} (Robertson, 1960). Hurwitz (1961) has discussed the problem of excitation-contraction coupling in smooth muscle.

muscles, which remain relaxed when they cease to respond; if the fast muscle (sartorius) is treated with high-Ca^{++} solution, its response to repeated stimulation now becomes a fatigue-contracture (Pauschinger & Brecht). The difference between fast and slow muscles thus seems to be connected with the amounts of Ca^{++} available to them to provide the coupling between excitation and contraction. It may well be that slow muscles depend to a greater extent on the extra influx of Ca^{++} during the spike for their coupling.*

Relaxing Factor. Activity of "relaxing factor" will clearly be important for rapid contraction, since the development of speed in contraction requires also speed in relaxation; it is interesting, therefore, that the efficiency of the Ca^{++}-pump of the sarcoplasmic reticulum of slow muscle is only one-third of that of fast muscle; furthermore, the yield of sarcoplasmic reticulum per gramme is only about a half, so that total relaxing activity of the slow muscle of rabbit is probably one-sixth of that of fast muscle.

ATPase Activity. ATPase activity of actomyosin and myosin is another parameter that should determine speed of contraction; according to Harigaya, Ogawa & Sugita (1968) these activities are much less in preparations from slow muscle; again Bárány *et al.* (1965) have shown that, although both types of muscle contain the same amounts of myosin per gramme, the ATPase activity of the myosin from slow muscle—whether activated by EDTA, actin or Ca^{++}—was 2–3 times less. In fact, when Bárány (1967) compared a wide variety of muscles, ranging from the fast and slow rabbit skeletal muscles, through rabbit uterus and *Mytilus* posterior adductor, with speeds varying from 24 muscle lengths/sec. to 0·1, the ATPase also ranged from 20 to 0·1 moles ATP split/moles myosin/sec. All mammalian foetal muscles are slow, and it is interesting that myosin ATPase from foetal rabbits was a half as active as the adult preparation.

Invertebrate Striated Muscles. Invertebrate striated muscles may also be characterized as twitch or slow; their anatomical and physiological characteristics have been described by Fahrenbach (1967); the twitch fibre gives a twitch in response to a brief stimulus although, unlike the corresponding vertebrate fibre, the membrane response may vary from a small graded depolarization to a full-sized spike. With sustained depolarization the twitch fibre relaxes rapidly. The slow fibres develop tension only in response to sustained depolarization, and relax very slowly. Structurally the most obvious difference between the two fibres is the short sarcomere length, namely 4·5 μ, in the twitch fibre compared with 12 μ in the slow one; in the electron-microscope the myofibrillar arrangements are different, one thick filament being surrounded by 6 thin in the fast and 12 in the slow. The fast fibres have a well ordered sarcoplasmic reticulum and the T-elements, derived from clefts in the plasma membrane, invade the spaces between the myofibrils.†

Cardiac Muscle

Here the maximum velocity of shortening and the maximal tension, P_o, are both less than those of the fast skeletal fibre; the large number of mitochondria in cardiac muscle presumably restricts the number of myofilaments per unit cross-sectional area, and this probably accounts for the smaller P_o. The long duration of the refractory period of cardiac muscle precludes the development

---

* The mechanical features of fast and slow muscles, including the *a* and *b* parameters of the force-velocity relationship (p. 1428) have been compared by Wells (1965).

† The microsome fraction of lobster muscle accumulates Ca^{++} and Sr^{++} in the presence of oxalate, and at a rate adequate to account easily for the relaxation-time (Van der Kloot & Glovsky, 1965); the authors point out that the overshoot in the action potential of barnacle muscle, induced by injections of K_2SO_4-EDTA, is unlikely to be due to the establishment of an inward gradient of Ca^{++}-concentration, since this gradient is very steep normally, due to this active uptake.

of a fused tetanus, so that strict comparison of maximal tensions, P_o, is not practicable.

Sarcotubular System. In the electron-microscope the myocardial fibre differs little from the skeletal fibre except in respect to the sarcotubular system, which is characteristically different (Forssman & Girardier, 1966); serial sections showed that the T-system was open to the extracellular space, a continuity that was confirmed by the accumulation of ferritin particles within the T-tubules;[*] there were no classical triads, the appearance being described by Forssman & Girardier as "triadoid", the characteristic membrane thickenings being absent. Rostgaard & Behnke (1965) described flattened cisterns just under the plasma membrane at the intercalated disc, where the plasma membrane ran parallel to the myofibrils (p. 1274); these probably belong to the sarcoplasmic reticulum rather than the T-system. Ferritin particles appeared within the L-system, indicating a connexion between this and the extracellular space, and tubules of the L-system were seen to open into this space at the intercalated discs.

Function of Ca^{++}. If this is true, a great deal more of the cardiac muscle fibre's Ca^{++} may be accounted as extracellular, and the heart will depend to a much greater extent on Ca^{++}-influx during the action potential for excitation-contraction coupling than skeletal muscle. This was, indeed, found by Winegrad (1961) when he plotted the relative uptake of ^{45}Ca per beat, at various frequencies of beat, against the developed tension, a straight-line relation being found, the greater the tension the greater the influx of ^{45}Ca. Again, Lüttgau (1963) showed that, if he replaced Ca^{++} by Mg^{++} in Ringer's solution, skeletal muscle could continue to contract for some time, presumably because it could employ intracellular stores; by contrast, the heart stopped at once.[†] In a similar way, Grossman & Furchgott (1964) showed that the amplitude of contraction of heart muscle increased with external concentration of Ca^{++} in precisely the same way as the increased influx of ^{45}Ca.[‡]

Ca^{++}-Fluxes

In a more elaborate study in which Ca^{++}-fluxes were measured when the heart muscle was caused to contract either by high-K^+ and low Na^+—contracture—or during normal beating, Niedergerke (1963) established that contraction was associated with an increased influx of ^{45}Ca; and this led, under conditions of prolonged contraction, to net uptake of Ca^{++} which was extruded on return-

---

[*] This continuity between extracellular space and T-system had also been described by Simpson & Oertelis (1962) and Nelson & Benson (1963). Rayns, Simpson & Bertaud (1967, 1968) have described the freeze-etched appearance of the transverse tubular system in cardiac muscle; they consider that their pictures represent inside and outside views of the system; from the outside, the tubules appear as an array of pits, and from the inside as the stumps of tubules arranged in rows corresponding to the space between myofibrils. We may note that the frog's ventricle has no transverse tubular system (Staley & Benson, 1968).

[†] Forssman & Girardier (1966) suggested that the mitochondria were responsible for maintaining a low intracellular concentration of Ca^{++} in cardiac fibres, i.e., for relaxing activity. This seems an unnecessary hypothesis, since Carsten (1964) and Briggs, Gertz & Hess (1966) have demonstrated uptake of Ca^{++} by sarcoplasmic vesicles comparable with that of skeletal muscle preparations. We may note that Sommer & Johnson (1968) have compared the essentially conducting Purkinje with the contracting ventricular VO fibre; the main difference is that there is no transverse T-system in the P-fibre; they attribute the transverse tubules, described by Page (1966) in the Purkinje fibre, to mere invaginations of the sarcolemma. As with skeletal muscle, ATPase activity seems to be restricted to the L-system so that only the terminal cisterns of triads show deposits of Pb phosphate in histochemical studies; when AMP was used as a substrate, precipitate was confined to the T-system, however (Rostgaard & Behnke, 1965).

[‡] These authors distinguished several pools of Ca^{++}, namely a minimum intracellular of some 0·5 mMoles/kg., an extracellular amount equivalent to an extracellular space of some 21 per cent. of the muscle weight; a pool of Ca^{++} associated with contraction, and a residue unaccounted for.

ing the muscle to a normal medium. Niedergerke suggested, as a working model, that depolarization caused an influx of Ca^{++} mediated by a carrier with which Na^+ and Ca^{++} competed, and this led to the constancy of the factor $[Ca^{++}]/[Na]^2$ described by Wilbrandt & Koller (1948) when measuring the effects of these ions on the strength of the heart beat. Immediately on leaving the membrane inside the cell, the Ca^{++} joined an "active pool" and contributed to activate the contractile machinery, a process that was counteracted by removal into an "inactive pool" whilst the level of Ca^{++} was ultimately held constant by active extrusion from the cell by Ca^{-++}pumps.

Staircase Phenomenon

Support for Niedergerke's model was provided by Orkand's study of the facilitation of contraction; thus the well-known staircase phenomenon is revealed by increased force of contraction with repeated stimulation, the action potential being relatively unaffected. Clearly, if some of the extra influx of Ca^{++} from the first stimulus is held over, *i.e.*, is neither extruded nor passed into the "inactive pool", the basis of facilitation is a gradual accumulation of active Ca^{++}. Orkand found the characteristic $[Ca^{++}]/[Na^+]^2$ antagonism, so far as the strength of the first contraction in a series was concerned; however, facilitation was less at low $[Ca^{++}]$ and $[Na^+]$ than at high, so that the ratio was not constant; and this accords with Niedergerke's finding that, under these conditions, the accumulation of ^{45}Ca was less. Again, at low external concentrations of Ca^{++} there was no facilitation,* and this corresponds with the finding that influx and efflux of Ca^{++} are the same so that there is no accumulation; when external Ca^{++} is raised, influx increases and transiently exceeds outflux, and so facilitation is observed. With prolonged stimulation, the ventricle comes into a steady state, with influx and efflux balanced, and the developed tension reaches a plateau. As with skeletal striated muscle, caffeine potentiates contraction; according to De Gubareff & Sleator (1965) this effect is strongly dependent on the concentration of Ca^{++} in the medium, and this suggests that influx of Ca^{++} is more important than release of internal stores.†

Dependence of Contraction on Duration of Action Potential. By contrast with the twitch system of skeletal muscle, where the contraction phase outlasts the action potential by many msec., the duration of the cardiac action potential is generally of the same order as that of its contraction, so that here the action potential may not only trigger the contractile process but also determine its duration. Morad & Trautwein (1968) voltage-clamped bundles of cardiac muscle fibres, using a sucrose-gap technique, and measured the development of tension when the spike was abruptly terminated at different phases; so long as the spike was allowed to last as long as its upstroke, development of tension was normal, but the maximum height and duration of tension were reduced if the spike-duration was reduced to less than some 200 msec. (normal duration 450–650 msec.). Thus only the first 200 msec. of the spike were obliga-

---

* Orkand discusses contrary findings that the staircase decreases with raised concentration of Ca^{++}, due, apparently, to failure to appreciate the very slow decay of facilitation in these preparations.

† The position of Ca^{++} in the excitation-concentration coupling is rather confused (Nayler, 1965; Daniel, 1965); there are definitely two main fractions of Ca^{++} in cardiac muscle differing in the rapidity with which they exchange with ^{45}Ca in the medium, and the same applies to the frog sartorius muscle, but it is not easy to decide whether the rapidly exchanging component is truly intracellular or whether it belongs to the sarcotubular system. According to Grossman & Furchgott, it is only the rapid phase of uptake of ^{45}Ca that is affected by frequency of stimulation. Kleinfeld & Stein (1968) have shown that a variety of divalent cations have a negative inotropic effect; since this is reversed by Ca^{++} it seems likely that, here again, we have an influence on excitation-contraction coupling, the ions probably competing with Ca^{++} for influx.

tory for the development of normal tension, so that when the amplitude of contraction was plotted against the duration of the spike this rose to a plateau at 200 msec.* With very brief pulses, tension failed to appear if these lasted less than 1 msec.; at 5 msec., the tension developed was 10–15 per cent. of normal. As with skeletal muscle, the tension developed as a result of a lengthy depolarization (2 sec.) increased with the degree of depolarization; with cardiac muscle the threshold membrane potential was −25 mV and maximum tension was reached with a positive potential of +75 mV. The importance of penetration of Ca^{++} in excitation-contraction coupling is shown by Fig. 787, where it is seen that, at high external Ca^{++}, a given duration of spike causes a much larger amplitude of contraction.

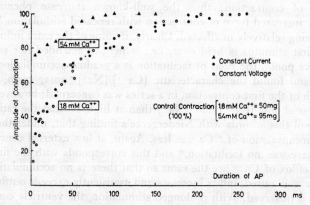

FIG. 787. Showing the effect of calcium on the relation between duration of action potential and amplitude of contraction of cardiac muscle. For details see text. Note that a given duration of action potential is more effective with the high concentration of Ca^{++}. (Morad & Trautwein. *Pflüg. Arch.*)

Cardiac Relaxing Factor. Microsomes of heart muscle are active in taking up Ca^{++} *in vitro* (Fanburg & Gergely, 1965); the features of this uptake are markedly different from those of mitochondria, so that it is very unlikely that the microsomal uptake could be due to contamination by mitochondria or their fragments. Patriarca & Carafoli (1968) have suggested, however, that the mitochondria of heart muscle do contribute to removal of sarcoplasmic Ca^{++}; they injected ^{45}Ca into rats and found the greatest specific activity in the mitochondria of heart muscle, rather than in the microsomes.

Energetics of Cardiac Contraction. The thermal aspects of cardiac contraction have been examined by Gibbs, Mommaerts & Ricchiuti (1967); the muscle differs little from skeletal muscle with respect, for example, to the relation between heat liberated and tension developed; when allowed to shorten, the muscle developed more heat but it was not practicable to identify the period of heat liberation with the actual period of shortening, so that this extra heat may not be strictly analogous with Hill's shortening heat. Efficiency, computed as the ratio of Work/Work + Heat was 0·12; this is low by comparison with *in vivo* estimates, but this was probably because of the experimental arrangement; for example, the heart does not naturally contract isotonically, but behaves like an ergometer; the authors considered that they may well have

* Rumberger (1968) showed that the second differential for development of tension, d^2p/dt^2, had a maximum at the end of the action potential of heart muscle, suggesting that the action potential continued to control the development of tension throughout its whole duration.

underestimated the efficiency by a half, in which case it would be comparable with Hill's figure of 0·20 for frog muscle.*

Myosin System. According to Bárány *et al.* (1964), the myosin extracted from rabbit cardiac muscle has the same molecular weight as that from skeletal muscle; it contains fewer cysteine residues, however, and this may be related to the lower nucleoside-phosphatase activity of the rabbit actomyosin.† Actomyosin extracted from the muscle is sensitive to small amounts of Ca^{++}, in so far as superprecipitation is concerned, and, as with skeletal muscle, the effects can be reversed by a relaxing factor or by Ca^{++}-chelating agents (Otsuka, Ebashi & Imai, 1964; Fanburg, Finkel & Martonosi, 1964).

Vertebrate Smooth Muscle

Myofilaments. This contractile tissue seems to be the only one in which a sliding filament mechanism may be unlikely, since it has been difficult, if not impossible, to demonstrate two types of filaments *in situ*. The filaments normally observed *in situ*, or in homogenates, are thin, of 50–70A diameter, and represent the actin component of smooth muscle. Those described by Panner & Honig (1967) had average dimensions of 40 × 60A in cross-section, and were similar to the actin filaments of striated muscle; they seemed to be made up of 30A diameter monomers on a double helix; with negative staining, they were 80A in cross-section and the globular subunits had 50A diameter; with 13 subunits per turn this gave a cross-over point every 350A.

Thick Filaments

Myosin may, indeed, be extracted from homogenates, and this will combine with actin to produce actomyosin, whilst under appropriate conditions an actomyosin may be directly extracted from smooth muscle, *e.g.*, tonoactomyosin from arterial smooth muscle.‡ Again, Hanson & Lowy (1963) separated a myosin from smooth muscle and by dialysing the preparation against a solution of low ionic strength and ATP concentration they obtained filaments some 250A in diameter and 0·5 μ long, thus demonstrating that the myosin can, under appropriate conditions, exist in the form of thick filaments. Kelly & Rice (1968) observed that thick filaments could, indeed, be isolated from chicken gizzard homogenates provided that the pH was held less than 6·6; these were 110 to 210A in diameter and 0·35 to 1·1 μ long, tapered at the ends. In thin sections of the muscle, too, they described fibrils of diameter up to 200A with a most common average diameter of 140A, and they concluded that the lability of myosin in neutral or alkaline media was responsible for failure to observe them in the usual preparations. Each thick fibril in the electron-micrographs was surrounded by a rosette of thin 50–70A fibrils, so that if this work can be substantiated, this arrangement provides the basis for a sliding filament mechanism of contraction.§ We may note that X-ray diffraction studies fail to reveal the

---

* Amytal reduces the force of contraction of heart muscle; this may be related to its inhibitory action on the uptake of Ca^{++} by sarcoplasmic vesicles since the effect is antagonized by ouabain, which likewise restores the strength of contraction (Briggs, Gertz & Hess, 1966).

† Carney & Brown (1966) have examined the molecule of cardiac myosin in the electron-microscope; like that of skeletal muscle it had the head-and-tail shape, the tail being some 1610A long with a cross-section of 15–20A and the head some 210A long with 35–40A cross-sectional diameter. We may note that Ebashi *et al.* (1966) have prepared an α-actinin from cardiac muscle.

‡ According to Needham & Williams (1963), skeletal muscle myosin will react with smooth muscle actin to give a highly viscous actomyosin solution.

§ Lane (1965) had described two types of filament with 30 and 80A diameters in intestinal smooth muscle cells; however, Panner & Honig considered the thicker filaments to represent artefacts.

presence of any reflexions that would correspond to an ordered arrangement of myosin molecules (Elliott, 1964; 1967).

Dense Bodies

A feature of the vertebrate smooth muscle cell is the dense body, which may well be the analogue of the Z-line of striated muscle; thus in molluscan "striated muscle" dense bodies, closely related to thin filaments, are regularly disposed in positions analogous to those of Z-lines (Fig. 788), and Panner & Honig showed that the dense bodies in vertebrate smooth muscle were probably densely packed thin filaments cemented together with some other protein. Contraction might

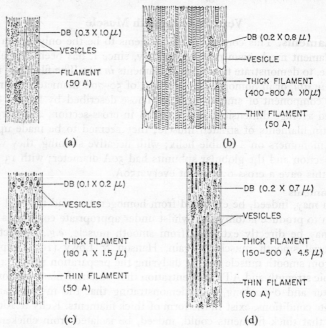

(a)

(b)

(c)

(d)

FIG. 788. Illustrating the morphological features of several types of muscle. (a) Transversely striated heart muscle fibres of *Archachatina* and *Sepia*. (b) Obliquely striated fast adductor fibres of *Crassostrea*. (c) Mammalian smooth muscle. (d) Molluscan smooth catch muscle. DB = dense body. (Twarog. *J. gen. Physiol.*)

occur by interaction between thin filaments anchored at dense bodies, and the interaction might be promoted by relatively small aggregates of myosin.*

Function of Ca++. As with other types of muscle examined, the activation of contraction seems to require Ca++, but a comparable sarcotubular system is absent, so that the immobilization of Ca++-ions within the cytoplasm is considered to occur on microcrystalline deposits analogous to those in mitochondria. The movement of Ca++ from the extracellular space into the fibre is probably important, and it would be for this reason that the completely depolarized smooth muscle contracts when Ca++ is added to the medium (Daniel, 1965).

---

* The myosin extracted from vertebrate smooth muscle, *e.g.*, chicken gizzard, has a comparable sedimentation coefficient with that of skeletal muscle, and in the electron-microscope the molecule has the typical rod-plus-head structure; the actin-activated Mg-ATPase activity is much less than that of skeletal muscle, and this is presumably related to the slow rate of contraction (Bárány *et al.*, 1966). Ebashi *et al.* (1966) have isolated both α-actinin and native tropomyosin from chicken gizzard.

Relation to Acetylcholine

Acetylcholine presumably increases permeability to Ca^{++} (Durbin & Jenkinson, 1961), thereby permitting its entry, although an additional factor may well be the release of Ca^{++} from a storage site within the fibre, as postulated by Evans & Schild; thus the internal concentration of free ions is of the order of $10^{-7}M$ (Van Breemen, Daniel & Van Breemen, 1966) compared with a Ca^{++}-content equivalent to $1\cdot0$ mM.* As Hurwitz *et al.* (1967) emphasize, the action of Ca^{++} is complicated by its effect on membrane permeability; thus it presumably binds to the surface to exert its effect on permeability, whilst another fraction is bound to internal structures and is available to activate the contractile machinery when released. Acetylcholine may merely act by competing with Ca^{++} for sites on the outside membrane. The effects of high external K^+ are different; it causes no increased influx of ^{45}Ca, as with acetylcholine (Van Breemen & Daniel, 1966), so that it presumably mobilizes intracellularly bound Ca^{++}.

Calcium Content and Exchange with Medium

The exchanges of Ca^{++} between the environment and smooth muscle have been studied in several tissues and, as Lüllmann & Siegfriedt (1968) have pointed out, the results are often contradictory. They found that the Ca^{++} content of longitudinal intestinal muscle depended on the concentration in the medium, varying from $1\cdot0$ μmole/g. in $0\cdot6$ μmolar medium to $3\cdot6$ μmole/g. in $2\cdot7$ μmolar medium; hence variations in the published values might well be due, at least in part, to variations in the composition of the bathing medium. ^{45}Ca exchanged readily with all the Ca^{++} in the tissue, except when Mg^{++} was absent from the medium. Wash-out of ^{45}Ca-loaded muscle exhibited two phases, in addition to that corresponding to loss of extracellular ion; this suggests that much of the intracellular Ca^{++} is sequestered. It is interesting that influx of ^{45}Ca was *accelerated* by increasing the external concentration of Ca^{++}.

Invertebrate Smooth Muscle

Catch-mechanism. The adductor muscles of the lamellibranchs such as the oyster or mussel are noted for their ability to remain contracted for a very long time when holding the shells closed; this is brought about by the expenditure of very little energy, so that a "catch-mechanism" has been invoked to explain the phenomenon, it being assumed that in some way the filaments are locked together. This may be an unnecessary hypothesis, since it has been shown by Bozler (1930) that the slower the response of a muscle the smaller is the energy required to maintain a given amount of tension, so that the adductor muscle of the snail, for example, is actually 30–180 times more economical than a frog's skeletal muscle (Bozler, 1930), with the result that a strong tension can be maintained without an impossible extra oxygen-consumption. Direct determination of the heat production by the retractor of the anterior byssus of *Mytilus*, carried out by Abbott & Lowy, showed that with this muscle, too, tonus could be effectively maintained with a heat production only one-twentieth of that required for a tetanus in frog's striated muscle. Nevertheless the question arises as to why, with a fast muscle, so much energy is consumed in doing what the slow muscle can do for so much less.

Structure. The muscles studied in respect to the catch-mechanism have been the anterior byssus retractor muscle (ABRM) of the mussel, *Mytilus edulis*, the smooth adductor of the clam, *Pecten*, and the adductor of the oyster, *Crassostrea angulata*; most of these muscles are made up of two parts, a translucent part

---

* Bauer, Goodford & Huter (1965) found some $2\cdot7$ mM/kg. wet wt. of Ca^{++} in the guinea-pig tænia coli; this increased by a statistically significant amount if the muscles were submitted to a sustained load.

that can close the shell rapidly and an opaque part that can hold the shell shut for a long time but only shortens slowly. The electron-microscopical studies of Philpott, Kahlbrock & A. G. Szent-Györgyi on a variety of smooth adductor muscles, and of Hanson & Lowy on that of the oyster, showed that these were essentially "striated" muscles, in the sense that they contained systems of thick and thin filaments made up of actin and myosin. The thin actin filaments were similar to those of vertebrate striated muscle, being some 50–80A in diameter, whilst the thick filaments were much thicker, with diameters varying from 300 to 1,500A.

Tropomyosin and Catch-mechanism

A feature of the thick filaments was the large amount of tropomyosin A* contained in them, in fact this protein dominated the X-ray picture and was described as *paramyosin*, the invertebrate muscles being described as "paramyosin muscles", their small-angle X-ray diffraction pattern being dominated by a fibrous protein with an axial repeat of 145A with accentuation of every fifth repeat to give a true periodicity of 725A (Bear; Bear & Selby). Since the thick fibres are discontinuous it is reasonable to postulate a sliding-filament mechanism for contraction. Because of the large amount of tropomyosin in these adductor muscles it was argued by Rüegg (1957, 1964) that this was in some way connected with the catch-mechanism (Rüegg, 1957), so that this type of smooth muscle might be expected to contain the orthodox actomyosin system, responsible for shortening, and a system of tropomyosin fibrils that shortened passively with the rest of the muscle but, when shortened, became "crystallized" in this shortened state and so took over the tension-maintaining task. That this type of muscle possessed the actomyosin system was shown by Rüegg, who extracted actomyosin from the smooth adductor of the clam, *Pecten*; this had high ATPase activity, in contrast with the tropomyosin which had none. Again, Rüegg showed that when the three regions of the clam's adductor were studied, namely, its striated portion and its translucent and opaque smooth portions, the proportions of actomyosin were 75, 47 and 20 per cent. respectively, whilst the proportions of tropomyosin were less than 1, 28 and 56 per cent. respectively. Thus the proportion of actomyosin increased with the degree to which the muscle showed phasic properties, whilst the tropomyosin proportion increased inversely with this characteristic.

Johnson *et al.* (1959) showed that the contraction of a glycerol-extracted ABRM muscle of *Mytilus* was sensitive to pH, being maximal at pH greater than pH 6·7. The solubility of paramyosin (from *Venus mercenaria*) was also pH-sensitive, so that at pH 6 it was insoluble and highly soluble at pH 7·5; thus it was argued that when the muscle showed powers of shortening the paramyosin was in the soluble condition; when only the "catch mechanism" was operating, the paramyosin would be in the crystallized state.

Subsidiary Function of Tropomyosin

As Lowy, Millman & Hanson (1964) and Hanson and Lowy (1964) have argued, there is little doubt that a sliding-filament mechanism operates in these catch-type muscles, the thick filaments having cross-bridges similar to those of striated skeletal muscles; during contraction their lengths and axial X-ray

---

* Bailey & Rüegg (1960) have described two forms of tropomyosin in the smooth muscle of lamellibranch adductors; the globulin type responsible for the paramyosin properties of the fibre is called tropomyosin A; tropomyosin B is considered to be a component in the actomyosin system and, according to Kominz, Saad & Laki (1957), it may be similar to the tropomyosin of vertebrates. These authors have examined the amino-acid composition of tropomyosins from a wide variety of classes of the vertebrate phylum and in different invertebrate phyla.

periodicities do not change. They point out, too, that the phasic parts of lamelli-branch muscles also contain large amounts of tropomyosin so that it is likely that tropomyosin is a supplementary protein, whose incorporation into the thick filament gives it the extra strength required of long sarcomeres; thus long sarcomeres, whilst they make for slowness of contraction, lead to the development of high tensions, so that extra strength becomes necessary; this is evident from Table LXXXIV, which compares some parameters in frog sartorius, the obliquely striated part of the oyster's adductor, and the ABRM of *Mytilus*.

TABLE LXXXIV

Some Parameters of Different Types of Muscle (Lowy, Millman & Hanson, 1964).

	Frog Sartorius	Oyster Adductor (Obliquely striated)		ABRM Mytilus	
Max Speed (Lengths/sec.)	6·0	1·5	(1/4)	0·25	(1/24)
Length of Thick Filaments (μ)	1·6	ca 8·0	($\times$5)	ca 30	($\times$19)
Tension per filament (10^{-8}g.)	5·2	60	($\times$11)	ca 450	($\times$86)
Area Thick Filament (A^2)	10^4	3.10^5	($\times$30)	ca 10^6	($\times$100)
Tension (10^{-8}g.) per 10^4 Cross-Sectional Area Thick Filament	5·2	ca 2		ca 4·5	

The magnitudes of the parameters of the oyster adductor and ABRM are indicated, in brackets, as fractions or multiples of those of the frog sartorius.

Slow Breakage of Bonds

In the light of this evidence, Lowy & Millman reject the simple morphological basis of the catch-mechanism and point to the alternative hypothesis, put forward by them earlier, that the essential basis for sustained contraction with low energy consumption is the slowness with which the actin-myosin links are broken down. Thus the energy for sustained contraction depends, on the basis of such models as that of A. F. Huxley, on the tendency for the bonds to break. If this happened very slowly much less energy would be required; this would favour tonic contraction but would, of course, be prejudicial to shortening, since this depends on the successive making and breaking of bonds. Thus the essence of the catch-mechanism would be the switch over to a process of slow breaking of bonds.

Electrical and Mechanical Features. The essential mechanical features of contraction by the anterior byssus retractor muscle of *Mytilus* were demonstrated in a classical paper by Winton (1937); he showed that if the muscle was stimulated repetitively the response was a phasic contraction, whilst if it was stimulated by direct current the response was the slow sustained "catch-type" contraction; in this condition the muscle exhibited high "viscosity", in the sense that the passive resistance to stretch was very high. Repetitive stimulation of the muscle during the phase of sustained contraction caused rapid *relaxation*.

Electrophysiology. The electrical events were first studied by Fletcher (1937). The muscle responded to direct electrical stimulation by an action potential which preceded the mechanical contraction; the threshold was high (1–2 V compared with a few mV in nerve), and the action potential was small, never greater than 6 mV, presumably because of short-circuiting. When one recording electrode was on an injured region, the action potential was definitely monophasic, indicating that the conduction was along fibres stretching the length of the muscle. In general, the total duration of the spike varied from 1·0 to 2·8 sec., the rising phase lasting 0·1 to 0·3 sec. It travelled without apparent decrement at about 18 cm./sec. at room temperature. One feature in particular, which is not observed in striated muscle, was the fusion of the action potentials at frequencies of stimulation greater than about five per sec. Apart from this feature, and the much slower time-scale of events, the smooth muscle of *Mytilus* behaved similarly to the striated muscle of the frog's sartorius.

Pharmacology

In later studies Twarog (1954, 1960, 1967) generally confirmed Fletcher's work and showed that agents such as KCl, acetylcholine and adrenaline all caused depolarization associated with contraction, the extent of depolarization being roughly proportional to the strength of contraction. Acetylcholine blocking agents, such as *d*-tubocurarine, inhibited the effects of acetylcholine. 5-Hydroxytryptamine (5-HT) caused a marked relaxation in muscle tonically contracted by acetylcholine, and Twarog considered that excitatory and inhibitory impulses to the smooth muscle were mediated by acetylcholine and 5-hydroxytryptamine respectively; at any rate, the muscle contained both these compounds.

Junction Potentials

The resting intracellularly recorded potential varied from 55 to 72 mV; submaximal stimulation gave rise to depolarizations that were described as junctional potentials with maximum amplitude of 20–25 mV; with repetitive stimulation successive junction potentials summated, and facilitation was observed; the spike arose from the junctional potential if this reached some 35–40 mV and this was always associated with contraction. Tetrodotoxin in concentrations up to 10^{-4} g/ml. had no effect on spike amplitude or contraction, and it is interesting that no overshoot of the spike was ever observed (Twarog, 1967, *b*). It would seem that junction potentials were the major component of electrical activity recorded by Fletcher, and this would account for the summation observed. The contraction caused by acetylcholine is of the tonic type, so that experimentally this aspect is often studied by applying this transmitter, whilst 5-HT is used to cause relaxation of tonic contraction.

Action Potentials

The catch-mechanism in its simplest formulation suggests that tension is maintained with no action potentials, and contrasts with the "tetanus mechanism" that postulates a continuous tonic nervous discharge. Support for the "catch-mechanism" theory was given by the failure of early investigators to detect action potentials during tonic contraction of the adductor muscles; however, in the intact animal, Lowy showed that tonus was definitely associated with bursts of action potentials occurring at regular intervals; between the bursts, much smaller and less frequent discharges occurred. In the scallop, *Pecten*, the adductor consists of a smooth and a striated part, the one operating in tonic contraction and the other in phasic swimming movements. Lowy showed that when the animal was undisturbed its shell closed; under these conditions bursts of action potentials were recorded from the smooth muscle, and none from the striated portion. When swimming movements occurred, action potentials were recorded from the striated portion only; these seemed to be the result of synchronized discharges of all-or-none fibres, by contrast with the very asynchronous discharge in the smooth muscle, whose tension may well have been maintained by graded responses similar to those in crustacean striated muscle. Bowden & Lowy showed that the smooth adductor muscle contained a rich nerve supply, including ganglia, so that the tonus, which is maintained when the extrinsic nerve supply is cut, might well be due to local nervous activity. At any rate, under these conditions, action potentials could be recorded, so that the tonus was not a simple contracture.

Inhibition

The possibility of inhibiting tonus in molluscan muscles was first shown by Pavlov, and has been confirmed by a number of more recent studies. For example, Benson, Hays & Lewis were able to induce relaxation in the slow muscle of *Pecten* by stimulation of the mouth, the relaxation being a reflex response mediated by a centre in the cerebral ganglion. By suitably cutting parts of the nervous supply to the slow muscle, they were able to obtain a preparation that responded by relaxation when the ganglion or the peripheral end of the cut nerve was stimulated. In *Mytilus*, it will be recalled, Twarog found that 5-hydroxy-tryptamine caused relaxation, possibly by mimicking the action of an inhibitory nerve, and in the intact animal Lowy was able to induce relaxation by appropriate stimulation of the cerebral ganglion.*

Some Mechanical Features. The mechanical responses to stimulation that reveal the peculiar features of the catch-mechanism have been studied in several laboratories.† In general, the results agree in showing that, in this type of muscle, we must distinguish an *active* state, during which the muscle can develop tension, which is measured by the return of tension after a quick release, and the "*frozen*" state, in which the muscle holds a tension, if this is established either by previous tonic stimulation (direct current or acetylcholine) or by imposing a stretch. Thus, as Johnson & Twarog (1960) showed, if a muscle is stretched after a d.c.-stimulus it develops a much larger passive tension than if it is stretched after a phasic, a.c.-stimulus. Again, they showed that, if a muscle is stimulated with d.c. at short lengths, very little tension is developed, but in spite of this there is a large passive tension developed on stretch, showing that the d.c.-stimulus can turn on the "stiffness" without necessarily turning on the tension mechanism.

Decay of Contractile State

Fig. 789, *a*, from Lowy & Millman (1963) illustrates the isometric response of the ARBM of *Mytilus* to a single stimulus; here the response is tonic, in the sense that tension is maintained for over 20 minutes. Curve (2) shows the decline of the active state, determined by quick release (p. 1436), and Curve (3) represents the decay of the "contractile state", determined by subtracting from the tension a fixed fraction of the twitch-tension held at the time of release, which allows for the parallel elastic element that necessarily develops tension when the muscle is stretched.‡ The curves show that the muscle's tension outlasts its active state, or contractile activity, by many minutes. The curves of Fig. 789, *b*, were obtained in the presence of 5-HT, and now the behaviour is characteristic of the twitch type of muscle. When the muscle was stimulated by acetylcholine or direct current, the peak tension and shortening speed were

* Hoyle & Lowy have discussed the differing responses of *Mytilus* muscle to alternating and direct current stimuli in terms of the excitatory and inhibitory innervation.

† The interested reader may be referred to Jewell (1959), Johnson & Twarog (1960), Johnson (1962), Lowy & Millman (1963) and Millman (1964); the last mentioned author isolated the opaque portion of the oyster's adductor and showed that this had all the features attributed to the muscle as a whole.

‡ Lowy & Millman (1963, p. 139) justify the use of this concept because, with catch-type muscle, the redevelopment of tension after quick release reflects not only the development of active processes but also the effects of inert elastic material. When contractile activity has ceased, all the tension present—passive tension—can be considered to be due to the stress exerted by inert elastic elements on linkages that still have to be broken to complete the cycle.

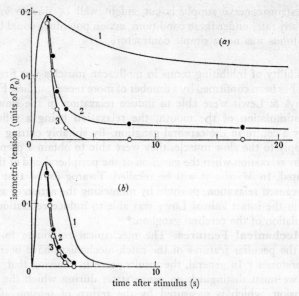

FIG. 789. Responses of the ABRM of *Mytilus* to single stimuli: (a) in sea-water, (b) in sea-water with 10^{-5} g/ml. 5 HT. Curves (1): isometric twitch. Curves (2) (●): "active-state", determined by measuring the tension re-developed following releases of 0·04 cm at a speed of 1·0 cm/sec. at different times after the stimulus. Curves (3) (○): decay of contractile activity, determined by subtracting from curves (2) an amount of tension equal to 0·14 of the twitch-tension held at the time of release. (Lowy & Millman. *Phil. Trans.*)

very similar to the phasic response, obtained by a repetitive stimulation, but in the latter case the exponential decay of tension was very much more rapid, the time-constant being 3 sec. compared with 31 min. for acetylcholine stimulation.

Passive Tension

Extension of the unstimulated muscle beyond resting length causes the development of a tension that decays slowly, to fall to ½-peak in about 30 min.; Lowy & Millman (1963) call this the *apparent resting tension*, to distinguish it from the true resting tension, which is obtained by treating the muscle with 5-HT, when the tension drops rapidly to a lower level and is now unaffected by a.c.-stimulation of the muscle; it is now analogous to the resting tension of frog sartorius. Thus it is the part of the tension abolished by 5-HT that is called by Lowy & Millman the *passive tension* at any moment; in *Pecten*, Bozler (1930) showed that such passive tension, induced by stretch, was similar to that caused by tonic stimulation; both decayed at the same rate and both were reduced by repetitive stimulation. Also, if the muscle is made inexcitable by treatment with $MgCl_2$, resistance to passive stretch is greatly reduced. Thus passive tension may be developed by previous contractile activity, or by passive stretch, whilst active tension is that which follows stimulation and gives the muscle power to shorten or develop tension actively.

TENSIONS DURING STRETCH

The tensions developed during stretch will obviously vary according to whether or not the muscle is being stimulated, and according to the type of isometric stimulus. Thus, when stretched from rest, it was 22 per cent. of P_o, the maximal tetanic tension; when stretched during repetitive stimulation, it was 46 per cent. of P_o; stretched during d.c.-stimulation it was 68 per cent., and *after* d.c.-stimulation it was 64 per cent. Similarly, during acetylcholine stimulation it was much higher than during repetitive stimulation or during repetitive stimulation in the presence of 5-HT, or with acetylcholine plus 5-HT.

Rapid and Slow Breakage of Linkages

All these results indicate a definite change in the state of muscle in the tonically contracted, or passively stretched, condition, a state that might well be associated with the presence of bridges between filaments that have become linked and are subject to only a very slow breakage. The magnitude of the peak stretch tension developed by a muscle would, on this basis, depend on the number of linkages remaining unbroken, and so should be greater, the greater the degree of passive tension following a tonic contraction; this is true. Thus the results are consistent with the hypothesis that contraction is produced by linkages of only one type, with one rate of formation, but with two different rates of breakage, one constant and the other variable. During active shortening, most of the linkages are broken at the constant rate,* whereas during lengthening, or isometric contraction, the linkages are broken at the variable rate, which is controlled by the concentration of relaxant (5-HT). In the absence of relaxant, linkages break at the slow rate and the muscle gives a tonic contraction; in the presence of relaxant the linkages break rapidly and a phasic contraction occurs.†

Chemical Aspects. In molluscan muscle the ADP, formed during metabolism, may be converted to ATP by phosphoryl arginine; and the breakdown of phosphoryl arginine has been used as a measure of ATP utilization in this type of muscle. Nauss & Davies (1966) showed that arginine liberation depended only on the work done by *Mytilus* ABRM, not on the mode of stimulation, so that when a tonically stimulated muscle was allowed to support a load there was no net increase in breakdown, in fact during this period 63 per cent. of the liberated inorganic phosphate was resynthesized to phosphoryl arginine. During relaxation from a tonic contraction, we might expect some ATP to be used up if the relaxation process involves sequestration of Ca^{++} by an ATP-activated pump; moreover, it can be (p. 1448) that ATP is involved in the filament linkage, so that breakage of the link results from a cleavage of this link by a myosin-ATPase to produce ADP; reconversion of this to ATP would involve breakdown of phosphoryl arginine. Nauss & Davies estimated that some 0·10 μmole of arginine would be liberated, and this compared with 0·2 μ/mole during relaxation. This study supports the contention of Lowy & Millman that the tonic contraction is only a modified phasic contraction.‡

O_2 Consumption

Nevertheless, more accurate experimental techniques, in which O_2 consumption was measured with an O_2-electrode during the development and maintenance of tension, revealed a very definite tension-related consumption during the slow process of relaxation from a tonic stimulus; release of the tension

---

* Lowy & Millman quote experiments of Howarth showing that the thermo-elastic response of ABRM, tonically contracted with acetylcholine, is the same as that of actively contracting amphibian striated muscle; when the acetylcholine was washed out, the thermo-elastic heat was initially unchanged, but it gradually reverted to that characteristic of resting muscle as the tension fell, suggesting that the links were still present but were gradually breaking.

† Thus the shortening speed at zero load is the same with tonic and phasic contractions.

‡ Some features of the development of catch tension in *Mytilus* have been described by Twarog (1967, *a*); an interesting point is that catch tension does not occur at temperatures above 30° C. In general, all factors that reduce catch tension *increase* muscle excitability. This is exemplified in the action of 5-HT; under conditions where catch is prominent this prolongs junction potentials and lowers threshold for spike discharge and contraction (Hidaka, Osa & Twarog, 1967).

caused a fall of this to a low value, whilst re-establishment of the tension brought the consumption back to its pre-release value (Baguet & Gillis, 1968); the actual amount of O_2 consumed is greatly in excess of the energy required to break the links established during development of tension, estimated on the basis of the breakdown of phosphoryl arginine during fast relaxation, so that it would seem that breakage of links *per se* is not the main cause of the energy consumption, and we must postulate a slow turnover of linkages during tonic contraction, too slow to permit recovery of tension from a quick release (Jewell, 1959).*

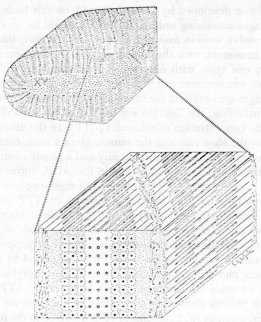

FIG. 790. Illustrating the ultrastructural basis of oblique striation in *Ascaris* muscle. In the XZ plane the myofilaments are staggered, with the result that the striations are oblique rather than transverse. A second consequence of the stagger is that the adjacent rows of myofilaments do not reach the XY plane in phase, resulting in the appearance of striation in this plane also. The YZ plane shows cross-striation. (Rosenbluth. *J. Cell Biol.*)

Obliquely Striated Muscle. The somatic muscle fibre of the parasitic nematode worm, *Ascaris lumbricoides*, reveals a complex series of oblique striations that have been examined by Rosenbluth (1965). The fibres contain two sets of myofilaments, thick and thin, exhibiting A- and I-bands, but because of the staggering of filaments in the direction of the long axis, a longitudinal section in the XZ-plane (Fig. 790) reveals an oblique striation; furthermore, in transverse sections the adjacent myofibrils do not reach the XY-plane, and this gives the appearance of striations in this plane, the section passing through I, A, H, A and I bands in this order. In the A-zone, each thick filament is surrounded by about 10–12 thin filaments; delicate cross-links seem to connect the thick with the thin filaments. At the lateral edges of the I-zones, the thin filaments become clumped together into small aggregates called Z-bundles; in the same dense zone that contains these Z-bundles there are dense bodies, which give the characteristic beaded appearance

* The oxygen consumption during phasic contraction of ABRM has been measured by Baguet & Gillis (1967); during the tetanus the cost of maintaining tension is remarkably low, about 1/250 that of frog sartorius, whilst the cost of establishing tension is about the same for the two muscles. An interesting feature of ABRM is the reduction in rate of O_2 consumption on stretching the muscle, the opposite of the Feng effect.

to the zone in the light-microscope. Finger-like extensions of the sarcolemma, extending deep into the fibre, probably represent the T-system; these come into relation with shallow intracellular membranous sacs that are probably the equivalent of the longitudinal system in vertebrate striated muscle. The musculature of the polychæte worm, *Glycera*, is also obliquely striated; it is physiologically faster, and this seems to be correlated with a much more highly developed sarcoplasmic reticulum (Rosenbluth, 1968).

REFERENCES

ABBOTT, B. C. & LOWY, J. (1958). Contraction in molluscan smooth muscle. *J. Physiol.*, **141**, 385–397.

ADRIAN, R. H. & PEACHEY, L. D. (1965). The membrane capacity of frog twitch and slow muscle fibres. *J. Physiol.*, **181**, 324–336.

BAGUET, F. & GILLIS, J. M. (1967). The respiration of the anterior byssus retractor muscle of *Mytilus edulis* (ABRM) after a phasic contraction. *J. Physiol.*, **188**, 67–82.

BAGUET, F. & GILLIS, J. M. (1968). Energy cost of tonic contraction in a lamellibranch catch muscle. *J. Physiol.*, **198**, 127–143.

BAILEY, K. & RÜEGG, J. C. (1960). Further chemical studies on the tropomyosins of lamellibranch muscle with special reference to *Pecten maximus*. *Biochim. biophys. Acta*, **38**, 239–245.

BÁRÁNY, M. (1967). ATPase activity of myosin correlated with speed of shortening. *J. gen. Physiol.*, **50**, No. 6, Pt. 2, 197–216.

BÁRÁNY, M., BÁRÁNY, K., GAETJENS, E. & BAILIN, G. (1966). Chicken gizzard myosin. *Arch. Biochem. Biophys.*, **113**, 205–221.

BÁRÁNY, M., BÁRÁNY, K., RECKARD, T. & VOLPE, A. (1965). Myosin of fast and slow muscles of the rabbit. *Arch. Biochem. Biophys.*, **109**, 185–191.

BÁRÁNY, M., GAETJENS, E., BÁRÁNY, K. & KARP, E. (1964). Comparative studies of rabbit cardiac and skeletal myosins. *Arch. Biochem. Biophys.*, **106**, 280–293.

BAUER, H., GOODFORD, P. J. & HÜTER, J. (1965). The calcium content and ^{45}calcium uptake of the smooth muscle of the guinea-pig taenia coli. *J. Physiol.*, **176**, 163–179.

BEAR, R. S. (1944). X-ray diffraction studies on protein fibres. II. Feather rachis, porcupine quill tip and clam muscle. *J. Amer. Chem. Soc.*, **66**, 2043–2050.

BEAR, R. S. & SELBY, C. C. (1956). The structure of paramyosin fibrils according to X-ray diffraction. *J. biophys. biochem. Cytol.*, **2**, 55–69.

BENSON, A. A., HAYS, J. T. & LEWIS, R. N. (1942). Inhibition in the slow muscle of the scallop, *Pecten circularis æquisulcatus* Carpenter. *Proc. Soc. exp. Biol., N.Y.*, **49**, 289–291.

BOWDEN, J. & LOWY, J. (1955). Innervation. *Nature*, **176**, 346–347.

BOZLER, E. (1930). The heat production of smooth muscle. *J. Physiol.*, **69**, 442–462.

BRIGGS, E. N., GERTZ, E. W. & HESS, M. L. (1966). Calcium uptake by cardiac vesicles: inhibition by amytal and reversal by ouabain. *Biochem. Z.*, **345**, 122–131.

BULLER, A. J. & LEWIS, D. M. (1965). The rate of tension development in isometric tetanic contractions of mammalian fast and slow skeletal muscles. *J. Physiol.*, **176**, 337–354.

CARNEY, J. A. & BROWN, A. L. (1966). An electron microscope study of canine cardiac myosin and some of its aggregates. *J. Cell Biol.*, **28**, 375–389.

CARSTEN, M. E. (1964). The cardiac calcium pump. *Proc. Nat. Acad. Sci. Wash.*, **52**, 1456–1462.

DANIEL, E. E. (1965). Attempted synthesis of data regarding divalent cations in muscle function. In *Muscle*. Ed. W. M. Paul *et al.* Oxford: Pergamon, pp. 295–313.

DE GUBAREFF, T. & SLEATOR, W. (1965). Effects of caffeine on mammalian atrial muscle, and its interaction with adenosine and calcium. *J. Pharmacol.*, **148**, 202–214.

DURBIN, R. P. & JENKINSON, D. H. (1961). The calcium dependence of tension development in depolarized smooth muscle. *J. Physiol.*, **157**, 90–96.

EBASHI, S., IWAKURA, H., NAKAJIMA, H., NAKAMURA, R. & OOI, Y. (1966). New structural proteins from dog heart and chicken gizzard. *Biochem. Z.*, **345**, 201–211.

ELLIOTT, G. F. (1964). X-ray diffraction studies on striated and smooth muscles. *Proc. Roy. Soc., B*, **160**, 467–472.

ELLIOTT, G. F. (1967). Variations of the contractile apparatus in smooth and striated muscles. X-ray diffraction studies at rest and in contraction. *J. gen. Physiol.*, **50**, No. 6, Pt. 2, 171–184.

EVANS, D. L., SCHILD, H. O. & THESLEFF, S. (1958). Effects of drugs on depolarized plain muscle. *J. Physiol.*, **143**, 474–485.

FAHRENBACH, W. H. (1967). The fine structure of fast and slow crustacean muscles. *J. Cell Biol.*, **35**, 69–79.

FANBURG, B., FINKEL, R. M. & MARTONOSI, A. (1964). The role of calcium in the mechanism of relaxation of cardiac muscle. *J. biol. Chem.*, **239**, 2298–2306.

FANBURG, B. & GERGELY, J. (1965). Studies on adenosine triphosphate-supported calcium accumulation by cardiac subcellular particles. *J. biol. Chem.*, **240**, 2721–2728.

FLETCHER, C. M. (1937). Action potentials recorded from an unstriated muscle of simple structure. *J. Physiol.*, **90**, 233–253.

FLETCHER, C. M. (1937). Excitation of the action potential of a molluscan unstriated muscle. *J. Physiol.*, **90**, 415–428.

FLETCHER, C. M. (1937). The relation between the mechanical and electrical activity of a molluscan unstriated muscle. *J. Physiol.*, **91**, 172–185.

FORSSMAN, W. G. & GIRARDIER, L. (1966). Untersuchungen zur Ultrastruktur des Rattenherzmuskels mit besonderer Berücksichtigung des Sarcoplasmatischen Retikulums. *Z. Zellforsch.*, **72**, 249–275.

GIBBS, C. L., MOMMAERTS, W. F. H. M. & RICCHIUTI, N. V. (1967). Energetics of cardiac contractions. *J. Physiol.*, **191**, 25–46.

GROSSMAN, A. & FURCHGOTT, R. F. (1964). The effects of frequency of stimulation and calcium concentration on Ca^{45} exchange and contractility in the isolated guinea-pig auricle. *J. Pharmacol.*, **143**, 120–130.

HANSON. J. & LOWY, J. (1961). The structure of the muscle fibres in the translucent part of the adductor of the oyster *Crassostrea angulata*. *Proc. Roy. Soc., B*, **154**, 173–196.

HANSON, J. & LOWY, J. (1963). The structure of F-actin and of actin-filaments isolated from muscle. *J. mol. Biol.*, **6**, 46–60.

HANSON, J. & LOWY, J. (1964). The structure of actin filaments and the origin of the axial periodicity in the I-substance of vertebrate striated muscle. *Proc. Roy. Soc., B*, **160**, 449–458.

HANSON, J. & LOWY, J. (1964). In *Biochemistry of Muscle Contraction*. Ed. J. Gergely. Boston: Little Brown, pp. 400–411.

HARIGAYA, S., OGAWA, Y. & SUGITA, H. (1968). Calcium binding activity of microsomal fraction of rabbit red muscle. *J. Biochem.*, (*Tokyo*), **63**, 324–331.

HESS, A. (1965). The sarcoplasmic reticulum, the T-system, and the motor terminals of slow and twitch muscle fibers in the garter snake. *J. Cell Biol.*, **26**, 467–476.

HIDAKA, T., OSA, T. & TWAROG, B. M. (1967). The action of 5-hydroxytryptamine on *Mytilus* smooth muscle. *J. Physiol.*, **192**, 869–877.

HOYLE, G. & LOWY, J. (1956). The paradox of *Mytilus* muscle. A new interpretation. *J. exp. Biol.*, **33**, 295–311.

HURWITZ, L. (1961). Electrochemistry of smooth muscle and its relation to contraction. In *Biophysics of Physiological and Pharmacological Actions*. Ed. Shanes. Washington: Amer. Ass. Adv. Sci., pp. 563–577.

HURWITZ, L., VON HAGEN, S. & JOINER, P. D. (1967). Acetylcholine and calcium in membrane permeability and contraction of intestinal smooth muscle. *J. gen. Physiol.*, **50**, 1157–1172.

JENKINSON, D. H. & NICHOLLS, J. G. (1961). Contractures and permeability change produced by acetylcholine in depolarized denervated muscle. *J. Physiol.*, **159**, 111–127.

JEWELL, B. R. (1959). The nature of the phasic and the tonic responses of the anterior byssus retractor muscle of *Mytilus*. *J. Physiol.*, **149**, 154–177.

JOHNSON, W. H. (1962). Tonic mechanisms in smooth muscle. *Physiol. Rev.*, **42**, Suppl. 5, 113–159.

JOHNSON, W. H., KAHN, J. S. & SZENT-GYÖRGYI, A. G. (1959). Paramyosin and contraction of "catch muscles". *Science*, **130**, 160–161.

JOHNSON, W. H. & TWAROG, B. M. (1960). The basis for prolonged contractions in molluscan muscles. *J. gen. Physiol.*, **43**, 941–960.

KELLY, R. E. & RICE, R. V. (1968). Localization of myosin filaments in smooth muscle. *J. Cell Biol.*, **37**, 105–116.

KLEINFELD, M. & STEIN, E. (1968). Action of divalent cations on membrane potentials and contractility in rat atrium. *Amer. J. Physiol.*, **215**, 593–599.

KOMINZ, D. R., SAAD, F. & LAKI, K. (1957). Vertebrate and invertebrate tropomyosins. *Nature*, **179**, 206–207.

KUTSCHA, W. (1961). Die Wirkung verschiedener Kontrakturstoffe auf den Verkurzungsvorgang des Skeletmuskels in Verbindung mit Calcium- und anderen Erdalkali-Ionen. *Pflüg. Arch.*, **273**, 409–412.

LANE, B. P. (1965). Alterations in the cytologic detail of intestinal smooth muscle cells in various stages of contraction. *J. Cell Biol.*, **27**, 199–213.

LOWY, J. (1953). Contraction and relaxation in the adductor muscles of *Mytilus edulis*. *J. Physiol.*, **120**, 129–140.

LOWY, J. (1954). Contraction and relaxation in the adductor muscles of *Pecten maximus*. *J. Physiol.*, **124**, 100–105.

LOWY, J. & MILLMAN, B. M. (1959). Contraction and relaxation in smooth muscle of lamellibranch molluscs. *Nature*, **183**, 1730–1731.

LOWY, J. & MILLMAN, B. M. (1963). The contractile mechanism of the anterior byssus retractor muscle of *Mytilus edulis*. *Phil. Trans.*, **246**, 105–148.

LOWY, J., MILLMAN, B. M. & HANSON, J. (1964). Structure and function in smooth tonic muscles of lamellibranch molluscs. *Proc. Roy. Soc.*, B, **160**, 525–536.

LÜLLMANN, H. & SIEGFRIEDT, A. (1968). Über den Calcium-Gehalt und den ^{45}Calcium austausch in Längsmuskulatur des Meerschweinchendünndarms. *Pflüg. Arch. ges. Physiol*, **300**, 108–119.

LÜTTGAU, H. C. (1963). The action of calcium ions on potassium contractures of single muscle fibres. *J. Physiol.*, **168**, 679–697.

MILLMAN, B. M. (1964). Contraction in the opaque part of the adductor muscle of the oyster (*Crassostrea angulata*). *J. Physiol.*, **173**, 238–262.

MORAD, M. & TRAUTWEIN, W. (1968). The effect of duration of the action potential on contraction in the mammalian heart muscle. *Pflüg. Arch. ges. Physiol.*, **299**, 66–82.

NAUSS, K. M. & DAVIES, R. E. (1966). Changes in inorganic phosphate and arginine during the development, maintenance and loss of tension in the anterior byssus retractor muscle of *Mytilus edulis*. *Biochem. Z.*, **345**, 173–187.

NEEDHAM, D. M. & WILLIAMS, J. M. (1963). The proteins of the dilution precipitate obtained from salt extracts of pregnant and non-pregnant uterus. *Biochem. J.*, **89**, 534–545.

NELSON, D. A. & BENSON, E. S. (1963). On the structural continuities of the transverse tubular system of rabbit and human myocardial cells. *J. Cells Biol.*, **16**, 297–313.

NIEDERGERKE, R. (1963). Movements of Ca in frog heart ventricles at rest and during contractures. *J. Physiol.*, **167**, 515–550.

ORKAND, R. K. (1968). Facilitation of heart muscle contraction and its dependence on external calcium and sodium. *J. Physiol.*, **196**, 311–325.

OTSUKA, M., EBASHI, F. & IMAI, S. (1964). Cardiac myosin B and calcium ions. *J. Biochem.*, (*Tokyo*), **55**, 192–194.

PAGE, E. (1966). Tubular systems in the cat heart. *J. Ultrastr. Res.*, **17**, 72–83.

PANNER, B. J. & HONIG, C. R. (1967). Filament ultrastructure and organization in vertebrate smooth muscle. *J. Cell Biol.*, **35**, 303–321.

PATRIARCA, P. & CARAFOLI, E. (1968). A study of the intracellular transport of calcium in rat heart. *J. cell. Physiol.*, **72**, 29–38.

PAUSCHINGER, P. & BRECHT, K. (1961). Influence of calcium on the potassium-contracture of "slow" and "fast" skeletal muscle fibres of the frog. *Nature*, **189**, 583–584.

PHILPOTT, D. E., KAHLBROCK, M. & SZENT-GYÖRGYI, A. G. (1960). Filamentous organization of molluscan muscles. *J. Ultrastr. Res.*, **3**, 254–269.

RAYNS, D. G., SIMPSON, F. O. & BERTAUD, W. S. (1967). Transverse tubule apertures in mammalian myocardial cells: surface array. *Science*, **156**, 656–657.

RAYNS, D. G., SIMPSON, F. O. & BERTAUD, W. S. (1968). Surface features of striated muscle. I. and II. *J. Cell Sci.*, **3**, 467–474; 475–482.

ROBERTSON, P. A. (1960). Calcium and contractility in depolarized smooth muscle. *Nature*, **186**, 316–317.

ROSENBLUTH, J. (1965). Ultrastructural organization of obliquely striated muscle fibers in *Ascaris lumbricoides*. *J. Cell Biol.*, **25**, 495–515.

ROSENBLUTH, J. (1968). Sarcoplasmic reticulum, contractile apparatus, and endomysium of the body muscle of a polychaete, *Glycera*, in relation to its speed. *J. Cell Biol.*, **36**, 245–259.

ROSTGAARD, J. & BEHNKE, O. (1965). Fine structural localization of adenine nucleoside phosphatase activity in the sarcoplasmic reticulum and the T-system of rat myocardium. *J. Ultrastr. Res.*, **12**, 579–591.

RÜEGG, J. C. (1957). Die Reinigung der Myosin-ATPase eines glatten Muskels. *Helv. Physiol. Pharmacol. Acta*, **15**, C 33–35.

RÜEGG, J. C. (1961). On the tropomyosin-paramyosin system in relation to the viscous tone of lamellibranch "catch" muscle. *Proc. Roy. Soc.*, B, **154**, 224–249.

RÜEGG, J. C. (1964). Tropomyosin-paramyosin system and "prolonged contraction" in a molluscan smooth muscle. *Proc. Roy. Soc.*, B., **160**, 536–542.

RUMBERGER, E. (1968). Uber Korrelationen Zwischen der Aktionspotentialdauer und dem zeitlichen Verlauf der Erschlaffung beim Herzmuskel des Warmund Kaltbluters. *Pflüg. Arch. ges. Physiol.*, **301**, 70–75.

SIMPSON, F. O. & OERTELIS, S. J. (1962). The fine structure of sheep myocardial cells: sarcolemmal invaginations and the transverse tubular system. *J. Cell Biol.*, **12**, 91–100.

SOMMER, J. R. & JOHNSON, E. A. (1968). Cardiac muscle. A comparative study of Purkinje fibres and ventricular fibres. *J. Cell Biol.*, **36**, 497–526.

STALEY, N. A. & BENSON, E. S. (1968). The ultrastructure of frog ventricular cardiac muscle and its relationship to mechanisms of excitation-contraction coupling. *J. Cell Biol.*, **38**, 99–114.

TWAROG, B. M. (1954). Response of a molluscan smooth muscle to acetylcholine and 5-hydroxytryptamine. *J. cell. comp. Physiol.*, **44**, 141–163.

TWAROG, B. M. (1960). Innervation and activity of a moluscan smooth muscle. *J. Physiol.*, **152**, 220–235.

TWAROG, B. M. (1967, a). Factors influencing contraction and catch in *Mytilus* smooth muscle. *J. Physiol.*, **192**, 847–856.

TWAROG, B. M. (1967, b). Excitation of *Mytilus* smooth muscle. *J. Physiol.*, **192**, 857–868.

TWAROG, B. M. (1967, c). The regulation of catch in molluscan muscle. *J. gen. Physiol.*, **50**, Pt. 2 of No. 6, 157–168.

VAN BREEMEN, C. & DANIEL, E. E. (1966). The influence of high potassium depolarization and acetylcholine on calcium exchange in the rat uterus. *J. gen. Physiol.*, **49**, 1299–1317.

VAN BREEMEN, C., DANIEL, E. E. & VAN BREEMEN, D. (1966). Calcium distribution and exchange in the rat uterus. *J. gen. Physiol.*, **49**, 1265–1297.

VAN DER KLOOT, W. G. & GLOVSKY, J. (1965). The uptake of Ca^{++} and Sr^{++} by fractions from lobster muscle. *Comp. Biochem. Physiol.*, **15**, 547–565.

WELLS, J. B. (1965). Comparison of mechanical properties between slow and fast mammalian muscles. *J. Physiol.*, **178**, 252–269.

WILBRANDT, W. & KOLLER, H. (1948). Die Calciumwirkung am Froschherzen als Funktion des Ionengleichgewichts zwischen Zellmembran und Umgebung. *Helv. Physiol. Acta*, **6**, 208–221.

WINEGRAD, S. (1961). The possible role of calcium in excitation-contraction coupling of heart muscle. *Circulation*, **24**, 523–529.

WINTON, F. R. (1937). The changes in viscosity of an unstriated muscle (*Mytilus edulis*) during and after stimulation with alternating, interrupted and uninterrupted direct currents. *J. Physiol.*, **88**, 492–511.

6. *Light: its Effect on, and its Emission by, the Organism*

CHAPTER XXIII

PHOTOSYNTHESIS, PHOTODYNAMIC ACTION, AND THE EFFECTS OF ULTRA-VIOLET LIGHT

LIGHT is the primary source of energy for green plants and certain coloured bacteria; the energy so absorbed is used to build up organic compounds of high energy-content which may be subsequently used in the metabolism of the cell; the process of absorption of energy is thus one of *photosynthesis*, and is described as *assimilation* in contrast to the process of *respiration* which involves the utilization of the energy stored by photosynthesis.* In the green plants and algæ the photosynthetic reaction consists in the conversion of CO_2 and H_2O to carbohydrate, it being generally agreed that other organic compounds within the plant, such as fats and proteins, are formed as a result of secondary chemical reactions. Stoichiometrically it is easy enough to write down a reaction of CO_2 with H_2O to give a sugar, thus:—

$$6 CO_2 + 6 H_2O \rightarrow C_6H_{12}O_6 + 6 O_2$$

As an overall description of photosynthesis it is probably correct, since it demands that for each volume of CO_2 absorbed, one volume of O_2 should be liberated, *i.e.*, that the *photosynthetic quotient*, $Q_P = \dfrac{\Delta O_2}{- \Delta CO_2}$, should be equal to unity. Numerous studies on green plants and algæ have shown that this quotient is, indeed, very close to unity. The reaction, as written above, appears simple; nevertheless, in common with the metabolic process of respiration, that of photosynthesis is, as we shall see, highly complex. At present, only one aspect of photosynthesis, namely the chemical reactions involved in the conversion of CO_2 to carbohydrate, is at all well understood. The most important aspect, namely the mechanism of absorption of light-energy, and of its transfer to the reacting molecules, has so far not been clarified, and only the broad nature of the process can be surmised from studies on the kinetics of photosynthesis; it is beyond the scope of this book to enter into a detailed analysis of these studies, nor yet to present in detail the recent conclusions of the prominent investigators in this field. It must suffice to present only a few of the main findings, and to indicate the general trend of opinion; to understand these findings we must, however, have a clear picture of the mechanism of chemical reactions, of which photosynthesis is a special case.

* It will be remembered that certain bacteria utilize the energy of chemical reactions, such as the oxidation of sulphur, in order to synthesize organic chemical compounds from CO_2; here assimilation consists of *chemosynthesis*. In the higher organisms assimilation consists largely in the absorption of organic compounds of high energy-content, "ready-made".

THE MECHANISM OF CHEMICAL REACTIONS

Photosynthesis is a special case of a more general type of chemical process —the *photochemical reaction*; and the understanding of the methods and results of modern studies is only possible with some knowledge of the fundamentals of reaction kinetics. Let us consider briefly, therefore, some of the elementary principles of this branch of physical chemistry. We have seen, in the second section of this book, that chemical reactions involve a redistribution of energy, which is generally manifest in the absorption or liberation of heat. The first problem of reaction kinetics is why, under a given set of conditions, certain reactions take place and others do not. The failure of certain reactions to occur spontaneously, such as the decomposition of water:—

$$2 H_2O \rightarrow 2 H_2 + O_2 - 68 \cdot 3 \text{ Cal.}$$

may reasonably be attributed to the absence of the necessary energy, at ordinary temperatures, to make such an endothermic reaction proceed, the absorption of the 68·3 Calories for each mole of water vapour decomposed placing an impossible drain on the thermal capacity of the system. However, the failure to observe the opposite reaction:—

$$2 H_2 + O_2 \rightarrow H_2O + 68 \cdot 3 \text{ Cal.}$$

at ordinary temperatures cannot possibly be due to this cause, so that it is not necessarily the exothermy or endothermy of a reaction that determines whether it will occur spontaneously at a measurable rate or not. If a flame is applied to a mixture of H_2 and O_2, the reaction proceeds violently, with the liberation of a great deal of energy as heat and light; the reaction is set off by the transfer of a relatively small amount of energy, and it is this transfer that gives the clue to the mechanism of chemical reaction. The molecules of H_2 and O_2 are very stable, the atoms being linked by bonds of high energy-content. The first step in the reaction between H_2 and O_2 must be a weakening, or rupture, of the bonds that hold the atoms together; and, because of the stability of these bonds, the energy required is high, 103 Calories in the case of H_2 and 117 Calories in the case of O_2. The final result of the reaction between H_2 and O_2 is to produce a compound with a smaller energy-content than that of the reactants, so that the excess energy is liberated as heat and light during the process, and the reaction is said to be exothermic; however, a considerable amount of energy is first required to weaken the bonds in the H_2 and O_2 molecules, and it is this requirement that makes necessary some preliminary "detonating action". Once a given pair of molecules has reacted, the water molecule may pass its extra energy to other H_2 and O_2 molecules,* and allow them to react to give more energy; and so the process goes on, gathering speed as a result of the liberation of energy until it may attain explosive proportions.

Activation Energy

An exothermic reaction will therefore proceed spontaneously at a measurable speed if there are a suitable number of molecules with the necessary *activation energy*; we may now ask what form does this activation energy take. The molecules of a gas may possess *kinetic*—or *translational*—energy, the average

---

* Actually the H_2O molecule *must* pass its energy on at the instant of formation, otherwise the atoms fly apart because of the high energy-content of the newly-formed molecule ; a *third body collision* is therefore a prerequisite for many reactions that are strongly exothermic.

for a perfect gas being given by $\frac{1}{2} Nm\bar{c}^2$ per mole, which* is equal to $3/2$ RT. In addition, a molecule may possess *internal energy* consisting of *rotations* about various axes, *vibrations* of the constituent atoms, and—finally—energy associated with the transition of the orbital electrons of the constituent atoms from one energy-state to another. Rotational and vibrational energy are associated with the emission or absorption of infra-red rays, whilst changes in electronic energy are accompanied by the emission or absorption of ultra-violet or visible light. We may thus speak of a variety of *energy-levels* of a given molecule, and, to be precise, we must indicate the state of the molecule in respect to these four types of energy—translational, rotational, vibrational and electronic. The last three types have a number of levels which are *quantized*, i.e., their energies are given by: $nh\nu$, where n is an integer, h is Planck's constant ($6\cdot5.10^{-27}$ erg. sec.) and ν is the frequency of the vibration, the energy corresponding to the case where n equals unity is the *quantum*, and is the smallest amount of energy that can be acquired or emitted by an oscillator of a given frequency.

Collision Theory

The energy of one molecule may be transmitted to another on collision; and this interchange of energy is not confined to the translational variety, a molecule with high electronic energy being able to transfer high vibrational energy by collision with another; moreover, the electronic energy of an excited molecule may be re-distributed into vibrational and rotational energies.

In an ideal gas, the number of collisions occurring per sec. in 1 ml. is given by kinetic theory as:—

$$z = 2 \, n^2\sigma^2(\pi RT/M)^{\frac{1}{2}}$$

(n = no. of molecules/ml. σ = collision diameter. R = $8\cdot3.10^7$ ergs. deg.$^{-1}$ mole^{-1}); and the number of molecules having more than a minimum energy, E, is given by:—

$$N = N_o \, e^{-E/RT}$$

By combining these equations one may compute the number of molecules colliding per sec. with more than this minimum energy, so that if E were the energy required for a collision to result in reaction, the rate of reaction would be given theoretically by:—

$$dx/dt = ze^{-E/RT}$$

assuming that each collision with the necessary amount of energy did, indeed, result in reaction. The activation energy, i.e., the energy required to induce the necessary instability in the reactants to permit the chemical change to occur, may be computed from the effect of temperature on the rate of reaction.

Temperature Coefficient

The relationship between activation energy, reaction-rate, and temperature is given by the Arrhenius equation:—

$$k = Ce^{-E/RT}$$

where k is the velocity constant of the reaction, and, on the basis of the "collision theory" of reaction kinetics, the constant, C, may be equated to z. Thus, if $\log k$ is plotted against $1/T$, a straight line is generally obtained, the

* N is the number of molecules in one mole, i.e., Avogadro's Number, $6\cdot023.10^{23}$; m is the weight of the molecule; $\bar{c}^2$ is the "mean square velocity" i.e., the mean of the squares of the velocities of the molecules; R is the gas constant, $8\cdot314.10^7$ ergs/°C/mole, or $1\cdot987$ cal./°C/mole, or $0\cdot082$ litre-atm./°C/mole.

slope being equal to $-E/R$, whence E may be obtained. The differential form of the Arrhenius equation is:—

$$\frac{d \ln k}{dT} = \frac{E}{RT^2}$$

from which it is clear that the increase in k with temperature will be greater the higher the activation energy, E, *i.e.*, the temperature coefficient, Q_{10}, equal to the ratio of the reaction rates at T and T + 10, will be higher, the higher the activation energy of the process. For most reactions studied, the Q_{10} lies between 2 and 3.* A Q_{10} in the region of 3 means that raising the temperature 100° C will increase the reaction rate about $3^{10} = 59,000$-fold; and this striking effect of temperature serves to emphasize the significance of activation energies. Thus, raising the temperature increases both the *frequency* and *energy* of the collisions; if all collisions resulted in reaction, the increased rate with a rise in temperature would be attributable only to the increased frequency of collision (which varies only as the square root of T). When, however, only a certain fraction of the collisions result in reaction, a fraction that is dependent on temperature, the reaction rate will increase on both counts: increased *frequency* of collision, and increased *average energy* of collision. Moreover, the higher the energy required (activation energy), the more will the process benefit on the second count, *i.e.*, the higher the activation energy the higher the Q_{10}. Finally, it will be obvious that the rate of reaction will vary inversely with the activation energy; consequently, the more rapid the reaction, the lower will be its Q_{10} (that is, of course, provided that factors other than collision-energy are not significant).

Order and Molecularity of Reactions

The mechanism of a reaction is elucidated by a study of the way it proceeds with time, the equation describing this determining its *"order"*. Thus, if the reaction:—

$$A + B \rightarrow AB$$

requires that A and B collide with a minimal energy, the equation governing the reaction may take the form:—

$$\text{Rate} = \text{Constant} \times [A] [B]$$

Such a reaction would be described as *bimolecular*, and its order of reaction would be the *second order*, because the equation takes the form:—

$$dx/dt = k(a-x)^2$$

where x is the amount of AB formed at time t, a is the original concentration of A and B, assumed equal, and k is the *velocity constant*.

First Order Reaction. A large number of reactions follow a simpler equation, of the form:—

$$dx/dt = k(a-x)$$

when they are called *first order reactions*, the rate of reaction at any time, t, being proportional to the number of unreacted molecules $(a-x)$. An example of this is the decomposition of nitrogen pentoxide:—

$$N_2O_5 \rightarrow N_2O_4 + \tfrac{1}{2}(O_2)$$

If the reaction depended simply on the collision of two molecules of N_2O_5, the decomposition taking place on collision, we should expect a second order

---

* The Q_{10} is not an accurate method of indicating the effect of temperature, since it varies with the temperature.

reaction, the rate being dependent on the square of the concentration of N_2O_5. If, on the other hand, the molecule of N_2O_5 spontaneously disintegrated, *i.e.*, without acquiring activation energy from collision, the reaction would be expected to follow a first order type of equation, just as the spontaneous radio-active decay of a substance does, since here the rate, *i.e.*, the number of molecules decomposed in a given time, is simply determined by the amount of substance remaining at any moment $(a-x)$. It could be argued, therefore, that, because it was a first order reaction, the decomposition of N_2O_5 occurred spontaneously, without activation. However, a study of the temperature coefficient of the reaction, and of many other first order reactions of a similar type, has shown that the activation energy is high, often as high as 40–60 Calories per mole. To overcome this difficulty, Lindemann suggested that the reacting molecules do indeed obtain their activation energy by collision, but that the rate at which this activation occurs is rapid compared with the subsequent decomposition. Thus, whereas the rate of activation is dependent on the square of the con-centration $(a-x)^2$, the actual rate of decomposition is determined, in ordinary circumstances, by the time required for the activated molecules to react; consequently the number of molecules reacting in a given time is proportional to the number of molecules present, *i.e.*, the state of affairs is analogous to spontaneous decomposition. When the rate of activation is sufficiently decreased —by lowering the pressure of the gas—we may expect the reaction rate to become dependent on this factor and to change from first to second order, a state of affairs that is actually found. This point reveals the importance of an exact analysis of the mechanism of a chemical reaction; a mere study of the overall kinetics suggested a non-activated type of reaction, similar to radio-active decay, whereas a further investigation revealed the existence of an activation process, the rate of which, however, was not *rate determining*. The first order reaction considered above is also *unimolecular*, in the sense that only one molecule decomposes at a time—*i.e.*, it is not necessary that two molecules of N_2O_5 should decompose simultaneously—so that the condition for activation is only that a collision should make *one* of the colliding molecules unstable, in contrast to the bimolecular reaction in which two molecules, A and B, had to react simultaneously on collision. A great many bimolecular reactions are first order for varying reasons; for example, the hydrolysis of cane sugar involves the reaction of a water molecule with a sugar molecule, so that it is definitely bimolecular; however, the concentration of water remains effectively constant during the reaction; hence the rate is determined by the product of a constant quantity with the concentration of sugar, *i.e.*, it is first order.* Many photochemical, and catalysed reactions, to be discussed later, are also first order, but for different reasons.†

* Thus $dx/dt = k.$ [Sugar] [Water]
$$= k\,(a-x) \times \text{Constant}$$
$$= k'\,(a-x)$$

The reaction is described as *pseudo-monomolecular*.

† It is important that the difference between *molecularity* and *order* of a chemical reaction should be appreciated, especially by biologists. A large number of reactions are first order, but by no means unimolecular, yet only too frequently does one find in the biological literature claims that a certain process involves the splitting of one molecule into two because the kinetics of the process are first order; or, again, that a process such as the regeneration of visual purple involves the interaction of two molecules because the kinetics follow a second order equation. The order of the reaction defines its kinetics, *i.e.*, how it proceeds with time, whilst the molecularity states how many molecules are involved in the actual reaction. The decomposition of N_2O_5 is unimolecular and first order, *i.e.*, although two molecules of N_2O_5 must collide to provide the activation energy, it is not necessary that both molecules decompose together.

Catalysed Reactions. Very many reactions take place at an interface, for instance on the surface of a solid, and are called *heterogeneous*, in contrast to the *homogeneous* reaction which takes place in a single phase; these reactions exhibit a variety of orders, the exponent, n, in the kinetic equation:—

$$\text{Rate} = k \, (a-x)^n$$

being not necessarily an integer and sometimes zero. A zero-order reaction would be represented by:—

$$\text{Rate} = k \, (a-x)^0$$
$$= k$$

i.e., the reaction occurs at a constant rate until completed. The conditions for such a zero-order reaction would be the following. The reaction takes place at a limited surface only, the surface being fully occupied at any moment by the reacting molecules; as soon as one molecule reacts and leaves the surface, another takes its place.* Thus, so long as the number of molecules on the surface is independent of the total concentration of the molecules, the rate of reaction will be independent of this concentration, and hence of zero order. If, on the other hand, only a small area of the available surface is occupied at any moment, the rate may be determined by the rate at which the molecules from the bulk phase reach and leave the surface, in which case, if the reaction is unimolecular—as for example the decomposition of phosphine on glass —the equation will be of the first order. With intermediate conditions of surface saturation it is clear that intermediate exponents will be obtained. The heterogeneous reactions mentioned here are examples of catalysed reactions, in which the catalyst represents a separate phase from that in which the bulk of the reactants are distributed, in contrast to the catalysis of the hydrolysis of sugar by hydrogen ions, in which catalyst and reactants occupy the same phase. In homogeneous catalysis, the acceleration of the reaction is undoubtedly brought about by the participation of the catalyst in the chemical reaction, a participation that permits the reaction to proceed in steps involving less activation energy than in its absence. Thus the conversion of SO_2 to SO_3 by O_2:—

$$SO_2 + \tfrac{1}{2} O_2 \rightarrow SO_3$$

is catalysed by nitric oxide, NO. A simple explanation would be that NO underwent oxidation first and was then reduced by SO_2:—

$$NO + \tfrac{1}{2} O_2 \rightarrow NO_2$$
$$NO_2 + SO_2 \rightarrow NO + SO_3$$

the catalyst behaving as an intermediary; although this may be too simple a view of the actual process, there is little doubt that NO permits the reaction to proceed along paths involving, at each stage, less activation energy than that involved in the simple reaction:—

$$SO_2 + \tfrac{1}{2} O_2 \rightarrow SO_3$$

Specificity

Catalysis taking place at solid surfaces is often not very specific, a variety of different reactions being catalysed by the same solid, *e.g.*, glass. Presumably the acceleration in reaction-rate is effected by a decrease in stability of the reacting molecule, resulting from its adsorption on the surface; in this unstable condition it may decompose directly, or as a result of a collision with another

* The surface is said to be saturated with the reacting molecules. An example of a heterogeneous zero-order reaction is the decomposition of gaseous HI at a gold surface.

molecule, the collision-energy being smaller than that necessary under homogeneous conditions. Many catalytic processes are highly specific, a given surface promoting one type of reaction and not another; if the catalyst behaves as an intermediary in the chemical reaction this is understandable, since it must be able to enter into chemical combination with at least one of the reactants; where an adsorption process is concerned it may very well be that, even though a variety of substances can adsorb a reactant, it will be only the substance that has a particular molecular, or crystalline, configuration at its surface that can adsorb the reactant in such a way as to decrease its stability, in fact some substances may increase the stability of a reactant and thus behave as inhibitors, or *negative catalysts*.* The modern view of adsorption, moreover, regards it as frequently differing little, if at all, from true chemical combination, the linkage between adsorbent and adsorbate varying from covalent, ionic, hydrogen-bond to simple van der Waals forces of attraction. Consequently the difference between the heterogeneous catalyst and the homogeneous one tends to disappear, both permitting a reaction to proceed more readily by lowering the minimal activation energy necessary. In biological systems the catalyst is described as an *enzyme*, and is very frequently a protein. The enzymes are highly specific, in that they accelerate chemical reactions involving only certain specific groupings; thus, although starch and cellulose are built up of glucose units, they are not hydrolysed by the same enzyme, the difference between the maltose and cellobiose type of linkage being sufficient to require a difference in enzymes. This high degree of specificity can only be explained on the assumption that the enzyme possesses definite chemical groupings at its surface with which the substrate must unite in some way; the union must be of such a character as to induce instability in the molecule, thereby permitting it to react. Thus, although the chemical groupings in the enzyme may allow union with a large variety of substrates, it is not difficult to envisage a state of affairs in which only one type of substrate molecule will be permitted to react. The poisoning of enzymes results, presumably, from the combination of the poison with the specific chemical groupings; this is doubtless true of such poisonings as that of cholinesterase by eserine, the poison having a similar chemical grouping to that of the substrate, acetylcholine; whether such non-specific poisons as the heavy metals and narcotics act in the same way is doubtful. The inactivation of enzymes by heat, for example, is probably due to a denaturation of the protein, and it may well be that the action of a heavy metal follows from a similar cause.

Photochemical Reactions

We have regarded activation as a loosening or breaking of bonds between the atoms of a molecule. The acquisition of this activation energy, in the cases so far presented, has resulted from thermal collisions, so that raising the temperature markedly accelerated the reaction-rate. Thermal energy of a molecule is manifest, as we have seen, not only in translatory motion, but also in rotation and vibration, such rotations and vibrations being reflected in the emission of infra-red radiation. As the temperature rises, a point is reached where the radiation becomes visible; it can be calculated that the energy of a quantum of this visible light, hv, corresponds to transitions from one electronic

---

* Negative catalysts, however, are most frequently those that tend to break up a chain reaction, *i.e.*, they tend to inactivate active molecules or atoms resulting from a primary reaction, preventing them from handing on their energy to other reactants; alternatively, they poison a catalyst already operating, by being themselves adsorbed on the active regions of the catalytic surface.

energy-level to another. Since any molecule emitting energy of a given wave-length must be able to absorb it, it follows that activation energy can be supplied in the form of visible or ultra-violet light. Reactions activated in this way are called *photochemical reactions*; the energy may be acquired directly by the reacting molecules, as in the decomposition of HI, or vicariously from some light-absorbing molecule; in the latter case the reaction is said to be *photosensitized*, an example being the photographic emulsion sensitized to react to certain colours, or infra-red light, by the incorporation of dye-stuffs.

Quantum Efficiency and Fluorescence. Once the possibility of absorbed light-energy acting as activation energy is admitted, the general principles of the photochemical reaction are simple and straightforward. According to the Einstein equivalence law, a molecule will react each time it has absorbed a quantum of light, provided that the energy of the quantum is equal to, or greater than, the activation energy.* A simple interpretation of the Einstein equivalence law would indicate that the *quantum efficiency, i.e.,* the number of molecules reacting per quantum absorbed, would be equal to unity; actually very few reactions exhibit this efficiency; some, such as the decomposition of HI by ultra-violet light, show an efficiency of 2, whilst many others have values of less than unity. The high efficiency in the decomposition of HI is due to the occurrence of secondary reactions following the primary reaction, which consists in the splitting of the molecule into two atoms:—

$$HI + h\nu \rightarrow H + I$$

following this the atoms may react in a variety of ways, the principal ones being:—

$$H + HI \rightarrow H_2 + I$$
$$I + I \rightarrow I_2$$

Thus, one quantum causes the splitting of two molecules. A quantum efficiency of less than unity is given by *inelastic collisions* of the activated molecule with normal molecules, the absorbed light-energy being dissipated as heat; alternatively, some of the light may be lost by *fluorescence*, the excited molecule retaining its energy for some 10^{-7} to 10^{-8} sec. and then re-emitting most of it at a longer wavelength.† The influence of inelastic collisions is well exemplified in the *quenching of fluorescence*, the activated molecule losing its energy by collision before it has time to fluoresce (the time between collisions in a gas at ordinary pressures and temperatures is of the order of 10^{-10} sec.); thus hydrogen is able to reduce the resonance fluorescence of mercury, the light-energy absorbed by the mercury being converted by inelastic collisions to heat. Studies of the quenching of fluorescence by various gases have shown, moreover, that the transfer of energy by these *collisions of the second kind* does not take place indiscriminately, since some molecules are able to quench more effectively than others. Fluorescence is a common feature of photosensitizing substances, substances that absorb light-energy and pass it on to the reactants;

---

* The quantum is given by $h\nu$; ν, the frequency of the light, is related to the wavelength, λ, by: $\nu\lambda = c$, where c is the velocity of light, so that the quantum is: hc/λ. If each molecule must absorb one quantum the activation energy for one mole is: Nhc/λ, where N is Avogadro's number. N is $6 \cdot 023.10^{23}$, h is $6 \cdot 624.10^{-27}$ and c is $2 \cdot 9977.10^{10}$ cm./sec. The energy will be given in ergs, so that on dividing by $4 \cdot 184.10^{10}$, the result comes out in Calories.

Thus $E = 2 \cdot 859.10^{10}/\lambda$ Cal./mole.

† If the same wavelength is emitted (*resonance fluorescence*) *all* of the absorbed energy is necessarily given out; generally, however, some of the absorbed energy is retained as vibrational and rotational energy, consequently the quantum for re-emission is smaller, and the wavelength therefore longer, than that absorbed. Atoms, which cannot exhibit vibrational or rotational energy, give only resonance fluorescence.

in the absence of these reactants fluorescence should be—and usually is—more pronounced, since the reactants can no longer quench it; a study of the quenching of fluorescence in photosensitized reactions may therefore lead to conclusions regarding the identity of the participating molecules; if a certain molecule does not quench fluorescence, it is unlikely to accept energy from the photosensitizer.

The photochemical reactions very frequently follow a first order equation; this is understandable since the rate of reaction will be determined by the rate of absorption of light and by the subsequent rate of reaction; both of these quantities will frequently depend on the first power of the concentration, thus giving a first-order type of equation. We may expect, and do indeed find, that the effect of temperature is small, the Q_{10} of many photochemical reactions being about unity; where, however, the photochemical process is only the first step in a series of reactions, it may well be that the overall reaction is considerably increased by a rise in temperature, since the subsequent steps may involve high activation energies.

Absorption Spectrum. The interpretation of the nature of the primary process in a photochemical reaction—for example, the decision as to whether a molecule on absorbing light immediately splits up into atoms, or remains only in an excited electronic state—is helped by a study of the *absorption spectra* of the reactants. Isolated atoms, *i.e.*, in the vapour state, emit light in limited regions of the visible and ultra-violet spectrum; the light so emitted appears as discrete lines in the spectroscope; the emission of this light is induced by transferring to the atoms the necessary energy which they re-emit; *e.g.*, NaCl, heated in a Bunsen flame, emits its characteristic yellow light. The emission of light is due to the fact that during activation, *e.g.*, the application of a flame, the orbital electrons of the atom assume new energy-levels; on returning to their lower, or *ground*, states they re-emit the energy, the frequency of the emitted light being given by the quantum expression:—

$$h\nu = E_1 - E_2$$

where E_1 and E_2 are two energy-levels.*

Rotation and Vibration

The spectrum of a molecule is not so simple as that of an atom, for the obvious reason that there are more possibilities of emission of energy (more *degrees of freedom*), since rotational and vibrational levels of energy are possible. By rotational levels of energy are meant the changes of energy associated with the rotation of the molecule about an axis; by vibrational levels, the oscillation of the atoms about a mean position. As with the electronic levels, these are quantized.† On the basis of the quantum theory these energy-levels may be calculated, and hence the wavelengths with which they correspond; rotational wavelengths are of the order of 2.10^6A and vibrational $1-23.10^4$A, they thus represent radiation in the far infra-red. This does not explain, however, the fact that molecules show band spectra, consisting of many lines close together, in the ultra-violet and visible region. The explanation for this is that changes of electronic level in molecules also induce changes in the vibrational and

---

* If, during the activation process, the atoms ionize, the energy-levels are altered so that the wavelength of the emitted light alters; the spectrum of the ionized atom is called the *spark spectrum* to distinguish it from the *flame* or *arc spectrum* of the normal atom.

† The rotational frequencies are given by: $\nu_r = \dfrac{h}{8\pi^2 K}(2n - 1)$, where K is the moment of inertia and n is the quantum number, which must be an integer.

rotational levels. Thus let us suppose that there is a certain electronic transition possible, corresponding to a frequency ν_2, so that the energy-change is equal to $h\nu_2$; if, now, the same amount of energy must be used to change the electronic level, and also to change the vibrational and rotational energies, it is clear that the electronic energy must be reduced, *i.e.*, it becomes $h\nu$, where ν is a new frequency, whilst the remaining energy is devoted to raising the vibrational and rotational energy-levels. The possible changes in these last-mentioned levels are numerous, and since the quantum of energy must remain the same, this means that $h\nu$ must vary, becoming less as the vibrational and rotational energies increase. Thus the electronic levels can take a large number of values, and this corresponds to a large number of values of the wavelength of emission or absorption in the ultra-violet or visible, thereby giving rise to bonds in the place of discrete lines.*

Finally, in solution, the molecules interact with each other to such an extent that a great many electronic levels are possible, and the bands are no longer resolvable into lines, they become continuous.

Ultra-Violet and Visible Absorption

We may note that the transition from one electronic level to another usually involves the emission, or absorption, of energy in the ultra-violet region; this is generally true of compounds in which the electrons are tightly bound in simple covalent bonds; with such bonds as $C = C$, and $N = O$, certain electrons are more loosely held, so that transitions corresponding to lower energy-changes are possible, with the result that absorption or emission is at longer wavelengths.

Activated States. In complex molecules, as in atoms, the absorption of radiation energy may lead to excited states in which an electron passes from one orbital to another; the electron is, as a result, in a condition of higher energy than that prevailing in the most stable, or ground, state. Modern developments of the quantum theory have defined a number of "rules" that determine what transitions are possible and what are "forbidden", the latter term being used to indicate rather a probability of existence than an absolute veto.† Thus, if a highly forbidden state exists we mean that it will occur very rarely; moreover, once achieved it will last a long time before reverting to the ground state. The states of a molecule are called "singlet" if the pairing of electrons is maintained with respect to spin, even though in the excited state one member of the pair is in a new, higher-energy orbit. This is a highly probable state and therefore the most common; the return to the ground singlet state is correspondingly rapid and may or may not give rise to fluorescence, depending on the likelihood of quenching collisions by other molecules or the possibility of internal conversion of energy leading to radiationless transfer.

Triplet State

The *triplet state* is a forbidden one and occurs when the pairing of electrons is lost; it is an excited state and lasts much longer than the singlet state; moreover it is the lowest excited state of the molecule. Because of its long life, being

* The possible frequencies corresponding to a change of electronic energy, $h\nu_2$, *i.e.*, a pure electronic energy-jump, are given by the equation:

$$\nu = \nu_1 + \nu_2 + \nu_0 n^2 - \nu_0{}' (n \pm 1)^2$$

Where ν is the frequency emitted, ν_1 is the vibrational frequency, $\nu_0{}'$ is the new value of the rotational frequency caused by the change in electronic level, whilst ν_0 is its original value. It is clear that ν can have a number of values.

† Factors determining "forbiddenness" have been recounted by Reid (1957); they include the conservation of electron spin; space forbiddenness, *i.e.*, the requirement that transition orbitals be not too remote from each other; orbitally-forbidden transitions involving large changes in angular momentum about a given axis; and finally various symmetry forbiddennesses.

greater than 10^{-4} sec. and lasting sometimes for a second or more, the triplet state of the molecule is responsible for the delayed emission of light called phosphorescence. The excitation of the molecule into this triplet state may be envisaged to occur as in Fig. 791, the allowed singlet state being achieved first, and then, by a radiationless transition, the lower triplet level is reached. In this way we are able to obtain a large number of molecules in the forbidden triplet state, whereas by direct excitation the number would be negligible. So efficient may this transition be that emission may occur from the triplet state with a quantum yield of unity. Fluorescence results from the passage from one singlet state to another, whilst phosphorescence results from the passage of the triplet state to the ground state.

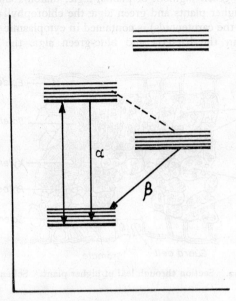

FIG. 791. Illustrating the mechanism of phosphorescence. The molecule passes by the α-transition first to the allowed singlet state, and thence by a radiationless β-transition to the triplet state. Passage from the triplet to the ground state is accompanied by phosphorescence. (Reid. *Excited States in Chemistry and Biology*. Butterworth.)

Fluorescence and Phosphorescence

The same molecule may be made to fluoresce and phosphoresce, and the absorption spectrum of the fluorescing and phosphorescing molecules may be measured; as we should expect, these are different, the phosphorescing molecule absorbing more strongly at the longer wavelength. By the same token the emitted light of phosphorescence has the longer wavelength; thus naphthalene fluoresces in the ultra-violet and phosphoresces in the green.

Beer's Law. Before passing to a consideration of the details of photosynthesis and other reactions involving light, we may briefly consider the experimental facts regarding the absorption of light. Beer's Law states that, if I_0 is the intensity of the light incident on a solution, I, the intensity of the transmitted light, is given by:—

$$I = I_0 . 10^{-\epsilon cd}$$

where c is the concentration of the solution in moles/l., d is the thickness through which the light is transmitted, and ϵ is called the *molar extinction* *coefficient* for the solute for a particular wavelength.* By plotting extinction coefficient against wavelength an absorption curve or *absorption spectrum* is obtained, the regions of peak absorption being called *absorption bands*.

THE STRUCTURE OF THE PHOTOSYNTHETIC APPARATUS

The Chloroplast

CO_2 and H_2O, the main reactants in photosynthesis, are colourless substances and therefore do not absorb strongly in the visible spectrum; the absorption of the necessary light-energy requires, therefore, a photosensitizer —*chlorophyll*—the green pigment of plants, algæ, diatoms and photosynthetic bacteria. In the higher plants and green algæ the chlorophyll (as well as other pigments, such as the carotenoids) is contained in cytoplasmic inclusion bodies, the *chloroplasts*; in the bacteria and blue-green algæ the photosensitizing

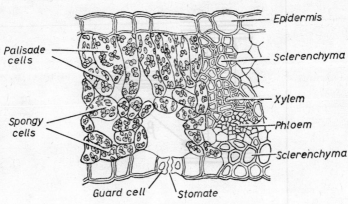

Epidermis

Palisade cells

Sclerenchyma

Xylem

Phloem

Spongy cells

Sclerenchyma

Guard cell   Stomate

FIG. 792.   Section through leaf of higher plant.   Schematic.

pigment is apparently localized on much smaller bodies that have been called *chromatophores* by Pardee, Schachman & Stanier.† Fig. 792 illustrates a section through a leaf of a higher plant; and it will be seen that the chloroplasts are contained mainly in the palisade and spongy cells. They are disk-shaped or flat ellipsoids, about 5 μ across; under the high power of the microscope they appear composite, being made up of dark *grana* some 0·3–2 μ across, more or less uniformly distributed through a lighter coloured stroma. Chloroplasts have been isolated from spinach leaves by the ultracentrifugal methods employed in the separation of inclusion bodies described earlier (p. 15).

Composition. The chemical analysis of the chloroplast has been discussed by Menke (1966) and some values for the major components are given in Table LXXXV. Chlorophyll is contained in the methanol-soluble fraction.

Birefringence. On the basis of the birefringence of the chloroplast, Menke

---

* If the law is used in the form:
$$I = I_0 e^{-\alpha cd}$$
is called the *molar absorption coefficient*; for a pure liquid the law takes the form:
$$I = I_0 10^{-kd}$$
where k is called the *extinction coefficient*.

† The term *chromoplast* is used to describe the form of differentiated plastids, containing carotenoid pigments, that are responsible for the yellow, orange and red colours of many plant structures; it develops from the chloroplast or amyloplast and its ulstrastructure has been described recently by Kirk & Juniper (1967).

PLATE LXX

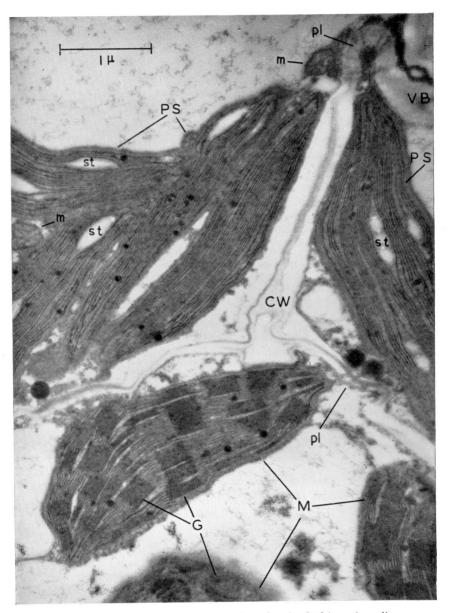

Fig. 793. Electron-micrograph of thin section of maize leaf in region adjacent to a vascular bundle (VB), showing typical parenchyma sheath (PS) and mesophyll (M) chloroplasts. Note the lamellation of both types, the presence of well-defined grana (G) in the mesophyll chloroplasts and their absence in those of the parenchyma sheath. The mesophyll chloroplast at the bottom of the figure has been sectioned in a plane essentially parallel to the plane of the lamellæ; under these conditions, the grana appear as circular dense areas. *st* = starch granules; *m* = mitochondrion; *pl* = plasmodesmatum in cell wall (CW). Osmium fixation. × 27,800. (Hodge McLean & Mercer. *J. biophys. biochem. Cytol.*)

PLATE LXXI

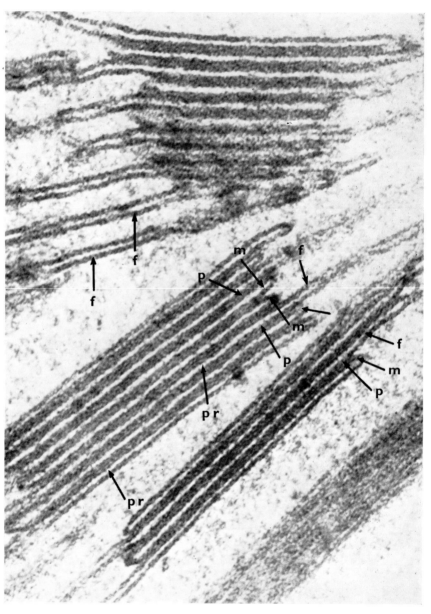

FIG. 796. Three small grana in a chloroplast of *Phaseolus vulgaris*. The union of margins (*m*), and frets (*f*) to form partitions (*p*) is well illustrated. A regular arrangement of globular units may be seen at *pr*. × 120,000. (Weier, Engelbrecht, Harrison & Risley. *J. Ultrastr. Res.*)

and later Frey-Wyssling concluded that it was built up of layers perpendicular to the optic axis, thus giving rise to a negative form birefringence, the optic axis being at right-angles to the surface. When this form birefringence was neutralized by a suitable solvent, a weak positive micellar birefringence was observed, presumably due to a radial arrangement of lipoid molecules. The picture derived from the birefringence of the chloroplast is therefore very similar to that of the nerve myelin sheath, namely layers of lipid separated by leaflets of protein (Fig. 257, p. 478).

TABLE LXXXV

Composition of Antirrhinum *Chloroplasts* (*Nickel, quoted by Menke,* 1966).

	Amount/Chloroplast $(10^{-12}$ g.)	Percentage
Structural protein	7·9	32
Lipids	5·5	22
Stroma Proteins	4·3	18
Methanol Soluble Material	5·0	20
Dry Weight of One Chloroplast	24·6	—

Electron Microscopy. Modern high-resolution electron microscopy of thin sections of chloroplasts has thrown some light on the nature of the membrane systems within the stroma of the organelle, although the relation of the granum to the other membranous components is still not completely clear. As Fig. 793, Pl. LXX shows, within the chloroplast there are layers of paired membranes, and the granum seems to be a region of denser packing of these membranes, appearing as a stack of discs or flattened cisterns. In the leaf of maize there are two types of chloroplast; in the parenchyma cells surrounding each vascular bundle these are simple laminated structures without grana, whilst in the mesophyll chloroplasts there are numerous grana, some 0.3–0.4 μ in diameter; the difference between the two types of chloroplast is well illustrated in Fig. 793, Pl. LXX.

Thylakoids or Frets

We may regard the fundamental basis of the membrane system in the chloroplast as the *thylakoid*, of Menke, or the *fret* of Weier; this is a flattened enclosed sac embedded in the stroma, and from this thylakoid the granum may arise, perhaps as a result of a bifurcation, as suggested by Fig. 794 from Mühlethaler

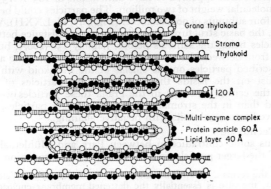

FIG. 794. Hypothetical arrangement of membranes to give a granum. The dark particles are attached to the outer surface; the white ones, to the inner side of the membranes. (Mühlethaler. *Biochemistry of Chloroplasts*. Academic Press.)

(1966). Viewed in this way the granum may be said to be made up of small thylakoids or compartments. Fig. 795 illustrates the different terminologies of Mühlethaler and Weier when describing the membranous structures, whilst Fig. 796, Pl. LXXI shows a granum at high magnification; the relation to the large thylakoid, or fret, is apparent.*

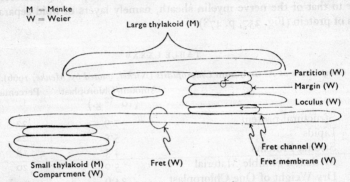

FIG. 795. Illustrating the nomenclature employed in describing the membrane structures in chloroplasts. (Park. *Int. Rev. Cytol.*)

Subunits

With $KMnO_4$ fixation, the thylakoid membranes have the classical unit structure described by Robertson; however, since the chloroplast is the photosynthetic unit, containing the absorbing pigments as well as the enzymatic apparatus required for converting CO_2 and H_2O to carbohydrate, we may expect to be able to resolve rather more detail than a simple bimolecular lipid-protein leaflet comparable with the myelin sheath. In fact, by the employment of shadowing, freeze-etching and negative-staining techniques, the presence of subunits, resting on, or embedded in, the substance of the membrane skeleton has been demonstrated, but the interpretation of these pictures, in terms of the known chemical composition of the chloroplast, is far from clear. Park & Pon (1961), studying sonicated chloroplasts in the electron microscope, showed that the lamellæ constituting the grana were sandwiches, on the inside of which were oblate granules that were probably particles embedded in the two layers. These units, some 200 × 100A, were considered to be the fundamental units of photosynthesis, analogous with the phosphorylation particles of the mitochondrion, and were called *quantasomes*. In a later study Park & Biggins and Park (1965) described the paracrystalline array of these particles, which appeared to have dimensions of 185 × 155 × 100A with an estimated molecular weight of two million. The particles could be clearly seen to be made up of four subunits of 90A diameter (Fig. 797, Pl. LXXII). When the lipids were removed, the basic structure was still present, the distance between the linear arrays of particles being the same as before, namely 150–200A, but material had been removed from between the arrays. Mühlethaler (1966), using a freeze-etching technique, described particles on the *outside* of the thylakoid with dimensions of 60 × 120 whilst on the inside there were smaller particles of diameter 60A embedded in the central stratum; in the grana these particles were much more densely packed than in the stromal lamellæ. The larger particles, on the inside, could be resolved into 4 subunits with negative staining.

Freeze-Etch Technique

There is thus some contradiction between Park and Mühlethaler in so far as the position of the larger particles is concerned. This contradiction may have been

---

* The use of the terms *discs* and *lamellæ* has been discussed by Weier, Bisalputra & Harrison (1966); the disc is essentially the flattened membrane-enclosed sac, and thus consists of two "unit membranes" closely apposed. It is equivalent to the thylakoid of Menke. When the discs are piled on each other they form a *band*, composed of lamellæ alternating with a light space; the lamellæ are composed of either two discs or one according as they are on the inside, or on the outside, of the band.

resolved by Branton & Park's (1967) re-examination of the freeze-etching technique in the light of Branton's claim that the tangentially fractured surfaces revealed by this method are the inner surfaces rather than the outer ones as postulated by Mühlethaler *et al.* (Fig. 798). Thus one membrane of a thylakoid, if split tangentially, should create four surfaces, some of which will be seen and others not. The most prominent surface actually seen is face B, consisting of 175A diameter particles on a relatively smooth background; these would be Park's quantasomes. Face A is only revealed when a granum is fractured and represents, presumably,

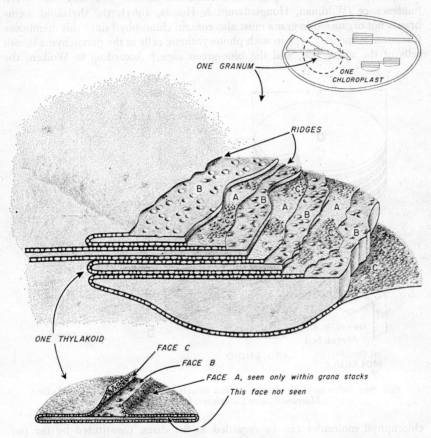

FIG. 798. Illustrating the interpretation of the appearances of freeze-etched preparations of the granum. The diagram relates the postulated structure of the membranes (cross-sectioned) to the various faces (stippled) actually seen in freeze-etched preparations. (Branton & Park. *J. Ultrastr. Res.*)

the outer surface of a thylakoid which is split because, in the granum, these faces are held together by hydrophobic bonds similar to the forces holding the two internal surfaces of a single membrane, *i.e.*, faces B and C. As Fig. 798 suggests, face C, which is normally seen at the edges of a fracture, is the other inside surface, and corresponds to the surface seen in the freeze-etched preparations with a large number of 110A particles, which presumably act as a matrix surrounding the larger quantasomes on the apposing face B. In air-dried and fixed specimens face A would be the only one seen and, because of differential drying, the large 175A particles would bulge through the surface, suggesting the presence of these on the *outer* surface. Whatever the final outcome of these structural studies, the general scheme portrayed by Fig. 794, according to which the membrane system of the chloroplast is made up of particles embedded in a matrix, will probably be sustained.* However, the quantasome concept, according to which the basic photochemical reactions are carried out on discrete particles, rather than more uniformly over the surface of the chloroplast membranes, may well have to be dropped, in view of the study of Howell & Moudrianakis (1967) on the photochemical activity of the whole membranes, and of particles similar to those described and isolated by Park and his colleagues (p. 1529).

Arrangement of the Molecules. Chlorophyll is undoubtedly present in the grana, since they are green in a colourless stroma, and they have a red fluorescence (Wildman, Hongladarom & Honda, 1964); the thylakoid membranes not organized in grana must also contain chlorophyll since this membrane formation is the only type in such photosynthetic cells as the parenchyma sheath cells of the maize leaf, and the blue-green algæ.† According to Wolken, the

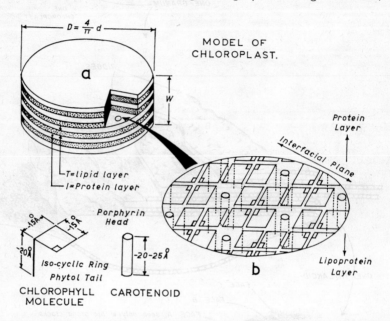

FIG. 799. Schematic illustration of the structure of the chloroplast. (Wolken. *Macromolecular Complexes*. Ronald Press Co.)

chlorophyll molecules can be regarded as flat discs, constituted by the porphyrin heads (p. 1498), connected to long hydrocarbon tails made up of the phytyl residues. The square laminæ of the porphyrin heads would fit closely to form a plane sheet, with the more reactive iso-cyclic ring of four molecules in close contact (Fig. 799). The phytol tail would be surrounded by lipid molecules, in particular the carotenoids which, as we shall see, enter closely into the photosynthetic process. Wolken has calculated the area available for the porphyrin heads from the known concentration of chlorophyll in the chloroplast and measurements of the number and size of the laminæ. It came out at

---

* The physical and chemical changes taking place during permanganate and osmium fixation, leading to the simple three-layered structure, have been discussed by Mühlethaler (1966). Weier *et al.* (1965) observed subunits in permanganate-fixed chloroplast fret-membranes; they had a light core of 37A diameter and a dark rim 28A wide.

† Lloyd (1924) demonstrated the red fluorescence of chloroplasts and Spencer & Wildman (1962) localized this to the grana; Lintilhac & Park (1966) compared pictures obtained by fluorescence-microscopy and the electron microscope and showed that non-grana regions also contained chlorophyll, presumably the thylakoids not organized as grana.

$225-246A^2$ comparing with the area of the porphyrin head of $225-242A^2$. The arrangement of the structural protein, lipid, and other more soluble constituents is a matter of speculation, and the interested reader should consult, for example, Weier & Benson (1966) for a reasoned argument in favour of sheets of lipoprotein subunits with the chlorophyll molecules mainly concentrated in the "partition" between each pair of sheets in the granum.

The Chloroplast as a Living Unit. In phase-contrast, the shape of the chloroplast is not fixed, but protuberances or pseudopods project and are withdrawn (Wildman, 1967); sometimes these become detached from the surface and join the mitochondria and other organelles during streaming movements.

Reproduction

It is now well established that the chloroplast is capable of reproduction, containing its own DNA and ribosomes. Thus Lyttleton (1962) isolated ribosomes from spinach chloroplasts; these had a sedimentation coefficient of 66 and contained 45 per cent. RNA; again Clark, Matthews & Ralph (1964) were able to separate cytoplasmic from chloroplast ribosomes, and they showed that the latter were of the bacterial 68 S type by contrast with the cytoplasmic particles, which were of the 83 S type.*

Ribosomes

Again, Wildman, by osmotically disrupting isolated chloroplasts obtained, besides Fraction -I protein (p. 1511), 70 S ribosomes with which protein synthesis was associated. These ribosomes are remarkably similar to those isolated from bacteria and blue-green algæ and are sharply distinguished from the 80 S ribosomes of erythrocytes and the cytoplasm of leaves.

This is shown by the differing Mg^{++} concentrations required for dissociation into subunits and for optimal synthetic activity, whilst in the electron microscope the chloroplast ribosomes have the cleft observed by Huxley & Zubay (1960) in *E. coli* ribosomes and not in cytoplasmic ribosomes of higher organisms (Miller, Karlsson & Boardman, 1966).

Chloroplast DNA

Again, Eisenstadt & Brawerman (1964) have shown that, not only are the ribosomes of *Euglena* chloroplasts different from those of the rest of the cell, but also the cell-free amino acid-incorporating system is different.

The chloroplast is weakly Feulgen-positive indicating the presence of DNA, and this is confirmed by radioautography after incorporation of ^{3}H-thymidine and by extraction and separation in a CsCl gradient (Kisley, Swift & Bogorad, 1965).† Synthesis of DNA was shown by Gibor & Izawa who removed the nuclei from *Acetabularia* by cutting off the basal region of the cell; the ^{14}C of $^{14}CO_2$ appeared

---

* Clark *et al.* (1964) showed that the 68 S ribosomes were not the product of breakdown of the 83 S ribosomes; in growing tips of cabbage stems there were very few chloroplasts and no 68 S ribosomes could be extracted. The authors described an interesting cycle of change from polyribosome to monomer and *vice versa* induced by light; at the end of the night the bulk of both types are in the monomeric form; from first light onwards the proportion declines until after several hours of sun the amounts of the monomeric form are very small or undetectable. Perhaps the best demonstration of the heredity of chloroplasts is that of Lyman, Epstein & Schiff (1961) on *Euglena*; exposure to ultraviolet light produced a colourless colony of cells, and these remained colourless through several hundred generations; the action spectrum indicated that the light had been absorbed by a nucleoprotein, presumably the DNA of the chloroplasts rather than that of the nucleus since there was no lethality. *Euglena* can be grown on a carbon-containing medium so that the loss of photosynthetic ability was not injurious. The multicellular green alga, *Ulva mutabilis*, has a single chloroplast which, during cell division, occupies a position such that it is cut in two by the advancing cleavage furrow; the process has been described by Løvlie & Bråten (1968).

† Gunning (1965) has described regions in the stroma of the chloroplast resembling the nucleoplasm of bacteria, networks of fine fibrils some 25–30A thick which may well be DNA. He also describes "stromacentres", aggregates of 80–85A thick fibrils about 1 μ in diameter which, in three dimensions, may be like balls of wool; they had no limiting membrane. Bisalputra & Bisalputra described electron-transparent regions in the chloroplast of a brown alga containing 15–25A fibrils that disappeared with DNAse treatment. Schiff & Zeldin (1968) have summarized the base compositions and the characteristics of the DNA's from nucleus, chloroplast and mitochondrion of *Euglena gracilis*.

in the isolated DNA which in a CsCl gradient had a specific gravity of $1 \cdot 695$ g./ml., the same as that of the "satellite" DNA isolated by Chun, Vaughan & Rich (1963) from the chloroplasts of *Chlamydomonas* and *Chlorella*.* The last named authors showed that both chloroplast and nuclear DNA's were double-stranded.

RNA Turnover

During light-stimulated chloroplast development there is an increased turnover of both cytoplasmic and chloroplast RNA, indicating the intervention of the cytoplasmic protein synthetic mechanism; this may well serve to provide the growing chloroplast with metabolic precursors necessary for its syntheses; thus the plastid is an auxotrophic resident within the cell during development (Zeldin & Schiff, 1967; Schiff & Zeldin, 1968), so that the chloroplast DNA is not the only DNA coding for the synthesis of its constituents; this accords with genetic evidence indicating interaction of the nuclear and chloroplast genomes. We may presume that the light-dependent conversion of protochlorophyll(ide) to chlorophyll(ide) (p. 1509) is intimately connected with the de-repression of synthesis of chloroplast proteins.

As Schiff & Zeldin (1968) have suggested, the effects of light in bringing about the manifold changes in synthetic activity of the proplastid, leading to the development of the mature chloroplast, are analogous with the phenomena of enzyme induction through substrates (p. 225); and we may presume that light unblocks the transcriptive activity of DNA.

Algæ. In lower forms of plant cell the regular stacking of double-membraned discs to form grana is rare, but the discs may be paired, as in the chloroplast of the alga *Chlamydomonas* described by Sager & Palade, or may occur in triplets, as in the alga *Ochromonas danica*, studied by Gibbs. We may treat the disc, then, as the fundamental unit of chloroplast structure, whilst we may describe the organization of discs into multiple layers as *lamination*, the highly orientated lamination of higher plants into grana being a special case.

Blue-Green Algæ

The blue-green algæ are an interesting group that share many features of the bacteria in so far as they have no membrane-bound intracellular organelles, such as nucleus, mitochondrion or chloroplast, so that they would be classed as prokaryonts, yet their photosynthetic pigment is chlorophyll-*a* and not *bacteriochlorophyll*, as in the bacteria. Thus their photosynthetic activity is similar to that of higher plants although their organization is bacterial. As described by Jost, in *Oscillatoria rubescens*, the membrane-system consists of thylakoids piled in stacks but separated by some 1,000A; both sides of the membranes had spherical bodies packed on to them; on the inner side they were some 100–200A in diameter, and probably corresponded with the quantasomes of Park & Pon. When the membrane was torn, the quantosomes disappeared during the freeze-etching process.

Chlorophyll-Deficient Mutant. A chlorophyll-deficient mutant of the tobacco plant, which was of the colour of pale parchment all over, was capable of blooming; Schmid & Gaffron (1967) showed that this was because light-saturation (p. 1511) occurred at much higher light-intensities than in the normal plant (50,000 erg/sec. cm.^{2} compared with 5,000–12,000); the plants could thus profit by very high light intensities and would be described as "sun plants". Under these conditions the efficiency of CO_2-fixation was some $1 \cdot 3$ to $2 \cdot 2$ times normal when measured on the basis of chlorophyll content. Quantum efficiency was the same as in normal plants, namely a minimum of 10 quanta per O_2-molecule evolved. In the electron-microscope the membrane structure was characteristically different, the organiza-

---

* Kisley, Swift & Bogorad (1965) have reviewed the evidence bearing on the multiplication of DNA in mitochondria and chloroplasts. Another interesting review on the genetic control of multiplication of chloroplasts is that by Gibor & Granick (1964), whilst a paper by Schiff & Zeldin (1968) discusses, in the light of experimental studies on RNA turnover, the relative contributions of nuclear and chloroplast DNA in the synthesis of the enzymatic apparatus of the chloroplast.

tion into grana being absent, although the single thylakoids were folded over at the ends. When a variety of chlorophyll-deficient plants were examined it seemed that this curling over at the end, to form primitive grana, was essential for photosynthesis (Schmid, 1967).

Bacterial System

It was originally considered that photosynthetic bacteria contained no specialized bodies for photosynthesis, their chlorophyll being distributed throughout the cytoplasm of the cell. However, Pardee, Schachman & Tanier disrupted the photosynthetic bacterium, *Rhodospirillum rubrum*, by grinding with alumina; an extract in NaCl was centrifuged at 20,000 r.p.m. to give a pellet of particles that, when examined as a dried preparation in the electron-microscope, appeared as discs; if these had been spheres it was estimated that their diameter would have been some 600A. This agrees reasonably well with their molecular weight of 30.10^6 as determined from their sedimentation coefficient. The absorption spectrum was characteristic of bacteriochlorophyll.

Photosynthetic Particle. It was shown by Frenkel in 1954 that subcellular particles of *Rhodospirillum rubrum* were capable of carrying out photosynthetic activity (photophosphorylation of ADP to ATP, p. 1519), and the more elaborate studies of Bergeron & Fuller on another photosynthetic bacterium, *Chromatium*, have shown that the "photosynthetic particle" is a sphere of some 320A diameter and a molecular weight of 13.10^6. Chemical analysis by Newton & Newton showed that the particle could be considered as made up of smaller units consisting of 1 molecule of carotenoid, 2 of chlorophyll, 10 of phospholipid, and protein equivalent to 220 amino-acids. In the electron-microscope the chromatophores were seen to fill the cytoplasm of the bacterium (Vatter & Wolfe). When grown in the dark no chromatophores were present.

Chemistry of Chromatophore

The chemical composition of the material isolated by Gorchein, Neuberger & Tait (1968) from *R. spheroides* is shown in Table LXXXVI. Apart from the fact that the material isolated from aerobically grown bacteria contained no chlorophyll or carotenoids, it was sufficiently similar to the chromatophores to suggest that they were both derived from similar membrane material.

TABLE LXXXVI

Chemical Composition of Chromatophore Material from Rhodopseudomonas spheroides
(Gorchein, Neuberger & Tait, 1968).

Component	Per cent. Dry Weight
Protein	61–63
Total lipid including pigments	27
Nucleic acid	0·28
Bacteriochlorophyll	6·8
Carotenoids	1·75

Internal Membranes

The electron microscopical appearance of another photosynthetic bacterium, *Rhodospirillum molischianum*, revealed a system of internal membranes rather than one of individual vesicles (Gibbs, Sistrom & Worden, 1965), formed obviously by invagination of the plasma membrane (Hickman & Frenkel, 1965); and it has been argued that, not only the chromatophores isolated from this bacterium, but also those from *R. rubrum* and *Chromatium*, represented fragmented cytoplasmic membranes analogous with the microsomes of the cells of higher organisms; the remarkable uniformity of size could be accounted for by the presence of weak points, as suggested by Cohen-Bazaire & Kunisawa (1963). Support for this view of the origin of the chromatophore was provided by Holt & Marr (1965), who showed that osmotic disruption of the lysozyme-treated cells of *R. rubrum* did not necessarily release chromatophores into the medium. It may be, as Gorchein (1968b) has claimed, that the isolated chromatophores are genuine units of structure that are normally attached to the cytoplasmic membranes; at any rate, by very gentle fragmentation of *R. spheroides*, material could be isolated, and examined with negative staining, that gave the appearance of membrane in which was embedded the characteristic chromatophore.* Fig. 800 illustrates Holt & Marr's interpretation of their stereoscopic electron micrographs, suggesting that the system is made up of a network of tubules attached to the cell envelope, originating as invaginations of this.

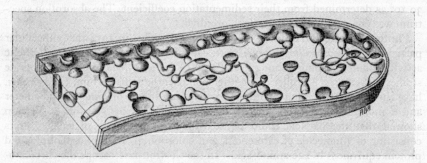

FIG. 800. Three-dimensional appearance of the membrane system in *Rhodospirillum*. (Holt & Marr. *J. Bact.*)

Development of the Chloroplast. The development of the chloroplast may be followed either by examining meristematic cells of rapidly growing tissue, or by first growing the plant in the absence of light, in which case the *etiolated* leaves have no chloroplasts; exposure to light causes the development of chlorophyll and at the same time the organized chloroplast of the mature leaf appears. Finally, chlorophyll-deficient mutants may be examined (v. Wettstein). According to the study of Hodge, McLean & Mercer, the first sign of a chloroplast in a growing plant cell is a membrane-bound vesicle with a few laminæ within.

Prolamellar Body

These laminæ appear to be derived from what these authors described as a *prolamellar body*, a mass of vesicles within the chloroplast; from this mass, lamellæ appeared to be growing out by fusion of the vesicles into flattened cisternæ. In the fully etiolated plant there were no lamellæ, but the same compact mass of vesicles was present; on exposure to light, lamellæ formed apparently by the same process of fusion of vesicles, whilst the prolamellar bodies disappeared. Formation of further lamellæ, after the disappearance of the prolamellar body, appeared to result from the fusion of vesicles that had been formed in a zone immediately under the limiting membrane of the chloroplast. These vesicles are probably

* Photosynthetic bacteria such as *R. rubrum* will grow aerobically in the dark, in which case they do not contain bacteriochlorophyll; cultures with bacteriochlorophyll contents varying by a factor of four may be prepared by exposing the bacteria to different light intensities. In general, the amount of chlorophyll present does not correlate well with the amount of membrane material within the cell; however, the total amount of membrane—plasmalemma plus cytoplasmic—was proportional to the bacteriochlorophyll content (Gibbs *et al.*, 1965). The pigment content of the isolated chromatophores did correlate with that of the cells from which they were isolated in the study of Cohen-Bazaire & Kunisawa (1963) but this might have been due to the greater contamination with non-chromatophore membrane in the dark-grown bacteria, since Gorchein (1968a) was able to separate material with relatively constant pigment composition from cells with varying bacteriochlorophyll content.

PLATE LXXII

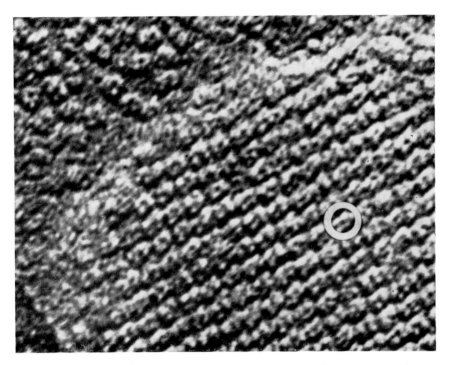

FIG. 797. Shadowed paracrystalline array of quantasomes. A quantasome with contained subunits is circled. × 330,000. Park. *J. Cell Biol.*)

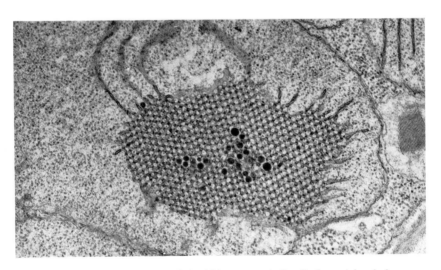

FIG. 801. Low magnification of plastid in a mesophyll cell of an etiolated *Avena* leaf containing highly crystalline prolamellar body. × 32,000 approx. (Gunning. *Protoplasma.*)

PLATE LXXIII

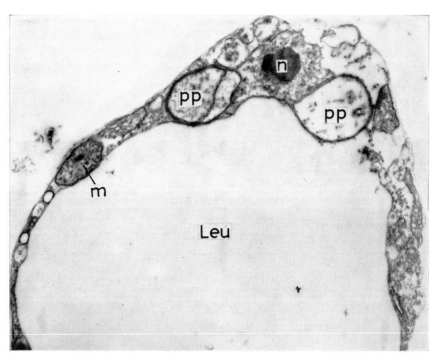

FIG. 804. *Top.* Dark-grown cell of *Ochromonas danica* containing a large leucosin vacuole (leu) and a small shield-shaped proplastid (pp) which partly encircles the nucleus (n). m = mitochondrion. × 8,450.

Bottom. Light-grown cell containing a much reduced leucosin vacuole and a large lamellate chloroplast (c). × 19,200. (Gibbs. *J. Cell Biol.*)

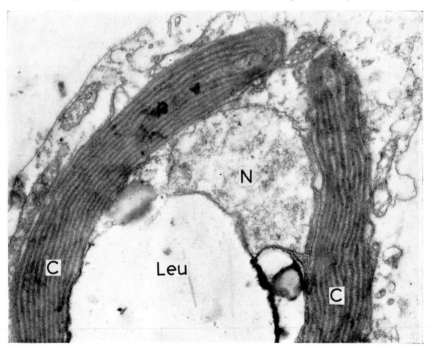

formed by an invagination of this outer limiting membrane. Mühlethaler & Frey-Wyssling came to essentially the same conclusions; the smallest bodies they found were vesicles some 200A in diameter; these grew in size and, when some 0·6 to 2 μ in diameter, their limiting membranes invaginated to form internal laminæ which, if development was normal, stacked up to form grana. In the absence of light, prolamellar bodies full of vesicles appeared, which, on exposure to light, arranged to form ordered laminæ.

Subsequent electron-microscopy of the prolamellar body by Gunning (1965, 1967) and Wehrmeyer (1965) has shown that the remarkable crystalline arrangement of what Hodge *et al.* described as vesicles probably represented sections through a lattice arrangement of tubules; the crystalline appearance is shown in Fig. 801, Pl. LXXII, whilst Fig. 802 is a schematic illustration of the postulated arrangement.*

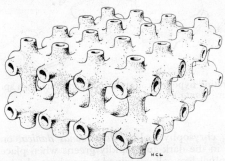

Fig. 802. Three-dimensional representation of the membrane system of the *Avena* prolamellar body. (Gunning. *Protoplasma.*)

Analogy with Mitochondrion

Mühlethaler & Frey-Wyssling emphasized the analogy with the mitochondrion; since this organelle employs oxygen in its metabolic activities it was presumably preceded in evolution by the chloroplast, since the atmosphere was originally a reducing one, the oxygen now present having been derived from photosynthesis. The formation of lamellæ in the mitochondrion leads to cristæ and this is the terminal point of their development; by contrast, in the chloroplast this is an initial stage which then proceeds to the development of organized stacks of discs.

Development of the Granum. The relation of the granum to the stromal lamellæ or large thylakoids is a vexed question, depending as it does on the interpretation in three dimensions of the two-dimensional pictures of the electron microscope; bound up with the final relation between grana and stromal lamellæ is the manner in which this relationship grew. In general, there are two main theories, namely the one that postulates a primary formation of the granum from which the stroma membranes grow, and the other postulating a growth of the granum from the stromal membranes. Since in phylogenetic development the granum appears in the higher forms, one might intuitively adopt the latter hypothesis; thus Fig. 803 from Wehrmeyer (1964) illustrates the growth of a granum from a growing membrane by what Wehrmeyer calls a spirocyclic process; an analogy would be the spiral growth of membrane in the formation of the myelin sheath. On this basis the grana would maintain connection with the stromal lamellæ, and presumably there would be continuity of the lumens of the respective membrane-bound sacs. A simpler process is suggested by other studies, in which the individual smaller discs are simply packed on top of each other, coming into

---

* Wehrmeyer (1965) has established a model of the probable crystal structure of the prolamellar body that seems to fit the electron-microscopical appearances of thin sections with some accuracy. Like Gunning's model, it is based on a system of tubules; these are branched at a constant angle of 109° and have a constant length from one branch to the next. These tubules are organized round a pentagonal dodecahedron, *i.e.*, a body with twelve faces made up of pentagons. The analogy with crystals of zinc-blende and wurtzite was established, so that a combination of the two is compatible with the final picture.

closer and closer approximation until the tightly packed grana are formed (Ohad, Siekevitz & Palade, 1967).

Relation to Chlorophyll Synthesis. The development of the membrane system in the chloroplast is closely bound up with the synthesis of chlorophyll; this has been demonstrated by the quantitative studies of Gibbs (1962) and Ohad *et al.* (1967).

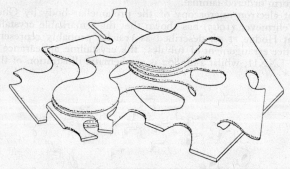

FIG. 803. Illustrating hypothetical growth of a granum by a spirocyclic process. (Wehrmeyer. *Planta.*)

Ochromonas

Gibbs chose the chrysophyte alga, *Ochromonas danica*, one that is yellowish white when grown in the dark and rapidly greens when placed in the light. The dark-grown chlorophyll-less cell contained a membrane-bound proplastid encircling the nucleus; after exposure to light this became larger and filled with piles of characteristic lamellæ (Fig. 804, Pl. LXXIII), which when fully developed were made up of three closely apposed discs. As earlier workers had found, the lamellæ were made up by the fusion of small vesicles, although there was no evidence that these had been formed by invagination of the protoplastid membrane. The lamella appeared as a three-layered structure, built up of three discs each apparently made by the flattening of vesicles; thus, during the early stages of development, lamellæ made up of one, two and finally three discs were observed. After 48 hours of exposure to light the algæ have about 50 per cent. of their normal chlorophyll content whilst the chloroplast has grown to its full size and has the characteristic lamellar structure, all lamellæ having three discs; further increase in chlorophyll content seems to be associated with an increasingly compact apposition of the discs constituting the lamellæ.

Chlamydomonas

Ohad *et al.* chose a mutant of another alga, *Chlamydomonas reinhardi*, that, unlike the wild species, was unable to synthesize chlorophyll in the dark; in the light it could do so and, on transfer to the dark, it continued to grow but did not synthesize chlorophyll, so that the amount in the cell decreased progressively by dilution of that previously synthesized. It was possible to follow development (greening) and loss (degreening) of chloroplast structure. Degreening was associated with breakdown of the membrane system, so that finally the cell contained only a few scattered vesicles and discs with few or no stacks (grana). Greening was associated with an increase in the number of membrane-bound vesicular and disc profiles, with an increase in length of the flattened disc-profiles, indicating a two-dimensional growth of the discs followed by fusion of neighbouring discs to form larger ones. Although new discs were found frequently in the neighbourhood of the chloroplast envelope, there was no evidence of budding off from this. By 5 hr. there was a remarkable tendency for the large flat discs to pair off, and the pairs were either closely applied to form a 2-disc granum or there was a space between. By 6·5 hr most discs appeared fused into stacks, and by 9–10 hr after the beginning of illumination the appearance was the same as that of light-grown algæ. Gibbs' study suggested that the final process of apposition and fusion into grana was dependent on chlorophyll synthesis, but Ohad *et al.* found that, if illumination was stopped at 5 hr, the paired discs, although separated by quite large spaces, would proceed to fuse in the dark.

Chlorophyll Synthesis and Membrane Assembly

In general, then, the synthesis of chlorophyll is closely linked with the synthesis and assembly of membrane material; it may be that chlorophyll accelerates the removal of synthesized polypeptide chains from their polyribosomes (Ohad *et al.*). Studies of the incorporation of labelled protein and lipid precursors into the cell suggested that growth occurred by a one-step assembly of completely photosynthetically capable components; moreover, there were no "growing points" on the membrane, radioautography indicating growth over the whole surface. The mode of formation of new membrane quite obviously varies with the particular system being studied; in the algæ studied by Gibbs and Ohad *et al.* this was not derived from chloroplast membrane but occurred within the chloroplast; in other algæ* and higher plants the chloroplast membrane is involved. In bacteria, Gibbs, Sistrom & Worden (1965) described the stacks of flattened sacs constituting the chromatophore material as being connected to the plasma membrane.

The Pigments

Photosynthetic organisms and tissues have a variety of colours; red, yellow, green, purple, brown, etc. Common to all the pigment systems associated with photosynthesis is the green pigment, chlorophyll, the actual colour of the organism being determined by the absolute amount of this substance and the amounts, in relation to this, of a variety of accessory pigments. Thus the various shades of the green leaf, commonly observed, are due to varying mixtures of yellow *carotenoids*;† the red colour, especially obvious in growing shoots, is due to the presence of water-soluble *flavones*, present in the vacuoles of the cells, which may be converted into red *anthocyanins*. The blue-green, red, purple and brown algæ contain *phycobilins*,‡ besides carotenoids, and it is the strong absorption by these substances in the green and yellow that robs the organism of the green colour due to chlorophyll, and leaves it red, purple, or blue.

Chlorophylls. The chlorophyll of green plants is a mixture of two compounds, *chlorophyll-a* and *chlorophyll-b*, consisting of structures built up of four pyrrole nuclei co-ordinated with an atom of Mg. They are esters of dibasic acids, the *chlorophyllins*, the esterifying alcohols being methyl alcohol and *phytol* ($C_{20}H_{39}OH$), so that the chlorophylls may be described as *methyl-phytyl-chlorophyllides*. Leaves contain an enzyme which splits the alcohol groups from the molecule; and with its aid a number of artificial chlorophyllides may be prepared. *Bacteriochlorophyll*, extracted from photosynthetic bacteria, is classified by Fischer as a relation of chlorophyll-*a*; Fischer considers bacteriochlorophyll as the most primitive of the photosynthetic pigments and indicates how, by a process of hydrogenation and subsequent splitting off of water,

---

* In the blue-green alga, *Oscillatoria rubescens*, Jost described invagination of the plasmalemma in quite a complicated fashion; this led to the formation of small vesicles which joined up to form thylakoids. He points out that although, during cell division, the thylakoids of the mother cell are split, and appear in the daughter cells, this cannot be the only source of thylakoids since in the course of development the thylakoids completely disappear yet new thylakoids appear in the daughter cells after division.

† Carotene is one of the yellow pigments found in green leaves; the remaining pigments are sometimes referred to as xanthophylls; they are generally oxidation products of carotene, containing OH- or CO-groups in the ring structure, and may best be described as carotenols; thus luteol, the most abundant of the xanthophylls in green leaves, is $C_{40}H_{54}(OH)_2$. *Fucoxanthol* is found in brown algæ and diatoms, but not in higher plants; other carotenoids peculiar to algæ are *myxoxanthone* and *myxoxanthophyll* of blue-green algæ. Purple bacteria contain *violascin*, *rhodopin*, *rhodopurpurin*, etc.

‡ Phycobilins are chromoproteins of high molecular weight; thus the red *phycoerythrin* and the blue *phycocyanin* have molecular weights in the region of 280,000, some 2 per cent. of the weight being due to the prosthetic pigment group and the remainder to a globulin type of protein. The prosthetic groups are built up of pyrrole units. The compounds are given the generic name of biliproteins and their distribution in algæ, and preparation, have been described by O'hEoacha (1966).

chlorophyll-*a* may be derived from it. Chlorophyll-*b*, he considers, was developed later. Hæmin, the prosthetic group of the respiratory pigments, is built up on the same pyrrole basis and could, according to Fischer, give rise to bacteriochlorophyll.

BACTERIOCHLOROPHYLL

PROTOCHLOROPHYLL   **CHLOROPHYLL a (b)**

Chlorophyll-c

Chlorophyll of the *a*-type (bacteriochlorophyll) was considered, up till recently, to be the only green pigment of bacteria, algæ and diatoms, although it had been suggested that a substance, *chlorofucin* (also called chlorophyll-*γ*, or chlorophyll-*c*),* found in diatoms and brown algæ in addition to chlorophyll-*a*, was also a photosynthetically active pigment. Wilstätter & Page concluded

---

* Not to be confused with the chlorophyll-*c* alleged to be present in the higher plants, the claim regarding which was subsequently withdrawn, it having been shown to be a mixture of chlorophyll-*a* and pheophytin-*a*, *i.e.*, of chlorophyll and the product obtained by removing Mg from the molecule.

that it was an artificial post-mortem product, but in 1942 Strain & Manning isolated chlorofucin by partition chromatography from extracts of diatoms (*Nitzschia closterium*), and showed that its absorption spectrum differed from those of both chlorophyll-*a* and -*b*; these authors suggested the term *chlorophyll-c* for the compound and emphasized the importance of this substance in the carbohydrate economy of the earth, since diatoms are the most abundant autotrophic organisms over much of the earth's surface.

Absorption Spectra. Chlorophyll-*a* and -*b* are soluble in alcohol, acetone, etc., and may be extracted from leaves by vigorous shaking with these solvents. The absorption spectra in ether solution are shown in Fig. 805; the green colour is due to the prominent absorption bands in the red, and blue.

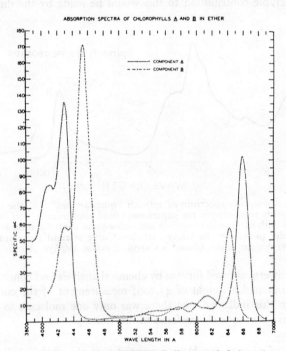

FIG. 805. Absorption spectra of chlorophylls *a* and *b* in ether. (Zscheile. *Bot. Gaz.*)

Chlorophyll-*a* has principal bands at 6,620, 4,300 and 4,100A whilst chlorophyll-*b* has main bands at 6,440, 4,550 and 4,300A. Thus, with light of the longer wave-lengths, chlorophyll-*a* would be the principal absorber of light. Bacterio-chlorophyll has its red absorption maximum at longer wavelengths, namely in the infra-red at 8,000A.

There is every reason to believe that the chlorophylls *in vivo* are linked to a protein, or lipid-protein complex; as a result of this linkage, their absorption spectra *in vivo* may be expected to be different from those of the extracted pigments in an organic solvent. In fact, measurements of the absorption spectrum of intact chloroplasts, carried out with microspectrophotometric techniques by Wolken, for example, or on clarified aqueous suspensions of homogenized chloroplasts in which the chlorophyll-protein link is probably retained, show absorption spectra in which the main red bands are shifted towards longer wavelengths, so that the main band of chlorophyll-*a*, for example, is close to

6,800A instead of 6,620A, and that of chlorophyll-*b* is at 6,500A. Fig. 806 is an absorption spectrum measured on a preparation of spinach chloroplasts that had been fragmented by ultrasonic vibrations.

Molecular Weight of Chlorophyll Complex. It has already been indicated that the absorption spectra of chlorophyll preparations suggest that the pigment is attached to protein, so that the chlorophyll molecule may be regarded as a chromophore group, like hæm in hæmoglobin or retinene in rhodopsin. Aqueous "solutions" of chlorophyll are made by treating the colloidal suspension of fragmented chloroplasts with a surface-active substance such as digitonin; this forms a complex with the natural particle and makes it water-soluble. Smith found that such preparations had a molecular weight of some 265,000 but a considerable contribution to this would be made by the digitonin itself.

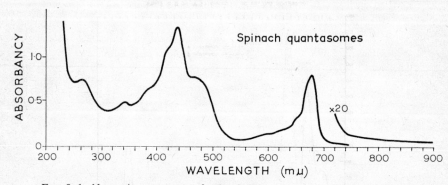

FIG. 806. Absorption spectrum of spinach "quantasomes", *i.e.*, the very small fragments remaining in the supernatant fluid after treatment of the chloroplasts with ultrasonic vibrations and subsequent centrifugation at 20,000 *g* in order to remove the larger particles. Curve at right represents 20-fold greater concentration. (Sauer & Calvin. *Biochim. biophys. Acta.*)

Wolken & Schwertz allowed for this by chemical analysis and concluded that the chlorophyll unit had a weight of 43,500; measurement of the contribution of chlorophyll to this indicated that there was only one molecule to each 43,500 unit. They gave the name *chloroplastin* to this unit.*

Inhomogeneity

A number of phenomena have suggested that the chlorophyll-*a in vivo* is really a mixture of pigments with slightly different absorption spectra; since, however, organic extractants that break down the link with protein leave only chlorophyll-*a* and chlorophyll-*b* in green plant extracts, the small differences in absorption are presumably due to linkage with different proteins, just as the variety of visual pigments is brought about by the combination of two organic molecules or chromophores—retinene$_1$ and retinene$_2$—with a variety of proteins called opsins (p. 1644). Thus Brown & French examined the absorption spectrum of homogenized chloroplasts of the green alga *Chlorella*, and were able to resolve it into the separate absorptions of probably four different components with slightly different wavelengths of maximum absorption (λ_{max}) in the far-red. The different pigments were indicated by these values of λ_{max}, as follows:—

C_a6,840, C_a6,730, C_a6,950 and C_a7,070, although the last might have been

---

* There have been numerous other preparations of chlorophyll-protein complexes, notably that of Kahn (1964) using Triton-X as solubilizer (see, Criddle, 1966, for a review). The chlorophyll will be attached to the structural protein of the chloroplast which, according to Criddle, has a molecular weight of 25,000 and is similar to that of mitochondria.

due to a pheophytin complex. Chlorophyll-b was represented by only a single band in the far-red at 6,500A, and was indicated by $C_b6,500$.* By growing the photosynthetic protozoan, *Euglena*, in light of low intensity, the proportion of $C_a6,950$ could be increased considerably, namely to some 20 per cent. of the total; this chlorophyll bleached more readily than the others (Brown & French). In a similar way, Duysens resolved the bacterial chlorophyll-a into four separate pigments B 8,000, B 8,600, B 8,700, B 8,900.

Films of Chlorophyll

Although linkage to different proteins is the most likely explanation for the various types of chlorophyll-a, or of bacteriochlorophyll, the mere orientation of chlorophyll in a monomolecular film may be responsible for an alteration in its absorption characteristics; at any rate Trurnit & Colmano found a shift in the λ_{max} of pure chlorophyll when this was spread as an interfacial film.

Effects of Solvent

Again, Aghion (1963) has shown that treatment of a detergent-solubilized preparation of chlorophyll with aqueous acetone or alcohol, at an appropriate concentration, will cause a shift of maximal absorption from 6,680 to 7,400A; the material absorbing at this long wavelength is still particulate, since it can be centrifuged down at 20,000 g; addition of excess of solvent reverses the change, which Aghion attributes to the crystallization of the chlorophyll molecules on the protein particles.

Thomas & Bartels (1966) also obtained reversible shifts of absorption on treatment of chloroplast fragments with aqueous acetone solutions, and they attributed these to dissociation of one or more lipoprotein-chlorophyll complexes, which were brought into solution by the acetone.

Action Spectra. The special feature of the chlorophylls is their absorption in the blue and red regions of the spectrum; if these pigments are responsible for the photosensitization of the reaction of CO_2 and H_2O to form carbohydrate, clearly we may expect the blue and red regions of the spectrum to be much more effective than the intermediate ones in bringing about this reaction. By plotting the amount of photosynthesis,† resulting from the supply of a given number of quanta of light-energy, against the wavelength of the light, we will obtain what is called an *action spectrum*, the region of maximal activity corresponding to the region of maximal absorption by the photosynthetic pigment. The two curves, action spectrum and absorption spectrum, will only be similar in outline provided that (a), one quantum of absorbed light is just as effective whatever its wavelength, and (b), either that the only pigment present is chlorophyll or that other absorbing pigments, such as the carotenoids, contribute their absorbed light-energy just as efficiently to photosynthesis as does chlorophyll.

* Smith & French (1963) quote some unpublished work of Allen, Murchio, Jeffrey & Bendix describing the separation of $C_a6,700$ from $C_a6,830$ without destruction of the natural complexes; they thus occur on separate particles and would therefore not exhibit the phenomenon of energy transfer. As Smith & French point out, the credit for demonstrating the inhomogeneity of chlorophyll-a *in vivo* is due to Albers & Knorr, who in 1936 measured the absorption spectra of single chloroplasts.

† In early studies of photosynthesis, the amount was measured by the absorption of CO_2; a great improvement in technique was made by Haxo & Blinks who determined the evolution of O_2 by the polarigraphic technique. The suspension of algæ is in contact with a platinum electrode which is the cathode of a battery; it attracts H^+-ions liberated by hydrolysis but becomes rapidly polarized when these ions are discharged. As soon as the H-atoms so formed combine with O_2 more current can flow, so that as long as O_2 is present in the suspension there will be a flow of current. In a closed vessel O_2 is used up by the electrode and the tissue; when photosynthesis occurs there will be an upward slope of the plot of current versus time, and by a suitable calibration the difference between this slope, and the slow decline that occurs before photosynthesis starts, becomes an accurate measure of the rate of evolution of O_2.

The first assumption is usually considered to follow from quantum theory, so that if it were found that the "quantum efficiency" varied with wavelength, the explanation would be sought in the interfering effects of subsidiary pigments. So far as the second assumption is concerned, it is known that chlorophyll is not the only pigment in the green leaf absorbing light, the subsidiary pigments, carotenoids, absorbing strongly in the blue region. Thus, unless these accessory pigments are indeed capable of acting as photosensitizers as efficiently as chlorophyll, we may expect the absorption spectrum in the blue region to be different from the action spectrum obtained by plotting the rate of photosynthesis against wavelength. Hence, to assess the function of chlorophyll in the green leaf or green alga, it is necessary to choose a region of the spectrum where interference by carotenoid absorption is negligible, namely in the red; as Fig. 807 shows, the action spectrum for photosynthesis, as measured by the rate of oxygen evolution of a suspension of *Chlorella*, corresponds remarkably well with the absorption of chlorophyll-*a* in the red region.

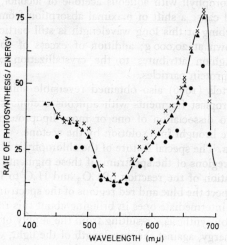

FIG. 807. Action spectrum of photosynthesis of *Chlorella*.
(Myers & French. *J. gen. Physiol.*)

Accessory Pigments

The overlap of the carotenoid and chlorophyll absorption in green plants and algæ would suggest that, if the carotenoids were ineffective in photosynthesis, the blue regions of the spectrum would be less effective than would be suggested by the absorption spectrum of chlorophyll, the light-energy absorbed by the carotenoids in this region being wasted. The question of the fate of this absorbed energy, and its value in photosynthesis, has been investigated in some detail. That photosynthesis may be carried out without the aid of the auxiliary pigments was shown by Warburg & Negelein, who demonstrated photosynthesis in red light, which is not absorbed by carotenoids. Emerson & Lewis have tackled the same problem in the blue-green alga, *Chröococcus*. From the absorption spectra of the separately extracted pigments—chlorophyll, carotenoids and phycocyanin—it was possible to compute, with fair accuracy, the percentage of the total incident light absorbed by a given pigment. Computed curves of this sort are shown in Fig. 808; they indicate that in the red region, at 6,800A, most of the absorption is due to chlorophyll, whereas a little further off, in the orange and yellow, absorption is predominantly by the blue pigment, phycocyanin. Clearly, if the light

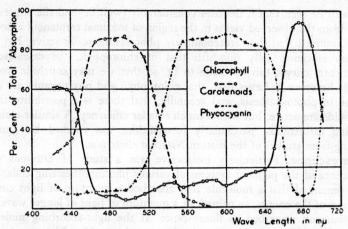

FIG. 808. Computed percentage absorption of the different wavelengths of visible light by the chlorophyll, carotenoids and phycocyanin of *Chröococcus*. Note that in the red region absorption is due predominantly to chlorophyll, whilst in the orange and yellow it is mainly due to phycocyanin. (Emerson & Lewis. *J. gen. Physiol.*)

absorbed by phycocyanin is useless for photosynthesis, a given number of quanta of yellow light at 6,000A should be far less efficient than the same number of red quanta. The quantum yield should therefore fall precipitately on passing from 6,760 to 6,000A. Fig. 809 shows the quantum yield plotted against wavelength; it will be seen that it is nearly constant over the range 6,900–5,700A, a result which can only be explained on the assumption that both phycocyanin and chlorophyll are photosynthetically active and, also, are equally efficient, *i.e.*, have a quantum yield of 0·08. The sudden fall in quantum yield beyond 5,700A is clearly associated with absorption of light by the carotenoids; this suggests that they are not photosynthetically active. However, the dotted curve in Fig. 809 has been computed on this assumption,

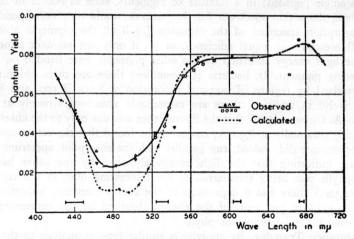

FIG. 809. The quantum yield of *Chröococcus* photosynthesis. The solid line is drawn through the experimental points. The dotted curve shows the expected dependence of the quantum yield on wavelength, on the assumption that the yield for light absorbed by chlorophyll and phycocyanin is 0·08 at all wavelengths, and that the light absorbed by the carotenoids is not available for photosynthesis. (Emerson & Lewis. *J. gen. Physiol.*)

and it will be seen that it deviates considerably (well beyond the experimental errors) from the observed yields in the region of maximal carotenoid absorption. If it is assumed that the carotenoids are photosynthetically active, but with a quantum efficiency only one-fifth that of chlorophyll, the calculated and experimental curves can be made to fit, so that we may conclude from this study of Emerson & Lewis first, that chlorophyll and phycocyanin are equally effective in photosynthesis, and secondly, that there is a possibility that the carotenoids are active, but with a much smaller efficiency. A similar conclusion regarding carotenoids, particularly fucoxanthin, was reached by Dutton & Manning from studies of the diatom *Nitzchia closterium.*

Fluorescence. Particularly instructive was a study of Duysens on the fluorescence of the pigment systems of various photosynthesizing cells. Earlier it was remarked that a molecule that absorbed a quantum of light could re-emit most of the energy, so gained, as a quantum of light of longer wavelength, a phenomenon described as fluorescence. If the light-absorbing molecule is involved in a simple reaction, *e.g.,* the decomposition of I_2 which, on absorption of a quantum, rapidly (within the vibration period of the atoms) splits into two I-atoms, fluorescence will not occur. In a complex photosensitized reaction, on the other hand, where the absorbed energy must be passed on to other reactants, the chances of a re-emission are much greater, and it is not surprising that photosensitizers are, in general, fluorescent. Chlorophyll is no exception, either in the intact chloroplast, where it may be recognized by fluorescence-microscopy, or in extracts, the emitted light being red of about the wavelength corresponding to the maximal absorption. Under optimal conditions for photosynthesis, the loss of energy in this way is remarkably small, only one quantum in about a thousand being re-emitted.

Action Spectrum of Fluorescence

Since fluorescence depends on the absorption of light, we may expect the action-spectrum for fluorescence, *i.e.,* the curve obtained by plotting intensity of fluorescence against wavelength, to correspond with the absorption spectrum of the compound emitting the light. This would be true of a system containing only a single pigment; in a mixture of pigments, such as occurs in bacteria and algæ, the action spectrum for fluorescence would only correspond with the absorption spectrum of the organism (*a*) if all the pigments absorbing light fluoresced with equal efficiency, or (*b*) if only one pigment fluoresced, but the light energy absorbed by the other pigments were handed on to the fluorescing pigment. In bacteria (*Chromatium*) there are three chlorophylls, characterized by regions of maximal absorption at 8,000, 8,500, and 8,900A respectively; in addition, there are carotenoids absorbing strongly at 5,500 to 4,500A. Duysens found that the fluorescence was due only to the chlorophyll absorbing maximally at 8,900A; moreover, he found that the action-spectrum for fluorescence did, indeed, run parallel with the absorption spectrum of the bacteria, indicating that the light-energy absorbed by the other bacterio-chlorophylls was being transferred to the fluorescing one; in the region of 5,500-4,500A there was a depression of the action-spectrum, indicating that only some 35-40 per cent. of the energy absorbed by the carotenoids was transferred to the bacteriochlorophyll.

Resonance Transfer. By applying a similar type of analysis to the green alga, *Chlorella,* Duysens concluded that chlorophyll-*a* alone was directly concerned in photosynthesis; nevertheless, the energy absorbed by chlorophyll-*b* seemed to be transferred to chlorophyll-*a* almost completely. A remarkable finding, confirmed later by Yocum & Blinks' studies of photochemical efficiency, was that although the quanta absorbed by phycobilins of red algæ were probably transferred to chlorophyll-*a*, they were more efficient in photosynthesis than quanta absorbed directly by chlorophyll-*a*. The only possible explanation for this anomaly is that chlorophyll-*a* is really a mixture of two molecular types, one of which is not involved in photosynthesis; the further assumption must be made that energy absorbed by phycobilin is transferred only to the chlorophyll involved in photosynthesis, presumably by the favourable location of the pigments within the cell. This point will be discussed further in a later section (p. 1532).

This "resonance transfer" is a well recognized phenomenon; thus, the fluorescence of naphthacine in a mixture of this with anthracene is excited by light absorbed by anthracene only. The conditions necessary for such a transfer are that the two molecules should be close to each other—less than one wavelength apart—and that the two molecules have common resonance frequencies. The existence of common resonance frequencies is indicated by an overlap of the fluorescence bands of the sensitizing dye molecule with the absorption band of the receiving molecule. A study of chlorophyll and phycoerythrin from this aspect suggests that the overlap is insufficient to permit resonance transfer; on the other hand, an overlap between phycoerythrin and phycocyanin of sufficient size does occur, and again between phycocyanin and chlorophyll. Thus, theoretically, it is likely that light absorbed by phycoerythrin will be transferred to chlorophyll by way of phycocyanin (Arnold & Oppenheimer, 1950; French & Young, 1952). These theoretical considerations are in accord with Duysens' (1951) experimental findings on the fluorescence of red algæ, from which he concluded that absorbed light-energy was transferred from phycoerythrin to chlorophyll-*a* by way of phycocyanin.

Structural Requirement

If the accessory pigments are to transfer energy to chlorophyll molecules, they must be closely associated with them on the chloroplast lamella or on the chromatophore of bacteria; Fuller, Bergeron & Anderson showed that the effectiveness of carotenoids, as determined by the relative effectiveness of green and red light in causing photosynthetic phosphorylation, could be greatly decreased by osmotic rupture of the chromatophores of *Chromatium*.*

Funnelling the Light-Energy. Hence we may picture the accessory pigments, including chlorophyll-*b*, as funnelling energy into chlorophyll-*a* by resonance transfer, so that on this basis the only photosensitizing pigment involved in the transfer of energy to CO_2 and H_2O is chlorophyll-*a*. This simple picture has been called into question, mainly by French and Rabinowitch, on the basis of several anomalies and the observed inhomogeneity of chlorophyll-*a* itself already described (p. 1500). Thus, we may ask which, if only one pigment is directly involved in photosynthesis, of the chlorophyll-*a*'s is the one. According to French, $C_a 6,840$ may be the main functional pigment, whilst $C_a 6,730$ might function as a chlorophyll-*b*, absorbing light and transmitting it to $C_a 6,840$. Similarly, $C_a 6,950$ might well be an accessory pigment and responsible for the

---

* The transfer of energy from one chlorophyll molecule to another, or from an accessory pigment to chlorophyll, may also be envisaged as a movement of an excited electron, and the "hole" it leaves behind, in opposite directions along the excited molecule, and from the excited molecule to neighbours. Such light-induced electron transfer gives rise to the phenomenon of semiconduction; a film of such light-sensitive material behaving like a metallic conductor. This aspect of energy-transfer has been reviewed by Clayton (1963), whilst experimental studies in which the photoconductivity of chlorophyll has been studied are those of Calvin (1959; 1961), and Commoner (1961). The appearance of "loose" or unpaired electrons in this way is revealed by the absorption of light energy under the influence of an appropriate magnetic field (Calvin, 1959), *i.e.*, by the so-called *spin resonance*.

Emerson effects that we shall now describe, but it must be appreciated that there is a possibility that the chemical events following absorption of light might be different according to which of two or more pigments absorbed the light, *i.e.*, we may picture, with French and Duysens, the presence of, say, two primary photosensitizers, perhaps $C_a6,840$ and $C_a6,950$, which sensitize different photochemical reactions, but both leading to synthesis of carbohydrate and oxygen evolution.*

Emerson Effect and Chromatic Transients. Emerson studied the efficiency of a quantum of light at different wavelengths and found that at long wavelengths, greater than 6,800A for green algæ, a single quantum became less and less effective. Since at these longer wavelengths it is only chlorophyll-*a* that is absorbing light to any marked extent, it appears that chlorophyll-*a* is in some way

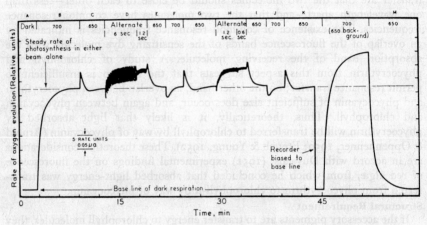

FIG. 810. Rate of O₂ evolution by *Chlorella pyrenoidosa* for light of 650 or 700 mμ adjusted in intensity to give equal steady-state rates. The chromatic transient effect of Blinks is shown between B and C, E and F, F and G, I and J and J and K. The Emerson enhancement, when both wavelengths are given together, is shown at L by the height of the record above the dot-dashed line. Enhancement by short period alternation of the beams, as contrasted with simultaneous presentation, is shown at D and H. (French. *Light and Life*. Johns Hopkins Press.)

less effective by itself than when acting together with chlorophyll-*b* and other pigments. Emerson, Chalmers & Cederstrand found, moreover, that the efficacy of a long wavelength light, *e.g.*, 7,000A, could be restored by simultaneous application of a background beam of shorter wavelength. Thus, if the rate of O₂-evolution of *Chlorella* is measured at 6,500A and at 7,000A, we may find that the longer wavelength is less effective, so that in order to obtain equal rates we must increase the intensity of the 7,000A light. Having adjusted the intensities

* The carotenoids serve the role, also, of protecting chlorophyll from being irreversibly bleached in the presence of light (photo-oxidation, p. 1536); thus Fuller & Anderson (1958) obtained normal photophosphorylation of carotene-deficient *Chromatium* bacteria so long as oxygen was excluded; in the presence of oxygen the process fell off, owing to the photo-oxidation of the chlorophyll. The protective effect of carotenoids was first demonstrated by Griffiths *et al.* (1955) in a purple bacterium. Calvin in an addendum to their paper suggests that the carotenoids may in effect be diverting light-energy from the long-lived activated state of chlorophyll that is necessary for inducing photo-oxidation. In higher plants, too, the absence of carotenoids leads to the photo-oxidative destruction of chlorophyll (Koski & Smith, 1951). A possible mechanism for this protective action has been discussed by Krinsky (1966). A carotenoid such as zeaxanthin being "epoxidized" to antheraxanthin by the photoactivated chlorophyll-O₂ complex; de-epoxidation of the antheraxanthin converts it back to zeaxanthin and permits the cycle to proceed again.

in such a way that during steady illumination we have equal rates of O_2-evolution, we may start with the longer wavelength and then superimpose the shorter wavelength; the rate of evolution is found to be more than twice, so that the combined effects of the two wavelengths are greater than their separate effects summed. We may say that the effect of the longer wavelength has been *enhanced* by the shorter wavelength. This effect, called the *Second Emerson Effect,** is shown in Fig. 810, where the rate of oxygen evolution, measured polarigraphically, *i.e.*, by the effects of the O_2 on the current passing through a platinum electrode, has been plotted against time. At K the suspension has been exposed to the shorter wavelength and a steady state of photosynthesis is achieved; the record is now biased to the base-line so that the rate of O_2-evolution is made to read zero; next the suspension is exposed to the 7,000A light while the 6,500A light is still shining. As indicated, the intensities of both beams had been adjusted to give equal rates of O_2-evolution, so that if there were no enhancement the record should return to the steady-state level indicated by the horizontal line. In fact the curve rises well above this (L).

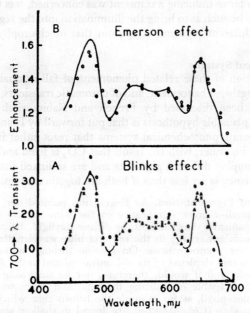

Fig. 811. Action spectra for "Emerson-enhancement" and "Blinks-transient" effects. (Myers & French. *J. gen. Physiol.*)

Action Spectrum of Enhancement

If we assume that the enhancement due to adding a shorter wavelength is due to the fact that absorption by a second pigment, besides chlorophyll-*a*, is necessary for efficient photosynthesis, then we may obtain a clue to the nature of this pigment by measuring an action spectrum for this enhancement effect, *i.e.*, we may measure the effectiveness of different wavelengths in enhancement. Myers & French measured the action spectrum in *Chlorella* and found that its maximum in the red corresponded with the maximum absorption of chlorophyll-*b*, namely 6,500A (Fig. 811, top). A more accurate study of the action spectrum in *Chlorella* by Govindjee & Rabinowitch showed that, besides chlorophyll-*b*, a form of

---

* By the first Emerson effect is meant the burst of evolution of CO_2 by plants during the first minutes of illumination.

chlorophyll-*a* absorbing maximally at about 6,700A was involved in the Emerson effect.

Chromatic Transient

Myers & French showed, moreover, that the action spectrum of the Emerson effect also corresponded with the action spectrum for another effect, discovered by Blinks—the *chromatic transient*. Blinks showed that when a suspension of algæ was exposed to a steady intensity of light of a given wavelength, changing the wavelength caused a transient increase or decrease in rate of O_2-evolution, in spite of the circumstance that the intensities of both wavelengths had been adjusted to give equal steady illumination. The change in rate of photosynthesis was thus only transient and is illustrated in Fig. 810. Blinks found that the action spectrum for this transient effect corresponded with the absorption spectrum of the accessory pigment; *e.g.*, in the green alga, *Chlorella*, it corresponded with chlorophyll-*b*; and this was confirmed by Myers & French (Fig. 811, bottom); in the brown alga maxima were found at 6,300 and 5,800A corresponding to chlorophyll-*c* and one at 5,200–5,400A corresponding to fucoxanthine. The essential point, so far as inducing a transient was concerned, was that the shift of wavelength had to be such as to bring the illumination into the region of maximal absorption of a different pigment, *e.g.*, from that of chlorophyll-*a* to that of chlorophyll-*b*.

Two Photochemical Systems

The interpretation of these related phenomena of fall in quantum efficiency with long wavelengths, enhancement, and of chromatic transients, is by no means obvious and has been discussed by French and Rabinowitch & Govindjee. Perhaps the most plausible hypothesis is that put forward by Franck, namely that there are two separate photochemical systems that react either in sequence* or in a more complex manner, with the result that CO_2 is fixed and O_2 is evolved. These reactions employ different pigments and are so linked that if only one is activated the efficiency is far less than if both are together (see p. 1524).†

Variety of Leaf Pigmentation. As Engelmann pointed out, the transfer of energy to chlorophyll-*a* gives significance to the wide variety of pigments in photosynthetic organisms; thus sunlight, or diffuse daylight, falling on terrestrial plants, contains sufficient energy in the red and blue wavelengths to provide an adequate amount for photosynthesis. On passing through water, on the other hand, the light becomes depleted of its red, infra-red and violet wavelengths, the light turning bluish; and it would, therefore, not be surprising if marine algæ developed pigments capable of utilizing the blue and green wavelengths more efficiently than chlorophyll, as for example the brown algæ which absorb in the green. Thus green algæ (*Chlorophyceæ*) are found in shallow water, whilst the red (*Floridæ*) are found in deep water, and the brown algæ occupy an intermediate position. Oltmanns emphasized, however, that the *intensity* of the light, as well as quality, also plays a part in the development of pigments, the total concentration of pigment tending to increase with depth of habitat as well as the proportion of phycoerythrin to chlorophyll (in the red algæ). In land plants chromatic adaptation is not so evident for the reason given above; nevertheless the adaptation to differences in the intensity of the prevailing illumination is very real, it being customary by botanists to classify plants as *ombrophilic* and *heliophilic*, the ombrophilic being characterized by their high density of chlorophyll (*e.g.*, *Aspidistra eliator*)‡;

* Govindjee, Rabinowitch & Thomas have described an inhibitory effect of extreme red light of wavelength greater than 7,200A on photosynthesis by red light of shorter wavelength, up to 7,200A.

† The sequential nature of these reactions is suggested by the finding by Myers & French that enhancement could be obtained when the two wavelengths of light were employed alternately as well as simultaneously; this is indicated in Fig. 810.

‡ Thus *Theobroma cacao*, an ombrophilic species, has a chlorophyll content of 0·79 per cent. of the fresh weight, compared with 0·11 per cent. for *Pinus sylvestris*, a typical heliophilic tree.

moreover, the relative proportions of chlorophyll-*a* to chlorophyll-*b* show striking changes, the relative amount of -*b* increasing with increasing ombrophilia. The absorption spectrum of chlorophyll-*b* differs from that of -*a* in that there is stronger absorption between 4,500 and 4,800A, *i.e.*, in the blue region, and this promotes the more efficient use of the light in shady regions, which has an abundance of the blue wavelengths.

Development of Chlorophyll. It is a common observation that a seedling allowed to develop in the absence of light, is very pale and deficient in chlorophyll (it is said to be *etiolated*); on exposure to light it becomes green. Clearly the chlorophyll is formed directly by the photochemical reaction of a precursor. Pringsheim, in 1874, first recognized this precursor by its absorption bands at 6,200–6,400A in an extract of etiolated leaves, and he called it *etiolin*, whilst Monteverde, in 1893, called it *protochlorophyll*; Scharfnagel, in 1931, recognized the characteristic absorption bands in etiolated *Zea* seedlings. Fischer identified the substance as chlorophyll-*a*, minus two H-atoms in the 7–8 positions. A variety of studies, notably those of Frank on the action spectrum for the formation of

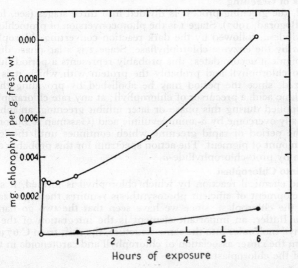

Fig. 812. The formation of chlorophyll-*a* (upper curve) and chlorophyll-*b* in dark-grown corn seedlings as a result of exposure to light. (Koski. *Arch. Biochem.*)

chlorophyll by etiolated leaves, and of Koski on the changes in relative amounts of protochlorophyll and chlorophyll, leave little doubt that protochlorophyll is, indeed, the precursor of chlorophyll-*a*, and probably also of chlorophyll-*b*. Thus Fig. 812 from Koski shows the course of formation of chlorophyll-*a* and chlorophyll-*b* in seedlings of *Zea*; most of the protochlorophyll was transformed to chlorophyll-*a* in the first five minutes of exposure to light, and this accounts for the sudden rise in chlorophyll-*a* concentration; subsequent formation of chlorophyll is largely determined by the chemical synthesis of protochlorophyll. The relationship of chlorophyll-*b* to chlorophyll-*a* in etiolated seedlings was investigated by Blaauw-Jansen, Komen & Thomas who found that the capacity for photosynthesis increased as the proportion of *b* to *a* increased, until the "normal" proportions were reached when, although the total concentration of chlorophyll increased with further exposure to light, the capacity for photosynthesis remained constant.*

* Albino mutants of corn are able to synthesize protochlorophyll and convert this to chlorophyll normally (Koski & Smith, 1951); the albinism is therefore due to an instability of the chlorophyll after formation; Smith, Durham & Wurster (1959) have been unable to associate any single factor, such as carotene deficiency or defective chlorophyllide esterification, with albinism in all mutants.

Thermal Reaction

The isolation of the thermal and photochemical components in the formation of chlorophyll was made by Yocum, who showed that 0·001 M cyanide inhibited the O_2-uptake of leaves from dark-grown bean seedlings by about 66 per cent., whilst there was no effect on the rate of formation of chlorophyll during the first hour, during which time, however, the concentration of protochlorophyll fell continuously. After this induction period, cyanide completely blocked the formation of chlorophyll. Thus cyanide apparently inhibits the synthesis of protochlorophyll, which is a thermal reaction, but has no effect on the photochemical conversion of protochlorophyll to chlorophyll.

Protochlorophyllide-Chlorophyllide Reaction

Studies of the change of absorption spectrum during greening were at first confusing, until it was appreciated that the material originally present in the etiolated leaf is not protochlorophyll but *protochlorophyllide*, lacking the phytol residue; this is converted by light to chlorophyllide, which is then converted by a dark reaction to chlorophyll-*a*.

Three Stages of Greening

In general, the greening process is divided into three stages (see, for example, Gassman & Bogorad, 1967). Stage 1 is the photoconversion of protochlorphyllide-*a* to chlorophyllide-*a*, followed by the dark reaction converting chlorophyllide-*a* to chlorophyll-*a* by the enzyme chlorophyllase. Stage 2 is a lag phase during which little or no pigment accumulates; this probably represents a period during which precursors to chlorophyll and probably the protein with which it is associated, are synthesized, since the period may be abolished by providing the leaf with δ-aminolevulinic acid, a precursor of chlorophyll; at any rate chloramphenicol and puromycin applied during this phase or later inhibit greening, an inhibition that may be partially overcome by δ-aminolevulinic acid (Gassman & Bogorad, 1967).* Stage 3 is the period of rapid greening which continues until the leaf contains the normal amount of pigment. The action spectrum for this probably corresponds to absorption by protochlorophyllide-*a*.

Integration into Chloroplast

Besides the chemical reaction by which chlorophyll is formed, it seems likely that the development of efficient photosynthesis requires the organization of the pigments in the chloroplast, since we have seen that the two go hand in hand. According to Butler, an important element is the integration of the carotenoid and chlorophyll molecules into the same structure, the shift from C 6730 to C 6770 resulting from the closer association of chlorophyll and carotenoids in the lamellar framework of the chloroplast.†

Dark-Grown Leaves

The process of grana formation in dark-grown leaves exposed to light has been described by Von Wettstein as proceeding through three light-dependent steps; first the prolamellar body consists of a system of tubules; these, during Stage 1, break up into vesicles; this is associated with the conversion of protochlorophyll to chlorophyll-*a*, and protochlorophyllide to chlorophyllide-*a*. Stage 2 consists of vesicle dispersal leading to the arrangement in regularly spaced primary layers; this apparently has different spectral light requirements from the first (Henningsen, 1967), and the pigment responsible for this has not been identified. The third step, formation of discs and aggregation into grana, occurs in strict correlation with the rapid phase of chlorophyll synthesis. The action-spectrum for Stage 2 indicates that only blue light of wavelength about 4,500A is effective, and this rules out chlorophyll-*a*, phytochrome and riboflavin as the effective pigments, whilst carotenoids are ruled out since mutants lacking these can carry out vesicle dispersal.

---

* According to Margulies (1962) although the inhibitor of bacterial protein synthesis, chloramphenicol, blocks photosynthesis in etiolated leaves it only partially blocks chlorophyll synthesis, so that the blocking of photosynthesis, which includes the Hill reaction, is due to other factors.

† Withrow, Wolff & Price showed that pre-treatment of etiolated leaves with red light reduced the latent period of synthesis of chlorophyll on subsequent illumination; as Price & Klein (1961) have shown, this effect may be inhibited by irradiation with a longer wavelength, *i.e.*, far-red light. The effect therefore belongs to the more general "red-far-red response" of plants, and represents an influence on synthetic processes as opposed to the photoconversion of protochlorophyll.

Protochlorophyll Holochromes. We may note that Smith and his colleagues (1957) have extracted photochlorophyll in association with its protein, by means of glycerol, from dark-grown plants; irradiation of the extract resulted in the formation of chlorophyll-a—which was presumably associated with protein—showing the characteristic absorption-maximum in the region of 6,800A. Smith gave the name "holochrome" to the pigment extract, to indicate that the essential principle contained both protein and prosthetic residues. Protochlorophyll holochromes from different sources had different absorption maxima and, corresponding with these, the chlorophyll holochromes derived from them had different absorption maxima. Thus barley leaf extracts gave an absorption maximum at 6,500A, corresponding to the protochlorophyll, which shifted to 6,800A on irradiation; extracts from bean leaves gave a maximum at 6,400A, shifting to 6,750A, whilst squash cotyledons gave a maximum at 6,450A, shifting to 6,800A. Subsequently, following on the pioneering work of the Russian investigators Krasnovskii & Kosobutskaya (1952), Smith *et al.*, prepared aqueous extracts of the protochlorophyll holochrome which they partially purified by differential centrifugation; the protochlorophyll activity was associated with material of molecular weight about 500,000. Later Boardman (1962) separated a protein complex of molecular weight 600,000, containing one protochlorophyll molecule; he emphasized that this might not be pure since Fraction I protein, or carboxy-dismutase, has a similar sedimentation constant.*

DYNAMICS OF PHOTOSYNTHESIS
Light- and Dark-Reactions

The overall photochemical synthesis of green plants may be described as:—

$$CO_2 + H_2O \xrightarrow{nh\nu} (CH_2O) + O_2 - 112 \text{ Cal.}$$

where the expression (CH_2O) is used to indicate a carbohydrate unit:—

$$H - \overset{|}{\underset{|}{C}} -- OH$$

which may be built up successively to a sugar; $nh\nu$ represents the number of quanta of light absorbed during the process. A simple experiment suffices to prove that photosynthesis can be broadly divided into a "light-process" followed by a "dark-reaction". Thus if the absorption of CO_2—or liberation of O_2—by a plant is measured with different intensities of illumination, it is found that the rate increases up to a point†; above this level of illumination, further increases have no effect, the system being *light-saturated*. This suggests strongly that with high light-intensities a secondary, or dark-reaction, determines the speed of photosynthesis; the supposition is confirmed, moreover, by the observation that with low intensities of illumination, where the light-process presumably dominates the picture, photosynthesis is almost independent of temperature, as we should expect of a purely photochemical reaction; at high light-intensities, on the other hand, temperature coefficients of the order of magnitude found with non-photochemical reactions are observed.

Intermittent Light. The more precise separation of photosynthesis into a light- and a dark-reaction was made possible by Warburg, who studied the effects of intermittent light on the absorption of CO_2 by plants; if the dark-

---

* Haselkorn *et al.* (1965) have examined Fraction I protein in the electron-microscope and biochemically; it certainly contains the carboxy dismutase activity (ribulose-1,5-diphosphate carboxylase), but they could not state definitely that it did not contain anything else; the molecules were built up of subunits, probably 24 with molecular weight of 22,500, to give a particle weight of 540,000.

† 20,000–40,000 lux for the alga *Chlorella*.

reaction requires a measurable time, we may expect to be able to use light more efficiently by illuminating the plant intermittently, allowing time between the flashes for the dark-reaction to proceed.* This was actually found by Warburg who, by using equal periods of light and dark, and varying the frequency of alternation (the flash-frequency), was able to double the efficiency of photosynthesis, i.e., a given amount of light-energy presented at a flash-frequency of some 133 per sec., produced twice the amount of O_2 as the same amount of light-energy presented as continuous illumination. Warburg observed that certain concentrations of cyanide—and other poisons that attacked enzymes containing the Fe-porphyrin type of catalyst, such as hydroxylamine, H_2S and azide—had no effect on the rate of photosynthesis when

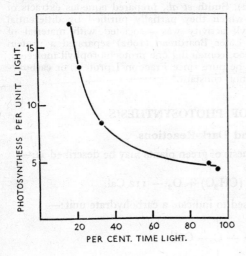

FIG. 813. Variation in photosynthetic yield, during flash illumination, with the bright period in the cycle. Note how the yield increases as the light-period decreases. (Emerson & Arnold. *J. gen. Physiol.*)

the intensity of illumination was low, i.e., when the overall rate was determined by the light-reaction; when the intensity was sufficiently high to produce light-saturation, on the other hand, the same concentration of cyanide, for example, markedly inhibited photosynthesis. It was concluded therefore that cyanide acted only on the slower dark-reaction and not on the initial, rapid, light-process. Emerson & Arnold developed the flash technique, varying not only the rate of flashing, but also the relative periods of light and darkness; by making the flashes very short (10^{-4} sec.) and the intervals long ($0.01-0.1$ sec.) they were able to effect a separation of the two main components in photosynthesis.

Fig. 813 shows the effect of varying the length of the light-period, in relation to the dark-period, from some 85 per cent. to 17 per cent.; it is seen that, as the dark-period lengthens, the efficiency of photosynthesis increases, so that a 400 per cent. improvement over the efficiency during steady illumination is obtained with only 17 per cent. of the cycle light; the frequency of flashing was 50/sec.; each light-period was therefore 0.0034 sec. and the dark-period 0.0166. Fig. 814 shows the variation in the photosynthetic yield-per-flash with the length of the dark-interval; it is seen that the yield-per-flash falls off when the interval becomes shorter than about 0.1 to 0.08 sec., so that, if the separation of the photosynthetic process into a light- and dark-reaction is

---

* Thus, if the result of the absorption of light were to produce a compound that must subsequently react with another compound, or else revert to its former state, light will be wasted if the newly-formed "light-product" has no compound to react with.

justified, the dark-reaction probably takes place within some 0·1 to 0·08 sec. at 6° C. The shape of the curve in Fig. 814 thus enables us to deduce the approximate length of the dark-, or *Blackman, period* as it is called, the length of the dark-period at which the yield-per-flash begins to decline being a measure of this; thus the lower curve represents the same type of experiment with a much lower concentration of CO_2; it is seen that the yield-per-flash falls off at about the same length of dark-period, a fact suggesting that the availability of CO_2 is not a decisive factor for the speed of the dark-reaction. The observation that the actual *yields* are consistently lower with the low concentration of CO_2 indicates, on the other hand, that the fast reaction, or primary photochemical process, is affected by the availability of CO_2.

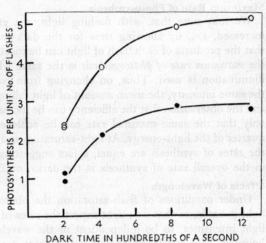

FIG. 814. Variation of photo-synthetic yield per flash with length of dark period between flashes at 6° C. Open circles, CO_2 concentration of $7·1.10^{-5}$ moles per l. Solid circles, CO_2 concentration of $4·1.10^{-6}$ moles per l. (Emerson & Arnold. *J. gen. Physiol.*)

Flash Saturation

Application of this form of analysis to the effects of cyanide and narcotics, such as urethane, showed that the photochemical process was apparently insensitive to cyanide whilst the dark-reaction was; narcotics, on the other hand, slowed the light-reaction, but not the dark. Emerson & Arnold showed that, by increasing the intensity of the flashes, an increase in yield-per-flash could be obtained—as one would expect—but that a maximum was reached, *flash saturation*, long before the intensity had been raised to such a height that every chlorophyll molecule was activated by each flash.

Photosynthetic Unit

By varying the concentration of chlorophyll, *i.e.*, the density of the algal suspension, Emerson & Arnold showed that it required some 2,480 molecules of chlorophyll to reduce one molecule of CO_2 per flash under conditions of flash-saturation. Since it is likely that some 10 quanta of light must be absorbed to reduce one molecule of CO_2 or produce one molecule of O_2, it would appear that some 248 molecules of chlorophyll are concerned in the useful absorption of a quantum, so that we may speak of the photosynthetic unit as a group of chlorophyll molecules able to funnel their quanta into the same synthetic unit, *i.e.*, into the same group of enzymes concerned in reducing a molecule of CO_2. Kok, by improving the flash technique still further, estimated the unit as lying between 150 and 400 chlorophyll molecules. If this concept is valid, it might well be as Kok & Hoch have suggested, that light absorbed by these molecules of chlorophyll is funnelled into a single photosensitizing molecule, different from those constituting the unit and which is the final light absorber in the photosynthetic process, transferring its energy to the compounds concerned in CO_2 fixation. As Rabinowitch (1959) expresses it: plant cells require a high concentration of pigment to be able to absorb sufficient light-energy, but they lack the space to provide a separate enzymatic conveyor belt for each chlorophyll molecule. A high chlorophyll : enzyme ratio is possible because even in direct sunlight a chlorophyll molecule absorbs quanta only about once every tenth of a second comparing with a turnover time of the enzyme reactions of only a fraction of this. Thus the migration of excitation energy through the photosynthetic unit provides an elegant solution to the problem of how to use the light-absorbing capacity of several hundred pigment molecules to feed a single conveyor belt.*

Maximum Rate of Photosynthesis

We may note that, with flashing light, the efficiency of the light can be increased, i.e., by allowing time for the dark-reaction to complete itself, so that the products of each flash of light can be most effectively utilized; however, the *maximum rate of photosynthesis* is the same whether continuous or flashing illumination is used. Thus, on changing from continuous to flashing light of the same intensity, the mean amount of light falling on to the system is reduced, and the observation that the efficiency can be increased by 400 per cent. indicates only that the same maximal rate can be achieved with the expenditure of a quarter of the light-energy. At light-saturation and at flash-saturation, therefore, the rates of synthesis are equal, a fact suggesting that the determining factor in the overall rate of synthesis is the dark-, or Blackman, series of reactions.

Effects of Wavelength

Under conditions of flash-saturation the photosynthetic system has all the light it needs, so that we might expect the rates of photosynthesis, at saturating light intensities, to be independent of the wavelength of light used. In fact, however, McLeod found that the rates were not the same in the green alga, *Chlorella*, a peak occurring at 6,500A, i.e., in the region of maximal absorption of chlorophyll-*b*. In a red alga the peak occurred at the λ_{max} of phycoerythrin. This "anomaly", together with the Emerson effects already described, is probably best explained on the assumption that there are two photochemical reactions involved, driven by different pigment systems with overlapping absorption spectra.†

Photosynthesis by Bacteria

The photosynthetic bacteria differ from the algæ and higher plants in not producing O_2; they are divided roughly into three classes, although differences within a class may be as great as those between classes.

The *green sulphur bacteria* reduce CO_2 and oxidize H_2S as follows:—

$$CO_2 + 2H_2S \rightarrow CH_2O + H_2O + 2S$$

The *red sulphur bacteria* (*Thiorhodaceæ*) grow autotrophically in the presence of a number of inorganic sulphur compounds such as S, H_2S, Na_2SO_3,

---

* The problem of the transfer of energy within the photosynthetic unit has been discussed by Duysens (1964) and Clayton (1965); it would seem that the singlet energy of one excited chlorophyll molecule may be transmitted to its neighbours, and thence to the final reaction centre, if the geometrical arrangement of the molecules is adequate to permit the dipole interactions on which the process depends.

† Pickett & Myers (1966) were unable to confirm McLeod's work; up to a wavelength of 7070A the maximum rate of photosynthesis, measured by O_2-evolution in *Chlorella*, was constant; with a wavelength of 7100A the same maximum was not achieved, but this was because they were unable to increase the intensity of this long-wavelength light sufficiently.

$Na_2S_2O_3$; they can also use organic compounds, and some species oxidize gaseous H_2; a typical reaction is:—

$$2CO_2 + H_2S + 2H_2O \rightarrow 2CH_2O + H_2SO_4$$

or, when H_2 is used in place of H_2S:—

$$CO_2 + 2H_2 \rightarrow CH_2O + H_2O$$

The *purple bacteria* (*Athiorhodaceæ*) require organic material, *e.g.*, aliphatic acids, which are oxidized. A recent example of this type of reaction has been provided by Foster, who grew bacteria on a secondary alcohol, to give the reaction:—*

$$\underset{\text{(Isopropanol)}}{2 (CH_3)_2CHOH} + CO_2 \rightarrow \underset{\text{(Acetone)}}{2 (CH_3)_2CO} + CH_2O + H_2O$$

Van Niel Hypothesis. As Van Niel pointed out, the reactions may be treated more generally as:—

$$CO_2 + 2 H_2A \rightarrow (CH_2O) + 2A + H_2O$$

the substance H_2A being, essentially, a reducing agent, or *hydrogen donor*, reducing CO_2 to CH_2O by the addition of hydrogen atoms, and being oxidized during the process to A and H_2O. The important point brought out by Van Niel is that the two processes, reduction of CO_2 and oxidation of substrate, are *linked reactions*, in the sense that the reduction of CO_2 necessarily demands the simultaneous oxidation of some substrate. Kluyver & Donker, in 1926, had suggested that H_2O might be considered as a reducing agent, or *hydrogen donor*, during the photosynthetic process, reducing CO_2 and becoming oxidized to O_2. Van Niel adopted this suggestion, and incorporated the photosynthesis of green plants into the general scheme by writing the equation:—

$$CO_2 + 2H_2O \rightarrow (CH_2O) + O_2 + H_2O$$

Hydrogen Donor

Van Niel postulated a primary photochemical process, common to all photo-synthetic reactions, the differences being due to the differing fates of the products of this primary reaction. He suggested that the common factor was the photo-chemical decomposition of water into a "reduced" and "oxidized" product; the reduced product eventually reducing CO_2 to give carbohydrate, whilst the oxidized product reacts with its characteristic hydrogen donor, *e.g.*, H_2S in the case of the green sulphur bacteria, fatty acid in the case of the *Athiorhodaceæ*, and with H_2O in the case of the green plants; in this last case producing H_2O_2 which finally gives gaseous O_2.

As we shall see, the common process is more likely to be the synthesis of ATP and $NADPH_2$ through the reduction of ferredoxin, the decomposition of H_2O being an incidental process peculiar to the green plants and algæ.

BIOCHEMISTRY OF PHOTOSYNTHESIS

Fixation of CO_2

Studies with Isotopic Carbon. The pioneering study in this field was that of Ruben and his colleagues with the aid of radioactive carbon, [11]C. This isotope

---

* Generally the organic substance oxidized in the photosynthesis is completely used up, so that it is difficult to formulate the primary photosynthetic process, distinct from subsequent metabolic reactions; in this case, however, acetone, formed from isopropanol, is not used by the bacteria. We may note that the presence of O_2 inhibits photosynthesis by the red (*Thiorhodaceæ*) and purple (*Athiorhodaceæ*) bacteria; they are thus anaerobic organisms; nevertheless many purple bacteria may be cultivated aerobically on suitable organic media; photosynthesis is abandoned and the bacteria adopt a chemosynthetic metabolism.

has a half-life of only 21·5 minutes, so that the effects of only brief exposures of algæ to $^{11}CO_2$ could be measured. The results showed that exposure to $^{11}CO_2$ in the dark caused a distinct uptake of ^{11}C which could be extracted as a water-soluble compound, the amount taken up being independent of the concentration of chlorophyll in the algæ. When *Chlorella* was exposed to $^{11}CO_2$ in the light for a few minutes, fixation of ^{11}C occurred, as we should expect; a rather smaller amount was found in the form of − COOH, but it is interesting that no ^{11}C was found in the extractable sugars. Formaldehyde has been suggested as an intermediate in the synthesis of carbohydrate, but no evidence of the presence of volatile aldehydes could be obtained. It would appear from Ruben's studies, then, that the fixation of CO_2 is a dark-reaction resulting from the accumulation, during exposure to light, of some "CO_2-acceptor".

Ruben suggested that there were two phases of sugar synthesis in the dark; a

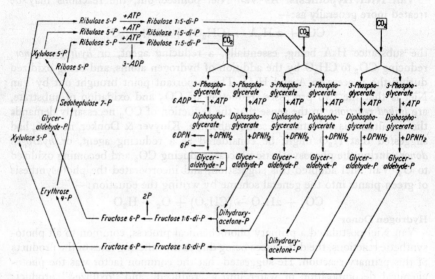

FIG. 815. The path of carbon in photosynthesis as deduced from the studies of Calvin and his associates. (Krebs & Kornberg. *Ergebn. Physiol.*)

carboxylation phase, dependent for its energy supply on ATP only and in which CO_2 carboxylates an acceptor molecule; and a *reductive phase*, in which the carboxyl group is reduced by a pyridine nucleotide with the aid of ATP.

The CO_2-Acceptor. This view was confirmed by the subsequent studies of Calvin and his associates, who employed the long-lived ^{14}C-isotope and examined the products of photosynthesis by a combination of paper-chromatography and radioautography. They showed that $^{14}CO_2$ appeared first in phosphoglyceric acid, which was later reduced to glyceraldehyde phosphate and thence converted to hexose by a reversal of the glycolytic process. The search for the primary CO_2-acceptor which reacted to give phospho-glyceric acid led to the identification of ribulose and sedoheptulose phosphate esters as normal constituents of green plants; and the kinetic studies of Calvin & Massini on the steady-state levels of phosphoglyceric acid and sugar diphosphates in *Scenedesmus*, when illumination was interrupted, indicated that ribulose diphosphate was the actual acceptor, interruption of illumination causing a fall in the concentration of this and a rise in that of phosphoglyceric acid. Addition of ribose diphosphate to an extract of ultrasonically treated

Chlorella showed that $^{14}CO_2$ could be converted to phosphoglyceric acid, *i.e.*, that the necessary enzyme system was present in the cells (Quayle, *et al.*, 1954). Further studies of the distribution of ^{14}C in ribulose and sedoheptulose indicated that these were formed during photosynthesis from glyceraldehyde and hexose phosphates; and this finding led to a proposed cycle (Bassham, Benson, Kay, Harris, Wilson & Calvin, 1954) that remains, in essence, the basis of the present-day view on the path of carbon in photosynthesis.

Photosynthetic Cycle. The scheme is illustrated in Fig. 815; as with the mechanism of aerobic respiration, it is cyclical, in the sense that a series of compounds are continuously being broken down and resynthesized. In the case of the photosynthetic cycle, ribulose 5-phosphate reacts with ATP to give ribulose diphosphate, the acceptor of CO_2; with one turn of the cycle 3 molecules of CO_2 react to give 6 molecules of 3-phosphoglycerate which, with ATP, become diphosphoglycerate, and are then reduced with $NADH_2$ to glyceraldehyde phosphate (triose phosphate). One of these six molecules leaves the cycle to be incorporated in a hexose and thence into starch, whilst the remaining five become involved in reactions leading to the resynthesis of ribulose 5-phosphate. According to this scheme, then, the net reaction is:—

$$3CO_2 + 9ATP + 5H_2O + 6NADH_2 \rightarrow \text{Glyceraldehyde 3 phosphate} \\ + 9ADP + 6NAD + 8 \text{ Phosphate}$$

The synthesis of 1 molecule of glucose from 2 molecules of glyceraldehyde phosphate is achieved by the following series of reactions which, with the exception of: Fructose 1:6-diphosphate → Fructose 6-phosphate are the reactions of glycolysis in reverse:—

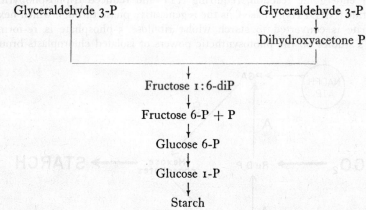

Glyceraldehyde 3-P   Glyceraldehyde 3-P
↓
Dihydroxyacetone P

Fructose 1:6-diP
↓
Fructose 6-P + P
↓
Glucose 6-P
↓
Glucose 1-P
↓
Starch

The reactions peculiar to photosynthesis are just two, namely the conversion of ribulose 5-phosphate to ribulose 1:5-diphosphate with the enzyme phosphoribulokinase:

and the cleavage of ribulose 1:5-diphosphate by CO_2 and H_2O, with the enzyme carboxydismutase

$$
\begin{array}{ccccc}
 & & & & CH_2OPO_3H_2 \\
 & & CH_2OPO_3H_2 & & | \\
 & & | & & HCOH \\
 & OH & C{=}O & & | \\
 & | & | & & HOC{=}O \\
 O{=}C & + & HCOH & \longrightarrow & + \\
 | & & | & & \\
 C & & HCOH & & O{=}COH \\
 HO & & | & & | \\
 & & CH_2OPO_3H_2 & & HCOH \\
 & & & & | \\
 & & & & CH_2OPO_3H_2
\end{array}
$$

The remaining reactions belong either to glycolysis or the pentose cycle (p. 291) working, of course, in reverse.

Assimilatory Power. The scheme for the biochemical events taking place during assimilation of CO_2 may be illustrated by Fig. 816 from Losada, Trebst & Arnon, which shows that it may be regarded as consisting of three phases. Phase A consists of carboxylation and is brought about by the conversion of ribulose 5-phosphate into ribulose diphosphate, an endergonic reaction requiring the participation of ATP; the ribulose diphosphate accepts CO_2 and splits to phosphoglyceric acid. Phase B consists in the reduction of phosphoglyceric acid, another endergonic reaction, requiring ATP and reduced triphosphopyridine nucleotide ($NADPH_2$). Phase C is the regenerative phase through which hexose phosphate is converted to starch whilst ribulose 5-phosphate is re-formed. Arnon's studies on the photosynthetic powers of isolated chloroplasts brought

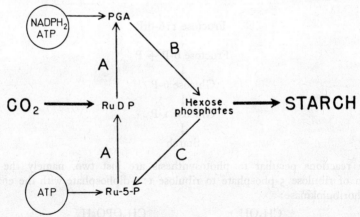

FIG. 816. Carbohydrate synthesis by isolated chloroplasts. The cycle consists of three phases. In the carboxylative phase (A), ribulose-5-phosphate (Ru-5-P) is phosphorylated to ribulose diphosphate (RuDP), which then accepts a molecule of CO_2 and is split to 2 molecules of phosphoglyceric acid (PGA); in the reductive phase (B) PGA is reduced and converted to hexose phosphates; in the regenerative phase (C) hexose phosphate is converted into storage carbohydrates (starch) and into the pentose monophosphate needed for the carboxylative phase. All the reactions of the cycle occur in the dark. The reactions of the carboxylative and reductive phases are driven by ATP and $NADPH_2$ formed in the light. (Arnon. *Light and Life.* John Hopkins Press.)

out the "dark" nature of these reactions, so that the essential feature of the light reaction is not assimilation of CO_2 but the formation of ATP and reduced NADP, *i.e.*, the formation of compounds that will participate in biochemical reactions that are not by any means peculiar to photosynthesis.* He called these "high-energy compounds" *assimilatory power*, and emphasized that the problem peculiar to photosynthesis was the mechanism of ATP formation and reduction of NADP. Thus light-energy is trapped as ATP and $NADPH_2$, but in order that it may be stored it must be transferred to CO_2 by "dark reactions". There is now no doubt that ATP is formed in the leaf from light-energy, and not as a result of a dark reaction by the mitochondria, *i.e.*, it is formed by what Arnon called *photosynthetic phosphorylation*. The experimental basis for Arnon's views consists of a series of studies on isolated chloroplasts, and these may be introduced by a description of earlier *in vitro* studies on the so-called Hill reaction.

Evolution of Oxygen by Isolated Chloroplasts. The Hill Reaction

Early studies on dried leaves or isolated chloroplasts failed to demonstrate any photosynthesis, nor yet could any considerable evolution of O_2 be obtained on exposure to light; that there is *some* evolution of O_2 associated with the illumination of these systems seems, however, to be generally proved, *e.g.*, by Molisch, but the methods necessary to demonstrate this evolution have had to be very sensitive,† and it was considered by Inman that it probably represented the decomposition of some easily reduced compound in the dried leaf or chloroplast. In 1939 R. Hill showed, with the aid of a spectroscopic test,‡ that O_2 was indeed evolved consistently by isolated chloroplasts, provided that aqueous extracts of acetone-extracted leaves, or yeast extracts, were added; the efficiency of the yeast extracts seemed to be related to the presence of ferric salts (although the leaf extracts contained none), and it was found that potassium ferric oxalate could replace these extracts, the evolution of O_2 following the stoichiometric equation:—

$$4K_3Fe(C_2O_4)_3 + 2H_2O + 4K^+ \rightarrow 4K_4Fe(C_2O_4)_3 + 4H^+ + O_2$$

The activity, in terms of output of O_2 per gramme of chlorophyll, was, however, only about one-tenth that of the intact leaf. Later, Hill & Scarisbrick showed that the liberation of O_2 was more vigorous, and comparable in extent with that taking place in the intact leaf, if potassium ferricyanide was also added, the ferricyanide oxidizing the ferrous oxalate back to the ferric form and thus preventing the evolved O_2 from doing so. In effect, therefore, the Hill-reaction consists of the decomposition of H_2O:—

$$2H_2O \rightarrow O_2 + 4H$$

or, as it is written:—

$$2H_2O \rightarrow O_2 + 4H^+ + 4e$$

the electrons being given to the ferric oxalate to convert it to the ferrous form:—

$$4Fe^{+++} + 4e \rightarrow 4Fe^{++}$$

---

* The only peculiar feature is that although both NAD- and NADP-linked triosephosphate dehydrogenases are present in the plant cell, only the NADP-linked enzyme is active under normal conditions.

† Luminous bacteria (p. 1588) were employed; they will luminesce in a concentration of O_2 corresponding to a partial pressure of 0·0007 mm. Hg; Beijerinck, in 1899, was the first to use this test on illuminated emulsions of clover-leaf chloroplasts.

‡ The conversion of added hæmoglobin to oxyhæmoglobin was the first test used; Holt & French, in following up this work, made use of the fact that the reaction results in the liberation of hydrogen ions, adding dilute alkali to the medium at a rate required to keep the *p*H constant; the same authors have also followed the reaction by the reduction of the dye, 2-6-dichlorophenol-indophenol.

Electron-Acceptor

Essentially it is the presence of the electron-acceptor (oxidizing agent) that permits the reaction to proceed. CO_2 is unnecessary for the reaction, so that the Hill-reaction permits an artificial resolution of the photosynthetic process into two stages. In the absence of any electron-acceptor, the decomposition of H_2O to O_2 and H_2 is immediately reversed, and a measurable evolution of O_2 is not observed; in the presence of ferric oxalate, on the other hand, the reduced product, whatever it is, may be oxidized and hence the recombination with O_2 may be prevented. In the intact plant CO_2 is ultimately the electron-acceptor; the disorganization of the photosynthetic system involved in the preparation of the chloroplasts presumably interferes with the enzymatic mechanisms. Thus the concept of H_2O as the source of O_2, already made use of in the earlier description of photosynthesis, fits in well with this scheme, and it is therefore of great interest that Ruben and his colleagues have shown, with the aid of ^{18}O, that the O_2 evolved in photosynthesis does, indeed, come from H_2O, the proportion of ^{18}O in the evolved gas being the same as that in the isotopic H_2O employed. According to the most recent study, the quantum number for the Hill reaction is of the order of 6–8, i.e., of the same order as that involved in natural photosynthesis (Lumry, Wayrynen & Spikes, 1957).*

Action Spectrum

That the evolution of O_2 by isolated chloroplasts is a part of the photosynthetic process is confirmed by the strong similarity between the action spectrum for oxygen evolution and that for absorption by chlorophyll suspensions; Chen found identical absorption and action maxima at 4,360 and 6,750A. Some 26 quanta of light were required to produce one molecule of oxygen.

Photosynthetic Phosphorylation

The Hill reaction, by which O_2 is evolved without CO_2 fixation, presumably fails as a photosynthetic reaction because in some way the formation of ATP has come uncoupled from the photolysis of water. However, Arnon, Whatley & Allen showed that isolated chloroplasts could, indeed, synthesize ATP from ADP and inorganic phosphate. The process was independent of CO_2-fixation, in that ATP accumulation was unimpaired by removing all CO_2. In the presence of CO_2, Allen et al. showed that CO_2 was fixed with a photosynthetic quotient of unity. If the chloroplasts were fragmented osmotically by placing in water, they lost their power of phosphorylation and CO_2-fixation, but this was restored on adding certain "catalysts" which presumably had been washed away during the preparation of the chloroplast suspension. The action spectrum for photophosphorylation showed maxima at 6,800 and 4,750A agreeing well with the absorption spectrum of chlorophyll-a; the high efficiency of the blue light indicated that the carotenoids were quite efficient in transferring their absorbed energy or otherwise utilizing it in the phosphorylating process (Jagendorf et al.).

Electron-Transport Chain. The catalysts required were ascorbic acid, flavine mononucleotide (FMN), and vitamin K; and it was suggested that the formation of ATP resulted from the linkage of ATP synthesizing reactions with an electron-transfer chain, in a similar way to the linkage with the electron-transport chain in the mitochondrion; in the latter case the energy is derived from the transfer of electrons from a hydrogen donor to oxygen, by way of flavine enzymes and the cytochromes, whilst in the photophosphorylating

---

* We may note the Hill reaction shows the second Emerson effect.

system Whatley *et al.* assumed that the electron donor would be the hydrogen derived from splitting of water, and that this electron would be transported through FMN, vitamin K, etc., to cytochrome and ultimately to the oxygen derived from the photolysis. This view of FMN, etc., acting in an electron-transport chain had to be modified when it was found that both FMN and vitamin K were not necessary, whilst the function of ascorbic acid seemed rather to protect the enzyme system from oxidation than to take part in a transport chain. Furthermore, an artificial electron acceptor, phenazine methasulphate, (PMS), was very effective in catalysing photophosphorylation (Jagendorf & Avron). Nevertheless, the general concept of electron-transport was retained.

Reduction of NADP. An important further step in defining the steps in assimilation was taken by Arnon, Whatley & Allen in 1958 when they showed that isolated chloroplasts were able to couple the reduction of NADP with synthesis of ATP, provided that a water-soluble factor from aqueous extracts of chloroplasts was added. This turned out to be the "photosynthetic TPNH-reductase" described by San Pietro & Lang (1958) equivalent, as we shall see, to ferredoxin. Thus the isolated chloroplasts were capable of carrying out the reaction:—

$$2NADP + 2ADP + 2P_i + 4H_2O \rightarrow 2NADPH_2 + 2ATP + O_2 + 2H_2O$$

thereby providing, in the light, all the assimilatory power needed for the fixation of CO_2. The chloroplast was thus shown to be, indeed, the "complete photosynthetic unit".

Cyclic and Non-Cyclic Photophosphorylation. Arnon described the photophosphorylation taking place in the absence of NADP reduction as cyclic ATP formation; here all the light-energy is devoted to synthesis of ATP, whereas in the non-cyclic photophosphorylation a part is used in reduction of NADP and the evolution of oxygen. By adding FMN, vitamin K, or PMS, the formation of $NADPH_2$ could be short-circuited so that the reaction was exclusively cyclic photophosphorylation. If the view of Arnon that the reaction indicated by the above equation constitutes the essential light process, then carbohydrate synthesis from CO_2 should be possible in isolated chloroplasts in the dark, if they were supplied with the assimilatory power artificially. In fact, Trebst, Tsujimoto & Arnon were able to demonstrate this either by supplying the chloroplasts with an external source of ATP and $NADPH_2$, or by exposing them to light in the absence of CO_2, and then exposing them to $^{14}CO_2$ in the dark; the assimilatory power synthesized in the light enabled the chloroplasts to assimilate the $^{14}CO_2$, the products being identified by chromatography.

Location of Enzymes

Because the process of cyclic photophosphorylation occurs in disrupted chloroplasts*, we may conclude that the enzymes responsible for this are structurally connected with the light-absorbing system; by contrast, the relative ease with which the enzymes for CO_2 assimilation are lost, suggests that these are much more loosely connected. From an evolutionary point of view this

---

* Kahn & v. Wettstein (1961) have examined chloroplasts from spinach in the electron-microscope; some appeared almost normal whilst others, broken chloroplasts, lacked outer membranes and seemed to consist of pieces of lamellæ. We may assume that the requirement of cofactors to obtain complete photosynthesis follows from the rupture of the chloroplast envelope which allows enzymes and substrates to leak out (Park & Pon, 1961; Lyttleton & Ts'o, 1958). Gee *et al.* (1965) showed that chloroplasts, homogenized in a medium containing NaCl, lost their outer membranes and were obviously disorganized; if leaves were homogenized without addition, and then diluted with 0·05 M Tris buffer, they obtained a CO_2-incorporation some 11 per cent. of that of intact leaves; addition of RuDP, ADP and ATP increased this to 44 per cent.

might suggest that production of ATP was the first step in photosynthesis, and only later did this become linked with fixation of CO_2 (Arnon, 1961). In bacteria, where evolution of O_2 does not occur, the photosynthetic phosphorylation is of the cyclic type.

Cytochromes

In 1959 Arnon suggested the scheme illustrated by Fig. 817; according to this, during non-cyclic electron-transport, the OH^--ions from water supplied the electron to the excited chloroplast molecule through the cytochrome chain, whilst ATP was formed by coupling reactions associated with the electron-transfer in a manner analogous with that associated with oxidative phosphorylation (p. 16). In the Hill reaction some other acceptor for the electron is present,

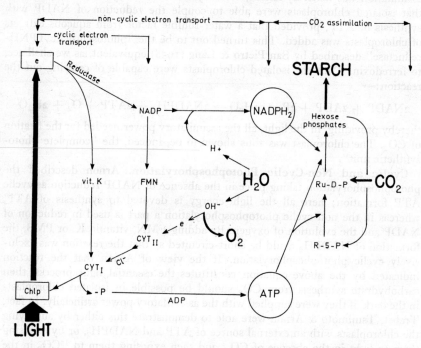

FIG. 817. Arnon's earlier scheme for photosynthesis. For details see text. (Arnon. *Nature*.)

and the light-energy is used solely to produce O_2 and to reduce the oxidizing agent (ferricyanide). Thus the difference between cyclic and non-cyclic phosphorylation resides in the pathway of the electron released by the excited chlorophyll and in the source of its replacement. In cyclic phosphorylation it returns through a redox system to the chlorophyll molecule and so no electrons are removed from the photosynthetic apparatus; in non-cyclic phosphorylation, the electron is transferred to NADP and used in the reduction of CO_2, being replaced by an electron from H_2O through the intermediary of the cytochrome system. Subsequent work has required successive modifications of this scheme, to take into account the role of *ferredoxin* as a primary electron acceptor, and also the Emerson and Blinks effects, which, together with the study of Amesz & Duysens on fluorescence, indicate the participation of two photochemical steps instead of one.

Ferredoxin

In the cyclic phosphorylation illustrated by Fig. 817, Vitamin K_3 and FMN, or an artificial substance such as phenazine methosulphate (PMS) were found effective as electron-acceptors, but these were probably taking the place of ferredoxin, the naturally occurring electron-acceptor that had been lost during the preparation of the chloroplasts. The same is true of the non-cyclic process; according to the early scheme illustrated by Fig. 817, the chlorophyll was supposed to reduce NADP directly, whereas more recent evidence indicates that here, too, the primary electron-acceptor is ferredoxin, whilst the reduction of NADP takes place by a dark-reaction catalysed by an enzyme, *ferredoxin-NADP reductase* (Shin & Arnon, 1965). Ferredoxin is the name originally given by Mortenson, Valentine & Carnahan (1962) to an iron-containing protein that contained neither a hæm nor a flavine prosthetic group; it was isolated from an anaerobic non-photosynthetic bacterium, *Clostridium pasteurianum* and was involved as a link between an enzyme dehydrogenase and different electron donors and acceptors. It became clear, subsequently, that this substance was not confined to non-photosynthetic systems, and it is chemically similar to, if not identical with, several substances isolated from chloroplasts and given such names as "methæmoglobin reducing factor", "TPN-reducing factor" and the "photosynthetic TPN-reductase" of San Pietro & Lang (p. 1521).* The special feature of ferredoxin is the extremely negative redox potential ($E_0' = -417$ mV at pH 7·55) close to that of hydrogen gas; and this permits it to reduce NADP easily and also to act as an electron-acceptor for the OH^- of water. The steps in which NADP is reduced involve photochemical reduction of ferredoxin, reoxidation of ferredoxin by a flavoprotein enzyme—ferredoxin-NADP reductase—and reoxidation of reduced ferredoxin-NADP reductase by NADP (Shin & Arnon, 1965). Thus the scheme for electron-flow under physiological conditions is:

Electron Donor System $\longrightarrow$ Ferredoxin

$\downarrow$

Ferredoxin-NADP Reductase

$\downarrow$

Free NADP

On this basis it should be possible to demonstrate the reduction of ferredoxin and simultaneous evolution of O_2, provided that the evolved O_2 were prevented from reoxidizing the reduced ferredoxin; by rigidly excluding extraneous O_2 from the system, it was possible to show that the O_2 evolved in response to illumination, measured polarigraphically, was equivalent to the ferredoxin reduced, as estimated spectroscopically (Arnon, Tsujimoto & McSwain, 1964), the stoicheiometry being 4 molecules of added ferredoxin to 1 molecule of O_2 produced, and thus the reduction of ferredoxin involved the transfer of one electron.

ATP Formation

In non-cyclic photophosphorylation, Arnon *et al.* (1964) found that one molecule of ATP was produced by 2 ferredoxin molecules, so that the process may be summarized by:—

$$4Fd_{ox} + 2ADP + 2P_1 + 2H_2O \rightarrow 4Fd_{red} + 2ATP + O_2 + 4H^+$$

* Ferredoxins isolated from five species of *Clostridium* were purified by Lovenberg, Buchanan & Rabinowitz (1963); they had the same molecular weight of 6,000, but crystallized in different forms, indicating differences in composition. The similarities between plant and bacterial ferredoxins have been described by Tagawa & Arnon (1968).

Cyclic and Non-Cyclic Phosphorylation

It was originally considered that the two processes of phosphorylation would occur together, but more recent work has shown that they are mutually exclusive, so that the cyclic process only occurs if the non-cyclic process is blocked by an inhibitor that prevents electron flow from H_2O. Alternatively, and very significantly, the two may be separated by the use of far-red light of wavelength greater than 7,000A; with this, cyclic phosphorylation is achieved without evolution of O_2 (Tagawa, Tsujimoto & Arnon, 1963). In general, it seems unlikely that the cyclic process occurs to any great extent when light of wavelengths less that 7,000A is available.

The Two Light-Reactions of Photosynthesis

We have already seen some of the evidence indicating the participation of two light-reactions in photosynthesis, namely the Emerson enhancement effect and the chromatic transients of Blinks.* Additional support for this concept was provided by Duysens & Amesz (1962) in their study of the red alga *Porphyridium cruentum*.

Short and Long Wavelengths. Duysens & Amesz found that the time-course of the change in degree of oxidation of cytochrome in the red alga, *Porphyridium cruentum*, measured by the change in absorption spectrum of the alga, was fundamentally different according as wavelengths of 6,800 and 5,600A were employed, *i.e.*, according as light was preferentially absorbed by the main bulk of chlorophyll-*a*, or by phycoerythrin, the accessory pigment of this red alga. With exposure to 6,800A there was a decrease in absorption of light at 5,200A, indicating oxidation of cytochrome; if, while the light was on, a second light of 5,600A was switched on, there was an increase in absorption at 5,200A, indicating reduction of cytochrome; increase in intensity of the 6,800A beam caused a decrease of absorption. A study of the interaction of the two wavelengths, and of the action spectra for inducing these effects, led Duysens & Amesz to postulate the presence of two linked systems, both containing chlorophyll-*a* and phycoerythrin, but in different proportions. One system, System I, contained mainly chlorophyll-*a* and was preferentially activated by light of 6,800A; this was supposed to oxidize a cytochrome and reduce CO_2 by way of $NADPH_2$; System II contained mainly phycoerythrin, and it reduced cytochrome when it absorbed light, the most effective wavelength being 5,600A; this reduction of cytochrome was associated with the simultaneous oxidation of water to O_2. In a study of a blue-green alga, *Anacystis nidulans*, Amesz & Duysens showed that the action spectrum for cytochrome paralleled that for NADP reduction; the action spectra indicated the participation of both chlorophyll-*a* and of phycocyanin, the accessory pigment of this alga, but chlorophyll-*a* was more active in reducing NADP than in the overall process of photosynthesis, as we should expect were chlorophyll-*a* concerned largely with only one phase of this process.

Systems I and II and Photophosphorylation

In general, the resolution of the photochemical processes into a System I, operating preferentially at longer wavelengths and accompanied by the reduction of CO_2, and a System II, operating at shorter wavelengths and accompanied by oxidation of H_2O to O_2, has been generally accepted (see, for example, Goedheer, 1965, who refers to the systems as q and p respectively). The scheme

---

* Warburg, Krippahl & Schroeder (1955) showed that the maximum efficiency of photosynthesis by *Chlorella*, illuminated with red light, was increased by very small amounts of blue light, an effect they attributed to the photo-activation of an enzyme.

illustrated by Fig. 817, based on a single photochemical step, meets difficulties, since a cytochrome is supposed to mediate the oxidation of H_2O, yet cytochromes sufficiently electropositive to remove the electron from OH^- are not known (requiring an E_o' of more than $+ 0.82$ V); this difficulty is overcome if a separate light-mediated step is invoked. According to Arnon, Tsujimoto & McSwain (1965), phosphorylation of ADP to ATP occurs in both cyclic and non-cyclic processes by separate light-activated reactions as indicated by Figs. 818 a and b. In the cyclic process, cytochromes-b_6 and -f are oxidized; and the dashed arrow indicates how a dye can short-circuit the cycle and make it insensitive to inhibition by antimycin A; thus cytochromes are confined to the cyclic system by Arnon because of its sensitivity to antimycin A and DNP, well-known

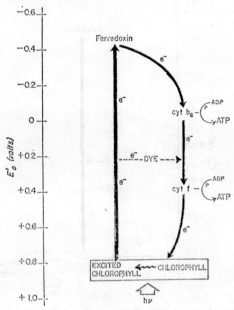

FIG. 818a. Scheme for cyclic photophosphorylation in chloroplasts. The dashed arrow indicates how a redoc dye can short-circuit the process. (Arnon. *Physiol. Rev.*)

inhibitors of electron-flow in oxidative phosphorylation by mitochondria. Non-cyclic phosphorylation is indicated by Fig. 818b; provision of a non-physiological electron-acceptor, such as ferricyanide, as in the Hill reaction, or by benzoquinone, gives a diminution of the light-generated reducing potential, and thus prevents reduction of NADP. ATP generation is coupled to the photo-oxidation of H_2O; and the rate of evolution of O_2 is greatly increased when this is accompanied by phosphorylation (Arnon, 1967).

"Separate Package Model"

It will be noted that the two photochemical events occur in parallel rather than in series ("separate package model"), so that in non-cyclic phosphorylation the transfer of an electron from H_2O *via* chlorophyll to ferredoxin involves a single photochemical event and not two in series as earlier postulated.* The steps involved in the electron-transfer are by no means clearly understood; the

---

* Clayton (1965) has discussed the relation between the two photosynthetic processes in terms of series models.

process requires Cl^- (Bové *et al.*, 1963), Mn^{++} (Kessler, 1955)* and plasta-quinone† (Arnon & Horton, 1963; Bishop, 1959), and is probably specifically inhibited by DCMU.

Three Light Reactions. Quite recently Knaff & Arnon (1969) have suggested that System II is really composite and they have modified the scheme illustrated by Figs. 818a, b by subdividing System II into two light reactions, IIa and IIb as illustrated in Fig. 819. This envisages System I, as before, operating in parallel with System II, which is now resolved into two short-wavelength photochemical steps, Systems IIa and IIb operating in series and joined by a "dark-reacting" electron-chain. Systems IIa and IIb are linked through C_{550}, the provisional name given to a new chloroplast component which, upon illumination with short wavelength light, shows a decrease in absorbance with a maximum at 5500 A.

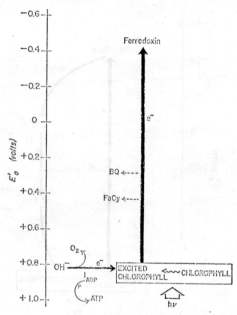

Fig. 818b. Scheme for non-cyclic photophosphorylation. (Arnon. *Physiol. Rev.*)

Separation of Systems

If, as is generally supposed, the evolution of O_2 from water, mediated through System II, requires shorter wavelengths of light than the System I reactions, it might be possible to separate the two processes by choosing a wavelength of light that the pigments of System II were unable to absorb; in fact, Tagawa, Tsujimoto & Arnon (1963) were able to obtain cyclic photosynthetic phosphorylation without evolution of O_2 by using light of 7,080A.

---

* Kessler (1955) showed that, although photosynthesis with evolution of O_2 was reduced to about a quarter in the Mn^{++}-deficient green alga, *Ankistrodesmus braunii*, it had no effect on the photoreduction of CO_2 in algæ adapted to carry out this process by maintaining in an atmosphere of H_2, so that the deficiency affects System II activity. Cheniae & Martin (1968) observed the same in *Scenedesmus*. We may note that Bishop & Gaffron (1962) showed that short-wavelength lights had no enhancement of this photo-reduction, once again indicating that System II activity requires relatively short wavelengths.

† A number of quinones are present in the photosynthetic apparatus, and they doubtless serve, in addition to the cytochromes, as electron carriers; of these, the *plastaquinones* have been the subject of much research, summarized by Crane, Henninger, Wood & Barr (1966). *Plastocyanin* is a copper-containing redox substance that is widely distributed in plant cells (Katoh *et al.*, 1961).

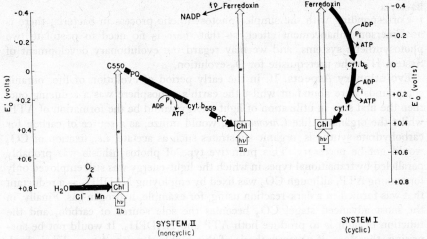

FIG. 819. Scheme for three light reactions in plant photosynthesis. System II consists of two "short wavelength" light reactions (IIb and IIa) operating in series and linked by a "dark" electron-transport chain associated with noncyclic phosphorylation. Parallel to System II is System I, consisting of a "long wavelength" light reaction linked to another dark electron transport chain associated with cyclic phosphorylation. (Knaff and Arnon. *Proc. Nat. Acad. Sci.*)

Switch Mechanism

The switch from non-cyclic to cyclic phosphorylation, if it occurs normally, might well be controlled by the supply of oxidized NADP. As long as this is available, electrons flow from H_2O to ferredoxin to NADP; if CO_2 assimilation ceased, for lack of ATP, reduced NADP would accumulate and electrons from reduced ferredoxin would begin to cycle to give cyclic phosphorylation. The resulting generation of ATP would promote CO_2 assimilation, so that non-cyclic phosphorylation would be re-established.*

Relation to Van Niel Hypothesis. On the basis of Arnon's scheme, the photolysis of water is not an obligatory prerequisite to the trapping of light-energy, so that the Van Niel hypothesis, which treats this as the universal step, may well be incorrect. Thus cyclic photophosphorylation represents a trapping of light-energy without evolution of O_2, and presumably without photolysis of water. In bacteria the reduction of NADP would be associated with oxidation of some substrate. In support of this view Arnon cites the photosynthesis of the bacterium *Chromatium* in an atmosphere of H_2 (Losada *et al.*, 1960). According to this, the only necessary light-reaction is the formation of ATP:—

$$Light: nADP + nP_i \rightarrow nATP$$

In the dark NADP would be reduced by H_2:—

$$Dark: 2NADP + 2H_2 \rightarrow 2NADPH_2$$

$$CO_2 + 2NADPH_2 + nATP \rightarrow (CH_2O) + H_2O + nADP + nP_i + 2NADP$$

Adding these equations we get.—

$$CO_2 + 2H_2 \xrightarrow{Light} (CH_2O) + H_2O$$

A process that Gaffron called photoreduction. Thus we have no photolysis of water, if this scheme is correct, yet photosynthesis of carbohydrate takes place.†

* A number of the biochemical features of photophosphorylation have been reviewed critically in a recent article by Avron & Neumann (1968).

† The assimilation of acetate in *Chromatium* is another example; the only function of

Bacteria

Corresponding with the simpler photosynthetic process in bacteria, there is no Emerson enhancement effect, so that there is no need to postulate two photosynthetic systems, and we may regard the evolutionary development of System II as the prerequisite for O_2-evolution.

Evolutionary Aspects. If, in the early period of evolution of life, organic compounds were abundant whilst the earth's atmosphere was a reducing one, then the first step in utilization of light-energy would be the formation of ATP, whilst the organism, like *Chromatium*, would utilize, as a source of carbon for carbohydrate synthesis, organic substrates such as acetate, *i.e.*, fixation of CO_2 would not be necessary. This primitive type of photosynthesis was probably paralleled by transitional types in which the light-energy was still employed only for making ATP, although CO_2 was fixed by employing $NADPH_2$ as a reductant that was formed in a dark-reaction using, for example, hydrogen gas. Finally, in the most advanced stage, CO_2 becomes the sole source of carbon, and the function of light is to produce both ATP and $NADPH_2$. It would not be surprising, therefore, if photosynthesis of this highly developed type did, indeed, require two sets of pigments (Losada *et al.*, 1960; Arnon, 1961). Expressed in terms of Arnon's electron-transport scheme (p. 1523), we would say that the green plants have acquired an "open", non-cyclic, type of electron-transport for reducing NADP by a photochemical reaction, whilst retaining the cyclic photophosphorylation common to plants and photosynthetic bacteria.

Quantum Efficiency

The change of free energy in the reaction:

$$CO_2 + 2H_2O \rightarrow 1/6(C_6H_{12}O_6) + H_2O + O_2$$

is some 117 Calories. The energy taken up by a gramme-mole of chlorophyll when each molecule absorbs one quantum of red light is some 41 Calories, hence $117/41 = ca.$ 3 might be expected to be the absolute minimum of quanta necessary to convert one molecule of CO_2 to carbohydrate, and the quantum efficiency of the reaction would then be said to be $1/2\cdot8 = 0\cdot36$.

In fact, of course, because photosynthesis involves a large series of reactions and the simultaneous utilization of several quanta, it seems very likely that many more than three quanta will be required.

According to Warburg's most recent studies (Warburg & Ostendorf, 1963), some $5\cdot5$ quanta are absorbed per molecule of CO_2 assimilated under optimal conditions for photosynthesis. In general, the non-cyclic process, with H_2O as the electron-donor, requires the greater amount of energy, and it may be for this reason that light of longer wavelength than 7,000 fails to bring about non-cyclic phosphorylation; furthermore, the threshold wavelength for obtaining maximum efficiency in photosynthesis is 6,850A, and this is the wavelength at which the peak of fluorescence occurs in chloroplasts at room temperature (Fig. 820).*

---

light here is to form ATP, so that, if this is provided, assimilation occurs in the dark (Losada, *et al.*, 1960).

* The quantum requirements for photophosphorylation have been discussed by Avron & Neumann (1968) who have described their own estimates, which depended on a variety of experimental parameters, such as wavelength of light, nature of substrate, pH, etc. For non-cyclic photophosphorylation with ferricyanide as redox substrate, they obtained a value of 3–4 quanta per ATP molecule in red light, and 12–17 quanta in far-red light. For cyclic photophosphorylation, using PMS as redox substrate, they obtained 3–5 per ATP molecule in the red, and 7–15 in the far-red.

Quenching of Fluorescence

Fluorescence is the loss of absorbed energy by re-emission of light, and we may expect the amount of this loss to depend in some measure on the ease with which the absorbed energy can be usefully employed chemically; thus we find that, whereas fluorescence in green leaves is only about 1 per cent. of the absorbed light at most, the fluorescence of chlorophyll in organic solvents may amount to some 30 per cent., and this is, presumably, because the activated chlorophyll molecules are unable to pass their energy on rapidly; the components of the photochemical system in the chloroplast may thus be described as *quenchers* of fluorescence, and it is conceivable that addition of components that favour increased photosynthesis will increase quenching; in fact Arnon, Tsujimoto & McSwain (1965) demonstrated a striking reduction of fluorescence of intact chloroplasts by addition of menadione or phenazine methosulphate, catalysts of cyclic photophosphorylation.*

The Photosynthetic Unit

The complex of chlorophyll and protein described as chloroplastin (p. 1500) had an estimated molecular weight of 43,500, and contained a single molecule of chlorophyll; photochemical studies indicate that the "unit of photosynthesis", in the sense of the smallest particle capable of photosynthesis, might contain perhaps 2,500 chlorophyll molecules, made up of some 10 subunits containing 250 molecules. On this basis, the smallest unit capable of photosynthesis, the quantasome of Park, would have a molecular weight of some 2 million. The difficulty in estimating this smallest unit by sonication, or otherwise breaking up the chloroplast, is the circumstance that photosynthesis is a double event, and it is possible to prepare particles that predominantly carry out System I or System II reactions; since this fractionation occurs, it can happen that the smaller particles are more active, per molecule of chlorophyll, than the whole chloroplast (Katoh & San Pietro, 1966), probably because they are released from the phosphorylation which may control the overall rate of photosynthesis (Izawa & Good, 1965).

Separated Particles. The smallest particle able to sustain the Hill reaction was found by Gross *et al.* (1964), by sonic rupture, to have a sedimentation coefficient of 38 S and an estimated molecular weight of $3 \cdot 5.10^6$; it contained 600 chlorophyll molecules. Four of these would cooperate to reduce one molecule of CO_2 (*i.e.* transport 4 electrons), and thus they might be equated with Emerson & Arnold's unit. The 38 S particles would contain some 2–3 quantasomes of Park. Another particle is that described by Becker, Shefner & Gross (1965), sedimenting at 20,000 to 50,000 g with a molecular weight of some $14 \cdot 4.10^7$; it was highly active in the Hill reaction and very stable, being little affected by tonicity, pH, etc. In the electron microscope it appeared as a disc of some 1,500A diameter and 360A high; it could probably accommodate 16,000 chlorophyll molecules on its surface and could be equated to some 88 quantasome particles.†

Non-Participation of the Quantasome

Howell & Moudrianakis (1967) have questioned the validity of the quantasome concept; their electron microscopical studies have confirmed the existence of discrete particles on the chloroplast membranes, built up of subunits of comparable size with those described by Park & Biggins, but they have shown that chloroplast membranes may be prepared, by treatment with EDTA, that, in spite of having no observable particles on them, are highly efficient in the Hill reaction. Hill-activity was measured by the use of an electron-dense tetrazolium salt, which

* In etiolated leaves immediately after illumination there is only one form of chlorophyll, and photosynthetic activity is not present; associated with this, the fluorescence is similar to that of chlorophyll in organic solution; with further illumination, photosynthesis becomes possible, and this is associated with the appearance of a more complex fluorescence spectrum at low temperatures indicative of the appearance of new pigments (Goedheer, 1965).

† Biggins & Park (1965) prepared what they considered to be the structural protein of the chloroplast by treatment with detergent; the average molecular weight was 22,000, and the material might well be analogous to Criddle's structural protein of the mitochondrion.

yielded an insoluble formazan that was deposited at the site of reduction on the chloroplast membranes. Observation of the precipitated material, when chloroplast membranes were allowed to react, showed that it was not localized to the "quantasome particles" but was uniformly distributed over the membranes. Again, preparations of the isolated particles failed to show Hill activity. It could be argued that the EDTA-treated membranes, lacking resolvable particles, really contained these, but that the treatment had flattened them so as to make them unresolvable; however, the extraction liberated protein material which was similar to, but not identical with, the Fraction-I particles described, for example, by Haselkorn et al. (1965); it could well be, therefore, that the quantasome particles are enzymes involved in the synthesis of carbohydrate, rather than chlorophyll-containing units in which the primary photochemical events leading to the Hill reaction take place.

FURTHER PHOTOCHEMICAL ASPECTS

A coherent theory of photosynthesis must be able to reconcile the physical changes in the pigment system with the chemical events; and it is just at this point that uncertainty prevails most strongly. The understanding of the physical events involves the elucidation of the absorption and fluorescence characteristics of the pigment system, and their interpretation in terms of the excited states of complex molecules. For example, it is easy enough to postulate, with Arnon, the tearing out of an electron from the chlorophyll molecule which then passes into an electron-transport-chain, but to show that the physical change in light absorption, or the fluorescence of the chlorophyll molecule, is consistent with this hypothesis is another matter. To examine the more physical features of the process in detail would take us beyond the permitted scope of this book* and we must confine our attention to just a few phenomena that have led to a radical change of thinking in the last few years.

Fluorescence. We have already referred to the fluorescence of chlorophyll, and seen how studies of this phenomenon have helped to show that light-energy, absorbed by other pigments, may be transferred by resonance transfer to chlorophyll-a. In various photosynthetic organisms the action spectrum for fluorescence is very close to that for photosynthesis and this seemed to confirm the position of chlorophyll-a as the photosensitizer in the primary light-reaction. Nevertheless, as we have seen, to speak of chlorophyll-a is somewhat ambiguous since there are probably several closely related pigments whose massed absorption spectrum corresponds to that of what is classically called chlorophyll-a.

Fluorescence Spectra. The fluorescent light† emitted by the chloroplast pigments in response to illumination has a characteristic distribution of energies in the different wavelengths of the spectrum, so that we may plot a *fluorescence emission spectrum* as in Fig. 820, and we may expect this to give a clue to the nature of the emitting molecule or molecules; if it is a single molecule the peaks of emission will be expected to occur close to the peaks of absorption, but the peaks of emission will be at rather longer wavelengths since this is a general feature of fluorescence (p. 1485). The emission spectrum of Fig. 820, measured at room temperature, shows a large peak in the red at 6,850A, corresponding to the red band of chlorophyll-a, and a much smaller peak in the far red around 7,300A; this is a general feature of the fluorescence of green plants and algæ, and it is considered, on the basis of experiments of different types, that the pigment responsible is C_a 6,700; thus, to employ the current terminology, F6,850 is equivalent to C_a 6,700. When the system is cooled to the temperature of liquid nitrogen ($-196°$ C), the spectrum is transformed to a more complex one; the intensity of the 6,850A band is greatly reduced whilst new bands appear at 6,960 and around 7,200A; this last band varies in position with the particular system, and is in the region of 7,300–7,400A in green leaves. Goedheer (1964)

* The interested reader may be referred to the Brookhaven Symposium No. 11 (1959) and a later one entitled *Light and Life* (Ed. McElroy & Glass, 1961), for some interesting articles on this aspect of photosynthesis.

† The intensity of fluorescence increases with time of exposure to light, reaching a steady-state value several times the initial value; this is considered to be due to the photoreduction of an electron acceptor, which thus reduces the electron-flow and so decreases quenching of the photosynthetic process (see, for example, Malkin & Kok, 1966).

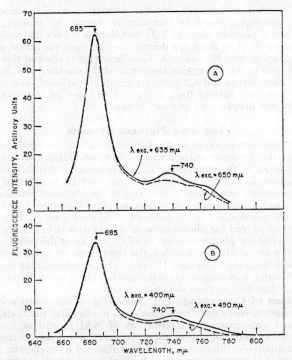

FIG. 820. Emission spectra of chloroplast fragments when excited by long
    wavelength light (A) and short wavelength light (B). (Govindjee & Yang.
    J. gen. Physiol.)

has suggested that these emission bands are associated with three forms of chlorophyll, the equivalence being indicated as follows:—

$$C_a\ 6{,}700\ \ ..\ \ F\ 6{,}860$$
$$C_a\ 6{,}800\ \ ..\ \ F\ 6960$$
$$C_a\ 6{,}950\ \ ..\ \ F\ 7{,}200^*$$

According to Goedheer, the shrinking of the chloroplast at low temperature
permits the transfer of energy from the chlorophyll molecule, normally emitting
at room temperature, F6,860, to the other species. As Robinson (1964) has argued,
the emission at room temperature is probably due to emission by the bulk chlorophyll, whilst the molecules actually transferring to the trap, or the trapping
molecules themselves (p. 1533), are responsible for the emission appearing at very
low temperatures. An indication as to the homogeneity or otherwise of the pigment
emitting the fluorescent radiation is given by studying the emission spectrum when
different wavelengths of light are used to excite the emission; thus the small
differences in the spectra of Fig. 820 are due to using 6,350 and 6,500A wavelengths
for excitation, and a so-called matrix analysis of the differences evoked in this
way indicates the involvement of two forms of chlorophyll-*a* in this emission at
room temperature, although the emission of one is extremely weak and normally
obscured by that of the other (Govindjee & Yang, 1966).

Action Spectra

Fig. 820 was obtained by using two selected wavelengths of light for stimulation
of fluorescence; if a wide range of wavelengths is employed, and the intensity of
fluorescence is plotted against the wavelength of the incident light, we have the
action spectrum for fluorescence; and a study of this can be useful in determining
the nature of the absorbing pigment or pigments leading to fluorescence. The

* The "accessory pigment" chlorophyll-*b* is, of course, present, and is preferentially
excited by the short-wavelength light, but it transfers its energy so efficiently to chloro-
phyll-*a* that fluorescence by chlorophyll-*b* is not normally observed.

information is limited, however, by the phenomenon of resonance transfer, whereby the absorbing molecule transfers its energy to another, fluorescing, molecule. When a pigment absorbs light but does not pass the energy on to the fluorescing molecule, then there is a discrepancy between the action spectrum and the absorption spectrum; for example, Goedheer (1965) showed that the xanthophylls contributed to the absorption spectrum of the etiolated leaf, but the action spectrum for fluorescence did not show a corresponding band, indicating that the xanthophylls were not passing their energy on to the C_a 6,700 considered to be responsible for the fluorescence at room temperature.

The Two Pigment Systems

The actual pigments involved in the two systems of photosynthesis, postulated to account for the Emerson enhancement effect, are difficult to determine with any precision since, in the chloroplast, light is necessarily absorbed to some extent by all the pigments, and the energy from one may be transferred to the final pigment, or pigments, most closely involved in the chemical reactions of photosynthesis. A variety of techniques have been employed, no one of which is capable of allocating unequivocally a given pigment to System I or II. When the wavelength of light employed for photosynthesis is varied, it is found that the shorter wavelengths favour the photoreduction of H_2O, and since the peaks of absorption of chlorophyll-*b* are at shorter wavelengths than those of chlorophyll-*a*, chlorophyll-*b* has been assigned to System II. Since photosynthesis may occur normally in mutants of barley that contain no chlorophyll-*b* (Highkin & Frenkel, 1962), we must conclude that this pigment is not essential.

Modifications of Quenching. The quenching of fluorescence occurs when the absorbed light is more effectively employed in chemical reactions, so that if a given treatment accentuates, say, reduction of NADP, and at the same time quenches the fluorescence of a given pigment, then we may assign this pigment to System I. Alternatively, fluorescence will be enhanced by metabolic poisons that inhibit the flow of electrons along one or both of the metabolic pathways involved in Systems I and II; for example, 3 (3,4-dichlorophenyl)-1,1-dimethylurea (DCMU) probably selectively inhibits the reduction of cytochrome by System II; and an observed increase in emission at 6,920A, caused by the inhibitor, suggested to Krey & Govindjee (1966) that a pigment emitting at this wavelength was involved in System II.

Modifications of O_2-uptake. Another approach is to measure the action spectra for photosynthetic activity under conditions where only one of the Systems is active, if we may suppose such a separation is really possible. Vidaver (1966) considered that the *uptake* of O_2, immediately following irradiation of the green alga, *Ulva*, represented System I activity, the uptake of O_2 being due to oxidation of a product, *e.g.*, NaDPH, formed by the light in System I. He considered that this was true because DCMU inhibits O_2-evolution but does not affect light-induced O_2-uptake. The action spectrum for this uptake had a maximum at 6,750A, corresponding to chlorophyll-*a* absorption, with a small shoulder suggesting the participation of chlorophyll-*b*; activity extended into the far-red at 7,400A, thus supporting the notion that System I can operate at long wavelengths. The action spectrum for System II was measured after exposure of the leaves to 50 per cent. O_2 in the dark for 14 hours. On return to air, the response to light was a continuous *uptake* of O_2, and this was saturated by far-red light. By imposing different wavelengths of light on the algæ exposed to the continuous far-red light, a reduction in O_2-consumption was obtained, the extent varying with the wavelength of the imposed light; this reduction was presumably due to the evolution of O_2 by the action of System II, so that the action spectrum for reduction of uptake is the action spectrum for System II. The main peak was at 6,500A, corresponding to chlorophyll-*b*; a shoulder at 6,750A indicated the participation of chlorophyll-*a*.

Photoreduction of CO_2. Again, the green alga, *Scenedesmus*, can be adapted to utilize H_2 instead of H_2O as an electron donor; the reversion to the use of H_2O can be prevented by DCMU and hydroxylamine, both inhibitors considered to block System II. Thus this photoreduction of CO_2 by H_2 is driven by System I, and the wavelength dependence shows the greater effectiveness of the long wavelengths by comparison with normal photosynthesis, whose rate is limited by the low absorption of System II in the far red (Bishop & Gaffron, 1962).

Heavy and Light Particles. Probably the most valuable evidence is derived from the fractionation of the pigment systems by breaking up the chloroplast into smaller units with the aid of detergents, since it has been possible to separate heavy and light particles with differing chlorophyll-*a*/chlorophyll-*b* contents, which exhibited different photosynthetic and fluorescent characteristics. Boardman, Thorne & Anderson (1966) obtained heavy particles sedimenting at 1,000–10,000 g with a low *a*/*b*-ratio compared with that of the intact chloroplasts; these were active in the Hill-reaction, indicating System II activity. The light particles, sedimenting at 50,000 to 144,000 g, had a low *a*/*b*-ratio, contained very little Mn^{++}, and exhibited very high reduction of NADP, and so were active in System I. Study of the fluorescence of these particles was instructive; the System I particles were only weakly fluorescent so that the normal fluorescence is due to a pigment or pigments of System II. It will be recalled that at very low temperatures the chloroplast emits at three wavelengths, 6,850, 6,960 and 7,350A. The System I particles emitted 97 per cent. of their light at 7,350A, so it was concluded that, at low temperatures, System I was responsible for the long wavelength emission whilst System II was responsible for the 6,860 and 6,960A.

In an essentially similar but more elaborate study, Ke & Vernon (1967), using Triton-X as a detergent, obtained heavy particles that were active in the Hill reaction, *e.g.*, the photoreduction of ferricyanide or dichlorophenolindophenol (DPIP), and thus in System II; these fluoresced at room temperature in the same way as whole chloroplasts, with a maximum at 6,830A, once again suggesting that the emission at 6,830 to 6,860A is due to a System II pigment. Hill reagents, such as ferricyanide, promote System II reactions and so might be expected to quench emission by System II pigments; and Ke & Vernon found that they did, indeed, quench emissions at 6,850 and 6,950A. The light particles with high *a*/*b*-ratio and β-carotene content, show high System I metabolic activity (photoreduction of NADP and oxidation of plastaquinone); these had only weak fluorescence at room temperature and at low temperature there was only one peak at 7,350A.*

To summarize, then, it would appear that the long wavelength band of fluorescence, F 7,350, appearing only in cooled chloroplasts, is associated with System I, whilst F 6,850 and 6,960 are associated with System II. The question now arises as to whether the pigments emitting the fluorescence are those involved in collecting the incident light-energy—chlorophylls and accessory pigments—or whether they are the energy traps into which the collected energy is funnelled.†

Energy Traps. Kok & Hoch have raised the question as to whether the light-absorbing characteristics of the pigment most intimately concerned in photosynthesis are altered significantly during the absorption of light. It is well established that the great bulk of the chlorophyll is photostable, retaining its absorption characteristics during exposure to light, in marked contrast to the photolabile pigments concerned in vision (p. 1628). Nevertheless, by employing the difference spectrum technique, *i.e.*, by measuring the absorption spectrum of the photosynthetic system before and after exposure to light, Kok & Hoch demonstrated a pronounced decrease in absorption at 7,000A after green algæ had been exposed to a flash of white light; this was attributed to the presence of a photolabile pigment, P7,000, which is reversibly converted to a non-absorbing product by light. In the dark the pigment recovered its absorbing power with a half-time of some 7 msec. The action spectrum for bleaching P7,000 indicated that chlorophyll-*a* and the carotenoids were concerned, so that Kok & Hoch suggested that chlorophyll-*a* and the carotenoids funnelled their absorbed light-energy into P7,000 which was

---

* Anderson & Vernon (1967) pointed out that if chloroplasts were treated with digitonin in a low-salt medium the resulting particles showed no fractionation of Systems I and II, having the same chlorophyll-*a*/*b* ratios; in the electron microscope it was seen that in the low salt medium the granum structure disintegrated, so that it would seem that if digitonin is to separate the two components it must have an intact granum-system to work on.

† More recently Wessels (1968) has described the separation of digitonin-soluble pigment-protein complexes; the heavy blue-green material had high *a*/*b*-ratio and P7,000 activity, and was enriched in α- and β-carotenes; it reduced NADP in light in the presence of ascorbate-dichlorophenolindophenol, ferredoxin, ferredoxin-NADP reductase and plastocyanin, and thus had high System I activity; the lighter yellow-green material had no NADP-reducing activity, or P 7000, and was strongly fluorescent at 6,800A at room temperature as well as at 77° K; the material could be separated into light and heavier material, the latter having high chlorophyll-*b* and xanthophyll contents.

the active pigment in photosynthesis,* acting as the final energy trap. Studies on fluorescence suggested, furthermore, that P7,000 might be responsible for the long-wavelength emission that had been attributed to System I, since particles with low a/b ratio exhibited high P7,000 activity.

In the light of the two-systems hypothesis, we might expect a second trap for System II, and Krey & Govindjee's (1966) study, based on the action of DCMU on the fluorescence of red algæ, suggests that this trap emits maximally at 6,920A, emission at this wavelength being enhanced by DCMU. If there is a difference of 150A in the wavelengths of emission and absorption, this would suggest a pigment absorbing maximally at about 6,800A, which may be characterized tentatively as P6,800.†

Morphological Correlates of Systems I and II. We have mentioned earlier (p. 1492) that photosynthetic activity per unit of chlorophyll, measured by O_2 evolution, is remarkably high in chlorophyll-deficient mutants, and it seemed that an essential structural requirement for this was that the thylakoids should be folded over at the ends to give a vestigial type of granum; in the absence of these foldings no O_2-evolution could be obtained. Homann & Schmid (1967) considered that System I was the rate limiting factor in photosynthesis at light-saturation, and suspected that Hill activity would be very high in the deficient mutant; and this turned out to be true, suggesting that System II, at any rate, required a folded-over thylakoid. Chloroplasts taken from the very chlorophyll-deficient areas of a variegated tobacco leaf, NC95, were found to be inactive in the Hill reaction, but these chloroplasts, although incapable of photosynthesis, were able to reduce NADP in the light with ascorbate-DPIP as an electron donor, and thus exhibited System I activity. In the electron-microscope these chloroplasts had no grana and no foldings of the thylakoids. Perhaps, because the evolution of O_2 requires the local accumulation of four oxidizing equivalents, it is necessary to have a "niche" equivalent to the granum, in which they can accumulate to be simultaneously effective.

Manganese Deficiency

Manganese-deficiency causes failure of System II (Kessler, 1955), and this is associated with severe disorganization of the membrane system (Mercer, Nittim & Possingham, 1962); and once again System I activity is unimpaired (Homann, 1967).

Luminescence of Chlorophyll. Fluorescence is a rapid process so that within less than a millionth of a second it is complete; Strehler & Arnold observed emission of light lasting for as long as 30 sec. after illumination; this was apparently a chemiluminescence, *i.e.*, the reversal of the primary photochemical reaction led to the excitation of the chlorophyll molecules; at any rate, the emission spectrum of this delayed luminescence was identical with that for fluorescence (Arnold & Davidson, 1954); moreover, it exhibited an optimal temperature of about $35°$ C, in common with the normal photosynthetic and Hill reactions.‡

It may be argued, however, that this "afterglow" is a re-emission of absorbed

---

* The order of photolability of the various chlorophylls identified by Brown & French (1959) was: Ca 650 < Ca 675 < Ca 683 < Ca 694. Sauer & Calvin (1962) have recently examined the photolability of chloroplast fragments.

† A corresponding energy trap associated with bacterial chlorophyll was demonstrated by Duysens in 1952, and is described as P8,700 or P8,900 according to the species; by selectively destroying the major bacteriochlorophyll complement by photooxidation in the presence of a detergent, the reactive substance was isolated; the material turned out to be bacteriochlorophyll, and owes its light-sensitivity to an association, perhaps with a cytochrome; at any rate, P8,700 activity is inversely related to light-induced cytochrome oxidation as though the light-oxidation of P8,700 were closely coupled with a dark-oxidation of cytochrome whereby the light-oxidized P8,700 is reduced. A mutant of *Rhodopseudomonas spheroides*, containing bacteriochlorophyll but not P8,700, was unable to support photosynthesis nor did it reveal any of the transient changes of fluorescence associated normally with light exposure (Sistrom & Clayton, 1964).

‡ Goedheer (1962) has described a "quenching" of the luminescence induced by a short wavelength by simultaneous exposure to a longer wavelength; he thus speaks of a promoting or "p" system favouring luminescence and a "q" system favouring quenching.

energy. This viewpoint has been supported by Tollin & Calvin who emphasize the analogy with the phosphorescence of inorganic crystals; if, therefore, the chlorophyll molecules are rigidly orientated as in a crystal, a similar type of emission is feasible. According to Tollin & Calvin, a great deal of trapped light-energy may pass into the excited triplet state from which electrons escape to various "electron-traps"; in these trapped states they are sufficiently stable that they require thermal energy to release them; when released they return to the "holes" in the chlorophyll molecule, giving rise to the excited triplet state again, and from here light emission may take place, which appears as the after-glow or luminescence. It is the stable state of the electrons in their traps that allows the process of emission to take place over such a long period.

Shape Changes of Chloroplasts

As with mitochondria, the size and shape of the chloroplast are influenced not only by osmotic movements of water but also by metabolic events that may be induced by illumination; in this respect the chloroplast is more favourable for study than the mitochondrion, since the reactions of photosynthesis can be so much more easily switched on and off than the thermal reactions of the mito-chondrion; nevertheless, the situation is still very confused; it has been reviewed by Packer & Siegenthaler (1966).

Rapid Reversible Changes. Light induces a rapid and reversible shrinkage of isolated spinach chloroplasts, as indicated by increased light-scattering; and the effects are very precisely correlated with conditions for non-cyclic and cyclic photophosphorylation (p. 1524); for example, it is stimulated by ascorbate-DPIP or ferricyanide which, by short-circuiting some of the steps, cause enhanced electron-transport; inhibitors of electron-transport, such as DCMU or NH_4^+, inhibit the change. In the electron microscope, Kushida et al. (1963) and Packer, Barnard & Deamer (1967), demonstrated remarkable increases in the axial ratio (from 1:7 to 2:6), so that the chloroplasts became longer and thinner.

Large-Amplitude Swelling. To confuse the issue, however, a much slower, and irreversible, light-induced swelling was also observed; this had a much greater amplitude, so that increases in volume up to 100 per cent. could occur; it was stimulated by co-factors favouring electron flow such as PMS, FMN and ferri-cyanide, but inhibited by NADP and ferredoxin; inorganic phosphate was a very strong inhibitor. This high-amplitude swelling was associated with deterioration of the ultra-structure (Packer, Siegenthal & Nobel, 1965).*

Contractile Protein. As to whether the rapid and reversible shrinkage is due to a contractile protein cannot be unequivocally stated; Packer & Marchant (1964) extracted a protein with ATPase activity, and Packer (1966) described a contraction of glycerinated chloroplasts in response to ATP.

Permeability of Chloroplast. The dependence of shape-changes on normal permeability of the chloroplast membrane was emphasized by Deamer & Crofts (1967); they caused chloroplasts to swell to more than three times their normal volume by treating with Triton-X, and this was due presumably to colloid-osmotic swelling comparable with that described in the erythrocyte (p. 553), since it could be prevented by increasing the colloid osmotic pressure of the medium; the swelling paralleled the blocking of light-induced shape-changes, so that the authors suggested that these depended on the establishment of ionic gradients by the light-induced reactions, gradients that could not be sustained with an abnormally highly permeable membrane.

---

* Nishida & Koshii (1964) found that, whereas chloroplasts retained their normal volume in 0·5 M sucrose, addition of 0·02 M salts caused swelling; the individual ions could be arranged in order of effectiveness corresponding to the lyotropic series, $SO_4^- > Cl^-$; $Ca^{++} > K^+$. Izawa & Good (1966) have described remarkable changes of organization of the membrane system in chloroplasts as a result of salt treatment and the uncoupling of phosphorylation.

PHOTO-OXIDATION AND PHOTODYNAMIC ACTION

Solarization

It has been recognized by botanists for a long time that excessive light may injure a green plant, the word "solarization" being used to describe this deleterious influence. A more particular investigation into the phenomenon has shown that it may be regarded as the result of a transition from one predominant type of photochemical reaction to another. The damage is more easily produced in an atmosphere deficient in CO_2, and it is observed that during the excessive illumination the evolution of O_2 ceases and gives place to an *absorption* of this gas. The amount of O_2 taken up depends on the partial pressure of O_2 in the atmosphere; and an essentially similar photo-oxidation is observed in plants killed by boiling, or in plant juices. The fact that the concentration of CO_2 is critical in determining the ability of a given light-intensity to cause solarization suggests that, in the absence of suitable concentrations of substrate, the light-energy, absorbed by chlorophyll, is used in a less specific manner to oxidize the constituents of the cell; if this oxidation proceeds unchecked, the structural components of the leaf—protein and carbohydrate—will become involved and the damage will be irreparable. According to Franck & French, the non-specific photo-oxidation normally runs parallel with photosynthesis, a certain amount of the intermediate products of photosynthesis in the plant being thereby wasted, but not sufficient to prejudice the normal metabolism; during CO_2-deficiency, however, the concentration of intermediates is reduced, and the photo-oxidative process becomes dangerous, attacking the vital structures.

We have already seen how the coloured carotenoids, presumably by directing light-energy from this destructive process, may protect photosynthetic plants and bacteria (Sistrom, Griffiths & Stanier).

Photodynamic Action

This non-specific photo-sensitized oxidation is not peculiar to the photosynthetic system; for many years a variety of phenomena have been classed together under the name of *photodynamic action*; and it would seem that they have the fundamental characteristic in common of being photosensitized oxidations. Thus a variety of unicellular organisms and viruses may be killed by treating them with a dye-stuff, *e.g.*, eosin, and subsequently exposing them to light. A frog muscle, treated in the same way, exhibits a series of twitches and eventually goes into a contracture; enzymes, such as invertase, peroxidase and catalase, may be inactivated; the proteins of plasma may be oxidized and denatured so that the fibrinogen, for example, fails to clot on the addition of thrombin; red blood cells may be made to hæmolyse, and so on. In all the cases examined, the effect may be completely prevented by the removal of O_2 from the system, either by anaerobiosis or by the addition of reducing agents such as sulphite; consequently photodynamic action may be described as photo-oxidation. The dye-stuffs most commonly employed in studying photodynamic phenomena belong to the fluorescein series, *e.g.*, eosin, rose Bengal, erythrosine, but many other pigments, such as chlorophyll and hæmatoporphyrin, are effective. Photodynamic dyes are more or less fluorescent in visible light, and this is symptomatic of their suitability as photosensitizers, since a fluorescent substance is one that can hold its quantum of absorbed energy for about 10^{-7} to 10^{-8} sec., in contrast to non-fluorescent substances which lose their absorbed energy by collision much more rapidly; the excited molecule has thus sufficient time to pass its quantum on to energy-absorbing reactants.

Activated Molecule

The general problems presented by photodynamic action are the nature of the oxidized substance, *i.e.*, whether it is protein, carbohydrate or fat ; and the nature of the activated molecule, *i.e.*, whether O_2 is activated and subsequently oxidizes substrate, or whether the substrate is activated, and reacts with normal O_2. The former hypothesis has been supported by Kautsky, who suggested that O_2 is activated to the $^1\Sigma$ state with a relatively long life of 10^{-3} to 10^{-2} sec. The energy of this state, however, requires a wavelength of less than 7,623A, whereas photo-oxidation with a wavelength as long as 8,000A has been observed. Studies of the quenching of fluorescence by substrates, moreover, indicate that O_2 need not necessarily be the energy-acceptor, and now opinion favours the activation of the oxidizable substrate as the primary process:—

$$Dye + h\nu \rightarrow Dye^*$$
$$Dye^* + X \rightarrow X^* + Dye$$
$$X^* + O_2 \rightarrow XO_2$$

where the asterisk denotes the activated condition. The fact that a frozen suspension of bacteria could be inactivated photodynamically led Heinmetz, Vinegar & Taylor to believe that the dye, oxygen and bacteria formed a complex, initially, and that it was this complex that absorbed the light; however, Blum & Kauzmann have shown that, so far as photodynamic hæmolysis is concerned at any rate, this is an unlikely mechanism.

Oxidized Substrate

With regard to the nature of the oxidizable substrate, protein is favoured by most workers; thus blood plasma, treated with hæmatoporphyrin and irradiated with visible light, exhibits a rapid consumption of O_2; Smetana has shown that this O_2-consumption is due predominantly to the proteins albumin and globulin; fats, glucose, and urea being responsible for only a negligible proportion. Again, Howell has shown that the clotting of plasma can be inhibited by photodynamic action, the effect being to alter the fibrinogen molecule, which no longer coagulates with heat. Thus, where photodynamic action is manifested in cell damage, there is little reason to doubt that structural proteins are attacked; according to Weil & Buchert's studies, the amino-acids most vulnerable to attack are those with aromatic nuclei, namely histidine and tryptophan; when these have disappeared, but only then, tyrosine begins to be destroyed.

Photodynamic Hæmolysis. One of the most exhaustively studied cases of photodynamic action is hæmolysis—*photodynamic hæmolysis*. Suspension

Rose Bengal

of red cells in saline containing a low concentration (say 1.10^{-6} M) of rose Bengal, for example, and exposure of this suspension to light from a 100-watt lamp, causes hæmolysis after an induction period (Fig. 821). According to the concentration of the dye, and the intensity of the light, the induction period varies from a matter of seconds to many minutes. If the light is switched off before hæmolysis is complete, the latter does not immediately cease, but continues for some time in the dark, the final degree of hæmolysis attained being determined by the amount that had taken place before extinguishing the light. Photodynamic hæmolysis consists, therefore, of a primary damage to the cells, followed by a dark or "after-light" hæmolytic process.* Blum showed that the dye is adsorbed on to the surface of the cell; consequently a prime factor in the relative efficacies of different dyes is the extent to which they are adsorbed; thus, from a $2–10.10^{-5}$ M solution, 80 per cent. of the rose Bengal is taken up

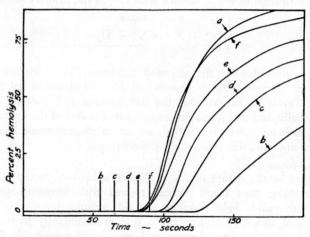

FIG. 821. Photodynamic hæmolysis. Percentage hæmolysis is plotted against time after illumination began. Curve *a* was obtained when light was continuous throughout the run; in curves *b* to *f* the light was discontinued after the point on the time abscissæ indicated by the vertical line bearing the corresponding letter. (Blum & Morgan. *J. cell. comp. Physiol.*)

by the cells, whilst only 30 per cent. of erythrosine is adsorbed. In terms of the actual concentrations of dyes required to cause a given degree of hæmolysis, rose Bengal is the more effective agent; on the basis of the amounts actually on the cell surface, however, the two dyes are equally effective.† Moreover, Blum found that the number of quanta of light required to hæmolyse a single cell was approximately 1.10^{10}, independently of the dye or its concentration. There is therefore little doubt that the first step in hæmolysis is the photosensitized destruction of a constituent of the cell membrane. The presence of reducing agents completely inhibits the light-hæmolysis; consequently this destructive action is oxidative.

* Photodynamic dyes are generally hæmolytic in the absence of light, but in much higher concentration than that required to cause photodynamic hæmolysis; such hæmolysis is described as *dark hæmolysis*, so that the process succeeding irradiation must be given another name; Davson & Ponder have used the term *after-light hæmolysis* to describe this phase.

† Small, Mantel & Epstein (1967) have examined the uptake by *Tetrahymena* of various polycyclic photodynamic compounds; the amount of uptake varied inversely with the molecular size, *i.e.*, the number of fused rings; however, activity was not determined exclusively by the amount of uptake.

Colloid-Osmotic Hæmolysis

The hæmolysis taking place after irradiation, the after-light hæmolysis, was investigated by Davson & Ponder; they showed that it was not due merely to a continuation of the oxidative process initiated in the light, since it was quite unaffected by reducing agents; furthermore, it was not due to the presence of lytic substances formed during the light-phase since dilution of a suspension of irradiated cells with isotonic NaCl had no effect on the course of after-light hæmolysis. It was shown that, before any hæmolysis occurred, the cells lost K^+ (Fig. 822) and that, as hæmolysis proceeded, the remaining cells continuously lost this ion. This escape of K^+ continued during the after-light period, and it

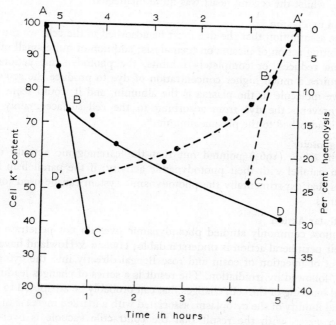

FIG. 822. Illustrating the escape of potassium from the erythrocyte accompanying photodynamic hæmolysis. Curve ABC shows the escape of K during illumination; curve BD shows escape of K continuing after cutting off illumination at B. Curves A'B'C' and A'B'D' represent the concomitant hæmolysis. (Davson & Ponder. *J. cell. comp. Physiol.*)

was concluded that the after-light hæmolysis was due, essentially, to the cation permeability induced during the light-phase, *i.e.*, that the cells, during the light-phase, had been made permeable to both Na^+ and K^+, a state of affairs that gives rise to the so-called colloid-osmotic hæmolysis described by Davson & Danielli (p. 553). Hence the persistence of the hæmolytic process after irradiation is due to secondary changes—colloid-osmotic swelling—resulting from the primary oxidative damage to the cell membrane.

This view of the process was confirmed by Green, Blum & Parpart, who found that the escape of K^+ was associated with penetration of Na^+, the longer the irradiation the more the gain of Na^+ exceeded the loss of K^+. The rates of cation loss indicated by Fig. 822 are too large to be due to a mere poisoning of the active-transport pump; they must represent an increase in passive permeability. At any rate, Borgese & Green have shown that glycolysis was barely affected.

Absence of Intermediate Substances

From the study of this type of hæmolysis we may conclude that photodynamic action results from the transfer of light-energy, by an adsorbed molecule of dye, to the structural proteins; as a result of this activation, they combine with molecular O_2. That the damage is not due to the formation of toxic substances in solution, which then damage the cell, is made very probable by the studies of Blum on the red cell, and is confirmed by the neat experiment of Rask & Howell in which two turtle hearts "in tandem" were perfused with a Ringer solution containing hæmatoporphyrin. Irradiation of the first heart (when the second was in darkness) caused irregularity in the beat and eventual stoppage, whilst the second heart was quite unaffected.

Inhibition by Plasma

On the assumption that the dye must be adsorbed to the cell, we can explain the inhibiting action of plasma on hæmolysis; addition of quite small quantities of plasma reduces, or completely inhibits, the photodynamic action, which now requires a much higher concentration of dye to produce the same effect; the active molecule in the plasma is the albumin, and it would seem that the plasma prevents the dye from adsorbing to the cell surface mainly because the dye is adsorbed to the plasma albumin.*

Carcinogenicity

Epstein et al. (1964) pointed out that the carcinogenic activity of many dyes ran parallel with their photodynamic activity; if the parallelism is exact, it means that experimentally the photodynamic system can be used for testing compounds.

Intracellular Action

The most commonly studied photodynamic dyes do not penetrate cells, so that their peripheral action is understandable; Hyman & Howland have studied the effect of injection of eosin and rose Bengal directly into the cytoplasm of *Amœba*, followed by irradiation. The result is a series of changes leading to the eventual bursting of the cell, changes that can probably be attributed to a greatly increased fluidity of the cytoplasm, associated with a marked increase in permeability to water, with the result that the contractile vacuole is over-worked. Internally, therefore, the photodynamic action is significant; and since a dye applied externally tends to destroy the cell membrane, it must favour its own penetration; the photodynamic lysis of protozoa, e.g., *Paramœcium*, is most probably due to combined internal and external actions.†

Again, the dye Janus green has a specific affinity for mitochondria which become stained when the living organism is treated with it; if kept in the dark, the motility of spermatozoa exposed to the dye is unaffected, but after illumination their mean velocity is decreased. Van Duijn considers that this effect is due to an attack on the mitochondria, although an influence on the cell membrane was not excluded.‡ Thus the great toxicity of Janus green to cells in tissue-culture may well be due entirely to this photosensitizing action since such cells would, during their examination, be exposed to a high intensity of illumination.

---

* This effect is not peculiar to photodynamic hæmolysis; saponin and many other lysins are inhibited by plasma.

† Giese has described the photodynamic killing of a protozoan, *Blepharisma*, the photosensitizer being the red pigment normally present in the cytoplasm. A similar action with other pigmented cells has not been described.

‡ Bull spermatozoa are actually sensitive to light in the absence of added dyes; this is due to the presence of a natural photosensitizer and is a true photodynamic effect since adding catalase, which breaks up peroxide as soon as formed, inhibits it (Norman & Goldberg, 1959).

That a muscle may be caused to twitch after photosensitizing it with a dye such as rose Bengal has been known for a long time; since end-plate blocking agents can inhibit this, the primary action is on the nerve or nerve-muscle junction (Rosenblum). The contracture that is also observed takes place even when the muscle is depolarized by KCl, and therefore represents an action on the contractile machinery.

Photodynamic Action in Living Animals. Since visible light cannot penetrate to any great depth of tissue, we may expect photodynamic action, resulting from the injection of a dye and subsequent irradiation of an animal, to be confined to the surface structures. This appears to be the case, white rats and mice so treated exhibiting symptoms attributable to a generalized stimulation of the sensory nerve endings in the skin. Later, necrotic lesions develop with loss of hair. The phenomena are by no means confined to experimental studies; it has been known for a long time* that cattle and sheep develop skin diseases as a result of eating certain weeds, and that these diseases attack a white animal in preference to a coloured one. Thus St. John's wort (*Hypericum*) is a weed that may become a serious menace to sheep in some parts of the world; that the disease is due to a photodynamically active pigment in the plant is made probable by a number of investigations; for example Horsley has extracted a fluorescent pigment, *hypericin*, from the whole plant; injection of the product into white rats produces typical photodynamic lesions. The disease of geeldikkop (yellow thick head) in sheep has the characteristic signs of photodynamic action, and is also apparently due to the eating of certain plants, such as *Lippia* sp., *Tribulus*, sp., etc. The toxic principle in the plant, in this case, is, however, not photodynamically active; it damages the liver and causes various breakdown products of chlorophyll (which are normally excreted in the bile) to appear in the bloodstream; and it is one of these products, *phylloerythrin*, that is photodynamically active. So long as the animal does not eat chlorophyll the signs do not occur. Finally we may mention the sensitization of animals to buckwheat (*Fagopyrum esculentum*), the condition of "fagopyrism"; Chick & Ellinger have shown that the flower and husk of the seed cause an inflammatory condition in albino rats when they are exposed to light in the range 5,400–6,100A; these authors extracted an active principle with acetic acid and methyl alcohol which gave a reddish fluorescence.

ULTRA-VIOLET LIGHT
Nuclear and Cytoplasmic Damage

In humans the effects of sensitization to hæmatoporphyrin, for example, are very similar to those of sun-burn, but in the latter case the signs result from a direct action of the radiation, without the mediation of a photosensitizer. The two phenomena—photodynamic and ultra-violet damage—differ also in that the effects of ultra-violet light are apparently not oxidation reactions, the hæmolysis, stimulation of muscle, destruction of bacteria, etc., being unaffected by anaerobiosis. In this connection we may remember that the study of cell structure was greatly facilitated by the characteristic absorption of the nucleic

---

* Blum quotes probably the earliest recorded case of a photodynamic disease in cattle from an article by J. Lambert in the Philosophical Transaction of the Royal Society, 1776, **66**, 493. "I shall now inform you of a very extraordinary and singular effect of lightning on a bullock in this neighbourhood, which happened about a fortnight since. The bullock is pyed, white and red. The lightning, as supposed, stripped off all the white hair from his back, but left the red hair without the least injury. . . ." Doubtless the loss of hair was due to photodynamic action, since only the white parts were affected.

acids in the region of 2,600A, whilst proteins absorbed most strongly at the longer wavelength region of about 2,800A; the strong absorption of nucleic acids in the ultra-violet might suggest that the lethal action of ultra-violet rays was due to nuclear damage, and this would appear to be true where bacteria, yeast and dermatophytes are concerned.*

Transforming Principle. The influence of ultra-violet light on nuclear material is best revealed, and most easily studied quantitatively, by measuring the behaviour of transforming DNA; for example, we may extract the DNA from streptomycin-resistant *Hemophilus influenzæ*, and measure its power to transform wild strains into resistant organisms. The effectiveness may be measured in terms of the number of "transforms" in unit time. After irradiation with ultra-violet light, the transforming effectiveness decreases in proportion to the dosage.

Cleavage Delay. Giese showed that irradiation of *Paramœcium* delayed its cleavage, the amount of delay being proportional to the dose of radiation; the action spectrum for this effect resembled fairly closely the absorption spectrum of nucleoprotein, the maximal effect being at 2,650A.

Subsequent studies on the delayed cleavage of *Arbacia* and of other echino-derm zygotes, carried out by Marshak, Blum, Cook and their collaborators, have left little doubt that the main point of attack, when cleavage is delayed, is the nuclear DNA. Thus, Marshak found that the dose of 2,537A radiation necessary to delay cleavage was very much less if the sperm, rather than the egg, was irradiated, indicating that absorption by the egg's cytoplasm was ineffective in causing delay. Again, Harding & Thomas centrifuged eggs so that the nuclei were all in the top zone; under these conditions, irradiation was much more effective from above than from below.

When sand-dollar zygotes were exposed to ultra-violet light at different stages after fertilization, it was found that, if irradiation was carried out after the replication of DNA, it had no delaying effect on division of the zygote, but it did delay that of the daughter cells; incorporation of the thymidine analogue, bromodeoxyuridine, increased the sensitivity to ultraviolet light and extended this sensitivity to longer wavelengths, 3,100–3,700A, characteristic of the absorption of the modified DNA. The interesting point emerging from Cook's (1968) study is that it is only lesions in the DNA that has not been replicated prior to the division that are capable of influencing this particular cleavage. Another point emerging was that, although a delay in replication of DNA is the prime event in ultraviolet damage, the delay *per se* is not a part of the mechanism, which is reflected in a delay in occurrence of prophase. The delay does not involve transcription of DNA so that the impaired function of DNA must be in some other process at the onset of mitosis with which DNA is associated, *e.g.*, separa-tion of centrioles or condensation of chromosomes.†

* Giese *et al.* (1956) found that flashing exposures of ultra-violet light were more effective than continuous ones, for the same total light-dosage, when studying the induc-tion of a delay in mitosis of the protozoan, *Didinium*. This suggested the existence of "dark-reactions". By contrast, Cook & Rieck (1962) found no evidence of this in the sand-dollar egg, the effect of dosage being independent of the way in which it was administered.

† The evidence implicating the replication of DNA as the sensitive step in delayed cleavage or mitosis is very strong; it is the S-phase of mitosis that is highly sensitive to ultraviolet light, the sensitivity decreasing during the latter half of the intermitotic period and becoming zero in early prophase in the mould *Physarum polycephalum* (Devi, Guttes & Guttes, 1968). With high doses, however, sensitivity remains constant during the period of rapidly declining DNA synthesis, so that other processes are involved at these doses; the effects of inhibitors of protein and RNA synthesis are similar to those of ultraviolet irradiation, in that they inhibit mitosis in *P. polycephalum*, and the latest times for application for both treatments to be effective was the same, namely 13 min. prior to metaphase (Cummins, Blomquist & Rusch, 1966).

Cytoplasmic Effects. It must be emphasized that most studies on the effects of ultra-violet irradiation are studies on what are essentially nuclear phenomena—time for division of eggs, viability of bacteria, etc.—and it could be that other effects would depend on cytoplasmic absorption, for example, the cytolysis of the egg following heavy doses, or the lifting of the fertilization membrane and the disappearance of the jelly coat (Blum, Cook & Loos, 1954), or the acceleration and later blocking of ciliary movement in *Paramœcium* (Saier & Giese, 1966). This is obviously true of the erythrocyte, which has no nucleus; ultra-violet light causes hæmolysis which is of the colloid-osmotic type similar to that described by Davson & Danielli and later workers.

Hæmolysis

This ultra-violet induced hæmolysis has been examined in some detail by Cook (1965), who has shown that the losses of K^+ and gains of Na^+ are quite consistent with the colloid-osmotic hypothesis, the dynamics being explicable in terms of the relative permeabilities to these ions induced by the damage, and the probable transmembrane potentials.

Yeast Cell

The escape of organic phosphates from the yeast cell irradiated with ultra-violet light seems to be a complex phenomenon, involving interference with nucleic acid metabolism as well as permeability changes (Swenson, 1960).

Nerve Conduction

Finally the effects of ultraviolet light in blocking nerve conduction have been known for some time; the literature has been summarized recently by Lieberman (1967) whose own studies indicate two sites for action. Thus light of 2,850A seems to block specifically the Na^+-activating mechanism, decreasing the amplitude and rate of rise of the spike, whilst the shorter wavelength 2,550A affects the resting membrane characteristics required for firing the spike. In this latter instance the amount of energy required per unit area is the same as that required for enzyme inhibition.

Ribonucleic Acid Absorption. Goldman & Setlow profited by the superficial cleavage of *Drosophila* eggs to effect a separation between nuclear and cytoplasmic damage. Immediately after fertilization, the zygote nucleus migrates to the centre of the egg where it divides several times; at this stage, then, it is protected by the surrounding cytoplasm from ultra-violet damage. At a later stage, most of the daughter nuclei migrate to the cortical cytoplasm, where subsequent development occurs; at this stage the nuclei are subject to damage. Ultra-violet damage to the cytoplasm led to faulty morphological development, *e.g.*, herniation of the intestine, whereas damage to the nuclei completely arrested development. Goldman & Setlow controlled the wavelength of their radiations with some precision with a monochromator, and were thus able to plot an action-spectrum for cytoplasmic damage; as Fig. 823 shows, maximum effectiveness is at 2,500–2,700A, corresponding to the absorption of purines; presumably, then, the cytoplasmic damage is due to absorption by ribonucleic acid. This view of the importance of ribonucleic acid absorption is strengthened by studies on the power of yeast to synthesize enzymes; Swenson & Giese showed that ultra-violet light inhibited this, the action spectrum being that for purine absorption; later Halvorson & Jackson* showed that there was an excellent correlation between the interference with ribonucleic acid metabolism, induced by ultra-violet light, and the interference with the synthesis of enzyme, hence the lethal effect of ultra-violet light on yeast, and probably on other microorganisms, would appear to be the

* Quoted by Spiegelman (1956).

consequence of an interference with nucleic acid metabolism, leading to a loss of synthetic power, especially manifest with respect to the production of enzymes.*

DNA. Because, then, both cytoplasmic and nuclear nucleic acids apparently absorb the damaging ultra-violet radiation, we may expect some competition between the two; since nuclear damage is more serious in a growing cell the damage caused by a given dose will be greater, the greater the proportion of DNA to RNA. This may account for the resistance of starved yeast to ultra-violet damage observed by Giese, Iverson & Sanders; a similar effect of starvation has

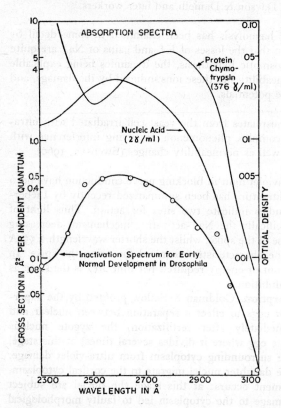

FIG. 823. Action spectrum for inhibition of development in *Drosophila* (lower curve) compared with absorption spectra of protein and nucleic acid. (Goldman & Setlow. *Exp. Cell Res.*)

also been observed in *Shigella sonnei* (Ramage & Nakamura); at any rate, if the DNA/RNA ratio of *E. coli* is doubled—and this may be induced by copper—the resistance to ultra-violet damage is increased (Weed & Longfellow). Since, therefore, so much of the damage caused by ultra-violet radiation can be attributed to nuclear events, the effects of light on the DNA† molecule are of extreme interest and have in consequence been studied at several levels, namely the chemical, the physico-chemical, and the biological. As indicated above, the biological change may be studied quantitatively by examining the influence of

---

* According to Weissman & Dingle (1961) ultra-violet light causes the release of proteases from lysosomes, and this may be responsible in some measure for cytoplasmic damage.

† The nucleic acid of the tobacco mosaic virus is of the ribose-type; the lesions are sensitive to ultra-violet irradiation. It is interesting that if infection is induced by the nucleic acid alone there is no lag period in sensitivity as occurs when the normal virus is used; thus the lag period probably represents the time during which the virus is shedding its protective coat of protein (Siegel, Ginoza & Wildman, 1957).

ultra-violet light on genetic markers, such as the transforming activity of bacterial DNA.

Formation of Dimers

At the chemical level, the studies of Wang, of Beukers, and of Wacker have implicated the thymine residues, which are altered so that they react with each other to form dimers; at any rate extraction of the labelled DNA from irradiated bacteria gives a labelled thymine derivative that is separable chromatographically from thymine. A possible interpretation of the effects of ultra-violet light, therefore, could be that it induced the formation of chemical cross-linkages, through the thymine residues, between the two DNA chains of the double helix. This is not the only factor, however, since the studies of Marmur *et al.* have shown that ultra-violet light has two effects, as revealed by the physical chemistry of the DNA molecule. The first is discovered as a partial dissociation of the double helix, *i.e.*, a denaturation, so that the temperature required to induce a given degree of separation of chains, as revealed by the hyperchroic effect (p. 171), is lowered. The opposite tendency, namely the formation of cross-linkages, is shown by a proportion of the molecules which become insensitive to the normal strand-separation procedures; the higher the adenine + thymine content the greater the proportion of this insensitive material.* We may presume, therefore, that the first effect of ultra-violet light is to cause some separation of strands by breakage of hydrogen-bonds; this permits the strong covalent cross-linking in localized regions following on the altered chemical constitution of the thymine residues; such a linkage between non-specific base-pairs presumably prevents replication in the regions where it happens. Since the single-stranded phage ϕX 174 DNA can be inactivated, we must assume, further, that linkage between thymine residues on the same strand can occur; in fact Bollum & Setlow (1963) found that native DNA was less sensitive to inhibition of its primer activity by ultra-violet light than was denatured, *i.e.*, single-stranded, DNA.

Sensitive and Insensitive Strains

The importance of dimer formation in inhibition of growth of bacteria has been shown by Swenson & Setlow (1966), who compared radiation-sensitive and -insensitive strains of *E. coli*. The essential difference between the two was that, whereas high doses would completely block DNA synthesis in the sensitive strain, the same doses would only block this temporarily in the insensitive strain, and the recovery of synthesis ran parallel with the splitting off of dimers from the DNA molecule, *i.e.*, the cell could, in some way, cut out the points at which replication of DNA ceased and presumably replace them with monomeric thymine. It must be emphasized that this form of reactivation of the bacterium is different from *photoreactivation*, which is revealed in both sensitive and insensitive strains.†

Photoreactivation

Of considerable interest is the phenomenon of photoreactivation. Kelner noticed, during the course of an investigation into spontaneous recovery of microorganisms from ultra-violet irradiation, that a suspension of irradiated spores of *Streptomyces griseus* showed a 100,000-fold recovery by comparison with another suspension, treated apparently in exactly the same manner; it

---

* Thus the DNA from *Streptomyces viridochromogenes*, which has some 26 per cent. adenine + thymine, gives less cross-linked material than that from *D. pneumoniæ* with 62 per cent.

† The possible significance of faulty repair work in the genesis of mutations has been discussed by Setlow (1964). Experimentally the dimers can be isolated because of their resistance to acid hydrolysis, and they can be labelled heavily with tritium.

turned out, however, that the suspension showing recovery had been exposed to a large amount of visible light; and later work showed that the destructive effect of ultra-violet light could be largely reversed by subsequent illumination of the organisms, the effect being definitely due to the visible portion of the spectrum—*i.e.,* not to the infra-red—and increasing with the duration of radiation. Irradiation of the organisms with visible light before treatment with ultra-violet was quite without effect.*

Importance of Cytoplasm

Studies on other systems, for example on delayed division of *Arbacia* or *Paramœcium*, have shown that photoreactivation is a general (but not a universal)† phenomenon; of special interest is the finding that, although ultra-violet action is essentially a nuclear phenomenon, reactivation seems to depend on the cytoplasm. For example, Marshak showed that the delayed division of the zygote,

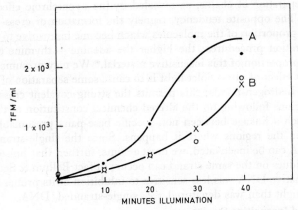

FIG. 824. Competitive inhibition of photoreactivation of transforming DNA that had been inactivated by ultra-violet light. Curve B shows the course of photoreactivation in the presence of irradiated unmarked DNA; this is to be compared with curve A which shows the recovery of marked (transforming) DNA in the absence of this addition. (Rupert. *J. gen. Physiol.*)

obtained by fertilizing a normal egg with an ultra-violet-irradiated sperm, could not be reversed by visible light; irradiation of the zygote, however, caused photoreversal.‡ Blum, Robinson & Loos (1951) observed the same phenomenon and commented on its analogy with the photorecovery from ultra-violet inactivation of phage; Dulbecco observed that this recovery could only be obtained when the phage was mixed with bacteria; exposure of the ultra-violet-inactivated phage to visible light in the absence of bacteria being useless.

Again, Cook & Rieck showed that, in the sand-dollar, the following systems could be photoreactivated: (*a*) egg exposed to ultra-violet and fertilized with

* At about the same time Dulbecco described the reactivation of phage by visible light; he incubated equal quantities of ultra-violet-inactivated phage with sensitive bacteria, the one in the light and the other in darkness. The light ones showed 1,000 times as many plaques as those in the dark. Wavelengths longer than 4,400A were ineffective in reactivation.

† Jagger (1958) has categorized the instances of non-photoreactiveness; many bacteria show it, *e.g.,* in strains of *B. cereus, B. megatherium, S. fæcalis, Diplococcus pneumoniæ,* and so on.

‡ The amount of delay in division depends on the phase of the mitotic cycle during which the egg is irradiated; the maximal effect occurs when the egg is irradiated before anaphase (Blum, Kauzmann & Chapman, 1954); photoreactivation is most effective during prophase (Marshak, 1949).

normal sperm; (b) eggs exposed to ultra-violet and fertilized with irradiated sperm; (c) zygotes exposed to ultra-violet. The fourth system examined, namely normal eggs fertilized with ultra-violet irradiated sperm that had been treated with longer-wavelength light, showed more delayed cleavages than if no attempt had been made to reactivate the sperm. Thus, so long as photoreactivation was attempted in the presence of cytoplasm it was successful.*

Reactivating Enzyme. The clue to this cytoplasmic effect was found by Rupert, who extracted from baker's yeast a material that could, when added to inactivated transforming DNA, permit photoreactivation. In the absence of the principle, the long-wavelength light was ineffective. Not all bacterial extracts were effective, however; thus one from *Hemophilus influenzæ* was useless, although the DNA transforming principle was actually derived from this bacterium, whilst an extract of *E. coli* was effective. This corresponds with the fact that intact baker's yeast and *E. coli* can be photoreactivated whereas *H. influenzæ* cannot. The extracted material had the characteristics of an enzyme, being heat-labile and non-dialysable; moreover, it could be shown to form a complex with the inactivated DNA, but not with normal DNA. The mechanism of repair would therefore seem to be first the formation of an enzyme-DNA complex; while complexed in this way the DNA would repair itself, and the effect of light would be to release the repaired molecule from its complex. Thus, on centrifuging the solution containing ultra-violet irradiated DNA and enzyme, the latter is found at the bottom of the tube with the DNA; only after this complex has been irradiated with longer-wavelength light, *i.e.*, after it has been photoreactivated, does the enzyme remain in the supernatant fluid. The photoreactivation process may be competitively inhibited by adding another DNA so long as this has been inactivated by ultra-violet light; this is illustrated by Fig. 824 which compares the increase in "transforms" with time of illumination in photoreactivating light, in the presence and absence of inactivated DNA. The inactivated DNA competitively inhibits the enzyme, reducing its activity to about one half, indicating that the DNA's compete on equal terms.

Condition of DNA

The power of inactivated DNA to inhibit the enzyme provides a delicate means of assessing the "state of repair" of a given DNA preparation. In this way we can prove, for example, that when an intact cell has been photoreactivated the cause of this reactivation is, indeed, the restoration of the DNA to normal. To do this it is only necessary to extract the DNA and determine whether it is capable of competing with the enzyme in photoreactivation of some standard system. Again, when a cell or phage has been damaged, the contribution of DNA damage may be assessed by extracting the DNA and once again determining its competitive power; in this way it was proved that the single-stranded phage, ϕX 174, is damaged by irradiation.

Mechanism of Repair. We have seen that the repair of ultra-violet induced damage occurs in a radiation-resistant strain of bacteria by virtue of the excision of dimers from the DNA molecule, rather than by reversal of the dimerization, since there is no reduction in the number of dimers that can be extracted from the cells when these radiation-resistant cells are allowed to recover in the dark (Setlow, Swenson & Carrier, 1963). There seems little doubt that photoreactivation, on the other hand, depends on an enzymatic depolymerization; thus Cook (1967) caused photoreactivation of ^{3}H-thymine-labelled DNA by exposing the ultra-violet inactivated material to yeast-photoreactivating enzyme and ultra-

---

* Dulbecco (1950) observed that phage could only be reactivated when adsorbed on to its bacterial host.

violet light of wavelengths greater than 3,200A. Reactivation was allowed to proceed for increasing periods, the DNA was precipitated and hydrolysed, and the ³H-labelled thymine and its dimers were separated chromatographically. There was a very close correspondence between the disappearance of activity in the dimer spot on the chromatogram and the increase in the activity of the thymine spot.

Significance of Photoreactivation

The need for some cytoplasmic mechanism for DNA repair was emphasized by Rupert (1961), who pointed out that exposure for one hour to bright sunlight would otherwise almost completely inactivate the DNA of cells exposed to this. The sun's rays contain both the damaging radiation—λ_{max} 2,540A—and the reactivating wavelength of λ_{max} 3,500A, so that inactivation of DNA is opposed by photoreactivation. Thus, we may expose DNA in a quartz flask to sunlight until its transforming activity has been reduced by a factor of about a thousand; the reactivating enzyme may now be added and the exposure continued, in which event transforming activity returns, to give some 90 per cent. recovery.

Nuclear Repair

The striking effects of photoreactivation on DNA repair suggest that photoreactivation is purely a repair of nuclear damage*; this fits in with Borstel & Wolff's work on the wasp egg. Here the nucleus is located superficially when laid, and it requires no sperm for normal development; the nucleus can thus be irradiated preferentially. When this is done, the damage, indicated by reduced hatchability, could be reversed by light of 3,600A. A much larger cytoplasmic

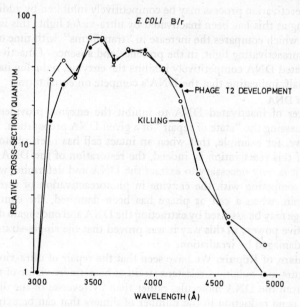

FIG. 825. Illustrating the action spectra for photoreactivation in *E. coli* suspensions. The relative effectiveness per quantum in producing one per cent reactivation has been plotted against wavelength. (Jagger. *Bact. Rev.*)

* Some cytoplasmic damage seems to be photoreactivable, however; for example Skreb & Errera (1957) found that the ultra-violet damage to anucleate *Amœba proteus* could be reversed; again Pierce & Giese (1957) found reversal of ultra-violet damage to frog and crab nerve fibres, *i.e.*, to part of the neurone that had no nucleus. Cabrera-Juarez (1963) obtained some reactivation of DNA from *H. influenzæ* by treatment with nitrous acid.

dose was required to produce an equivalent damage, and this damage could not be reversed by the longer wavelength light.

Action Spectrum. Action spectra for reactivation of ultra-violet damage indicate that the most effective wavelengths are in the ultra-violet and violet; the spectrum for reactivation of *E. coli* and Phage T 2 are shown in Fig. 825 from Jagger & Latarjet; the falling off in effect at short wavelengths is presumably due to the overlap with the DNA-inactivation region of the ultra-violet light.* Since the spectrum is not the same for all organisms, it may be that different absorbing "chromophores" are concerned, although the effectiveness of the enzyme extracted from one organism—baker's yeast—in reactivating the DNA from all the organisms tried, would suggest that the absorbing molecule is the DNA-enzyme complex, which may well be the same for all organisms.

* The pitfalls to be avoided in the determination of an action spectrum for photo-reactivation have been discussed in detail by Jagger (1958) in an exhaustive review of the whole subject. Giese *et al.* (1953) have claimed that wavelengths between 4,050 and 3,660A are equally effective in reactivating *Colpidium*.

REFERENCES

AGHION, J. (1963). Propriétés des extraits aqueux de chloroplastes ayant un maximum d'absorption à 740 mμ. *Biochim. biophys. Acta*, **66**, 212–217.

AMESZ, J., & DUYSENS, L. N. M. (1962). Action spectrum, kinetics and quantum requirement of phosphopyridine nucleotide reduction and cytochrome oxidation in the blue-green alga *Anacystis nidulans*. *Biochim. biophys. Acta*, **64**, 261–278.

ANDERSON, J. M. & VERNON, L. P. (1967). Digitonin incubation of spinach chloroplasts in tris (hydroxymethyl) methylglycine solutions of varying ionic strengths. *Biochim. biophys. Acta*, **143**, 363–376.

ARNOLD, W., & DAVIDSON, J. B. (1954). "The Identity of the Fluorescent and Delayed Light Emission Spectra in *Chlorella*." *J. gen. Physiol.*, **37**, 677–684.

ARNOLD, W., & OPPENHEIMER, J. R. (1950). "Internal Conversion in the Photosynthetic Mechanism of Blue-green Algae." *J. gen. Physiol.*, **33**, 423.

ARNON, D. I. (1955). "The Chloroplast as a complete Photosynthetic Unit." *Science*, **122**, 9–16.

ARNON, D. I. (1959). Conversion of light into chemical energy in photosynthesis. *Nature*, **184**, 10–21.

ARNON, D. I. (1961). Cell-free photosynthesis and the energy conversion process. In *Light and Life*. Ed. McElroy & Glass. Baltimore: Johns Hopkins Press, pp. 489–565.

ARNON, D. I. (1967). Photosynthetic activity of isolated chloroplasts. *Physiol. Rev.*, **47**, 317–358.

ARNON, D. I. & HORTON, A. A. (1963). Site of action of plastaquinone in the electron transport chain of photosynthesis. *Acta chem. scand.*, **17**, S135–S139.

ARNON, D. I., TSUJIMOTO, H. Y. & McSWAIN, B. D. (1964). Role of ferridoxin in photosynthetic production of oxygen and phosphorylation by chloroplasts. *Proc. Nat. Acad. Sci., Wash.*, **51**, 1274–1282.

ARNON, D. I., TSUJIMOTO, H. Y. & McSWAIN, B. D. (1965). Quenching of chloroplast fluorescence by photosynthetic phosphorylation and electron transfer. *Proc. Nat. Acad. Sci., Wash.*, **54**, 927–934.

ARNON, D. I., WHATLEY, F. R., & ALLEN, M. B. (1958). Assimilatory power in photosynthesis. *Science*, **127**, 1026–1034.

ARNON, D. I., WHATLEY, F. R., & ALLEN, M. B. (1959). Photosynthetic phosphorylation and the generation of assimilatory power. *Biochim. biophys. Acta*, **32**, 47–57.

AVRON, M. & NEUMANN, J. (1968). Photophosphorylation in chloroplasts. *Ann. Rev. Plant Physiol.*, **19**, 137–166.

BASSHAM, J. A., BENSON, A. A., KAY, L. D., HARRIS, A. Z., WILSON, A. T., & CALVIN, M. (1954) "The Path of Carbon in Photosynthesis." *J. Amer. Chem. Soc.*, **76**, 1760–1770

BECKER, M. J., SHEFNER, A. M. & GROSS, J. A. (1965). Comparative studies of the Hill activity of differentially centrifuged chloroplast fractions. *Plant Physiol.*, **40**, 243–250.

BERGERON, J. A., & FULLER, R. C. (1961). The submicroscopic basis of bacterial photosynthesis: the chromatophore. In *Macromolecular Complexes*. Ed. Edds. New York: Ronald Press Co., pp. 179–202.

BEUKERS, R., & BERENDS, W. (1960). Isolation and identification of the irradiation product of thymine. *Biochim. biophys. Acta*, **41**, 550–551.

BIGGINS, J. & PARK, R. B. (1965). Physical properties of spinach chloroplast lamellar proteins. *Plant Physiol.*, **40**, 1109–1115.

BISALPUTRA, T. & BISALPUTRA, A. A. (1967). Chloroplast and mitochondrial DNA in brown alga *Egregia menziesii*. *J. Cell Biol.*, **33**, 511–520.

BISHOP, N. I. (1959). The reactivity of a naturally occurring quinone (Q 255) in photo-chemical reactions of isolated chloroplasts. *Proc. Nat. Acad. Sci., Wash.*, **45**, 1696–1702.

BISHOP, N. I. & GAFFRON, H. (1962). Photoreduction at 705 mμ in adapted algae. *Biochem. Biophys. Res. Comm.*, **8**, 471–476.

BLAAUW-JANSEN, G., KOMEN, J. G., & THOMAS, J. B. (1950). "On the Relation between the Formation of Assimilatory Pigments and the Rate of Photosynthesis." *Biochim. biophys. Acta*, **5**, 179–185.

BLINKS, L. R. (1960). Action spectra of chromatic transients and the Emerson effect in marine algæ. *Proc. Nat. Acad. Sci., Wash.*, **46**, 327–333.

BLUM, H. F. (1941). *Photodynamic Action and Diseases caused by Light*. Reinhold., New York.

BLUM, H. F., COOK, J. S., & LOOS, G. M. (1954). "A Comparison of Five Effects of Ultraviolet Light on the *Arbacia* Egg." *J. gen. Physiol.*, **37**, 313–324.

BLUM, H. & KAUZMANN, E. F. (1954). Photodynamic haemolysis at low temperature. *J. gen. Physiol.*, **37**, 301–311.

BLUM, H. F., KAUZMANN, E. F., & CHAPMAN, G. B. (1954). "Ultraviolet Light and the Mitotic Cycle in the Sea Urchin's Egg." *J. gen. Physiol.*, **37**, 325–333.

BLUM, H. F., & MORGAN, J. L. (1939). "Photodynamic Hæmolysis. III." *J. cell. comp. Physiol.*, **13**, 269.

BOARDMAN, N. K. (1962). Studies on a protochlorophyll-protein complex. I. *Biochim. biophys. Acta*, **62**, 63–79.

BOARDMAN, N. K., THORNE, S. W. & ANDERSON, J. M. (1966). Fluorescence properties of particles obtained by digitonin fragmentation of spinach chloroplasts. *Proc. Nat. Acad. Sci., Wash.*, **56**, 586–593.

BOLLUM, F. J. & SETLOW, R. B. (1963). Ultraviolet inactivation of DNA primer activity. I. Effects of different wavelengths and doses. *Biochim. biophys. Acta*, **68**, 599–607.

BORGESE, T. A., & GREEN, J. W. (1962). Cation exchanges and glycolytic inhibition by photosensitized rabbit erythrocytes. *J. cell. comp. Physiol.*, **59**, 215–222.

v. BORSTEL, R. C. & WOLFF, S. (1955). Photoreactivation experiments on the nucleus and cytoplasm of the *Habrobracon* egg. *Proc. Nat. Acad. Sci., Wash.*, **41**, 1004–1009.

BOVÉ, J. M., BOVÉ, C., WHATLEY, F. R. & ARNON, D. I. (1963). Chloride requirement for oxygen evolution in photosynthesis. *Z. Naturf.*, **18 b**, 683–688.

BRANTON, D. & PARK, R. B. (1967). Subunits in chloroplast lamellae. *J. Ultrastr. Res.*, **19**, 283–303.

BROWN, J. S., & FRENCH, C. S. (1959). Absorption spectra and relative photostability of the different forms of chlorophyll in *Chlorella*. *Plant Physiol.*, **34**, 305–309.

BROWN, J. S., & FRENCH, C. S. (1961). The long wavelength forms of chlorophyll-*a*. *Biophys. J.*, **1**, 539–550.

BUTLER, W. L. (1961). Chloroplast development: energy transfer and structure. *Arch. Biochem. Biophys.*, **92**, 287–295.

BUTLER, W. L. (1961). A far-red absorbing form of chlorophyll, *in vivo*. *Arch. Biochem. Biophys.*, **93**, 413–422.

BUTLER, W. L. (1962). Effects of red and far-red light on the fluorescent yield of chlorophyll *in vivo*. *Biochim. biophys. Acta*, **64**, 309–317.

CALVIN, M. (1959). From microstructure to macrostructure and function in the photo-chemical apparatus. *Brookhaven Symp. Biol.*, No. 11, 160–179.

CALVIN, M. (1961). Some photochemical and photophysical reactions of chlorophyll and its derivatives. In *Light and Life*. Ed. McElroy & Glass. Baltimore: Johns Hopkins Press, pp. 317–355.

CALVIN, M., & MASSINI, P. (1952). "The Path of Carbon in Photosynthesis. XX. The Steady State." *Experientia*, **8**, 445–457.

CHEN, S. L. (1952). The action spectrum for the evolution of oxygen by isolated chloro-plasts. *Plant Physiol.*, **27**, 35–48.

CHENIAE, G. M. & MARTIN, I. F. (1968). Site of manganese function in photosynthesis. *Biochim. biophys. Acta*, **153**, 819–837.

CHICK, H., & ELLINGER, P. (1941). "The Photo-sensitising Action of Buckwheat (*Fagopyrum esculentum*)." *J. Physiol.*, **100**, 212.

CHUN, E. H. L., VAUGHAN, M. H. & RICH, A. (1963). The isolation and characterization of DNA associated with chloroplast preparations. *J. mol. Biol.*, **7**, 130–141.

CLARK, M. F., MATTHEWS, R. E. F. & RALPH, R. K. (1964). Ribosomes and polyribosomes in *Brassica Pekinensis*. *Biochim. biophys. Acta*, **91**, 289–304.

CLAYTON, R. K. (1963). Photosynthesis: primary physical and chemical processes. *Ann. Rev. Plant Physiol.*, **14**, 159–180.

CLAYTON, R. K. (1965). Biophysical problems of photosynthesis. *Science*, **149**, 1346–1354.

COHEN-BAZAIRE, G. & KUNISAWA, R. (1963). The fine structure of *Rhodospirillum rubrum*. *J. Cell Biol.*, **16**, 401–419.

COMMONER, B. (1961). Electron spin resonance studies of photosynthetic systems. In *Light and Life*. Ed. McElroy & Glass. Baltimore: Johns Hopkins Press, pp. 356–377.

COOK, J. S. (1956). Some characteristics of hæmolysis by ultraviolet light. *J. cell. comp. Physiol.*, **47**, 55–84.

COOK, J. S. (1965). The quantitative interrelationships between ion fluxes, cell swelling, and radiation dose in ultraviolet hemolysis. *J. gen. Physiol.*, **48**, 719–734.

COOK, J. S. (1967). Direct demonstration of the monomerization of thymine-containing dimers in u.v.-irradiated DNA by yeast photoreactivating enzyme and light. *Photochem. Photobiol.*, **6**, 97–101.

COOK, J. S. (1968). On the role of DNA in the ultraviolet-sensitivity of cell division in sand-dollar zygotes. *Exp. Cell Res.*, **50**, 627–638.

COOK, J. S., & RIECK, A. F. (1962). Studies on photoreactivation in gametes and zygotes of the sand-dollar, *Echinarachnius parma*. *J. cell. comp. Physiol.*, **59**, 77–84.

CRANE, F. L., HENNINGER, M. D., WOOD, P. M. & BARR, R. (1966). Quinones in chloroplasts. In *Biochemistry of Chloroplasts*. Vol. I. London: Academic Press, pp. 133–151.

CRIDDLE, R. S. (1966). Protein and lipoprotein organization in the chloroplast. In *Biochemistry of Chloroplasts*. Vol. I. Ed. T. W. Goodwin. London: Academic Press, pp. 203–231.

CUMMINS, J. E., BLOMQUIST, J. C. & RUSCH, H. P. (1966). Anaphase delay after inhibition of protein synthesis between late prophase and prometaphase. *Science*, **154**, 1343–1344.

DAVSON, H., & PONDER, E. (1940). "Photodynamically Induced Cation Permeability and its relation to Hæmolysis." *J. cell. comp. Physiol.*, **15**, 67.

DEAMER, D. W. & CROFTS, A. (1967). Action of Triton X-100 on chloroplast membranes. *J. Cell Biol.*, **33**, 395–410.

DEVI, V. R., GUTTES, E. & GUTTES, S. (1968). Effects of ultraviolet light on mitosis in *Physarum polycephalum*. *Exp. Cell Res.*, **50**, 589–598.

DULBECCO, R. (1949). "Reactivation of Ultra-violet-inactivated Bacteriophage by Visible Light." *Nature*, **163**, 949–950.

DULBECCO, R. (1950). Experiments on photoreactivation of bacteriophages inactivated with ultraviolet radiation. *J. Bact.*, **59**, 329–347.

DUTTON, H. J., & MANNING, W. M. (1941). "Evidence for Carotenoid-sensitised Photosynthesis in the Diatom *Nitzschia closterium*." *Amer. J. Bot.*, **28**, 516.

DUYSENS, L. N. M. (1951). "Transfer of Light Energy within the Pigment Systems present in Photosynthesizing Cells." *Nature*, **168**, 548–550.

DUYSENS, L. N. M. (1952). *Transfer of Excitation Energy In Photosynthesis*. Thesis. Univ. of Utrecht. (Quoted by Arnon, 1967).

DUYSENS, L. N. M. (1956). "Energy Transformations in Photosynthesis." *Ann. Rev. Plant Physiol.*, **7**, 25–50.

DUYSENS, L. N. M. (1964). Photosynthesis. *Progr. Biophys.*, **14**, 3–104.

DUYSENS, L. N. M., & AMESZ, J. (1962). Function and identification of two photochemical systems in photosynthesis. *Biochim. biophys. Acta*, **64**, 243–260.

EISENSTADT, J. M. & BRAWERMAN, G. (1964). The protein-synthesizing systems from the cytoplasm and the chloroplasts of *Euglena gracilis*. *J. mol. Biol.*, **10**, 392–402.

EMERSON, R. (1958). Yield of photosynthesis from simultaneous illumination with pairs of wavelengths. *Science*, **127**, 1059–1060.

EMERSON, R., & ARNOLD, W. (1932). The photochemical reaction in photosynthesis. *J. gen. Physiol.*, **16**, 191–205.

EMERSON, R., & ARNOLD, W. (1932). "A Separation of the Reactions in Photosynthesis by means of Intermittent Light." *J. gen. Physiol.*, **15**, 391.

EMERSON, R., CHALMERS, R., & CEDERSTRAND, C. (1957). Some factors influencing the long-wave limit of photosynthesis. *Proc. Nat. Acad. Sci., Wash.*, **43**, 133–143.

EMERSON, R., & LEWIS, C. M. (1939). "Factors influencing the Efficiency of Photosynthesis." *Amer. J. Bot.*, **26**, 808.

EMERSON, R., & LEWIS, C. M. (1942). "The Photosynthetic Efficiency of Phycocyanin in *Chröococcus*, and the Problem of Carotenoid Participation in Photosynthesis." *J. gen. Physiol.*, **25**, 579.

EMERSON, R., & LEWIS, C. M. (1943). "The Dependence of the Quantum Yield of *Chlorella* Photosynthesis on Wavelength of Light." *Amer. J. Bot.*, **30**, 165.

ENGELMANN, T. W. (1883; 1884). *Bot. Z.*, **41**, 18; **42**, 81, 97. (Quoted by Rabinowitch.)

EPSTEIN, S. S., SMALL, M., FALK, H. L. & MANTEL, N. (1964). Association between photodynamic and carcinogenic activities in polycyclic compounds. *Cancer Res.*, **24**, 855–862.

FISCHER, H., LAMBRECHT, R., & MITTENZWEI, H. (1938). "Uber Bacterio-chlorophyll." *Z. physiol. Chem.*, **253**, 1.

FOSTER, J. W. (1940). Role of organic substrates in photosynthesis of purple bacteria. *J. gen. Physiol.*, **24**, 123–134.

FRANCK, J. & FRENCH, C. S. (1941). Photo-oxidation processes in plants. *J. gen. Physiol.*, **25**, 309.

FRANK, S. R. (1946). "The Effectiveness of the Spectrum in Chlorophyll Formation." *J. gen. Physiol.*, **29**, 157.

FRENCH, C. S. (1961). Light pigments and photosynthesis. In *Light and Life*. Ed. McElroy & Glass. Baltimore: Johns Hopkins Press, pp. 447–471.

FRENCH, C. S., & YOUNG, V. K. (1952). "The Fluorescence Spectra of Red Algae and the transfer of Energy from Phycoerythrin to Phycocyanin and Chlorophyll." *J. gen. Physiol.*, **35**, 873–890.

FREY-WYSSLING, A., & STEINMANN, E. (1948). "Die Schichtendoppelbrechung grosser Chloroplasten." *Biochim. biophys. Acta*, **2**, 254.

FULLER, R. C., & ANDERSON, I. C. (1958). Suppression of carotenoid synthesis and its effect on activity of photosynthetic bacterial chromatophores. *Nature*, **181**, 252–254.

FULLER, R. C., BERGERON, J. A., & ANDERSON, I. C. (1961). Relation of photosynthetic activity to carotenoid-bacteriochlorophyll interaction in *Chromatium*. *Arch. Biochem. Biophys.*, **92**, 273–279.

GASSMAN, M. & BOGORAD, L. (1967). Control of chlorophyll production in rapidly greening bean leaves. *Plant Physiol.*, **42**, 774–780.

GEE, R., JOSHI, G. BILS, R. F. & SALTMAN, P. (1965). Light and dark C^{14} O_2 fixation by spinach leaf systems. *Plant Physiol.*, **40**, 89–96.

GIBBS, S. P. (1962). Chloroplast development in *Ochromonas danica*. *J. biophys. biochem. Cytol.*, **15**, 343–361.

GIBBS, S. P., SISTROM, W. R. & WORDEN, P. B. (1965). The photosynthetic apparatus of *Rhodospirillum molischianum*. *J. Cell Biol.*, **26**, 395–412.

GIBOR, A. & GRANICK, S. (1964). Plastids and mitochondria: inheritable systems. *Science*, **145**, 890–897.

GIBOR, A. & IZAWA, M. (1963). The DNA content of the chloroplasts of *Acetabularia*. *Proc. Nat. Acad. Sci. Wash.*, **50**, 1164–1169.

GIESE, A. C. (1939). "Effects of 2654 and 2804A on *Paramœcium caudatum*." *J. cell. comp. Physiol.*, **13**, 139.

GIESE, A. C. (1945). "The Ultraviolet Action Spectrum for retardation of Division of *Paramœcium*." *J. cell. comp. Physiol.*, **26**, 47.

GIESE, A. C. (1946). "An Intracellular Photodynamic Sensitiser in *Blepharisma*." *J. cell. comp. Physiol.*, **28**, 119.

GIESE, A. C., IVERSON, R. M., & SANDERS, R. T. (1957). Effect of nutritional state and other conditions on ultraviolet resistance and photoreactivation in yeast. *J. Bact.*, **74**, 271–279.

GIESE, A. C., IVERSON, R. M., SHEPARD, D. C., JACOBSON, C. & BRANDT, C. L. (1953). Quantum relations in photoreactivation in *Colpidium*. *J. gen. Physiol.*, **37**, 249–258.

GIESE, A. C., SHEPARD, D. C., BENNETT, J., FARMANFARMAIAN, A., & BRANDT, C. L. (1956). Evidence for thermal reactions following exposure of *Didinium* to intermittent ultraviolet radiations. *J. gen. Physiol.*, **40**, 311–325.

GOEDHEER, J. C. (1962). Afterglow of chlorophyll *in vivo* and photosynthesis. *Biochim. biophys. Acta*, **64**, 294–308.

GOEDHEER, J. C. (1964). Fluorescence bands and chlorophyll *a* forms. *Biochim. biophys. Acta*, **88**, 304–317.

GOEDHEER, J. C. (1965). Fluorescence action spectra of algae and bean leaves at room and at liquid nitrogen temperatures. *Biochim. biophys. Acta*, **102**, 73–89.

GOLDMAN, A. S., & SETLOW, R. B. (1956). "The Effects of Monochromatic Ultraviolet Light on the Egg of *Drosophila*." *Exp. Cell Res.*, **11**, 146–159.

GOOD, N. E. (1962). Uncoupling of the Hill reaction from photophosphorylation by anions. *Arch. Biochem. Biophys.*, **96**, 653–661.

GORCHEIN, A. (1968, a). Relation between pigment content of isolated chromatophores and that of whole cell in *Rhodopseudomonas spheroides*. *Proc. Roy. Soc., B*, **170**, 247–254.

GORCHEIN, A. (1968, b). The nature of the internal fine structure of *Rhodopseudomonas spheroides*. *Proc. Roy. Soc., B*, **170**, 255–263.

GORCHEIN, A., NEUBERGER, A. & TAIT, G. H. (1968). The isolation and characterization of subcellular fractions from pigmented and unpigmented cells of *Rhodopseudomonas spheroides*. *Proc. Roy. Soc., B*, **170**, 229–246.

GOVINDJEE, & RABINOWITCH, E. (1960). Action-spectrum of the "second Emerson effect". *Biophys. J.*, **1**, 73–89.

GOVINDJEE, RABINOWITCH, E., & THOMAS, J. B. (1960). Inhibition of photosynthesis in certain algæ by extreme red light. *Biophys. J.*, **1**, 91–97.

GOVINDJEE & YANG, L. (1966). Structure of the red fluorescence band in chloroplasts. *J. gen. Physiol.*, **49**, 763–780.

GREEN, J. W., BLUM, H. F., & PARPART, A. K. (1959). Prehæmolytic studies of photosensitized rabbit erythrocytes. *J. cell. comp. Physiol.*, **54**, 5–10.

GRIFFITHS, M., SISTROM, W. R., COHEN-BAZIRE, G., & STANIER, R. Y. (1955). Function of carotenoids in photosynthesis. *Nature*, **176**, 1211–1214.

GROSS, J. A., BECKER, M. J. & SHEFNER, A. M. (1964). Some properties of differentially centrifuged particulate fractions from chloroplasts. *Nature*, **203**, 1263–1265.

GUNNING, B. E. S. (1965). The fine structure of chloroplast stroma following aldehyde osmium-tetroxide fixation. *J. Cell Biol.*, **24**, 79–93.

GUNNING B. E. S. (1965). The greening process in plastids. I. *Protoplasma*, **60**, 111–130.

GUNNING B. E. S. & JAGOE, M. P. (1967). The prolemellar body. In *Biochemistry of Chloroplasts*. Vol. II. Ed. T. W. Goodwin. London: Academic Press, pp. 655–676.

HARDING, C. V., THOMAS, L. J. (1950) "Ultraviolet Light induced Delay in Cleavage of Centrifuged *Arbacia* Eggs." *J. cell. comp. Physiol.*, **35**, 403–411.

HASELKORN, R., FERNÁNDEZ-MORÁN, H., KIERAS, F. J. & VAN BRUGGENS, E. F. J. (1965). Electron microscopic and biochemical characterization of fraction I protein. *Science*, **150**, 1598–1601.

HAXO F. T., & BLINKS, L. R. (1946). "Photosynthetic Action Spectra in Red Algae." *Amer. J. Bot.*, **33**, 836.

HAXO, F. T., & BLINKS, L. R. (1950). "Photosynthetic Action Spectra of Marine Algae." *J. gen. Physiol.*, **33**, 389.

HEINMETZ, F., VINEGAR, R., & TAYLOR, W. W. (1952). "Studies on the Mechanism of the Photosensitized Inactivation of *E. coli* and Reactivation Phenomenon." *J. gen. Physiol.*, **36**, 207–226.

HENNINGSEN, K. W. (1967). An action spectrum for vesicle dispersal in bean plastids. In *Biochemistry of Chloroplasts*. Vol. II. Ed. T. W. Goodwin. London: Academic Press, pp. 453–457.

HICKMAN, D. H. & FRENKEL, A. W. (1965). Observations on the structure of *Rhodospirillum molischianum*. *J. Cell Biol.*, **25**, 261–278.

HIGHKIN, H. R. & FRENKEL, A. W. (1962). Studies of growth and metabolism of a barley mutant lacking chlorophyll *b*. *Plant Physiol.*, **37**, 814–820.

HILL, R. (1939). "O_2 Produced by Isolated Chloroplasts." *Proc. Roy. Soc., B*, **127**, 192.

HILL, R. & SCARISBRICK, R. (1940). Production of O_2 by isolated chloroplasts. *Proc. Roy. Soc., B*, **129**, 238–255.

HODGE, A. J., McLEAN, J. D., & MERCER, F. V. (1955). "Ultrastructure of the Lamellae and Grana in the Chloroplasts of *Zea mays* L." *J. biophys. biochem. Cytol.*, **1**, 605–614.

HODGE, A. J., McLEAN, J. D. & MERCER, F. V. (1956). Possible mechanism for morphogenesis of lamellar systems in plant cells. *J. biophys. biochem. Cytol.*, **2**, 597–607.

HOLT, A. S., & FRENCH, C. S. (1946). "Evolution of O_2 by Isolated Chloroplasts immersed in Solutions of various Oxidising Agents." *Amer. J. Bot.*, **33**, 836.

HOLT, S. C. & MARR, A. G. (1965). Location of chlorophyll in *Rhodospirillum rubrum*. *J. Bact.*, **89**, 1402–1412.

HOMANN, P. H. (1967). Studies on manganese of chloroplast. *Plant Physiol.*, **42**, 997–1007.

HOMANN, P. H. & SCHMID, G. H. (1967). Photosynthetic reactions of chloroplasts with unusual structures. *Plant Physiol.*, **42**, 1619–1632.

HORSLEY, C. H. (1934). Investigation of Action of St. Johns Wort. *J. Pharmacol.*, **50**, 310.

HOWELL, S. H. & MOUDRIANAKIS, E. N. (1967). Hill reaction site in chloroplast membranes: non-participation of quantasome particle in photoreduction. *J. mol. Biol.*, **27**, 323–333.

HOWELL, W. H. (1921). "L'Action Photodynamique de l'Hématoporphyrine sur le Fibrinogène." *Arch. int. Physiol.*, **18**, 269.

HUXLEY, H. E. & ZUBAY, G. (1960). Electron microscope observations on the structure of microsomal particles from *Escherichia coli*. *J. mol. Biol.*, **2**, 10–18.

HYMAN, C., & HOWLAND, R. B. (1940). "Intracellular Photodynamic Action." *J. cell. comp. Physiol.*, **16**, 207.

IZAWA, S. & GOOD, N. E. (1965). Hill reaction rates and chloroplast fragment size. *Biochim. biophys. Acta*, **109**, 372–381.

IZAWA, S. & GOOD, N. E. (1966). Effect of salts and electron transport on the conformation of isolated chloroplasts I. and II. *Plant Physiol.*, **41**, 533–543; 544–552.

JAGENDORF, A. T., & AVRON, M. (1958). Cofactors and rates of photosynthetic phosphorylation by spinach chloroplasts. *J. biol. Chem.*, **231**, 277–290.

JAGENDORF, A. T., HENDRICKS, S. B., AVRON, M. & EVANS, M. B. (1958). Action spectrum for photosynthetic phosphorylation by spinach chloroplasts. *Plant Physiol.*, **33**, 72–73.

JAGGER, J. (1958). Photoreactivation. *Bact. Rev.*, **22**, 99–142.

JOST, M. (1965). Die Ultrastruktur von *Oscillatoria rubescens* D.C. *Arch. f. Microbiol.*, **50**, 211–245.

KAHN, A. & V. WETTSTEIN, D. (1961). Structure of isolated spinach chloroplasts. *J. Ultrastr. Res.*, **5**, 557–574.

KAHN, J. S. (1964). A soluble protein-chlorophyll complex from spinach chloroplasts. I. *Biochim. biophys. Acta*, **79**, 234–240.

KATOH, S. & SAN PIETRO, A. (1966). Activities of chloroplast fragments. I. Hill reaction and ascorbate-indophenol photoreductions. *J. biol. Chem.*, **241**, 3575–3581.

KATOH, S., SUGA, I., SHIRATORI, I., & TAKAMIYA, A. (1961). Distribution of plastocyanin in plants, with special reference to its localization in chloroplasts. *Arch. Biochem. Biophys.*, **94**, 136–141.

KAUTSKY, H. (1936). "Chlorophyllfluoreszenz und Kohlensäureassimilation." *Biochem. Z.*, **284**, 412.

KE, B. & VERNON, L. P. (1967). Fluorescence of the subchloroplast particles obtained by the action of Triton X-100. *Biochem.* **7**, 2221–2226.

KELNER, A. (1949). Effect of visible light on the recovery of *Streptomyces griseus* Conidia from ultraviolet irradiation injury. *Proc. Nat. Acad. Sci., Wash.*, **35**, 73–79.

KESSLER, E. (1955). On the role of manganese in the oxygen-evolving system of photosynthesis. *Arch. Biochem. Biophys.*, **59**, 527–529.

KIRK, J. T. O. & JUNIPER, B. E. (1967). The ultrastructure of the chromoplasts of different colour varieties of *Capsicum*. In *Biochemistry of Chloroplasts*. Ed. T. W. Goodwin. London: Academic Press, pp. 691–701.

KISLEY, N., SWIFT, H. & BOGORAD, L. (1965). Nucleic acids of chloroplasts and mitochondria in Swiss chard. *J. Cell Biol.*, **25**, 327–344.

KLUYVER, A. J., & DONKER, H. J. L. (1926). *Chem. Zelle Gewebe*, **13**, 134. (Quoted by Van Niel.)

KNAFF, D. B., & ARNON, D. I. (1969). A concept of three light reactions in photosynthesis by green plants. *Proc. Nat. Acad. Sci., Wash.*, **64**, 715–722.

KOK, B. (1956). Photosynthesis in flashing light. *Biochim. biophys. Acta*, **21**, 245–258.

KOK, B., & HOCH, G. (1961). Spectral changes in photosynthesis. In *Light and Life*. Ed. McElroy & Glass. Baltimore: Johns Hopkins Press, pp. 397–416.

KOSKI, V. M. (1950). "Chlorophyll Formation in Seedlings of *Zea mays* L." *Arch. Biochem.*, **29**, 339–343.

KOSKI, V. M., & SMITH, J. H. C. (1951). "Chlorophyll Formation in a Mutant, White Seedling-3." *Arch. Biochem.*, **34**, 189–195.

KRASNOVSKII, A. A., & KOSOBUTSKAYA, L. M. (1952). Quoted by Smith *et al.* (1957).

KREBS, H. A., & KORNBERG, H. L. (1957). "A Survey of the Energy Transformations in Living Matter." *Ergebn. Physiol.*, **49**, 212–298.

KREY, A. & GOVINDJEE. (1966). Fluorescence studies on a red alga. *Porphyridium cruentum*. *Biochim. biophys. Acta*, **120**, 1–18.

KRINSKY, N. I. (1966). The role of carotenoid pigments as protective agents against photosensitized oxidations in chloroplasts. In *Biochemistry of Chloroplasts*. Vol. I. Ed. T. W. Goodwin. London: Academic Press, pp. 423–430.

KUSHIDA, H., ITOH, M., IZAWA, S. & SHIBATA, K. (1963). Deformation of chloroplasts on illumination in intact spinach leaves. *Biochim. biophys. Acta*, **79**, 201–203.

LIEBERMAN, E. M. (1967). Structural and functional sites of action of ultraviolet radiations in crab nerves. I., II. & III. *Exp. Cell Res.*, **47**, 489–507; 508–517; 518–535.

LINTILHAC, P. M. & PARK, R. B. (1966). Localization of chlorophyll in spinach chloroplast lamellae by fluorescence microscopy. *J. Cell Biol.*, **28**, 582–585.

LLOYD. (1924). *Science*, **59**, 241.

LOSADA, M., TREBST, A. V., & ARNON, D. I. (1960). CO_2 assimilation in a reconstituted chloroplast system. *J. biol. Chem.*, **235**, 832–839.

LOSADA, M., TREBST, A. V., OGATA, S., & ARNON, D. I. (1960). Equivalence of light and adenosine triphosphate in bacterial photosynthesis. *Nature*, **186**, 753–760.

LOVENBERG, W., BUCHANAN, B. B. & RABINOWITZ, J. C. (1963). Studies on the chemical nature of clostridial ferredoxin. *J. biol. Chem.*, **238**, 3899–3913.

LØVLIE, A. & BRÅTEN, T. (1968). The division of cytoplasm and chloroplast in the multi-cellular green alga *Ulva mutabilis* Føyn. *Exp. Cell Res.*, **51**, 211–220.

LUMRY, R., WAYRYNEN, R. E., & SPIKES, J. D. (1956). "The Mechanism of the Photo-chemical Activity of Isolated Chloroplasts. II. Quantum Requirements." *Arch. Biochem.*, **67**, 453–465.

LYMAN, H., EPSTEIN, H. T. & SCHIFF, J. A. (1961). Studies of chloroplast development in *Euglena*. I. Inactivation of green colony formation by u.v. light. *Biochim. biophys. Acta*, **50**, 301–309.

LYTTLETON, J. W. (1962). Isolation of ribosomes from spinach chloroplasts. *Exp. Cell Res.*, **26**, 312–317.

LYTTLETON, J. W. & Ts'o, P. O. P. (1958). The localization of Fraction I protein of green leaves in the chloroplasts. *Arch. Biochem. Biophys.*, **73**, 120–126.

MALKIN, S. & KOK, B. (1966). Fluorescence induction studies in isolated chloroplasts. I. and II. *Biochim. biophys. Acta*, **126**, 413–432; 433–442.

MARGULIES, M. M. (1962). Effect of chloramphenicol on light dependent development of seedlings of *Phaseolus vulgaris*. *Plant Physiol.*, **37**, 473–480.

MARMUR, J., ANDERSON, W. F., MATTHEWS, L., BERNS, K., GAJEWSKA, E., LANE, D., & DOTY, P. (1961). Effects of ultraviolet light on biological and physical chemical properties of deoxyribonucleic acids. *J. cell. comp. Physiol.*, **58**, Suppl. 33–55.

MARSHAK, A. (1949). "Recovery from Ultra-violet Light induced Delay in Cleavage of *Arbacia* Eggs by Irradiation with Visible Light." *Biol. Bull. Wood's Hole*, **97**, 315–322.

MCLEOD, G. C. (1961). Action spectra of light-saturated photosynthesis. *Plant Physiol.*, **36**, 114–117.

MENKE, W. (1966). The structure of the chloroplasts. In *Biochemistry of Chloroplasts*. Vol. I. Ed. T. W. Goodwin. London: Academic Press, pp. 3–18.

MENKE, W., & KOYDL, E. (1939). "Direkter Nachweis des lamellaren Feinbaues der Chloroplasten." *Naturwiss.*, **27**, 29.

MERCER, F. V., NITTIM, M. & POSSINGHAM, J. V. (1962). The effect of manganese deficiency on the structure of spinach chloroplasts. *J. Cell Biol.*, **15**, 379–381.

MILLER, A., KARLSSON, U. & BOARDMAN, N. K. (1966). Appendix to paper by Boardman, Francki & Wildman. *J. mol. Biol.*, **17**, 487–489.

MORTENSON, L. E., VALENTINE, R. C. & CARNAHAN, J. E. (1962). An electron transport factor from *Clostridium pasteurianum*. *Biochem. Biophys. Res. Comm.*, **7**, 448–452.

MÜHLETHALER, K. (1966). The ultrastructure of the plastid lamellae. In *Biochemistry of Chloroplasts*. Ed. T. W. Goodwin. London: Academic Press, pp. 49–64.

MÜHLETHALER, K., & FREY-WYSSLING, A. (1959). Entwicklung und Struktur der Proto-plastiden. *J. biophys. biochem. Cytol.*, **6**, 507–512.

MYERS, J., & FRENCH, C. S. (1960). Evidence from action spectra for a specific partici-pation of chlorophyll-*b* in photosynthesis. *J. gen. Physiol.*, **43**, 723–736.

NEWTON, J. W., & NEWTON, G. A. (1957). Composition of the photoactive subcellular particles from *Chromatium*. *Arch. Biochem. Biophys.*, **71**, 250–265.

NISHIDA, K. & KOSHII, K. (1964). On the volume changes of isolated spinach chloroplasts caused by electrolytes and sugars. *Physiol. Plant*, **17**, 846–854.

NORMAN, C. & GOLDBERG, E. (1959). Effect of light on motility, life-span, and respiration of bovine spermatozoa. *Science*, **130**, 624–625.

OHAD, I., SIEKEVITZ, P. & PALADE, G. E. (1967). Biogenesis of chloroplast membranes. I. and II. *J. Cell Biol.*, **35**, 521–552; 553–584.

Ó'HEOCHA, C. (1966). Biliproteins. In *Biochemistry of Chloroplasts*. Vol. I. Ed. T. W. Goodwin. London: Academic Press, pp. 407–421.

OLTMANNS, F. (1893). *Jahb. f. wiss. Bot.*, **16**, 1. (Quoted by Rabinowitch.)

PACKER, L. (1966). Evidence of contractility in chloroplasts. In *Biochemistry of Chloro-plasts*. Vol. I. Ed. T. W. Goodwin. London: Academic Press, pp. 233–242.

PACKER, L., BARNARD, A. C. & DEAMER, D. W. (1967). Ultrastructural and photometric evidence for light-induced changes in chloroplast structure *in vivo*. *Plant Physiol.*, **42**, 283–293.

PACKER, L. & MARCHANT, R. H. (1964). Action of adenosine triphosphate on chloroplast structure. *J. biol. Chem.*, **239**, 2061–2064.

PACKER, L., SIEGENTHALER, P. A. & NOBEL, P. S. (1965). Light-induced volume changes in spinach chloroplasts. *J. Cell Biol.*, **26**, 593–599.

PACKER, L. & SIEGENTHALER, P. A. (1966). Control of chloroplast structure by light. *Int. Rev. Cytol.*, **20**, 97-124.

PARDEE, A. B., SCHACHMAN, H. K., & STANIER, R. Y. (1952). Chromatophores of *Rhodospirillum rubrum*. *Nature*, **169**, 282-283.

PARK, R. B. (1965). Substructure of chloroplast lamellae. *J. Cell Biol.*, **27**, 151-161.

PARK, R. B. (1966). Subunits of chloroplast structure and quantum conversion in photosynthesis. *Int. Rev. Cytol.*, **20**, 67-95.

PARK, R. B. & BIGGINS, J. (1964). Quantasome: size and composition. *Science*, **144**, 1009-1011.

PARK, R. B., & PON, N. G. (1961). Correlation of structure and function in *Spinacea oberacea* chloroplasts. *J. mol. Biol.*, **3**, 1-10.

PIERCE, S., & GIESE, A. C. (1956). Photoreversal of ultraviolet injury to frog and crab nerves. *J. cell. comp. Physiol.*, **49**, 303-317.

PICKETT, J. M. & MYERS, J. (1966). Monochromatic light saturation curves for photosynthesis in *Chlorella*. *Plant Physiol.*, **41**, 90-98.

PRICE, L., & KLEIN, W. H. (1961). Red, far-red response and chlorophyll synthesis. *Plant Physiol.*, **36**, 733-735.

QUAYLE, J. H., FULLER, R. C., BENSON, A. A., & CALVIN, M. (1954). "Enzymatic Carboxylation of Ribulose Diphosphate." *J. Amer. Chem. Soc.*, **76**, 3610-3611.

RABINOWITCH, E. (1959). Primary photochemical and photophysical processes in photosynthesis. *Plant Physiol.*, **34**, 213-218.

RABINOWITCH, E., & GOVINDJEE (1961). Different forms of chlorophyll A *in vivo* and their photochemical functions. In *Light and Life*. Ed. McElroy & Glass. Baltimore: Johns Hopkins Press, pp. 378-391.

RAMAGE, C. M. & NAKAMURA, M. (1962). Alterations in ultraviolet sensitivity of *Shigella sonnei* starved at 8° and 37° C for varying periods of time. *Exp. Cell Res.*, **27**, 147-149.

RASK, E. N., & HOWELL, W. H. (1928). "The Photodynamic Action of Hematoporphyrin." *Amer. J. Physiol.*, **84**, 363.

REID, C. (1957). *Excited States in Chemistry and Biology*. London: Butterworth.

ROBINSON, W. (1965). Quantum processes in photosynthesis. *Ann. Rev. Phys. Chem.*, **15**, 311-348.

ROSENBLUM, W. I. (1960). The stimulation of frog skeletal muscle by light and dye. *J. cell. comp. Physiol.*, **55**, 73-79.

RUBEN, S., HASSID, W. Z., & KAMEN, M. D. (1939). "Radioactive Carbon in the Study of Photosynthesis." *J. Amer. Chem. Soc.*, **61**, 661.

RUBEN, S., KAMEN, M. D. & HASSID, W. Z. (1940). Photosynthesis with radioactive carbon. II. Chemical properties of the intermediates. *J. Amer. Chem. Soc.*, **62**, 3443.

RUPERT, C. S. (1960). Photoreactivation of transforming DNA by an enzyme from baker's yeast. *J. gen. Physiol.*, **43**, 573-595.

RUPERT, C. S. (1961). Repair of ultraviolet damage in cellular DNA. *J. cell. comp. Physiol.*, **58**, Suppl. 57-68.

RUPERT, C. S. (1962). Photoenzymatic repair of ultraviolet damage in DNA. I and II. *J. gen. Physiol.*, **45**, 703-724; 725-741.

SAGER, R. & PALADE, G. E. (1957). Structure and devlopment of the chloroplast in *Chlamydomonas*. *J. biophys. biochem. Cytol.*, **3**, 463-487.

SAIER, F. L. & GIESE, A. C. (1966). Action of ultraviolet radiation upon ciliary movement in *Paramecium*. *Exp. Cell Res.*, **44**, 321-331.

SAN PIETRO, A., & LANG, H. M. (1958). Photosynthetic pyridine nucleotide reductase. *J. biol. Chem.*, **231**, 211-229.

SAUER, K., & CALVIN, M. (1962). Absorption spectra of spinach quantosomes and bleaching of the pigments. *Biochim. biophys. Acta*, **64**, 324-339.

SCHIFF, J. A. & ZELDIN, M. H. (1968). The developmental aspect of chloroplast continuity in *Euglena*. *J. Cell Biol.*, **72**, Suppl. 1. 103-127.

SCHMID, G. H. (1967). Photosynthetic capacity and lamellar structure in various chlorophyll-deficient plants. *J. de Microscopie*, **6**, 485-498.

SCHMID, G. H. & GAFFRON, H. (1967). Light metabolism and chloroplast structure in chlorophyll-deficient tobacco mutants. *J. gen. Physiol.*, **50**, 563-582.

SCHMID, G. H. & GAFFRON, H. (1967). Quantum requirements for photosynthesis in chlorophyll-deficient plants with unusual lamellar structures. *J. gen. Physiol.*, **50**, 2131-2144.

SETLOW, R. B. (1964). Physical changes and mutagenesis. *J. cell comp. Physiol.*, **64**, Suppl., pp. 51-68.

SETLOW, R. B., SWENSON, P. A. & CARRIER, W. L. (1963). Thymine dimers and inhibition of DNA synthesis by ultraviolet irradiation of cells. *Science*, **142**, 1464-1466.

SHIN, M. & ARNON, D. I. (1965). Enzymic mechanisms of pyridine nucleotide reduction in chloroplasts. *J. biol. Chem.*, **240**, 1405–1411.

SIEGEL, A., GINOZA, W., & WILDMAN, S. G. (1957). The early events of infection with tobacco mosaic virus nucleic acid. *Virology*, **3**, 554–559.

SISTROM, W. R. & CLAYTON, R. K. (1964). Studies on a mutant of *Rhodopseudomonas spheroides* unable to grow photosynthetically. *Biochim. biophys. Acta*, **88**, 61–73.

SISTROM, W. R., GRIFFITHS, M., & STANIER, R. Y. (1956). The biology of a photosynthetic bacterium which lacks coloured carotenoids. *J. cell. comp. Physiol.*, **48**, 473–515.

SKREB, Y., & ERRERA, M. (1957). Action des rayons u.v. sur des fragments nucléés et anucléés d'amibes. *Exp. Cell Res.*, **12**, 649–656.

SMALL, M., MANTEL, N. & EPSTEIN, S. (1967). Role of cell-uptake of polycyclic compounds in photodynamic injury of *Tetrahymena pyriformis*. *Exp. Cell Res.*, **45**, 206–217.

SMETANA, H. (1938).Photo-oxidation of body fluids. *J. biol. Chem.*, **124**, 667–691.

SMITH, J. H. C., DURHAM, L. J., & WURSTER, C. F. (1959). Formation and bleaching of chlorophyll in albino corn seedlings. *Plant Physiol.*, **34**, 340–345.

SMITH, J. H. C., & FRENCH, C. S. (1963). The major and accessory pigments in photosynthesis. *Ann. Rev. Plant Physiol.*, **14**, 181–224.

SMITH, J. H. C., KUPKE, D. W., LOEFFLER, J. E., BENITEZ, A., AHME, I., & GIESS, A. T. (1957). "The Natural State of Protochlorophyll." *Research in Photosynthesis*. Interscience. N.Y., pp. 464–474.

SPENCER, D. & WILDMAN, S. G. (1962). Observations on the structure of grana-containing chloroplasts and a proposed model of chloroplast structure. *Austral. J. Biol. Sci.*, **15**, 599–610.

SPIEGELMAN, S. (1956). "On the Nature of the Enzyme-forming System." *Enzymes*. Ed. Gaebler. Acad. Press. N.Y., pp. 67–89.

STRAIN, H. H. & MANNING, W. M. (1942). Chlorofucine (chlorophyll-γ), a green pigment of diatoms and brown algae. *J. biol. Chem.*, **144**, 625–636.

STREHLER, B. L., & ARNOLD, W. (1951). "Light Production by Green Plants." *J. gen. Physiol.*, **34**, 809–820.

SWENSON, P. A. (1960). Leakage of phosphorus compounds from ultraviolet-irradiated yeast cells. *J. cell. comp. Physiol.*, **56**, 77–91.

SWENSON, P. A., & GIESE, A. C. (1950). "Photoreactivation of Galactozymase Formation in Yeast." *J. cell. comp. Physiol.*, **36**, 369–380.

SWENSON, P. A. & SETLOW, R. B. (1966). Effects of ultraviolet radiation on macromolecular synthesis in *Escherichia coli*. *J. mol. Biol.*, **15**, 201–219.

TAGAWA, K. & ARNON, D. I. (1962). Ferredoxins as electron carriers in photosynthesis and in biological production and consumption of hydrogen. *Nature*, **195**, 537–543.

TAGAWA, K. & ARNON, D. I. (1968). Oxidation-reduction potentials and stoichiometry of electron transfer in ferredoxins. *Biochim. biophys. Acta*, **153**, 602–613.

TAGAWA, K., TSUJIMOTO, H. Y. & ARNON, D. I. (1963). Separation by monochromatic light of photosynthetic phosphorylation from oxygen evolution. *Proc. Nat. Acad. Sci., Wash.*, **50**, 544–549.

THOMAS, J. B. & BARTELS, C. T. (1966). Studies on chlorophyll complexes *in vitro*. In *Biochemistry of Chloroplasts*. Vol. I. Ed. T. W. Goodwin. London: Academic Press, pp. 257–267.

TOLLIN, G., & CALVIN, M. (1957). The luminescence of chlorophyll-containing plant material. *Proc. Nat. Acad. Sci., Wash.*, **43**, 895–908.

TREBST, A. V., TSUJIMOTO, H. Y., & ARNON, D. I. (1958). Separation of light and dark phases in the photosynthesis of isolated chloroplasts. *Nature*, **182**, 351–355.

TRURNIT, H. J., & COLMANO, G. (1959). Absorption spectra of chlorophyll monolayers at liquid interfaces. *Biochim. biophys. Acta*, **31**, 434–447.

VAN DUIJN, C. (1961). Photodynamic effects of vital staining with diazine green (Janus green) on living bull spermatozoa. *Exp. Cell Res.*, **25**, 120–130.

VAN NIEL, C. B. (1941). "The Bacterial Photosyntheses and their Importance for the General Problem of Photosynthesis." *Adv. in Enzym.*, **1**, 263.

VATTER, A. E., & WOLFE, R. S. (1958). The structure of photosynthetic bacteria. *J. Bact.*, **75**, 480–483.

VIDAVER, W. (1966). Separate action spectra for the two photochemical systems of photosynthesis. *Plant Physiol.*, **41**, 87–89.

WACKER, A., DELLWEG, H., & WEINBLUM, D. (1960). Strahlenchemische Veränderung der Bakterien-Desoxyribonucleinsäure *in vivo*. *Naturwiss.*, **47**, 477.

WANG, S. Y. (1961). Photochemical reactions in frozen solutions. *Nature*, **190**, 690–694.

WARBURG, O. (1928). "U. d. katalytischen Wirkungen der lebendigen Substanz." Berlin, Springer. (Quoted by Franck & Gaffron.)

WARBURG, O. (1949). *Heavy Metal Prosthetic Groups and Enzyme Action*. Oxford: O.U.P.
WARBURG, O., KRIPPAHL, G. & SCHRÖDER, W. (1955). Wirkungsspektrum eines photosynthese-Ferments. *Z. Naturf.*, **10b**, 631–639.
WARBURG O., & NEGELEIN, E. (1923). "U. d. Einfluss der Wellenlänge auf den Energieumsatz bei der Kohlensäureassimilation." *Z. physik. Chem.*, **106**, 191.
WARBURG, O & OSTENDORF, P. (1963). Quantenbedarf der Photosynthese in Blättern. *Z. Naturf.*, **18 b**, 933–936.
WEED, L. L., & LONGFELLOW, D. (1954). Morphological and biochemical changes induced by copper in a population of *Escherichia coli*. *J. Bact.*, **67**, 27–33.
WEHRMEYER, W. (1964). Zur Entstehung der Grana durch Membranuberschiebung. *Planta*, **63**, 13–30.
WEHRMEYER, W. (1965). Zur Kristalgitterstruktur der sogenannten Prolamellarkörper in Proplastiden etiolierter Bohnen. I.–III. *Z. Naturf.* **20b**. 1270–1278; 1278–1288; 1288–1296.
WEIER, T. E. & BENSON, A. A. (1966). The molecular layer of chloroplast membranes. In *Biochemistry of Chloroplasts*. Ed. T. W. Goodwin. London: Academic Press.
WEIER, T. E., BISALPUTRA, T. & HARRISON, A. (1966). Subunits in chloroplast membranes of *Scenedesmus quadricauda*. *J. Ultrastr. Res.*, **15**, 38–56.
WEIER, T. E., ENGELBRECHT, A. H. P., HARRISON, A. & RISLEY, E. B. (1965). Subunits in the membranes of chloroplasts of *Phaseolus vulgaris*, *Pisum sativum*, and *Aspidistra sp.* *J. Ultrastr. Res.*, **13**, 92–111.
WEIL, L., & BUCHERT, A. R. (1951). "Photoöxidation of Crystalline β-Lactoglobulin in the Presence of Methylene Blue." *Arch. Biochem.*, **34**, 1–15.
WEISSMAN, G., & DINGLE, J. (1961). Release of lysosomal protease by ultraviolet irradiation and inhibition by hydrocortisone. *Exp. Cell Res.*, **25**, 207–210.
WESSELS, J. S. C. (1968). Isolation and properties of two digitonin-soluble pigment-protein complexes from spinach. *Biochim. biophys. Acta*, **153**, 497–499.
v. WETTSTEIN, D. (1959). The formation of plastid structures. *Brookhaven Symp. Biol.*, No. 11, 138–157.
v. WETTSTEIN, D. (1966). On the physiology of chloroplast structures. In *Biochemistry of Chloroplasts*. Vol. I. Ed. T. W. Goodwin. London: Academic Press, pp. 19–22.
WHATLEY, F. R., ALLEN, M. B., ROSENBERG, L. L., CAFINDALE, J. B., & ARNON, D. I. (1956). "Photosynthesis by Isolated Chloroplasts. V. Phosphorylation and Carbon Dioxide Fixation by Broken Chloroplasts." *Biochim. biophys. Acta*, **20**, 462–468.
WILDMAN, S. G. (1967). The organization of grana-containing chloroplasts in relation to location of some enzymatic systems concerned with photosynthesis, protein synthesis, and ribonucleic acid synthesis. In *Biochemistry of Chloroplasts*. Vol. II. Ed. T. W. Goodwin. London: Academic Press, pp. 295–319.
WITHROW, R. B., WOLFF, J. B. & PRICE, I. (1956). Elimination of the lag phase of chlorophyll synthesis in dark-grown bean leaves by a pretreatment with low irradiances o monochromatic energy. *Plant Physiol.*, **31**, Suppl. xiii–xiv.
WOLKEN, J. J. (1961). The chloroplast: its lamellar structure and molecular organization. In *Macromolecular Complexes*. Ed. Edds. N.Y. Ronald Press Co., pp. 85–112.
WOLKEN J. J., & SCHWERTZ, F. A. (1953). "Chlorophyll Monolayers in Chloroplasts." *J. gen. Physiol.*, **37**, 111–119.
WOLKEN, J. J., & SCHWERTZ, F. A. (1956). Molecular weight of algal chloroplastin. *Nature*, **177**, 136–138.
YOCUM, C. S. (1946). Relation between respiration and chlorophyll formation. *Amer. J. Bot.*, **33**, 828.
YOCUM, C. S., & BLINKS, L. R. (1954). "Photosynthetic Efficiency of Marine Plants." *J. gen. Physiol.*, **38**, 1–16.
ZELDIN, M. H. & SCHIFF, J. A. (1967). RNA metabolism during light-induced chloroplast development in *Euglena*. *Plant Physiol.*, **42**, 922–932.
ZSCHEILE, F. P., & COMAR, C. L. (1941). "Influence of Preparative Procedure on the Purity of Chlorophyll Components as shown by Absorption Spectra." *Bot. Gazette*, **102**, 463.

CHAPTER XXIV

THE PIGMENTARY RESPONSE TO LIGHT

THE proverbial colour changes of the chameleon, the paling and darkening of frogs, crabs and fishes, and certain adaptive colorations of some insect larvæ, are all well-recognized instances of what we may call the *pigmentary response to light.**

Chromatophores

The colours of animals are determined by the nature and variety of their pigmented cells, generally situated immediately under the epidermis, although in insects the pigment is usually extracellular, being distributed in granular form throughout the cuticle. The *changes* in colour, with which we shall be concerned here, follow from alterations in the degree of dispersion of the pigment within the coloured cells, the name *chromatophore* being given to those pigment cells capable of dispersing or concentrating their pigment. The most common, and certainly the most intensively studied, type of chromatophore is the *melanophore*, containing the brownish-black pigment, *melanin*, in the form of granules. Other types are distinguished by the nature of their coloured inclusions, *e.g.*, *xanthophores* contain predominantly yellow pigments and *erythrophores* predominantly red ones; those containing highly refractile purine crystals, *e.g.*, guanine, are called *iridophores.*†

Pigments and Ultrastructure. The melanophores contain granules of the black pigment, melanin (Fig. 826); and darkening of the skin results from the migration of these granules from the body of the cell into the dendritic, finger-like processes (Fig. 828, Pl. LXXV).

Melanin Granules

These particles have been examined in the electron microscope by Drochmans (1960) and more recently by Bikle, Tilney & Porter (1966) and by Shearer (1969); according to the last author they are ovoid bodies ranging in size from 0.2μ diameter in the smallest near-spherical units to $2.2 \times 0.6 \mu$ in the large ovoid particles; they are surrounded by a unit membrane. According to Drochmans they seem to have a crystalline structure and are surrounded by a more diffuse layer, but the high electron-density makes resolution of detail difficult. Birbeck (1963) has examined the growth of melanin granules in melanocytes; as with secretory granules, they are apparently synthesized in the Golgi apparatus, beginning as small vesicles, some 0.05μ in diameter and becoming large vesicles of the size of the mature granule; the outer membrane of the vesicle is the typical

---

* The winter dress of many mammals and the breeding dress of some fishes represent colour changes with which we shall not be concerned here; they are the result of the breakdown or build-up of pigment in the tissues, not a redistribution. They have been discussed by Parker (1955). In the same way, prolonged exposure of the animal to conditions that produce a colour change lead to increased amounts of a given pigment in the chromatophore. It is customary to distinguish the rapid changes as "physiological" and the slower changes as "morphological".

† Bagnara (1966) has discussed the terminology of chromatophores, and I have adopted his recommendations; it was common to name the cell after the pigment it was supposed to contain so that iridophores were called *guanophores*, whilst xanthophores were called *lipophores* because they contained the lipid-soluble carotenoids. However, most chromatophores contain a mixture of pigments, so that typical xanthophores and erythrophores contain both carotenoids and pteridines, and iridophores contain adenine besides hypoxanthine, whilst riboflavine is a constituent of some reflecting cells.

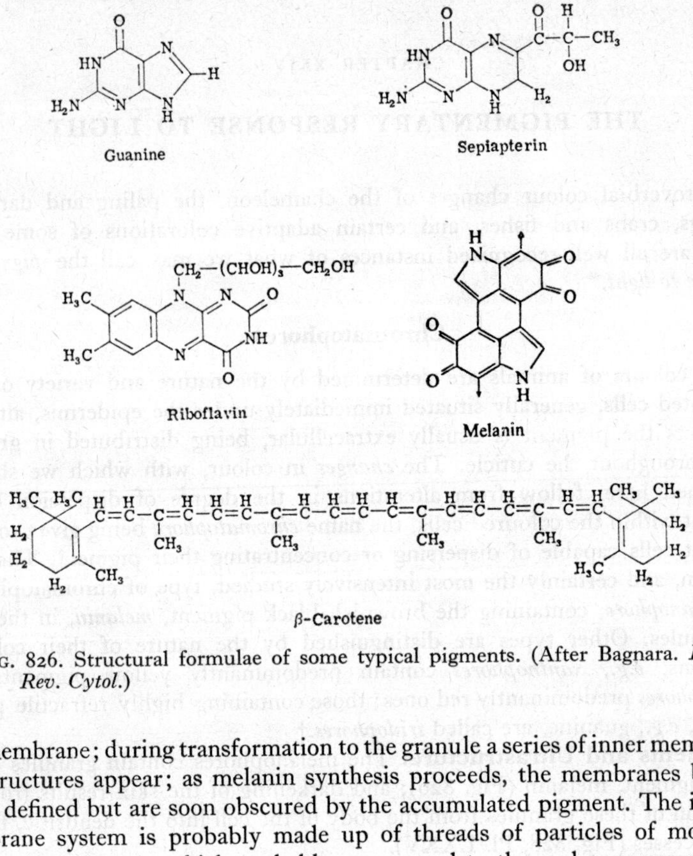

FIG. 826. Structural formulae of some typical pigments. (After Bagnara. *Int. Rev. Cytol.*)

unit membrane; during transformation to the granule a series of inner membrane-like structures appear; as melanin synthesis proceeds, the membranes become better defined but are soon obscured by the accumulated pigment. The internal membrane system is probably made up of threads of particles of molecular weight about 80,000, which probably correspond to the melanosomes of Seiji, Fitzpatrick & Birbeck (1961), namely particles with tyrosinase activity and containing melanin.

Iridophores

The great majority of the iridophores contain purines; in amphibia Bagnara (1966) showed that guanine, adenine and hypoxanthine were present in equal amounts. During preparation for electron-microscopy the crystals are usually lost and their places are revealed as empty needle-like spaces, as in Fig. 827, Pl. LXXIV, surrounded by membranes. According to Setoguti (1967), the irregularity of shape depends on the direction of cutting, so that the reflecting bodies probably exist as flat plates parallel to the cell surface with spaces some 0·1 μ between.

Xanthophores and Erythrophores

The bright yellow and red colorations typically seen in amphibia and lizards are due to the xanthophores and erythrophores; Obika (1963) analysed the coloured skin of larval salamanders at different stages of development and showed that, contrary to belief, the development of yellow xanthophores was associated with the appearance of pteridines,* mainly the yellow-fluorescent sepiapterin (Fig. 826), rather than carotenoids, and histologically the amount of

* The pteridines are both coloured and colourless; sepiapterins are yellow and drosopterins are red; the colourless give blue or violet fluorescence in ultra-violet light (Hama, 1963).

PLATE LXXIV

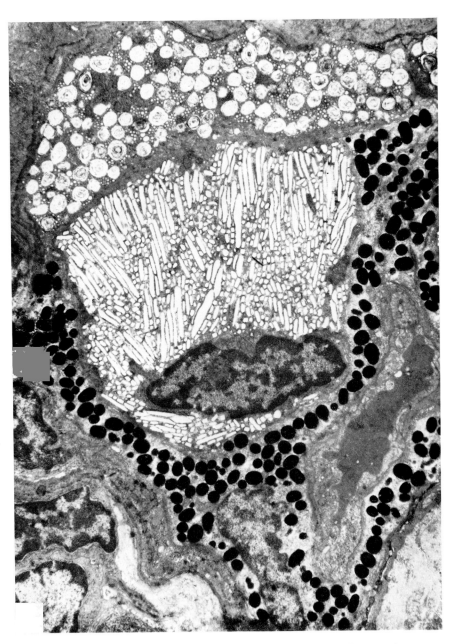

FIG. 827. Transverse section of *H. cineria* skin showing dermal chromatophore unit in white background-adapted state. Note that the melanosomes are uniformly distributed in arms around the sides of the iridophore but do not enter the "fingers" over the iridophore. × 8100. (Bagnara, Taylor & Hadley. *J. Cell Biol.*)

PLATE LXXV

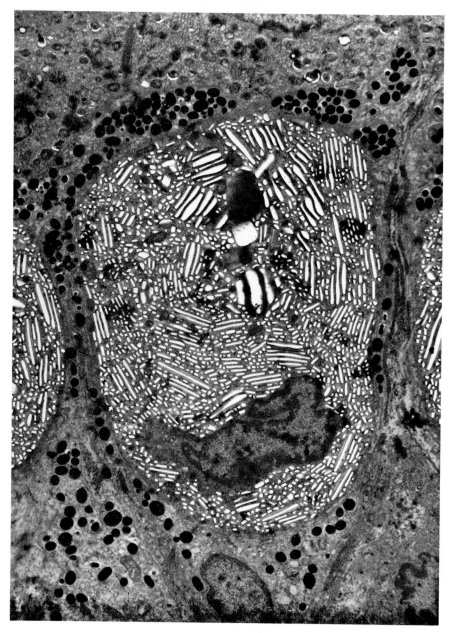

FIG. 828. Similar to Fig. 827 except that the animal was treated with intermedin, so that the melanosomes now fill the "fingers" over the iridophore at the expense of the arms and body of the melanophore. × 6000. (Bagnara, Taylor & Hadley. *J. Cell Biol.*)

water-soluble granules in the cells correlated with the chemically extracted pterins. In the larval stages there were no carotenoids, which appeared in the adult. In a more extensive study of different species, Obika & Bagnara (1964) showed that the red erythrophores of amphibian skin contained principally drosopterin, in addition to carotenoids, and the xanthophores mainly sepia-pterin. In many species the bright colours were due only to pteridines.*

PTERINOSOMES

The erythrophores of the swordtail, *Xiphophorus*, have been studied in the electron microscope by Matsumoto; they contain carotenoids, diffusely distri-buted and responsible mainly for the central yellow spot, and red organelles, some $0.7-0.5\,\mu$ in diameter, eluted by 1 per cent. aqueous NH_3 which contain a variety of pteridines, *e.g.*, the orange-fluorescent drosopterin, the blue-fluorescent biopterin, and so on. The granules, called *pterinosomes*, were sur-rounded by unit membranes and were made up of some 13–20 concentrically arranged membranes, packed rather like myelin. Removal of the pteridines caused breakdown of structure.

CAROTENOIDS

We may presume that the lipid-soluble carotenoids are present as fine drop-lets in the cytoplasm of the chromatophore; histochemically they are identified by their extraction with lipid-solvents, so that, as indicated earlier, it was common to describe the coloured chromatophores as *lipophores*.

Iridescent Colours. The iridescent colours of many fish and insects may be largely due to interference of light, rather than to the presence of pigments; this is brought about by the arrangement of alternate layers of material with high refractive index, *e.g.*, guanine crystals, with material of low refractive index, namely cytoplasm. The optics of these interference colours have been described by Land (1966), with special reference to the reflector in the eye of the scallop, *Pecten maximus*, and by Denton & Nicol (1965) with reference to the colours of fish scales. The colour of the reflected light is determined by the constructive interference at preferential wavelengths; and the highest reflectivity at a given wavelength, λ_0, together with the widest band of wavelengths of high reflectivity, is given when the layers of reflecting plates all have an optical thickness of $\frac{1}{4}\lambda_0$. Denton & Land (1967) have used the interference-microscope to measure the path-difference between reflecting plates in the scales of several fish; this path-difference is given by:—

$$\text{Object thickness} \times \text{Refractive index difference}$$

and should be equal to $\frac{1}{4}\lambda_0$. They found that the computed values of λ_0 corres-ponded remarkably well with the colours of reflected light from the scales, so that the plates seem to be laid down close to the ideal thickness necessary to give the actual reflected colours. On theoretical grounds the colour of the reflected light could be changed by altering the spacing between the plates; experimentally Land (1966) showed that such changes could be brought about by changing the osmolality of the medium surrounding the reflecting cells of *Pecten*.

* Ortiz & Williams-Ashman (1963) and Ortiz *et al.* (1963) have reviewed the findings of red-, blue-, violet-, green- and yellow-fluorescing pteridines in several poikilothermic vertebrates; their own work on lizards (anoles) has identified many of these compounds in the coloured dewlap of *Anolis*. Hama (1963) has reviewed the changes in the pteridine composition of amphibia and fishes during ontogenetic development. The distribution of the various pteridines in the integuments of some fifty amphibian species has been described by Bagnara & Obika (1965) in an effort to relate these to amphibian phylogeny and to the evolution of pteridines and pteridine metabolism.

Organization of Chromatophores

According to the variety and nature of the colour changes that the animal may undergo, the organization of the chromatophores varies from a simple arrangement of melanophores to a highly complex pattern of many types. As an example of a relatively complex organization we may quote the studies of W. J. Schmidt on the tree-frog, *Hyla arborea*; this animal may change from a pale green through dark green, lemon-yellow to grey. Fig. 829 illustrates the ways in which these changes of colour are brought about. Thus (*a*) represents the bright-green stage; a layer of yellow xanthophores occupies the most superficial position; behind them are the reflecting iridophores, whilst behind

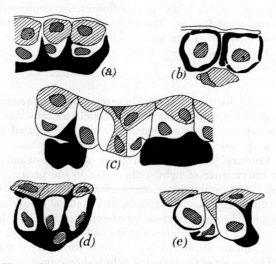

FIG. 829. Illustrating the behaviour of different cell-types during the colour-changes of the tree-frog. Iridophores, white; xanthophores, shaded; melanophores, black. (*a*) Bright green condition. Typical xantholeucosomes on the dark background of melanophores. (*b*) Grey, black spotted condition. Iridophores pressed against collagen boundary layer; under these, xanthophores. (*c*) Lemon-yellow condition. Xanthophores pushed between iridophores. (*d*) Dark green condition. Processes of melanophores extend round iridophores. (*e*) Grey (green) condition. Iridophores completely surrounded by melanophores; xanthophores flattened over them or pushed between. (After W. J. Schmidt. *Arch. Mikr. Anat.*)

these, and partially enveloping them with their processes, are the melanophores. In the lemon-yellow stage (*c*), there has been a concentration of the melanophores so that the dark pigment does not surround the iridophores. In the grey condition with black spots, the reflecting iridophores are virtually surrounded by the processes of the melanophores, and the xanthophores may be squeezed out of the surface (*b*).

In the simplest case, changes of colour may be brought about by alterations in the degree of dispersion of pigment granules in the melanophore alone, even though other types of chromatophore are present. As an example, we may quote the case of the lizard, *Anolis*, studied particularly by Kleinholz; this animal exhibits a change from green to brown. The chromatic system consists of a layer of yellow oil-droplets and xanthophores just beneath the epidermis; internal to this is a much thicker layer of iridophores composed of a number of irregularly placed blocks or plates. Finally, below this is the melanophore layer

with the cell bodies partly embedded in the iridophores and partly in the underlying connective tissue of the dermis. In the brown state, the pigment in the
melanophores is dispersed throughout the long processes, which stretch up
between the iridophores and form a thin layer of pigment just under the
epidermis. In the green condition, the pigment is concentrated in the body and
proximal branches of the melanophores, an arrangement permitting the reflection of light by the iridophores through the yellow layer. The iridophores and
xanthophores appear to play only a passive role in this chromatic response.

Chromatophore Unit. In general, the organization of melanophore, iridophore and xanthophore, or erythrophore, is such as to permit co-operation in
the production of colour change, and this has led to the concept of the *chromatophore unit*. Fig. 827, Pl. LXXIV illustrates the situation in the frog *Hyla cinerea;*
the melanophore, with its granule-containing processes, surrounds the
reflecting iridophore immediately above it; above the iridophore is the xanthophore that imparts the yellow-green colour to the frog. The skin is in the white-
background state with the animal a bright yellow-green. Fig. 828 shows the
situation when the animal was darkened by the injection of pituitary hormone,
intermedin. Now the melanin granules have moved into the cytoplasmic fingers,
or dendrites, of the melanophore, thus shading the iridophore and reducing
the light reflected through the overlying xanthophore. In this colour change, as
in *Anolis*, the main mechanism consists in the dispersion of pigment in the
melanophore, but in other species, *e.g.*, in the frog *Rana pipiens*, and in fishes,
the dispersion and concentration in the iridophores are well established (see, for
example, Fries, 1958) and the two phenomena of melanophore dispersion and
iridophore concentration are brought about by the same hormonal influences
(Bagnara, 1968), in fact when various hormone preparations were examined
their melanophore-dispersing activity ran parallel with the iridophore-concen-
trating activity (Bagnara, 1964). The xanthophores and erythrophores of
vertebrates probably play mainly a passive role since it has been difficult to
establish the role of hormones or nervous influences on their state.*

Melanocyte. In the skin of mammals and cold-blooded vertebrates the more
permanent coloration is due to the presence of melanin-bearing cells, *melanocytes*, of neural crest origin, which come into intimate relation, through their
dendritic processes, with neighbouring Malpighian cells of the epidermis. In
response to a stimulus, *e.g.*, ultra-violet irradiation, the melanocyte releases
pigment into the neighbouring Malpighian cells, whilst synthesis of melanin by
the melanocyte is likewise stimulated. This is the basis for the darkening of
human skin, and is to be distinguished from the rapid changes of coloration due
to readily reversible migration of granules into the dendritic processes of
melanophores; and it is common to speak of the *physiological* and *morphological*
colour changes respectively. The difference between the two is not absolute,
however; thus Lerner & McGuire (1961) showed that movement of pigment into
the dendritic processes of the melanocytes *per se* must be responsible for the
rapid darkening of human skin in response to injection of melanocyte-stimulating
hormone, whilst the more prolonged effects are due to transport of pigment
to neighbouring cells. In the frog, the distinction between the melanophore and
melanocyte is confounded; Hadley & Quevedo showed that the "epidermal
melanocytes" behaved like melanophores, in that their pigment was concentrated or dispersed in response to appropriate stimuli, but also they behaved

---

* Bagnara, Taylor & Hadley (1968) have described a dispersion of pigment in the
xanthophores of the tree-frog *Hyla arenicolor* in response to intermedin; as they point
out, this is the first description of such a process.

like melanocytes in giving up pigment to the neighbouring cells of what may be called the *melanin unit*. It seems likely that the dispersion of the pigment is the prime cause of the accelerated synthesis of melanin that occurs in response to prolonged stimuli.

MECHANISM OF DISPERSION AND CONCENTRATION

With the exception of the cephalopods it would seem that the dispersion of pigment in the "expanded state" of the chromatophore is due to the movement of pigment granules into pre-existing dendritic processes, rather than due to the extension of pigment-containing processes from the body of the cell.

Cephalopoda

In the cephalopod, *e.g.*, the squid or octopus, we find indisputable evidence for a change in shape of the melanophore, brought about by the contraction of radially disposed muscle fibres. The phenomenon has been studied by Bozler in some detail, and Fig. 830 illustrates his conception of the expansion and contraction of the pigment. The melanophore, in its relaxed condition,

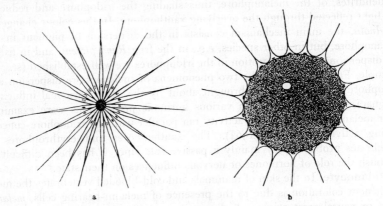

FIG. 830. Chromatophore, with attached muscle cells, of the squid, *Loligo*. (a) Muscle fibres are relaxed and the pigment is concentrated, giving the light condition. (b) The muscle fibres are maximally contracted, giving dispersion of pigment. (Bozler. *Z. vgl. Physiol.*)

is a small spherical cell, so that if all the melanophores in the skin are in this *concentrated* condition the animal is pale; the muscle fibres pull radially and, when contracted, draw the sphere out into a large thin disc; in this condition the pigment is said to be *dispersed*, and the animal is dark.

Vertebrates

In vertebrates the melanophores are cells with many branching processes; Fig. 831, from Hogben & Slome, illustrates various conditions of the melanophores of the horned toad, *Xenopus*, the different states being characterized by the varying degrees of dispersion of the pigment along the fine ramifying processes. The stages of dispersal of pigment are given numbers, from 1 to 5; maximal dispersal being represented by 5, and maximal concentration by 1. This forms the basis for most quantitative studies on pigment dispersal.

In the light-microscope it is difficult to determine whether the changes in appearance are due to changes in the morphology of the cell or to movement of pigment along existing structures. Conclusive proof that the morphology remained the same was given by Matthews' studies of the migration of pigment in cultured cells; grown in tissue culture, the outlines of melanophores remain visible even when the pigment is in the concentrated state. By treating the cells with adrenaline the pigment was concentrated, and the outline remained the same; on treatment with atropine the pigment migrated peripherally to re-fill the contours.

Again, Bikle, Tilney & Porter were able to follow the movements of granules into and out of empty dendritic processes by the aid of the phase-contrast microscope.

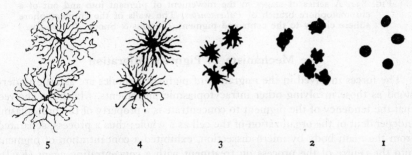

5   4   3   2   1

FIG. 831. Different degrees of dispersion and concentration of pigment in the melanophores of the horned toad, *Xenopus*. (Hogben & Slome. *Proc. Roy. Soc.*)

In the electron microscope Bagnara, Taylor & Hadley (1968) observed membranes between the basement membrane and the iridophore when the skin was from a bright-background frog with melanophores concentrated; these membranes were probably the empty dendritic processes of the melanophores; in *Rana pipiens* there was considerable expansion of the iridophores under these conditions, and their cytoplasm showed indentations as though expansion were being hindered by the melanophore dendrites; with *Hyla cinerea* or *A. dachnicolor*, whose iridophores do not expand, there were no deep indentations. These observations thus suggest that, whereas the melanophore retains the same morphology, the iridophore does show physical expansion.*

Arthropoda

With respect to the red chromatophores of the shrimp, *Palæmonetes*, Perkins & Snook have arrived at generally similar conclusions to those of Matthews, in that the processes are a part of the cell and not pre-existing intercellular spaces. Their microscopical observations, however, led them to believe that the dispersion of pigment was a filling up of collapsed tubes, as illustrated in Fig. 832. If this is true, dispersion must be accompanied by a considerable increase in cell volume, presumably brought about by osmosis from outside.†

---

* According to Novales (1962) the melanophores of the salamander in tissue-culture contract and expand; only when they become fixed in the adult tissue do they lose this amœboid movement.

† True amœboid movements of pigment cells are found in the Culicidæ, of which *Corethra* has been studied by Martini & Achundow; thus larvæ, if kept on a white background, become thoroughly transparent; and on a black background become black, a colour change brought about by the rapid movement of the black pigment cells on the air-sacs.

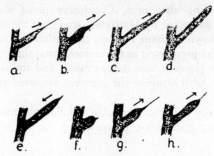

FIG. 832. A series of stages in the movement of pigment into and out of a chromatophore branch of *Palæmonetes*. The walls of the chromatophore adhere closely to the contained pigment. (Perkins & Snook. *J. exp. Zool.*)

The Mechanism of Pigment Migration

The forces involved in the migration of pigment granules are as little understood as those involving other intracytoplasmic movements. Matthews showed that the tendency of the pigment to concentrate is a property of the protoplasm, independent of the organization of the cell as a whole; thus a process, separated from the main body by micro-dissection, exhibited a concentration of pigment into the centre of the process on treatment with a concentrating agent (KCl); replacement of KCl with NaCl caused a subsequent dispersion in both the isolated process and the remaining cell. That a change in protoplasmic structure is associated with the migration of pigment is shown by Matthews' observation that, in the concentrated condition, the cell was brittle and not easily distorted, in marked contrast with the expanded state.

Again, Bikle *et al.*, in their study of isolated fish scales, noted that, when the scales were placed in a solution stimulating migration of the granules, these began a rapid vibrating movement which seemed to be directed predominantly along the longitudinal axis of the process; following this, the pigment began to move *en masse* at a uniform rate; some independence of movement was observed, in the sense that a file of granules, in the centre of a process, might be seen moving in the opposite sense to the main migration. The general impression obtained by Bikle *et al.* was that the granules were confined to definite channels, and their electron microscope study suggested that the microtubules, which radiated from the centriole and were arranged parallel with the long axes of the processes, guided the granules.

Effects of Pressure

Marsland showed that hydrostatic pressures of 7,000–8,000 lb./in.² caused the concentrated chromatophore in the isolated scale of *Fundulus* to expand, a fact suggesting that the dispersed state of the pigment resulted from a sol-like condition of the cytoplasm, whilst concentration was the consequence of a gelation. This view was supported by his observation that centrifuging chromatophores at 70,000 g. or more had no influence on the position of the pigment when they were in the concentrated condition, whereas it caused a fairly ready displacement in the expanded chromatophores. If, as Marsland supposes, the plasmagel has the power of contracting (p. 124), then concentration of pigment may be the consequence of, first, a gelation of the plasmasol, followed by shrinkage of the gel drawing the attached pigment granules into the centre of the cell, their place in the processes being occupied by sol driven out of the central region. Observation under the microscope suggested that the pigment granules, moving towards the centre during concentration, were moving against a current, exhibiting a "peculiar bobbing movement" distinctly different from Brownian movement. It is possible, then, that the granules are carried in the meshwork of a contracting gel. The opposite process of expansion is not a purely passive migration by Brownian movement, since it occurs far too rapidly; it is possible that the contracted gel expands, carrying the pigment granules with it, and finally solates.*

Intracytoplasmic Potential Gradient

Kinosita (1963) has suggested that migration of melanin granules is determined by a gradient of electrical potential between cell body and cell process; thus the inside of the melanophore is negative with respect to the medium, but the degree of negativity at the centre of the cell is different from that in the process, and the direction of this difference is correlated with concentration or dispersal of the pigment. In the expanded state the centre was more negative than the periphery, and the gradient would cause the negatively charged granules to move out from the centre of the cell; in the contracted state the gradient of potential was reversed. With iridophores, which respond in the reverse manner to melanophores, the gradient of potential varied in the opposite sense. More recently Martin & Snell (1968) have measured the resting-potential of frog melanophores; this was of the order of 70–80 mV, but was quite unaffected by the degree of dispersion of pigment induced by α-MSH, showing that major changes of ionic permeability were probably not at the basis of the altered dispersion. Again, Freeman, Connell & Fingerman (1968) measured an average resting potential of the red chromatophores of the prawn, *Palæmonetes vulgaris*, of 55 ± 15 mV; the concentrating hormone caused a hyperpolarization; and they suggested that, if this was confined to the body of the cell, it would create a gradient of potential along the processes favouring migration of granules.

Effects of Osmolality and Sodium

Novales (1959) showed that in a hypotonic medium the melanophores of the frog's skin were reversibly dispersed; in a hypertonic medium the pigment was concentrated and now intermedin was ineffective in causing dispersion, perhaps because of a collapse of the dendritic processes. In a Na^+-free medium intermedin was ineffective, and it was suggested that the hormone, like pitressin, exerted its influence primarily by increasing the influx of Na^+ into the cell, thereby triggering a movement of the granules.

Similar effects were observed with isolated dogfish skin (Novales & Novales, 1966). The effect was remarkably specific for Na^+, so that its replacement by other monovalent cations was similar to using a sucrose-medium. A Ca^{++}-free medium enhanced the action of intermedin (Novales *et al.*, 1962). Essentially similar effects were described by Fingerman, Miyawaki & Oguro (1963) on the isolated legs of *Uca*, hypotonic sea-water causing reversible dispersion and 200–300 per cent. sea-water blocking the action of the sinus gland; unlike the vertebrate situation, K^+ and Li^+ were as effective as Na^+. In a later study of the prawn, *Palæmonetes vulgaris*, Fingerman, Connell & Yoshioka (1967) found that the action of the concentrating hormone for the red chromatophores had a requirement for Na^+, and that of the dispersal hormone a requirement for Ca^{++}. Ouabain inhibited the concentrating hormone by 75 per cent. and tetrodotoxin enhanced the response, and they suggested that the action of the concentrating hormone depended on a high gradient of concentration of Na^{++}

---

* The effects of D_2O, which, as we have seen, enhances gelation, are consistent with the effects of pressure (Marsland & Meisner, 1967). The tissue-culture study of *Ambystoma* by Zimmerman & Dalton (1961) adds some interesting details to the microscopically observable processes of concentration and dispersion of pigment.

across the cell; thus ouabain would cause this to be reduced whilst tetrodotoxin, by supposedly blocking passive influx of Na^+, would increase the gradient.

THE CONTROL OF PIGMENT MIGRATION

Changes in pigment distribution result most frequently, but not invariably, from a change in the conditions of illumination of the animal; thus a frog placed in darkness becomes black, and in a bright environment becomes pale; the shrimp, *Palæmonetes*, becomes pale in darkness and dark in daylight; the flat fishes not only adapt themselves to the colour of their background, but also to the pattern, as Mast has shown. In some instances, however, stimuli, other than light, play an important role; for example, the frog, *Rana temporaria*, is very sensitive to humidity and temperature, becoming dark at low temperature and high humidity, and pale under the reverse conditions. Again, the fiddler crab, *Uca*, exhibits a definite 24-hourly rhythm of lightness and darkness, becoming pale at night and dark during the day; Brown & Sandeen have shown that keeping this crab in darkness for weeks fails to modify the rhythm, which is thus inherent. Finally, the amphipod *Hyperia* becomes whitish whenever it attaches itself to a jellyfish, but when free is a dark reddish brown. Apart from these, and a few other instances, however, the light-stimulus is undoubtedly the most important.

Responses to Light

In general we may distinguish two types of response to light, as follows: (*a*) The *primary response* which is a direct reaction to illumination of the skin and occurs in the blinded animal; it is generally manifest as a dispersion of the pigment in those chromatophores that principally determine the colour of the animal.* (*b*) The *secondary response*, which depends on the eyes, represents the dispersion of dark pigment when the animal is placed—in the light—on a dark background, and the concentration of pigment when on a bright background. These two types of secondary response are often described as *black-background* and *white-background* responses, and are determined by the ratio of the incident and reflected intensities of light—the so-called *albedo*.

The Primary Response. This is seen in its most pronounced form in the chameleon; thus Zoond & Eyre state that, if a letter Y is cut out of a piece of copper sheet and placed on the skin of a chameleon in bright sunlight, an exact reproduction is obtained on the skin like a photographic negative. Yoshida studied the effects of very small patches of illumination (ca. 3μ diameter) on the dispersal of pigment in individual chromatophores of the sea urchin *Diadema*. In general, by illuminating a portion of a fully dispersed chromatophore, he was able to prevent the concentration that occurred in neighbouring chromatophores when the animal was in darkness; the illuminated part of the chromatophore thus retained its outline. Again, when the pigment-free part of a concentrated chromatophore was illuminated, the pigment migrated into it, showing that migration occurred in the direction of the light, and suggesting that the chromatophore was directly susceptible to the action of light.

Secondary Response. That the black-background response is not due to a diminished illumination of the eye, *per se*, was shown by Hogben & Slome, who found that a blinded South African clawed toad, *Xenopus*, adopted an intermediate colour between the extreme pallor obtained in white surroundings

* The *Natantia*, e.g., the shrimp, are said, by Hogben, to prove an exception to this rule.

and the dark colour when placed on a black background; the same was true if normal animals were kept in the dark for a long time, their colour being that of blind animals. Hence an animal in daylight, but on a black background, shows the most intense dispersion of its pigment. The same state of affairs was found with the isopod crustacean, *Ligia oceanica*, by H. G. Smith. Hogben & Slome concluded that the eye must be broadly differentiated into two regions, so far as the pigmentary response is concerned. Thus, with an aquatic vertebrate, on a black background, a limited ventral region of the retina will be illuminated, since only rays within a cone, whose half-angle is the critical angle for an air-water interface, can enter the eye. On the black background, only this region is stimulated, and the result is a maximal expansion of the melanophores. On a white background the whole retina is illuminated; if stimulation of the dorsal region of the retina causes concentration of pigment, and if this region is prepotent over the ventral region, the white-background response can be explained. On this basis the blind animal is not so dark as the black-background animal because the important ventral region is unstimulated, nor yet is it so pale as the white-background animal because the dorsal region is unstimulated.

Retinal Differentiation

Evidence for such a differentiation in the gillyfish, *Fundulus heteroclitus*, has been provided by the histological and physiological studies of Butcher; the upper (dorsal) region of the retina contained rods and single and double cones, whilst the lower region, some 30 per cent. of the whole area, only contained rods and double cones; moreover, there was a specialized "crescentic ridge" in the lower retina. By putting a "blinker" over the fish's eye in such a way as to cut off different regions of the retina from illumination, Butcher showed that stimulation of the upper portion of the retina caused paleness, and that this area was prepotent over the lower region which caused, by itself, darkening. Further confirmation was obtained by making incisions through the eye to sever different regions of the retina from their central connections. Essentially the same state of affairs was found by H. G. Smith in the crustacean, *Ligia*; in this (compound) eye, it was the latero-ventral group of ommatidia that were concerned with the white-background response, whilst the dorsal group controlled the black-background response.

Control Mechanisms

The secondary responses are thus mediated by the eyes; we have yet to discuss the manner in which the effects of these visual stimuli are transmitted to the effectors, the chromatophores. In the cephalopods, nervous control over the pigmentary system is well established; the muscle fibres, causing the expansion of the melanophores, are activated by nerve fibres; cutting a mantle nerve in the squid leads to a blanching of the area innervated, a finding that suggests that the muscle fibres are maintained in a tonically contracted state. The detailed mechanism of the control is not understood; there seems no doubt that there are a number of chromatic centres in the ganglia, and that certain drugs may act on these centres, whilst others act more peripherally; of the latter group, adrenaline causes an expansion of the chromatophores, *i.e.*, a contraction of the muscle fibres, whilst acetylcholine has the opposite effect.

Pituitary Hormones. The colour changes effected by the cephalopods are extremely rapid; moreover, the expansion of the melanophores depends only on a muscular contraction, so that a predominantly nervous control is not surprising. In amphibians, on the other hand, the very sluggish responses (hours are required for the development of colour changes in frogs) suggest that the activation of the melanophores requires the release of a hormone into the blood.* The early work of Smelt in 1916, among others, showed that the pituitary gland was important, and this was confirmed by Hogben & Winton, who showed that an extract of beef pituitary caused a rapid blackening of a pale frog; removal of the pituitary made the animals pale.

Melanocyte Stimulating Hormone (MSH)

Subsequent work has shown that the factor responsible is different from other pituitary hormones and is described as *intermedin* or *melanocyte stimulating hormone* (MSH);† the factors responsible in the pituitary for this action are polypeptides; thus according to Pickering & Hao Li (1962) the frog's pituitary contains three principles, α-MSH, β-MSH and ACTH containing respectively 14, 18 and 19 amino acids, which have in common a heptapeptide moiety, namely His-Phe-Arg-Try-Gly-Ser-Pro; the smallest sequence showing activity is given by His-Phe-Arg-Try-Gly, and all synthetically active preparations contain this. So far as mammals are concerned, all species examined have α-MSH; β-MSH is not a single entity, some five different polypeptides having been isolated from pig, cattle, horse, monkey and human glands respectively, differing in the numbers of amino acids and their sequences (Lee, Lerner & Buettner-Janusch, 1963). Adrenocorticotrophic hormone (ACTH) from the anterior lobe of the pituitary has some chromophorotropic activity.

Nervous Mechanisms. It has frequently been denied that the melanophores of amphibians are innervated; recently, however, both histological and physiological evidence has been provided by Stoppani and by Vilter, among others, in favour of an active concentration of pigment by nervous means. Thus Stoppani cut the sciatic nerve of the toad, *Bufo arenarum*, and found that the leg darkened a little; on stimulation of the peripheral end, blanching occurred. Certainly the injection of adrenaline (*e.g.*, 1 ml. of 1:10,000) causes a blanching which begins in about ten minutes and lasts for hours; consequently Vilter's hypothesis that the sympathetic is antagonistic to the pituitary has some support.

In fishes, as we shall see, nervous influences are predominant, mediated presumably through vesicle-filled nerve terminals that make a synaptic type of contact with the melanophore (Bikle, Tilney & Porter, 1966).‡

Crustacean Hormones. In crustaceans there is no innervation of the chromatophores; Koller showed that perfusing a pale shrimp, *Crangon vulgaris*, with blood taken from a dark one (dark, because it had been kept on a dark background) caused the pale shrimp to become dark; the reverse experiment, namely, perfusing a dark shrimp with blood from a pale one, led to no change in colour. Later Perkins showed that within an hour after injection of an extract of the eye-stalk of a pale shrimp, *Palæmonetes*, into a blinded animal, whose chromatophores were expanded, the pigment had contracted. So far as the shrimp is concerned, therefore, the active hormone tends to concentrate the pigment.

Secretory Organs. In addition to the eye-stalks, the central nervous system

---

* The hormone, once secreted into the blood, may act quickly, but unless it is rapidly removed a change in colour will take a long time, since what is required is a significant change in its concentration in the blood. As Hogben points out, however, we cannot immediately deduce from the sluggishness of the responses that the control is humoral; the effectors themselves may have a very long latent period. Thus the reaction time in some teleosts is less than one minute, whilst in *Xenopus* it is 100 minutes.

† According to Fingerman (1965), the American Society of Zoologists prefer to retain the name intermedin for the class of hormones.

‡ Snell & Kulovich (1967) were unable to detect any dispersion of melanin granules in the hypophysectomized frog with sciatic nerve stimulation; furthermore, there was no influence of stimulation on the action of MSH.

of the crustacean contains active principles that, when extracted, will influence the chromatophores; thus Perkins, in the study alluded to, showed that extracts of the brain caused concentration of red chromatophores. Again, Brown & Ederstrom found that, whereas the eye-stalk extracts caused concentration of pigment in the black chromatophores of *Crago*, extracts of the œsophageal commissures, including the commissural ganglion, would cause dispersion of pigment, the two extracts antagonizing each other.

Sinus Gland

The eye-stalk is a complex structure containing, besides the eye and its associated structures, a *sinus gland* and what Hanström described as an *X-organ*. The sinus gland was believed by Hanström to contain cells that actually secreted the active hormones, but it is now generally conceded that it is merely the receptacle for hormones secreted by neurones which may be in the eye-stalk itself, or in other parts of the nervous system. The gland thus consists of swollen terminal expansions of neurones—*neurosecretory cells*—in which are concentrated the hormones in the form of granules.

Thus, the histological studies of Passano, of Bliss & Welsh, and of Knowles concur in showing that the sinus gland really consists of a connective-tissue bag containing the swollen terminal expansions of neurones whose cell bodies may be in the X-organ or in other parts of the nervous system, including the brain and perhaps the thoracic ganglion. The relationship between the neurones and the sinus gland is therefore similar to the neurosecretory relationship of the hypothalamus and the posterior pituitary of vertebrates. Consequently, extirpation of the sinus gland does not abolish hormonal control in the crab, whether it is measured by the tendency to moult or by pigment dispersion and concentration (Passano); removal of both X-organ and sinus gland, on the other hand, is just as effective as eye-stalk removal. Excision of the sinus gland amounts, on this basis, to cutting off the terminals of neurosecretory nerves; regeneration of these endings should lead to the formation of a new gland, and this does, indeed, happen (Bliss & Welsh), extracts of this showing hormonal activity. The term "gland" is therefore somewhat of a misnomer.*

X-Organ

Passano examined the cell-bodies and axons of nerves leading to the sinus gland, and concluded that active secretion of the sinus-gland material occurs in a group of neurosecretory cells in the nearby medulla terminalis, the secreted material passing along the axons. The secretory material consisted of granules of diameter $0.3\,\mu$. In all the eleven species of brachyurans examined by Passano, the X-organ consisted of a dozen or so large, faintly blue, cell-bodies with axons that were also faintly blue, and which passed to the sinus gland as the *sinus gland nerve*. The relationship of the X-organ to the sinus gland is indicated schematically in Fig. 833; here several X-organs, characterized by their whitish-blue appearance, the large size and granular nature of their cell-bodies, and their connexion with the sinus gland, are shown (Kleinholz *et al.*). In this context we may note that the X-organs described by Passano, Bliss, and others, are different in structure and location from that originally described by Hanström, which apparently represented the transformed sensory cells of a rudimentary eye papilla or sensory pore. Hanström considered this to be a secretory organ; whether or not this is true, it should be distinguished from the X-organ of the medulla terminalis and other similar regions of the eye-stalk. Thus the Hanström

---

* Enami (1951) observed granular material in the axons supplying the sinus gland and this appeared to migrate from cell-body to gland. Nevertheless Enami concluded that the sinus gland did actually elaborate its own secretion.

X-organ might well be called the PDX (*pars distalis X-organi*) by contrast with the PGX (*pars ganglionis X-organi*) belonging to the medulla terminalis (Knowles & Carlisle). The Hanström X-organ, or PDX, contains structures of several sorts including swollen terminals of neurones, some of which originate in the medulla terminalis of the eye-stalk, and others in the brain. Thus, in part at least, this X-organ is a site of release of neurosecretion and therefore partakes of the character of the sinus gland so that, as Knowles & Carlisle point out, it is

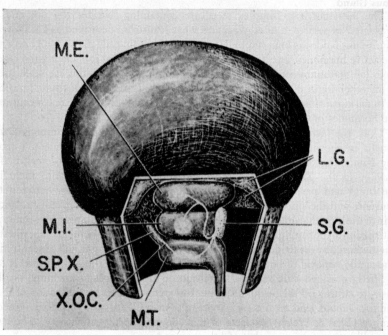

FIG. 833. Dissection of a left eyestalk of *Pandalus borealis*, approaching some-what obliquely from the dorso-abaxial aspect, so that the mid-dorsal line is towards the right-hand side of the drawing. Only the nervous tissues are shown. Ganglionic X-organs show in their natural appearance as whitish patches against their corresponding medullæ; from each of them a neuro-secretory tract runs to the sinus gland. A tract also runs from the brain to the sinus gland. In life these tracts also appear as white lines against the nervous tissue. Dissection by D. B. Carlisle. LG, lamina ganglionaris; ME, medulla externa, or second optic ganglion; MI, medulla interna; MT, medulla termi-nalis; SG, sinus gland; SPX, sensory pore X-organ (Hanström's X-organ); XOC, X-organ connective. (Kleinholz *et al. Biol. Bull.*)

unfortunate that the name X-organ has been given to two bodies so different in structure and function.

Granules

That the granules in the swollen nerve terminals constituting the sinus gland do actually contain hormones was made likely by Pérez-González who homo-genized the sinus glands and separated the granules by differential centrifugation; if these had been prepared in sucrose they had little activity when injected into the crab but if they were prepared in distilled water, or if after preparation they were treated with a detergent, then they had a powerful dispersing activity on the black chromatophores of the crab, *Uca pugilator*.* A possible objection to

---

* Boiling the various extracts of eye-stalk or brain-tissue has very little effect on activity (Kleinholz *et al.*, 1962); the small increase that occurs, for example, with the

the interpretation of the effects of distilled water is that hypotonicity, *per se*, will cause expansion of chromatophores (p. 1567).

Direct Secretion into Blood

There is little doubt, both on anatomical and physiological grounds, that neurosecretions may find their way into the blood directly from the secretory neurones as well as indirectly by way of the sinus gland. For example, in the prawn, *Penæus*, the post-commissure lamellæ are actually fused to the wall of a blood sinus from the gills (Knowles), whilst Fingerman, Sandeen & Lowe have shown that the sinus gland of the prawn *Palæmonetes* did not contain any pigment dispersal hormone whereas the nearby optic ganglia did, a finding that

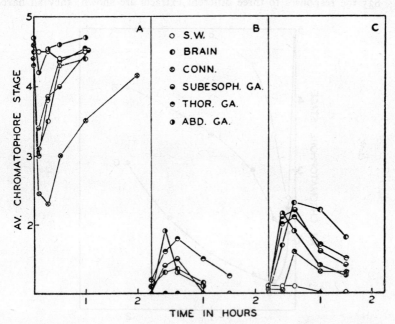

FIG. 834. Responses of red chromatophores of prawns to extracts of various parts of the central nervous system. Ordinates represent degree of dispersion of pigment (maximum = 5). A. Eye-stalkless recipients. B. Normal. C. One-eye-stalked recipient on a white background. S.W. = sea-water control; Conn. = tritocerebral commissure; Subesoph. Ga. = subœsophageal ganglion; Thor. Ga. = thoracic ganglion; Abd. Ga. = abdominal ganglion. (Brown, Webb & Sandeen. *J. exp. Zool.*)

suggests that the neurosecretory cells of these ganglia release their hormone directly into the circulation.

Dispersal and Concentrating Hormones. Some of the complexities in the hormonal control of *Palæmonetes* are illustrated by the study of Brown, Webb & Sandeen. Fig. 834 shows the response of this shrimp to extracts of various parts of the central nervous system. In A, eye-stalkless animals were injected; initially their pigment was completely dispersed, owing to the absence of eye-stalk hormones which normally have a net pigment-concentrating action. All of the extracts caused a concentration of pigment, which tended to reverse with time, presumably as the injected hormone was destroyed. The extracts of the

---

red-pigment dispersing hormone of *Palæmonetes* (Fingerman *et al.*, 1959) may be due to breakdown of granules to smaller components.

central nervous system therefore had a concentrating hormone, the circum-œsophageal connectives and tritocerebral commissure (Conn. in Fig. 834) apparently containing most. When injected into normal shrimps, adapted to a white background and therefore pale, the effect was one of pigment dispersion, the extract causing least dispersion being the one that caused most concentration in the blinded animals. It appears, then, that all the extracts contained two hormones, concentrating and dispersing, the proportions of the two being different according to the nervous tissue extracted. The extent to which the concentrating hormone was "contaminated" by dispersal hormone could be assessed roughly by the time-course of change in the chromatophores; thus, in Fig. 835 the responses to three different extracts are shown; they all have a

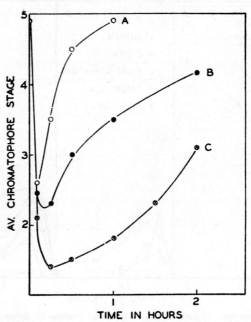

Fig. 835. Responses of red chromatophores of eye-stalkless prawns to different extracts of the central nervous system. A. Circumœsophageal connectives without tritocerebral commissure. B. Connectives with commissure. C. Commissure alone. Ordinates represent degree of dispersion of pigment (maximum = 5). All the preparations have concentrating activity, but the rapid reversal shown by extracts A and B suggests the presence of dispersal hormone as well. (Brown, Webb & Sandeen. *J. exp. Zool.*)

pronounced concentrating action but in A the return to the dispersed state is rapid whilst in C it is slow. A represents an extract of circumœsophageal connectives alone, which presumably contained a large amount of dispersal hormone; C represents an extract of tritocerebral commissure alone, which contained apparently very little or no dispersal factor, whilst B represents an extract of both tissues. Practically, then, the extract of circumœsophageal connectives could be treated as pure concentrating hormone, whilst abdominal extracts could be treated as pure dispersal hormone.

Isolation of Single Factor

A similar mode of procedure with the eye-stalks showed that the predominantly concentrating effect of the extract was contaminated by some dispersing activity; by comparing extracts of whole glands with extracts of sinus glands, it appeared that the latter contained little or no dispersal hormone. The dose-response curves of extracts of sinus gland and extracts of tritocerebral commissures could be made to overlap perfectly, so that it would appear that the red pigment-concentrating hormone in both extracts was the same. Having established to their satisfaction that they could prepare extracts containing one factor uncontaminated by the other, Brown *et al.* then studied the antagonism between the two hormones; for example, injection of abdominal cord extract into animals that had concentrated chromatophores, as a result of injection of an extract of tritocerebral commissure, led to a rapid dispersal of pigment. In general, it was concluded that a massive liberation of one hormone would suppress the activity of the other, so that rapid responses to environmental changes could be brought about in this way. By contrast, final steady-state conditions would be determined by a delicate balance between the opposing effects of the two hormones.*

Electrophoretic Separation

Electrophoresis of eye-stalk extracts on filter-paper has permitted a separation of the red-pigment concentrating hormone of *Palæmonetes*, which passed to the cathode, from the dispersing hormone which passed to the anode; an examination of the behaviour of the dispersing hormone indicated the presence of two dispersal hormones, migrating in opposite directions† (Fingerman, Sandeen & Lowe).

Temperature. The response to a change of temperature, inasmuch as it is independent of the eyes, can be classified as a primary response (Fingerman & Tinkle, 1956). Its physiological significance is not always clear; in the frog it has a thermoregulatory function, in the sense that high temperature is usually associated with a bright light; a concentration of pigment will thus favour reflection of radiant heat. In crabs, Brown & Sandeen considered that the blanching of crabs at high temperatures had a thermoregulatory role, and this was neatly demonstrated by Wilkens & Fingerman (1965) who showed that pale crabs (*Uca pugilator*) maintained their temperatures some 2° C cooler than dark crabs at the same ambient temperature. They pointed out that these crabs enter their burrows every 20 minutes to cool themselves. As Barnwell (1968) has emphasized in his study of several species of *Uca* with varying habitat, the state of the chromatophores under any given conditions will depend on a variety of reactions so that in one species a given environmental influence may exert a much stronger effect than in another; for example, the necessity to match the environment may take precedence over thermoregulation.‡

Retinal Pigment. The compound eyes of crustaceans exhibit characteristic movements of pigment on exposure to light—the light-adaptation response;

---

* More recent perfusion studies of Fingerman on the crayfish *Cambarellus shufeldtii* and the crab, *Uca*, suggest that, in these species too, the state of the chromatophores is determined by a balance between dispersal and concentrating hormones. Fingerman draws attention to a tendency for the red chromatophores of *Cambarellus* to disperse spontaneously, and the white chromatophores to concentrate, when the carapace is isolated. Attention should be drawn to a recent paper by Shibley (1968) showing that in *Cancer*, removal of the X-organ-sinus gland complex does not abolish the animal's responses to illumination, although complete removal of the eye-stalks does.

† Knowles, Carlisle & Dupont-Raabe (1955) have also separated, by paper electrophoresis, two principles in extracts of sinus gland and post-commissure organ. Sinus gland contained a Substance A, which caused concentration of pigment in all the red chromatophores of *Leander*; the post-commissure organ contained this and also Substance B, which caused concentration in the large red chromatophores and expansion in the small red chromatophores.

‡ Crane (1944) has pointed to the role of blanching of the carapace in display; in the most terrestrial subgroups this has developed to the point where the carapace becomes dazzlingly white.

essentially this consists of a dispersal of pigment granules in chromatophores surrounding the distal part of the individual ommatidia (distal pigment migration), together with a movement of the proximal pigment in the pigment cells surrounding the rhabdom (Fig. 836). In the dark, these changes are reversed —dark-adaptation. By simply examining the eye in the microscope the diameter of the black region gives a measure of the degree of dispersion—the so-called distal retinal index (Sandeen & Brown). Fingerman, Lowe & Sundaraj have shown that the distal pigment is under the same pigment-concentrating and pigment-dispersal hormones as those concerned with the chromatophores of the body surface. Thus, an injection of whole eye-stalk extract gave a characteristic light-adapting response followed by a dark-adapting response, indicating the presence of two antagonistic hormones in the eye-stalk. The sinus gland was

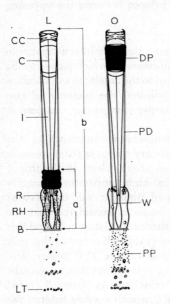

Fig. 836. Ommatidium in the light- (L) and dark-adapted (O) conditions. The change in position of the distal and proximal pigments is depicted; the position of the reflecting pigment is not shown. The measurement a/b represents the distal pigment index.

CC, cells of crystalline cone; C, cone; I, intermediate crystalline tract; R, retinular cell; RH, rhabdom; B, basement membrane; LT, lamina ganglionaris; DP, Distal pigment; PD, Proximal extension of distal pigment cell; W, distal extension of reflecting pigment cell; PP, proximal pigment. (Knowles. *Biol. Bull.*)

just as effective as the eye-stalk. Extracts of the supraœsophageal ganglion plus circumœsophageal connectives, which as we have seen contain mainly dispersal hormone, gave a pure light-adaptation response. By electrophoresis of eye-stalk extracts at pH 9 Fingerman & Mobberly were able to separate the two hormones, the anodal material containing only dark-adapting hormone. Both hormones could be inactivated by tryptic digestion, so that it appears that they are polypeptides.*

Tapetum Lucidum. The shining eyes of a cat or dog at night are due to the reflexion of incident light by a cellular layer behind the retina known as the *tapetum lucidum.* An important function of this layer is to increase the sensitivity of the eye by causing light to pass twice through the light-absorbing pigment of the retinal receptors; and it is interesting, as Denton & Nicol (1964) have

---

* Kleinholz, Burgess, Carlisle & Pflueger (1962) have examined the activity of extracts from eye-stalks, eye-stalk minus sinus glands, eye-stalks minus optic ganglia, and so on, in a number of decapod crustaceans. In all cases, removal of the sinus gland had no measurable effect on activity; on the other hand, extracts of sinus gland were measurably active. Details of the method of assay of hormone by the effect on the distal retinal pigment have been described by Kleinholz, Esper & Kimball (1962). Fingerman, Nagabhushanam & Philpott (1961) have described the physiology of the melanophores of the crab, *Sesarma,* as part of a study designed to extend our knowledge to different species.

shown, that when a variety of fishes are compared, with and without tapeta, those without tapeta have rods approximately twice as long as those with. A typical fish tapetum is illustrated by Fig. 837, the reflecting surface being made up of a series of flattened cells, containing guanine crystals, or plates; in many fish the degree of reflexion can be reduced by the migration of pigment into the processes of melanophores inserted between the reflecting plates; this movement of pigment thus reduces the amount of light reflected back along the retinal receptors and subserves the function of a pupil, preventing dazzle when illumination is high. In *Scyliorhinus* there is no migration of pigment, and the fish

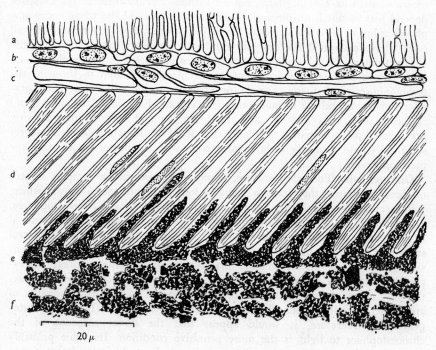

FIG. 837. A section through the tapetum lucidum in the dorsal region of the eye of *Scyliorhinus canicula.* The centre of the eye, with reference to this drawing, is towards the right. *a,* rods; *b,* pigment epithelium; *c.* choriocapillaris; *d,* reflecting cells (plates); *e,* pigment cells of tapetum, with processes; *f,* underlying pigmented chorioid. (Denton & Nicol. *J. Mar. Biol. Ass. U.K.*)

has a very efficient pupil; by contrast, in *Squalus* the pupil is inert and movements of pigment occur in the tapetum. Modifications of the reflecting powers of the tapetum also modify the amount of eye-shine, and according to Denton & Nicol this aspect may be more important in *Squalus* than any modifications of light-sensitivity of the eye; and it is interesting that the reflecting plates of *Squalus* are never completely covered by pigment; if they were, the eyes would become conspicuous, not because of eye-shine, but because they were black against the grey of the fish.

Species Variations. It is not possible to speak of a typical crustacean behaviour so far as the response to hormones is concerned. Brown (1948) has classified them into three types according to their response to removal of the eye-stalks. Group I is represented by *Palæmonetes,* which responds to removal by dispersion of its dark (red) pigment. Group III, containing all the studied brachyurans (*e.g.*, *Uca*) except *Sesarma*, responds in the reverse manner, *i.e.*, the animals become pale on removal of the eye-stalks; whilst *Crago*, the single example of Group II, responds in an intermediate fashion. Again, we must note that the chromatophores usually studied are those whose dispersal or concentration results in the most obvious colour change, *e.g.*, the red chromatophores of *Palæmonetes* or the black ones of the crab, *Uca*. As Abramowitz has shown, however, the white, black, red and yellow chromatophores of the crab, *Portunus*, all show responses to illumination, responses that are largely independent of each other.* Of obvious physiological significance are the white chromatophores of crabs, such as *Uca pugilator*, since lightness and darkness of the carapace depend just as much on dispersion and concentration of the white chromatophores as on the reverse changes in the melanophores. Rao, Fingerman & Bartell (1967) have examined the daily rhythm and hormonal control of the white chromatophores of a Florida specimen of *Uca pugilator*; as one might expect, their responses to the environment are reciprocally related to those of the melanophores so that, for example, on exposure to light the white chromatophores disperse and the melanophores concentrate. Extracts of the sinus gland and central nervous system had both dispersing and concentrating activity, presumably due to separate hormones that acted on the white chromatophores specifically, and probably without interaction with the hormones affecting the melanophores.†

Ontogenesis. The adult shrimp, *Crangon crangon*, has a complex system of chromatophores; they contain probably four pigments—sepia-brown, white, yellow, and red—which may appear alone in monochromatic chromatophores, or together in bi-, tri-, and tetrachromatic cells. With this apparatus, the animal is very clever at adapting itself to its background. In the zoea—the first larval stage—on the other hand, the chromatophore system is much simpler, consisting mainly of monochromatic blackish-brown chromatophores; corresponding with this simple system, the zoea shows no background response, the only effect of light being the primary dispersal of pigment; injection of an extract of adult sinus gland, moreover, has no effect (Pautsch). Ontogenetically, therefore, it would appear that the direct response of the chromatophore to light is the more primitive condition. Its value probably varies with the particular organism; according to Pautsch, in the larvæ it may act as a protection of the internal organs against the harmful effect of light, rather than as a camouflage.

Diurnal Rhythm

Persistence. We have referred briefly to the diurnal rhythmic change in pigment dispersal of the crab *Uca*, the animal being dark by day and pale by night; this is of general interest in interpreting not only the physiology of the chromatophoric responses, but also the rhythmicity associated with cosmic events, and is therefore worth considering a little farther. The diurnal rhythm is persistent, in the sense that keeping the animals in constant dim illumination

---

* Fingerman & Mobberley found hormones in the eye-stalks and other regions of the blind cave crayfish, *Orconectes pellucidus australis*; these caused pigment concentration and retinal light-adaptation in *Cambarellus shufeldtii* although *Orconectes* itself has no retinal pigment or chromatophores. Again, Sandeen & Costlow (1961) have found maximal pigment dispersal of the black chromatophores of *Uca* by extracts from the barnacle's central nervous system, although the barnacle has no compound eyes and no chromatophores. These hormones may be vestigial or else may subserve different functions in the animals of origin.

† The authors quote studies on other crabs demonstrating control of white chromatophores by two hormones, *e.g.* Powell (1962) in *Carcinus*.

in the laboratory for several weeks does not alter the precise synchronism with the solar changes. Brown & Webb showed that it was possible to set the rhythm out of phase by cooling the crab to around 0° C for several hours; on re-warming, the changes of colour took place every 24 hours, as before, but they were out of phase by the number of hours the crab was kept cold.

Tidal Influence. In investigating the diurnal rhythm more precisely, Brown,

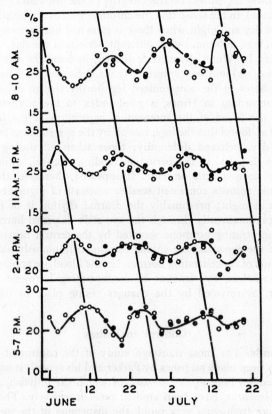

FIG. 838. The relationship between the day of the month and the degree of pigment dispersal shown by *Uca pugilator* at different times of the day. Note the tidal maxima passing over the daily periods at the tidal rate of approximately 50 minutes per day. There is also a 14·8 day cycle, the interval between each of the parallel diagonal lines. Circles and dots represent crabs collected on June 1st and June 15th respectively. Ordinates give the percentage of the daily dispersion taking place during the indicated period. (Fingerman. *Biol. Bull. Wood's Hole.*)

Fingerman, Sandeen & Webb noticed a certain skewness in the curve of diurnal variation of pigment dispersal, suggesting that another rhythm was super-imposing itself, and this turned out to be a tidal rhythm of 12·4 hours' periodicity, low tide favouring dispersal. A tidal rhythm of 12·4 hours' periodicity imposed on a diurnal rhythm, should lead to a semi-lunar rhythm of 14·8 days, in the sense that once every 14·8 days we may expect the maximal tidal effect to coincide with the maximal diurnal effect. This becomes clear in Fig. 838, taken from a more recent study of Fingerman. Crabs collected from beaches with their peaks of high tide differing by 4 hours showed, in the laboratory, tidal rhythms correspondingly out of phase. The natural rhythm could be altered in phase temporarily by exposing the crabs to a bright light during successive nights; in this way the time for maximal pigment dispersal could be shifted from 12 midday to 8 a.m., for example. It is interesting that the tidal rhythm was correspondingly shifted, so that the mechanisms for the two rhythms are somehow related.

Hormonal Control. Brown & Sandeen (1948) showed that the diurnal rhythm of *Uca* was dependent on the integrity of the eye-stalks (which secrete a dispersal hormone), in the sense that the chromatophores remained concentrated throughout the day and night when these organs had been removed. Nevertheless, studies on legs, autotomized at different periods of the day, carried out by Hines, and later by Webb, Bennett & Brown, indicate the influence of both dispersing (eye-stalk) and concentrating (c.n.s.?) hormones. The behaviour of the chromatophores in the autotomized leg during the 30 minutes following autotomy is, according to Hines, a good index to the amount of hormone circulating in the blood at the moment of separation of the limb from the body. Webb *et al.* found that the legs, cast off by the eye-stalkless crab at different times of the day, behaved differently; those taken off during the morning showed a large dispersal in the period succeeding removal, whilst those taken in the evening showed little or none. It appeared, therefore, that in the day the blood of the animals contained smaller amounts of pigment-concentrating hormone than at night; presumably the diurnal rhythm is of nervous origin and operates predominantly through the eye-stalk dispersal hormone, but also through a concentrating hormone secreted by the central nervous system. In the absence of the eye-stalk, the animal remains continuously pale because the smallest amounts of concentrating hormone in the blood are apparently adequate for this; nevertheless, the amount of concentrating hormone secreted does vary diurnally, as revealed by the changes taking place in the autotomized legs.

Control in Fishes

Caudal Bands. The most thorough study of the control of the migration of pigment has been made on fishes by Parker and his school; it will be sufficient to confine attention here to the teleosts, of which the gillyfish, *Fundulus*, and the cat-fish, *Ameiurus*, have been studied most intensively. The responses of teleost fishes are frequently very rapid, the dispersion of the melanophores in *Fundulus* requiring only 45 seconds. Such rapid changes demand a direct nervous action on the effectors. This innervation is best proven by the formation of *caudal bands*; thus, by a localized incision in the tail of *Fundulus*, the nerve supply to a limited region can be severed without interference with the blood supply to the same region. The first effect of denervation is a dark band, appearing within about half a minute and taking five minutes to attain its maximum density. Within six hours it has begun to fade, and after two to three days it is indistinguishable from the rest of the animal. When this has happened, the region is still functionally different from the rest since, on placing the fish on a black background, the caudal band appears bright against the darkened animal. If the animal is maintained for a day or two on the black background, however, the caudal band eventually darkens. Thus the caudal band is distinguished by a very much delayed response to visual stimuli, a fact suggesting that slow-acting humoral mechanisms come into play over long periods. Certainly fishes, like amphibia, respond to pituitary extracts by darkening, and Kleinholz has shown that injection of intermedin into a *Fundulus* with a blanched caudal band causes the latter to darken. However, this does not explain the blanching of the caudal band when the animal is maintained on a white background for some time.

Cholinergic and Adrenergic Control

According to Parker, the production and disappearance of caudal bands may best be explained on the basis of a double innervation of the melanophores, an adrenergic—pigment-concentrating—and a cholinergic—pigment-dispersing —innervation. The effect of an incision is thought not to be due to the cutting off of tonic concentrating impulses but, on the contrary, to an irritative excitation of the cut nerve, the dispersing fibres being more sensitive to the irritation than the concentrating fibres. Evidence for this rather unusual effect of nerve section is the following: if a fish with a caudal band is kept in a white vessel, as we have seen, the dark band fades. On making a new incision, distal to the original cut, a second dark band appears within the region where the first band was. Regeneration of the nerve fibres cannot be invoked to explain this finding, since experiments have shown that this takes many days. It would seem, therefore, that the cut fibres, which had subsided into quiescence, were re-excited by the new cut. Support for this view is given by the effects of cold blocks applied distally to the cut; if one is applied before the second incision, the new band extends only as far as the block; if the second cut has already been made, a cold block causes the new band to fade distally to the block.

Diffusion of Neurohumours

Electrical excitation of the nerves of a fish has long been known to cause a blanching of the region supplied, so that, if there really are dispersing and concentrating fibres, they differ radically in their responses to electrical excitation and cutting; thus the concentrating fibres are activated by electrical stimulation and the dispersing fibres by section. This rather *ad hoc* proposal is not easy to accept, but some studies by Parker & Rosenblueth on the cat-fish have shown that high-frequency stimulation causes concentration of pigment whilst low-frequency stimulation of the same nerve, with pulses of long duration, causes dispersion. That the two antagonistic nervous mechanisms are equivalent to the sympathetic and parasympathetic systems of warm-blooded animals is suggested by the action of adrenaline, which blanches, and of acetylcholine, which darkens, an eserinized cat-fish. The gradual blanching of a dark caudal band on a fish kept on a white background, and the gradual darkening of a pale band of a fish on a dark background, have been explained by Mills and by Parker as being due to the slow diffusion of adrenaline or of acetylcholine, liberated at the nerve endings in the adjacent innervated regions. Thus Mills observed that the fading of a band did not occur evenly, but progressively from without inwards, as though the fading depended on the diffusion of some principle. Parker showed, further, that the time required for fading or darkening of a band depended on the width, as the following table shows:—

TABLE LXXXVII

Times Required for Caudal Bands and Body of Fish to Change Colour Under Different Conditions. (Parker, 1934.)

Loss of Initial stripe . . .	1 mm. stripe	30	hour
	2 mm. ,,	78	,,
Change from Light to Dark .	Body	1·8	,
	1 mm. ,,	20·5	,
	2 mm. ,,	37	.,
Change from Dark to Light	Body	4·6	,,
	1 mm. ,,	26·4	,,
	2 mm. ,,	51·6	,,

On this basis, then, the paling of a caudal band is due to the slow diffusion of adrenaline, liberated in neighbouring regions of the pale fish, whilst the darkening of a pale band is due to diffusion of acetylcholine, liberated by the active dispersing fibres of the dark fish. The diffusion, according to Parker, is trans-cellular, similar to that which must occur in the nutritive processes of coelenterates where the more superficial ectodermal elements receive their nutriment by diffusion of material from the gastrovascular cavity.

Hormonal Control in Teleosts. Teleosts possess, besides a nervous, a humoral control over their melanophores, although the latter is generally subservient to the former. Injection of extracts of mammalian or fish pituitary causes pigment dispersion in the cat-fish, *Ameiurus*, for example; in *Fundulus* it seems doubtful whether pituitary extract has much influence on the innervated melanophores.* In the cat-fish the importance of hormonal control is well brought out by a study of Osborn, who showed that a hypophysectomized fish fails to exhibit the maximal contraction of melanophores obtained in normal animals; in this fish, therefore, both hormonal and nervous factors are involved in dispersion.

Primary Response

The primary response of the skin to light, obtainable in blinded animals, has been briefly mentioned; it is seen in its most characteristic form in lizards, such as the chameleon, the "horned toad", *Phrynosoma*, and *Anolis*. The control of chromatic behaviour of *Anolis* has been well worked out by Kleinholz, who has shown that a direct nervous influence on the melanophores is most probably not present in this species, denervated areas responding to changes in background in exactly the same way as normal areas. The pituitary plays a dominant, if not exclusive, role in chromatic reactions in *Anolis*, a hypophysectomized animal becoming light green and showing no response to background. Not only does the hypophysectomized *Anolis* fail to respond to backgrounds, but it also loses its primary response to light, remaining a pale green in sunlight, whilst a normal, but blinded, animal goes brown under the same conditions. Clearly, the primary response is not a simple effect of light on the chromatophores, but apparently depends on the presence of dermal photoreceptors which reflexly transmit impulses to the pituitary. We may note, however, that Kleinholz' conclusions have been questioned by Parker, as also those of Zoond & Eyre on the chameleon; these workers denervated one side of an eviscerated chameleon by a longitudinal incision beside the spinal cord; they found that the denervated side failed to respond directly to changes in illumination, but remained permanently dark. The control of the melanophores of the chameleon is largely nervous (in contrast to that of *Anolis*), so that Zoond & Eyre postulated a local reflex mechanism, mediated by nervous paths, in this primary response. Hogben has questioned the likelihood of such an accurate reflex response, and is inclined to agree with Parker that the melanophores are, indeed, directly susceptible to light. Certainly the highly localized primary responses cannot be easily interpreted on the basis of a reflex activation of the pituitary, as it is difficult to see how, on this basis, the response could be confined to a definite region; Kleinholz' finding, that hypophysectomized animals fail to show a direct response, is best interpreted on the assumption that the primary response is more effective when there is some intermedin in the blood.

---

* As indicated earlier (p. 1580), when the melanophores are released from nervous control, the extract causes pigment dispersion.

This short review of the main findings concerned with the mechanisms of chromatic responses has revealed an unexpected degree of complexity, even when attention is concentrated only on those chromatophores that contribute most to the colour changes, namely the melanophores; when it is realized that colour changes are rarely effected by only one type of chromatophore, and that the different types may sometimes respond in the same way, and sometimes differently, to the same stimulus, it is not surprising that we are still far from a complete elucidation of the processes involved in any one organism. The value of these studies, interesting as they are as steps towards the elucidation of problems that are fascinating in themselves, extends beyond the immediate phenomena of chromatic behaviour, since their understanding demands an extension of our still limited knowledge of the nervous and humoral mechanisms in cold-blooded vertebrates and invertebrates.

REFERENCES

ABRAMOWITZ, A. A. (1935). "Colour Changes in Cancroid Crabs of Bermuda." *Proc. Nat. Acad. Sci., Wash.*, **21**, 677.

BAGNARA, J. T. (1964). Stimulation of melanophores and guanophores by melanophore-stimulating hormone peptides. *Gen. comp. Endocrin.*, **4**, 290–294.

BAGNARA, J. T. (1966). Cytology and cytophysiology of non-melanophore pigment cells. *Int. Rev. Cytol.*, **20**, 173–205.

BAGNARA, J. T. (1968). Hypophyseal control of guanophores in anuran larvae. *J. exp. Zool.*, **137**, 265–279.

BAGNARA, J. T. & OBIKA, M. (1965). Comparative aspects of integumental pteridine distribution among amphibians. *Comp. Biochem. Physiol.*, **15**, 33–49.

BAGNARA, J. T., TAYLOR, J. D. & HADLEY, M. E. (1968). The dermal chromatophore unit. *J. Cell Biol.*, **38**, 67–79.

BARNWELL, F. H. (1968). Comparative aspects of the chromatophoric responses to light and temperature in fiddler crabs of the genus *Uca*. *Biol. Bull.*, **134**, 221–234.

BIKLE, D., TILNEY, L. G. & PORTER, K. R. (1966). Microtubules and pigment migration in the melanophores of *Fundulus heteroclitus* L. *Protoplasma*, **61**, 322–345.

BIRBECK, M. S. C. (1963). Electron microscopy of melanocytes: the fine structure of hair-bulb premelanosomes. *Ann. N.Y. Acad. Sci.*, **100**, 540–547.

BLISS, D. E., & WELSH, J. H. (1952). The neurosecretory system of brachyuran crustacea. *Biol. Bull.*, **103**, 157–169.

BOZLER, E. (1928; 1931). "Die Chromatophoren der Cephalopoden." *Z. vgl. Physiol.*, 1928, **7**, 379; 1931. **13**, 762.

BROWN, F. A. (1948). Hormones in crustaceans. *The Hormones*. Vol. I. Chap. V. N.Y.: Academic Press.

BROWN, F. A., & EDERSTROM, H. E. (1940). Dual control of certain black chromatophores of *Crago*. *J. exp. Zool.*, **85**, 53–69.

BROWN, F. A., FINGERMAN M., SANDEEN, M. I., & WEBB, H. M. (1953). "Persistent Diurnal and Tidal Rhythms of Colour Change in the Fiddler Crab, *Uca pugnax*." *J. exp. Zool.*, **123**, 29–60.

BROWN, F. A., & SANDEEN, M. I. (1948). "Responses of the Chromatophores of the Fiddler Crab, *Uca*, to Light and Temperature." *Physiol. Zool.*, **21**, 361.

BROWN, F. A., & WEBB, H. M. (1948). "Temperature Relations of an Endogenous Rhythmicity in the Fiddler Crab, *Uca*." *Physiol. Zool.*, **21**, 371.

BROWN, F. A., WEBB, H. M., & SANDEEN, M. I. (1952). "The Action of Two Hormones regulating the Red Chromatophores of *Palæmonetes*." *J. exp. Zool.*, **120**, 391–420.

BUTCHER, E. O. (1938). "Structure of the Retina of *Fundulus heteroclitus* and the Regions of the Retina associated with the different Chromatophoric Responses." *J. exp. Zool.*, **79**, 275.

CRANE, J. (1944). On the color changes in fiddler crabs (genus *Uca*) in the field. *Zoologica*, **29**, 161–168. (Quoted by Barnwell.)

DENTON, E. J. & LAND, M. F. (1967). Optical properties of the lamellae causing interference colours in animal reflectors. *J. Physiol.*, **191**, 23–24 P.

DENTON, E. J. & NICOL, J. A. C. (1964). The choroidal tapeta of some cartilaginous fishes (*Chondrichthyes*). *J. mar. biol. Ass. U.K.*, **44**, 219–258.

DENTON, E. J. & NICOL, J. A. C. (1965). Reflexion of light by external surfaces of the herring, *Clupea harengus*. *J. mar. biol. Ass. U.K.*, **45**, 711–738.

DROCHMANS, P. (1960). Electron microscope studies of epidermal melanocytes, and the fine structure of melanin granules. *J. Cell Biol.*, **8**, 165–180.

ENAMI, M. (1951). The sources and activities of two chromophorotropic hormones in crabs of the genus *Sesarma*. *Biol. Bull.*, **101**, 241–258.

FINGERMAN, M. (1956). "Phase Difference in the Tidal Rhythms of Colour Change of Two Species of Fiddler Crab." *Biol. Bull. Wood's Hole*, **110**, 270–290.

FINGERMAN, M. (1956). "Black Pigment Concentrating Factor in the Fiddler Crab." *Science*, **123**, 585–586.

FINGERMAN, M. (1957). "Physiology of the Red and White Chromatophores of the Dwarf Crayfish *Cambarellus shufeldtii*." *Physiol. Zool.*, **30**, 142–154.

FINGERMAN, M. (1965). Chromatophores. *Physiol. Rev.*, **45**, 296–339.

FINGERMAN, M., CONNELL, P. M. & YOSHIOKA, P. (1967). A possible mechanism for the actions of the red pigment-concentrating and red pigment-dispersing hormones in the prawn, *Palaemonetes vulgaris*. *Biol. Bull.*, **133**, 463–464.

FINGERMAN, M., LOWE, M. E., & SUNDARAJ, B. I. (1959). Dark-adapting and light-adapting hormones controlling the distal retinal pigment of the prawn. *Biol. Bull.*, **116**, 30–36.

FINGERMAN, M., MIYAWAKI, M. & OGURO, C. (1963). Effects of osmotic pressure and cations on the response of the melanophores in the fiddler crab, *Uca pugnax*, to the melanin-dispersing principle from the sinus gland. *Gen. comp. Endocrin.*, **3**, 496–504.

FINGERMAN, M., & MOBBERLY, W. C. (1957). Trophic substances in a blind cave crayfish. *Science*, **132**, 44–45.

FINGERMAN, M., & MOBBERLY, W. C. (1960). Investigation of the hormones controlling the distal retinal pigment of the prawn *Palæmonetes*. *Biol. Bull.*, **118**, 393–406.

FINGERMAN, M., NAGABHUSHANAM, R., & PHILPOTT, L. (1961). Physiology of the melanophores in the crab *Sesarma reticulatum*. *Biol. Bull.*, **120**, 337–347.

FINGERMAN, M., SANDEEN, M. I., & LOWE, M. E. (1959). Experimental analysis of the red chromatophore system of the prawn. *Physiol. Zool.*, **32**, 128–149.

FINGERMAN, M., & TINKLE, D. W. (1956). "Responses of the White Chromatophores of two species of Prawns (*Palæmonetes*) to Light and Temperature." *Biol. Bull. Wood's Hole*, **110**, 144–152.

FREEMAN, A. R., CONNELL, P. M. & FINGERMAN, M. (1968). An electrophysiological study of the red chromatophore of the prawn, *Palaemonetes*: observations on the action of the red pigment-concentrating hormone. *Comp. Biochem. Physiol.*, **26**, 1015–1029.

FRIES, E. F. B. (1958). Iridescent white reflecting chromatophores (antaugophores, leucophores) in certain teleost fishes, particularly *Bathygobius*. *J. Morph.*, **103**, 203–253.

HADLEY, M. E. & QUEVEDO, W. C. (1966). Vertebrate epidermal melanin unit. *Nature*, **209**, 1334–1335.

HAMA, T. (1963). The relation between the chromatophores and pterin compounds. *Ann. N.Y. Acad. Sci.*, **100**, 977–986.

HANSTRÖM, B. (1939). *Hormones in Invertebrates*. O.U.P. Oxford.

HINES, M. N. (1954). "A Tidal Rhythm of Behaviour of Melanophores in Autotomized Legs of *Uca pugnax*." *Biol. Bull. Wood's Hole*, **107**, 386–396.

HOGBEN L. (1942). "Chromatic Behaviour." Croonian Lecture. *Proc. Roy. Soc.*, B, **131**, 111.

HOGBEN, L., & SLOME, D. (1931). "Dual Character of Endocrine Co-ordination in Amphibian Colour Change." *Proc. Roy. Soc.*, B, **108**, 10.

HOGBEN, L., & SLOME, D. (1936). "Dual Receptive Mechanism of the Amphibian Background Response." *Proc. Roy. Soc.*, B, **120**, 158.

HOGBEN, L., & WINTON, F. R. (1922). "The Melanophore Stimulant in Posterior Lobe Extracts." *Biochem. J.*, **16**, 619.

KINOSITA, H. (1963). Electrophoretic theory of pigment migration within fish melanophore. *Ann. N.Y. Acad. Sci.*, **100**, 992–1003.

KLEINHOLZ, L. H. (1935). "The Melanophore Dispersing Principle in the Hypophysis of *Fundulus heteroclitus*." *Biol. Bull.*, **69**, 379.

KLEINHOLZ, L. H. (1938). "Pituitary and Adrenal Glands in the Regulation of the Melanophores of *Anolis corolinensis*." *J. exp. Biol.*, **15**, 474.

KLEINHOLZ, L. H. (1938). "Control of the Light Phase and Behaviour of Isolated Skin." *J. exp. Biol.*, **15**, 492.

KLEINHOLZ, L. H., BURGESS, P. R., CARLISLE, D. B., & PFLUEGER, O. (1962). Neuro-secretion and crustacean retinal pigment hormone: distribution of the light-adapting hormone. *Biol. Bull.*, **122**, 73–85.

KLEINHOLZ, L. H., ESPER, H., & KIMBALL, F. (1962). Neurosecretion and crustacean pigment hormone. *Biol. Bull.*, **123**, 317–329.

KNOWLES, F. G. W. (1950). The control of retinal pigment migration in *Leander serratus*. *Biol. Bull.*, **98**, 66–80.

KNOWLES, F. G. W. (1951). Hormone production within the nervous system of a crusta-cean. *Nature*, **167**, 564–565.

KNOWLES, F. G. W., & CARLISLE, D. B. (1956). Endocrine control in the crustacea. *Biol. Rev.*, **31**, 396–473.

KNOWLES, F. G. W., CARLISLE, D. B., & DUPONT-RAABE, M. (1955). The separation by paper electrophoresis of chromactivating substances in arthropods. *J. Mar. Biol. Ass.*, **34**, 611–635.

KOLLER, G. (1927). "Uber Chromatophorensystem, Farbensinn und Farbwechsel bei *Crangon vulgaris*." *Z. vgl. Physiol.*, **5**, 191–246.

LAND, M. F. (1966). A multilayer interference reflector in the eye of the scallop, *Pecten maximus*. *J. exp. Biol.*, **45**, 433–447.

LEE, T. H., LERNER, A. B. & BUETTNER-JANUSCH, V. (1963). Species differences and struc-tural requirements for melanocyte-stimulating activity of melanocyte-stimulating hormones. *Ann. N.Y. Acad. Sci.*, **100**, 658–668.

LERNER, A. B. & MCGUIRE, J. S. (1961). Effect of alpha- and beta-melanocyte stimulating hormones on the skin colour of man. *Nature*, **189**, 176–179.

MARSLAND, D. A. (1944). "Mechanism of Pigment Displacement in Unicellular Chromatophores." *Biol. Bull. Wood's Hole*, **87**, 252–261.

MARSLAND, D. & MEISNER, D. (1967). Effects of D₂O on the mechanism of pigment dispersal in the melanocytes of *Fundulus heteroclitus*: a pressure-temperature analysis. *J. cell. Physiol.*, **70**, 209–215.

MARTIN, A. R. & SNELL, R. S. (1968). A note on transmembrane potential in dermal melanophores of the frog and movement of melanin granules. *J. Physiol.*, **195**, 755–759.

MARTINI, E., & ACHUNDOW, I. (1929). "Versuche über Farbenanpassung bei Culiciden." *Zool. Anz.*, **81**, 25.

MATSUMOTO, J. (1965). Studies on fine structure and cytochemical properties of erythro-phores in swordtail, *Xiphophorus helleri*. *J. Cell Biol.*, **27**, 493–504.

MATTHEWS, S. A. (1931). "Pigment Migration within the Fish Melanophore." *J. exp. Zool.*, **58**, 471.

MILLS, S. M. (1932). "Evidence for a Neurohumoral Control of Fish Melanophores." *J. exp. Zool.*, **64**, 245.

NOVALES, R. R. (1959). The effects of osmotic pressure and sodium concentration on the responses of melanophores to intermedin. *Physiol. Zool.*, **32**, 15–28.

NOVALES, R. R. (1962). The role of ionic factors in hormone action on the vertebrate melanophore. *Amer. Zoöl.*, **2**, 337–352.

NOVALES, R. R. & NOVALES, B. J. (1966). Factors influencing the response of isolated dog-fish skin melanophores to melanocyte-stimulating hormone. *Biol. Bull.*, **131**, 470–478.

NOVALES, R. R., NOVALES, B. J., ZINNER, S. H. & STONER, J. A. (1962). The effects of sodium, chloride, and calcium concentration on the response of melanophores to melanocyte-stimulating hormone (MSH). *Gen. comp. Endocrin.*, **2**, 286–295.

OBIKA, M. (1963). Association of pteridines with amphibian larval pigmentation and their biosynthesis in developing chromatophores. *Develop. Biol.*, **6**, 99–112.

OBIKA, M. & BAGNARA, J. T. (1964). Pteridines as pigments in amphibians. *Science*, **143**, 485–487.

ORTIZ, E., BÄCHLI, E., PRICE, D. & WILLIAMS-ASHMAN, H. G. (1963). Red pteridine pigments in the dewlaps of some anoles. *Physiol. Zoöl.*, **36**, 97–103.

ORTIZ, E. & WILLIAMS-ASHMAN, H. G. (1963). Identification of skin pteridines in the pasture lizard *Anolis pulchellus*. *Comp. Biochem. Physiol.*, **10**, 181–190.

OSBORN, C. M. (1938). "The Role of the Melanophore Dispersing Principle of the Pituitary in the Colour Change of the Catfish." *Biol. Bull.*, **79**, 309.

OSBORN, C. M. (1939). "The Physiology of Colour Changes in Flatfishes." *J. exp. Zool.* **81**, 479.

PARKER, G. H. (1934). "Cellular Transfer of Substances, especially Neurohumours." *J. exp. Biol.*, **11**, 81.

PARKER, G. H. (1948). *Animal Colour Changes and their Neurohumours.* C.U.P., Cambridge.

PARKER, G. H. (1955). "Background Adaptations." *Quart. Rev. Biol.,* **30,** 105–115.

PARKER, G. H., & ROSENBLUETH, A. (1941). "Electric Stimulation of the Concentrating and Dispersing Nerve Fibres of the Melanophores of the Catfish." *Proc. Nat. Acad. Sci., Wash.,* **27,** 198.

PASSANO, L. M. (1951). The X organ-sinus gland neurosecretory system in crabs. *Anat. Rec.,* **111,** 502.

PASSANO, L. M. (1951). The X organ, a neurosecretory gland controlling moulting in crabs. *Anat. Rec.,* **111,** 559.

PAUTSCH, F. (1953). "The Colour Changes of the Zoea of the Shrimp *Crangon crangon* L." *Experientia,* **9,** 274–276.

PÉREZ-GONZÁLEZ, M. D. (1957). Evidence for hormone-containing granules in sinus glands of the fiddler crab *Uca pugilator. Biol. Bull.,* **113,** 426–441.

PERKINS, E. B. (1928). "Colour Changes in Crustaceans, especially in *Palæmonetes.*" *J. exp. Zool.,* **50,** 71.

PERKINS, E. B., & SNOOK, T. (1932). "Movement of Pigment within the Chromatophores of *Palæmonetes.*" *J. exp. Zool.,* **61,** 115.

PICKERING, B. T. & HAO LI, C. (1962). Sone aspects of the relationship between chemical structure and melanocyte-stimulating properties of several peptides related to adrenocorticotropin. *Biochim. biophys. Acta,* **62,** 475–482.

POWELL, B. L. (1962). Types distribution and rhythmical behaviour of the chromatophores of juvenile *Carcinus maenas* (L). *J. Anim. Ecol.,* **31,** 251–261. (Quoted by Rao *et al.*). 1967.

RAO, K. R., FINGERMAN, M. & BARTELL, C. K. (1967). Physiology of the white chromatophores in the fiddler crab, *Uca pugilator. Biol. Bull.,* **133,** 606–617.

SANDEEN, M. I., & BROWN, F. A. (1952). Responses of the distal retinal pigment of *Palæmonetes* to illumination. *Physiol. Zool.,* **25,** 222–230.

SANDEEN, M. I., & COSTLOW, J. D. (1961). The presence of decapod-pigment activating substances in the central nervous system of representative cirripedia. *Biol. Bull.,* **120,** 192–205.

SCHMIDT, W. J. (1920). "U. d. Verhalten der verschiedenartigen Chromatophoren beim Farbwechsel des Laubfrosches." *Arch. Mikr. Anat.,* **93,** 414.

SEIJI, M., FITZPATRICK, T. B. & BIRBECK, M. S. C. (1961). The melanosome: a distinctive subcellular particle of mammalian melanocytes and the site of melanogenesis. *J. invest. Derm.,* **36,** 243–252.

SETOGUTI, T. (1967). Ultrastructure of guanophores. *J. Ultrastr. Res.,* **18,** 324–332.

SHEARER, A. C. I. (1969). Morphology of the isolated pigment particle of the eye by scanning electron microscopy. *Exp. Eye Res.,* **8,** 134–138.

SHIBLEY, G. A. (1968). Eyestalk function in chromatophore control in a crab, *Cancer magister. Physiol. Zool.,* **41,** 268–279.

SMITH, H. G. (1938). "Receptor Mechanism of the Background Response in Chromatic Behaviour of Crustacea." *Proc. Roy. Soc., B,* **125,** 249.

SNELL, R. S. & KULOVICH, S. (1967). Nerve stimulation and the movement of melanin granules in the pigment cells of the frog's web. *J. invest. Derm.,* **48,** 438–443.

STOPPANI, A. O. M. (1942). "Neuroendocrine Mechanism of Colour Change in *Bufo arenarum.*" *Endocrinology,* **30,** 782.

VILTER, V. Quoted by Parker (1948).

WEBB, H. M., BENNETT, M. F., & BROWN, F. A. (1954). "A Persistent Rhythm of Chromatophoric Response in Eyestalkless *Uca pugilator.*" *Biol. Bull. Wood's Hole,* **106,** 371–377.

WILKENS, J. L. & FINGERMAN, M. (1965). Heat tolerance and temperature relationships of the fiddler crab, *Uca pugilator,* with reference to body coloration. *Biol. Bull.,* **128,** 133–141.

YOSHIDA, M. (1956). On the light response of the chromatophore of the sea-urchin *Diadema setosum* (leske), *J. exp. Biol.,* **33,** 119–123.

ZIMMERMAN, S. B., & DALTON, H. C. (1961). Physiological responses of amphibian melanophores. *Physiol. Zool.,* **34,** 21–33.

ZOOND, A., & EYRE, J. (1934). "Studies on Reptilian Colour Response." *Phil. Trans.,* **223,** 27.

CHAPTER XXV

BIOLUMINESCENCE

IN photosynthesis, absorbed light drives a reaction which, in the green plants at least, would not proceed in the absence of this supply of energy, first, because the activation energy of at least one step is too high to permit the reaction to proceed at a measurable speed at ordinary temperatures, and secondly, because the reaction is strongly endothermic. Just as an electric motor, driven in reverse, can produce electricity, so we may expect that certain exothermic chemical reactions can give up their energy of reaction as light. The emission of light during a chemical reaction is, of course, a familiar phenomenon, and represents the return of excited atoms and molecules to their ground states of energy; but most frequently such emission of light is associated with a high temperature of the reactants, for example, the light of a flame from coal gas. In other words, the emission of electronic energy is associated with the liberation of large amounts of vibrational, rotational and kinetic energy. Such luminous reactions take place in the gaseous phase, since the activated molecules must emit their energy before they lose it by inelastic collisions; and, in solution, this takes place very rapidly. If, however, an activated molecule can keep its energy for sufficient time without losing it by inelastic collisions, it may give it up as light; and if the great majority of the activated molecules possess this immunity from loss by collision, the liberation of energy need not be accompanied by any great production of heat, and, moreover, can take place in solution. We have seen that molecules that can retain their activated state for appreciable periods (of the order of 10^{-8} sec.) are fluorescent; we may therefore expect to find this liberation of "cold light" by molecules that react to form fluorescent compounds.

Chemiluminescence

The existence of chemiluminescence has been recognized for a long time; thus the "phosphorescence" of phosphorus* is due to its slow oxidation; and a cold flame below room temperature can be obtained by boiling a solution of phosphorus in chloroform under reduced pressure and mixing the vapour with air. A very large number of organic compounds luminesce on slow oxidation in solution; we need only mention here the Grignard-reagents which, on oxidation in ethereal solution, may be brightly luminescent; luminol, aminophthalylhydrazide, on oxidation gives a brilliant luminescence, visible light being detectable at dilutions of the reagent as low as one part in a hundred-million, whilst the quaternary salts of dimethyldiacridylium luminesce in a dilution a

---

* Phosphorescence, correctly defined, is a delayed fluorescence, the light emitted persisting for some time after the exciting radiation has been cut off (a matter of seconds or more). Many fluorescent compounds can be made to phosphoresce by taking steps to reduce the de-activating collisions, e.g., by dissolving in glycerol and cooling to form a rigid glass. A study of this type of phosphorescence exhibits the presence of two bands; one, characteristic of the normal fluorescence (the α-phosphorescence) and another, of longer wavelength, which is apparently due to the activated molecule having passed to a new activated state of lower energy (the P or phosphorescent state); on returning to its ground level it emits the longer-wavelength β-phosphorescence; alternatively, if it receives activation energy by collision, it may return to its F, or fluorescent, state before going to its ground level, in which case the phosphorescence has the shorter wavelength.

H
|
Si
/ \
HSi—O   SiH
| \
O
/
HSi—O   SiH
\ |
Si
|
H

Siloxene

hundred times greater than this (1 in 10^{10}). Kautsky showed that siloxene can be made to chemiluminesce with permanganate in solution; moreover, on addition of a fluorescent dye-stuff to the medium, *e.g.*, rhodamine, the spectral characteristics of the emitted light were identical with those of the fluorescence-spectrum of the dye; the emission of light by the dye, however, was not a simple fluorescence, resulting from the absorption of light previously emitted by siloxene, but was due to a direct activation of the dye-stuff molecules by the siloxene molecules.* Thus, the chemiluminescence of siloxene alone is equivalent to a reversed photochemical reaction whilst the *induced* chemiluminescence of the rhodamine is equivalent to a reversed photosensitized reaction.

Luminescent Organisms

The emission of light by living organisms has excited the interest of investigators since the time of Aristotle; the "phosphorescence" of the sea, the light of the glow-worm, and the flash of the firefly are familiar terms but, to the untravelled town-dweller in England, as remote from his experience as the aurora borealis or tropical storms. Luminescence, however, is a widespread phenomenon, quite apart from these three instances; it may be observed on a rotting tree-trunk in many woods in consequence of the growth of luminous fungi; on recently dead fishes on the beach; on meat,† and on many aquatic animals, of which the jelly-fish may be taken as an example. That O_2 was necessary for luminescence was shown by Boyle, as long ago as 1667, in his experiments with the vacuum, but the association of luminosity with microscopical organisms is largely the result of the work of Pflüger, who actually cultured luminous bacteria scraped from dead fish. Altogether, luminescence in some form or other has been described in species belonging to some 40 orders, so that a comprehensive description of the living phenomena is out of the question here; instead, we may consider some typical examples: the bacteria, a protozoon, a medusa, the much studied ostracod *Cypridina*, the firefly, and several types of fish.

Bacteria. The luminescence of many living marine forms is due to the presence of luminous bacteria, either on their surface or in their blood and tissues‡; moreover, the luminescence of dead fish, so commonly observed, is due to saprophytic organisms. The luminous bacteria do not form a definite taxonomic group of their own, but may be found under such diverse classifications as micrococcus, bacillus, pseudomonas, microspira, etc. Most are more or less elongated, exhibiting rapid motion in virtue of their flagella; they are most commonly found on dead fish, cast on the beach, in which decay has begun a short time before; in the laboratory many types can be cultured in 1·5 per cent. agar jelly in sea-water containing 2 per cent. peptone, 1 per cent. glycerol and powdered $CaCO_3$, the last being necessary to maintain the pH constant, otherwise the acids formed by metabolism eventually inhibit luminescence. Grown under these conditions, and with a suitable supply of

* Or rather by the activated oxidation product of siloxene.

† Dahlgren refers to reports of glowing sausages hanging up in butchers' shops; the meat had presumably been preserved with sea-salt containing luminous bacteria.

‡ Thus the amphipod crustacean, *Talitrus*, glows when it is infected with luminous bacteria; the infection can be passed from one individual to another, and apparently from one species of crustacean to another. On culturing in an artificial medium the organisms multiply, but do not glow; on transferring the cultured bacteria to a *Talitrus*, however, they at once glow.

O_2, the emission of light is continuous, the chemical reaction, leading to this, doubtless taking place within the organisms since filtrates from actively luminescent bacteria fail to luminesce. It is of interest to record that Beijerinck, towards the end of the last century, exploited the O_2 requirements of luminous bacteria to establish the liberation of this gas by illuminated chloroplasts; thus he mixed luminous bacteria with an emulsion of clover chloroplasts in the dark; when all the O_2 had been used up, the bacteria were extinguished, but immediately started to luminesce again after exposure of the system to light.*

Dinoflagellates. The "phosphorescence" of the sea is not ordinarily due to bacteria but to protozoa, of which the dinoflagellate, *Noctiluca miliaris*, is the most widely distributed. Although unicellular, it is about 0·5–1 mm. in diameter; Fig. 839 illustrates a specimen with a quadrant removed to reveal the interior, the bulk of which is occupied by the flotation vacuole. *Noctiluca* does not emit light unless stimulated, *e.g.*, by disturbing the sea-water; and, as with the bacteria, the luminiscence is intracellular in that filtrates show no luminescence.

According to the classical microscopical study of Pratje, the cytoplasm contains numerous droplets which, being soluble in alcohol and staining with Sudan III, are probably lipoid. Quatrefages, in 1850, observed that, under the microscope, the glowing of the organism was generally confined to certain regions; on increasing the magnification of the microscope he observed that the glow was not uniform but consisted of a large number of discrete bright points; if the glow or "blush" moved over the surface, this was due to the lighting up of new points and the extinguishing of others in a progressive fashion.

Nervous Control of *Noctiluca* Flash

This aspect of the luminiscence has been studied by Eckert and his colleagues, employing ingenious devices for recording light-emission and membrane electrical responses. The light-emission derives from the surface perivacuolar cytoplasm; this contains numerous strands and, between these, numerous small bodies, some $1·5\,\mu$ to less than $0·2\,\mu$ in diameter, revealed in the polarizing microscope by their strong phase-retardation or in the fluorescence microscope by their strong fluorescence in ultraviolet light.† On stimulating the organism to emit light, these bodies may be seen to brighten; and there is little doubt that they are the microsources of light that constitute the basis of the flash (Fig. 840, Pl. LXXVI); with an image-intensifying device, the flash can be amplified sufficiently to permit photography; the total number of emissions on a 500μ cell was $1·6–3·6.10^4$; since the macroflash contained $2·5.10^9$ photons, the number per microflash was of the order of 10^5.

ELECTRICAL STIMULATION

The organism could be stimulated electrically in one of two ways: locally, by passing current through a micropipette attached to the surface, or by passing current through an internal electrode in the vacuole. In the former instance the evoked action potential was propagated at some $60\,\mu$/sec. over the surface of the cell; accompanying the spike the microwaves were excited, so that the macroflash does, in fact, spread over the surface, although propagation is sufficiently rapid for the responses of the microflashes to be very nearly simultaneous. As

---

* Harvey used the same test to demonstrate the *absence* of evolution of O_2 by illuminated photosynthetic sulphur bacteria. The minimum tension of O_2 for luminescence of bacteria is of the order of 0·005 mm. Hg.

† The microsources of Eckert are probably analogous with the "scintillons" of De Sa, Hastings & Vatter (1963) but *Noctiluca* does not contain the guanine crystals that are a component of the scintillon (p. 1610).

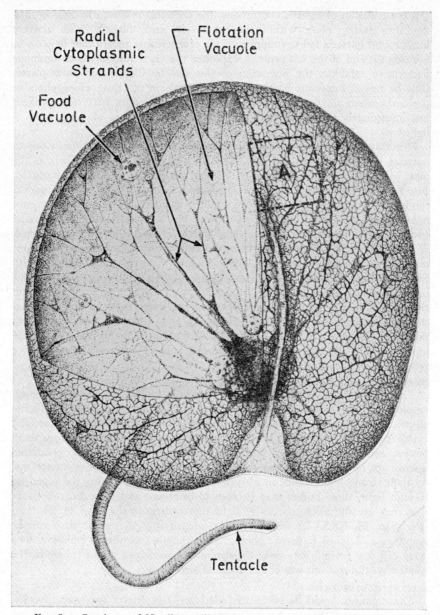

Radial Cytoplasmic Strands

Flotation Vacuole

Food Vacuole

Tentacle

FIG. 839. Specimen of *Noctiluca miliaris* drawn with a hemiquadrant removed to expose internal features. The flotation vacuole occupies the bulk of the cellular volume. (Eckert & Reynolds. *J. gen. Physiol.*)

Fig. 841 shows, the time-course of a microflash is very similar to that of the macroflash of the whole organism. With the internal electrode, drawing current through the whole surface of the cell, the action potential was synchronous over the whole cell, and the flash was correspondingly synchronous.

DIRECTION OF STIMULATING CURRENT

Eckert & Sibaoka (1968) drew attention to the "unorthodox" character of the electrical stimulus, namely the passage of positive current from vacuole to the

PLATE LXXVI

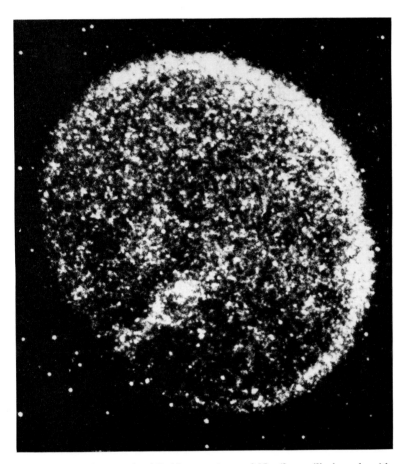

FIG. 840. Autophotograph of flashing specimen of *Noctiluca miliaris* made with the aid of an image-intensifier. White spots in the field are artefacts inherent in the image-intensifier. (Courtesy R. Eckert.)

[*To face p.* 1590.

PLATE LXXVII

FIG. 842. *Pelagia noctiluca*. (Dahlgren. *J. Franklin Inst.*)

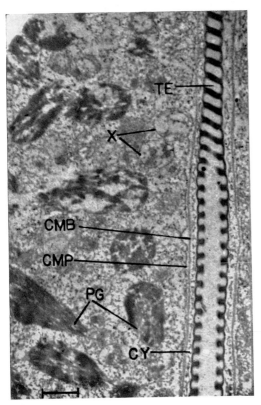

FIG. 845. Electron-micrograph of section through part of firefly (*Photinus pyralis*) light organ showing tracheole lying between photogenic cells. TE = tracheole; CMB = end bulb membrane; CMP = photogenic cell membrane; PG = photogenic granules; CY = end bulb cytoplasm investing tracheole. Osmium fixation. × 10,000. (Beams & Anderson. *Biol. Bull. Wood's Hole.*)

FIG. 846. Photograph of *Porichthys notatus* to show relative size and distribution of the photophores visible from the ventral view. (Greene & Greene. *Amer. J. Physiol.*)

outside of the cell, so that the action potential appears as a hyperpolarizing response. Nevertheless, because propagation is associated with decreased impedance similar to that described by Cole & Curtis (p. 1043), there is good reason to believe that the basis of the spike is similar to that in nerve, and it would appear that the active membrane is the one separating the flotation vacuole from the surrounding thin layer of protoplasm, rather than the plasma-lemma, as in *Valonia*. At rest, the cytoplasm is negative with respect to the vacuole,* the resting potential being some -20 to -40 mV, and during the spike this reverses sign so that the active phase of the spike depends on a flow of current through the vacuolar membrane from vacuole into cytoplasm. An imposed current from bath to vacuole causes a potential drop across each of the series of membranes in its path, and so inward positive current from bath to vacuole through pellicle, cytoplasm and vacuolar membrane increases cyto-plasmic positivity in relation to the vacuole. Since the cytoplasmic resting

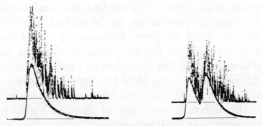

FIG. 841. Wave-shape of the microflash (upper wavy trace of each record) compared with that of the macroflash. Record at right demonstrates the ability of the microsource to flash in rapid sequence. (Eckert & Reynolds. *J. gen. Physiol.*)

potential is negative in respect to both bath and vacuole, the reaction is to depolarize the vacuolar membrane.

MECHANICAL STIMULUS

Eckert (1966) used a fine stylus to stimulate *Noctiluca* locally, and obtained graded negative-going potentials, recorded from the vacuole; these increased with intensity of mechanical stimulus until they fired off a spike; the latency between local current flow and the flash from the same region was some 2–3 msec.; this time is sufficiently long for diffusion of some hypothetical substance from the surface of the cell into the microsources. It is interesting that the duration of the microflash is not controlled by the availability of substrates since it is possible to obtain temporal summation of responses, a second action potential giving a bigger response than the first; we may postulate that the microsource contains the main substrates for the chemiluminescent reaction; activation could be due to the diffusion into it of some co-substrate that is consumed during the flash.

DIURNAL RHYTHM

Hastings & Sweeney (1958) have described a diurnal rhythm in the lumines-cent response of the dinoflagellate, *Gonyaulax polyedra*; when cultured under conditions of alternative light and dark 12-hour periods, the luminiscence was some 40 to 60 times greater during the dark period. The rhythm is inherent, since the organisms may be cultured for a year in strong light and yet, on transfer to dim light, exhibit the same diurnal rhythm. Nevertheless Sweeney, Haxo &

* The vacuole is usually negative with respect to the bath but less negative than the cytoplasm (Eckert & Siboaka, 1968).

Hastings (1959) have shown that an important feature in the rhythm is the photosensitizing action of light during the day that enhances the luminescence during the following night; presumably photosynthesis builds up the supply of metabolic material required for luminescence. Hastings & Bode (1962) found variations in biochemical activity of extracts according as they were taken during day or night, but the difference was only fourfold.

Coelenterates. The most typical luminous coelenterate is the jelly-fish, *Pelagia noctiluca*, common on the shores of the Mediterranean (Fig. 842, Pl. LXXVII). It has the typical jelly-fish form and swims by a series of rhythmic contractions of its umbrella. For the emission of light, a definite stimulus is necessary; for example, a slight touch with a glass rod produces at first a local glow which spreads by the same nervous influences as those that are responsible for the rhythmic contractions. Increasing the strength of the stimulus increases the area of the luminous region; and, with a sufficiently strong excitation, the whole animal glows brightly. It was concluded that in this, and other hydromedusæ, in contrast to the forms so far studied, the luminescence was, indeed, caused by the secretion of material on to the surface, because touching with the finger caused the latter to glow. However, it seems more likely from Davenport & Nicol's studies on the hydromedusa *Aequorea* that the luminescence is intracellular, and that the exudation of luminous material is the result of trauma, the damaged cells losing their photogenic material.

As with all coelenterates, the epithelial layer of the ex-umbrella consists of a variety of cells with different functions; of these, a highly granular type is considered to be responsible for luminescence.

In *Aequoria*, Davenport & Nicol observed localized accumulations—*oval masses*—of what appeared to be photogenic cells, which were analogous with the light-organ of the firefly (p. 1593). The cells were immediately above the endoderm in the mesogloea, and were not organized round ducts, so that the secretion of photogenic material on to the outside of the organism seemed most unlikely, *i.e.*, luminiscence was very probably intracellular.

Nervous Control

Pelagia exhibits a diurnal rhythm in its sensitivity to stimuli, responding in the early evening but not during the day; it would appear, therefore, that light acts as an inhibiting stimulus. The studies of Heymans & Moore on the effects of varying the salt content of the medium bring out the importance of the nervous mechanism in the control of luminescence; thus, the omission of Ca^{++} or K^+ from the sea-water confined the effects of a stimulus to a local glow, whilst the omission of Mg^{++} caused hyperirritability both in respect to contraction of the umbrella and to luminescence, flashes appearing spontaneously and the smallest stimulus causing the whole body to break into light. In *Aequoria*, Davenport & Nicol found the response to an electrical stimulus to be very highly localized, so that spread of excitation by the nervous net, which apparently occurs in the luminescent pennatulids (for example, the sea pansy; Nicol, 1955) seems not to take place in this medusa.

Crustaceans. Of the crustaceans, the ostracod *Cypridina* may be described since, owing to its abundance, its luminous secretion has been very thoroughly studied from the biochemical viewpoint. It is only one-eighth of an inch long, covered by a hinged chitinous shell which almost hides its swimming legs; the luminous organ may be likened to a gland which opens on to the surface of the body near the mouth by way of several ducts. According to Okada the gland of *Cypridina hilgendorfii* contains four types of cells, each group opening into a different duct. One type, common to all ostracods, contains large yellow granules some 10 μ in diameter; there are two other types of granular cell—the granules being very much smaller—and mucous cells. The gland or organ is invested with muscle fibres which, on contraction, apparently extrude the cellular secretions. According to Dahlgren, the luminous organ of the related ostracod, *Pyrocypris*, consists of an invaginated reservoir, opening through the upper lip by several ducts, whilst the gland proper, made up of two types of cell, secretes into this reservoir, the muscle expelling the secretion out of the reservoir along the ducts. Luminescence takes place when the secretion is mixed with sea-water.

The Firefly. The firefly is an example of an organism containing a definite *light-organ* which may be activated by nervous influences to emit a bright flash of light, the luminescence being intracellular. In the male, the organ occupies the whole ventral surface of the 6th and 7th segments, whilst in the female only about two-thirds of these segments are occupied (Fig. 843). The

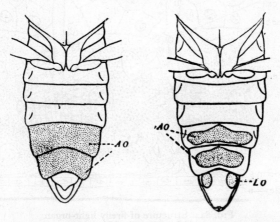

Fig. 843. Illustrating positions of light-organs in the firefly. Left: Adult male, ventral view of abdomen. *AO*, adult light-organ, located on the sixth and seventh abdominal segments. Right: Adult female taken immediately after emergence, ventral view of abdomen. *AO*, adult light-organ; *LO*, larval light-organ. (Hess. *J. Morph.*)

structure of the organ of *Photuris** is shown in Fig. 844; it consists of two layers, reflecting and photogenic. The cells of the former contain crystals of some urate which give the layer a reflecting property, whilst the latter is made up of large cells filled with yellow granules. An adequate supply of O_2 is ensured by the numerous tracheæ which penetrate both layers and branch profusely in the photogenic layer; each branch ends in a *tracheal end-cell* which gives rise to tracheal capillaries, or *tracheoles*, which were considered to enter the photogenic cells. However, Beams & Anderson, in their electron-microscopical study, were never able to see a tracheole enter the cell. The end-cell is peculiar to the tracheal system of the light-organ, and is remarkable for the large number of mitochondria it contains. The tracheolar twig is surrounded by the end-cell and, more distally, by a rounded body that has been called the *end-bulb* by Beams & Anderson (the *rounded body* of Dahlgren). Before emerging from this, it breaks into two tracheoles which pass between the photogenic cells, the membrane and cytoplasm of the end-bulb investing the outer layer of the tracheole (Fig. 845, Pl. LXXVII). The outer cytoplasmic zones of the photo-

* Buck (1948) has reviewed compendiously the anatomy of the light organ, which is a highly variable structure. He recognizes, on a histological basis, six different types.

genic cells are differentiated from the inner regions by being free of photogenic granules and containing numerous small mitochondria. The photogenic granules have a laminar basis (Fig. 845, Pl. LXXVII).

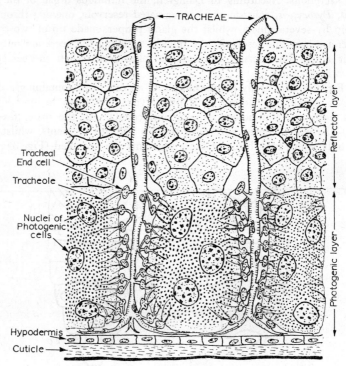

FIG. 844. Structure of firefly light-organ.
(After Hess. *J. Morph.*)

Control of Emission

It was thought that the control of the emission of light was exercised by regulation of the supply of O_2 to the photogenic cells through the tracheoles, the end-body acting as a muscular valve; however, the electron-microscope study of Beams & Anderson has shown that the structures that had been considered to be muscular elements were really mitochondria, and this simple view of the function of the tracheal end-cell has been abandoned. There is no doubt, however, that the end-cell is concerned in some way with the ability of the fly to emit a discrete flash, since Buck has shown that there is a good correlation between the ability to emit a flash and the presence of this cell, when different species are compared; it is not necessary for control *per se*, however, since glow-worms do exert control over their emission but have no end-cells.

Fishes. The Californian singing fish,* *Porichthys notatus*, is an example of a luminous vertebrate, its ventral surface containing some 840 small organs, *photophores*, which emit light to produce a definite pattern on the body (Fig. 846, Pl. LXXVII). The luminescence takes place within the photophores, a section through one being illustrated in Fig. 847. The epidermis of the fish has no scales and is well supplied with mucous cells; the dermis is a thick vascularized layer of connective tissue in which the organ is embedded. From within outwards

---

* This animal makes a noise by compressing its air-sac, the diaphragm separating its two compartments being set in vibration by the unequal pressures generated.

it consists of a *reflector*, a *gland*, and a *lens*. The reflector is made up of connective tissue, the matrix of which is modified into fine spicules which reflect
light strongly; blood vessels pass through on their way to the gland which is a
shallow cup of granular cells; the lens is avascular. Thus the photophore
exhibits the main features of an eye, but its function is to emit, instead of to
receive, light. Histological study failed to reveal any definite nerve supply to
the organ, and Greene concluded that it was controlled by a humoral mechanism.
Mechanical stimulation does not usually cause luminescence; with powerful
electrical stimulation, on the other hand, all the photophores light up and,
if the fish is held some 10 to 12 inches from the face in complete darkness,
the features of the person holding it may be recognized. Since, with a 2 sec.
tetanic stimulus, the latent period is about 8–10 sec., it is possible that the nervous

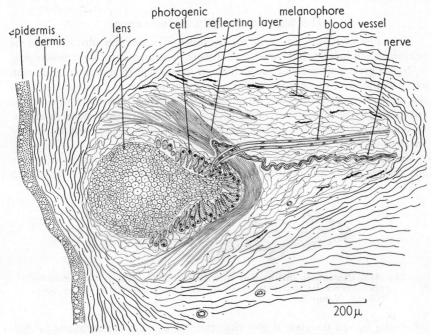

Fig. 847. Light-organ of *Porichthys myriaster*. (Courtesy J. A. C. Nicol.)

discharge liberates a hormone which subsequently acts on the photophores;
Greene found that a subcutaneous injection of adrenaline was followed by
luminescence, the number of activated photophores increasing progressively
until, in about 10 minutes, the whole animal was alight and remained thus for
longer than an hour. The view that the control was purely humoral stems from
Greene's failure to identify a nervous supply; however, Nicol (1957) has identified
nerve fibres supplying the organ and his experiments on the effects of cord
section would suggest that both humoral and nervous (adrenergic) mechanisms
are operative.*

* The photophore system of the teleost *Maurolicus* has been described in some detail
by Bassot (1960); the cells responsible for luminescence constitute a far greater proportion
of the bulk of the photophore than in *Porichthys*, and are grouped in what Bassot calls
the *projector* and, dorsal to this, *reservoir*; he considers that it is the most anterior cells
that emit light, but the exact mode by which the other cells contribute to the process is
by no means clear.

Subocular Organs

A perfect example of symbiosis is provided by the East Indian fishes *Photoblepharon palpebratus* and *Anomalops katoptron*; these have a large white organ just under each eye which may be concealed at will either by drawing a fold of black tissue over it, in the case of *Photoblepharon*, or by turning the whole organ over and downwards into a groove so that the white surface is not exposed. The organ consists of rows of tubes, well supplied with capillaries; the tubes are filled with a mass of bacteria which luminesce continuously so long as they are adequately supplied with O_2; these organisms may be grown in an artificial medium but fail to luminesce under these conditions. Unlike the majority of luminous fishes, *Photoblepharon* and *Anomalops* frequent shallow water, consequently their luminous organs are of value only at night and are presumably of use in recognition.*

The Significance of Luminescence

The value of luminescence to the organism is sometimes obvious from its behaviour and habitat, but sometimes obscure. The amount of light penetrating to depths greater than about 300 fathoms is insufficient to stimulate the human eye; nevertheless, a great many deep-sea fishes have well-developed eyes, so that it is very probable that luminescence plays an important role in the recognition of the sexes, in attracting prey, or blinding an enemy, at these depths. In nocturnal animals and insects the same functions are apparent in shallow water or on land.

Mesopelagic Organisms. In the mesopelagic zone of the sea, some 500 to 800 metres deep, bioluminescence is very frequent amongst the organisms

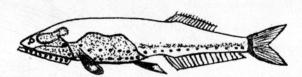

FIG. 848. Photophore distribution on a species of *Cyclothone*. Note that the photophores lie under the heavily pigmented structures in the body. (Clarke. *Nature*.)

inhabiting it, and the zone is also characterized by diurnal migrations to shallower waters, during the night, and deeper in the day. Clarke (1963) suggested that the convergent evolution that led to the development of luminescence in such widely separated groups as the squids, crustacea and fish could have been due to a common environmental factor, namely the requirement of countershading to avoid recognition by predators from below. He pointed out that light remains directional down to some 1,000 metres, so that a fish could avoid recognition from below if it emitted light from its ventral surface of sufficient intensity and of suitable wavelength to match the light falling from above. The aphausid shrimp varies from being translucent to nearly transparent, and the photophores are concentrated under regions of least transparence, namely the cephalothorax and gut. Again, the hatchet fish has its photophores tightly packed along its ventral "keel", whilst the lateral photophores shine down silvery channels along the lower sides of the body. As Fig. 848 shows, in

* The subocular organs of a number of stomiatoid fishes have been examined by Nicol (1961). Unlike those just described, they contain photocytes generating light intracellularly; these are under nervous control. The organ may be rotated into a pigmented pocket by contraction of a large muscle.

Cyclothone the photophores are much more densely packed over pigmented regions. Thus the distribution of photophores strongly supports Clarke's idea, and the positions of the eyes of many predators suggest that they do, indeed, hunt from below.

The light reaching the mesopelagic regions has a wavelength of maximum intensity of some 4,780A, and the wavelength of maximal emission of the ephausid shrimp, *Thysanoessa raschii*, is 4,760A (Boden & Kampa, 1959) and that of the lantern fish, *Myclophur punctatum*, is 4,700A (Nicol, 1960).

It might be argued that a group of bright point-sources of light would make the fish more obvious than otherwise, but fish are myopic so that at a distance these points would become diffuse circles of light merging into a uniform luminance.* The optical problems associated with the emission of light in such a way that the net result is to be a fairly uniform illumination of the ventral and lateral aspects of the fish have been discussed by Denton, Gilpin-Brown & Roberts (1969) with special reference to the photophores of *Argyropelecus*, and they conclude that, because of the half-silvered nature of the walls of the tube conducting light from the luminous cells to the surface of the body, the fish are able to match the external light-distribution for all possible angles from which they can be viewed.

As Clarke pointed out, this function of luminescence in countershading does not rule out other functions, such as sex recognition, blinding of predators, and so on.

Firefly. The significance of the flashing of the firefly, *Photinus pyralis*, has been elucidated by the studies of Buck, among others. The female is wingless,† and in the evening perches on a blade of grass whilst the males, who during the day remain quiescent, fly about emitting flashes. The female never flashes spontaneously, but only in response to the flash of a male within some 3–4 metres; if the female makes this response, the male immediately turns directly towards her and flashes again; the female flashes in response and after about five of these exchanges the insects mate. The ability of the male fly to distinguish a flash, emitted by a female, from one emitted by a male is not due to any difference in the spectral quality of the light emitted by the two sexes, but depends entirely on the interval elapsing between his own flash and the answering one. If this is precisely two seconds, the male responds to it, whereas he ignores any flashes not occurring at other intervals after his own; thus Buck was able to attract males to a torch by flashing it exactly two seconds after the appearance of a flash, or alternatively, by squeezing a male (which makes him emit a flash) just two seconds after seeing a flash.

Marine Worm. In the marine worm, *Odontosyllis enopia*, the sexual function is again well illustrated; the female suddenly becomes acutely phosphorescent, and swims rapidly through the water in small luminous circles; around each circle is a halo of luminescence, probably due to luminous eggs. If the male does not appear, the illumination ceases after some 10–20 sec., but the performance is repeated several times. The male appears as a delicate glint of light about 10–15 ft. from the luminous female, and when he has located her they rotate together in spirals, scattering sperm and eggs in the water. The luminosity of the polynoid worm, *Achola astericola*, is under control, and is used as a sacrificial lure; if the worm is cut in two parts, for example by a crab, the

---

* Hardy (1962) has shown that the photophores of the ephausid, *Meganyctiphanes norvegica*, will rotate so as to vary the direction of emitted light in accordance with the direction of illumination.

† The glow-worm is a wingless female or a larva.

posterior part lights up and wriggles, whilst the anterior part remains dark and seeks concealment.

Other Organisms. The use of light for protection, much as the squid uses its ink, is illustrated by the organ of the fish *Malæocephalus lævis*, studied by Hickley; the gland consists of an invagination of the surface epithelium, and it opens around the anus; muscle fibres, on contraction, squeeze out the secretion which, in contact with sea-water, luminesces brightly for a long time. Hickley thinks that the fish uses its light to blind an attacking fish.

Again, the deep-sea squid, *Heteroteuthis*, has only a rudimentary ink gland and it apparently uses its luminous secretion in the same way as the surface squid uses ink, emitting, instead of a cloud of ink, a burst of luminescence which, by dazzling a predator, enables it to get away (Harvey, 1956).* Finally, in many instances the luminescence is apparently quite useless; for example, the annelid worm, *Chætopterus*, lives in the sand and mud, between low water and 5 fathoms, in a tough parchment-like U-shaped tube secreted by its own integument; the ends of the tube project into the water, and a current—the respiratory stream— is forced through the tube by means of three flat paddles in its middle. It is one of the most luminescent of animals but, as it never comes out of its tube, it is very difficult to see why it should luminesce.

In a similar way it is difficult to see why bacteria and fungi should emit light continuously.

Spectral Distribution of Energy

Figure 849 shows the spectral characteristics of the light emitted by several bacteria and a fungus (*A. mellea*); the most commonly studied bacterium, *Ph. fischeri*, has a maximum at about 4,800A in the blue. The variations in the emission spectra from one bacterium to another may represent variations in

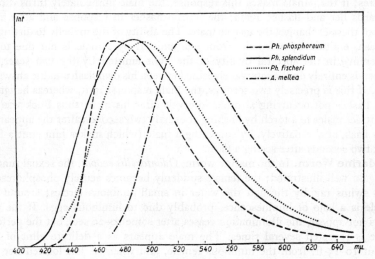

FIG. 849. Emission spectra of luminous bacteria. Ordinates: Relative energy. Abscissæ: Wavelength. (Van der Burg. *Biochim. biophys. Acta*.)

* The luminescence of the marine copepod, *Metridia lucens*, is regarded by David & Conover (1961) as an escape mechanism; as with other planktons, the luminescence only results from an external stimulus, and there was some evidence from their experiments, in which the flashes emitted when a predator was added to the sea-water were counted, that the flash was used in an avoidance technique. Certainly, when the copepod was stimulated electrically, it emitted a point of luminescence and immediately darted off.

the nature of the emitting molecule, but since, according to Cormier & Strehler, the chemistry of bacterial luminescence seems remarkably uniform, it is possible that the differences result from variations in the degree of self-absorption of the original luminescence.* The emission of the fungus is sufficiently different to warrant the assumption of a different emitting molecule.

The emission spectra for fireflies have maxima in the region of 5,500A (Fig. 853, p. 1606), corresponding with the much greener light emitted by these flies. *Cypridina*, according to Eymers & van Schouwenburg, has a maximum at about 4,640A.

BIOCHEMISTRY OF LUMINESCENCE

Luciferin-Luciferase System

As long ago as 1887 Dubois postulated that the essential feature of luminescence was the oxidation of a substance, *luciferin*—the constituent of the coloured granules of photogenic cells—by means of an enzyme, *luciferase*. Thus on boiling a luminous organ, and then adding fresh organ to it, the boiled organ was able to luminesce, the enzyme, luciferase, being destroyed by the boiling whilst the luciferin was not. More recent work has confirmed in general outlines this resolution of the reactants into an enzyme, of a protein nature—and therefore purified by prolonged dialysis in a collodion sac—and a substrate, luciferin, a substance of comparatively low molecular weight which is separated from the enzyme during dialysis. The fact that the secretions of *Cypridina*, and of many other organs, only luminesce when extruded into sea-water, and the fact that extracts always require the presence of O_2 to produce luminescence,† suggest that luciferin and luciferase are secreted as granules, their dissolution in water permitting the oxidative reaction.

Granules. That the breakdown of the granules is an important element in luminescence is shown by the fact that many cytolytic agents, such as saponin, increase luminescence; for example, the luminous slime from *Pholas*, allowed to stand until its light disappears, lights again on the addition of saponin which liberates luciferin from the remaining granules, the enzyme luciferase being still present. The observations of Hickley on the secretion of the fish *Malæocephalus lævis* are of some interest in this connection. Histological studies have shown that the secreting cells accumulate granules, and that secretion is accompanied by the complete breakdown of the cells, in marked contrast to the behaviour of secreting cells in other glands. The secretion is thus a mass of granules, but these are clumped together in globes some 30–40 μ in diameter and apparently covered with a cytoplasmic sheath. In sea-water these globes break up rapidly to give granules about 1–2 μ long, with a greenish fluorescence. It is a common observation that in distilled water the photogenic secretions are not luminous; and Hickley observed that, in the case of *M. lævis*, the globes failed to break up in this medium; the envelopes disintegrated,

---

* Spruit & Spruit-Van der Burg (1955) have analysed the spectral characteristics of bacterial luminescence in some detail, in the hope that they can be related to the fluorescence spectra of the emitting molecules. Harvey, Chase & McElroy have given accurate spectral energy curves for *Cypridina*, *Mnemiopsis* and *A. fischeri*.

† An exception to this rule is provided by extracts of the luminous ctenophore (comb-jelly) *Mnemiopsis Leidyi*; Harvey showed that the extract was luminous in the presence of nascent H_2, and concluded from his experiments that some "bound O_2" was present, which was liberated by the appropriate stimulus. The inhibitory action of light, observed in this organism, could then be due to photodynamic oxidation; the fact that light-exposed ctenophores cannot regain the power to luminesce in the absence of O_2 certainly indicates that storage of O_2, probably as a loosely bound complex similar to oxyhæmoglobin, is a characteristic feature of this luminous secretion.

but the granules remained clumped. If sea-water was added immediately, the luminescence appeared, but only if less than three minutes elapsed; it would seem that the deleterious effect of distilled water is due to its stabilization of the granular clumps, in much the same way that salt-free solutions cause the agglutination of many cells.

Variety of Mechanisms. It is impossible to generalize on the mechanism of luminescence, even if a luciferin-luciferase system is involved in all examples; with some glands it seems certain that the enzyme and substrate are secreted by separate types of cell, only the combined secretions possessing the potentiality for luminescence; in others it may well be that both are secreted by the same cell, in which case the dissolution of the granules, and the aerobic conditions in sea-water, allow the luminescence to proceed. With intracellular luminescence, as in bacteria and photophores, both enzyme and substrate must be present in the same cell; the bacteria luminesce continuously, but in the case of the photophore we must seek some mechanism that prevents the reaction from proceeding.

Bacterial System

Up to the late 'forties of this century, research into the chemical nature of luciferin and luciferase was concentrated largely on the ostracod, *Cypridina*, and the firefly, whose luminous organs permitted the resolution of well defined luciferin-luciferase systems. Early attempts to resolve the bacterial luminescent system into a luciferin and luciferase were almost uniformly unsuccessful, and it was not until 1953 that Strehler & Cormier showed that acetone-powders —obtained by precipitating the bacteria with acetone, washing with acetic acid, and drying—would, on addition to distilled water, give a weak lumines-

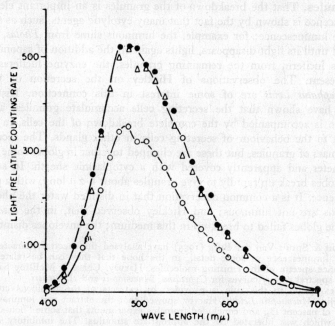

FIG. 850. Emission spectra of *A. fischeri* and luminescent extracts. Dots: Emission of bacteria. Open circles: Emission of extracts. Triangles: Extract with maximum emission equated to that of the bacteria. (Strehler & Cormier. *Arch. Biochem.*)

cence. They found, moreover, that when NAD was added to an aqueous extract of the powder, a bright luminescence was obtained with spectral characteristics similar to those of the bacterial luminescence (Fig. 850). If the extract was dialysed, NAD was only effective in the reduced form—$NADH_2$— unless a substrate, such as malate, was added; and it was concluded that $NADH_2$ was the actual luciferin, in the sense that it was the first substance to become rate-limiting. Other factors favouring emission were shown to be a long-chain aldehyde—plasmal or palmitic aldehyde, extracted from animal tissues such as kidney cortex—and flavine mononucleotide (FMN).

Flavine Mononucleotide

In the same year, McElroy, Hastings, Sonnenfeld & Coulombre showed the absolute importance of FMN. A thermolabile luciferase preparation was obtained by extracting bacteria with water, and fractional precipitation with ammonium sulphate; this was irradiated with ultra-violet light to remove flavines. Addition of $NADH_2$ had a negligible effect, whilst FMN gave a strong luminescence. It was concluded that the function of $NADH_2$ was to reduce FMN, which subsequently reacted with oxygen to give the chemiluminescent reaction. These authors noted, however, that some other factor was necessary, since addition of bacterial extracts, without luciferase activity, would allow more luminescence to occur, in spite of the system's containing plenty of FMN; this additional factor they called bacterial luciferin, and it was subsequently shown by Strehler & Cormier to be identical with the long-chain aldehyde that they had already denoted as "kidney-cortex factor".

Thus, the various factors that favour luminescence are malate, NAD, FMN, aldehyde, oxygen and luciferase.* By studying the rate of development of luminescence on adding one component to the remaining five, some idea of the order in which the various components were used could be obtained. For example, addition of malate to a mixture of NAD, FMN, aldehyde and O_2 gave a "half-rise time"† of 108 sec. This was the same as when NAD was added to a mixture of malate, FMN, aldehyde and O_2, so that it could be presumed that NAD and malate reacted with each other. If $NADH_2$ was added to a mixture of FMN, aldehyde and O_2, the time was now only 2·7 sec., indicating that the malate-NAD reaction occurred first. On the basis of these studies, the suggested order of reactions was:—

$$NAD + Malate \rightarrow NADH_2$$
$$NADH_2 + FMN \rightarrow FMNH_2$$
$$FMNH_2 + Aldehyde + O_2 \xrightarrow{\text{Luciferase}} FMN + h\nu$$

Utilization of Aldehyde

According to McElroy & Green, the aldehyde is not merely a catalyst, but is gradually used up during the chemiluminescent reaction; and they have suggested that it is essentially the energy derived from the oxidation of the aldehyde that is emitted as light, the emitting molecule being flavine mononucleotide. The objections to this hypothesis have been frankly considered by Strehler & McElroy; first, there is the established fact that the fluorescence spectrum of the flavines would lead one to expect a predominantly yellow emission, which contrasts with the blue light of bacterial luminescence. Secondly, there is the energetic requirement. An Einstein of light of 4,800A corresponds

---

* Luciferase is not in itself a $NADH_2$-oxidase; Cormier, Totter & Rostorfer have shown that a luciferase, purified by starch-column electrophoresis, would not give a luminescence on addition of $NADH_2$, FMN, aldehyde and O_2; only when $NADH_2$-oxidase, obtained from E. coli, was added was luminescence obtained.

† That is, the time required for the luminescence to reach half its maximum intensity.

to 60 Calories. The direct oxidation of an aldehyde would not give this amount of free-energy, so that McElroy & Green have suggested the intermediate formation of a peroxide of flavine mononucleotide that reacts with a compound formed by the reaction of another molecule of $FMNH_2$ and aldehyde:—

$$FMNH_2 + -CHO \rightarrow FMNCH + H_2O$$

$$FMNCH + FMNH_2O_2 \rightarrow FMN + CHCOOH + H_2O$$

a reaction that would lead, according to them, to the liberation of some 75 Calories per mole. On this basis, the emission of light would be an improbable event, being determined by the reaction between two FMN-derivatives, neither of which would be stable. This would fit in with the low quantum efficiency of emission; according to Eymers & van Schouwenbourg only 18 per cent. of the total O_2-consumption is concerned in the light-emitting reaction, so that at 16° C they compute that some 450 molecules of O_2 are consumed per quantum of light emitted.

Support for this point of view is provided by Cormier & Totter who have shown that the number of quanta of light-energy obtainable from a mixture of FMN, dodecyl aldehyde, luciferase and O_2 depends directly on the amount of aldehyde added; the rate of emission of energy fell linearly with time, and this could be shown to be due to the utilization of aldehyde by waiting till the rate had halved and then adding half the amount of aldehyde originally present —the result was always to bring the rate back to approximately its original value.

Chain Length

The effectiveness of different aliphatic aldehydes in causing light emission was studied by Rogers & McElroy on a dark mutant of *A. fischeri* that was normally unable to emit light because it lacked the power to synthesize aldehyde. Rogers & McElroy measured the light emission when exogenous aldehydes were added to the bacterial suspension and found that the intensity increased from C_7 to C_{14} and this was associated with an increased affinity for the enzyme, luciferase.*

Firefly Luminescence

The early studies of Dubois showed that the lanterns of fireflies contained material that could be resolved into a thermolabile enzyme—luciferase—and a thermostable luciferin. McElroy and his colleagues showed, however, that the "luciferin" really consisted of several substances: a faintly yellowish compound with a brilliant yellow-green fluorescence—luciferin proper—ATP, and an inorganic divalent cation, Mg^{++} or Mn^{++}.

Structure of Luciferin. The chemical identity of luciferin proper, *i.e.*, of the fluorescent molecule, was finally established by White, McCapra, Field & McElroy in 1961 who also described its synthesis. The structural formula is given below together with that of the reversibly oxidized product, oxyluciferin.

---

* Rogers & McElroy adapted this system to measure the rate of penetration of the aldehydes into the cell; surprisingly the rate decreased with increasing chain-length, perhaps because of the greater adsorption of the long-chain molecules at the membrane surface.

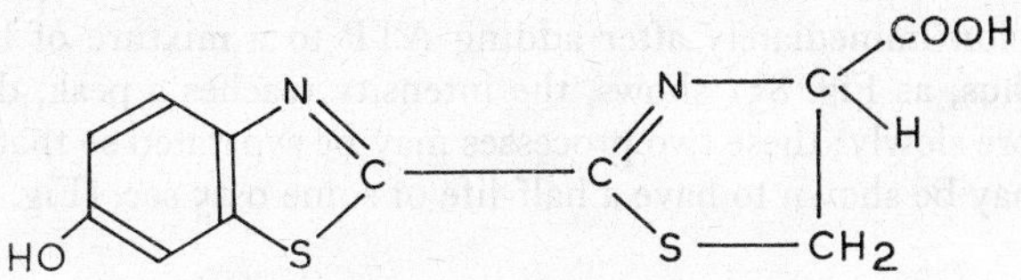

Firefly Luciferin

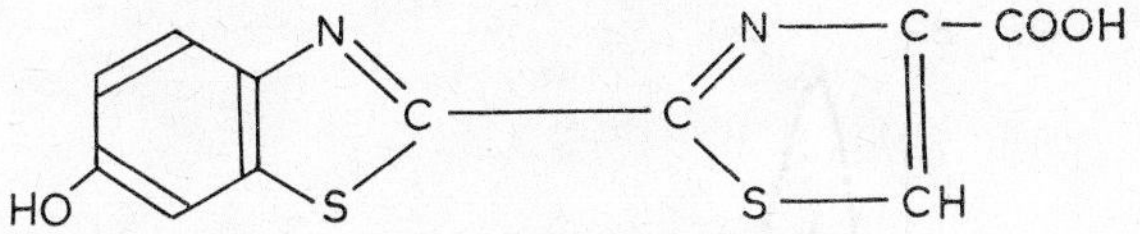

Firefly Oxyluciferin

Luciferin has an asymmetrical C-atom and may be synthesized in the D- or L-forms; only the natural, D-form, will participate in the luminescent reaction with luciferase and ATP (Seliger, McElroy, White & Field).

Luciferase. The enzyme, firefly luciferase, was prepared pure in the crystalline form by Green & McElroy by extraction of an acetone power of firefly lanterns and fractionation by precipitation with ammonium sulphate. It is a euglobulin of molecular weight approximately 100,000.

Chemical Mechanism. The chemical events taking place during luminescence in the firefly system have turned out to be remarkably complex but thanks to the astute investigations of McElroy and his colleagues, a clear picture of these reactions has emerged. They may be summarized as follows:

$$E + LH_2 - AMP$$
$$(1) \quad \updownarrow$$
$$E + LH_2 + ATP \rightleftharpoons E - LH_2 - AMP + PP$$
$$(2)\ O_2 \big| \rightarrow Light$$
$$(3) \qquad \downarrow$$
$$E + L + ATP \rightleftharpoons \quad E - L - AMP + PP$$
$$\Updownarrow$$
$$E + L - AMP$$

According to this scheme, we may mix the enzyme, luciferase (E), and the substrate, luciferin (LH_2), in an aerated solution and then add ATP; the emission of light that ensues is the consequence of the oxidation of a complex of enzyme, luciferin and adenylic acid, $E - LH_2AMP$. The rate-determining reaction, however, is that which leads to the formation of this complex:

$$E + LH_2 + ATP \rightleftharpoons E - LH_2AMP + PP$$

together with inorganic pyrophosphate.

Thus, the complex of luciferin and adenylate ($LH_2 - AMP$) may be synthesized by a direct reaction between luciferin and adenylic acid (Rhodes & McElroy), and this may produce light if it is added to luciferase in the presence of O_2.

Product Inhibition

The light-giving reaction consists in the oxidation of the ELH_2AMP complex to give a new complex containing oxidized luciferin, L, namely, $E - L - AMP$. Once again the compound $L - AMP$ may be synthesized directly from adenylate and oxidized luciferin. Because of the stability of the complex, $E - L - AMP$, the supply of enzyme is progressively reduced during maintained luminescence, and this accounts for the rapid diminution in the intensity of the emission immediately after adding ATP to a mixture of luciferase and luciferin. Thus, as Fig. 851 shows, the intensity reaches a peak, decays rapidly and then more slowly; these two processes may be separated so that the product-inhibition may be shown to have a half-life of some 0·25 sec. (Fig. 851).

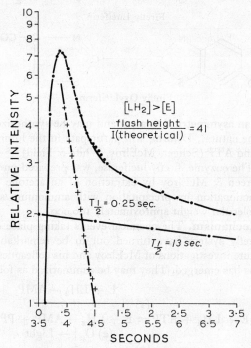

FIG. 851. Light emission response to ATP. The initial rapid decay with a half-time of 0·25 seconds represents product inhibition. The slower decay after the first few seconds, with a half-life of 13 seconds, represents a partial reversal of the inhibition by pyrophosphate and ATP. (McElroy & Seliger. *Light and Life.* Johns Hopkins Press.)

Importance of Pyrophosphate

The slower decay is due to a partial reversal of the inhibition corresponding to reaction (3), the accumulation of pyrophosphate pushing this over to the left thereby liberating free enzyme that may participate in the primary reaction (1). Addition of pyrophosphate to the reaction mixture after the slow decay phase has been reached leads to an immediate increase in emission. Addition of pyrophosphatase, which breaks up pyrophosphate, causes inhibition of the luminescent reaction. The importance of pyrophosphate in maintaining luminescence is further emphasized by observing the luminescence when, instead of mixing luciferin, luciferase and ATP, we begin with the synthetic product luciferin adenylate, L — AMP; in this case there is no formation of pyrophosphate and product-inhibition becomes more pronounced (Rhodes & McElroy).

Quantum Efficiency

The light emission is far more efficient than that from the bacterial reaction; according to Seliger *et al.*, on the average 0·88 ± 0·25 quanta are emitted per luciferin molecule oxidized; this is probably a lower limit, so that a quantum efficiency of unity may be accepted.

The Light-Emitting Molecule

We cannot, as yet, speak with any certainty regarding the light-emission process; the biochemical studies have shown that emission results from oxidation of the luciferin-adenylate-enzyme complex, E—LH_2—AMP, so that it is reasonable to assume that the emitting molecule is an activated condition of this complex or of its oxidized form, E—L—AMP. A clue to the nature of the emitting molecule should be given by a study of the fluorescence-spectrum of luciferin and oxyluciferin and its complexes, *i.e.*, by the spectral composition of the lights emitted by these compounds when they are excited by the absorption of light. The emission spectra for luciferin and oxyluciferin are shown in Fig. 852, and it is seen that their peaks are at 5,350 and 5,440A respectively; these are different from the maximum for light emission in the chemiluminescent reaction, at 5,620A, and hence we must examine the fluorescence of the complexes formed by

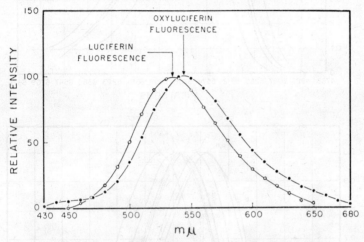

Fig. 852. Fluorescence spectra of luciferin and oxyluciferin. The peaks are at 535 and 544 millimicrons, respectively. The curves have been corrected for phototube spectral efficiency. (McElroy & Seliger. *Light and Life.* Johns Hopkins Press.)

luciferin and oxyluciferin. Here difficulties arise because of the instability of LH_2—AMP and L—AMP at physiological *p*H; in acid solution they are stable and luciferin adenylate, LH_2—AMP, has a peak at 5,700A, very close to that for light-emission in the chemiluminescent reaction. On the other hand, this luminescence undergoes a large shift to longer wavelengths in acid solution, so that it becomes a red glow; in consequence, the comparison of the emission spectrum of luciferin adenylate with that of the chemiluminescent reaction loses its significance.

SPECTRAL DISTRIBUTIONS

A more profitable approach is to study the variation in the emission spectra with species and to examine the effects of using the luciferase of one species with the luciferin of another. By stimulating the fly with ethyl acetate vapour a bright continuous emission was obtained and analysed spectrally (Seliger *et al.*, 1964, Biggley, Lloyd & Seliger, 1967). In general, the peak intensity of emission occurs over the range of 5,520 to 5,820A when different species are compared (Fig. 853); the spectra are quite symmetrical so that it is unlikely that the different colours are due to the presence of two emitting molecules, one in the

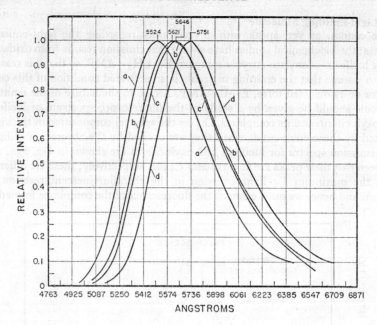

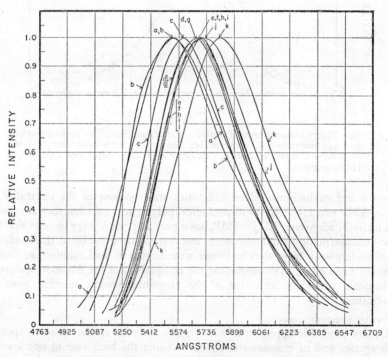

FIG. 853. Top: Bioluminescence emission spectra of some American fireflies.
(a) *Photuris pennsylvania*; (b) *Photinus pyralis* ♂ and ♀; (c) *Photinus marginellus*; (d) *Photinus scintillans* ♂ and ♀. Bottom: Jamaican fireflies.
(a) *Pyrophorus plagiophthalamus* (thoracic organ); (b) *Diphotus sp.*; (c) *Photinus pardalis*; (d) *Photinus xanthophotis*; (e) *Photinus leucopyge*; (f) *Photinus melanurus*; (g) *Photinus nothus*; (h) *Photinus evanescens*; (i) *Photinus ceratus* or *morbosus*; (j) *Photinus gracilobus*; (k) *Pyrophorus plagiophthalamus* (ventral organ). (Seliger, Buck, Fastie & McElroy. *J. gen. Physiol.*)

red and another in the green; thus experimentally, *in vitro*, quite large changes of colour can be achieved by altering pH or the concentration of Zn, and this is due to the red and green emissions of separate excited states; in this event the emission spectrum reveals great asymmetry with two shoulders, a situation never seen *in vivo*. When *Photuris pennsylvanica* luciferase was mixed with *P. pyralis* luciferin, the green emission of *P. pennsylvanica*, was obtained; when the luciferase of *P. pyralis* and the luciferin of *P. pennsylvanica* were mixed, the yellow-green emission of *P. pyralis* resulted, so that there is little doubt that it is the luciferase that determines the character of the emitted light. We may presume that small changes in structure of the enzyme account for the species variations.*

Energy Requirement

As with the bacterial reaction, the energy requirement is high, probably 73 kcal./mole (Seliger & McElroy); of this only some 12·6 kcal. could be provided by dissociation of the luciferase-oxyluciferyl adenylate complex, so that 60 kcal. are required from the biological oxidation.

Control of Emission. The luminescence of fireflies is of special interest since it takes place intracellularly, and it can be controlled by the organism. McElroy has studied the reactions in solution, in order to determine at what point in the chain this control could be exercised; for example, by the liberation of ATP, by the admission of O_2 to the system, and so on. At one time it was thought (*e.g.*, by Alexander) that the control was achieved by regulation of the supply of O_2 to the photogenic cells by a valve-like action of the end-cells of the tracheæ. The main evidence supporting this viewpoint was based on Snell's experiment. Snell examined the luminescence under low partial pressure of O_2; if flies were maintained at about 4 mm. Hg, the O_2-regulating mechanism seemed to fail and the flies became continuously luminescent—the so-called "hypoxic glow". Suddenly raising the O_2-tension caused a very brilliant flash —the "pseudo-flash"—whilst decreasing the O_2-tension below 4 mm. Hg extinguished the glow. It was considered that deficiency of O_2 caused the valve to open, giving the hypoxic glow which had a low intensity because of the low O_2-tension. Admitting O_2 increased the amount available to the cells, causing the pseudo-flash; this was followed by recovery of the end-cells, which then shut off the supply of O_2.†

Liberation of Pyrophosphate

However, the electron-microscope studies of Beams & Anderson have shown that the structures in the tracheolar system that had been considered to be muscular elements were really mitochondria, so that this simple view of the function of the tracheal end-cell had to be abandoned. In view of the importance of pyrophosphate in releasing luciferase from an inactive complex with oxy-luciferin, McElroy inclines to the view that the liberation of pyrophosphate, by a nervous impulse, is the trigger mechanism causing a flash, whilst the

---

* In the click beetle, *Pyrophorus*, the emission maxima for the abdominal and thoracic light organs are greatly different (Seliger *et al.* 1964).

† The influence of neural factors in the development of the hypoxic glow and subsequent pseudo-flash is shown by Carlson's observation that it is only in spontaneously flashing flies that these phenomena can be evoked appreciably, a quiescent non-flashing fly giving only a very dull hypoxic glow and a very small pseudo-flash. If the fly is stimulated into activity before inducing anoxia, then the anoxic glow and pseudo-flash become prominent. Carlson (1965) has examined the hypoxic glow and pseudo-flash in the larval firefly (glow worm) in the hope that a comparison with the situation in the adult firefly might throw further light on the mechanism of the pseudo-flash. The main difference is that the pseudo-flash can only be elicited from the glow worm if it is stimulated during hypoxia, and it may be that this reflects the absence of spontaneous nervous activity that definitely occurs in the adult, and may be responsible for triggering the pseudo-flash.

pyrophosphatase, which is so abundant in the lantern, would decompose the pyrophosphate and permit the re-formation of the inactive complex. The liberation of pyrophosphate could be caused by the reaction:

$$\text{Acetylcholine} \longrightarrow \text{Choline} + \text{Acetate} \xrightarrow[\text{CoA}]{\text{ATP}} \text{PP} + \text{AcCoA} + \text{Adenosine*}$$

Nervous Control

The nervous control is well established, so that spontaneous flashing can only be observed when the brain is connected to the cord; this aspect of the firefly flash has been elaborately investigated by Buck and his collaborators who have shown that the lantern exhibits most of the features of a neuro-effector system, *e.g.*, strength-duration curve, facilitation, and so on. Direct stimulation of the organ can reduce the latency, presumably through bypassing stages in the neuro-effector system, which is assumed to consist of nerve terminal, end-organ and photocyte; thus extremely strong stimuli reduce the latency to 1 msec., and this may correspond to direct photocyte stimulation. The central nervous aspects have also been examined; for example, volleys of action potentials, synchronized one for one with flashes, may be recorded from the posterior cord; stimulation of the eye can either inhibit or enhance flashing, and so on (Buck & Case, 1961; Case & Buck, 1963; Buck, Case & Hanson, 1963).

Effects of Drugs

Adrenergic drugs such as norepinephrine, dopamine, tyramine and isoproterenol all stimulated the *Photuris* flash, giving typical sigmoid dose-response curves characteristic of drug-receptor interaction, perhaps involving a direct influence on the photocyte membrane (Carlson, 1968, *a*). Synephrine and closely related monophenolic drugs were more effective than norepinephrine, and Carlson (1968, *b*) suggests that the chemical transmitter may well be monophenolic, an adaptation necessary in the interest of maintaining transparency of the cuticle in the region of the light-organ. If the neural control is, indeed, exerted through an adrenergic type of transmitter, then the release of pyrophosphate might well be through interaction with adenylcyclase and ATP to produce cyclic AMP and pyrophosphate.

Cypridina

The luciferin-luciferase system of *Cypridina hilgendorfii* has been examined intensively; it is probably more simple than the bacterial and firefly systems, the reaction consisting of a luciferase-catalysed oxidation that requires, apparently, no other substrate than luciferin.

Luciferase. The enzyme luciferase is a protein and has been isolated in pure form by Shimomura, Johnson & Saiga; on the basis of a sedimentation constant of 3·93 Svedberg units, they computed a molecular weight of 48,500 to 53,000; a more exact estimation was carried out by Tsuji & Sowinski who arrived at a molecular weight of 79,650 agreeing with the limit of Fedden & Chase ($\not> $ 80,000); this is rather larger than Chase & Langridge's estimate of 70,000.

Luciferin. The purification of luciferin was first undertaken in 1935 by Anderson, who extracted *Cypridina* with methyl alcohol in the absence of O_2 and submitted the extract to repeated benzoylation and hydrolysis. In this way a product some 2,000 times as active (per gramme of dry weight) as the original material was obtained. It was a yellow material but apparently not a carotenoid. It behaved as a compound of relatively low molecular weight, and the fact that

---

* The significance of coenzyme A in the luminescent reaction has been studied by Airth, Rhodes & McElroy (1958); apparently it stimulates light-emission by removing oxyluciferin from the enzyme surface, leaving the latter free to react with luciferin and ATP; a compound of coenzyme A and oxyluciferin, L-CoA, is formed.

it could be prepared by benzoylation suggested to Anderson that it was a polyhydroxybenzene derivative.

Subsequent studies have led to more active preparations but to contradictory conclusions as to the character of this compound (Mason & Davis; Tsuji, Chase & Harvey); according to Johnson & Sie, the probable structural formula is as follows:

Cypridina Luciferin

Consequently the view that luciferin was a polypeptide (Tsuji *et al.*; Marfey) on this basis is unsound.* As prepared most recently, the luciferin is yellow and gives a yellow fluorescence in ultra-violet light; on standing in solution exposed to air the colour changes to red, and three substances can be isolated chromatographically, namely, an inactive red substance plus two colourless blue-fluorescent substances called oxyluciferin A and B (Johnson & Sie). When a crystalline preparation was examined chromatographically, Haneda *et al.* found evidence for the presence of two components, but they concluded that this could well have been due to the effects of impurities.

Absence of Other Substrates. All attempts to demonstrate the necessity for other factors than luciferin and luciferase have failed; thus, McElroy & Chase purified luciferase 150 times and found it just as effective as a crude water extract, so that it must be concluded that, in *Cypridina*, luminescence results from the catalysed oxidation of luciferin which is, presumably, the light-emitting molecule.

Non-Luminescent Oxidation of Luciferin. The observation of Harvey that *Cypridina* extracts, which had finished luminescing, could be made to luminesce again after the addition of a reducing agent suggested that the chemiluminescent reaction was reversible; however, Anderson showed that it is only the non-luminous form of oxidation that produces a reversible oxidation product, the true luminescent reaction being irreversible; the fact that *Cypridina* extracts become luminescent on addition of reducing agents means simply that, in ordinary circumstances, a part of the luciferin is always oxidized reversibly, *i.e.*, without emission of light. Treating *Cypridina* luciferin with steadily rising concentrations of ferricyanide progressively modifies the response on mixing with luciferase. Without ferricyanide treatment, the emission rapidly develops its maximum intensity and then falls; but, as the concentration of ferricyanide is raised, the rapid emission becomes smaller whilst a slower emission follows. Presumably the slow emission is determined by the time taken for the reversibly oxidized luciferin to be reduced by the extract, in which condition it can be oxidized by the irreversible luciferin-luciferase reaction with the emission of light. Corresponding with the oxidative change there is a characteristic alteration in the absorption spectrum leading, eventually, to the loss of the yellow colour.†

* Marfey, Craig & Harvey (1961) were able to separate the principle responsible for luminescence from polypeptide material by dialysis.

† Chase, Hurst & Zeft have examined the effects of temperature on the luminescent reaction; there is an optimum at 22° C but the falling off in rate at higher temperatures is

Dinoflagellates

Relatively little is known of the biochemical reactions involved in emission by *Noctiluca* or related dinoflagellates; the requirements are similar to those of *Cypridina*, namely only luciferin and luciferase, no other small molecules being required. The system of *Gonyaulax polyedra* has been studied by Hastings and his colleagues; crystalline particles, probably corresponding to the microsources of Eckert, could be isolated from the cells by centrifugation; if isolated at pH 8·2 they could be made to flash by simply changing the pH to 5·7; and the intensity of the flash of any preparation was proportional to the number of the particles. In the electron microscope the scintillons had a characteristic crystalline structure due to the presence of guanine, although guanine itself apparently does not participate in the chemistry of the luminescence. Its function is unknown as it is unlikely to act as a reflector since no directional effect is required (De Sa & Hastings, 1968).

Æquorin

In general, some dozen luciferin-luciferase systems have been extracted from different organisms; these are biologically specific in the sense that the enzyme for one system will not cause luminescence with the substrate of another. The hydromedusa, *Æquorea*, like *Pelagia noctiluca*, requires no oxygen, and Shimomura, Johnson & Saiga have shown that, if an extract of the tissue is prepared in a calcium-free medium, with the aid of the chelating agent EDTA, then the only requirement for luminescence is the addition of calcium. The luminescent material is contained in yellowish masses of tissue on either side of the tentacular bulbs; on rupture of the photogenic cells, granules are released which luminesce on contact with water presumably by virtue of the calcium present. Intracellular luminescence, in response to a nervous stimulus, may well take place by the penetration of calcium into the cells or by its release from a complex in a similar way to that postulated for activation of the contractile machinery of muscle (p. 1398). The purified extract, which Shimomura *et al.* have called *æquorin*, is a protein, and gives an almost colourless solution; all efforts to separate components with light-emitting activity failed, so it must be assumed that the mere addition of Ca^{++} is adequate to induce electronic rearrangements that lead to the emission of light; a change in the chemistry of the æquorin molecule resulting from light emission is revealed by the appearance of a new peak of absorption in the ultra-violet at 3,330A, possibly indicative of the presence of a reduced pyrinium derivative combined with protein.

Luminous Fungi

Up till recently attempts to prepare luciferase and luciferin from fungi have failed; however Airth & McElroy have now achieved this, the past failure being due to the lability of luciferin and the small amounts of luciferase that are extracted; furthermore, they showed that reduced pyridine nucleotide was

---

apparently not due to the non-luminescent oxidation of luciferin, so that quenching effects are probably responsible. The effects of cyanide on luminescence of *Cypridina* extracts are confusing; Johnson, Shimomura & Saiga (1962), working on pure preparations, have concluded that the effect is to induce an oxidation of luciferin; the failure to observe inhibition in crude extracts is due to the presence of reducing materials that prevent oxidation. It is remarkable that a heavy metal like Co^{++} promotes the inhibitory action of cyanide, since this metal is employed as an antidote to the poison.

essential for the luminescent reaction. Aldehydes, flavines and bacterial luciferin were useless.*

EVOLUTION OF LUMINESCENCE

It may well be that the very uselessness of bioluminescence in lower forms such as bacteria and fungi gives the clue to their evolution. McElroy & Seliger (1962) have argued that in early biological time, because the atmosphere was a reducing one, the organisms extant would be anaerobes to which oxygen would be toxic in so far as it inhibited growth. The development of photosynthetic systems would lead to the evolution of oxygen so that eventually these anaerobic organisms would have to develop a means of "detoxicating" this gas; this could be done by developing specific oxidases—luciferases—that would catalyse the oxidation of a specific substrate and dissipate the energy as light.

* Some of the characteristics of fungal luminescence have been described by Airth & Foerster (1960); the luminous mould *Armillaria mellea* was studied. The wavelength for maximum emission was 5,300A in agreement with Van der Burg's finding (Fig. 849); ultra-violet light at first inhibited emission but, while irradiation continued, there was recovery.

REFERENCES

AIRTH, R. L., & McELROY, W. D. (1959). Light emission from extracts of luminous fungi. *J. Bact.*, **77**, 249–250.

AIRTH, R. L., & FOERSTER, G. E. (1960). Some aspects of fungal bioluminescence. *J. cell. comp. Physiol.*, **56**, 173–182.

AIRTH, R. L., RHODES, W. C., & McELROY, W. D. (1958). The function of coenzyme A in luminescence. *Biochim. biophys. Acta*, **27**, 519–532.

ALEXANDER, R. S. (1943). "Factors controlling Firefly Luminescence." *J. cell. comp. Physiol.*, **22**, 51–71.

ANDERSON, R. S. (1935). "The Partial Purification of *Cypridina* Luciferin." *J. gen. Physiol.*, **19**, 301.

ANDERSON, R. S. (1936). "The Reversible Reaction of *Cypridina* Luciferin; its relation to the Luminescent Reaction." *J. cell. comp. Physiol.*, **8**, 261.

BASSOT, J. M. (1960). Données histochimiques et cytologiques sur les photophores du teleostéen *Maurolicus pennanti*. *Arch. d'Anat. Micr.*, **49**, 23–71.

BEAMS, H. W., & ANDERSON, E. (1955). "Light and Electron Microscope Studies on the Light Organ of the Firefly (*Photinus pyralis*)." *Biol. Bull. Wood's Hole*, **109**, 375–393.

BIGGLEY, W. H., LLOYD, J. E. & SELIGER, H. H. (1967). The spectral distribution of firefly light. II. *J. gen. Physiol.*, **50**, 1681–1692.

BODEN, B. P. & KAMPA, E. M. (1959). Spectral composition of the luminescence of the ephausid *Thysanoessa raschii*. *Nature*, **184**, 1321–1322.

BUCK, J. B. (1948). "The Anatomy and Physiology of the Light Organ in Fireflies." *Ann. N.Y. Acad. Sci.*, **49**, 397–482.

BUCK, J., & CASE, J. F. (1961). Control of flashing in fireflies. I. The lantern as a neuro-effector organ. *Biol. Bull.*, **121**, 234–256.

BUCK, J., CASE, J. F. & HANSON, F. E. (1963). Control of flashing in fireflies. III. Peripheral excitation. *Biol Bull.*, **125**, 251–269.

CARLSON, A. D. (1961). Effects of neural activity on the firefly pseudoflash. *Biol. Bull.*, **121**, 265–276.

CARLSON, A. D. (1965). Factors affecting firefly larval luminescence. *Biol. Bull.*, **129**, 234–243.

CARLSON, A. D. (1968, a). Effect of adrenergic drugs on the lantern of the larval *Photuris* firefly. *J. exp. Biol.*, **48**, 381–387.

CARLSON, A. D. (1968, b). Effect of drugs on luminescence in larval fireflies. *J. exp. Biol.*, **49**, 195–199.

CASE, J. F. & BUCK, J. (1963). Control of flashing in fireflies. II. Role of the central nervous system. *Biol. Bull.*, **125**, 234–250.

CHASE, A. M., HURST, F. S., & ZEFT, H. J. (1959). The effect of temperature on the non-luminescent oxidation of *Cypridina* luciferin. *J. cell. comp. Physiol.*, **54**, 115–123.

CHASE, A. M., & LANGRIDGE, R. (1960). The sedimentation constant and molecular weight of *Cypridina* luciferase. *Arch. Biochem. Biophys.*, **88**, 294–297.

CLARKE, W. D. (1963). Function of bioluminescence in mesopelagic organisms. *Nature*, **198**, 1244–1246.

CORMIER, M. J. & STREHLER, B. L. (1954). Some comparative biochemical aspects of cell-free bacterial luminescence. *J. cell. comp. Physiol.*, **44**, 277–289.

CORMIER, M. J., & TOTTER, J. R. (1957). "Quantum Efficiency Determinations on Components of the Bacterial Luminescent System." *Biochim. biophys. Acta*, **25**, 229–237.

CORMIER, M. J., TOTTER, J. R., & ROSTORFER, H. H. (1956). "Comparative Studies on Different Bacterial Luciferase Preparations." *Arch. Biochem.*, **63**, 414–426.

DAHLGREN, U. (1915–17). "The Production of Light by Animals." *J. Franklin Inst.*, **180**, 513, 711; **181**, 109, 243, 377, 525, 658, 805; **183**, 79, 211, 323, 593, 735.

DAVENPORT, D., & NICOL, J. A. C. (1956). "Luminescence in Hydromedusæ." *Proc. Roy. Soc., B*, **144**, 399–411.

DAVID, C. N., & CONOVER, R. J. (1961). Preliminary investigations on the physiology and ecology of luminescence in the copepod, *Metridia lucens*. *Biol. Bull.*, **121**, 92–107.

DENTON, E. J., GILPIN-BROWN, J. B. & ROBERTS, B. L. (1969). On the organization and function of the photophores of *Argyropelecus*. *J. Physiol.*, **204**, 38–39 P.

DESA, R. & HASTINGS, J. W. (1968). The characterization of scintillons. *J. gen. Physiol.*, **51**, 105–122.

DESA, R., HASTINGS, J. W. & VATTER, A. E. (1963). Luminescent "crystalline" particles: an organized subcellular bioluminescent system. *Science*, **141**, 1269–1270.

ECKERT, R. (1966). Excitation and luminescence in *Noctiluca miliaris*. In *Bioluminescence in Progress*. Ed. F. H. Johnson & Y. Haneda. Princeton: University Press, pp. 269–300.

ECKERT, R. (1966). Subcellular sources of luminescence in *Noctiluca*. *Science*, **151**, 349–352.

ECKERT, R. (1967). The wave form of luminescence emitted by *Noctiluca*. *J. gen. Physiol.*, **50**, 2211–2237.

ECKERT, R. & REYNOLDS, G. T. (1967). The subcellular origin of bioluminescence in *Noctiluca miliaris*. *J. gen. Physiol.*, **50**, 1429–1458.

ECKERT, R. & SIBAOKA, T. (1968). The flash-triggering action potential of the luminescent dinoflagellate *Noctiluca*. *J. gen. Physiol.*, **52**, 258–282.

EYMERS, J. G., & VAN SCHOUWENBURG, K. L. (1936). "A Quantitative Study of the Spectrum of the Light emitted by *Photobacterium phosphoreum* and by some Chemiluminescent Reactions." *Enzym.*, **1**, 107.

EYMERS, J. G., & VAN SCHOUWENBURG, K. L. (1937). "Determination of O_2 consumed in the Light-emitting Process of *Photobacterium phosphoreum*." *Enzym.*, **1**, 328.

EYMERS, J. G., & VAN SCHOUWENBURG, K. L. (1937). "Further Quantitative Data regarding Spectra connected with Bioluminescence." *Enzym.*, **3**, 235.

FEDDEN, G. A., & CHASE, A. M. (1959). The diffusion constant of *Cypridina* luciferase. *Biochim. biophys. Acta*, **32**, 176–181.

GREEN, A. A., & McELROY, W. D. (1956). Crystalline firefly luciferase. *Biochim. biophys. Acta*, **20**, 170–176.

GREENE, C. W. (1899). "The Phosphorescent Organ in the Toad Fish, *Porichthys notatus*." *J. Morph.*, **15**, 684.

GREENE, C. W., & GREENE, H. H. (1924). "Phosphorescence of *Porichthys notatus*, the Californian Singing Fish." *Amer. J. Physiol.*, **70**, 500.

HANEDA, Y., JOHNSON, F. H., MASUDA, Y., SAIGA, Y., SHIMOMURA, O., SIE, H.-C., SUGIYAMA, N., & TAKATSUKI, I. (1961). Crystalline luciferin from live *Cypridina*. *J. cell. comp. Physiol.*, **57**, 55–62.

HARDY, M. G. (1962). Photophore and eye movement in the ephausid *Meganyctiphanes norvegica* (G. O. Sars.) *Nature*, **196**, 790–791.

HARVEY, E N. (1940). *Living Light*. University Press. Princeton.

HARVEY, E. N. (1952). *Bioluminescence*. Acad. Press. N.Y.

HARVEY, E. N. (1956). Evolution and bioluminescence. *Quart. Rev. Biol.*, **31**, 270–287.

HARVEY, E. N., CHASE, A. M., & McELROY, W. D. (1957). "The Spectral Energy Curve of Luminescence of the Ostracod Crustacean, *Cypridina* and other Luminous Organisms." *J. cell. comp. Physiol.*, **50**, 499–505.

HASTINGS, J. W, & BODE, V. C. (1962). Biochemistry of rhythmic systems. *Ann. N.Y. Acad. Sci.*, **98**, 876–889.

HASTINGS, J. W., & SWEENEY, B. M. (1958). A persistent diurnal rhythm of luminescence in *Gonyaulax polyedra*. *Biol. Bull.*, **115**, 440–458.

HESS, W. N. (1922). "Origin and Development of the Light Organs of *Photurus Pennsylvanica*." *J. Morph.*, **36**, 245.

HEYMANS, C., & MOORE, A. R. (1924). "Bioluminescence in *Pelagia noctiluca*." *J. gen. Physiol.*, **6**, 273.

HICKLEY, C. F. (1924; 1926). "A New Type of Luminescence in Fishes." *J. Mar. Biol. Assn.*, **13**, 914; **14**, 495.

JOHNSON, F. H., SHIMOMURA, O., & SAIGA, Y. (1962). Action of cyanide on *Cypridina* luciferin. *J. cell. comp. Physiol.*, **59**, 265–272.

JOHNSON, F. H., & SIE, E. H.-C. (1961). The luciferin-luciferase reaction. In *Light and Life*. Ed. McElroy & Glass. Baltimore: Johns Hopkins Press, pp. 206–218.

KAUTSKY, H., & ZOCHER, H. (1923). "U. d. Wesen der Chemilumineszenz." *Z. Elektrochem.*, **29**, 308.

MARFEY, S. P. (1958). Fractionation of *Cypridina* luciferase and its benzoyl derivative. *Biol. Bull.*, **115**, 339.

MARFEY, P., CRAIG, L. C., & HARVEY, E. N. (1961). Isolation studies with *Cypridina* luciferin. *Arch. Biochem. Biophys.*, **92**, 301–311.

MASON, H. S., & DAVIS, E. F. (1952). "*Cypridina* Luciferin. Partition Chromatography." *J. biol. Chem.*, **197**, 41–45.

McELROY, W. D. (1955/56). Chemistry and physiology of bioluminescence. *Harvey Lectures*, **51**, 240–265.

McELROY, W. D., & CHASE, A. M. (1951). "Purification of *Cypridina* Luciferase." *J. cell. comp. Physiol.*, **38**, 401–408.

McELROY, W. D., & GREEN, A. A. (1955). "Enzymatic Properties of Bacterial Luciferase." *Arch. Biochem.*, **56**, 240–255.

McELROY, W. D., & HASTINGS, J. W. (1955). "Biochemistry of Firefly Luminescence." *The Luminescence of Biological Systems*. Ed. Johnson. Amer. Ass. Adv. Sci., Washington, D.C., pp. 161–198.

McELROY, W. D., HASTINGS, J. W., SONNENFELD, V., & COULOMBRE, J. (1953). "The Requirement of Riboflavin Phosphate for Bacterial Luminescence." *Science*, **118**, 385–386.

McELROY, W. D., & SELIGER, H. H. (1961). Mechanisms of bioluminescent reactions. In *Light and Life*. Ed. McElroy & Glass. Baltimore: Johns Hopkins Press, pp. 219–257.

McELROY, W. D., & SELIGER, H. H. (1962). Origin and evolution of bioluminescence. In *Horizons in Biochemistry*. Ed. Kasha & Pullman. New York: Academic Press, pp. 91–101.

NICOL, J. A. C. (1955). Nervous regulation of luminescence in the sea-pansy *Renilla kollikeri*. *J. exp. Biol.*, **32**, 619–635.

NICOL, J. A. C. (1957). Observations on photophores and luminescence in the teleost. *Porichthys. Quart. Z. Micr. Sci.*, **98**, 179–188.

NICOL, J. A. C. (1960). Spectral composition of the light of the lantern fish, *Myctophum punctatum*. *J. Mar. Biol. Ass. U.K.*, **39**, 27–32.

NICOL, J. A. C. (1962). Animal luminescence. *Adv. Comp. Physiol. Biochem.*, **1**, 217–273.

OKADA, Y. K. (1926). "Luminescence et Organe Photogène des Ostracodes." *Bull. Soc. Zool. de France*, **51**, 478.

PRATJE, A. (1921). *Noctiluca miliaris*. Morphologie und Physiologie. *Archiv. f. Protistenk.*, **42**, 1–98.

RHODES, W. C., & McELROY, W. D. (1958). The synthesis and function of luciferyladenylate and oxyluciferyladenylate. *J. biol. Chem.*, **233**, 1528–1537.

ROGERS, P., & McELROY, W. D. (1958). Effect of chain length on light emission and penetration. *Arch. Biochem. Biophys.*, **75**, 87–105.

SELIGER, H. H., BUCK, J. B., FASTIE, W. G. & McELROY, W. D. (1964). The spectral distribution of firefly light. *J. gen. Physiol.*, **48**, 95–104.

SELIGER, H. H., & McELROY, W. D. (1960). Spectral emission and quantum yield of firefly luminescence. *Arch. Biochem. Biophys.*, **88**, 136–141.

SELIGER, H. H., McELROY, W. D., WHITE, E. H., & FIELD, G. F. (1961). Stereospecificity and firefly bioluminescence, a comparison of natural and synthetic luciferins. *Proc. Nat. Acad. Sci. Wash.*, **47**, 1129–1141.

SHIMOMURA, O., JOHNSON, F. H., & SAIGA, Y. (1961). Purification and properties of *Cypridina* luciferase. *J. cell. comp. Physiol.*, **58**, 113–123.

SHIMOMURA, O., JOHNSON, F. H., & SAIGA, Y. (1962). Extraction, purification and properties of æquorin, a bioluminescent protein from the luminous hydromedusan, *Æquorea. J. cell. comp. Physiol.*, **59**, 223–239.

SPRUIT, C. J. P., & SPRUIT VAN DER BURG, A. (1955). Spectroscopic investigation of luminescent systems. In *The Luminescence of Biological Systems*. Ed. Johnson. Washington: Amer. Ass. Adv. Sci., pp. 99–124.

STREHLER, B. L., & CORMIER, M. J. (1953). "Factors affecting the Luminescence of Cell-free Extracts of the Luminous Bacterium *Achromobacter fischeri*." *Arch. Biochem.*, **47**, 16–33.

STREHLER, B. L., & McELROY, W. D. (1949). "Purification of Firefly Luciferin." *J. cell. comp. Physiol.*, **34**, 457.

TSUJI F. I., CHASE, A. M., & HARVEY, E. N. (1955). "Recent Studies on the Chemistry of *Cypridina* Luciferin." *The Luminescence of Biological Systems*. Ed. Johnson. Amer. Ass. Adv. Sci. Washington, D.C., pp. 127–159.

TSUJI, F. I., & SOWINSKI, R. (1961). Purification and molecular weight of *Cypridina* luciferase. *J. cell. comp. Physiol.*, **58**, 125–129.

VAN DER BURG, A. S. (1950). "Emission Spectra of Luminous Bacteria." *Biochim. biophys. Acta*, **5**, 175.

WHITE, E. H., McCAPRA, F., FIELD, G. F., & McELROY, W. D. (1961). The structure and synthesis of firefly luciferin. *J. Amer. Chem. Soc.*, **83**, 2402–2403.

CHAPTER XXVI

PHOTOCHEMICAL ASPECTS OF VISION

Light as a Stimulus

IN an earlier chapter we have seen how the energy of light may be trapped for the purpose of photosynthesis, the prime source of energy in the living world; in the present chapter we shall be concerned with the use of light-energy in another type of photochemical reaction, the *light-stimulus*. In the higher organisms this takes place in highly differentiated organs—eyes—which are essentially aggregates of light-sensitive cells with some form of dioptric apparatus to permit an exact localization of the stimulus. The receptors make nervous connections with the brain so that the light-stimulus may govern or modify the motor activity of the animal. In lower forms, such as the earthworm, the light cells may be scattered in the superficial parts of the body, but their close relationship with the nervous system permits motor responses to these local light-stimuli.*

Action Spectra. It is reasonable to assume that the fundamental process involved in the response of an organism to light is a photochemical reaction

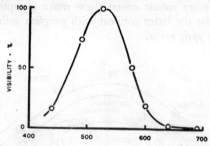

FIG. 854. The reciprocal of the energy, necessary to evoke a discharge of fixed magnitude from a single visual cell of *Limulus*, has been plotted against the wavelength of the stimulating light to give an action spectrum. (Hartline. *J. Opt. Soc. Amer.*)

taking place in the light-sensitive receptors. We can conclude, therefore, that these receptors contain a photosensitive substance, or at any rate a light-absorbing substance. We have already seen that the nature of a photochemically active substance may be inferred from the *action spectrum* of the tissue in which the reaction takes place, the argument being that, if the reaction depends on the absorption of light, the most efficient wavelength of the spectrum will be that which is absorbed most strongly. (It may be recalled that a comparison of the action spectrum with the absorption spectrum revealed the importance of accessory pigments in photosynthesis by certain algæ, p. 1502.) The action

---

* The unicellar organism, *Amœba*, responds to a source of light by moving away from it—it is *negatively phototropic*. The mechanism of this response has been elucidated by Mast; the whole surface of *Amœba* is probably uniformly sensitive to light, in that irradiation of any part favours gelation; since locomotion is achieved by the extension of a pseudopod, the tendency for one side of the organism to gelate will be reflected in movement *away* from this side, because the pseudopod will be formed in a region favouring solation (p. 99).

spectra of a number of invertebrates have been studied, use being made of some phototropic response—as in the case of the clam, *Mya*, studied exhaustively by Hecht—or, alternatively, by measuring the action potentials in the optic nerve, as Hartline has done with the horseshoe crab, *Limulus polyphemus*. Thus, if the intensities of different coloured lights, required to produce the same frequency of discharge in the optic nerve, are plotted against the wavelength, as in Fig. 854, a typical visual action spectrum, or *spectral sensitivity curve*, is obtained; a maximum in the region of 5,300A being found with this eye. In the clam, *Mya*, Hecht found that a blue-green light of about 5,000A was the most effective; and, in general, action spectra seem to fall into one of two classes, showing maxima at about 5,000A and 5,300–5,500A. With some reservations as to the absorption of light by non-photosensitive pigments, these action spectra should conform to the absorption spectra of the photosensitive pigments in the respective light-sensitive organs; the most exact correlation between the two has been established for the human eye, but before going into this in detail we must consider the nature of vision in man.

The Vertebrate Eye

Figure 855 is a diagrammatic representation of a section through the human eye, and Fig. 856 illustrates the structure of the retina, which consists essentially of three layers of cells—the receptors, the bipolar cells, and ganglion cells; the receptors of the human eye (as with most vertebrates) are of two types, long thin *rods* and more robust *cones*; these make synaptic connections with the *bipolar cells*, whilst the latter connect with *ganglion cells* whose axons lead out of the eye as the *optic nerve*.

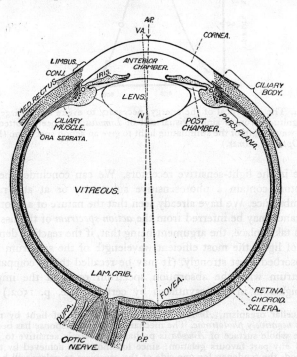

Fig. 855. Horizontal section of the eye. *P.P.*, posterior pole; *A.P.*, anterior pole; *V.A.*, visual axis. (Wolff. *Anatomy of the Eye and Orbit*. H. K. Lewis.)

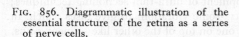

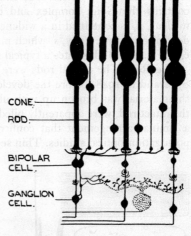

FIG. 856. Diagrammatic illustration of the essential structure of the retina as a series of nerve cells.

CONE.

ROD.

BIPOLAR CELL.

GANGLION CELL.

Structure of the Receptors

In most retinæ the receptors consist of a mixture of two types of cell, the *rods* and *cones*, but their basic structure is similar and is indicated by the generalized receptor of Fig. 857, A; the light-absorbing region, containing the visual pigment, is the *outer segment*, joined by a connecting structure, *c*, to the *inner segment*, which contains the organelles characteristic of a living cell; the ellipsoid, *e*, is

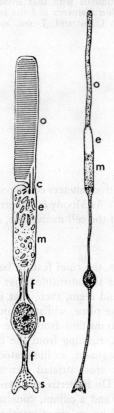

FIG. 857. Diagrammatic representation of retinal photo-receptor cells. A is a schematic drawing of a vertebrate photoreceptor, showing the compartmentation of organelles as revealed by the electron microscope. The following subdivisions of the cell may be distinguished: outer segment (*o*); connecting structure (*c*); ellipsoid (*e*) and myoid (*m*), which comprise the inner segment; fibre (*f*); nucleus (*n*); and synaptic body (*s*). The scleral (apical) end of the cell is at the top in this diagram. B depicts a photoreceptor cell (rod) from the rat. (Young. *J. Cell Biol.*)

characterized by a dense aggregation of mitochondria whilst the myoid, *m*, contains the Golgi complex and considerable quantities of ribosomes. The nucleus, *n*, is contained in a widened part of the fibre, *f*, which terminates in a dendritic synaptic body, *s*, which makes connexion with a neurone, the bipolar cell. Fig. 857 B illustrates a typical rod from the rat's retina.

The Rod. The retinal rods were first studied in the electron-microscope by Sjöstrand in 1948, before the development of ultra-thin sectioning techniques; the material was broken up with ultra-sonic vibrations to give discs which, in the intact rod, were apparently piled one on top of the other like so many coins, a laminar arrangement that confirmed the deductions from W. J. Schmidt's polarization-optical studies. Thin sections through the rod outer segment reveal

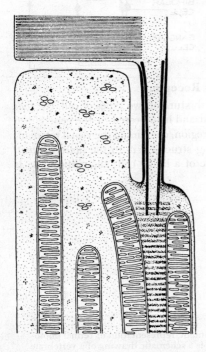

FIG. 859. Diagram illustrating the relationship of outer (above) to inner segment of a retinal rod; the membrane lining the outer segment is continuous with that covering the bundle of thin filaments and the inner segment proper. (Sjöstrand. *J. cell. comp. Physiol.*)

a series of flattened sacs made up of membrane pairs with diameters corresponding to that of the outer segment (Fig. 858, Pl. LXXVIII). As Moody & Robertson showed, these sacs arise through a deep invagination of the cell membrane, and they become separated by a process of nipping off.

Inner Segment

The structure of the inner segment is quite different, the chief feature being the long slender mitochondria packed tightly together to constitute what the light-microscopists have called the "ellipsoid". It would seem, then, that it is in the inner segments that metabolic events largely take place, whilst the outer segments contain the photosensitive pigment. The transition from outer to inner segment is made by a bundle of thin filaments, running from the base of the outer segment to the distal end of the inner segment, as illustrated in Fig. 859. This bundle of filaments connects with a cross-striated structure running through the apical half of the inner segment. De Robertis emphasized the remarkable analogy between this connecting fibril and a cilium, consisting as it does of some nine pairs of filaments surrounded by an outer membrane

PLATE LXXVIII

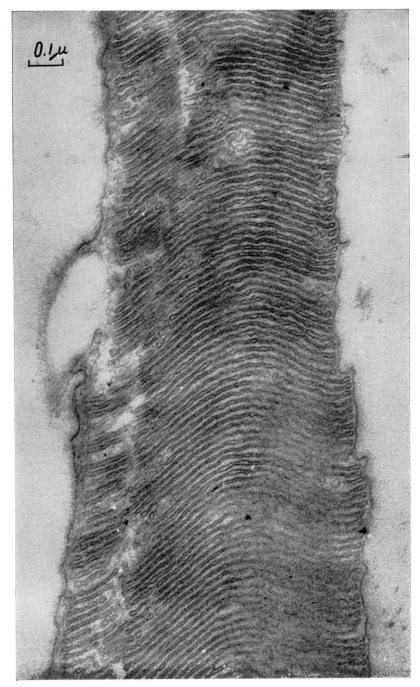

FIG. 858. Electron-micrograph of thin section of outer segment of retinal rod. Osmium fixation. × 90,000. (Courtesy F. S. Sjöstrand.)

PLATE LXXIX

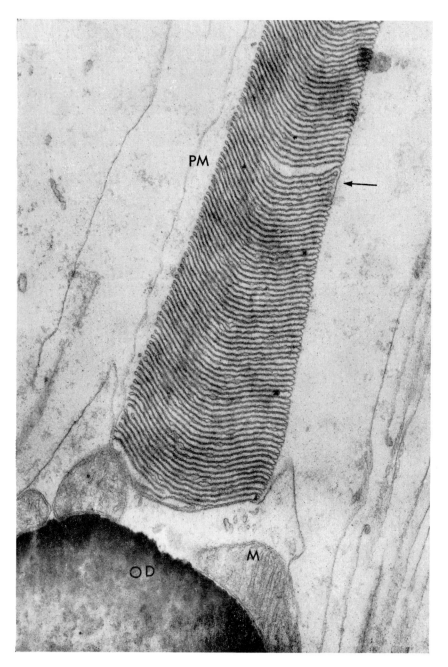

FIG. 861. Longitudinal section through base of the outer segment of a pigeon cone. With the exception of a few (arrow), most of the saccules are continuous with the plasma membrane on the right-hand side of the picture but only two basal saccules show this relationship on the left-hand side. PM, plasma membrane. M, mitochondrion. OD, oil-drop. Chrome-osmium fixed; stained lead. × 49,150. (Cohen. *Exp. Eye Res.*)

and coming into relation with a basal body. On this view, the rod would be regarded as a differentiated ciliated cell.

In fact the morphogenesis of the outer segment does, indeed, involve the development of a primitive cilium from the centriole, projecting from the inner segment primordium, followed by sac-like invaginations of the cell membrane to form five-layered discs or sacs.

Turnover of the Discs

The process of disc formation proceeds from the base, so that new discs are added successively from below, displacing the first-formed discs towards the apex of the segment (Nilsson, 1964). According to Young's (1967) radioautographic study of the incorporation of labelled protein into the outer segment,

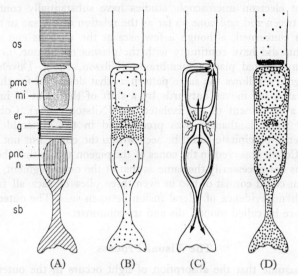

FIG. 860. Illustrating the spread of labelled protein in the rod from its region of synthesis in the myoid portion of the inner segment. (*A*) indicates the cell components that were analysed. *os*, outer segment; *pmc*, perimitochondrial cytoplasm; *mi*, mitochondria; *er*, ergastoplasm; *g*, Golgi complex; *pnc*, perinuclear cytoplasm; *n*, nucleus; *sb*, synaptic body. (*B*) indicates distribution of label after 10 min. (*C*) illustrates the probable pathways by which the label is distributed within the cell. (*D*) indicates distribution after 8 hr. At this time the concentration of labelling in the basal outer segment discs is nearly 20 times greater than that in the ergastoplasm. (Young & Droz. *J. Cell Biol.*)

the process of disc formation continues in the adult retina; at any rate, the newly incorporated material appears first in the inner segment and proceeds as a "reaction band" through the outer segment to reach, finally, the apex from which it disappears, presumably into the adjacent pigment epithelium.* Fig. 860, from a more recent study by Young & Droz (1968), illustrates schematically the progressive shift of labelled protein, as recognized autoradiographically, from its region of synthesis in the endoplasmic reticulum of the myoid portion of the inner segment of the rod to the Golgi apparatus, and thence anteriorly

* Absorption of light occurs in the outer segment; its effects must presumably be transmitted to the inner segment, and it is suggested that the primitive cilium and ciliated rootlet constitute a conducting pathway; the rootlet makes close contact with the mitochondria and endoplasmic reticulum of the inner segment, and may touch its plasma membrane (Matsusaka, 1967).

to the outer segment. From the most proximal part of the outer segment it passed forward, as before, indicating a regular turnover of the structure by repeated additions of new membranous discs at the base of the outer segment and their ultimate removal at the extremity. With the cones, however, the protein became diffusely distributed over the whole cell, and there was no evidence of a comparable turnover. Young & Droz estimated that the discs were replaced at a rate of some 25 to 36 per day, according to the type of rod.

The Cones. Most vertebrate retinæ contain both rods and cones; these last are built on essentially the same plan, the outer segments consisting of a pile of discs or saccules. In the rod, the saccules are clearly separate from the enveloping plasma membrane but in the cone this relationship is relatively rare, continuity between saccules and plasma membrane being common (Fig. 861, Pl. LXXIX).

Subsequent electron microscopic studies have substantially confirmed the difference between rod and cone so far as the relation of the sac to the plasma membrane is concerned, although a few sacs at the vitreous end of the rod outer segments did have continuity with the plasma membrane, *i.e.*, they remained as invaginated plasma membranes (Nilsson, 1965). Development of the outer segment follows the same pattern as that described for the rod, but each invaginated sac moves outwards by a shift of the point of invagination rather than a movement of the isolated sac (Nilsson, 1964). Cohen (1968) showed that when lanthanum was precipitated in the glutaraldehyde-fixed retina, the dense precipitate could be seen within the cone-, but not within the rod-, sacs.* Cohen observed in the cones of the pigeon retina that the diameter of the sac was not necessarily the same as that of the outer segment, so that the outer segment could consist of two or even three piles of sacs all in accurate register, with no evidence of lateral fusion between sacs. The outer segments could therefore be called mono-, di- and tri-columnar.

The Visual Process

We may assume that the absorption of light occurs in the outer segment, where the pigment is concentrated, and that the electrical change in the cell, induced by this photochemical event, is transmitted along the connecting filament to the inner segment. Possibly this change triggers off the liberation of energy by the mitochondria, leading to a more profound excitatory state in the proximal portion of the rod, which resembles a non-myelinated nerve fibre in this region. It will be seen, however, that the absorption of light leads to chemical changes that must be reversed in the dark; this reversal requires energy, and it may be that some of the metabolic events, connected with this, take place in the inner segment. In this connexion we must note that the tips of the outer segments lie in contact with the pigment epithelium which is certainly concerned in exchanges of material between the receptor and itself, and ultimately with the general circulation.

Duplicity Theory of Vision

Action and Absorption Spectra. The human fovea contains only cones, so that if the two types of receptor subserve different visual functions we may expect an essential difference according as an object is directly viewed, in which

---

* Preliminary fixation with glutaraldehyde is generally necessary for the success of the experiment, since exposure of the unfixed retina to lanthanum resulted in disintegration of the outer segments; when some survived, the same difference between rods and cones was observed.

case its image is formed on the fovea, or indirectly, when a more peripheral part of the retina, containing both rods and cones, is used. Such, indeed, is the case, a finding that represents the main factual basis for the *duplicity theory of vision*, which postulates that the rods are concerned with night vision and the cones with day vision. The main facts of night vision may be briefly enumerated. On placing a subject in a dark room, at first the visual threshold is high, a light-stimulus of relatively high intensity being required to stimulate the "light sense"; as he remains in the dark the threshold falls progressively, so that by the end of half an hour, when he is said to be fully dark-adapted, the sensitivity to light may be a thousand times greater than before. Vision under these conditions is peripheral, the test stimulus being perceived only when its image falls off the fovea; moreover, it is achromatic, in the sense that different wavelengths produce only a sensation of light and not of colour.*

Sensitivity to Different Wavelengths

This does not mean, however, that a series of stimuli of different wavelength, but the same energy-content, will produce the same sensation of brightness;

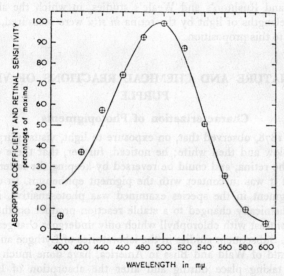

Fig. 862. Comparison of human retinal scotopic sensitivity at different wavelengths with the absorption spectrum of human visual purple. The curve represents the absorption spectrum whilst the plotted points represent human scotopic retinal sensitivity. (Crescitelli & Dartnell. *Nature*.)

on the contrary, under these conditions the different wavelengths appear differently bright, but not coloured. The action spectrum for the human dark-adapted eye, *i.e.*, for the rods, may be obtained, therefore, by plotting the number of quanta of the different wavelengths, required to stimulate the light-sense, against the wavelength. In general, the reciprocal of this number of quanta is plotted, so that the *efficiencies of the different wavelengths in threshold stimulation* are shown, as in Fig. 862. It will be seen that the most effective wavelength is in the blue-green, at about 5,000A, whilst red and violet lights

* On raising the intensity of a coloured stimulus a point is reached when a sensation of colour is obtained, *i.e.*, the *chromatic threshold* has been crossed; at about this level of intensity the cones are being stimulated.

are much less efficient. It is fair to deduce from this curve that the photo-pigment in the human rods has an absorption maximum at about 5,000A.

Rhodopsin

Extraction of the vertebrate eye with aqueous digitonin gives a magenta-coloured solution containing a light-sensitive pigment, called by Kühne *visual purple* or *rhodopsin*; the relative amounts of this substance obtainable from different eyes correlated fairly well with the relative numbers of rods (*e.g.*, the guinea-pig retina is almost a pure rod retina whilst that of the chicken has mainly cones; very little visual purple is extractable from the latter type of retina). The absorption spectrum of a purified solution of visual purple has been studied by a number of workers, with excellent agreement, the maximum occurring at about 5,000A, the exact value of the maximum varying with the species, being 5,020A in the frog and 4,970–4,980 in cattle, rat, dogfish and man. In Fig. 862 both rod sensitivity and visual purple absorption have been plotted together, the points being those for human scotopic sensitivity and the actual line representing the absorption spectrum of human visual purple. The evidence that visual purple is directly concerned with rod vision is therefore very strong, and Rushton's and Weale's studies, in which the absorption of different wavelengths of light by the retina *in situ* were measured, have added further proof to this proposition.

THE NATURE AND CHEMICAL REACTIONS OF VISUAL PURPLE

Characterization of Photopigments

Kühne, in 1878, observed that, on exposure to light, visual purple solutions went first yellow and then white; he noticed, further, that these changes also occurred in the retina, and could be reversed by keeping the latter in the dark provided that it was in contact with the pigment epithelium. In other words, the visual pigment in the species examined was photosensitive, in the sense that it was chemically changed to a stable reaction-product on the absorption of light, by contrast with chlorophyll which only undergoes changes to unstable excited states. Subsequent studies in the laboratories of Lythgoe and of Morton in England, and of Wald and Bliss in America, have done much to elucidate the changes taking place during and after the absorption of light. Before proceeding further, however, it would be profitable to consider the technique of study of photopigments; the amounts of these substances obtainable are so small that direct chemical studies are not easy, and the most useful mode of approach has been concerned with the study of their absorption spectra.

Absorption and Density Spectra. The absorption of light of wavelength, λ, by any solution obeying Beer's Law, is determined by the length of path through which the light travels, and the concentration of the pigment, in accordance with the equation:—

$$\log_e I_{Inc}/I_{Trans} = d_\lambda lc \quad . \quad . \quad . \quad . \quad . \quad . \quad (71)$$

where I_{Inc} and I_{Trans} are the intensities of the incident and transmitted beams respectively, c is the concentration, l the path-length, and d_λ is the *extinction coefficient* for this particular wavelength, and measures the tendency for the molecules to absorb light. Since the left-hand side of the equation is a ratio, without dimensions, the extinction coefficient is given by the reciprocal of concentration times length, *i.e.*, 1/moles per cm.^{3} $\times$ cm., or cm.^{2}/mole.

If logarithms to base 10 are used, and the concentration is expressed as moles *per litre*, the *optical density*, D_λ, is given by:—

$$D_\lambda = \log_{10} \frac{I_{Inc}}{I_{Trans}} = \epsilon_\lambda \, lc \quad . \quad . \quad . \quad . \quad . \quad (72)$$

where ϵ_λ is the *molar extinction coefficient*.

Density Spectrum

If, then, a given preparation is to be studied in solution, the density may be determined at different wavelengths, and it is clear from Equation 72 that this will be proportional to the extinction coefficient. Plotting this against

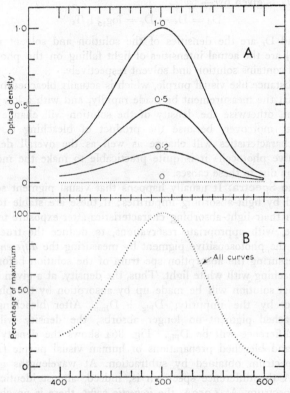

FIG. 863. A. Density spectra of different concentrations of visual purple. B. The same results plotted as percentages of their respective maxima, so that all curves come out the same. (Dartnall. *The Visual Pigments.* Methuen.)

wavelength gives a *density spectrum*, indicating the effectiveness of the molecule in absorbing light of different wavelengths. Thus Fig. 863, from Dartnall, gives the density spectrum for solutions of visual purple of three different concentrations. We may note that the shape varies with the concentration; if, however, the density at a particular wavelength is expressed as a percentage of the density at the wavelength of maximum absorption, λ_{max}, *i.e.*, as

$$\frac{D_\lambda}{D_{\lambda_{max}}} \times 100$$

the curve becomes independent of concentration, as Fig. 863, B, shows.

(Thus $D_\lambda = \epsilon_\lambda cl$ and $D_{\lambda_{max}} = \epsilon_{\lambda_{max}} cl$, so that c and l disappear from the ratio.*)

Measurement

To determine D_λ, the incident and transmitted intensities must be measured; this is done by passing a beam of light through a glass cell containing the solution, and allowing it to fall on a photoelectric cell which, with an appropriate calibration, indicates the intensity of the transmitted light. Some of the light lost will be caused by reflection from the interfaces, and also by absorption by the solvent. These are allowed for by measuring the light transmitted by an identical cell containing the solvent only. It may be shown that the density of the solute, D_d, is given by:—

$$D_d = D_s - D_r = \log_{10} I_r/I_s$$

where D_s and D_r are the densities of the solution and solvent respectively, and I_s and I_r are the actual intensities of light falling on the photocell when the glass cell contains solution and solvent respectively.

With a substance like visual purple, which is actually bleached by light, it is important that the measurement be made rapidly, and with a low intensity of incident light, otherwise the density of the solution will change during the measurement; moreover, because the product of bleaching is yellow, the absorption characteristics will change as well as the overall density. With highly sensitive photocells it is quite practicable to make the measurements without errors due to these causes.

Difference Spectra. It usually happens that visual pigment solutions are contaminated by light-absorbing impurities; if these are stable to light, *i.e.*, if they retain their light-absorbing characteristics after exposure to light, then it is possible, with appropriate reservations, to deduce the true absorption spectrum of the photosensitive pigment by measuring the *difference spectrum*, given by measuring the absorption spectrum of the solution before and after complete bleaching with white light. Thus, the density, at a given wavelength, of the original solution will be made up by absorption by the photosensitive pigment and by the impurity, $D_{Pig} + D_{Imp}$. After bleaching, provided that the bleached pigment no longer absorbs, the density will be D_{Imp}. Hence the difference will be D_{Pig}. Fig. 864 shows the density spectra for unbleached and bleached preparations of human visual purple (A), and the difference spectrum obtained by subtraction. At wavelengths greater than about 4,250A the difference spectrum is, indeed, almost identical with the absorption spectrum. At 4,250A, the *isobestic point*, there is no change in absorption, whilst at shorter wavelengths there is an *increase* of absorption after bleaching, due to the fact that bleaching leads to the formation of yellow pigments which absorb blue and violet light more strongly than visual purple. (By adding hydroxylamine to the solution the bleached product can be made to form retinene oxime whose λ_{max} is still farther towards the violet end of the spectrum, and thus the difference spectrum approximates more closely to that of visual purple.)

Characterization of Purity. The purest preparations of visual purple and other photosensitive pigments indicate that the minimum density on the blue side of the spectrum is some 20 per cent. of the maximum density;

---

* The *absorption spectrum* is given by plotting the percentage of light absorbed against wavelength, *i.e.*, $\dfrac{I_{Inc} - I_{Trans}}{I_{Trans}} \times 100$. Terminology in this respect is loose, so that a density spectrum is usually referred to as an absorption spectrum.

the presence of yellow impurities, by increasing absorption in the blue region, tends to increase this minimum density, and to shift the maximum towards shorter wavelengths. Thus, for a pure solution, D_{min}/D_{max} is about 0·2. If, in practice, this ratio were found to be, say, 0·7, it could be inferred that the preparation was impure and that the measured wavelength of maximum absorption, λ_{max}, was shorter than the true one. By plotting D_{min}/D_{max} against λ_{max} for a variety of preparations, Crescitelli & Dartnall established an empirical relationship between the two, so that for any given preparation

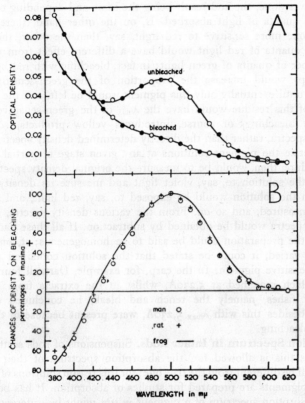

Fig. 864. Illustrating the difference spectrum. A. The optical densities of an extract of human visual purple before and after bleaching by white light. B. The difference spectrum obtained by plotting the changes of density on bleaching against wavelength. Difference spectra for rat and frog extracts are included. (Crescitelli & Dartnall. *Nature.*)

the true λ_{max} could be predicted from a knowledge of the ratio D_{min}/D_{max}. As an example, Saito in 1938 described a photopigment from carp retinæ with λ_{max} of 5,150A; this was impure because the ratio D_{min}/D_{max} was 0·70 and, according to Crescitelli & Dartnall's empirical relationship, the actual shift of λ_{max} due to the impurities was 50A, giving a true λ_{max} of 5,200A. This compares with a value of 5,230A, obtained from the purest solutions of Crescitelli & Dartnall. Again, Collins & Morton described an extract from the pike's retina with λ_{max} of 5,250A; this might be confused with the pigment from the carp, with λ_{max} of 5,230A, but the ratio, D_{min}/D_{max}, of 0·86 indicated that the preparation was grossly impure, and that the true λ_{max} was some 100A longer, *i.e.*, 5,350A, a figure agreeing with that determined from the much purer preparation of Dartnall, which had a λ_{max} of 5,300A.

Homogeneity. As we shall see, the recent work of Dartnall and others has shown that extracts from some retinæ contain mixtures of at least two photosensitive pigments; the λ_{max} for such preparations will therefore lie between the true values for the separate pigments. Dartnall showed that the effects of partial bleaching on the difference spectrum would vary according as the preparation were homogeneous or not. Thus, if there were only one absorbing pigment in the preparation, it should not matter how a given amount of bleaching was done, the end-result being the same and depending only on the number of quanta of light absorbed. If, on the other hand, there were two pigments, one more sensitive to red light, say, than the other, then a given number of quanta of red light would have a different effect from that of the same number of quanta of green light; in fact, bleaching with red light in successive steps would increase the proportion of the green-sensitive pigment remaining, until eventually only this pigment would be left, and the absorption spectrum of this residue would have the λ_{max} of the green-sensitive pigment. Progressive bleaching, of course, introduces yellow products, so that the difference spectra, rather than the directly determined density spectra, are used to characterize the pigment solutions at any given stage of partial bleaching. The procedure, then, would be to measure the original density spectrum; then to expose the solution to, say, violet light and measure its density spectrum again; then the solution would be exposed to, say, red light, and its density spectrum measured, and so on. From the various density spectra, a series of difference spectra would be obtained by subtraction. If all these had the same λ_{max}, then the preparation could be said to be homogeneous; if, on the other hand, λ_{max} varied, it could be stated that the solution contained more than one photosensitive pigment. In the carp, for example, Dartnall found no shift of λ_{max}, which remained at 5,230A, whilst in the extracts from two other fresh-water fishes, namely the tench and bleak, he concluded that other pigments, besides this with λ_{max} 5,230A, were present because of the shift of λ_{max} with bleaching.

Absorption Spectrum in Intact Rods. Suspensions of rods scatter light so that, unless this is allowed for, the absorption spectrum of their contained pigment will be different from the true one; for this reason solutions of rhodopsin and other pigments are prepared for studies of absorption. It has been argued that the absorption spectrum of a pigment *in situ* might be different from that in solution, and Wald & Brown did in fact find the difference spectrum for rhodopsin in the intact rods to be discrepant from that of a solution. Dartnall, however, showed that the absorption spectra are identical, the discrepancy between the difference spectra being due to the circumstance that the products of bleaching in the two situations are different.

Nature of the Visual Purple Molecule

Molecular Weight. Visual purple is a chromoprotein, which is separated from the rods by treatment with digitonin or other detergents. In the ultracentrifuge, Hecht & Pickels found that the material prepared with digitonin sedimented at a rate corresponding to a particle-size of 270,000, but since Smith & Pickels showed that digitonin itself forms micelles of molecular weight 75,000, it was likely that the material sedimented by Hecht & Pickels was a complex, containing the actual visual purple unit linked to one or more of these micelles. The answer to this problem was given by a study of the extinction coefficient of visual purple solutions.

It will be recalled that the molar extinction coefficient is given by the relationship:—

$$D_\lambda = \log_{10} \frac{I_{Inc}}{I_{Trans}} = \epsilon_\lambda l\, c$$

With a given solution, the density, $\log_{10} I_{Inc}/I_{Trans}$, of a known thickness, l, can be measured so that if the extinction coefficient is known, the concentration, c, can be obtained by substitution in the equation. If the weight of material in the solution is known, the molecular weight can be computed:—

$$\text{Concn. in Moles/litre} = \frac{\text{Grammes per litre of solution}}{\text{Molecular weight}}$$

Molar Extinction Coefficient

The problem therefore resolves itself into determining the molar extinction coefficient. This was computed indirectly by Dartnall, Goodeve & Lythgoe from kinetic studies of the rate of bleaching of visual purple solutions. Thus, the *photosensitivity* is defined as the product of the extinction coefficient, α_λ, and the quantum efficiency, γ. This product could be obtained from the kinetic studies, and came out at 9.10^{-17} cm.²/chromophore for a wavelength of 5,000A. It seemed very likely that, under the conditions of bleaching employed, the quantum efficiency was unity, hence the extinction coefficient was 9.10^{-17} cm.²/chromophore. To convert this to the molar extinction coefficient it must be multiplied by $6.1.10^{23}$ (Avogadro's number), by 0·43 to convert to base-10 logarithms, and by 10^{-3} to convert cm.³ to litres. This gives 24,000 as the value of ϵ_λ in the above equation, where the concentration is expressed as *moles of chromophore* per litre. On this basis, by measuring the optical densities of preparations of visual purple, Broda, Goodeve & Lythgoe obtained a molecular weight of 34,500 to 43,000 per chromophore. Wald & Brown measured the densities of visual purple solutions together with the concentrations of retinene in the solutions after bleaching; as we shall see, this substance retinene is the chromophore group detached from the visual purple molecule, so that the value of c, in terms of moles of chromophore per litre, was obtained; whence the molar extinction coefficient was given as cm.²/mole of chromophore. This came out at 40,600 for a wavelength of 5,000A and is accepted as nearer to the true value than the figure of 24,000 obtained by Dartnall *et al*. If this value is employed in Broda *et al.'s* calculation, a molecular weight of 50,000 per chromophore is obtained.

Chromophores per Molecule

We are still left to determine the actual molecular weight of the visual purple particle, however, since it may have several chromophores on it, in which case the molecular weight will be $40,000/n$, where n is the number of chromophores per molecule. Hubbard measured the molecular weight of the digitonin-visual purple complex in the ultra-centrifuge, and found it to be 260,000 to 290,000. By analysing the nitrogen-content of the complex, she computed that 14 per cent. of the particle was protein, *i.e.*, that the contribution to a particle by visual purple was 36,000 to 41,000. Thus, the molecular weight would be about 40,000 if one molecule of visual purple were contained in the complex, or 20,000 if there were two. She found that a preparation containing 7·1 mg./ml. of complex had a value of $\log_{10} I_{Inc}/I_{Trans}$ of unity when measured over a path-length of 1 cm. Such a solution would contain $7·1/10^3 \times 260,000–290,000$

$= 2 \cdot 4 - 2 \cdot 7 . 10^{-8}$ moles of micelles per ml. From the value of the molar extinction coefficient, namely 40,600, we can deduce that the same solution would have $1/40,600 \times 1,000 = $ ca. $2 \cdot 5 . 10^{-8}$ moles of retinene per ml. in it. Thus the number of moles of micelles is equal to the number of moles of retinene, whence we can conclude that each mole of visual purple contains one chromophore group only, so that the molecular weight is about 40,000.

According to Krinsky (1958), rhodopsin is a lipoprotein with a large portion of its weight constituted by lipid that is not separated from the protein moiety by treatment with bacterial phospholipase or fat solvents. His estimate of the molecular weight of the lipoprotein was 32,000.

Chemical Events on Exposure to Light

Bleaching. As mentioned earlier, Kühne observed that visual purple, either in solution or in the retina, went first yellow and then colourless on exposure to light, changes that, in the intact retina, could be reversed in the dark. Lythgoe & Quilliam, by working at $3°$ C, showed that the first product was apparently a substance, *transient orange*, which was rapidly transformed in the dark to a yellow substance, *indicator yellow*, which was given this name because, in alkaline solution, it was pale yellow but changed to a deep chrome yellow in acid solution; in the latter medium (*e.g.*, pH $4 \cdot 9$), indicator yellow is unstable and is transformed to colourless products. According to Lythgoe's results, light was necessary to form transient orange, whilst the other products resulted from thermal or dark-reactions. Wald extracted retinæ at different stages of exposure to light; he found that, after exposure just long enough to produce a yellow colour, a substance which he called *retinene*, and which he identified as a carotenoid related to vitamin A, could be extracted with mild reagents like petroleum ether; if exposure to light was continued, however, the amount of extractable retinene decreased whilst vitamin A made its appearance, and there was little doubt that retinene was being converted to this substance. Moreover, regeneration of visual purple from vitamin A took place in the retina on maintaining it in the dark. If the retina was not exposed to light at all, *i.e.*, if it was extracted in the dark-adapted state, no retinene was obtainable with petroleum ether, although a more vigorous extractive, such as chloroform, was effective. Wald therefore suggested that the action of light was to loosen retinene in the visual purple molecule (or possibly to break it off), thereby allowing its extraction with petroleum ether. Continued exposure to light—by preventing the regeneration of visual purple from retinene that would otherwise occur—allowed a further thermal process to proceed, namely the conversion of retinene to vitamin A.

Retinene and Vitamin A. Retinene was shown by Morton & Goodwin to be the aldehyde of vitamin A:—

$$\begin{array}{c} CH_3 \quad CH_3 \\ \diagdown \diagup \\ C \\ \diagup \diagdown \qquad\qquad CH_3 \qquad\qquad CH_3 \\ H_2C \quad C-C=C-C=C-C=C-C=C-CH_2OH \\ | \qquad\quad || \ H \ H \qquad H \ H \ H \qquad H \\ H_2C \quad C \\ \diagdown \diagup \diagdown \\ C \quad CH_3 \\ H_2 \end{array}$$

Vitamin A

so that the conversion of retinene to vitamin A in the retina represents a reduction, a process that is carried out, according to Bliss, by the enzyme, *alcohol dehydrogenase*, which catalyses the equilibrium between alcohol and acetaldehyde, together with cozymase (NAD). In the normal retina the products of bleaching—retinene and vitamin A—will be converted to rhodopsin, and there is a large bulk of evidence suggesting that an essential step in dark-adaptation, *i.e.*, the recovery of sensitivity to light that occurs, after light exposure, when the subject is placed in the dark, is the reconversion of these products to rhodopsin. Retinene is now usually described by the more correct name of *retinal*.

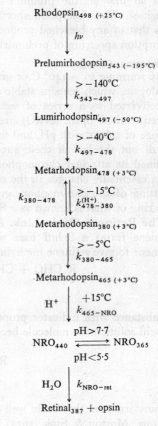

FIG. 865. Thermal reactions in the thermolysis of rhodopsin. (Abrahamson & Ostroy. *Progr. Biophys.*)

Intermediate Steps. Subsequent work on the bleaching of rhodopsin by light, carried out mainly by Wald and his collaborators, has revealed further steps, which may be identified by arresting the thermal changes at a given stage by cooling the reaction mixture sufficiently. The scheme proposed by Abrahamson & Ostroy (1967) is that shown in Fig. 865; the successive steps are recognized by the change in wavelength of maximum absorption (λ_{max}), and the new molecules so formed are indicated by their λ_{max} deduced from the difference spectra.

Prelumirhodopsin

According to this scheme, the first change, and the only one requiring absorption of light, is the conversion of rhodopsin to prelumirhodopsin; the thermal conversion of this to the next stage, lumirhodopsin, is prevented by carrying out the exposure to light in a glycerol-water mixture at the temperature of liquid nitrogen, $-195°$ C. Since light can reverse this transition, there is an equilibrium mixture of rhodopsin and prelumirhodopsin during continued exposure to light, and the equilibrium position is determined by the wavelength of the light, since a short wavelength favours the production of prelumirhodopsin and a long wavelength the production of rhodopsin.

The absorption spectrum of prelumirhodopsin* was established by Yoshizawa & Wald (1963) by exposing rhodopsin to short wavelength light (4,100A) at $-195°$ C and measuring the absorption spectrum of the resulting mixture, which was mainly prelumirhodopsin. The mixture was then warmed in the dark so that all the prelumirhodopsin was bleached whilst the rhodopsin remained. The absorption spectrum now corresponded to that of rhodopsin plus that of any bleached products, and after making allowance for these, the absorption spectrum of prelumirhodopsin was obtained by difference.

Subsequent Changes

By warming to $-140°$ C or greater, prelumirhodopsin is converted to lumirhodopsin, which remains stable at temperatures up to $-40°$ C, above which it is converted to a series of *metarhodopsins*, differing in their λ_{max}, then to N-retinylidene opsin (NRO) previously called *indicator yellow* because of its change of colour with pH, and finally NRO is hydrolysed to retinal and opsin. In all but the last of these successive changes, the rhodopsin molecule has retained its attached chromophore group, retinal, so that the alterations have been essentially changes in the configuration of the whole molecule. The final splitting of the chromophore group from the protein moiety represents the breaking of what is known as a Schiff-base link.

The Retinene-Opsin Link. According to the studies of Collins & Morton retinene forms a Schiff base with amines; thus, methylamine reacts with retinene to give retinene methylimine:—

$$R.CHO + CH_3.NH_2 \rightarrow R.C = N.CH_3$$
$$|$$
$$H$$

a substance with indicator properties analogous to those of indicator yellow; in acid solution the molecule becomes a cation:—

$$\overset{+}{R.C = N.CH_3}$$
$$| \quad |$$
$$H \quad H$$

a transformation that could well cause a shift in the absorption spectrum (Pitt, Collins, Morton & Stok, 1955). By analogy, then, the structure of indicator yellow would be:—

$$H$$
$$|$$
$$C_{19}H_{27} C = N — Opsin$$

i.e., N- retinylidene opsin†

* Until a few years ago, lumirhodopsin was considered to be the first product of the action of light on rhodopsin; Yoshizawa, Kito & Ishigami (1960) and Kito, Ishigami & Yoshizawa (1961) were the first to demonstrate the presence of a precursor at the temperature of liquid nitrogen. The flash-photolysis study of Grellman, Livingston & Pratt (1962) also indicated a precursor, which could be reversibly transformed back to rhodopsin by absorption of light.

† The scheme of decay of rhodopsin given in Fig. 865 includes the formation of N-retinylidene opsin before the final splitting off of retinal from the opsin moiety; this

In neutral solution there will be an equilibrium between the protonated and neutral forms:—

$$C_{19} H_{27} CH = N - R \rightleftharpoons C_{19} H_{27} CH = {}^+NHR$$

Alkaline   Acid

Finally, the conversion of N-retinylidene opsin to retinal plus opsin is a process of hydrolysis requiring water:—

$$C_{19} H_{27} CH = NR + H_2O \rightarrow C_{19} H_{27} CH = O + RNH_2$$

N-Retinylidene opsin   Retinal   Opsin

Regeneration of Rhodopsin. The recombination of retinal with opsin to form rhodopsin was not observed in the early experiments on isolated retinæ, separated from their pigment epithelium. Chase & Smith obtained some 15 per cent. regeneration in a solution of rhodopsin that had been bleached to retinal and opsin, and later Wald & Brown showed that, by adding an excess of retinal, a much larger recovery of rhodopsin could be achieved; in fact, by preparing retinal and opsin separately, they showed that the reaction was spontaneous.

Cis-Trans **Isomerism.** A very interesting finding, which helped to explain the inefficiency of recovery of rhodopsin, was the stereo-specificity of the reaction between retinal and opsin. Retinene and vitamin A contain five double bonds joining C-atoms, and the presence of these gives the possibility of numerous *cis-trans* isomers of the general form:—

It would seem, from Wald's studies, that, of the various possible isomers, only two are capable of condensing with opsin. Thus, ordinary crystalline vitamin A was useless for the synthesis of rhodopsin, whereas the vitamin A in fish-liver oil gave good yields.

An isomer named neoretinal-b, or 11-*cis*-retinal, on incubation with opsin in the dark yielded a rhodopsin identical with that extracted from the retina. Another, isoretinal-a, or 9-*cis*-retinal, yielded a photosensitive pigment very similar to rhodopsin but with λ_{max} at about 4,870A; this is iso-rhodopsin. The structural formulæ of these and other isomers are indicated in Fig. 866.

Isomerization

The reason why rhodopsin does not bleach reversibly to opsin and retinal is that the retinal that emerges from the bleaching is almost entirely the all-*trans* isomer, which is inactive for synthesis. Thus, retinal enters rhodopsin in one configuration and emerges in another; before the all-*trans* isomer can be used, it must be isomerized; this may be done by an enzyme, extracted from eye tissues and called *retinal-isomerase*, which, in the presence of dim light, catalyses the conversion to 11-*cis*-retinal. Thus, the cycle may be written after Hubbard & Wald:—

---

substance, the indicator yellow of Lythgoe, has always been treated as an artefact by Wald, being formed, according to him, from the retinal liberated from rhodopsin. Abrahamson & Ostroy (1967) point out that the failure to identify NRO during bleaching is accounted for by its ready hydrolysis at physiological pH.

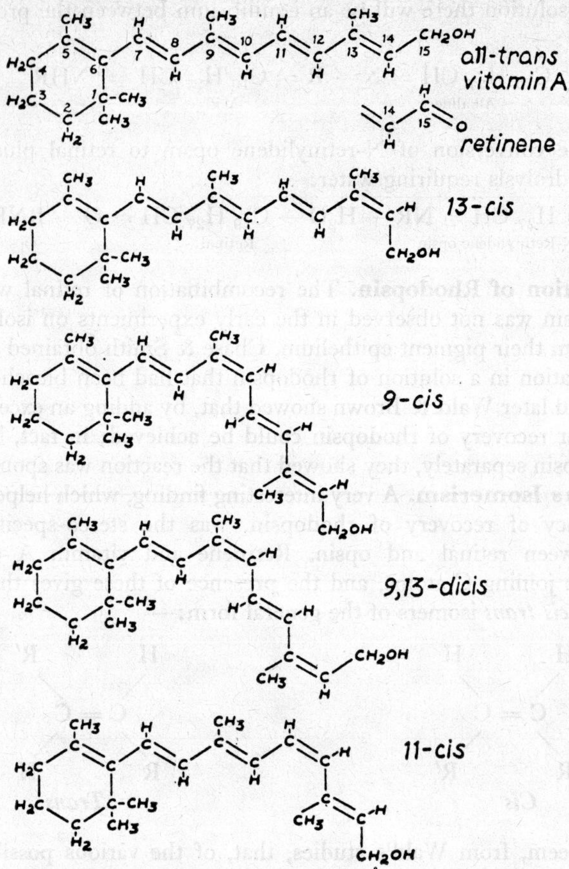

FIG. 866. Structural formulae of the more prevalent geometrical isomers of vitamin A and retinene (retinal). (Hubbard, Bownds & Yoshizawa. *Cold Spr. Harbor Symp. quant. Biol.*)

$$\text{Neoretinal-b + opsin} \rightleftharpoons \text{All-}trans \text{ retinal + opsin}$$

$$\Updownarrow \text{(Alcohol dehydrogenase, cozymase)} \Updownarrow$$

$$\text{Neovitamin Ab} \qquad \rightleftharpoons \qquad \text{All-}trans \text{ vitamin A}$$

Rhodopsin
↗   ↘ Light

Pigment Epithelium

The importance of the pigment epithelial cells, whose processes envelop the ends of the receptors, now becomes evident since these will provide the retinal and vitamin A in the appropriately isomerized 11-*cis*configuration.

Photoisomerization

The first stage in bleaching, the formation of prelumirhodopsin, consists in the conversion of 11-*cis* retinal to the all-*trans* form, and this gives rise to the change in λ_{max} from 4,980 to 5,430A; this may be called a *photoisomerization* process, and is reversible, *i.e.*, the absorption of a quantum of light by pre-lumirhodopsin will favour the return to rhodopsin; subsequent changes require a higher temperature, but some of these will still take place in the frozen condition; and we may assume that the changes are essentially alterations in the folding of the polypeptide chains that determines the tertiary structure of the opsin molecule.

As originally suggested by Kropf & Hubbard, the retinal moiety was held in place on the opsin molecule by virtue of the special shape of the 11-*cis* configuration; photoisomerization to the all-*trans* form would make a neat fit impossible, and this might then allow the opsin to alter its configuration, as illustrated by Fig. 867. The change in configuration would expose previously shielded reactive groups, such as the SH-group as demonstrated chemically by Wald & Brown; and in this way there would be a *molecular amplification* of the primary photochemical event. On this basis, the changes taking place constitute varying degrees of reversible denaturation of the opsin molecule, and the fact that the retinal moiety can be photoisomerized to the 11-*cis* condition while still *in situ* on the opsin molecule means that rhodopsin may be regenerated from the

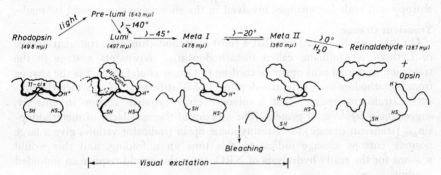

FIG. 867. Pictorial representation of stages in the bleaching of rhodopsin. Rhodopsin has, as chromophore, 11-*cis* retinal, which fits closely a section of the opsin structure. The only action of light is to isomerize retinal to the all-*trans* configuration (prelumirhodopsin). Then the structure of opsin opens in stages (lumirhodopsin, metarhodopsin I and II), until finally the retinaldehyde is hydrolysed away from opsin. Bleaching occurs in the transition from metarhodopsin I to II; and by this stage visual excitation must also have occurred. The opening of the opsin structure exposes new groups, including two -SH groups and one H^+- binding group. The absorption maxima shown are for prelumirhodopsin at $-190°C$, lumirhodopsin at $-65°$, and the other pigments at room temperature. (Wald. *Proc. Int. Congr. Physiol.*)

intermediate steps. Thus Yoshizawa & Wald (1963) demonstrated the photoconversion of lumirhodopsin to prelumirhodopsin at $-195°C$; the conversion of metarhodopsin to prelumirhodopsin required a higher temperature, so that the low temperature prevents the change in the opsin molecule in both directions.

Flash Photolysis Techniques. Considerable progress in the elucidation of the changes taking place in the rhodopsin molecule as a result of absorption of light has been made by exposing the retina or a solution of rhodopsin to a single brief flash and following the changes of absorption of light; thus Hagins exposed the excised rabbit eye to a 20 μsec. flash and observed a transient rise in absorption at 4,860A, due to the appearance of metarhodopsin, and in fact this is the first identifiable change taking place at ordinary temperatures, so that some of the intermediates described by Fig. 867 may well be the products of the special experimental conditions.

Metarhodopsins I and II

Exploiting this technique, Matthews *et al.* (1963) showed that the change of absorption following the decay of the so-called metarhodopsin, with λ_{max} of about 4,800A, involved two steps instead of a single one as previously thought, and they identified, as an intermediate, a yellow compound which they called *Metarhodopsin II* with λ_{max} 3,800. They considered that the next step consisted of the splitting off of retinal, to give retinene and opsin. Thus:—

$$\text{Metarhodopsin I} \longrightarrow \text{Metarhodopsin II} \longrightarrow \text{Retinal} + \text{Opsin}$$

Metarhodopsin II is essentially a tautomeric form of Metarhodopsin I, the tautomeric equilibrium being a thermal one and unaffected by light; the change from I to II involves binding of a proton and is accompanied by a large increase in entropy, signifying a considerable modification in the shape of the molecule.

Kinetics of Decay. Ostroy, Erhardt & Abrahamson (1966) have examined the kinetics of decay of metarhodopsin* in some detail and have come to the conclusion that the kinetics may be described by three first-order reactions in series. By the application of absolute reaction rate theory, estimates of the entropy and enthalpy changes involved in the successive steps could be made.

Transient Orange

On this basis they consider that a third metarhodopsin—the transient orange of Lythgoe & Quilliam, called metarhodopsin$_{465}$, represents a stage in the transition to retinal plus opsin; according to Ostroy *et al.* (1966) it is the change from metarhodopsin$_{380}$ (Metarhodopsin II of Matthews *et al.*) to metarhodopsin$_{465}$ (transient orange) that involves a large negative change in entropy, suggesting a *refolding* of protein. The kinetics of thermal decay of metarhodopsin$_{465}$ (transient orange) to N-retinylidene opsin (indicator yellow) give a large positive entropy change indicating this time an unfolding, and this would account for the ready hydrolysis of NRO, since this would require an unfolded position.

Exposure of SH-Groups. The deductions from kinetic parameters involving only measurements of absorption spectra are obviously precarious, and more specific chemical evidence regarding the changes in rhodopsin is obviously required. This point has been discussed recently by Abrahamson & Ostroy (1967) in the light of the kinetic data. Thus the appearance of reactive SH-groups has been indicated earlier (Fig. 867, p. 1633); this has been associated with the lumi$_{497}$ to meta$_{478}$ process, *i.e.*, the appearance of Wald's Metarhodopsin I, and has been considered to be the "trigger reaction" in vision, *i.e.*, it is at this point that sufficient chemical change has been induced to activate the rod (Wald, Brown & Gibbons, 1963). However, Erhardt *et al.* found no significant changes in the AgNO$_3$ titration or pH when rhodopsin was illuminated in 50 per cent. aqueous glycerol at $-29°$ C, so that SH-groups were not exposed up to the stage of the lumi- to meta-rhodopsin conversion; in their own experiments Ostroy, Rudney & Abrahamson (1966) indicated that the extra SH-groups were exposed in the meta$_{380}$ to meta$_{465}$ process, *i.e.*, the conversion to transient orange.

Borohydride Reduction. Again Bownds & Wald (1965) found that borohydride was only able to reduce the site of attachment of retinal with opsin when bleaching had proceeded to the meta$_{380}$ (Metarhodopsin II) stage; we must presume that this reduction requires the molecule to be in a strongly unfolded condition, and this conforms with the deductions from entropy changes, namely that meta$_{380}$ is the first unfolded stage. As we have seen, the kinetic results

---

* Abrahamson & Ostroy (1967) have reviewed comparable studies on the thermal decay of prelumirhodopsin and lumirhodopsin (Grellman, Livingston & Pratt, 1962; Pratt, Livingston & Grellman, 1964); the kinetics of the decay of prelumirhodopsin show three concurrent first-order reactions, all with very small enthalpy and entropy changes indicating a minimum of structural change in the molecule.

suggest that some degree of refolding occurs next, with a final unfolding that permits the hydrolysis of NRO to retinal and opsin.

Isorhodopsin. Rhodopsin is stable in the dark at body temperature presumably because of the special fit of the 11-*cis* isomer of retinene into the opsin molecule; the 9-*cis* isomer also fits comfortably, giving a compound with λ_{max} of 4,860A, nearer the violet than that of rhodopsin. Because light acts primarily to photoisomerize retinene while attached to the opsin molecule, the effect of light will not only be to produce the all-*trans* retinene, which initiates the steps of thermal decay shown in Fig. 864, but it will also cause the formation of 9-*cis* which will be thermally stable as isorhodopsin. Thus the effect of exposure to light is not to produce exclusively bleached products but a mixture of these with any compounds between opsin and retinal isomers that are thermostable, namely rhodopsin and isorhodopsin, and this occurs by virtue of the reversibility of the original isomerization of the retinal attached to rhodopsin. Thus, when rhodopsin is exposed exhaustively to white light at $-17°$ C the product is what has been called simply metarhodopsin or metarhodopsin$_{478}$. This is thermally unstable, so that when the frozen mixture is warmed, bleaching occurs to give retinene and vitamin A; this is only half the story, however, since there is some reconversion to rhodopsin, together with the formation of a new pigment, *isorhodopsin*. It was originally considered by Wald and his colleagues that the process represented a *thermal* conversion of metarhodopsin to rhodopsin and isorhodopsin, *i.e.*, the all-*trans* retinene on the opsin molecule was thought to be thermally isomerized to the two *cis*- forms, thus permitting them to adopt the appropriate position in the opsin "nest". In fact, however, Hubbard & Kropf showed that the isomerization had occurred during the exposure to light. This could be shown by making use of the fact that bleaching could also be induced thermally, in which case there was no isomerization; thus thermal bleaching of rhodopsin gives 11-*cis* retinal, and bleaching metarhodopsin gives all-*trans* retinal. On the other hand, when the retinene derived from this thermal bleaching is irradiated, then, because of the *photoisomerization* that takes place, a steady-state mixture of all the isomers will be formed.

Metarhodopsin as a Mixture

Thus, if we irradiate the retinene derived from thermal bleaching of rhodopsin there is a shift of λ_{max} indicating a change from *cis*- to *trans*- forms, whereas if the retinene derived from the thermal bleaching of metarhodopsin is irradiated, there is a shift from the *trans*- to the *cis*- forms of retinene. Hubbard & Kropf showed that the mixture obtained by irradiating rhodopsin at $-17°$ C, which has been called metarhodopsin, was in effect a mixture of chromoproteins formed from the isomerization of the retinene on the original opsin molecule. The first effect of light is to form all-*trans* lumirhodopsin; this goes thermally to all-*trans* metarhodopsin; further exposure to light now isomerizes the retinene on this all-*trans* chromoprotein to a mixture; and it is this mixture of chromo-

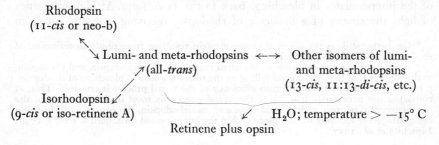

Rhodopsin
(11-*cis* or neo-b)

Lumi- and meta-rhodopsins ⟷ Other isomers of lumi-
(all-*trans*)   and meta-rhodopsins
             (13-*cis*, 11:13-*di-cis*, etc.)

Isorhodopsin
(9-*cis* or iso-retinene A) H_2O; temperature $> -15°$ C
Retinene plus opsin

proteins that in the past has been called "metarhodopsin". Of the mixture of chromoproteins formed by the continued action of light, rhodopsin and iso-rhodopsin are thermally stable, so that on warming the "metarhodopsin" mixture to room temperature these remain, whilst the others, all-*trans*-, 13-*cis*, 11:13-*di-cis*, and so on, decompose to bleached products.

KINETICS OF THERMAL BLEACHING

If this view is correct, it should be possible to demonstrate the inhomogeneity of the "metarhodopsin" formed at $-17°$ C; in fact Hubbard & Kropf proved this to be true, showing that the kinetics of thermal bleaching, at highly alkaline *p*H and at the low temperature at which the mixture was formed, indicated the presence of rhodopsin—bleaching with a half-life of 12 hours—isorhodopsin—with a half-life of 5 hours—and a third, rapidly bleaching component with a half-life of 15 minutes. This last component would presumably be the chromo-proteins formed by the remaining isomers of retinene, all-*trans*, 13-*cis*, etc., which, as we have seen, at higher temperatures are highly unstable and bleach rapidly. Metarhodopsin may now be redefined as the thermally unstable chromoproteins formed by irradiating rhodopsin or isorhodopsin at $-17°$ C, *i.e.*, the fraction that at temperatures greater than $-15°$ C hydrolyses to retinene plus opsin.*

To return to isorhodopsin, its presence in the retina, or in solutions, can be demonstrated by the technique of flash photolysis but it seems clear from the studies of Wulff *et al.* (1958), and Bridges (1961), that very high intensities of light are required to produce significant quantities; at the highest intensities the proportions of isorhodopsin and rhodopsin were equal.

In Vivo **Flash**

In rabbits, Hagins showed that, if sufficient quanta of light were delivered to the eye in the lifetime of lumi- and meta-rhodopsins, there was indeed a steady-state in the retina with only half the rhodopsin bleached.† Under ordinary conditions of illumination, however, it seems unlikely that photoisomerization, with the formation of isorhodopsin, will be a significant feature, and this accounts for the failure to isolate isorhodopsin from the retina; thus the enzymatic isomerization of retinene is unlike the photoisomerization process, in so far as it produces only the 11-*cis* form of retinene instead of an equilibrium mixture of all isomers.

Quantum Efficiency. Because of the photoreversal of bleaching imposed by the phenomenon of photoisomerization, there will clearly be an upper limit to the proportion of light falling on the retina that is employed in bleaching rhodopsin. Thus Williams has pointed out that the system may be equated with Fig. 868, according to which absorption of a quantum of light causes the formation of prelumirhodopsin, which decays thermally to lumirhodopsin and metarhodopsin. At any stage up to this, the absorption of a second quantum will cause photoreversal in so far as it will isomerize the all-*trans* retinal, or any of the intermediates in bleaching, back to the 11-*cis* form. At high intensities of light the chance of a molecule of rhodopsin receiving a second quantum

---

* Or, better still, as the all-*trans* chromoprotein resulting from photoisomerization of rhodopsin (Hubbard, Brown & Kropf, 1959).

† Because of the reconversion of bleached rhodopsin to isorhodopsin and rhodopsin, so that only half of the light quanta falling on the retina are used in bleaching of rhodopsin, it has been argued that the quantum efficiency of the visual process is one-half. This, of course, is not necessarily true, since the visual process is most likely the result of the conversion of rhodopsin to lumirhodopsin and metarhodopsin, the subsequent bleaching being altogether too slow to take place within the latent period (Hubbard & Kropf, 1959; Linschitz *et al.*, 1957).

will increase, and thus the quantum efficiency, $\gamma =$ No. of Molecules Bleached/ No. of Quanta Absorbed, will be less than one and will get smaller as the intensity increases. Fig. 869 shows how γ varies with the fraction of the rhodopsin bleached by a flash of light. By applying simple statistical theory, on the basis that bleaching will occur if odd numbers of quanta are absorbed and will fail if even numbers are, Williams (1964) showed that the fraction of rhodopsin bleached as a function of light intensity was predictable and agreed with the

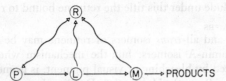

FIG. 868. Simplified bleaching scheme showing photoreversal. Wavy arrows are photochemical steps, straight arrows are thermal steps. Starting with rhodopsin, *R*, note that an odd number of quanta absorbed results in unstable *P*, *L*, or *M* species which decay to PRODUCTS. Even numbers of quanta absorbed result in regeneration of rhodopsin. (Williams. *Vision Res.*)

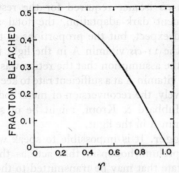

FIG. 869. Calculated values of the quantum efficiency, γ, for different values of the fraction bleached with short flashes. Note that γ approaches zero and unity as the fraction bleached approaches 0·5 and zero, respectively. (Williams. *Vision Res.*)

measured fraction. This was on the assumption that the quantum efficiency for the first step:—

$$\text{Rhodopsin} + h\nu \longrightarrow \text{Prelumirhodopsin}$$

was unity. Thus at very low intensities the quantum efficiency for the total reaction should approach unity, whilst at very high intensities it should be very low indeed.

Vitamin A. The conversion of retinene plus opsin to rhodopsin, and the reverse process, have been discussed earlier. We may now consider the relationship of vitamin A to the visual process. The equilibrium between retinene and vitamin A is catalysed by alcohol dehydrogenase, but it lies so far in favour of vitamin A that there must be some means of driving the reaction in the reverse direction. This is achieved, *in vitro*, by adding a trapping reagent that removes the retinene as soon as formed; Wald & Hubbard used hydroxylamine, which formed an oxime of retinene. *In vivo*, the rapid spontaneous condensation of retinene with opsin to form rhodopsin would be the equivalent of the trapping process. At any rate, *in vitro*, rhodopsin may be formed in good yield from a mixture of vitamin, cozymase A, alcohol dehydrogenase and opsin.

Exchanges with Pigment Epithelium

The *in vivo* relationships have been studied in several laboratories. Thus Jancsó & Jancsó profited by the fluorescence of vitamin A in ultra-violet light to demonstrate that after maximal light-adaptation there is a high concentration of the vitamin in the pigment epithelium; during full dark-adaptation none could be demonstrated. Again, the chemical studies of Hubbard & Coleman and of Dowling have confirmed this migration of vitamin A during light- and dark-adaptation, and have shown that the total amount of vitamin A in the eye remains constant, if we include under this title the retinene bound to rhodopsin.

THE VITAMIN A ISOMERS

Both the 11-*cis* and all-*trans* isomers of retinene may be converted to their corresponding vitamin-A isomers, but the mechanism whereby the all-*trans* vitamin A, produced by bleaching of visual pigment, is isomerized to the 11-*cis* form is not known. Recently Hubbard & Dowling have examined the relative concentrations of the isomers of vitamin A in the retina during light- and dark-adaptation. Remarkably, the store of the 11-*cis* isomer remains unchanged in spite of prolonged light-adaptation, a condition in which all-*trans* retinene is being formed in large amounts, and in which we may expect a considerable demand on the reserves of the 11-*cis* isomer required for the resynthesis of rhodopsin. Again, during subsequent dark-adaptation, the total amount of vitamin A decreases, as we should expect, but the proportions of the isomers remain the same. The sparing of the 11-*cis* vitamin A in the light is of great interest and could be explained on the assumption that the retina was capable of isomerizing the all-*trans* retinene or vitamin A at a sufficient rate to keep up with the bleaching of rhodopsin. Alternatively, the reconversion of metarhodopsin to rhodopsin by light, discovered by Hubbard & Kropf, might be of sufficient magnitude to spare the stores of vitamin A in the light.

The Chemical Trigger. It is impossible to speak with certainty as to the step in the decay of prelumirhodopsin that acts as the trigger for the rod, leading to an excited state that may be transmitted to the bipolar cell. It seems unlikely that the mere photoisomerization of 11-*cis* retinal to all-*trans* is sufficient, since this involves so little change in the structure of the opsin moiety; according to Abrahamson & Ostroy, it is most likely to be the substantial change in protein configuration involved in the conversion of metarhodopsin$_{478}$ to metarhodopsin$_{380}$ that by itself, or coupled to some enzymatic process, produces sufficient change in the rod membrane to be reflected in some change in membrane potential.

Electrical Events

As with other excitatory phenomena, we may expect to measure, at some stage in the process of visual excitation, a generator potential, *i.e.* a depolarization of the receptor cell's membrane.

Electroretinogram. That electrical events do occur in the rods and cones is now reasonably certain; classical studies of these events involved measuring the electroretinogram (ERG),* a succession of potential changes recorded from the surface of the cornea in response to a light stimulus; these consisted typically of an initial negative *a*-wave, followed by a large positive *b*-wave. Thanks to Brown & Watanabe's (1962) studies, in which a microelectrode was placed close to the surface of the retinal cells, the early negative wave, recorded now as a positive wave, can be attributed to the rods and cones, while the later *b*-wave

---

* A valuable review of the recent work on the electroretinogram is that by Brown (1968).

belongs to the bipolar cells and may well be an index to propagated activity in these neurones.

Receptor Potentials

The latency of the *a*-wave, or *receptor potential*, was 1·5 msec.; in a later study Brown & Murakami (1964) described a potential change of much shorter latency (less than 0·1 msec.), preceding the positive wave, and they called this the *early receptor potential* (ERP); the ERP could be resolved into two phases by lowering the temperature, an initial positive (1) and a later negative (2) wave. Thus the whole receptor potential complex consists of positive, negative and positive changes, the last being called the *late receptor potential* (3).

EARLY RECEPTOR POTENTIAL

The early receptor potential had an action spectrum corresponding to the absorption of rhodopsin, and its magnitude was linearly related to the degree of bleaching; according to Brown & Gage (1966), neither phase of the ERP was affected by ionic changes in the medium and stage 1 was not affected by very low temperatures; stage 3, the late receptor potential, was affected by ions in the way expected of a change across an electrically excitable membrane. Thus the electrical changes involved in the ERP are those of a different character, and much speculation has gone into the interpretation.

CONTRIBUTION OF PIGMENT EPITHELIUM

It must be emphasized, however, that essentially similar electrical changes occur in pigment cells that are quite clearly not concerned directly with vision,

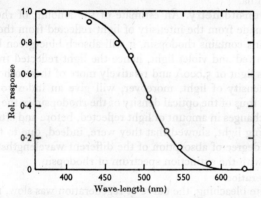

FIG. 870. Action spectrum for the response of the rod outer segments to light measured as a change in impedance of the suspension. Circles are experimental points; the line is the integrated optical density spectrum (difference spectrum) of frog rhodopsin. (Falk & Fatt. *J. Physiol.*)

namely the pigment epithelium of the retina; this may be demonstrated by measuring the ERG of the eye from which the retina has been removed, leaving the pigment epithelium intact. Under these conditions, both early and late receptor potential are recorded (Brown, 1965; Brown & Gage, 1966); the potentials can be recorded intracellularly from the pigment epithelial cells, and the flat action spectrum indicates that melanin is the pigment responsible. Thus the electrical events are the consequence of absorption of light but it seems doubtful whether they are in the direct sequence of events leading to transmission from receptor to bipolar cell.

Impedance of Rods. Perhaps the most fruitful approach to the effects of light on the receptors is that of Falk & Fatt (1968) in which the impedance of suspensions of rods was measured before and after exposure to light. The changes* could be resolved into two components, and their combined effects had an action spectrum corresponding to the absorption of rhodopsin (Fig. 870). Component II seemed most likely to be the change that might be physiologically significant; it was considered to represent a change in the conductance along the surface of the rod rather than across its membrane, or internally, and could have been due, say, to the release of ions into the surface. By contrast, Component I was probably due to a change in membrane conductance, but its magnitude was too small to cause any functionally significant change in membrane potential.

Significance of Dark-Adaptation

Rhodopsin is photosensitive; it not only absorbs light in the process leading to excitation of the rod, but undergoes chemical change. The phenomena of dark-adaptation may therefore be explicable on the basis of the necessity for the regeneration of the rhodopsin, bleached during previous exposure of the eye to light. Thus, if the sensitivity of the rods were a direct function of the concentration of rhodopsin in them, the rods, immediately after a period of light-adaptation, would be insensitive to light, the threshold in these circumstances being determined, in a mixed retina, by the cones. As the period in the dark increased, the amount of rhodopsin extractable from the retina would rise and the sensitivity of the rods would run a parallel course, so that, in the completely dark-adapted eye, sensitivity and concentration of rhodopsin would be at their maxima.

Reflexion Densitometry. An estimate of the amount of rhodopsin in the retina may be made from the intensity of light reflected from the fundus of the eye.† If the retina contains rhodopsin, it will absorb blue-green light of 5,000A in preference to red and violet light, hence the light reflected from the fundus will contain less light of 5,000A and relatively more of these other wavelengths. The loss of intensity of light, moreover, will give an index to the degree of absorption, and thus of the optical density of the rhodopsin in the retina. Careful analysis of the changes in amount of light reflected, before and after exposing the eye to a bleaching light, showed that they were, indeed, due to the presence of rhodopsin, the degree of absorption of the different wavelengths corresponding approximately with the extinction spectrum of rhodopsin.

Rates of Regeneration

After complete bleaching, the curve of regeneration was slow, requiring some 30 minutes for completion in man, and thus being equivalent to his time for complete dark-adaptation. In the cat, where dark-adaptation is considerably slower, the regeneration, measured in this way, took much longer (Weale). Again, in the alligator, which adapts relatively rapidly, the course of regeneration of pigment was correspondingly rapid (Wald *et al.*).

Rhodopsin Concentration and Sensitivity

The relationship between concentration of rhodopsin in the retina and its sensitivity to light is not so simple as at first thought. Thus, exposure of the

---

* Changes due to heat and alterations of pH also occurred, but these could be separated from the other changes discussed here.

† Brindley & Willmer (1952) initiated this technique, and it has been applied by both Rushton (Rushton & Cohen, 1954; Campbell & Rushton, 1955; Rushton, 1959) and Weale (1953–1962) to human and animal eyes. Weale has emphasized that the peripheral retina, although it contains a predominance of rods, nevertheless contains cones, the bleaching of whose pigments may influence the measurements of rhodopsin. His experiments certainly indicate the presence of photolabile material, in addition to rhodopsin, in the peripheral retina. In the fovea, of course, there are only cones and the labile pigments are different from rhodopsin (p. 1648). The limitations of the technique of fundus densitometry have been examined recently by Weale (1965).

dark-adapted eye to a flash that only bleached 1 per cent. of its rhodopsin caused an increase of tenfold in its threshold (Rushton & Cohen), and if the light-thresholds of human subjects were measured at different stages of dark-adaptation, it was found that the rods only became sensitive to light when some 92 per cent. of the rhodopsin had regenerated. Thus, the thousandfold improvement in retinal sensitivity that one obtains during complete dark-adaptation of the rods corresponds to the regeneration of only the last 8 per cent. of the retinal rhodopsin. Rushton's studies, and *in vitro* studies of Dowling, have shown that the relationship between retinal sensitivity and rhodopsin concentration is logarithmic.

An elegant demonstration of this has been made recently by Weinstein, Hobson & Dowling (1967), who measured the changes in concentration of rhodopsin in the isolated rat retina and correlated these with visual sensitivity,

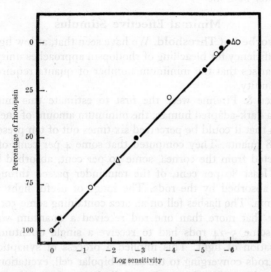

FIG. 871. Showing linear relation between concentration of visual pigment in the rat retina and the logarithm of its sensitivity to light. (Weinstein, Hobson & Dowling. *Nature.*)

as measured by the *b*-wave of the electroretinogram; the isolated retina remains functional in an artificial medium by virtue of its extreme thinness that permits adequate oxygenation, but it does not regenerate rhodopsin after bleaching since it has lost its attachment to the pigment epithelium; thus the preparation is ideal for making controlled changes in the amount of rhodopsin in the unbleached condition. Fig. 871 shows the linear relation between the concentration of rhodopsin and the logarithm of the sensitivity.

Saturation of Rods

The finding that so much of the rhodopsin in a receptor is apparently useless has excited some speculation (see, for example, Wald, 1954); whatever the mechanism, it means that when the eye is exposed to light of sufficient intensity to cause bleaching of only 8 per cent. of its rhodopsin, further increases in intensity of light cannot be appreciated since a visual stimulus must necessarily be the result of a change in rate of bleaching. This certainly corresponds with the finding of Aguilar & Stiles, based on purely psychophysical measurements, that the rods do become "saturated" at relatively low intensities of light, and with

Weale's (1961) measurements of the density of rhodopsin in the retina under these "saturating" conditions.

Summation Pool

According to Rushton (1965), variations in sensitivity of the retina, revealed by the effects of dark- or light-adaptation, are largely determined by changes in the degree of functional convergence of the receptors on higher-order neurones—bipolar cells and ganglion cells. In other words, the power of the retina to exhibit spatial summation of subliminal stimuli varies with the degree of adaptation, being high in the dark-adapted state and negligible in the fully light-adapted state. To account for the experimental relation between rhodopsin content of the retina and visual sensitivity, some sort of feed-back mechanism must be postulated, so that alterations in the rhodopsin content of a rod signal to a "summation pool", which determines how many receptors will converge on a higher-order neurone.*

Minimal Effective Stimulus

Quanta Absorbed at Threshold. We have seen that, at low light intensities, the quantum efficiency for bleaching of rhodopsin approaches unity; the suggestion naturally arises that the minimum number of quanta required to excite a rod is likewise unity.

Hecht, Shlaer & Pirenne were the first to estimate this minimum: they found that, in a dark-adapted human, the minimum amount of energy contained in a flash, such that it could be perceived six times out of ten presentations, was some 54 to 148 quanta. They computed that some 4 per cent. of the incident light was reflected from the cornea, some 50 per cent. absorbed by the ocular media, and at least 80 per cent. of the remainder passed through the retina without being absorbed by the rods. The range of useful light was therefore some 5–14 quanta. The flashes fell on an area containing some 500 rods, so that the probability that more than one rod received a quantum was negligible; consequently, some 5–14 rods had to receive a single quantum in order to evoke the sensation of light. This implied a process of synaptic summation, the stimulated rods converging to the same bipolar cell, excitation of the latter occurring only when a certain number of discharges (5–14) from the separate rods reached it.†

Quantum Fluctuations. Energy corresponding to five quanta is an extremely small amount, so small that fluctuations in the number of quanta in a flash, imposed by the uncertainty principle, become significant. Thus, in the region of the threshold, a flash that is presented once and recognized, is not necessarily seen at the next presentation; this variability in the threshold had been recognized for a long time, and ascribed to a variation in the physiological factors; Hecht, Shlaer & Pirenne, however, pointed out that the variability could be explained by fluctuations in the stimulus, since the number of quanta in a given flash could only be expressed as a probability, not a certainty. By making use of this variability, these authors were able to estimate, by an independent method, the minimum number of quanta required to excite vision. For instance, let us

---

* Donner & Reuter (1967) suggested that the concentration of Metarhodopsin II in the retina might determine the sensitivity of the rods in the initial stages of dark-adaptation; however, Frank & Dowling (1968), using their isolated retina technique, were unable to demonstrate any effects of photoproducts on the sensitivity to light; thus the loss of sensitivity on exposing the retina to light is immediate, whilst photoproducts may be detected in the retina for as long as 30 min. The problems of the mechanism and site of adaptation have been discussed by Dowling (1967).

† That such a summation can occur is beyond doubt, both on histological and physiological grounds.

suppose that n quanta are necessary; the chance of seeing a flash depends, therefore, on the chance that this flash has n or more quanta in it. This probability can be calculated from the Poisson distribution:—

$$P_n = a^n/e^a n$$

where P_n is the probability that the flash will yield n quanta, and a is the *average* number of quanta in a flash, *i.e.*, the number that all flashes would

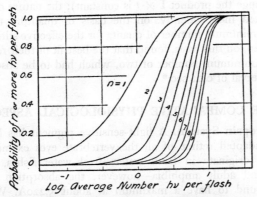

FIG. 872. Poisson probability distributions. For any average number of quanta (*hv*) per flash, the ordinates give the probabilities that the flash will deliver to the retina n or more quanta, depending on the value assumed for n. (Hecht, Shlaer & Pirenne. *J. gen. Physiol.*)

have if there were no quantum fluctuations. In Fig. 872 the probability that there will be n or more quanta in a flash has been plotted against the average number of quanta per flash; it will be seen that the curves are characteristically different according as n varies. If, now, the probability of seeing a flash depends

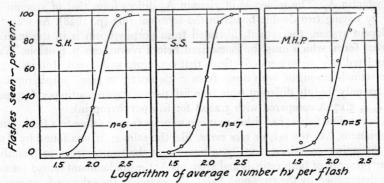

FIG. 873. Relationship between the average energy-content of a flash of light (in number of quanta) and the frequency with which it is seen by three observers. Each point represents 50 flashes, except for S.H. where the number is 35. The curves are the Poisson distributions of Fig. 872 for values of n equal to 5, 6, and 7. (Hecht, Shlaer & Pirenne. *J. gen. Physiol.*)

only on the probability that this flash has a certain number, n, or more, quanta in it, the curve obtained by plotting the *probability of seeing* against the average number of quanta per flash, a, should have a characteristic form, and its exact shape will be given by the *actual* number of quanta necessary. For example, if only one is necessary the rise in the curve will be gradual, whilst if n is 9 the rise will be very steep. In Fig. 873 these frequency-of-seeing curves have been plotted for three observers, the points being frequencies-of-seeing whilst the lines have been computed from the Poisson equation as in Fig. 872; it will be seen that, according to the subject, a number of 5–7 quanta represents the minimum.

The same statistical approach can be made to a number of other visual phenomena; thus, if both the intensity and time of a flash are varied, the threshold stimulus may be expressed as a function of these two variables (e.g., over a certain range the product $I \times t$ is constant); the nature of this function may be computed independently, on the basis of quantum theory, on the assumption of a minimum number of quanta for the effective stimulus. Bouman & Van der Velden, on this basis, found that the best fit of their results could be obtained with a minimum number of two, which had to be absorbed within a definite time-interval of 0·02 sec.*

SOME COMPARATIVE PHYSIOLOGICAL ASPECTS

The Porphyropsin System. A light-sensitive pigment has been extracted from the dark-adapted retinæ of all the vertebrate eyes examined; in those vertebrates that originate in fresh water—fresh-water fishes, lampreys and certain larval and adult amphibia—however, the absorption of the visual pigment was found to have a maximum at about 5,220A. Wald therefore classified the pigments into two main divisions—the *rhodopsins*, being pigments with λ_{max} about 5,000A, found in marine fishes and mammals, and the *porphyropsins*, pigments from fresh-water fishes having λ_{max} at about 5,220A. Wald showed that the porphyropsins went through a similar cycle to that of frog or ox rhodopsin, but the retinal and vitamin A extracted from the fresh-water fish were different, their absorption spectra being correspondingly shifted by 200 to 230A towards the red. The retinal and vitamin A were called retinal$_2$ and vitamin A$_2$, whilst those from the rhodopsin system were called retinal$_1$ and vitamin A$_1$. The molecule of vitamin A$_2$ differs from that of vitamin A$_1$ only by having two double-bonds in the terminal ring (p. 1628). As with the rhodopsin system, the retinal$_2$ liberated from porphyropsin is in the inactive all-*trans* form, which must be isomerized before resynthesis is possible. Wald has isolated two *cis*-isomers of retinal$_2$ that conjugate with opsin. Interestingly, these retinals conjugate with opsin from cattle retinæ to give "porphyropsins" that are only slightly different from true fish porphyropsin (cattle porphyropsin had λ_{max} 5,170A compared with 5,220A for fish porphyropsin).

In euryhaline fishes, *i.e.*, fishes that can migrate from a saline to a fresh-water environment, like the eel, or *vice versa*, like the salmon, it was found by Wald that the retinæ contained, usually, a mixture of the two systems, as identified by the presence of both retinals and vitamins A; the predominant retinal seemed to be determined by the spawning habit. For example, the salmonids, spawning in fresh water, had predominantly retinal$_2$, *i.e.*, porphyropsin, systems, whilst eels had predominantly retinal$_1$, or rhodopsin, systems.

Development of New Pigments. As Dartnall has emphasized, we must take care lest Wald's classification of the scotopic visual systems obscure

---

* Rushton, by measuring the light reflected from the human retina before and after bleaching, has confirmed Hecht, Shlaer & Pirenne's estimate that only 10 per cent. of the light falling on the cornea ($\lambda = 5,000A$) is absorbed by the retinal rods of the fully dark-adapted eye. He and Weale (1962) discuss in some detail estimates of the minimal stimulus and emphasize that the higher limit of the 5–14 quanta given by Hecht *et al.* is likely to be near the true value. Weale (1961) has computed, from the flux required to bleach a given percentage of the pigment, and the flux for threshold vision, in man, that the minimum number of quanta is 7–14.

differences within the classes. Rhodopsin, for example, has a different λ_{max} according as it is derived from frog (5,020A), cattle, rat or dogfish (4,980A), or the squid (4,930A). Again, according to Dartnall's studies, the "porphyropsin" of fresh-water fishes may have λ_{max} of 5,230A, if extracted from the carp, or of 5,330A, if extracted from the pike. In the euryhaline teleost *Gillichthys mirabilis* (mudsucker), Munz found a retinal extract with λ_{max} 5,120A, and by applying Dartnall's test for homogeneity (p. 1624) he showed that the extract was not a mixture of rhodopsin and porphyropsin, as would have been assumed on the basis of Wald's general findings. The chromophore isolated from this retina was retinal$_1$, so that, according to Wald's classification, it would be "rhodopsin". The mudsucker has thus developed a new pigment, with a λ_{max} intermediate between those of Wald's rhodopsin and porphyropsin, instead of using a mixture of these pigments, as apparently happens with many other euryhaline fishes. Presumably the shift of λ_{max} of the scotopic visual pigment towards the red in fresh-water fishes has some adaptive significance, although this is not clear. In the deep-sea fishes, the absorption maxima of the visual pigments in the retinæ apparently correspond with the predominant wavelength in very deep water,* namely about 4,750A. Thus, Denton & Warren found the λ_{max} of the visual pigments of four species of deep-sea fishes as 4,800A, whilst Walker obtained λ_{max} of 4,870 and 4,970A for retinal extracts of the conger eel and gurnard respectively. Subsequent studies, notably those of Wald, Brown & Brown and of Munz, have confirmed the presence of scotopic visual pigments with λ_{max} about 4,800 in deep-sea fishes, and suggest that there is a continuous gradation of λ_{max} with depth of habitat. All these pigments are based on retinal$_1$, so that they may be classed as rhodopsins; the retinal, moreover, has the same 11-*cis* configuration, so that Wald argues that the differences among pigments in the one or other class are due to differences in the nature of the opsin.

Spectral Clustering of Pigments. Dartnall & Lythgoe (1965) have examined the λ_{max} of a large number of visual pigments based on retinal$_1$, whilst Bridges (1964) has made a similar study of pigments based on retinal$_2$ or a mixture of this and retinal$_1$. As Fig. 874 shows, there is a tendency for the λ_{max} to cluster around preferred wavelengths. When a retina contains a mixture of retinals, we may expect the two pigments to make use of the same opsin, otherwise we should expect to find four pigments. In Fig. 875 the λ_{max} of the pigment based on retinal$_1$ has been plotted against the λ_{max} of the pigment based on retinal$_2$ for those fishes that have two pigments in their retinæ; it will be seen that there is a linear relation between the two and also that two artificial pigments, made by mixing the retinal of one retina with the opsin of another, also conform with this relationship. Thus, for a given opsin and retinal, we may predict the effect of mixing this opsin with another retinal; these predictions are incorporated into Fig. 874 for the retinal$_1$ pigments, and it is interesting that, with the four peaks at 4,940, 5,005, 5,065 and 5,120A, the predicted λ_{max} for the corresponding retinal$_2$ pigments coincide fairly closely with the humps on the retinal$_2$ λ_{max} distributions.

Temporal Variations. The pigment composition of a retina is not invariable; for example, immature (yellow) eels have a mixture of pigments based on retinal$_1$ and retinal$_2$, whereas the adult (silver) eel has only the pigment based on retinal$_1$ (Wald, 1960). Again, Dartnall, Lander & Munz found that, in a

---

* In coastal waters green light penetrates most effectively, whereas in such deep waters as the Sargasso Sea and Gulf Stream blue light reaches by far the greatest depth (Oster & Clarke, 1936; Clarke & James, 1939).

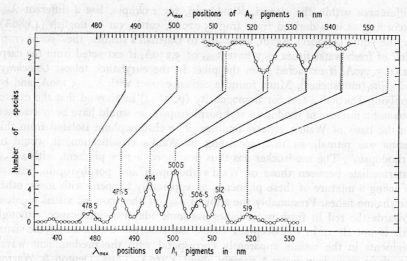

FIG. 874. The relationship between the spectral distributions of A_1 and A_2 pigments in teleost fishes. The data for the A_1 pigments are derived from those fishes that possess *only* A_1 pigments; the data for the A_2 pigments are not restricted: *i.e.* they include all fishes that possess A_2 pigments, some of which have A_1 pigments as well. Note that the cluster positions for the A_1 pigments, when translated (sloping continuous lines) by means of Fig. 875, indicate theoretical positions for A_2 pigments (upper set of vertical dotted lines), some of which agree closely with those actually found (vertical continuous lines). (Dartnall & Lythgoe. *Vision Res.*)

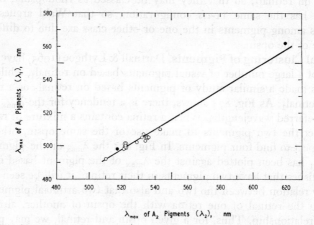

FIG. 875. The relationship between the A_2 pigments in identical-opsin pairs. Open circles represent the sixteen naturally-occurring pairs; filled circles, the data obtained in Wald, Brown and Smith's direct substitution experiments. (Dartnall & Lythgoe. *Vision Res.*)

given population of rudd, the proportions of the two pigments based on retinals 1 and 2 varied with the time of year; by artificially varying the environmental illumination of fish kept in an aquarium, Dartnall *et al.* were able to cause changes in the proportions of the two pigments.

Photopic Vision

Rod and Cone Thresholds. Vision in most vertebrates is duplex in nature, mediated by rods and cones; the latter are the less sensitive receptors, and this was until recently considered to be due to a lower concentration of pigment in them, which would decrease the probability of absorption of a quantum of incident light. According to the present view, based largely on Pirenne's work, the two types of receptor may actually have the same threshold, *i.e.*, a quantum of light has the same chance of inducing photochemical change in both, so that the difference in sensitivity resides in the different organization of the receptors. The rods are so organized that they permit the summation of their effects, so that when a patch of light falls on the peripheral retina, the effects of the light on individual rods will converge on to a single bipolar cell. When the size of the illuminated field on the retina is reduced, the rod-threshold becomes much higher and, according to Arden & Weale, approaches that of the cones.*

Iodopsin and Cyanopsin. The wavelength for maximum sensitivity of the eye to light under conditions of high illumination, *i.e.*, when only the cones are operative in determining visual sensation, is different from that of the eye under scotopic, or low-intensity conditions. As we have seen, the wavelength of maximal sensitivity under scotopic conditions corresponds with the wavelength of maximal absorption of rhodopsin, 4,970A. The wavelength of maximal sensitivity for cone vision is approximately 5,620, in the yellow-green, and investigators have been tempted to seek a pigment in the cones with a λ_{max} of this order. In 1937, Wald extracted from the chicken retina a pigment *iodopsin*, which had an absorption maximum at 5,620A. Subsequent studies showed it to be a chromoprotein, containing retinal$_1$ linked to a protein which Wald called *photopsin*, to distinguish it from the *scotopsins*, the protein moieties of the scotopic pigments. It would seem that it undergoes the same series of reactions as that undergone by rhodopsin, in the sense that neoretinene-b is isomerized during bleaching to the all-*trans* form; that this forms vitamin A, and so on (Hubbard, Brown & Kropf).

Cyanopsin

By analogy with the scotopic system, we may expect to produce a new pigment by condensing retinal$_2$ with the opsin from chick retina; in fact, Wald, Brown & Smith have done this, giving the name *cyanopsin* to the product which has a λ_{max} of 6,200A.

Pigeon Cone Pigment. The predominantly cone-retina of the pigeon has been extracted by Bridges who has identified a new pigment, in addition to the rhodopsin that this retina also contains, with λ_{max} 5,440A. Thus the cone-pigment of this bird is not iodopsin, and yet the wavelength for maximal visual sensitivity is in the region of 5,800 to 5,900A (Blough, 1957).

Colour Vision. It would be misleading, however, to suggest that rod and cone vision are comparable in respect to a relationship between visual sensitivity and pigment absorption maxima. Variations of wavelength of light incident on the rod-dominated retina evoke only variations in intensity of the sensation of light—they produce no qualitative changes; by contrast, in cone vision, similar variations produce changes in quality that we term changes in colour or hue. It is beyond the scope of this book to discuss the physiological basis for these variations in sensation; suffice it to say that a system of at least three types of receptor, containing three types of pigment or three different pigment mixtures, must be postulated. The sensation of light intensity may well be the result of the integrated action of light on all three types of cone, so that maximal sensitivity may occur at a wavelength that is identical with the maximal absorption of none of the individual pigments. The techniques of reflexion densitometry, as we have seen, permit the identification of rhodopsin in the living retina; when applied to the

* Pirenne (1962, p. 106) has criticized the conclusion that the thresholds are actually equal.

fovea by Rushton, and to both fovea and peripheral retina by Weale, the presence of non-rhodopsin photosensitive pigments has been demonstrated in human and animal eyes. These have been called erythrolabe and chlorolabe by Rushton, *i.e.*, red- and green-sensitive pigments with λ_{max} of 5,900A and 5,400A respectively. Weale has called pigments with similar λ_{max} secundochrome and primochrome respectively.

Absorption Spectra of Cones. A more precise characterization of the types of pigment in the cones is given by measuring the absorption spectra of individual cones exposed to a minute pencil of light on a microscope slide (Marks, Dobelle & MacNichol, 1964; Brown & Wald, 1964); as Fig. 876 shows, three pigments

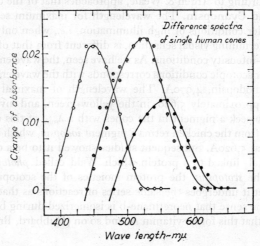

Fig. 876. Difference spectra of the visual pigments in single cones of the human parafovea. In each case the absorption spectrum was recorded in the dark, then again after bleaching with a flash of yellow light. The differences between these spectra are shown. They involve one blue-sensitive cone, two green-sensitive cones, and one red-sensitive cone. (Brown & Wald. *Science.*)

can be identified with λ_{max} in the blue, green and red as predicted by the three-colour theory; there are thus three types of cone differing in their contained pigment, and the sensation evoked by a given wavelength is determined by the extents to which the three cones absorb this wavelength.

Action Spectra of Cones. Again, Tomita *et al.* (1967) impaled single cones* of the carp retina; these had a resting potential of about 30 mV which exhibited a slow hyperpolarizing response to light. When different wavelengths of light were used, a typical action spectrum for the electrical response was obtained for a given cone, and when different cones were examined the action spectra fell into three classes, with λ_{max} at 4,620, 5,290 and 6,110A respectively, comparing with absorption maxima of single goldfish cones of 4,550, 5,300 and 6,250A measured by Marks (1965).

* The potentials recorded by Tomita *et al.* were definitely from impaled inner cone segments; the S-potentials described by earlier workers, *e.g.* Naka & Rushton (1966), are probably not derived from impaled cones, but from cells in the layers between the ganglion cell layer and the layer of cones. The character of the S-potential has been discussed by Tomita (1965).

REFERENCES

ABRAHAMSON, E. W. & OSTROY, S. E. (1967). The photochemical and macromolecular aspects of vision. *Progr. Biophys.*, **17**, 181–215.

AGUILAR, M., & STILES, W. S. (1954). Saturation of the rod mechanism of the retina at high levels of stimulation. *Optica Acta*, **1**, 59–65.

ARDEN, G. B., & WEALE, R. A. (1954). Nervous mechanisms and dark-adaptation. *J. Physiol.*, **125**, 417–426.

BLISS, A. F. (1948). "Mechanism of Retinal Vitamin A Formation." *J. biol. Chem.* **172**, 165.

BLOUGH, D. S. (1957). Spectral sensitivity in the pigeon. *J. Opt. Soc. Amer.*, **47**, 827–833.

BOUMAN, M. A., & VAN DER VELDEN, H. A. (1947). "The Two-quanta Explanation of the Dependence of the Threshold Values and Visual Acuity on the Visual Angle and the Time of Observation." *J. Opt. Soc. Amer.*, **37**, 908.

BOWNDS, D. & WALD, G. (1965). Reaction of the rhodopsin chromophore with sodium borohydride. *Nature*, **205**, 254–257.

BRIDGES, C. D. B. (1961). Studies on the flash photolysis of visual pigments. I and II. *Biochem. J.*, **79**, 128–134; 135–143.

BRIDGES, C. D. B. (1962). Visual pigments of the pigeon (*Columba livia*). *Vision Res.*, **2**, 125–137.

BRIDGES, C. D. B. (1964). The distribution of visual pigments in freshwater fishes. *Abstr. Fourth Internat. Congr. Photobiol.*, Oxford, p. 53. Bucks: Beacon Press.

BRINDLEY, G. S., & WILLMER, E. N. (1952). "The Reflexion of Light from the Macular and Peripheral Fundus Oculi in Man." *J. Physiol.*, **116**, 350–356.

BRODA, E. E., GOODEVE, C. F., & LYTHGOE, R. J. (1940). "The Weight of the Chromophore Carrier in the Visual Purple Molecule." *J. Physiol.*, **98**, 397.

BROWN, K. T. (1965). An early potential evoked by light from the pigment epithelium-choroid complex of the eye of the toad. *Nature*, **209**, 1249–1253.

BROWN, K. T. (1968). The electroretinogram: its components and their origins. *Vision Res.*, **8**, 633–677.

BROWN, K. T & CRAWFORD, J. M. (1967). Melanin and the rapid light-evoked responses from pigment epithelium cells of the frog eye. *Vision Res.*, **7**, 165–178.

BROWN, K. T. & CRAWFORD, J. M. (1967). Intracellular recording of rapid light-evoked responses from pigment epithelium cells of the frog eye. *Vision Res.*, **7**, 149–163.

BROWN, K. T. & GAGE, P. W. (1966). An earlier phase of the light-evoked electrical response from the pigment epithelium-choroid complex of the eye of the toad. *Nature*, **211**, 155–158.

BROWN, K. T. & MURAKAMI, M. (1964). A new receptor potential of the monkey retina with no detectable latency. *Nature*, **201**, 626–628.

BROWN, K. T. & WATANABE, K. (1962). Isolation and identification of a receptor potential from the pure cone fovea of the monkey retina. *Nature*, **193**, 958–960.

BROWN, P. K. & WALD, G. (1964). Visual pigments in single rods and cones of the human retina. *Science*, **144**, 45–52.

CAMPBELL, F. W., & RUSHTON, W. A. H. (1955). Measurement of the scotopic pigment in the living human eye. *J. Physiol.*, **130**, 131–147.

CHASE, A. M., & SMITH, E. L. (1940). "Regeneration of Visual Purple in Solution." *J. gen. Physiol.*, **23**, 21–39.

CLARKE, G. L., & JAMES, H. R. (1939). "Laboratory Analysis of the Selective Absorption of Light by Sea Water." *J. Opt. Soc. Amer.*, **29**, 43–55.

COHEN, A. I. (1963). The fine structure of the visual receptors of the pigeon. *Exp. Eye Res.*, **2**, 88–97.

COHEN, A. I. (1968). New evidence supporting the linkage to extracellular space of outer segment saccules of frog cones but not rods. *J. Cell Biol.*, **37**, 424–444.

COLLINS, F. D., & MORTON, R. A. (1950). "Studies on Rhodopsin. 1. Methods of Extraction and the Absorption Spectrum." *Biochem. J.*, **47**, 3.

COLLINS, F. D., & MORTON, R. A. (1950). "Studies on Rhodopsin. 2. Indicator Yellow." *Biochem. J.*, **47**, 10.

COLLINS, F. D., & MORTON, R. A. (1950). "Studies on Rhodopsin. 3. Rhodopsin and Transient Orange." *Biochem. J.*, **47**, 18.

CRESCITELLI, F., & DARTNALL, H. J. A. (1953). "Human Visual Purple." *Nature*, **172**, 195–196.

DARTNALL, H. J. A. (1957). *The Visual Pigments*. Methuen, London.

DARTNALL, H. J. A. (1961). Visual pigments before and after extraction from visual cells. *Proc. Roy. Soc., B,* **154,** 250–266.

DARTNALL, H. J. A. (1962). The photobiology of visual processes. In *The Eye.* Ed. Davson. Vol. II. New York: Academic Press, pp. 323–574.

DARTNALL, H. J. A., GOODEVE, C. F. & LYTHGOE, R. J. (1936). "Quantitative Analysis of the Photochemical Bleaching of Visual Purple Solutions in Monochromatic Light." *Proc. Roy. Soc., A,* **156,** 158.

DARTNALL, H. J. A., LANDER, M. R. & MUNZ, F. W. (1961). Periodic changes in the visual pigment of a fish. In *Progress in Photobiology.* Amsterdam: Elsevier, pp. 203–213.

DARTNALL, H. J. A. & LYTHGOE, J. N. (1965). The spectral clustering of visual pigments. *Vision Res.,* **5,** 81–100.

DENTON, E. J., & WARREN, F. J. (1956). "Visual Pigments of Deep-sea Fish." *Nature,* **178,** 1059.

DONNER, K. O. & REUTER, T. (1967). Dark-adaptation processes in the rhodopsin rods of the frog's retina. *Vision Res.,* **7,** 17–41.

DOWLING, J. E. (1960). Chemistry of visual adaptation in the rat. *Nature,* **188,** 114–118.

DOWLING, J. E. (1967). The site of visual adaptation. *Science,* **155,** 273–279.

ERHARDT, F., OSTROY, S. E. & ABRAHAMSON, E. W. (1966). Protein configuration changes in the photolysis of rhodopsin. I. The thermal decay of cattle rhodopsin *in vitro. Biochim. biophys. Acta,* **112,** 256–264.

FALK, G. & FATT, P. (1968). Conductance changes produced by light in rod outer segments. *J. Physiol.,* **198,** 647–699.

FRANK, R. N. & DOWLING, J. E. (1968). Rhodopsin photoproducts: effects on electroretinogram sensitivity in isolated perfused rat retina. *Science,* **161,** 487–489.

GRELLMAN, K.-H., LIVINGSTON, R., & PRATT, D. (1962). A flash photolytic investigation of rhodopsin at low temperatures. *Nature,* **193,** 1258–1260.

HAGINS, W. A. (1955). The quantum efficiency of bleaching of rhodopsin *in situ. J. Physiol.,* **129,** 22–23P.

HARTLINE, H. K. (1940). Nerve messages in the fibres of the visual pathway. *J. Opt. Soc. Amer.,* **30,** 239.

HECHT, S. (1921). "Relation between the Wavelength of Light and its Effect on the Photosensory Process." *J. gen. Physiol.,* **3,** 375.

HECHT, S., & PICKELS, E. G. (1938). "Sedimentation Constant of Visual Purple." *Proc. Nat. Acad. Sci., Wash.,* **24,** 172.

HECHT, S., SHLAER, S., & PIRENNE, M. H. (1942). "Energy, Quanta and Vision." *J. gen. Physiol.,* **25,** 819.

HUBBARD, R. (1954). "The Molecular Weight of Rhodopsin and the Nature of the Rhodopsin-digitonin Complex." *J. gen. Physiol.,* **37,** 373–379

HUBBARD, R., BOWNDS, G. & YOSHIZAWA, T. (1965). The chemistry of visual photoreception. *Cold Spr. Harb. Symp. quant. Biol.,* **30,** 301–315.

HUBBARD, R., BROWN, P. K., & KROPF, A. (1959). Action of light on visual pigments. *Nature,* **183,** 442–450.

HUBBARD, R., & COLMAN, A. D. (1959). Vitamin A content of the frog eye during light and dark adaptation. *Science,* **130,** 977–978.

HUBBARD, R., & DOWLING, J. E. (1962). Formation and utilization of 11-*cis* vitamin A by the eye tissues during light and dark adaptation. *Nature,* **193,** 341–343.

HUBBARD, R., & KROPF, A. (1958). The action of light on the eye. *Proc. Nat. Acad. Sci. Wash.,* **44,** 130–139.

HUBBARD, R., & KROPF, A. (1959). Molecular aspects of visual excitation. *Ann. N.Y. Acad. Sci.,* **81,** 388–398.

HUBBARD, R., & WALD, G. (1952). "Cis-trans Isomers of Vitamin A and Retinene in the Rhodopsin System." *J. gen. Physiol.,* **36,** 269–315.

JANCSÓ, N.v., & JANCSÓ, H.v. (1936). Fluoreszenzmikroskopische Beobachtung der reversiblen Vitamin-A Bildung in der Netzhaut wahrend des Sehaktes. *Biochem. Z.,* **287,** 289–290.

KITO, Y., ISHIGAMI, M., & YOSHIZAWA, T. (1961). On the labile intermediate of rhodopsin as demonstrated by low temperature illumination. *Biochim. biophys. Acta,* **48,** 287–298.

KRINSKY, N. I. (1958). The lipoprotein nature of rhodopsin. *Arch. Opthal.,* **60,** 688.

KROPF, A., & HUBBARD, R. (1958). The mechanism of bleaching rhodopsin. *Ann. N.Y. Acad. Sci.,* **74,** 266–280.

LINSCHITZ, H., WULFF, V. J., ADAMS, R. G., & ABRAHAMSON, E. W. (1957). Light-initiated changes of rhodopsin in solution. *Arch. Biochem. Biophys.,* **68,** 233–236.

LYTHGOE, R. J. (1937). "Absorption Spectra of Visual Purple and of Indicator Yellow." *J. Physiol.*, **89**, 331.

LYTHGOE, R. J., & QUILLIAM, J. P. (1938). "Thermal Decomposition of Visual Purple." *J. Physiol.*, **93**, 24.

LYTHGOE, R. J., & QUILLIAM, J. P. (1938). "Relation of Transient Orange to Visual Purple and Indicator Yellow." *J. Physiol.*, **94**, 399.

MARKS, W. B. (1965). Visual pigments of single goldfish cones. *J. Physiol.*, **178**, 14–32.

MARKS, W. B., DOBELLE, W. H. & MACNICHOL, E. F. (1964). Visual pigments of single primate cones. *Science*, **143**, 1181–1183.

MAST, S. O. (1938). "Factors involved in the Process of Orientation of Lower Organisms in Light." *Biol. Rev.*, **13**, 186.

MATSUSAKA, T. (1967). Lamellar bodies in the synaptic cytoplasm of the accessory cone from the chick retina as revealed by electron microscopy. *J. Ultrastr. Res.*, **18**, 55–70.

MATTHEWS, R. G., HUBBARD, R., BROWN, P. K. & WALD, G. (1963). Tautomeric forms of metarhodopsin. *J. gen. Physiol.*, **47**, 215–240.

MOODY, M. F. & ROBERTSON, J. D. (1960). The fine structure of some retinal photo-receptors. *J. biophys. biochem. Cytol.*, **7**, 87–92.

MORTON, R. A. & GOODWIN, T. W. (1944). Preparation of retinene *in vitro. Nature*, **153**, 405–406.

MUNZ, F. W. (1956). "A new Photosensitive Pigment of the Euryhaline Teleost, *Gillichthys mirabilis.*" *J. gen. Physiol.*, **40**, 233–249.

MUNZ, F. W. (1958). "Photosensitive Pigments from the Retinae of certain Deep-sea Fishes." *J. Physiol.*, **140**, 220–235.

NAKA, K. I. & RUSHTON, W. A. H. (1966). S-potentials from colour units in the retina of fish (*Cypridinae*). *J. Physiol.*, **185**, 536–555.

NILSSON, S. E. G. (1964). Receptor cell outer segment development and ultrastructure of the disk membranes in the retina of the tadpole (*Rana pipiens*). *J. Ultrastr. Res.*, **11**, 581–620.

NILSSON, S. E. G. (1965). The ultrastructure of the receptor outer segments in the retina of the leopard frog (*Rana pipiens*). *J. Ultrastr. Res.*, **12**, 207–231.

OSTER, R. H., & CLARKE, G. L. (1936). *J. Opt. Soc. Amer.*, **25**, 84. (Quoted by Wald, Brown & Brown, 1957.)

OSTROY, S. E., ERHARDT, F. & ABRAHAMSON, E. W. (1966). Protein configurational changes in the photolysis of rhodopsin. II. *Biochim. biophys. Acta*, **112**, 265–277.

OSTROY, S. E., RUDNEY, H. & ABRAHAMSON, E. W. (1966). The sulfhydryl groups of rhodopsin. *Biochim. biophys. Acta*, **126**, 409–412.

PIRENNE, M. H. (1962). Visual functions in man. In *The Eye*. Ed. Davson. Vol. II. New York: Academic Press.

PITT, G. A. J., COLLINS, F. D., MORTON, R. A., & STOK, P. (1955). "Retinylidenemethyl-amine an Indicator Yellow Analogue." *Biochem. J.*, **59**, 122–128.

PRATT, D. C., LIVINGSTON, R. & GRELLMAN, K. H. (1964). Flash photolysis of rod particle suspensions. *Photochemistry and Photobiology*, **3**, 121. (Quoted by Abrahamson & Ostroy.)

DE ROBERTIS, E. (1956). "Electron Microscope Observations on the Submicroscopic Organization of the Retinal Rods." *J. biophys. biochem Cytol.*, **2**, 310–330.

RUSHTON, W. A. H. (1956). "The Difference Spectrum and the Photosensitivity of Rhodopsin in the Living Human Eye." *J. Physiol.*, **134**, 11–29.

RUSHTON, W. A. H. (1956). "The Rhodopsin Density in the Human Rods." *J. Physiol.*, **134**, 30–46.

RUSHTON, W. A. H. (1959). Visual pigments in man and animals and their relation to seeing. *Progr. Biophys.*, **9**, 239–283.

RUSHTON, W. A. H. (1965). Visual adaptation. *Proc. Roy. Soc.*, B, **162**, 20–40.

RUSHTON, W. A. H., & COHEN, R. D. (1954). "Visual Purple Level and the Course of Dark Adaptation." *Nature*, **173**, 301–302.

SAITO, Z. (1938). Isolierung der Stäbchenaussenglieder und spektraler Untersuchung des deraus hergestellten Sehpurpurextraktes. *Tohoku J. exp. Med.*, **32**, 432–446.

SJÖSTRAND, F. S. (1953). "The Ultrastructure of the Inner Segments of the Retinal Rods of the Guinea Pig Eye as revealed by Electron Microscopy." *J. cell. comp. Physiol.*, **42**, 45–70.

SMITH, E. L. & PICKELS, E. G. (1940). Micelle formation in aqueous solutions of digitonin. *Proc. Nat. Acad. Sci., Wash.*, **26**, 272–277.

TOMITA, T. (1965). Electrophysiological study of the mechanisms subserving color coding in the fish retina. *Cold Spr. Harb. Symp. quant. Biol.*, **30**, 559–566.

TOMITA, T., KANEKO, A., MURAKAMI, M. & PAUTLER, E. L. (1967). Spectral response curves of single cones in the carp. *Vision Res.*, **7**, 519–531.

WALD, G. (1935). "Carotenoids and the Visual Cycle." *J. gen. Physiol.*, **19**, 351.

WALD, G. (1937). "Visual Purple System in Fresh-water Fishes." *Nature*, **139**, 1017.

WALD, G. (1937). "Photolabile Pigments of the Chicken Retina." *Nature*, **140**, 545.

WALD, G. (1938). "On Rhodopsin in Solution." *J. gen. Physiol.*, **21**, 795.

WALD, G. (1939). "The Porphyropsin Visual System." *J. gen. Physiol.*, **22**, 775.

WALD, G. (1945/46). "The Chemical Evolution of Vision." *Harvey Lectures*, **41**, 117

WALD, G. (1954). "On the Mechanism of the Visual Threshold and Visual Adaptation." *Science*, **119**, 887–892.

WALD, G. (1960). The significance of vertebrate metamorphosis. *Circulation*, **21**, 916–938.

WALD, G. (1965). Receptor mechanisms in human vision. *Proc. 13th. Int. Physiol. Congr.* Tokyo. Lectures & Symp, pp. 69–79.

WALD, G., & BROWN, P. K. (1952). "The Role of Sulphydryl Groups in the Bleaching and Synthesis of Rhodopsin." *J. gen. Physiol.*, **35**, 797–821.

WALD, G., BROWN, P. K., & BROWN, P. S. (1957). "Visual Pigments and Depth of Habitat of Marine Fishes." *Nature*, **180**, 969–971.

WALD, G., BROWN, P. K. & GIBBONS, I. R. (1963). The problem of visual excitation. *J. opt. Soc. Amer.*, **53**, 20–35.

WALD, G., BROWN, P. K., & KENNEDY, D. (1957). The visual system of the alligator. *J. gen. Physiol.*, **40**, 703–713.

WALD, G., BROWN, P. K., & SMITH, P. H. (1953). "Cyanopsin, a new Pigment of Cone Vision." *Science*, **118**, 505–508.

WALD, G., BROWN, P. K., & SMITH, P. H. (1955). "Iodopsin." *J. gen. Physiol.*, **38**, 623–681.

WALD, G., & HUBBARD, R. (1949). "The Reduction of Retinene to Vitamin-A *in vitro*." *J. gen. Physiol.*, **32**, 367.

WALKER, M. A. (1956). Homogeneity tests on visual pigment solutions from two sea fish. *J. Physiol.*, **133**, 56 P.

WEALE, R. A. (1953). Photochemical reactions in the living cat's retina. *J. Physiol.*, **122**, 322–331.

WEALE, R. A. (1955). Bleaching experiments on eyes of living grey squirrels (*Sciurus carolinensis leucotis*). *J. Physiol.*, **127**, 587–591.

WEALE, R. A. (1959). Photo-sensitive reactions in foveæ of normal and cone-monochromatic observers. *Optica Acta*, **6**, 158–174.

WEALE, R. A. (1961). Limits of human vision. *Nature*, **191**, 471–473.

WEALE, R. A. (1961). The duplicity theory of vision. *Ann. Roy. Coll. Surg.*, **28**, 16–35

WEALE, R. A. (1962). Further studies of photo-chemical reactions in living human eyes. *Vision Res.*, **1**, 354–378.

WEALE, R. A. (1965). Vision and fundus reflectometry: a review. *Photochemistry & Photobiology*, **4**, 67–87.

WEINSTEIN, G. W., HOBSON, R. R. & DOWLING, J. E. (1967). Light and dark adaptation in the isolated rat retina. *Nature*, **215**, 134–138.

WILLIAMS, T. P. (1964). Photoreversal of rhodopsin bleaching. *J. gen. Physiol.*, **47**, 679–689.

WILLIAMS, T. P. (1965). Rhodopsin bleaching: relative effectiveness of high and low intensity flashes. *Vision Res.*, **5**, 633–638.

WULFF, V. J., ADAMS, R. G., LINSCHITZ, H., & KENNEDY, D. (1958). The behaviour of flash-illuminated rhodopsin in solution. *Arch. Ophthal.*, **60**, 695–701.

YOSHIZAWA, T., KITO, Y., & ISHIGAMI, M. (1960). Studies on the metastable states in the rhodopsin cycle. *Biochim. biophys. Acta*, **43**, 329–334.

YOSHIZAWA, T. & WALD, G. (1963). Pre-lumirhodopsin and the bleaching of visual pigment. *Nature*, **197**, 1279–1286.

YOUNG, R. W. (1967). The renewal of receptor cell outer segments. *J. Cell Biol.*, **33**, 61–72.

YOUNG, R. W. & DROZ, B. (1968). The renewal of protein in retinal rods and cones. *J. Cell Biol.*, **39**, 169–184.

INDEX

α-actinin, 1379, 1461, 1462
α-adrenergic receptor, 1340
α-chains of collagen, 264
α-chitin, 249
α-collagen, 263
α-configuration, 85, 1354
α-disaccharide, 78
α-elevation of action potential, 1074
α-gelatin, 266
α-helix, 85, 87
  of membrane, 475
  relation to muscle contraction, 1447
α-keratin, 85
α-phosphorescence, 1587
A-band, 1347, 1355
  behaviour during contraction, 1366
  birefringence of, 1355
  effect of stretch on, 1367
  localization of myosin in, 1355
A-elevation of action potential, 1074
A-filaments, 1357
  behaviour during shortening, 1367
A substance, 1355
a-wave of ERG, 1639
Absolute refractory period, 1065, 1075, 1088
Absorption, by plant roots, 671
Absorption, intestinal, 751
  of amino-acids, 778
  of autogenous serum, 753
  of fats, 787
  of proteins, 798
  of sugars, 762
  of water and salts, 753, 775, 777
Absorption band, 1486
Absorption droplets, 35
Absorption spectrum, 1483, 1486, 1622, 1624
  of chlorophyll, 1499
  of cones, 1648
  of rhodopsin, 1621, 1626
Acceleration of heart, 1301
Acceptor site, 202
Accessory pigments, 1497
  role in photosynthesis, 1502
Acclimatization, 306, 347
  of poikilotherms, 351
  to cold, 306, 347
  to heat, 349
Accommodation, 1053, 1056
  in sensory nerve, 1244, 1246, 1266
  mechanism of, 1090
Accumulation, 373
  of ions by plant roots, 671
  of K+ by bacteria, 689
    by erythrocyte, 555, 610
    by muscle and nerve, 630
    by yeast, 686
Acetal phosphatide, 630
Acetamide, renal excretion of 857, 906, 907, 908
Acetazoleamide (Diamox) effect of on acidification of turtle urine, 609
  on c.s.f., 739
  on gastric secretion, 831
  on renal excretion, 911
  on salt gland, 962

Acetoacetate, renal excretion of, 880
Acetylcholine, 1134
  block, 1163
  chemical formula of, 1159
  contracture 1414, 1456
  effect of, on ABRM, 1466
    on autonomic ganglion, 1196, 1198
    on electric organ, 1228
    on end-plate, 1143
    on heart, 1282, 1300
    on inside of muscle fibre, 1147
    on nerve fibres, 1144
    on pigmentary response, 1581
    on pre- and post-synaptic neurones, 1269
    on Renshaw cells, 1216
    on slow muscle, 1175
    on smooth muscle, 1337, 1463
    nominal effective dose, 444, 1146
    muscarine action, 1135
    nicotine action, 1135
    release from ganglion, 1183
    sensitivity of muscle fibre to, 1157, 1180
    splitting by cholinesterase, 1137
    synthesis of, 1166
Acetyl coenzyme A, 286
Acetylglucosamine, 41, 249, 269
Acetylmuramic acid, 41
Acetylsalicylic acid, 878
Achromatic apparatus, 112
  origin of, 116
Acid, excretion by yeast, 686
Acid-base balance, 926
Acid-base equilibria of c.s.f., 735
Acid-fast bacteria, 40
Acid phosphatase, effect on nucleic acids, 167
  in autosomes, 36
  in Golgi zone, 34
  in monocytes, 32
Acidification of urine, 926
  effect of K+ on, 936
  sites of, 928
Acidosis, effect on c.s.f. secretion, 740
  gastric secretion, and, 832
Acrosome, structure of, 141
Acrosome filament, 140
Acrosome reaction, 140
Actin, 1354
  G-F transformation, 1378
  helix of, 1360, 1374
  linkage with myosin, 1379
  smooth muscle, 1461
Actin-like proteins, 107, 381
Actinomycin D, effect on active transport, 607
    on mitochondrial protein synthesis, 231
    on nucleolus 223
    on renal function, 947
    on RNA synthesis, 193
Action potential, 1038, 1046
  all-or-none character, 1053
  calculated, 1084, 1097, 1289
  cardiac, 1279, 1281, 1285
    in relation to contraction, 1459